Nanostructure Science and Technology

Series editor

David J. Lockwood, FRSC
National Research Council of Canada
Ottawa, Ontario, Canada

More information about this series at http://www.springer.com/series/6331

Naotoshi Nakashima
Editor

Nanocarbons for Energy Conversion: Supramolecular Approaches

Springer

Editor
Naotoshi Nakashima
International Institute for Carbon-Neutral
Energy Research (WPI-I2CNER)
Kyushu University
Fukuoka
Japan

ISSN 1571-5744 ISSN 2197-7976 (electronic)
Nanostructure Science and Technology
ISBN 978-3-030-06546-1 ISBN 978-3-319-92917-0 (eBook)
https://doi.org/10.1007/978-3-319-92917-0

Softcover re-print of the Hardcover 1st edition 2019

Printed on acid-free paper

This Springer imprint is published by the registered company Springer Nature Switzerland AG
The registered company address is: Gewerbestrasse 11, 6330 Cham, Switzerland

Preface

As seen in the periodic table, there are more than 110 elements on the earth. Various organic and polymer, inorganic, bio, and composite materials using these elements have been developed and enriched our lives. Among them, carbon is one of the most widely used common materials, especially new types of nanocarbons have attracted many researchers. Carbon nanotubes (discovered by S. Iijima in 1991) and graphene are representative examples of such nanomaterials that are expected to be key materials of science and technology in the twenty-first century in the world of nanoscience and technology. Carbon nanotubes (CNTs) and graphene are π-conjugated one-dimensional ($\sim$1–3 nm in diameter, several microns in length) and two-dimensional crystal nanostructural "macromolecules", respectively. In this sense, their nature is quite different from conventional carbon black which is an amorphous carbon. Such nanocarbons are made from $\sim$100% carbon, so it is very light. CNTs and graphene have remarkable electronic, electric, mechanical, thermal, thermoelectric, as well as unique optical properties, i.e., their thermal conductivities are comparable to silver and heat resistance is very high (up to 500 °C even in the air, and in vacuum, they reach 1000 °C). We can make films like plastic polymers. They are very stable in the air, acid, and alkaline solutions, and handling them is easy. Thus, these nanocarbons are exactly "dream" compounds with extreme functions.

Since oil, coal, and natural gas, as the main source of energy for long, will drain in the near future, there is a strong social demand to find an alternative sustainable green source of energy with a high power density. Promising alternative energy technologies contain batteries, fuel cells, supercapacitors, as well as solar energy in addition to natural energy such as wind power, geothermal power generation, hydraulic power generation, etc.

This book focuses on nanocarbons (carbon nanotubes, graphene, nanoporous carbon, and carbon black) and related materials for energy applications, and summarizes their recent progress in novel electrode catalysts design and performance. The targets contain the developments of polymer-wrapped carbon nanotube (or carbon black) based novel electrode catalysts for (high-temperature) polymer electrolyte fuels cells and direct methanol fuel cells with high durability and performance, (i) theoretical approaches for fuel cells and batteries design, (ii) graphene (graphene foam)-based catalysts for fuel cells and water splitting, (iii) polymer electrolyte alcohol electrosynthesis fuel cells, (iv) CO reduction for energy conversion, (v) non-precious metal for fuel cells and batteries, (vi) polymer electrolyte membrane design and synthesis in acid and alkaline media, (vii) carbon nanotube-based biofuel cells, (viii) carbon nanotube-based solar cells, (ix) carbon nanotube-based photocatalyst, (x) carbon nanotube-based up-conversion for new energy material design, and (xi) carbon nanotube-based thermoelectric devices. Now multi-walled carbon nanotubes are widely used in Li ion batteries for notebook computers, mobile phones, etc., while their use in industry is rather limited because of several reasons including cost problem.

Keywords in the chapters of this book are: carbon nanotubes, graphene, energy conversion, fuel cells, biofuel cells, solar cells, photocatalyst, capacitor, Li ion battery, zinc–air battery, non-precious metal catalysts, water splitting, computational design for catalysts, durable polymer electrolyte membranes, up-conversion for new energy material, and thermoelectric conversion.

This book summarizes recent progress in the design, fabrication, and characterization of nanocarbon-based energy and related areas catalyst. Many well-written chapters were made by professors/researches of expertise fields in the world. Notably, the authors present a new concept for such catalyst design for energy and energy conversion as well as and state-of-the-art data in this and related fields. Major drawbacks of catalysts for energy conversion such as fuel cells and batteries are low durability, sluggish mechanism and high cost. This book provides hints to solve such problem.

We believe that this book is highly useful and is of benefit for many readers in the areas of nanocarbons, energy materials including fuel cells, batteries, and capacitor as well as chemical/physical/polymer science, photo science, materials design, and supramolecular science.

Fukuoka, Japan Naotoshi Nakashima

Contents

Chapter 1
Carbon Nanotube-Based Fuel Cell Catalysts-Comparison with Carbon Black

Naotoshi Nakashima and Tsuyohiko Fujigaya

1.1 Introduction

1.1.1 Carbon Nanotubes

Carbon nanotubes (CNTs) discovered in 1991 by Iijima are made of rolled-up graphene sheets with one-dimensional extended π-conjugated structures [1], which are classified into three types, i.e., single-walled carbon nanotubes (SWNTs), double-walled carbon nanotubes (DWNTs) and multi-walled carbon nanotubes (MWNTs), having one, two and more than three walls, respectively (Fig. 1.1). CNTs have been central materials in the field of nanotechnology and nanoscience due to their remarkable physical, physicochemical, thermal, mechanical, electrical and thermoelectrical properties. Theoretical and experimental values (vary depending on purity) of their physical properties are summarized in Table 1.1. Solubilization/dispersion techniques can be categorized mainly into two methods, i.e., chemical modification and physical modification [2–9]. The authors of this chapter presented a concept (Fig. 1.2) to solubilize CNTs based on the physical adsorption of designed dispersant molecules [10–13].

It is a quite reasonable strategy to use conjugated polymers for the wrapping of the conjugated surfaces of the CNTs, in which effective interactions, such as π-π and/or CH-π are expected [2–9]. One of the advantages of the polymer wrapping is the synergetic functionalization of CNTs by combining the functions of the poly-

N. Nakashima (✉)
International Institute for Carbon-Neutral Energy Research (WPI-I2CNER), Kyushu University, 744 Motooka, Fukuoka 819-0395, Japan
e-mail: nakashima.naotoshi.614@m.kyushu-u.ac.jp

T. Fujigaya
Department of Applied Chemistry, Kyushu University, 744 Motooka, Fukuoka 819-0395, Japan
e-mail: fujigaya.tsuyohiko.948@m.kyushu-u.ac.jp

N. Nakashima (ed.), *Nanocarbons for Energy Conversion: Supramolecular Approaches*, Nanostructure Science and Technology,
https://doi.org/10.1007/978-3-319-92917-0_1

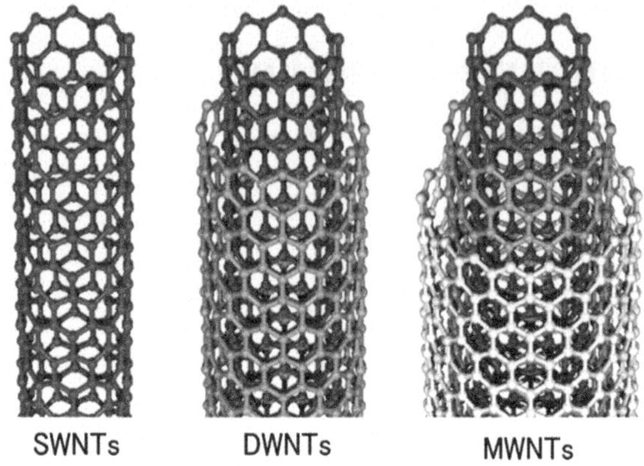

Fig. 1.1 Structures of SWNTs, DWNTs and MWNTs

Table 1.1 Physical properties of SWNTs, MWNTs and reference materials

	SWNTs	MWNTs	Metals
Tensile strength (GPa)	~100	~60	~1.3 (steel)
Young modulus (TPa)	~3.4	~1.3	~0.2 (steel)
Current density (A/cm^2)	~10^9	~10^9	~10^6 (cupper)
Carrier mobility (m^2/V s)	~80		~10 (amorphous silicone)
Thermal conductivity (W/mK)	~3000	~3000	~375 (cupper)

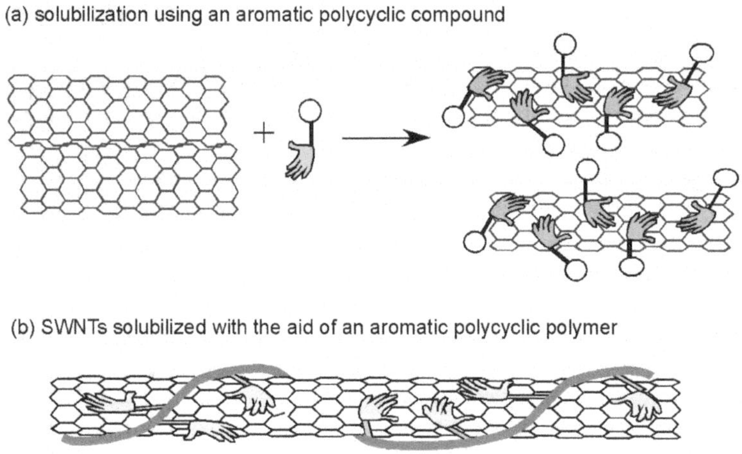

Fig. 1.2 Schematic drawing of bundled/individual CNTs (top) and a concept for CNT solubilization based on adsorption of polycyclic monomeric (**a**) and polymeric (**b**) compounds carrying a solvophilic moiety. Reproduction from [11] with permission of John Wiley and Sons, Inc

mers. Thus, the strategic design of polymers for wrapping is quite important. When the wrapped polymer possesses a sufficient affinity to the surfaces of the CNTs, such polymers should remain on the tubes even after vigorous rinsing to remove excess polymers. As a result, the CNT surfaces are decorated by the used polymer to provide a core-shell structure with an extremely thin polymer layer. Polybenzimidazole (PBI) is one of such polymer; i.e., the authors of this chapter discovered that PBI is a very good CNT solubilizer [14], and its structure was successfully visualized by SEM based on the difference in the efficiency of the electron scattering from polymer and CNTs, in which non-coated bare islands were also observed [15].

CNTs are expected to be an excellent supporting material for fuel cells due to their very high electrical conductivity [16, 17], high purity [18], and high durability [19–23] compared to the conventional supporting materials, such as carbon black (CB). However, due to the lack of binding sites, such as –COOH, C=O and –OH groups, on the tubes, deposition of metal catalysts on the surfaces of pristine CNTs are very difficult. For this reason, strong oxidation of the CNTs has been carried out in order to introduce such hydrophilic groups [16, 18, 22–30]. However, the oxidation introduces severely damages to the graphitic structure of CNTs which decrease the excellent electrochemical stability. Thus, a novel methodology to load the catalyst onto the non-oxidized CNTs has been demanded to utilize the intrinsic stability of the pristine CNTs. Toward the goal, the introduction of binding sites by polymer wrapping of the pristine CNTs is one of promising solutions. CNTs wrapped by poly(aromatic hydrocarbons) [31, 32], chitosan [33] polyanilines [34–36], poly(diallyldimethylammonium chloride) [37, 38], and polypyrrole [36, 39, 40] were successfully used to anchor metal nanoparticles onto the surfaces of the pristine CNTs.

1.1.2 Carbon Support for Fuel Cells

Fuel cells (FCs) are categorized into five different types including polymer electrolyte FCs (PEFC), alkaline FCs (AFC), phosphoric acid FCs (PAFC), molten carbonate FCs (MCFC), and solid oxide FCs (SOFC) [42–44]. Among them, PEFC has been extensively studied due to their compactness and simple fabrication process as well as high performance, and has been commercialized as automobile (from Toyota Motor Corp. and Honda Mortor Co. Ltd.) as well as stationary FC power systems (EneFirm). For more global commercialization, the reduction of cost is strongly requested from the industry side. PEFCs are electrochemical power source with high energy conversion efficiency, high power density, and low pollutant emission, and have the potential to solve major problems associated with the production and consumption of energy, considered as promising, attractive, reliable and clean energy generation for automotive and stationary applications. Currently, perfluorosulphonic acid (PFSA) polymers, such as Nafion®, carbon black (CB) and platinum (Pt) are commonly used as a polymer electrolyte, catalyst supporting

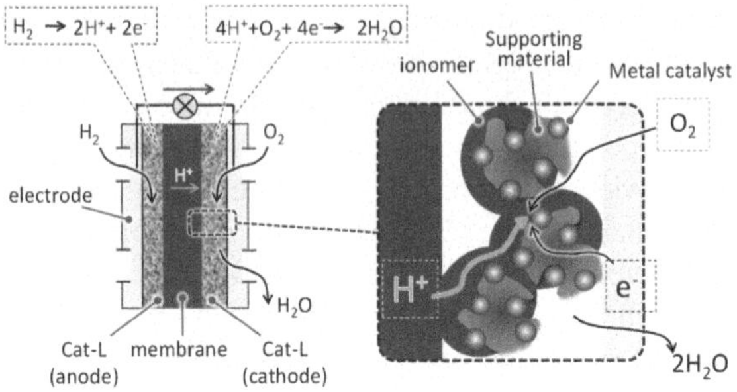

Fig. 1.3 Schematic illustration of a PEFC cell. Reproduction from [41] with permission of John Wiley and Sons, Inc

material, and metal catalyst, respectively (Fig. 1.3), and their operation conditions are temperature of 70–80 °C and high relative humidity ($\sim$100% RH). Under such conditions, cost of water manegement is needed. Another type of PEFC is a high-temperature PEFC (HT-PEFC) system, in which PEFC operations are carried out at temperatures over 100 °C (120–180 °C) [44–46]. In such systems, there are many benefits including (i) enhancement of reaction rates [45, 46], (ii) high CO tolerance [47, 48] and (iii) replacement of Pt to other less-expensive metals, leading to the reduction of the cost of PEFC systems. In addition, the PEFC operation at higher temperatures results in high power generation efficiency. While, Nafion® is not suitable as a electrolyte of high-temperature (HT) PEFCs since its proton conduction is very poor at temperatures over 100 °C. For HT-PEFCs, acid-doped polybenzimidazole derivatives (PBIs) have been used as an electrolyte membrane since the PBIs exhibits high proton conductivity at high temperature (>100 °C) even without humidification [49–60]. In this system, doped acid is responsible for the proton conduction sinse proton conduction via hopping mechanism through the hydrogen bonding network is occurred.

In HT-PEFCs, selection of carbon supporting materials is very important since acceleration of the oxidation and corrosion of carbon supports causes the decrease in the durability of the cells. As for the catalyst supporting material, use of carbon nanotube (CNTs) in place of CB is promising because CNTs possess excellent electrochemical durability due to the well-developed graphitic structures [16, 17, 20, 61].

1.2 Polymer-Coated Fuel Cell Catalyst for High-Temperature Polymer Electrolyte Fuel Cell (HT-PEFC)

For the enhancement of the durability of fuel cells, polymer-coated method has been applied [41]. Here we introduced performance of PBI (or PyPBI)-wrapped MWNTs as a carbon support in PEFCs. Chemical structures of PBI and PyPBI are shown in Fig. 1.4. MWNTs are preferable for practical applications because (i) they are much

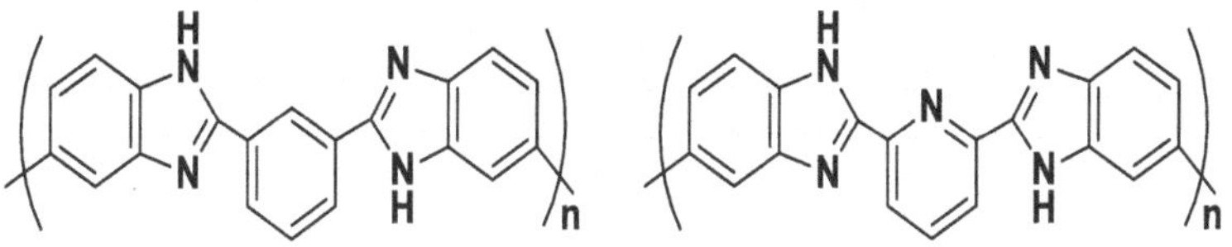

Fig. 1.4 Chemical structures of PBI (left) and PyPBI (right)

cheaper than SWNTs, (ii) their contamination with metal impurity is very low and (iii) they possess very high electrochemical and thermal resistivity.

Deposition of Pt nanoparticles (Pt-NPs) on the PBI-wrapped MWNTs is easy; i.e., by heating the material that is being dispersed in a 60% ethylene glycol aqueous solution at 120 °C in the presence of H_2PtCl_6 to provide MWNT/PBIs/Pt (Fig. 1.5). The thickness of the PBI layer based on thermogravimetric analysis (TGA) data is ~0.5–0.8 nm. The transmission electron microscope (TEM, Fig. 1.5) images of a typical MWNT/PBI(or PyPBI)/Pt show Pt-NPs with a diameter of 3–4 nm [62, 63]. Notably, the Pt-NPs are homogeneously loaded, while Pt-deposition without PBI (or PyPBI) caused aggregation of Pt on the tubes. Homogeneous Pt loading is due to the formation of sufficient binding sites by PBIs for anchoring the precursor metal ions or metal nanoparticles.

Such preparation method is also applicable to CB and graphene in place of MWNTs to provide CB/PBI/Pt and graphen/PBI/Pt. Electrochemical activity of fuel cell electrocatalysts is evaluated using cyclic voltammetry (CV). The electrochemically active surface area (ECSA) is calculated using the following Eq. (1.1) [64].

$$\mathrm{ECSA} = Q_H / (210 \times \text{ Pt loading on electrode}), \tag{1.1}$$

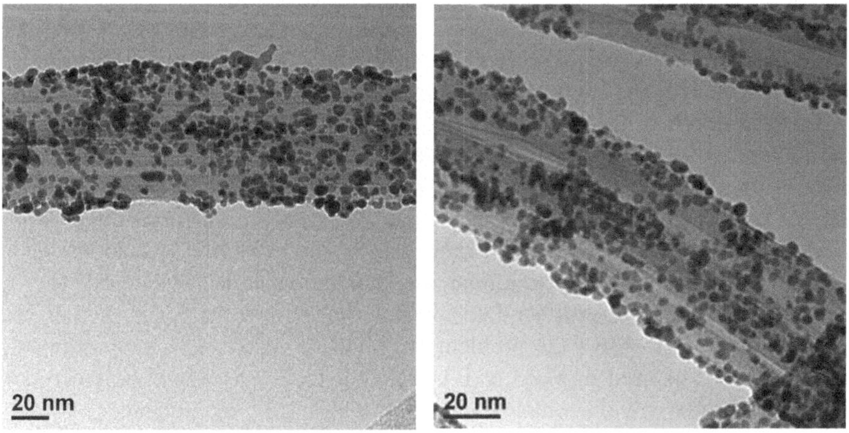

Fig. 1.5 Typical TEM images of (left) MWNT/PBI/Pt and (right) MWNT/PyPBI/Pt

where Q_H is the charge exchanged during the electroadsorption of H on Pt (from −0.2 to 0.1 V vs. Ag/AgCl). The ECSA values of the MWNT/PBIs/Pt are ~44 $m^2\ g^{-1}$ of Pt, which are comparable to the conventional commercially available electrocatalyst, Pt/CB (ECSA ~ 50 $m^2\ g^{-1}$). Such ECSA value clearly indicates that the dispersed Pt-NPs are well exposed to the electrolyte solution despite the polymer-coating.

A high resolution TEM (HRTEM) image of the MWNT/PyPBI/Pt shows a close contact between the Pt nanoparticles and MWNT surfaces, as well as the Pt exposure over the PBI layers. This interface structure is most likely the structure that has been widely recognized as the 'ideal' triple-phase boundary structure (TPB) that enables excellent fuel cell catalyst efficiency [65]. The formation mechanism of such structure is the *in situ* surface growth of the Pt-NPs on the surface of the PBI layers, followed by nucleation to form close contact structures with the MWNT surfaces.

PBIs have been used for a PEFC electrolyte membrane material since the early 1990s [66] due to their high proton conductivity after acid doping [49–53]. Inorganic acids, such as HCl, H_2SO_4, and H_3PO_4 (PA) are used as a dopant, especially PA is a typical acid dopant [67, 68]. Thus, MWNT/PyPBI/Pt was doped with PA, and the structure of the MWNT/PyPBI/Pt after PA doping (denoted MWNT/PA-PyPBI/Pt) was determined by IR, in which a new peak at 1,100 cm^{-1} that is assignable to the PO_2 band in the PA-PBIs [69–71].

Membrane electrode assemblies (MEAs) were fabricated using a PA-doped PBI membrane film and MWNT/PA-PyPBI/Pt (or MWNT/PyPBI/Pt) as an electrolyte membrane and catalysts (for both anode and cathode), respectively. One of advantages of the PBI-based PEFC over conventional Nafion-based PEFCs is in their operation at high temperature (>100 °C) under a non-humidified condition since an operation temperature below 80 °C with less than 100% relative humidity is required in a Nafion-based PEFC system. Higher operation temperatures without external humidification eliminate to equip a cooling and water management instrument [72].

Typical polarization curves of single cell tests of the MEAs fabricated using the MWNT/PA-PyPBI/Pt and MWNT/Pt performed at 120 °C under non-humidified hydrogen and air as the anode and cathode gases, respectively is shown in Fig. 1.6 [73]. The peak power density of the PA-doped MEA was over 100 mW cm^{-2}, while the non-doped MEA showed only ~50 mW cm^{-2} (Fig. 1.6b), indicating importance of proton conduction in the cell. On the other hand, for the MWNT/Pt, the absence of an ionomer limited the delivery of the protons, which results in lower fuel cell performance. The lower power density (~50 mW cm^{-2}) would be due to leaching of PA from electrolyte film as was pointed out in the literature [74].

PA is liquid and very difficult to avoid leakage from the MEA of fuel cell systems. In order to solve the problem, a polymeric acid, poly(vinylphosphonic acid) (PVPA) was used in place of PA [75, 76]. PVPA is an acid polymer, and enables to form a polyion-complex with PBIs which are basic polymers.

Namely, PVPA was expected to form a thin film on a PBI-wrapped MWNTs. PVPA-doped MWNT/PyPBI/Pt was prepared through a simple mixing of MWNT/

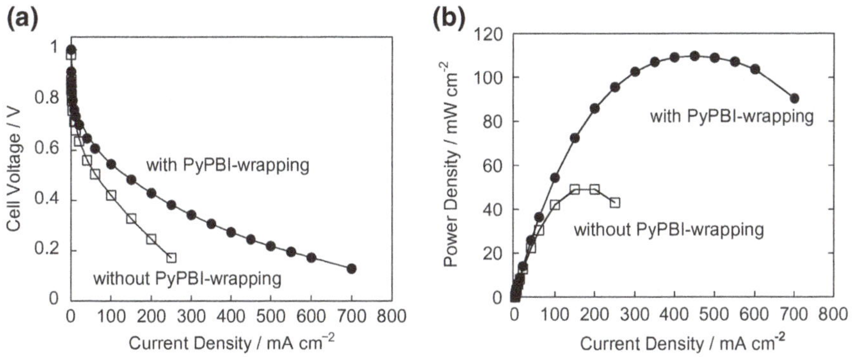

Fig. 1.6 **a** Polarization- and **b** power density curves of the MEAs fabricated with the MWNT/PA-PyPBI/Pt (circles) and MWNT/Pt (squares); the same modified electrodes were used as the anode and cathode electrodes (Pt loading: 0.45 mg cm^{-2}). The cell temperature was 120 °C, and anode and cathode gases were dry H_2 (flow rate: 100 mL min^{-1}) and dry air (flow rate: 200 mL min^{-1}), respectively. Reproduction from [73] with permission of Royal Society of Chemistry

PyPBI/Pt with PVPA in an aqueous solution followed by removing the unbound excess PVPA by vigorous water rinsing [75, 76]. Figure 1.7 displays an illustration for the preparation of the MWNT/PyPBI/Pt/PVPA together with the chemical structures of PyPBI and PVPA, as well as photographs MWNT/PyPBI/Pt (left) and MWNT/PyPBI/Pt/PVPA in water and HR-TEM image, in which good dispersion of MWNT/PyPBI/Pt/PVPA as well as a thin PVPA layer formation are observed.

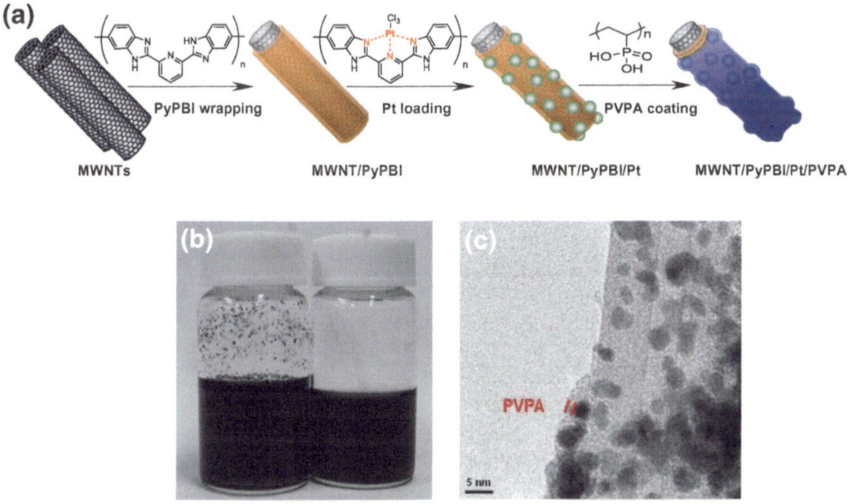

Fig. 1.7 **a** Schematic illustration for the preparation of the MWNT/PyPBI/Pt/PVPA together with the chemical structures of PyPBI and PVPA. **b** Photographs of the dispersity of the MWNT/PyPBI/Pt (left) and MWNT/PyPBI/Pt/PVPA in water. **c** HR-TEM image of the MWNT/PyPBI/Pt/PVPA. Reproduction from [75] with permission of Nature Publishing Group

In XPS of the MWNT/PyPBI/Pt/PVPA, Pt_{4f} doublet peaks observed at 71.1 and 74.4 eV which are attributable to $Pt^0_{4f7/2}$ and $Pt^0_{4f5/2}$, respectively. The ECSA value of the catalyst was ~40.2 m^2/g. A single cell tests using PVPA-doped MEA and PVPA-non-doped MEA were carried out at 120 °C without external humidification, in which as an electrolyte membrane, a PVPA-doped PBI film was used. The open-circuit voltages (OCV) of both MEAs were ~0.9 V, while the PVPA-doped MEA showed a higher cell voltage at all applied current densities (Fig. 1.8), i.e., the maximum power density of the PVPA-doped MEA reached 252 mW/cm^2 which is higher than that of the PVPA-non-doped MEA.

The most interesting feature of the PVPA-doped MEA is its durability evaluated based on the FCCJ protocol [77]. The polarization curves of the PVPA-doped MEA (Fig. 1.8a) together with the PA-non-doped MEA (Fig. 1.8b) measured after every 40,000 cycles of the potential sweeps between 1.0 and 1.5 V indicated very high durability of the PVPA-doped MEA (Fig. 1.8c) compare to the PA-non-doped

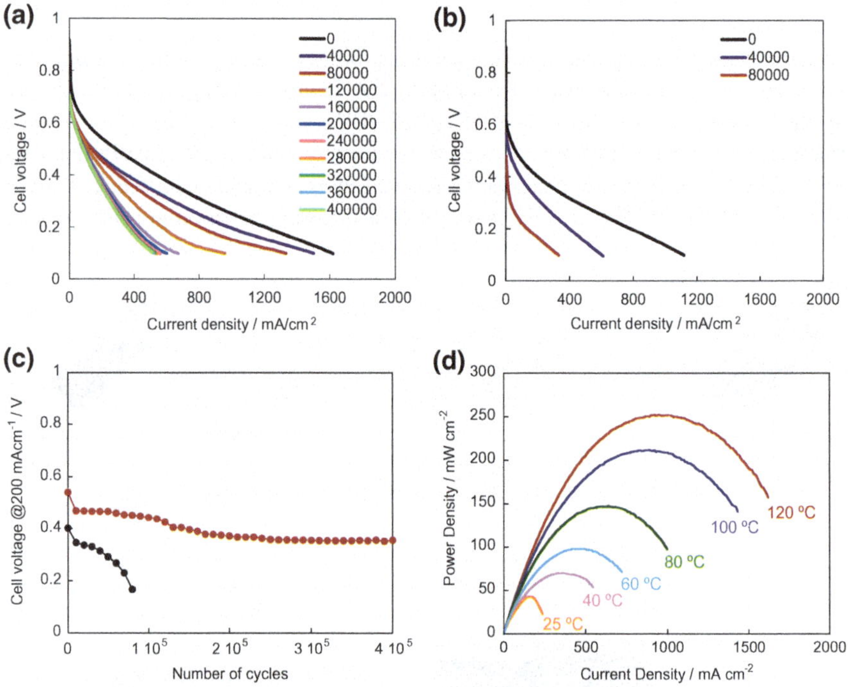

Fig. 1.8 Durability test of MEA. **a**, **b** Polarization curves of a durability test using PVPA-doped MEA (**a**) and PA-non-doped MEA (**b**). The polarization curves were measured after every 1,000 cycles of the potential sweep between 1.0 and 1.5 V. For convenience, the data were plotted every 40,000 cycles. **c** Plots of the cell voltage at 200 mW/cm^2 for the PVPA-doped MEA (red line) and PA-non-doped MEA (black line). **d** Power density curves of the PVPA-doped MEA measured at 25 °C (orange line), 40 °C (pink line), 60 °C (blue line), 80 °C (green line), 100 °C (purple line) and 120 °C (red line). Reproduction from [75] with permission of Nature Publishing Group

MEA (black circles in Fig. 1.8c). Evidently, that PVPA-doped MEA (red circles in Fig. 1.8c) is still alive even after 400,000 cycles although its performance decreased. Temperature dependent power density curves of PVPA-doped MEA is shown in Fig. 1.8d, in which, as expected, the increase in the operating temperature increased the output power density, while even at room temperature, the MEA worked.

In summary, PVPA can be used in place of PA, and PVPA act as a proton conducting layer which prevent leakage to the proton source, which is important for durability of fuel cell catalysts.

1.3 Comparison with Other Carbon Support

Here, comparison of durability/performance of the PBI-wrapped MWNT-based fuel cell catalyst with PBI-wrapped CB-based catalyst under non-humidified and 120 °C [76, 78–81], as well as PBI-Coated MWNT/CB composite fuel cell catalysts [79] are summarized. Study using graphene instead of MWNTs has also been reported [82].

Durability test of the CB/PyPBI/Pt-based MEA (MEA_{CB}; red) and CNT/PyPBI/Pt-based MEA (MEA_{CNT}; black) measured based on the FCCJ protocol [77] are shown in Fig. 1.9. Figure 1.9 shows the polarization curves of the MEA_{CB} (Fig. 1.9a) and the MEA_{CNT} (Fig. 1.9b). The initial power densities of the MEA_{CB} and MEA_{CNT} were ~135 mW/cm^2 and ~160 mW/cm^2, respectively. The Nyquist plots (EIS) of the MEA_{CB} and MEA_{CNT} in the frequency range of 100 kHz to 0.1 Hz at the current densities of 100 and 800 mA/cm^2 were analyzed using the equivalent circuit the circuit consisted of an ohmic resistance (R_Ω), charge transfer resistance (R_{ct}) and gas diffusion resistance (R_g) [83]. All the resistances (R_Ω, R_{ct} and R_g) of the MEA_{CNT} were lower than those of the MEA_{CB}. Agglomerative structure in the SEM image of MEA_{CB} was observed, while MEA_{CNT} showed a fibrous network structure. Such morphological difference might be related to the difference in the impedance results, and suggested that the fibrous network structures of the CNTs are important for a better proton conduction, charge transfer and gas diffusion pathways, which are important for improved FC performance.

The cell voltages of the MEA_{CB} and the MEA_{CNT} at 200 mA/cm^2 are plotted in Fig. 1.9c. The 200 mA/cm^2 region is chosen since this current density is often used for the stationary PEFCs [84]. The data shows that after 15,000 potential cycles of both MEAs, the cell voltage of the MEA_{CB} (red line) drastically decreased. In sharp contrast, the MEA_{CNT} (black line) shows a gradual decrease upon cycling. Such durability difference is due to the difference in the oxidation stability between the CB and CNTs.

It is important to understand the role of PBI (PyPBI) on the catalysts. A TEM image of the pristine CNTs/Pt formed large aggregates of the Pt metal because pristine CNTs lack the binding sites of Pt. The ECSA value of the CNTs/Pt was low as 16.6 m^2/g_{Pt} due to the Pt aggregation. Thus, the MEA of the CNTs/Pt showed a

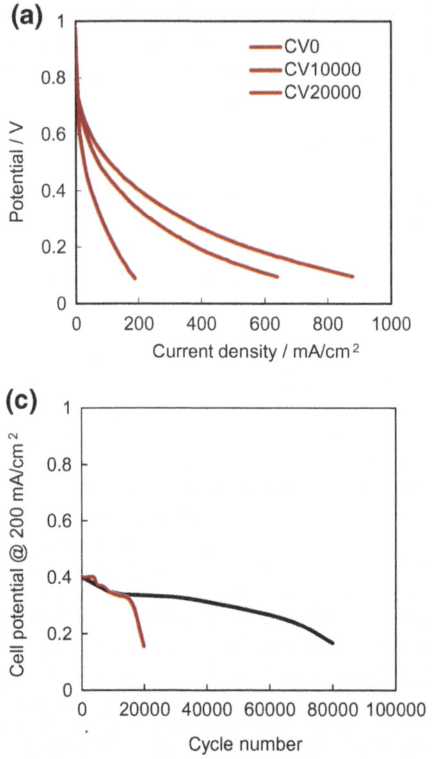

Fig. 1.9 Polarization curves of **a** MEA_{CB} and **b** MEA_{CNT} during durability testing. For simple understanding of the figures, the polarization curves were displayed every 10,000 cycles. **c** Plots of the cell voltage at 200 $mAcm^{-2}$ as a function of the number of potential cycling for the MEA_{CB} (red line) and MEA_{CNT} (black line) during the durability testing. Reproduction from [76] with permission of Royal Society of Chemistry

poor fuel cell performance due to the lack of the electrocatalyst homogeneity into the MEA, indication the importance of the PBI for the formation of homogeneous Pt-NPs in the catalyst. CNT-based and CB-based MEAs, MEA_{CNT} and MEA_{CB}, respectively, were delaminated during and after the durability testing, and then analyzed them using TEM and XRD. The TEM mages of the electrocatalysts of the cathode side of both MEAs after 12,000 potential cycles, the Pt particle sizes increased for both catalysts compared to their initial particle sizes, and the increase in the Pt particle size was more obvious for the CB-based catalyst where it reached 6.2 nm, which was greater than that of CNT-based one (5.6 nm), suggesting faster corrosion process of the CB-support material; namely, for CB-based catalyst, upon potential cycling, an accelerated fusion process of the neighboring Pt particles occurs, leading to the formation of bigger Pt particles.

In summary, the PBI (PyPBI)-wrapped CNT-based-MEA showed a remarkable durability compared to the CB-based MEAs. Such a high durability of the

CNT-based MEA was due to the inherent structural stability of the pristine CNTs and the polymer wrapping. The accelerated deterioration behavior of the MEA_{CB} was derived from the faster corrosion of the CB, leading to the formation of the Pt agglomeration, then resulting in the decrease in fuel cell performance.

1.4 Mechanism of Fuel Cell Durability of Carbon Black-Based Catalyst by Polybenzimidazole Coating

Detailed mechanism of the durability improvement of carbon blacks (CBs; Vulcan and Ketjen) by coating with PBI was reported [78]. The fabricated composites are Vulcan/PBI and Ketjen/PBI (they do not contain Pt). Figure 1.10 displays the CVs of (a) Vulcan, (b) Vulcan/PBI, (c) Ketjen and (d) Ketjen/PBI. The result clearly indicated that the oxidation durability was dramatically enhanced after coating with PBI. The N_2 adsorption and desorption profiles of the 4 different materials are

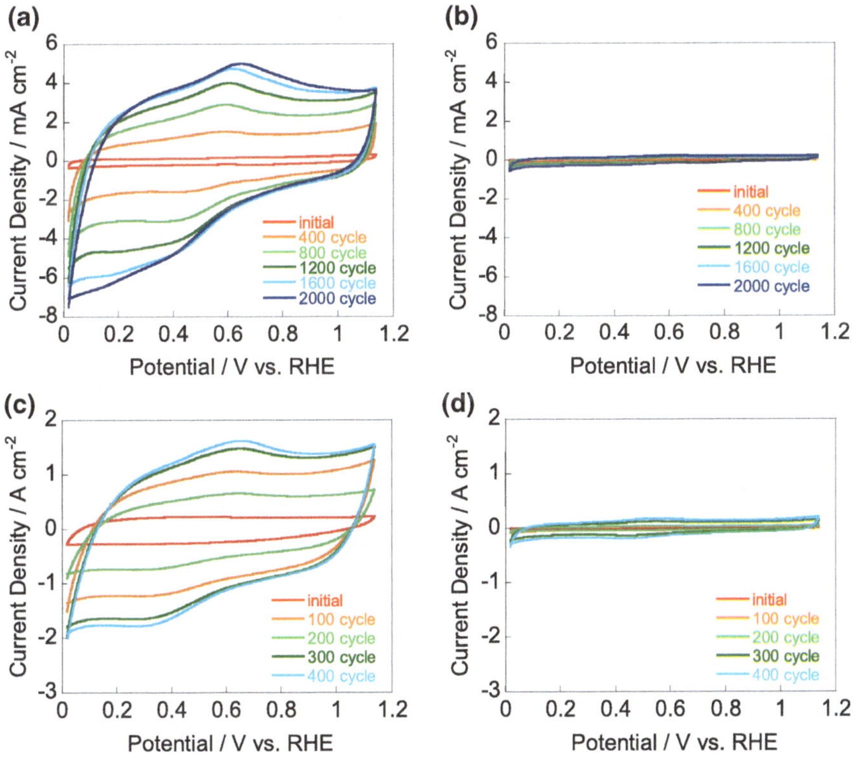

Fig. 1.10 CVs of the **a** Vulcan, **b** Vulcan/PBI, **c** Ketjen and **d** Ketjen/PBI measured in 0.1 M $HClO_4$ aqueous solutions. Reproduction from [78] with permission of The American Chemical Society

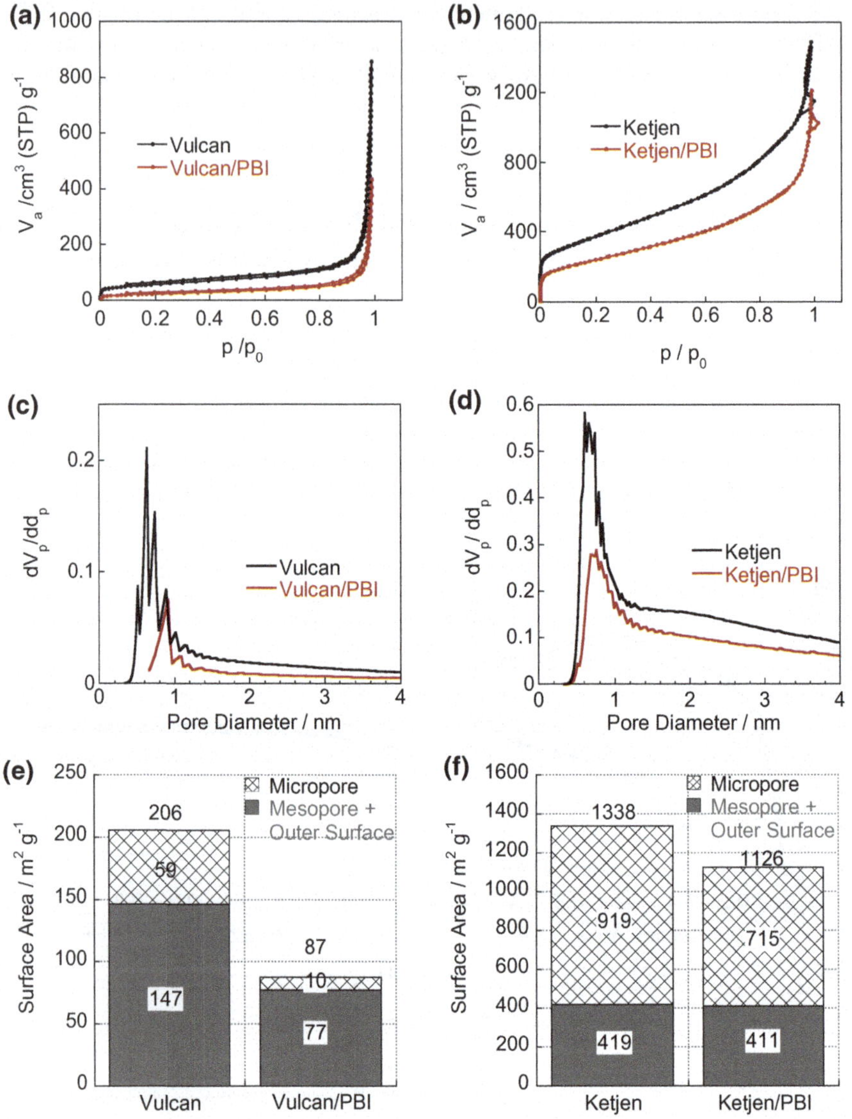

Fig. 1.11 **a**, **b** Absorption-desorption isotherms of N_2 for **a** Vulcan (black) and Vulcan/PBI (red) and **b** Ketjen (black) and Ketjen/PBI (red). **c**, **d** The pore diameter distribution evaluated based on the HK method for **c** Vulcan (black) and Vulcan/PBI (red) and **d** Ketjen (black) and Ketjen/PBI (red). **e**, **f** Bar graphs displaying the outer and micropore surface areas of **e** Vulcan (left) and Vulcan/PBI (right) and **f** Ketjen (left) and Ketjen/PBI (right) evaluated based on the t-plot method. Reproduction from [78] with permission of The American Chemical Society

shown in Fig. 1.11. The surface areas of the Vulcan (206 m^2/g) and Ketjen (1338 m^2/g) agreed well with the reported values [85], and their values of Vulcan/PBI and Ketjen/PBI decreased to 87 and 1126 m^2/g, respectively, due to the PBI coating. Based on the Horvath–Kawazoe (HK) method [86], the decrease in the surface areas after the PBI coating was mainly due to the decrease in the micropore areas, especially below 1.0 nm (Fig. 1.11c, d). By the quantitative analysis of the micropore area using the t-plots method [87], the micropore areas decreased after the PBI coating from 59 to 10 m^2/g and 919 to 715 m^2/g for the Vulcan and Ketjen, respectively, while the rest of surface (=mesopore + outer surfaces) decreased after the PBI coating from 147 to 77 m^2/g and 419 to 411 m^2/g for the Vulcan and Ketjen, respectively (Fig. 1.11e, f).

The thickness of the PBI layers on the Vulcan and Ketjen were ~0.23 and ~0.13 nm for the respectively, which were lower than the monolayer thickness of the PBI and thinner than those of PBI coated on the CNTs (1.2 nm) and graphene (1.1 nm) as previously reported [63, 82]. Based on this analysis, a schematic drawing for the PBI-coated CBs (Vulcan and Ketjen) is presented as shown in Fig. 1.12, in which the PBI preferentially coated the outer surfaces and the interior of mesopore surface of the CBs, but did not coat the micropore surface. As a result, the gate of the micropore is expected to be covered by PBI. This model explains the significant decrease in the micropore volume. The capping of the gate of the narrow pore has reported in the porous membrane systems [88].

The degree of defect on the surfaces of the inner micropores was reported to be higher than that of the outer ones [89], the capping is quite effective to protect the inner micropores from the penetration of the electrolyte and to prevent oxidation of the CBs.

Such a result would explain the enhanced durability of the MEA having Pt-NPs loaded on the Vulcan/PBI not only for Nafion-based MEA for low-temperature operation, but also PA-based MEA for high-temperature operation.

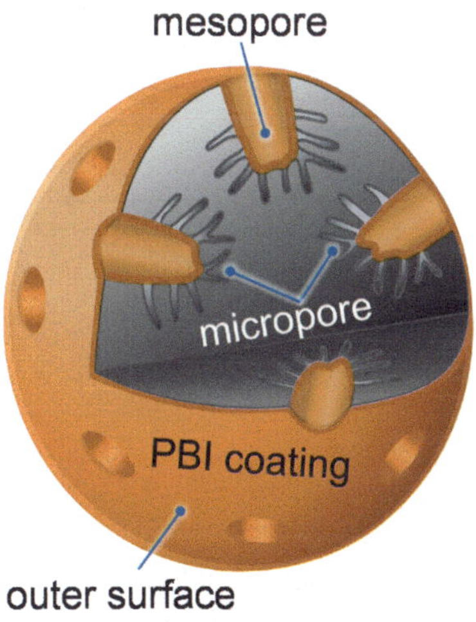

Fig. 1.12 Schematic drawing of the proposed structure of the PBI-coated CBs (Vulcan and Ketjen), showing a capping of the micropore gate by PBI coating. Reproduction from [78] with permission of The American Chemical Society

1.5 Reduction of Pt

Pt is a very expensive and rare; thus, it is highly important to reduce the amount of Pt in fuel cell electrocatalysts in order to reduce the cost of PEFCs [90–92]. For the goal, we need to maximize the mass activity of Pt in the catalysts. The strategy of increasing the mass activity of Pt is to form a Pt core-shell [93–96] or a Pt alloy in order to increase the surface area of the Pt per unit amount [97–102]. However, under a strong acidic condition, a dissolution process of the second metal occurs, resulting in a low fuel cell durability [103–107]. Another promising strategy is to reduce the particle size of Pt in order to increase the surface area of the Pt nanoparticles, leading to an increase in the Pt mass activity [108–110], and the simplest technique for this is to reduce the amount of Pt-ions fed into the Pt reduction process in order to regulate the growth of Pt nanoparticles. Successful examples of the Pt size reduction for carbon black (CB) [110] and graphene oxide [109] have already been reported.

The decrease of the feeding amount of the Pt ions to 1/2, 1/5, 1/10, and 1/20 (CNT/PyPBI/Pt_{25}, CNT/PyPBI/Pt_{10}, CNT/PyPBI/Pt_5 and CNT/PyPBI/$Pt_{2.5}$, respectively) provided catalysts with the Pt size of ~3.6, ~3.3, ~2.7, and ~2.3 nm for CNT/PyPBI/Pt_{25}, CNT/PyPBI/Pt_{10}, CNT/PyPBI/Pt_5 and CNT/PyPBI/$Pt_{2.5}$, respectively (Fig. 1.13) [15]. From the TEM images (Fig. 1.14; left images), the dispersions of the Pt-NPs on the CNTs were highly homogeneous, and the Pt size distributions for all the composites were very narrow (Fig. 1.14; right). Such results would be due to the rapid

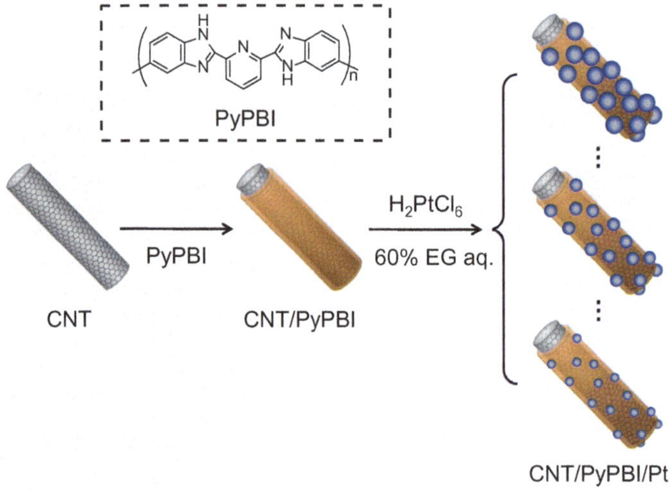

Fig. 1.13 Schematic drawing describing the preparation of the CNT/PyPBI/Pt composites. A serious of CNT/PyPBI/Pt composite was prepared by changing the feeding amount of Pt salt. Chemical structure of PyPBI is presented in the dotted frame. Reproduction from [15] with permission of Nature Publishing Group

nucleation of the Pt seed on the PyPBI-wrapped CNTs through the coordination of the Pt ions with the benzimidazole unit of the PyPBI [63].

The loading efficiency of Pt was estimated to be ~100% of the fed Pt ions, indicating an effective binding of Pt ion onto PyPBI that leads to the quantitative and homogeneous loading of Pt onto the CNT/PyPBI. Thus, excellent Pt-deposition controllability is realized.

The decrease in the Pt diameter resulted in a dramatic increase in the ECSA values; namely, the highest ECSA value, 114.5 $m^2/g_{Pt,}$ was obtained for the CNT/PyPBI/$Pt_{2.5}$, which is comparable to those of Pt with a similar size on graphene oxide [109] and CB [111]. The Pt utilization efficiency values are examined, and the reduction of the Pt size resulted in an increase in the Pt utilization efficiency; i.e., the value reached as high as 94% for the smallest Pt particles in the CNT/PyPBI/$Pt_{2.5}$ catalyst. Watanabe et al. reported that, when the Pt nanoparticles are too close to each other (roughly below 1.8 nm for oxygen gas and 2.0 nm for air), a poor supply of oxygen is observed since not all the Pt surface areas are available for the oxygen reduction reaction (ORR), thus not only the size of Pt, but also the Pt inter-particle distance plays an important role for the increase in the Pt mass activity [110]. For CNT/PyPBI/Pt catalysts, the distributions of the Pt inter-particle distances measured from the TEM images shifted to higher values (>18 nm) when the loading amounts of Pt were decreased, and the current density per unit weight of Pt increased with the decrease in the diameter of the Pt-NPs.

MEAs using CNT/PyPBI/Pt_{50}, CNT/PyPBI/Pt_{10} or CNT/PyPBI/$Pt_{2.5}$ for both the anode and cathode (denoted MEA_{50}, MEA_{10}, and $MEA_{2.5}$, respectively) were fabricated and their fuel cell performance was measured under dry conditions at 120 °C. As a result, small Pt loading values of MEA_{10} and $MEA_{2.5}$ (0.09 and 0.04 mg/cm^2) were lower than those of current PEFCs, in which the Pt amounts are 0.4–1.0 mg/cm^2. The I-V curves and their power density curves for the MEA_{50} (gray), MEA_{10} (blue) and $MEA_{2.5}$ (green) are shown in Fig. 1.15a. The I-V curves were re-plotted based on the total amounts of Pt in the anode and cathode to compare the Pt mass activity of the MEAs (Fig. 1.15b). Notably, the mass activity of $MEA_{2.5,}$ was around 8 times greater than that of MEA_{50}. The result indicates that fine control of the Pt size is important for the improvement of the mass activity [112, 113]. While we need to notice that several reports pointed out that smaller sized Pt-NPs (<2 nm) have a lower ORR activity, and these are not suitable for the fuel catalysts [114–116].

Durability of the MEA_{10} was compared to that of the MEA_{50} (Fig. 1.16), and found that the degradation rates of both MEAs were almost the same, indicating a good stability of the low Pt loading composites.

It is concluded that a fine control of the size of PtNPs in catalysts of CNT/PyPBI/Pt highly useful for the reduction of the Pt amount that is a strong demanded from industry.

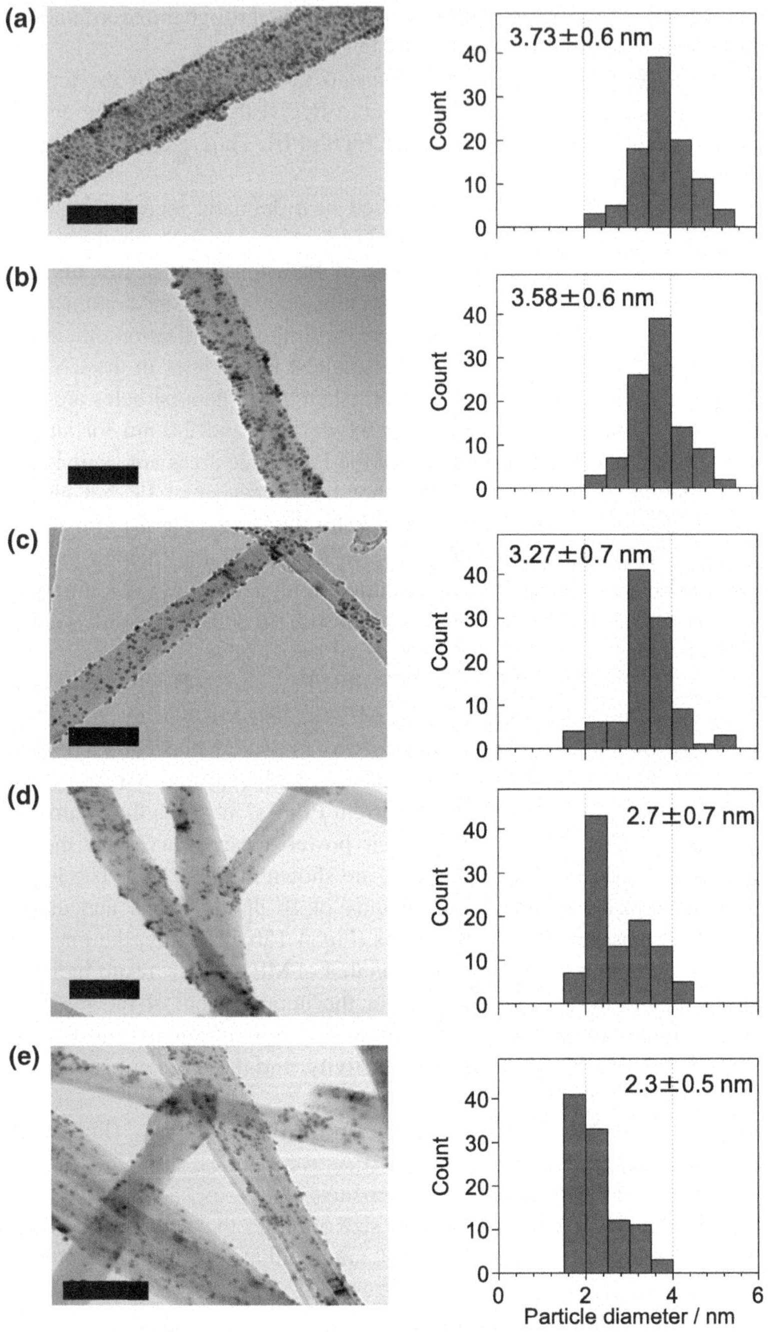
(a)
3.73±0.6 nm
(b)
3.58±0.6 nm
(c)
3.27±0.7 nm
(d)
2.7±0.7 nm
(e)
2.3±0.5 nm
Count
0
10
20
30
40
0
2
4
6
Particle diameter / nm

◂**Fig. 1.14** Pt diameter and distribution analyses of the CNT/PyPBI/Pt. The TEM images (left) and diameter histograms (right) of **a** CNT/PyPBI/Pt_{50}, **b** CNT/PyPBI/Pt_{25}, **c** CNT/PyPBI/Pt_{10}, **d** CNT/PyPBI/Pt_5 and e, CNT/PyPBI/$Pt_{2.5}$ revealed the diameter of Pt and density of Pt loading were decreased by decreasing the feeding amount of Pt salt with maintaining the homogeneous dispersion of Pt on the CNT/PyPBI. Scale bars, 50 nm. Reproduction from [15] with permission of Nature Publishing Group

1.6 Durable Carbon Nanotube-Based Polymer Electrolyte Fuel Cell Under Humidified Condition

Fabrication of an almost "immortal" fuel cell electrocatalyst using double layer-polymer-coated carbon nanotubes as the catalyst support was reported [63]. Here, the interesting features on this topic are summarized. The polymers being used for the double layer-coating are PBI (or PyPBI) and Nafion®.

As described at Sect. 1.2 in this chapter, PyPBI-wrapped MWNTs/Pt-NPs that were further coated with a proton conducting ionomer, polyvinylphosphonic acid (PVPA) through an acid-base reaction between PyPBI and PVPA to provide MWNT/PyPBI/Pt/PVPA as a fuel cell catalyst with a high durability under non-humidified and 120 °C operation.

This "double polymer-coating" technique was applied to the conventional fuel cell systems that are operated at low temperatures (~70 °C) under humidified conditions using Nafion® (Fig. 1.17), in which a Nafion® ionomer was used to further coat the PyPBI-coated MWNTs/Pt as the 2nd layer-polymer via an acid–base interaction to provide MWNT/PyPBI/Pt/Nafion [117]. For the comparison of MWNTs, an electrocatalyst using CB (CB/PyPBI/Pt/Nafion) was also synthesized. The ECSA values of these electrocatalysts were 59.5 and 62.0 m^2/g_{Pt} for the MWNT/PyPBI/Pt/Nafion and CB/PyPBI/Pt/Nafion, respectively (Fig. 1.18a), which are close to that (64.5 m^2/g_{Pt}) of the commercial CB/Pt (Nafion-treated).

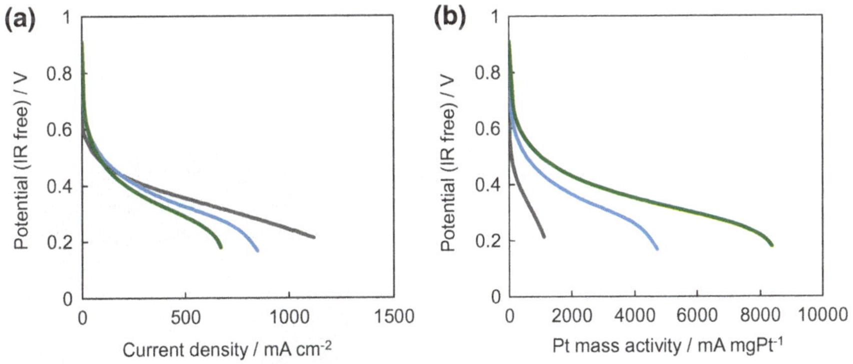

Fig. 1.15 **a** I-V curves and power density curves for MEA_{50} (gray), MEA_{10} (blue) and $MEA_{2.5}$ (green) plotted as a function of the current density. **b** I-V curves for MEA_{50} (gray), MEA_{10} (blue) and $MEA_{2.5}$ (green) plotted as a function of Pt mass activity. Reproduction from [15] with permission of Nature Publishing Group

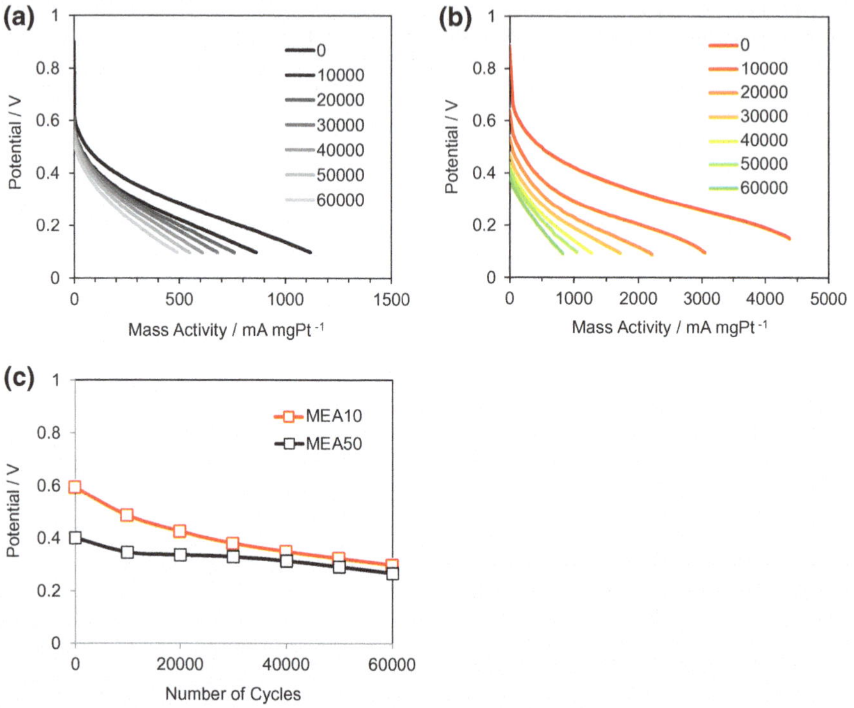

Fig. 1.16 **a**, **b** Polarization curves of a durability test using MEA_{50} (**a**) and MEA_{10} (**b**). The polarization curves were measured after every 1,000 cycles. For convenience, the data were plotted every 10,000 cycles. **c** Plots of the cell voltage at 200 mA $mgPt^{-1}$ for the MEA_{50} and MEA_{10}. Reproduction from [15] with permission of Nature Publishing Group

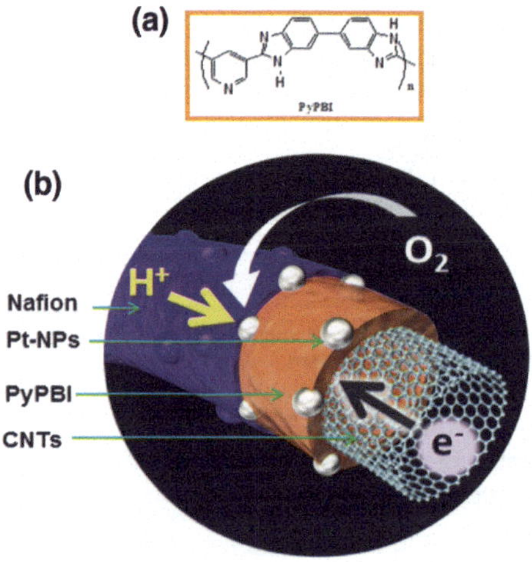

Fig. 1.17 **a** Chemical structure of PyPBI and **b** schematic illustration for the double layer-polymer-coated MWNTs. Reproduction from [117] with permission of Nature Publishing Group

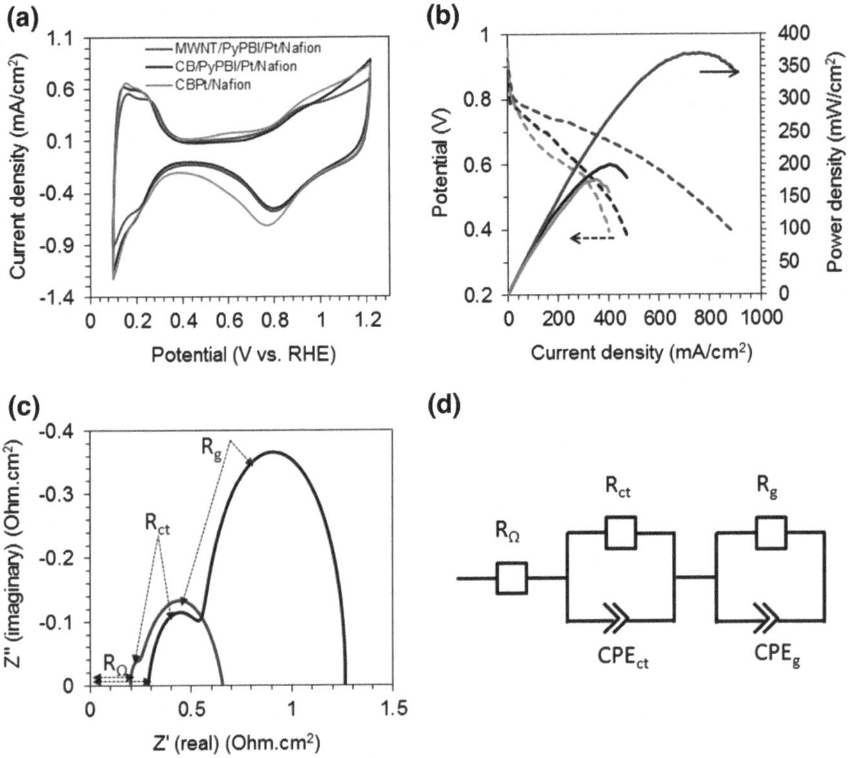

Fig. 1.18 **a** CVs of MWNT/PyPBI/Pt/Nafion (purple), CB/PyPBI/Pt/Nafion (black), and CB/Pt/Nafion-treated (gray). **b** Polarization (dashed) and power density (solid) curves of the MWNT/PyPBI/Pt/Nafion-based MEA (purple), CB/PyPBI/Pt/Nafion-based MEA (black), and CB/Pt/Nafion-treated (gray). **c** Nyquist plots of MWNT/PyPBI/Pt/Nafion-based MEA (purple) and CB/PyPBI/Pt/Nafion-based MEA (black). **d** Equivalent circuit used to fit the obtained impedance spectra. Reproduction from [117] with permission of Nature Publishing Group

The fuel cell performance of the assembled MEA using the catalyst measured at 70 °C under a relative humidity of ~50% was examined. Figure 1.18b shows the polarization (dashed) and power density (solid) curves of MWNT/PyPBI/Pt/Nafion-based MEA (purple) and CB/PyPBI/Pt/Nafion-based MEA (black), in which the OCV of the two different MEAs were almost 1.0 V. It is evident that the MWNT/PyPBI/Pt/Nafion-based MEA exhibited a higher performance compared to that of CB/PyPBI/Pt/Nafion-based MEA. EIS data is shown in Fig. 1.18c, d (the Nyquist plots of the MWNT/PyPBI/Pt/Nafion-based MEA (purple) and CB/PyPBI/Pt/Nafion-based MEA (black)). In the equivalent circuit used to fit the obtained impedance spectra, each spectrum shows two overlapped loops corresponding to charge transfer resistance (R_{ct}) and gas diffusion resistance (R_g) in the high and low frequency regions, respectively, and the high frequency intercept at the real impedance axis represents the ohmic resistance of the cell (R_Ω) [83], in which a greater decrease in the total impedance of the MWNT/PyPBI/Pt/Nafion-based MEA compared to that of the CB/PyPBI/Pt/Nafion-based MEA is obtained. Such result reflected the fuel cell performance (Fig. 1.18b). The R_{ct} of the MWNT/PyPBI/Pt/Nafion-based MEA showed a lower value compared to that of the CB/PyPBI/Pt/Nafion-based MEA, indicating a smoother charge transfer during the fuel cell operation of the MWNT/PyPBI/Pt/Nafion-based MEA. A remarkable increase in the R_g value for the CB/PyPBI/Pt/Nafion-based MEA compared to that of the MWNT/PyPBI/Pt/Nafion-based MEA is due to the difference in the morphology of the carbon supports that affects the gas diffusion during the fuel cell operation; i.e., the MWNTs have a fibrous network structure, which is an advantage in such gas diffusion, while CB has an agglomerative structure that might work as a slight barrier to fuel gas access [76]. The Nafion ionomer homogeneously distributed around the PyPBI-wrapped MWNTs is important to provide a good proton-conducting path as well as the formation of an ideal triple phase boundary structure that is suitable for smooth electrochemical reactions for the fuel cell operation.

Accelerated stress durability test was carried out. The polarization (**A**) and power density (**B**) curves of the MWNT/PyPBI/Pt/Nafion-based MEA and the CB/PyPBI/Pt/Nafion-based MEA are shown in Fig. 1.19 and Fig. 1.20, respectively. It is evident that the FC performance of MWNT/PyPBI/Pt/Nafion-based.

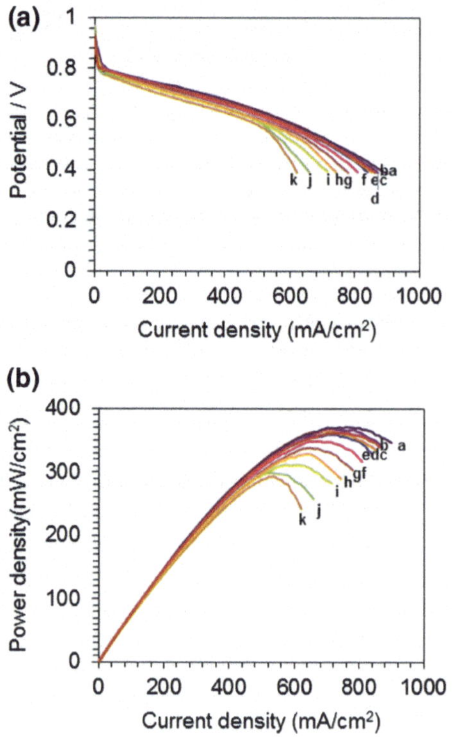

Fig. 1.19 Polarization curves (**A**) and power density curves (**B**) during durability tests of the CB/PyPBI/Pt/Nafion-based MEA. For simplicity, some representative curves are selected for display. The Arabic letters, a, b, c, d, e, f, g, h, I, j and k, represent the performance after 1,000, 50,000, 100,000, 150,000, 200,000, 250,000, 300,000, 350,000, 400,000, 450,000, and 500,000 potential cycles, respectively. Reproduction from [117] with permission of Nature Publishing Group

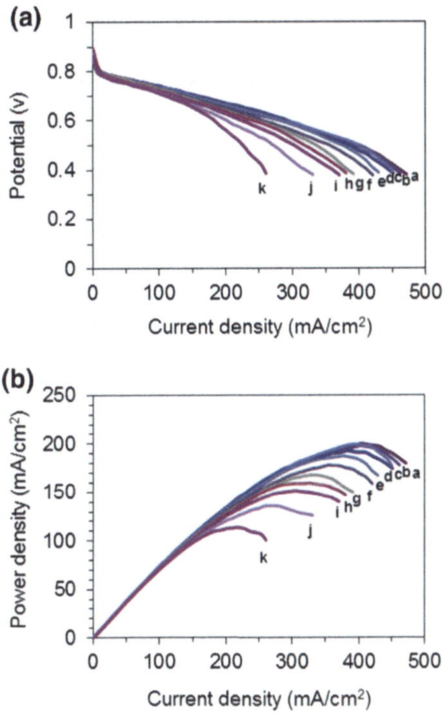

Fig. 1.20 Polarization curves (**A**) and power density curves (**B**) during durability tests of the MWNT/PyPBI/Pt/Nafion-based MEA. For simplicity, some representative curves are selected for display. The Arabic letters, a, b, c, d, e, f, g, h, I, j and k, represent the performance after 1,000, 50,000, 100,000, 150,000, 200,000, 250,000, 300,000, 350,000, 400,000, 450,000, and 500,000 potential cycles, respectively. Reproduction from [117] with permission of Nature Publishing Group

MEA was still active even after the 500,000 cycle-durability test (~5% loss of the initial voltage at 200 mA cm^{-2} was observed, and the maximum power density (initial value 375 mW cm^{-2}) of the MEA showed only a 20% decrease even after such high number of potential-dynamic cycles. The CB/PyPBI/Pt/Nafion-MEA is also durable. Such a high durability has not been recognized for the commercial CB/Pt-based MEA, which shows a 50% loss of the initial FC voltage at 200 mA cm^{-2} and 65% loss of the maximum power density after only 10,000 potential-dynamic cycles.

TEM studies indicated that for the MWNT/PyPBI/Pt/Nafion-based catalyst, CNT nanostructures were not very different from the original one, and the CB/PyPBI/Pt/Nafion-based catalyst also showed a similar behavior. Such behavior is due to the polymer wrapping. For the commercial CB/Pt-based catalyst, the TEM image after the durability test showed very few Pt-NPs with a large bare carbon support area, indicating a severe corrosion on the CB, leading to the loss of the Pt-NPs. Raman spectra of the MWNT/PyPBI/Pt/Nafion before and after the durability test indicated that a small change in the (I_D/I_G) ratio (from 0.62 before the durability test to ~0.95 after the durability test) which reflects the morphology change of the MWNTs.

In summary, the MEA fabricated using the MWNT/PyPBI/Pt/Nafion showed a high maximum power density of 375 mW/cm^2 and a remarkably high durability with only a 5% loss of the initial FC potential at 200 mA/cm^2 and 20% loss of the maximum power density after 500,000 potential cycles at 80 °C under fully humidified condition. The replacement of the CB by MWNT is important for the improvement of the durability of the MWNT-based MEA compared to the CB-based MEA due to their crystalline structure of the pristine MWNT compared to CB. In addition, use of a Nafion® ionomer as an outer layer of the MWNT/PyPBI/Pt was crucial in the reduction of the total ohmic resistance and the charge transfer resistance, leading to smoother electrochemical reactions, which resulted in such a high fuel cell performance and durability.

The above study provides a promising material design for the next-generation PEFC since such a high durability is strongly demanded from industry. The replacement of the conventional CB-based fuel cell catalyst that has been currently used in homes and automobiles with the presented catalyst would be an advantageous step.

References

1. Iijima S (1991) Helical microtubules of graphitic carbon. Nature 354:56–58
2. Nakashima N (2005) Soluble carbon nanotubes: fundamentals and applications. Int J Nanosci 4:119–137
3. Murakami H, Nakashima N (2006) Soluble carbon nanotubes and their applications. J Nanosci Nanotechnol 6:16–27
4. Nakashima N, Fujigaya T (2007) Fundamentals and applications of soluble carbon nanotubes. Chem Lett 36:692–697
5. Fujigaya T, Nakashima N (2012) Soluble carbon nanotubes and nanotube-polymer composites. J Nanosci Nanotechnol 12:1717–1738
6. Fujigaya T, Nakashima N (2015) Non-covalent polymer wrapping of carbon nanotubes and the role of wrapped polymers as functional dispersants. Sci Technol Adv Mater 16:Art No 024802
7. Dyke CA, Tour JM (2004) Covalent functionalization of single-walled carbon nanotubes for materials applications. J Phys Chem A 108:11151–11159
8. Tasis D, Tagmatarchis N, Bianco A, Prato M (2006) Chemistry of carbon nanotubes. Chem Rev 106:1105–1136
9. Sgobba V, Guldi DM (2009) Carbon nanotubes—electronic/electrochemical properties and application for nanoelectronics and photonics. Chem Soc Rev 38:165–184
10. Nakashima N, Tomonari Y, Murakami H (2002) Water-soluble single-walled carbon nanotubes via noncovalent sidewall-functionalization with a pyrene-carrying ammonium ion. Chem Lett: 638–639
11. Tomonari Y, Murakami H, Nakashima N (2006) Solubilizaton of single-walled carbon nanotubes by using polycyclic aromatic ammonium amphiphiles in water—strategy for the design of high-performance solubilizers. Chem Eur J 12:4027–4034
12. Murakami H, Nomura T, Nakashima N (2003) Noncovalent porphyrin-functionalized single-walled carbon nanotubes in solution and the formation of porphyrin-nanotube nanocomposites. Chem Phys Lett 378:481–485
13. Nakashima N, Okuzono S, Murakami H, Nakai T, Yoshikawa K (2003) DNA dissolves single-walled carbon nanotubes in water [Erratum to document cited in CA139:186348]. Chem Lett 32:782

14. Okamoto M, Fujigaya T, Nakashima N (2008) Individual dissolution of single-walled carbon nanotubes (SWNTs) using polybenzimidazole (PBI) and high effective reinforcement of SWNTs/PBI composite films. Adv Funct Mater 18:1776–1782
15. Hafez IH, Berber MR, Fujigaya T, Nakashima N (2014) Enhancement of platinum mass activity on the surface of polymer-wrapped carbon nanotube-based fuel cell electrocatalysts. Sci Rep 4:Article number: 6295
16. Li W, Liang C, Zhou W, Qiu J, Zhou Z, Sun G, Xin Q (2003) Preparation and characterization of multiwalled carbon nanotube-supported platinum for cathode catalysts of direct methanol fuel cells. J Phys Chem B 107:6292–6299
17. Tian ZQ, Jiang SP, Liang YM, Shen PK (2006) Synthesis and characterization of platinum catalysts on multiwalled carbon nanotubes by intermittent microwave irradiation for fuel cell applications. J Phys Chem B 110:5343–5350
18. Li W, Liang C, Qiu J, Zhou W, Han H, Wei Z, Sun G, Xin Q (2002) Carbon nanotubes as support for cathode catalyst of a direct methanol fuel cell. Carbon 40:791–794
19. Li L, Xing Y (2008) Electrochemical durability of carbon nanotubes at 80 °C. J Power Sources 178:75–79
20. Li L, Xing Y (2006) Electrochemical durability of carbon nanotubes in noncatalyzed and catalyzed oxidations. J Electrochem Soc 153:A1823–A1828
21. Kongkanand A, Kuwabata S, Girishkumar G, Kamat P (2006) Single-wall carbon nanotubes supported platinum nanoparticles with improved electrocatalytic activity for oxygen reduction reaction. Langmuir 22:2392–2396
22. Shao Y, Yin G, Gao Y, Shi P (2006) Durability study of Pt/C and Pt/CNTs catalysts under simulated PEM fuel cell conditions. J Electrochem Soc 153:A1093–A1097
23. Wang X, Li W, Chen Z, Waje M, Yan Y (2006) Durability investigation of carbon nanotube as catalyst support for proton exchange membrane fuel cell. J Power Sources 158:154–159
24. Ebbesen TW (1996) Decoration of carbon nanotubes. Adv Mater 8:155–157
25. Xu H, Zeng L, Xing S, Shi G, Xian Y, Jin L (2008) Microwave-radiated synthesis of gold nanoparticles/carbon nanotubes composites and its application to voltammetric detection of trace mercury(II). Electrochem Commun 10:1839–1843
26. de Paula CC, Garcia Ramos A, da Silva AC, Cocchieri Botelho E, Rezende MC (2002) Fabrication of glassy carbon spools for utilization in fiber optic gyroscopes. Carbon 40:787–788
27. Guo D-J, Li H-L (2005) High dispersion and electrocatalytic properties of palladium nanoparticles on single-walled carbon nanotubes. J Colloid Interface Sci 286:274–279
28. Chao G, Wenwen L, Yi Zheng J, Hao K (2006) Facile and large-scale synthesis and characterization of carbon nanotube/silver nanocrystal nanohybrids. Nanotechnology 17:2882
29. Wu G, Chen YS, Xu BQ (2005) Remarkable support effect of SWNTs in Pt catalyst for methanol electrooxidation. Electrochem Commun 7:1237–1243
30. Che G, Lakshmi BB, Martin CR, Fisher ER (1999) Metal-nanocluster-filled carbon nanotubes: catalytic properties and possible applications in electrochemical energy storage and production. Langmuir 15:750–758
31. Zhang S, Shao Y, Yin G, Lin Y (2010) Carbon nanotubes decorated with Pt nanoparticles via electrostatic self-assembly: a highly active oxygen reduction electrocatalyst. J Mater Chem 20:2826–2830
32. Grzelczak M, Correa-Duarte MA, Salgueiriño-Maceira V, Rodríguez-González B, Rivas J, Liz-Marzán LM (2007) Pt-catalyzed formation of Ni nanoshells on carbon nanotubes. Angew Chem Int Ed 46:7026–7030
33. Wu B, Zhang Y, Kuang Y, Yu Y, Zhang X, Chen J (2012) Chitosan-functionalized carbon nanotubes as support for the high dispersion of PtRu nanoparticles and their electrocatalytic oxidation of methanol. Chem Asian J 7:190–195
34. Hsu C-H, Liao H-Y, Kuo P-L (2010) Aniline as a dispersant and stabilizer for the preparation of Pt nanoparticles deposited on carbon nanotubes. J Phys Chem C 114:7933–7939

35. He D, Zeng C, Xu C, Cheng N, Li H, Mu S, Pan M (2011) Polyaniline-functionalized carbon nanotube supported platinum catalysts. Langmuir 27:5582–5588
36. Zhao Y, Yang X, Tian J, Wang F, Zhan L (2010) A facile and novel approach toward synthetic polypyrrole oligomers functionalization of multi-walled carbon nanotubes as PtRu catalyst support for methanol electro-oxidation. J Power Sources 195:4634–4640
37. Wang S, Jiang SP, White TJ, Guo J, Wang X (2009) Electrocatalytic activity and interconnectivity of Pt nanoparticles on multiwalled carbon nanotubes for fuel cells. J Phys Chem C 113:18935–18945
38. Zhang S, Shao Y, Yin G, Lin Y (2011) Self-assembly of Pt nanoparticles on highly graphitized carbon nanotubes as an excellent oxygen-reduction catalyst. Appl Catal B 102:372–377
39. Selvaraj V, Alagar M, Kumar KS (2007) Synthesis and characterization of metal nanoparticles-decorated PPY-CNT composite and their electrocatalytic oxidation of formic acid and formaldehyde for fuel cell applications. Appl Catal B 75:129–138
40. Oh H-S, Kim K, Kim H (2011) Polypyrrole-modified hydrophobic carbon nanotubes as promising electrocatalyst supports in polymer electrolyte membrane fuel cells. Int J Hydrogen Energy 36:11564–11571
41. Fujigaya T, Nakashima N (2013) Fuel cell electrocatalyst using polybenzimidazole-modified carbon nanotubes as support materials. Adv Mater 25:1666–1681
42. Steele BCH, Heinzel A (2001) Materials for fuel-cell technologies. Nature 414:345–352
43. Borup R, Meyers J, Pivovar B, Kim YS, Mukundan R, Garland N, Myers D, Wilson M, Garzon F, Wood D, Zelenay P, More K, Stroh K, Zawodzinski T, Boncella J, McGrath JE, Inaba M, Miyatake K, Hori M, Ota K, Ogumi Z, Miyata S, Nishikata A, Siroma Z, Uchimoto Y, Yasuda K, Kimijima KI, Iwashita N (2007) Scientific aspects of polymer electrolyte fuel cell durability and degradation. Chem Rev 107:3904–3951
44. Asensio JA, Sanchez EM, Gomez-Romero P (2010) Proton-conducting membranes based on benzimidazole polymers for high-temperature PEM fuel cells. A chemical quest. Chem Soc Rev 39:3210–3239
45. Zhang J, Xie Z, Zhang J, Tang Y, Song C, Navessin T, Shi Z, Song D, Wang H, Wilkinson DP, Liu Z-S, Holdcroft S (2006) High temperature PEM fuel cells. J Power Sources 160:872–891
46. Parthasarathy A, Srinivasan S, Appleby AJ, Martin CR (1992) Temperature dependence of the electrode kinetics of oxygen reduction at the Platinum/Nafion® interface—a microelectrode investigation. J Electrochem Soc 139:2530–2537
47. Yang C, Costamagna P, Srinivasan S, Benziger J, Bocarsly AB (2001) Approaches and technical challenges to high temperature operation of proton exchange membrane fuel cells. J Power Sources 103:1–9
48. Li Q, He R, Gao J-A, Jensen JO, Bjerrum NJ (2003) The CO Poisoning Effect in PEMFCs Operational at Temperatures up to 200 °C. J Electrochem Soc 150:A1599–A1605
49. Li Q, He R, Jensen JO, Bjerrum NJ (2004) PBI-based polymer membranes for high temperature fuel cells - preparation, characterization and fuel cell demonstration. Fuel Cells 4:147–159
50. Li Q, Jensen JO, Savinell RF, Bjerrum NJ (2009) High temperature proton exchange membranes based on polybenzimidazoles for fuel cells. Prog Polym Sci 34:449–477
51. Xiao L, Zhang H, Jana T, Scanlon E, Chen R, Choe EW, Ramanathan LS, Yu S, Benicewicz BC (2005) Synthesis and characterization of pyridine-based polybenzimidazoles for high temperature polymer electrolyte membrane fuel cell applicationsx. Fuel Cells 5:287–295
52. Wang JT, Savinell RF, Wainright J, Litt M, Yu H (1996) A H2/O2 fuel cell using acid doped polybenzimidazole as polymer electrolyte. Electrochim Acta 41:193–197
53. Weber J, Kreuer K-D, Maier J, Thomas A (2008) Proton conductivity enhancement by nanostructural control of poly(benzimidazole)-phosphoric acid adducts. Adv Mater 20:2595–2598

54. Seland F, Berning T, Børresen B, Tunold R (2006) Improving the performance of high-temperature PEM fuel cells based on PBI electrolyte. J Power Sources 160:27–36
55. Pan C, Li Q, Jensen JO, He R, Cleemann LN, Nilsson MS, Bjerrum NJ, Zeng Q (2007) Preparation and operation of gas diffusion electrodes for high-temperature proton exchange membrane fuel cells. J Power Sources 172:278–286
56. Zhai Y, Zhang H, Liu G, Hu J, Yi B (2007) Degradation Study on MEA in H3PO4/PBI High-Temperature PEMFC Life Test. J Electrochem Soc 154:B72–B76
57. Kim H-J, An SJ, Kim J-Y, Jin KM, Cho SY, Eun YC, Yoon H-K, Park Y, Kweon H-J, Shin E-M (2004) Polybenzimidazoles for high temperature fuel cell applications. Macromol Rapid Commun 25:1410–1413
58. Schuster M, Rager T, Noda A, Kreuer KD, Maier J (2005) About the choice of the protogenic group in PEM separator materials for intermediate temperature, low humidity operation: a critical comparison of sulfonic acid, phosphonic acid and imidazole functionalized model compounds. Fuel Cells 5:355–365
59. Paddison SJ, Kreuer K-D, Maier J (2006) About the choice of the protogenic group in polymer electrolyte membranes: ab initio modelling of sulfonic acid, phosphonic acid, and imidazole functionalized alkanes. Phys Chem Chem Phys 8:4530–4542
60. Steininger H, Schuster M, Kreuer KD, Kaltbeitzel A, Bingöl B, Meyer WH, Schauff S, Brunklaus G, Maier J, Spiess HW (2007) Intermediate temperature proton conductors for PEM fuel cells based on phosphonic acid as protogenic group: a progress report. Phys Chem Chem Phys 9:1764–1773
61. Lee K, Zhang J, Wang H, Wilkinson DP (2006) Progress in the synthesis of carbon nanotube- and nanofiber-supported Pt electrocatalysts for PEM fuel cell catalysis. J Appl Electrochem 36:507–522
62. Okamoto M, Fujigaya T, Nakashima N (2009) Design of an assembly of poly(benzimidazole), carbon nanotubes, and Pt nanoparticles for a fuel-cell electrocatalyst with an ideal interfacial nanostructure. Small 5:735–740
63. Fujigaya T, Okamoto M, Nakashima N (2009) Design of an assembly of pyridine-containing polybenzimidazole, carbon nanotubes and Pt nanoparticles for a fuel cell electrocatalyst with a high electrochemically active surface area. Carbon 47:3227–3232
64. Reddy ALM, Ramaprabhu S (2007) Pt/SWNT-Pt/C nanocomposite electrocatalysts for proton-exchange membrane fuel cells. J Phys Chem C 111:16138–16146
65. Munakata H, Ishida T, Kanamura K (2007) Electrophoretic deposition for nanostructural design of catalyst layers on Nafion membrane. J Electrochem Soc 154:B1368–B1372
66. Wainright JS, Wang JT, Weng D, Savinell RF, Litt M (1995) Acid-doped polybenzimidazoles: a new polymer electrolyte. J Electrochem Soc 142:L121–L123
67. Jones DJ, Rozie`re J (2001) Recent advances in the functionalisation of polybenzimidazole and polyetherketone for fuel cell applications. J Membr Sci 185:41–58
68. Xing B, Savadogo O (1999) The effect of acid doping on the conductivity of polybenzimidazole (PBI). J New Mater Electrochem Syst 2:95–101
69. Bouchet R, Siebert E (1999) Proton conduction in acid doped polybenzimidazole. Solid State Ionics 118:287–299
70. Asensio JA, Borros S, Gomez-Romero P (2002) Proton-conducting polymers based on benzimidazoles and sulfonated benzimidazoles. J Polym Sci Part A: Polym Chem 40:3703–3710
71. Glipa X, Bonnet B, Mula B, Jones DJ, Rozière J (1999) Investigation of the conduction properties of phosphoric and sulfuric acid doped polybenzimidazole. J Mater Chem 9:3045–3049
72. Li Q, He R, Jensen JO, Bjerrum NJ (2003) Approaches and recent development of polymer electrolyte membranes for fuel cells operating above 100 °C. Chem Mater 15:4896–4915
73. Matsumoto K, Fujigaya T, Sasaki K, Nakashima N (2011) Bottom-up design of carbon nanotube-based electrocatalysts and their application in high temperature operating polymer electrolyte fuel cells. J Mater Chem 21:1187–1190

74. Oono Y, Sounai A, Hori M (2009) Influence of the phosphoric acid-doping level in a polybenzimidazole membrane on the cell performance of high-temperature proton exchange membrane fuel cells. J Power Sources 189:943–949
75. Berber MR, Fujigaya T, Sasaki K, Nakashima N (2013) Remarkably durable high temperature polymer electrolyte fuel cell based on poly(vinylphosphonic acid)-doped polybenzimidazole. Sci Rep 3:1764
76. Berber MR, Hafez IH, Fujigaya T, Nakashima N (2014) Durability analysis of polymer-coated pristine carbon nanotube-based fuel cell electrocatalysts under non-humidified conditions. J Mater Chem A 2:19053–19059
77. Ohma A, Shinohara K, Iiyama A, Yoshida T, Daimaru A (2011) Membrane and catalyst performance targets for automotive fuel cells by FCCJ membrane, catalyst, MEA WG. ECS Trans 41:775–784
78. Fujigaya T, Berber MR, Nakashima N (2016) Improved durability of electrocatalyst based on coating of carbon black with polybenzimidazole and their application in polymer electrolyte fuel cells. ACS Appl Mater Interfaces 8:14494–14502
79. Yang Z, Berber MR, Nakashima N (2015) Design of polymer-coated multi-walled carbon nanotube/carbon black-based fuel cell catalysts with high durability and performance under non-humidified condition. Electrochim Acta 170:1–8
80. Nasef MM, Fujigaya T, Abouzari-Lotf E, Nakashima N, Yang Z (2016) Enhancement of performance of pyridine modified polybenzimidazole fuel cell membranes using zirconium oxide nanoclusters and optimized phosphoric acid doping level. Int J Hydrogen Energy 41:6842–6854
81. Yang Z, Fujigaya T, Nakashima N (2016) NaOH-aided platinum nanoparticle size regulation on polybenzimidazole-wrapped carbon nanotubes for use as non-humidified polymer electrolyte fuel cell catalyst. ChemCatChem 8:268–275
82. Fujigaya T, Hirata S, Nakashima N (2014) A highly durable fuel cell electrocatalyst based on polybenzimidazole-coated stacked graphene. J Mater Chem A 2:3888–3893
83. Yuan XZ, Wang HJ, Sun JC, Zhang JJ (2007) AC impedance technique in PEM fuel cell diagnosis—a review. Int J Hydrogen Energy 32:4365–4380
84. Wu J, Yuan XZ, Martin JJ, Wang H, Zhang J, Shen J, Wu S, Merida W (2008) A review of PEM fuel cell durability: degradation mechanisms and mitigation strategies. J Power Sources 184:104–119
85. Uchida M, Fukuoka Y, Sugawara Y, Eda N, Ohta A (1996) Effects of microstructure of carbon support in the catalyst layer on the performance of polymer-electrolyte fuel cells. J Electrochem Soc 143:2245–2252
86. Horv Aacute, Th G, Eacute Za, Kawazoe K (1983) Method for the calculation of effective pore size distribution in molecular sieve carbon. J Chem Eng Jpn 16:470–475
87. de Boer JH, Lippens BC, Linsen BG, Broekhoff JCP, van den Heuvel A, Osinga TJ (1966) Thet-curve of multimolecular N_2-adsorption. J Colloid Interface Sci 21:405–414
88. Phillip WA, Mika Dorin R, Werner J, Hoek EMV, Wiesner U, Elimelech M (2011) Tuning structure and properties of graded triblock terpolymer-based mesoporous and hybrid films. Nano Lett 11:2892–2900
89. M-x Wang, Liu Q, H-f Sun, Ogbeifun N, Xu F, Stach EA, Xie J (2010) Investigation of carbon corrosion in polymer electrolyte fuel cells using steam etching. Mater Chem Phys 123:761–766
90. Debe MK (2012) Electrocatalyst approaches and challenges for automotive fuel cells. Nature 486:43–51
91. Sun X, Song P, Zhang Y, Liu C, Xu W, Xing W (2013) A class of high performance metal-free oxygen reduction electrocatalysts based on cheap carbon blacks. Sci Rep 3:2505
92. Jaouen F, Proietti E, Lefevre M, Chenitz R, Dodelet J-P, Wu G, Chung HT, Johnston CM, Zelenay P (2011) Recent advances in non-precious metal catalysis for oxygen-reduction reaction in polymer electrolyte fuel cells. Energy Environ Sci 4:114–130

93. Beard KD, Borrelli D, Cramer AM, Blom D, Van Zee JW, Monnier JR (2009) Preparation and structural analysis of carbon-supported Co core/Pt shell electrocatalysts using electroless deposition methods. ACS Nano 3:2841–2853
94. Sasaki K, Naohara H, Cai Y, Choi YM, Liu P, Vukmirovic MB, Wang JX, Adzic RR (2010) Core-protected platinum monolayer shell high-stability electrocatalysts for fuel-cell cathodes. Angew Chem Int Ed 49:8602–8607
95. Wang JX, Inada H, Wu L, Zhu Y, Choi Y, Liu P, Zhou W-P, Adzic RR (2009) Oxygen reduction on well-defined core-shell nanocatalysts: particle size, facet, and Pt shell thickness effects. J Am Chem Soc 131:17298–17302
96. Wang D, Xin HL, Wang H, Yu Y, Rus E, Muller DA, DiSalvo FJ, Abruña HD (2012) Facile synthesis of carbon-supported Pd–Co core-shell nanoparticles as oxygen reduction electrocatalysts and their enhanced activity and stability with monolayer Pt decoration. Chem Mater 24:2274–2281
97. Wang C, Daimon H, Sun SH (2009) Dumbbell-like Pt-Fe_3O_4 nanoparticles and their enhanced catalysis for oxygen reduction reaction. Nano Lett 9:1493–1496
98. Lim B, Jiang M, Camargo PH, Cho EC, Tao J, Lu X, Zhu Y, Xia Y (2009) Pd-Pt bimetallic nanodendrites with high activity for oxygen reduction. Science 324:1302–1305
99. Wu J, Zhang J, Peng Z, Yang S, Wagner FT, Yang H (2010) Truncated octahedral Pt3Ni oxygen reduction reaction electrocatalysts. J Am Chem Soc 132:4984–4985
100. Li Y, Qi W, Huang B, Ji W, Wang M (2013) Size- and composition-dependent structural stability of core-shell and alloy Pd–Pt and Au–Ag nanoparticles. J Phys Chem C 117:15394–15401
101. Gasteiger HA, Kocha SS, Sompalli B, Wagner FT (2005) Activity benchmarks and requirements for Pt, Pt-alloy, and non-Pt oxygen reduction catalysts for PEMFCs. Appl Catal B 56:9–35
102. Zhang J, Mo Y, Vukmirovic MB, Klie R, Sasaki K, Adzic RR (2004) Platinum monolayer electrocatalysts for O_2 reduction: Pt monolayer on Pd(111) and on carbon-supported Pd nanoparticles. J Phys Chem B 108:10955–10964
103. Oezaslan M, Hasché F, Strasser P (2013) Pt-based core-shell catalyst architectures for oxygen fuel cell electrodes. J Phys Chem Lett 4:3273–3291
104. Colon-Mercado HR, Popov BN (2006) Stability of platinum based alloy cathode catalysts in PEM fuel cells. J Power Sources 155:253–263
105. Stamenkovic VR, Mun BS, Mayrhofer KJJ, Ross PN, Markovic NM (2006) Effect of surface composition on electronic structure, stability, and electrocatalytic properties of Pt-transition metal alloys: Pt-skin versus Pt-skeleton surfaces. J Am Chem Soc 128:8813–8819
106. Stephens IEL, Bondarenko AS, Grønbjerg U, Rossmeisl J, Chorkendorff I (2012) Understanding the electrocatalysis of oxygen reduction on platinum and its alloys. Energy Environ Sci 5:6744–6762
107. Hwang SJ, Kim S-K, Lee J-G, Lee S-C, Jang JH, Kim P, Lim T-H, Sung Y-E, Yoo SJ (2012) Role of electronic perturbation in stability and activity of Pt-based alloy nanocatalysts for oxygen reduction. J Am Chem Soc 134:19508–19511
108. Fang B, Chaudhari NK, Kim M-S, Kim JH, Yu J-S (2009) Homogeneous deposition of platinum nanoparticles on carbon black for proton exchange membrane fuel cell. J Am Chem Soc 131:15330–15338
109. Siburian R, Kondo T, Nakamura J (2013) Size control to a sub-nanometer scale in platinum catalysts on graphene. J Phys Chem C 117:3635–3645
110. Watanabe M, Sei H, Stonehart P (1989) The influence of platinum crystallite size on the electroreduction of oxygen. J Electroanal Chem Interfacial Electrochem 261:375–387
111. Zhang S, Chen S (2013) Enhanced-electrocatalytic activity of Pt nanoparticles supported on nitrogen-doped carbon for the oxygen reduction reaction. J Power Sources 240:60–65
112. Imaoka T, Kitazawa H, Chun W-J, Omura S, Albrecht K, Yamamoto K (2013) Magic number Pt13 and misshapen Pt12 clusters: which one is the better catalyst? J Am Chem Soc 135:13089–13095

113. Yamamoto K, Imaoka T (2014) Precision synthesis of subnanoparticles using dendrimers as a superatom synthesizer. Acc Chem Res 47:1127–1136
114. Kinoshita K (1990) Particle size effects for oxygen reduction on highly dispersed platinum in acid electrolytes. J Electrochem Soc 137:845–848
115. Nesselberger M, Ashton S, Meier JC, Katsounaros I, Mayrhofer KJJ, Arenz M (2011) The particle size effect on the oxygen reduction reaction activity of Pt catalysts: influence of electrolyte and relation to single crystal models. J Am Chem Soc 133:17428–17433
116. Mayrhofer KJJ, Blizanac BB, Arenz M, Stamenkovic VR, Ross PN, Markovic NM (2005) The impact of geometric and surface electronic properties of Pt-catalysts on the particle size effect in electrocatalysis. J Phys Chem B 109:14433–14440
117. Berber MR, Hafez IH, Fujigaya T, Nakashima N (2015) A highly durable fuel cell electrocatalyst based on double-polymer-coated carbon nanotubes. Sci Rep 5: art no 16711

Chapter 2
Carbon Nanotube-Based Direct Methanol Fuel Cell Catalysts

Naotoshi Nakashima

2.1 Introduction

Development of high efficient, clean, low lost and renewable energy sources is a strong social demand because we are faced with rapidly growing energy consumption and massive combustion of fossil fuels. Direct methanol fuel cells (DMFCs) (Fig. 2.1) have been recognized as an ideal power source for mobile applications due to the widely available fuel (methanol), high energy density (5500 W h kg^{-1}), and easy storage and transportation compared to hydrogen. Thus, they have received considerable attention. While, DMFC anode electrocatalysts still suffer from the following problems which hinder the DMFC for wide-spread commercialization; namely, (i) the carbon monoxide (CO) poisoning and low durability caused by the incomplete oxidation of the methanol ($Pt(CH_3OH)_{ads} \rightarrow Pt(CO)_{ads} + 4H^+ + 4e^-$), (ii) serious carbon corrosion ($C + 2H_2O \rightarrow CO_2 + 4H^+ + 4e^-$, 0.207 V vs. RHE) and (iii) the sluggish kinetics of the methanol oxidation reaction (MOR) of the DMFC anode electrocatalysts compared to the oxidation of the hydrogen, which degrade the fuel cell performance during the operation.

So far, several review articles on this topic have been published [1–6], in which the authors focused on catalyst preparation methods, catalyst carbon supports, catalyst nanostructures, methanol crossover, methanol concentration, stability/durability, catalyst degradation mechanism, proton conductivity, electrolyte membrane, power density, water management, Pt particle size, CO oxidation, methanol oxidation reduction reaction (MOR), CO tolerance, etc.

Catalyst design is the most important. Xing et al. summarized advances in catalysts for DMFC [3]. Although Pt-based catalysts are most commonly used, Pt is very expensive metal, so the reduction of Pt amount has been a strong demand from

N. Nakashima (✉)
International Institute for Carbon-Neutral Energy Research (WPI-I2CNER), Kyushu University, Motooka, Nishi-ku, Fukuoka 819-0395, Japan
e-mail: nakashima.naotoshi.614@m.kyushu-u.ac.jp

N. Nakashima (ed.), *Nanocarbons for Energy Conversion: Supramolecular Approaches*, Nanostructure Science and Technology,
https://doi.org/10.1007/978-3-319-92917-0_2

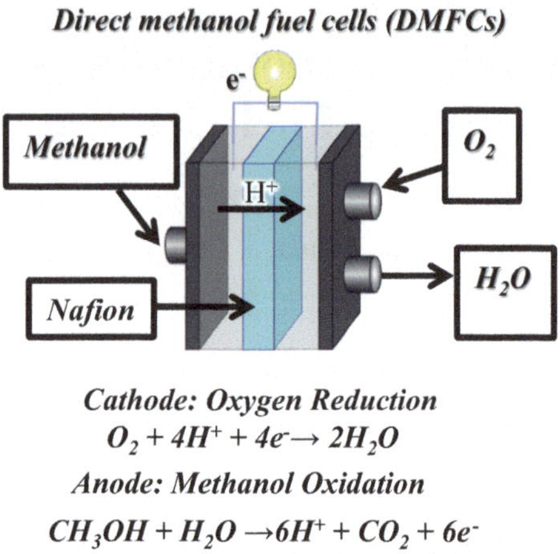

Fig. 2.1 Schematic drawing of direct methanol fuel cells (DMFC)

industry. For this line, metal alloys including Ru-decorated Pt, Pt-decorated Ru, Pt-Ni, Pt-Co, Pt-Mo, and Pt-Sn, and Pt-based core-shell catalysts have been intensively investigated. The developments of non-precious metal catalysts are also highly important in this DMFC study. For such a catalyst design, Chaps. 11–13 in this book should provide a hint.

Most widely used carbon-based catalyst in DMFC is carbon black (CB). Besides CB, mesoporous carbon, carbon fibers, carbon nanotubes, graphene, carbon aerogel and carbon paper have been used. Wang et al. summarized recent progress on this topic [5]. Very recently, Thotiyl et al. [7] reported a shingle chamber DMFC encompassing a Pt-free cathode half-cell with a freely diffusing electron acceptor delivering ~6 times higher power and current densities than conventional Pt-based DMFC, in which they used sodium persulfate as a reducing agent, and obtained a power density of 60 mW/cm^2.

Methanol crossover is one main issue that affects the efficiency and power density of DMFCs because methanol can easily go through a Nafion electrolyte membrane which is often used electrolyte membrane, and be oxidized at the cathode, poisoning the electrocatalyst, resulting in the decrease of the FC voltage and power density, especially when a high concentration of methanol (8–15 M) is fed to the anode side [8]. Recently, many studies have focused on the design and fabrication of a methanol tolerant electrocatalyst. Along this line, transition metals, such as Pd/Ag [9, 10], Ru/Se [11, 12], Pd/Co [13, 14], Ru/Mo/Se [15, 16] etc., were tested. Such electrocatalysts showed a high methanol tolerance due to the suppression of MOR. These transition metal electrocatalysts, however, are limited to practical DMFC applications because these transition metals show very low oxygen reduction reaction (ORR) activities compared to that of platinum (Pt), which is the most effective metal for ORR [17, 18]. Accordingly, Pt alloys with transition metals, such as Pt/Au [19, 20], Pt/Pd/Cu [21, 22], Pt/Cr [23, 24], Pt/Ni [25, 26], Pt/Co [27, 28] etc., were used, while the durability of such electrocatalysts was not high due to dissolution of the metals during real operating conditions [29–32]. Yang et al. [33] reported a selective electrocatalyst-based DMFC operated at high concentrations (10–15 M) of methanol, in which they used Au@Ag2S@Pt nanocomposites at the anode and core-shell Au@Pd nanoparticles at the cathode, resulting in the power density of 89.7 mW/cm^2.

The development of a durable and methanol tolerant electrocatalyst with a high ORR activity is still a significant and important challenge. In this chapter, we focus on the use of carbon nanotubes and carbon black (Vulcan and Ketjen Black) as well as mesoporous (nanoporous) carbon for comparison as a carbon support for DMFC [34–38]. Here, design of a new electrocatalyst with a high durability as well as high methanol and CO tolerance is highly important.

2.2 Polymer-Coated Carbon Nanotube-Based Direct Methanol Fuel Cells

Preparation of catalysts of polymer-coated nanocarbons is simple. As a carbon support, MWNTs, CB (Vulcan or Ketjen Black) were used. The method is similar to that of the PEFC catalysts as described in Chap. 1 (see Fig. 1.13), in which a nanocarbon is wrapped with PyPBI to which Pt nanoparticles were deposited on which PVPA is further coated to provide electrocatalysts (MWNTs (or NanoPC or CB)/PyPBI/Pt/PVPA). For comparison, catalysts (nanocarbon/PyPBI/Pt) without PVPA coating are used (see Sect. 2.2). Figure 2.2 shows the ORR measured in the presence of a given concentration of methanol, in which in the absence of methanol, the mass activities of the CB/Pt (commercial available), MWNT/PyPBI/Pt, and MWNT/PyPBI/Pt/PVPA calculated using the Levich-Koutecky equation, $1/i = 1/i_k + 1/i_d$ (where i is the experimentally measured current, and i_d is the diffusion-limited current) are 48.8, 187.0, and 157.7 mA/mg$_{Pt}$ at 0.85 V versus RHE, respectively [39–42]. The PVPA-coated electrocatalyst (MWNT/PyPBI/Pt/PVPA) shows a decrease by ~15.6% in the ORR activity compared to that of the MWNT/PyPBI/Pt due to the polymer-coating of the catalysts. Meanwhile, the diffusion-limited current density of the PVPA-coated electrocatalyst (−5.2 mA/cm^2) was almost identical to that of the MWNT/PyPBI/Pt (−5.3 mA/cm^2), suggesting that the PVPA coating showed a negligible effect on the O_2 accessibility [43]. For the MWNT/PyPBI/Pt, a small methanol oxidation peak is observed at 0.9 V versus RHE in the presence of methanol in the electrolyte. On the contrary, no such peak is observed for MWNT/PyPBI/Pt/PVPA. This result suggests that the polymer coating suppressed methanol oxidation due to methanol absorption [38, 44]. While, clear methanol oxidation peak is observed for the conventional CB/Pt (see Fig. 2.2a) in the presence of methanol in the electrolyte solution. The methanol oxidation reaction (MOR) peak showed a positive shift and the peak current increased with the increased methanol concentration (Fig. 2.2a). With the increase in the methanol concentration, the $E_{1/2}$ of the three electrocatalysts decreases due to increased coverage of CO on the Pt nanoparticles (Pt-NPs) generated during the MOR. In the presence of a high concentration of methanol, the $E_{1/2}$ of the CB/Pt and MWNT/PyPBI/Pt sharply decreased by 150 mV and 200 mV, respectively, while for the MWNT/PyPBI/Pt/PVPA, the value decreased only by 80 mV. It is evident that the coating of the catalyst with PVPA plays an important role in the ORR activity.

Compared to the MWNT/PyPBI/Pt/PVPA (see red and green lines in Fig. 2.2c, 1.0 mA/cm^2), the ORR curves of the CB/Pt (see red and green lines in Fig. 2.2a) and MWNT/PyPBI/Pt (see red and green lines in Fig. 2.2b) show high negative

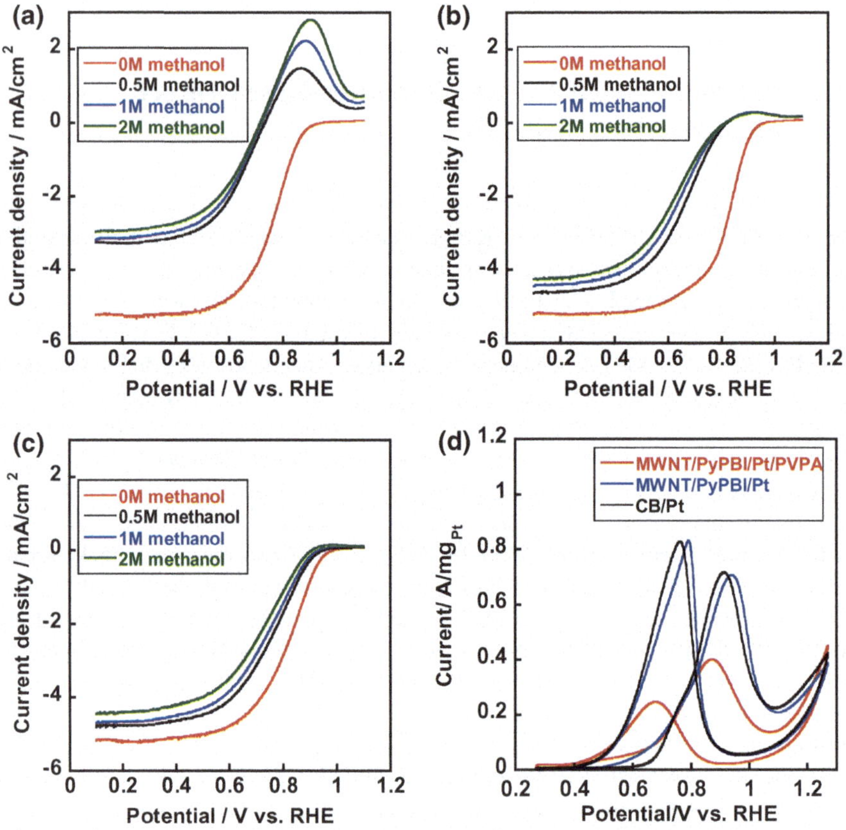

Fig. 2.2 Electrochemical measurements of the conventional CB/Pt, MWNT/PyPBI/Pt and MWNT/PyPBI/Pt/PVA. ORR polarization curves for the CB/Pt (**a**), MWNT/PyPBI/Pt (**b**) and MWNT/PyPBI/Pt/PVPA (**c**) in O_2-saturated 0.1 M $HClO_4$ and varying concentrations of methanol at 25 °C, rotation rate of 1600 rpm, and sweep rate of 10 mV/s. In all figures, methanol concentrations are: 0 M (red), 0.5 M (black), 1 M (blue) and 2 M (green). **d** Methanol oxidation reaction (MOR) curves were recorded in 0.1 M $HClO_4$ and 1 M methanol at a scan rate of 50 mV/s for the CB/Pt (black line), MWNT/PyPBI/Pt (blue line), and MWNT/PyPBI/Pt/PVPA (red line) before durability test. Reproduction from [37] with permission of Nature Publishing Group

shifts (3.0 and 1.9 mA/cm^2 at 0.85 V vs. RHE for the CB/Pt and MWNT/PyPBI/Pt, respectively) when 2 M methanol was added to the electrolyte, indicating that the CB/Pt and MWNT/PyPBI/Pt heavily suffered from methanol poisoning [45]. The diffusion-limited current density of the MWNT/PyPBI/Pt/PVPA shows a decrease only by 0.7 mA/cm^2 (see Fig. 2.2c) after adding 2 M methanol to the electrolyte, indicating a slight effect on O_2 accessibility due to the prevention of methanol absorption on the Pt-NPs by the PVPA coating, while the MWNT/PyPBI/Pt and CB/Pt showed a 1.0 and 2.3 mA/cm^2-loss (see Fig. 2.2a, b) in the diffusion-limited current densities, respectively. It is reported that during the ORR, the oxygen molecules were supplied by spherical diffusion to the individual Pt-NPs [46, 47]. Based on the forementioned results, after coating with PVPA, the methanol absorption would be partly blocked as shown in Fig. 2.2d, in which the mass current density of the MWNT/PyPBI/Pt/PVPA decreased by ~50% compared to that of the MWNT/PyPBI/Pt due to the coverage of the Pt-NP surfaces. Meanwhile, PVPA shows almost no effect on the O_2 diffusion due to the smaller size of the oxygen molecule as well as their easy diffusion in the membrane. Such event is schematically drawn in Fig. 2.3.

Although the H_2O_2 generation of the three forementioned electrocatalysts in the absence of methanol is very low due to the four electrons transferred during the ORR. However, under 2 M methanol concentration, the H_2O_2 generations of the CB/Pt and MWNT/PyPBI/Pt dramatically increased to 20.5% and 7.5%, respectively, which are much higher than that (0.8%) of the MWNT/PyPBI/Pt/PVPA. The lower H_2O_2 generation on MWNT/PyPBI/Pt/PVPA would be due to the PVPA polymer layer that prevents the H_2O_2 detaching from the electrocatalyst, which is very important for the reduction of H_2O_2 to H_2O on the Pt-NPs. Such higher H_2O_2 generations for the CB/Pt and MWNT/PyPBI/Pt should accelerate the corrosion of the carbon supporting materials, resulting in the decreased Pt-NPs stability and low performance.

Durability of the above three electrocatalysts were tested according to the protocol from the Fuel Cell Conference of Japan (FCCJ), in which the carbon corrosion ($C + 2H_2O \rightarrow CO_2 + 4H^+ + 4e^-$, 0.207 V vs. RHE) is accelerated. The loss in carbon supporting material leads to degradation of fuel cell performance due to the loss of the Pt-NPs. The ECSA values of the CB/Pt lost ~46% after 10,000 cycles

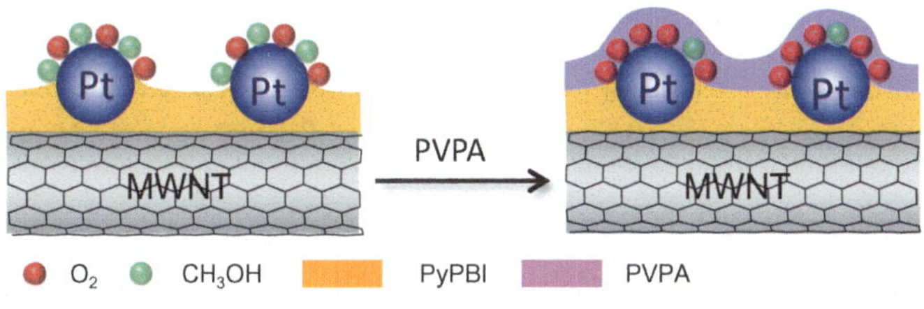

Fig. 2.3 Possible mechanism showing a low methanol tolerance of the MWNT/PyPBI/Pt (left) and high methanol tolerance (right) of the MWNT/PyPBI/Pt/PVPA. Reproduction from [37] with permission of Nature Publishing Group

(1.0–1.5 V vs. RHE) are due to lower carbon corrosion resistance. For MWNT/PyPBI/Pt, ECSA loss was only ~10% even after the cycling, meanwhile the MWNT/PyPBI/Pt/PVPA shows the highest durability among the three electrocatalysts (~6% loss in the ECSA). Such high resistance towards carbon corrosion is derived from the PVPA coating, which plays an important role in the reduction of the carbon corrosion and Pt agglomeration.

The membrane electrode assemblies (MEAs) were fabricated for practical application in DMFC system. Figure 2.4 shows the polarization curves of the MEAs measured under 100% relative humidity (RH) at 70 °C. The maximum power density of the MEA fabricated from MWNT/PyPBI/Pt/PVPA was 187 mW/cm^2, which was ~2.3 times higher than that of the commercial CB/Pt-based MEA (81 mW/cm^2). The MEA fabricated from MWNT/PyPBI/Pt was 165 mW/cm^2, which was somewhat lower than that of the MWNT/PyPBI/Pt/PVPA, suggesting that homogeneous PVPA coating might be important for a better proton conduction during the fuel cell measurement.

High CO-tolerance and durability of anodic fuel cell electrocatalysts is one of the important factors for commercialization of DMFCs. Another type of double-polymer-coated CNT-based catalyst was tested; namely, fabricated catalyst is MWNT/PyPBI/PVPA/Pt, in which MWNTs are coated with PyPBI and then with PVPA on which Pt-NPs are deposited (Fig. 2.5). In this catalyst, Pt-NPs positioned at outermost coat, which is differ from MWNT/PyPBI/Pt/PVPA, where PVPA is outermost coat. Durability tested by cycled ECSA has revealed that MWNT/PyPBI/

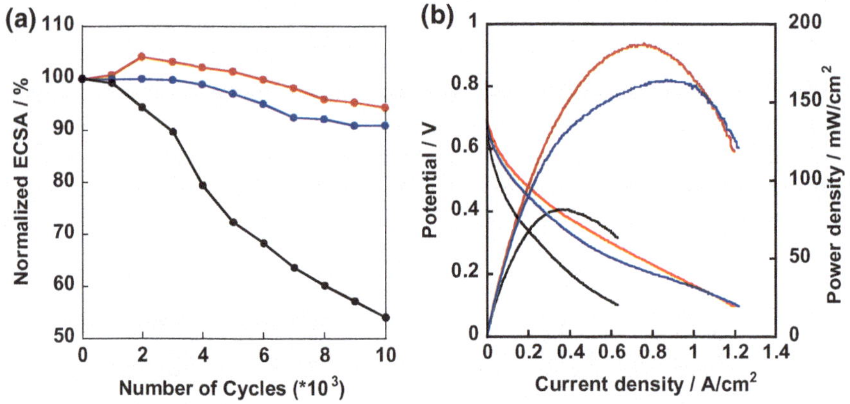

Fig. 2.4 Durability test and single cell performance of three electrocatalysts. **a** Plots of normalized ECSAs of the CB/Pt (black line), MWNT/PyPBI/Pt (Blue line) and MWNT/PyPBI/Pt/PVPA (red line) as a function of the numbers of potential cycles from 1.0 to 1.5 V versus RHE. **b** Polarization I-V and power density curves of MEAs fabricated from the CB/Pt (black line), MWNT/PyPBI/Pt (Blue line) and MWNT/PyPBI/Pt/PVPA (red line) under 70 °C with 8 M methanol (9 mL/min) and 100%RH humidified air (200 mL/min) for anode and cathode, respectively. Reproduction from [37] with permission of Nature Publishing Group

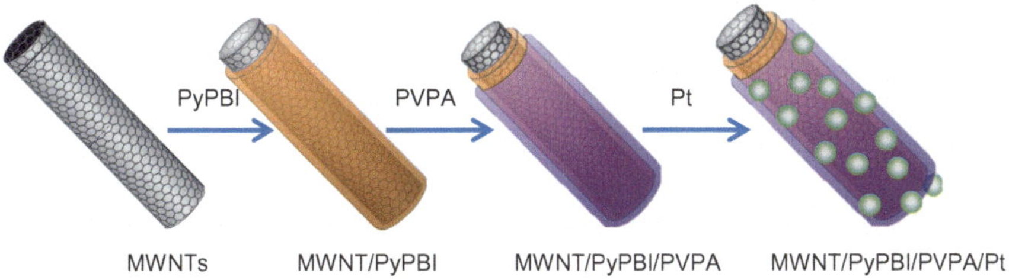

Fig. 2.5 Schematic illustration for the preparation of the MWNT/PyPBI/PVPA/Pt electrocatalyst. Reproduction from [35] with permission of Elsevier

PVPA/Pt and MWNT/PyPBI/Pt/PVPA shows a similar durability; namely, the ECSA values did not change even after 1,000 cyclings [35].

During DMFC operation, the methanol is oxidized at the anodic side, while incomplete methanol oxidation generates carbon monoxide (CO), which adsorbs onto the Pt surfaces, and poisons the Pt-NPs, resulting in deterioration of the fuel cell performance. Thus, the CO anti-poisoning is very important for the anodic electrocatalysts in DMFCs. For MOR measurements, we observe two peaks, i.e., anodic peak (I_f) and reverse anodic peak (I_b), in which I_f peak is the oxidation of the methanol to CO_2, CO, and HCOOH and I_b peak is the oxidation of the CO and HCOOH [48, 49]. MWNT/PyPBI/PVPA/Pt was found to be more active during the MOR compared to the CB/Pt and MWNT/PyPBI/Pt/PVPA. The CO anti-poisoning can be estimated from the ratio of I_f/I_b because the abundant by-product is CO. Here, higher I_f/I_b ratio indicates less generation of by-product, which denotes a higher CO anti-poisoning [50, 51]. The CO-tolerance of MWNT/PyPBI/PVPA/Pt was 1.5 times higher than that of the commercial CB/Pt before and after the durability test. Stable I_f/I_b ratio on MWNT/PyPBI/PVPA/Pt suggested that the PVPA was stable during the potential cycling.

The CO stripping voltammetry is also an important method to directly prove the CO-tolerance, in which CO was oxidized by the electrocatalyst. A sharp peak appeared at 0.9 V versus RHE in the CO stripping voltammetry is due to the oxidation of the absorbed CO species [35]. The peak of the MWNT/PyPBI/PVPA/Pt shows a negative shift compared to that of the commercial CB/Pt, which indicates that the CO species were more easily removed from the Pt-NP surfaces. Thus, MWNT/PyPBI/PVPA/Pt shows a higher CO tolerance than the commercial CB/Pt.

High CO tolerance was observed when using poly(vinylpyrrolidone) (PVP) instead of PVPA to provide MWNT/PyPBI/Pt/PVP (Fig. 2.6) [34]. PVP-coating is important for the formation of multipoint hydrogen bonding interactions between the NH groups on PBI (or PyPBI) and C=O groups on PVP [52]. In this catalyst (MWNT/PyPBI/Pt/PVP), ECSA loss was only ~8% even after 5000-potential cycling, while MWNT/PyPBI/Pt lost ~30% of the initial ECSA [49]. The I_f/I_b ratios for MOR measured in N_2-saturated 0.1 M $HClO_4$ in the presence of 4 M methanol before and after the durability test were 8.25 and 6.86, respectively, which were much higher than those of MWNT/PyPBI/Pt. Such result

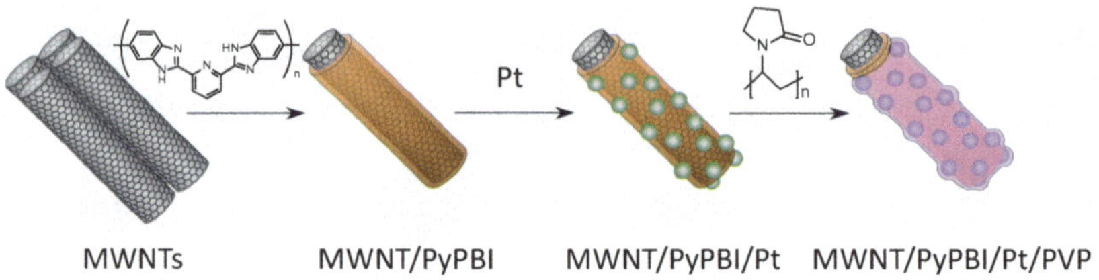

Fig. 2.6 Schematic illustration of the preparation of MWNT/PyPBI/Pt/PVP as well as chemical structures of PyPBI and PVP. Reproduction from [34] with permission of Wiley and Sons, Inc

clearly demonstrated the importance of the PVP-coating. The maximum power density of the MEA of MWNT/PyPBI/Pt/PVP was 166 mW/cm^2, which was comparable to that of the MWNT/PyPBI/Pt (165 mW/cm^2), meanwhile, 2 times higher than that (81 mW/cm^2) of the previously reported CB/Pt [37]. This result shows that the PVP layer of MWNT/PyPBI/Pt/PVP showed a negligible effect on the performance of DMFC.

In summary, double-polymer-coated electrocatalysts for DMFCs including MWNT/PyPBI/Pt/PVPA, MWNT/PyPBI/PVPA/Pt and MWNT/PyPBI/Pt/PVP are a promising candidate for future DMFC electrocatalysts with high performance and durability. Carbon nanotube-based polymer-coating method is simple and scalable, thus provides a new strategy for the design a practically applicable DMFCs.

2.3 Carbon Black-Based DMFC Catalysts–Comparison Study

Polymer-coating method for fabrication of DMFC electrocatalysts can be readily applicable to Carbon Black (CB). In this section, three different Carbon Black, Vulcan, Ketjen Black (KB) and nanoporous CB (NanoPC), which have different specific surface areas are compared [53]. Specific surface area of KB is higher than that of Vulcan, thus the amounts of PyPBI and PVPA on the carbon are higher, suggesting enhanced DMFC performance.

The specific surface areas calculated from the Brunauer–Emmett–Teller (BET) method for pristine KB and Vulcan are 1,232 and 235 m^2 g^{-1}, respectively. After wrapping with the PyPBI, the obtained KB/PyPBI (805 m^2 g^{-1}) still has ~4.5 times higher specific surface area than that of the Vulcan/PyPBI (176 m^2 g^{-1}). XPS study revealed that the intensities of the N_{1s} and P_{2p} peaks of KB/PyPBI/Pt/PVPA are higher than those of the Vulcan/PyPBI/Pt/PVPA, indicating greater amounts of PyPBI and PVPA in KB/PyPBI/Pt/PVPA. By analysis of the surface elemental ratios, the P and N in KB/PyPBI/Pt/PVPA are 2 and 3.5 times higher than those in the Vulcan/PyPBI/Pt/PVPA, respectively, which would be due to their larger specific surface area. Thermogravimetry analysis (TGA) measurements revealed that the PVPA amount in KB/PyPBI/Pt/PVPA was 2 times higher than that in Vulcan/

PyPBI/Pt/PVPA, which agreed with the XPS data. The obtained ESCA values of KB/PyPBI/Pt/PVPA and Vulcan/PyPBI/Pt/PVPA were 60.0 and 41.6 $m^2\ g^{-1}$, respectively.

Durability of these catalysts was different; namely, KB/PyPBI/Pt/PVPA lost only by 7% of the initial ECSA after 10,000 cycles, while the VC/PyPBI/Pt/PVPA showed 20%-loss. It is evident that KB/PyPBI/Pt/PVPA has a higher durability compared to Vulcan/PyPBI/Pt/PVPA. Pt-NPs directly supported on KB were reported to be more durable than those on the Vulcan because the Vulcan has a higher density of the surface defects, such as edges and corners of basal planes, leading to carbon corrosion [54]. The average particle size of Pt-NPs in KB/PyPBI/Pt/PVPA was still ~2.8 nm even after the durability test, which was still smaller than that in Vulcan/PyPBI/Pt/PVPA (~4.2 nm), which agreed with higher durability in KB/PyPBI/Pt/PVPA (Table 2.1).

Methanol concentration effect on the catalysts was examined [54]. MOR data measured before and after durability test by continuously adding the methanol up to 8 M to the electrolyte is shown in Fig. 2.7 [53], in which, as predicted, the current densities of both I_f and I_b increased with the increase in methanol concentrations, and before the durability test, the I_f/I_b ratios decreased by 19% and 31% with the increase in the methanol concentrations (from 0.5 to 8 M) for the KB/PyPBI/Pt/PVPA and Vulcan/PyPBI/Pt/PVPA, respectively. Such results indicate that the methanol concentration strongly affects the CO tolerance of the Vulcan/PyPBI/Pt/PVPA compared to that of the KB/PyPBI/Pt/PVPA. Under 8 M methanol concentration, which is close to the practical application of the DMFCs, the I_f/I_b ratio of the KB/PyPBI/Pt/PVPA was almost 2 times higher than those of the Vulcan/PyPBI/Pt/PVPA.

Such obtained higher CO tolerance would be derived from the higher PVPA amount in KB/PyPBI/Pt/PVPA, which facilitating the water adsorption on the Pt-NP surfaces to accelerate the formation of $Pt(OH)_{ads}$, which consumes the $Pt(CO)_{ads}$ and promoted the removal of the CO species from the catalyst. Such

Table 2.1 Comparisons of the KB/PyPBI/Pt/PVPA and Vulcan/PyPBI/Pt/PVPA regarding CO tolerance before and after durability test. Reproduction from [53] with permission of American Chemical Society

Electrocatalyst	0.5 M methanol	1 M methanol	2 M methanol	4 M methanol	8 M methanol
KB/PyPBI/Pt/PVPA (before durability)	2.45	2.11	2.08	2.02	2.00
KB/PyPBI/Pt/PVPA (after durability)	2.39	1.96	1.69	1.67	1.63
Vulcan/PyPBI/Pt/PVPA (before durability)	1.56	1.31	1.25	1.19	1.02
Vulcan/PyPBI/Pt/PVPA (after durability)	1.38	1.21	1.2	1.03	0.87

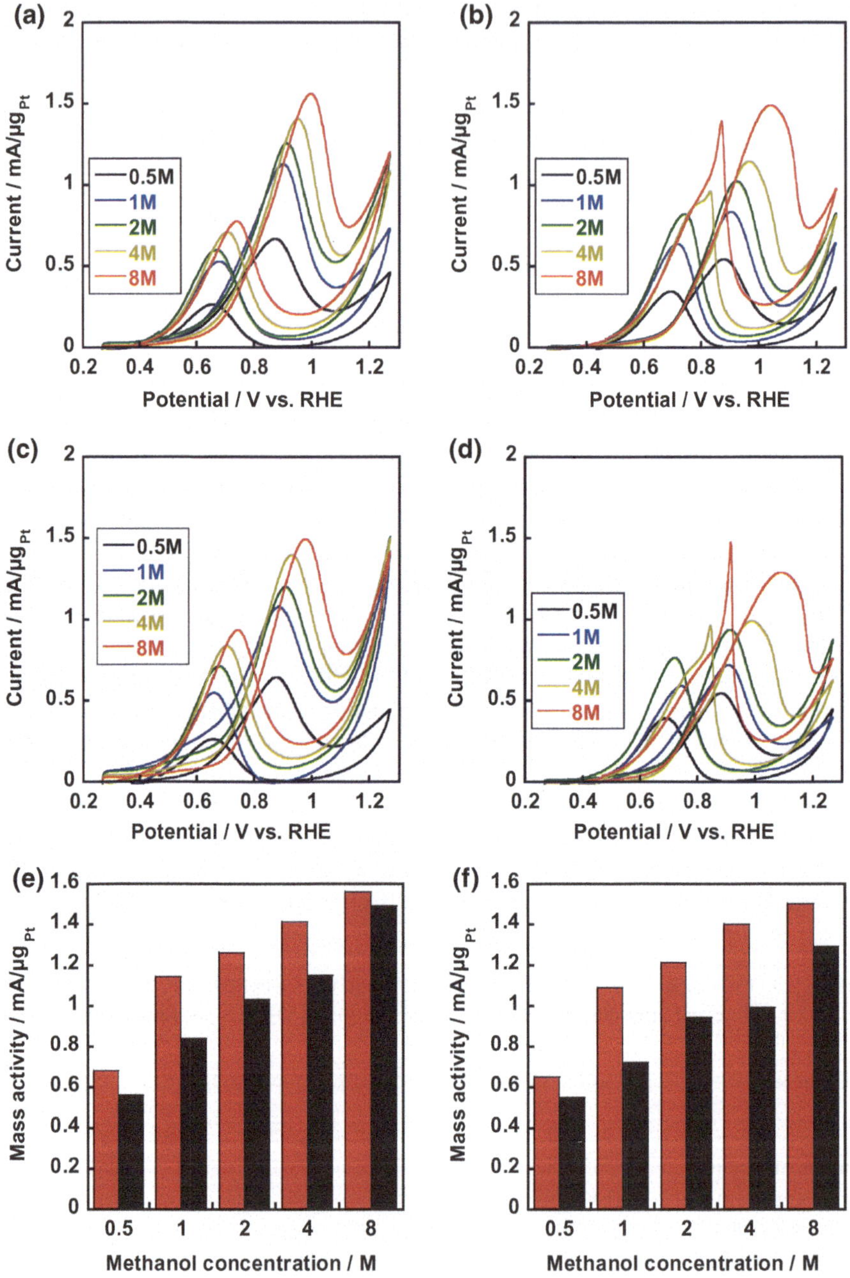

process was confirmed by the CO stripping voltammograms. As shown in Fig. 2.8, the CO oxidation peak of KB/PyPBI/Pt/PVPA exhibits a negative shift compared to that of the Vulcan/PyPBI/Pt/PVPA. After the durability test, the I_f/I_b ratios for KB/

◂**Fig. 2.7** MOR curves measured in an N_2-saturated 0.1 M $HClO_4$ and specified methanol concentration of the KB/PyPBI/Pt/PVPA (**a**, **c**) and Vulcan/PyPBI/Pt/PVPA (**b**, **d**) before and after durability test. Mass activities of the KB/PyPBI/Pt/PVPA (red column) and Vulcan/PyPBI/Pt/PVPA (black column) as a function of methanol concentration before (**e**) and after (**f**) durability test. Reproduction from [53] with permission of American Chemical Society

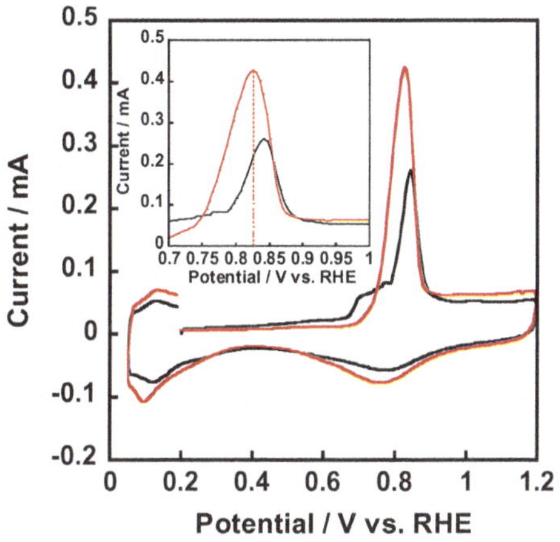

Fig. 2.8 First cycle of the CO stripping voltammograms of the Vulcan/PyPBI/Pt/PVPA (black line) and KB/PyPBI/Pt/PVPA (red line) at the scan rate of 50 mV s^{-1} at 25 °C. Zoom of CO stripping profile in 0.7–1.0 V versus RHE is shown as the inset. Reproduction from [53] with permission of American Chemical Society

PyPBI/Pt/PVPA and Vulcan/PyPBI/Pt/PVPA decreased by 32% and 37%, respectively, with the increase in the methanol concentration from 0.5 to 8 M, and after the durability test, the I_f/I_b ratios of the KB/PyPBI/Pt/PVPA were also higher than those of the Vulcan/PyPBI/Pt/PVPA, especially under 8 M methanol concentration, in which the value was still 2 times higher than that of Vulcan/PyPBI/Pt/PVPA. Such a high CO tolerance both before and after durability test indicated that KB/PyPBI/Pt/PVPA would be practically utilized in the DMFC anode side, in which high methanol concentration is fed. It is important to address that KB/PyPBI/Pt/PVPA showed almost no loss in the Pt mass activity even after durability test, while for Vulcan/PyPBI/Pt/PVPA, lost was ~10% of the initial mass activity.

The size of Pt-NPs is important for the performance of DMFC. Pt-size control of the CB(Vulcan)/PyPBI/Pt/PVPA was carried out by changing the weight ratios between the Pt feeding and polymer (PyPBI)-wrapped carbon support to give four catalysts (CB/PyPBI/Pt_1/PVPA, CB/PyPBI/$Pt_{0.5}$/PVPA, CB/PyPBI/$Pt_{0.2}$/PVPA, and CB/PyPBI/$Pt_{0.1}$/PVPA), in which the Pt-loading was 44.1 wt%, 27.5 wt%, 10.3 wt%, and 7.2 wt%, respectively [36]. All the electrocatalysts showed narrowed Pt-NPs distributions (~10% deviation; 3.2 ± 0.2, 3.0 ± 0.3, 2.6 ± 0.1, and 2.4 ± 0.2 nm for CB/PyPBI/Pt_1/PVPA, CB/PyPBI/$Pt_{0.5}$/PVPA, CB/PyPBI/$Pt_{0.2}$/PVPA, and CB/PyPBI/$Pt_{0.1}$/PVPA, respectively). Notably, CB/PyPBI/$Pt_{0.1}$/PVPA having the smallest particle size presented the highest ECSA (120.8 ± 12.0 m^2/g_{Pt}). The ECSA of other three electrocatalysts, CB/PyPBI/Pt_1/PVPA, CB/PyPBI/$Pt_{0.5}$/PVPA, and CB/PyPBI/$Pt_{0.2}$/PVPA were 48.0 ± 3.2, 56.8±2.8, and 95.3 ± 7.7 m^2/g_{Pt}, respectively. The high ECSA value was due to the high utilization efficiency of the Pt-NPs synthesized based on our preparation method by which almost all the surfaces of the Pt-NPs were exposed to the electrolyte. It is reported that high ECSA values for lower weight ratios was derived from the high utilization efficiency of the Pt-NPs [55].

Durability test based on the Fuel Cell Commercialization Conference of Japan (FCCJ) was carried out, and normalized ECSA values after every 10,000 potential cycles were 50%, 58%, 60%, and 62% of the initial ECSA values remained for the CB/PyPBI/Pt_1/PVPA, CB/PyPBI/$Pt_{0.5}$/PVPA, CB/PyPBI/$Pt_{0.2}$/PVPA, and CB/PyPBI/$Pt_{0.1}$/PVPA, respectively. Compared to carbon nanotube-based catalysts described in Sect. 2.1, CB-based durability was weak. In contrast, CB/PyPBI/$Pt_{0.1}$/PVPA electrocatalyst exhibited ~10 times higher durability compared to the commercially available CB/Pt [44]. The CO tolerance estimated from the I_f/I_b values of the four electrocatalysts measured in a methanol solution [56, 57] before and after the durability test were 1.31, 1.36, 1.38, and 2.69, and 0.95, 1.04, 1.22, and 1.84, respectively. Such data indicate that the CO anti-poisoning of the CB/PyPBI/$Pt_{0.1}$/PVPA is higher than those of the other three electrocatalysts.

All results indicate that the PVPA-coated electrocatalysts are attractive materials to fabricate a state-of-the-art DMFC electrocatalyst with high CO and methanol tolerance.

2.4 Concluding Remarks

In this chapter, we summarized the preparation of polymer-coated carbon nanotube (carbon black)-based catalysts for DMFCs. Similar to the H_2-based PEFC systems described in Chap. 1, polymer-coating method was powerful to enhance the durability of DMFCs. The polybenzimidazole-wrapped carbon nanotube (or carbon black)/Pt catalysts shown here exhibit high methanol tolerance, high CO-tolerance in methanol oxidation reaction, suggesting that the catalysts are important for an anode material for use in DMFCs. Such a polymer-coating method would be applicable to direct ethanol fuel cell systems (DEFCs). However, compared to DMFCs, DEFCs have many drawbacks including incomplete oxidation, slow kinetics, ethanol tolerance, low durability, etc., thus further studies in both operations in acidic and alkaline media are suggested to conquer such drawbacks.

References

1. Liu H, Song C, Zhang L, Zhang J, Wang H, Wilkinson DP (2006) A review of anode catalysis in the direct methanol fuel cell. J Power Sources 155:95–110
2. Neburchilov V, Martin J, Wang H, Zhang J (2007) A review of polymer electrolyte membranes for direct methanol fuel cells. J Power Sources 169:221–238
3. Zhao X, Yin M, Liang M, Liang L, Liu C, Liao J, Luc T, Xing W (2011) Recent advances in catalysts for direct methanol fuel cells. Energy Environ Sci 4:2736–2753

4. Li X, Faghri A (2013) Review and advances of direct methanol fuel cells (DMFCs) Part I: Design, fabrication, and testing with high concentration methanol solutions. J Power Sources 226:223–240
5. Huang H, Wang X (2014) Recent progress on carbon-based support materials for electrocatalysts of direct methanol fuel cells. J Mater Chem A 2:6626–6291
6. Mehmood A, Scibioh MA, Prabhuram J, An M-G, Ha HY (2015) A review on durability issues and restoration techniques in long-term operations of direct methanol fuel cells. J Power Sources 297:224–241
7. Thimmappa R, Aralekallu S, Devendrachari MC, Kottaichamy AR, Bhat ZM, Shafi SP, Lokesh KS, Thotiyl MO (2017) Adv Mater Interfaces 4:no 1700321
8. Franceschini EA, Bruno MM, Viva FA, Williams FJ, Jobbágy M, Corti HR (2012) Mesoporous Pt electrocatalyst for methanol tolerant cathodes of DMFC. Electrochim Acta 71:173–180
9. Liu M, Lu Y, Chen W (2013) PdAg nanorings supported on graphene nanosheets: highly methanol-tolerant cathode electrocatalyst for alkaline fuel cells. Adv Funct Mater 23:1289–1296
10. Sekol RC, Li X, Cohen P, Doubek G, Carmo M, Taylor AD (2013) Silver palladium core-shell electrocatalyst supported on MWNTs for ORR in alkaline media. Appl Catal B 138–139:285–293
11. Colmenares L, Jusys Z, Behm RJ (2007) Activity, selectivity, and methanol tolerance of Se-modified Ru/C cathode catalysts. J Phys Chem C 111:1273–1283
12. Choi JH, Johnston CM, Cao D, Babu PK, Zelenay P (2011) Se-modified Ru nanoparticles as ORR catalysts: Part 2: Evaluation for use as DMFC cathodes. J Electroanal Chem 662:267–273
13. Gharibi H, Golmohammadi F, Kheirmand M (2013) Fabrication of MEA based on optimum amount of Co in PdxCo/C alloy nanoparticles as a new cathode for oxygen reduction reaction in passive direct methanol fuel cells. Electrochim Acta 89:212–221
14. Li X, Huang Q, Zou Z, Xia B, Yang H (2008) Low temperature preparation of carbon-supported Pd-Co alloy electrocatalysts for methanol-tolerant oxygen reduction reaction. Electrochim Acta 53:6662–6667
15. Yu JS, Kim MS, Kim JH (2010) Combinatorial discovery of new methanol-tolerant non-noble metal cathode electrocatalysts for direct methanol fuel cells. Phys Chem Chem Phys 12:15274–15281
16. Schmidt TJ, Paulus UA, Gasteiger HA, Alonso-Vante N, Behm RJ (2000) Oxygen reduction on Ru1.92Mo0.08SeO4, Ru/carbon, and Pt/carbon in pure and methanol-containing electrolytes. J Electrochem Soc 147:2620–2624
17. Gao MR, Gao Q, Jiang J, Cui CH, Yao WT, Yu SH (2011) A methanol-tolerant Pt/$CoSe_2$ nanobelt cathode catalyst for direct methanol fuel cells. Angew Chem Int Ed 50:4905–4908
18. Selvarani G, Maheswari S, Sridhar P, Pitchumani S, Shukla AK (2009) Carbon-supported Pt-TiO_2 as a methanol-tolerant oxygen-reduction catalyst for DMFCs. J Electrochem Soc 156: B1354–B1360
19. Selvarani GS, Selvaganesh V, Krishnamurthy S, Kiruthika GVM, Sridhar P, Pitchumani S, A. Shukla AK (2009) A methanol-tolerant carbon-supported Pt-Au alloy cathode catalyst for direct methanol fuel cells and its evaluation by DFT. J Phys Chem C 113:7461–7468
20. Wang J, Yin G, Liu H, Li R, Flemming RL, Sun X (2009) Carbon nanotubes supported Pt–Au catalysts for methanol-tolerant oxygen reduction reaction: a comparison between Pt/Au and PtAu nanoparticles. J Power Sources 194:668–673
21. Cochell T, Li W, Manthiram A (2013) Effects of Pt coverage in Pt@PdCu5/C core-shell electrocatalysts on the oxygen reduction reaction and tolerance. J Phys Chem C 117:3865–3873
22. Nishanth KG, Sridhar P, Pitchumani S, Shukla AK (2011) A DMFC with methanol-tolerant-carbon-supported-Pt-Pd-alloy cathode. J Electrochem Soc 158:B871–B876

23. Yang H, Alonso-Vante N, Léger JM, Lamy C (2004) Tailoring, structure, and activity of carbon-supported nanosized Pt-Cr alloy electrocatalysts for oxygen reduction in pure and methanol-containing electrolytes. J Phys Chem B 108:1938–1947
24. Maillard F, Martin M, Gloaguen F, Léger JM (2002) Oxygen electroreduction on carbon-supported platinum catalysts. Particle-size effect on the tolerance to methanol competition. Electrochim Acta 47:3431–3440
25. Drillet JF, Ee A, Friedemann J, Kötz R, Schnyder B, Schmidt VM (2002) Oxygen reduction at Pt and Pt70Ni30 in H_2SO_4/CH_3OH solution. Electrochim Acta 47:1983–1988
26. Yang H, Coutanceau C, Léger JM, Alonso-Vante N, Lamy C (2005) Methanol tolerant oxygen reduction on carbon-supported Pt-Ni alloy nanoparticles. J Electroanal Chem 576:305–313
27. Lima FHB, Lizcano-Valbuena WH, Teixeira-Neto E, Nart FC, Gonzalez ER, Ticianelli EA (2006) Pt-Co/C nanoparticles as electrocatalysts for oxygen reduction in H_2SO_4 and H_2SO_4/CH_3OH electrolytes. Electrochim Acta 52:385–393
28. Salgado JRC, Antolini E, Gonzalez ER (2005) Carbon supported Pt–Co alloys as methanol-resistant oxygen-reduction electrocatalysts for direct methanol fuel cells. Appl Catal B 57:283–290
29. Stephens IEL, Bondarenko AS, Gronbjerg U, Rossmeisl J, Chorkendorff I (2012) Understanding the electrocatalysis of oxygen reduction on platinum and its alloys. Energy Environ Sci 5:6744–6762
30. Hwang SJ, Hwangn SJ, Kim S-K, Lee J-G, Lee S-C, Jang JH, Kim P, Lim T-H, Sung Y-E, Yoo SJ (2012) Role of electronic perturbation in stability and activity of Pt-based alloy nanocatalysts for oxygen reduction. J Am Chem Soc 134:19508–19511
31. Stamenkovic VR, Mun BS, Mayrhofer KJJ, Ross PN, Markovic NM (2006) Effect of surface composition on electronic structure, stability, and electrocatalytic properties of Pt-transition metal alloys: Pt-skin versus Pt-skeleton surfaces. J Am Chem Soc 128:8813–8819
32. Wang S, Cochell T, Manthiram A (2012) Boron-doped carbon nanotube-supported Pt nanoparticles with improved CO tolerance for methanol electro-oxidation. Phys Chem Chem Phys 14:13910–13913
33. Feng Y, Liu H, Yang J (2017) A selective electrocatalyst-based direct methanol fuel cell operated at high concentrations of methanol. Sci Adv 3:e1700580
34. Yang Z, Nakashima N (2016) An electrocatalyst based on carbon nanotubes coated with poly (vinylpyrrolidone) shows a high tolerance to carbon monoxide (CO) in a direct methanol fuel cell. ChemCatChem 8:600–606
35. Yang Z, Nagashima A, Fujigaya T, Nakashima N (2016) Electrocatalyst composed of platinum nanoparticles deposited on doubly polymer-coated carbon nanotubes shows a high CO-tolerance in methanol oxidation reaction. Internat J Hydrogen Energy 41:19182–19190
36. Yang Z, Hafez IH, Berber MR, Nakashima N (2015) An enhanced anode based on polymer-coated carbon black for use as a direct methanol fuel cell electrocatalyst. ChemCatChem 7:808–813
37. Yang Z, Nakashima N (2015) A simple preparation of very high methanol tolerant cathode electrocatalyst for direct methanol fuel cell based on polymer-coated carbon nanotube/ platinum. Sci Rep 5:no 12236
38. Yang Z, Berber MR, N. Nakashima N (2014) Polymer-coated carbon black-based fuel cell electrocatalyst with high CO-tolerance and durability in direct methanol oxidation. J Mater Chem A 2:18875–18880
39. Li Y, Zhu E, McLouth T, Chiu CY, Huang X, Huang Y (2012) Stabilization of high-performance oxygen reduction reaction Pt electrocatalyst supported on reduced graphene oxide/carbon black composite. J Am Chem Soc 134:12326–12329
40. Wang D, Wang D, Xin HL, Hovden R, Wang H, Yu Y, Muller DA, DiSalvo FJ, Abruña HD (2013) Structurally ordered intermetallic platinum-cobalt core-shell nanoparticles with enhanced activity and stability as oxygen reduction electrocatalysts. Nat Mater 12:81–87
41. He G, Song Y, Liu K, Walter A, Chen S, Chen S (2013) Oxygen reduction catalyzed by platinum nanoparticles supported on graphene quantum dots. ACS Catal 3:831–838

42. Guo S, Sun S (2012) FePt nanoparticles assembled on graphene as enhanced catalyst for oxygen reduction reaction. J Am Chem Soc 134:2492–2495
43. Matsumoto K, Fujigaya T, Yanagi H, Nakashima N (2011) Very high performance alkali anion-exchange membrane fuel cells. Adv Funct Mater 21:1089–1094
44. Fujigaya T, Okamoto M, Matsumoto K, Kaneko K, Nakashima N (2013) Interfacial engineering of platinum catalysts for fuel cells: methanol oxidation is dramatically improved by polymer coating on a platinum catalyst. ChemCatChem 5:1701–1704
45. Liu M, Lu Y, Chen W (2013) PdAg nanorings supported on graphene nanosheets: highly methanol-tolerant cathode electrocatalyst for alkaline fuel cells. Adv Funct Mater 23:1289–1296
46. Lee M, Uchida M, Okaya K, Uchida H, Watanabe M (2011) Durability of Pt/graphitized carbon catalyst prepared by the nanocapsule method for the start/stop operating condition of polymer electrolyte fuel cells. Electrochemistry 79:381–387
47. Yano H, Akiyama T, Bele P, Uchida H, Watanabe M (2010) Durability of Pt/graphitized carbon catalysts for the oxygen reduction reaction prepared by the nanocapsule method. Phys Chem Chem Phys 12:3806–3814
48. Gu Y-J, Wong W-T (2006) Nanostructure PtRu/MWNTs as anode catalysts prepared in a vacuum for direct methanol oxidation. Langmuir 22:11447–11452
49. Koenigsmann C, Wong SS (2013) Tailoring chemical composition to achieve enhanced methanol oxidation reaction and methanol-tolerant oxygen reduction reaction performance in palladium-based nanowire systems. ACS Catal 3:2031–2040
50. Yu LH, Kuo CH, Yeh CT (2007) Poly(vinylpyrrolidone)-modified graphite carbon nanofibers as promising supports for PtRu catalysts in direct methanol fuel cells. J Am Chem Soc 129:9999–10010
51. Ding L-X, Wang A-L, Li G-R, Liu Z-Q, Zhao W-X, Su C-Y et al (2012) Porous Pt-Ni-P composite nanotube arrays: highly electroactive and durable catalysts for methanol electrooxidation. J Am Chem Soc 134:5730–5733
52. Pu H, Liu Q, Qiao L, Z. Yang Z (2005) Studies on proton conductivity of acid doped polybenzimidazole/polyimide and polybenzimidazole/polyvinylpyrrolidone blends. Polym Eng Sci 45:1395–1400
53. Yang Z, Kim C, Hirata S, Fujigaya T, Nakashima N (2015) Facile enhancement in CO-tolerance of a polymer-coated Pt electrocatalyst supported on carbon black-comparison between Vulcan and Ketjenblack. ACS Appl Mater Interfaces 7:15885–15891
54. Speder J, Zana Spanos AI, Kirkensgaard JJK, Mortensen K, Hanzlik M, Arenz M (2014) Comparative degradation study of carbon supported proton exchange membrane fuel cell electrocatalysts—the influence of the platinum to carbon ratio on the degradation rate. J Power Sources 261:14–22
55. Hafez IH, Berber MR, Fujigaya T, Nakashima N (2014) Enhancement of platinum mass activity on the surface of polymer-wrapped carbon nanotube-based fuel cell. Electrocatalysts Sci Rep 4:no 6295
56. Sun X, Li D, Ding Y, Zhu W, Guo S, Wang ZL, Sun S (2014) Core/shell Au/CuPt nanoparticles and their dual electrocatalysis for both reduction and oxidation reaction. J Am Chem Soc 136:5745–5749
57. Guo S, Dong S, Wang E (2010) Three-dimensional Pt-on-Pd bimetallic nanodendrites supported on graphene nanosheet: facile synthesis and used as an advanced nanoelectrocatalyst for methanol oxidation. ACS Nano 4:547–555

Chapter 3
High-Temperature Polymer Electrolyte Membrane Fuel Cells

Jens Oluf Jensen, David Aili, Yang Hu, Lars N. Cleemann and Qingfeng Li

3.1 The Concept of Higher Working Temperature

Polymer electrolyte membrane fuel cells or proton exchange membrane fuel cells (both names abbreviated PEMFC) is the kind of fuel cells that have reached the largest degree of prevalence. Among the attractive features is the low working temperature from sub-zero to slightly below 100 °C. The low working temperature makes startup fast and allows for the use of a wide range of materials including fluorinated materials for tailoring surface properties and elastomers for sealing. Adding the general experience that chemical degradation processes become more severe at higher temperature, it may alltogether seem not to serve a purpose to try to increase the working temperature of a PEMFC. However, as shall be explained shortly, there are still a number of good reasons to aim for a higher working temperature in relation to certain applications and high-temperature PEMFC (HT-PEMFC) has been an active field of research and development for more than two decades.

In this context, *high temperature* refers to the temperature range from 100 to 200 °C and is meant relative to the working temperature of usual PEMFC, in this context labeled low-temperature PEMFC (LT-PEMFC). On a broader fuel cell perspective both cells belong to the low-temperature family of fuel cell (up to 200 °C) in contrast to the (real) high-temperature fuel cells, the molten carbonate fuel cell (MCFC) and the solid oxide fuel cell (SOFC), working at 650 °C and above.

In terms of materials and cell components, the HT-PEMFC is not much different from the LT-PEMFC. The main difference is the electrolyte membrane, which has to be a proton conductor at temperatures where the usual sulfonic acid-based membrane is not functioning well due to the drying effect.

J. O. Jensen (✉) · D. Aili · Y. Hu · L. N. Cleemann · Q. Li
Department of Energy Conversion and Storage, Technical University of Denmark, Building 375, 2800 Kgs. Lyngby, Denmark
e-mail: jojen@dtu.dk

N. Nakashima (ed.), *Nanocarbons for Energy Conversion: Supramolecular Approaches*, Nanostructure Science and Technology,
https://doi.org/10.1007/978-3-319-92917-0_3

Nanocarbons, which is the common theme of this book, are used very much the same way in HT-PEMFC as in LT-PEMFC and some degree of redundancy with other chapters is inevitable.

3.1.1 Why Complicating Things with a Higher Working Temperature?

Along with all the advantages, there are also a number of shortcomings of the traditional PEMFC, the LT-PEMFC. These are in brief that (1) the platinum catalyst is very sensitive to fuel impurities, (2) a strict humidity control (water management) is crucial, (3) cooling the cells can be challenging due to the limited temperature difference to that of the surroundings, (4) the utilization of the waste heat is limited due to the low temperature and finally, (5) the search for a more abundant and cost-efficient catalyst material than platinum or platinum group metals (PGM) has not yet proven successful. With a significantly higher working temperature, the four first shortcomings are much less pronounced and there is a chance that the elevated temperature can be the parameter that makes alternative catalyst materials active enough to replace the PGMs. Let us elaborate on the five points.

3.1.1.1 Poisoning of the Platinum Catalyst

The platinum catalyst is very sensitive to traces of carbon monoxide (CO) and sulfur-containing species. LT-PEMFC are normally operated on very pure hydrogen, but due to the poorly developed hydrogen infrastructure and the bulky storage of gaseous hydrogen, it is in many cases desirable to power the fuel cell by hydrogen produced locally from common carbonaceous fuels by reforming. Reforming is a chemical process where hydrogen is extracted from the carbonaceous fuel (normally hydrocarbons or alcohols) by a thermal process. Water, and in some cases air, is added to provide enough oxygen for producing carbon dioxide, the final oxidation product of the carbon atoms. Unfortunately, the chemical equilibria dictate that CO is formed along with carbon dioxide as a byproduct and the content can be from several hundred ppm to a few percents on a molar basis. Such a high content of CO is lethal for the platinum catalyst of the LT-PEMFC. CO adsorbs strongly to platinum rendering the catalytic sites unavailable to the hydrogen fuel. Already at concentrations of 20–50 ppm CO, the poisoning effect is severe. If a reformer system is to be used with an LT-PEMFC, an advanced CO removal unit must be inserted, but this adds dramatically to the system complexity, volume, weight, and cost. In large stationary hydrogen producing plant, this is not a big problem and even though by far the most hydrogen is produced by natural gas reforming, CO poisoning from commercial technical quality hydrogen is not an issue.

The CO adsorption on platinum is temperature dependent and at temperatures of 150–200 °C, it has been proven with HT-PEMFC that the CO tolerance is increased by about three orders of magnitude. At 160 °C, the most common operational temperature of HT-PEMFC, even one percent of CO is in most cases not a problem at all [1]. The tolerance to sulfur compounds, especially hydrogen sulfide, is also increased markedly as the temperature is raised, but this has less practical relevance since sulfur must be removed upstream of the reforming system to protect the reformer catalyst.

3.1.1.2 Humidity Control

An LT-PEMFC uses a perfluorosulfonic acid (PFSA) membrane as electrolyte. It requires a high water content to maintain a high proton conductivity, and a careful water management is mandatory. It is carried out by humidification of especially the cathode air that would otherwise dry out the membrane. A fuel cell with a PFSA membrane cannot easily be operated above 80–90 °C because of the difficulty keeping the membrane humid. Another related complication of the water management of the LT-PEMFC is that excessive water addition results in condensation and flooding of the electrodes and gas channels.

An HT-PEMFC will have to rely on a membrane conduction mechanism depending less on high humidity and water management will be less critical or even unnecessary.

3.1.1.3 Cooling of PEMFC

Cooling of PEMFC is important for the maintenance of the correct temperature. With 50–60% electrical efficiency (lower heating value), 40–50% of the fuel energy is released as heat. If part of the water formed is condensed in the cell, additional heat is formed. Cell cooling can be performed by air or a liquid, but in both cases, the cooling fluid will ultimately transfer the heat elsewhere, typically to the surroundings. The driving force of heat transfer is the temperature difference and with an LT-EPMFC and with the ambient air as surroundings, this temperature difference can be as low as 40 °C (fuel cell at 80 °C and surroundings at 40 °C). In case of a closed cooling system, this difference has moreover to be split between the two interfaces, fuel cell/coolant, and coolant/air. It was realized by the car manufacturers that a fuel cell vehicle may need a larger cooling radiator than a combustion engine vehicle does despite the higher efficiency or less heat release from the fuel cell vehicle. This is because the engine block temperature and thus the temperature of the cooling water are higher than that of a fuel cell. Besides, a large fraction of the heat from a combustion engine is emitted via the exhaust and will not be a burden to the cooler. A higher working temperature of the fuel cell would make for a more compact cooling system in fuel cell vehicles. Nevertheless, it appears that the cooling problem has found a solution in modern fuel cell vehicles without the use of HT-PEMFC.

3.1.1.4 Waste Heat Quality

When about half of the energy of the fuel converted in a fuel cell is released as heat, it is a noble wish to make this heat useful. The amount of heat produced at a given cell voltage is the same regardless of the working temperature, but the quality is not. Heat at about 70 °C can be used for domestic heating in a house or cabin heating in a vehicle, but apart from that, applications are very limited. A fuel cell working at 160 °C may deliver heat at say 140–150 °C and then it is possible to make steam and evaporate alcoholic fuels for a reformer [2, 3]. The ultimate quality of a heat flow can be expressed as the exergy content, i.e., the maximum work it can produce in the ideal case. The exergy, $\dot{E}x$, of a heat flow, $\dot{Q}$, is calculated from the temperature, T_h, and the temperature of the surroundings, T_s, (both are absolute temperatures) as

$$\dot{E}x = \dot{Q}\frac{T_h - T_s}{T_h} \tag{3.1}$$

For the heat flow at 70 °C, and with the temperature of the surrounding set to 25 °C, the exergy flow is 13% of the thermal energy, while for a 150 °C flow it is 30%. Note that this is purely a thermodynamic consideration and that in both cases the work extracted in real systems will be much less. Actually, both temperatures are so low, that it is questionable whether it is worthwhile to extract work.

3.1.1.5 The Search for Platinum-Free Catalysts

The strong dependence on platinum or PGM as the active catalyst material for both the hydrogen and the oxygen electrode has not really been a decisive cost driver yet. This is because other components have been expensive too, especially the bipolar plates and the PFSA membrane, but the cost of these is continuously reduced and with a larger production volume, the cost is expected to reduce further. The cost of the platinum catalyst has been reduced too due to better utilization and therefore lower loading used, but this will reach its limit and increased production volume will not reduce materials cost but instead increase it as a direct result of the growing demand. The annual production and the known resources of platinum are limited and not only the cost but also availability should be taken into account. The global annual production of platinum is around 200 tons [4]. Assuming (optimistically) a use of 0.2 g of platinum per $kW_{electrical}$ for the automotive fuel cell stacks, a total replacement of the annually produced 70 million new combustion engine vehicles, at average rated at 50 kW, would require an annual platinum supply of 700 tons. It is apparent that a very high recycling rate of platinum is needed to make such a technology sustainable. In brief, it is highly desirable to find a less expensive catalyst that can be manufactured from abundant elements.

What has that to do with the working temperature if both systems use platinum? Actually, the HT-PEMFC typically uses more platinum per electrode area and per unit power, so the temperature alone is not the key to a solution. The advantage of the higher temperature is still more speculative. There are two main criteria for new catalysts besides cost; they must be active and durable like platinum in an acidic environment. All new candidates have so far failed to reach either. It can be expected that the increased temperature will increase the electrocatalytic activity, anything else unchanged. A very rough rule of thumb states that for every 10 °C the temperature is increased, the reaction rate of a chemical reaction is doubled. Naturally, this rule cannot be used quantitatively without a risk of seriously misleading predictions, but it illustrates the power of thermal activation of processes. It is possible that the increased temperature can promote an alternative catalyst in the competition with platinum. The activity of platinum is of course also increased, but perhaps not to the same extent. Catalysts are often ranked in a volcano plot by means of Sabatier's principle. This principle says that the reactants shall adsorb on the catalytic site with a bond strong enough to secure a sufficiently large coverage and that the reactants remain bonded until the reaction has taken place. The products, on the other hand, must be bonded loosely enough to desorb fast and leave the site for new reactant molecules. If reactants and products are assumed comparable with respect to binding energy, an optimum bond strength must exist for an optimum balance between adsorption of reactants and desorption of products. This optimum bond strength will depend on temperature since desorption is thermally activated. A species that bonds slightly too strongly at one temperature may move closer to the optimum at a higher temperature. Experiments suggest, as an example, that tungsten carbide (WC) increase the hydrogen evolution activity faster with temperature than platinum [5].

Another effect on platinum in the HT-PEMFC, as we know it today, is related to the presence of phosphoric acid in the catalyst layer of the cell (we shall discuss the role of this acid later). Phosphoric acid adsorbs on the platinum catalyst and reduces its activity significantly. An alternative catalyst that is not inhibited by phosphoric acid will have a strong competing edge in phosphoric acid containing cells.

3.1.2 How to Increase the Working Temperature

The component that limits the working temperature of a PEMFC to below 100 °C is the proton conducting membrane. Proton conductivity is brought about by immobilized acids in the polymeric material. These acid moieties are sulfonic acids, which possess very high acid strength, i.e., they have a proton that is easily liberated. Mobile protons is naturally a prerequisite for proton conduction. The sulfonic acid groups are fixed to side chains of the polymer backbone, but even though they are close, the distance is still too long for direct proton transfer between the acid groups. Moreover, in order to receive a proton, an adjacent acidic moiety must let go of the proton that is already filling its basic site. For both these reasons, another proton acceptor must be present in the membrane, and this is water. Water absorbed in the membrane accepts the protons from the highly acidic side chains and forms domains with high concentrations of hydronium ions, H_3O^+, which are the real proton conducting species. The aqueous domains percolate into proton conducting pathways across the membrane while the polymer backbone forms a structural element. Conduction of the protons is a mix of vehicular transport where mobile hydronium ions carry the charge by migration in the electrical field within the membrane, and structural diffusion in which protons are transferred between hydronium ions and water molecules as a chain reaction facilitated by hydrogen bonding between the species (the Grotthuss mechanism).

Regardless of the detailed conduction mechanism, a high water content in the membrane is mandatory for high proton conductivity. The higher the temperature, the higher the volatility of the water in the membrane and even though the cathodic air is humidified to mitigate the drying effect (water management), there is a practical upper limit to the working temperature, which is 80–90 °C depending on the operating pressure and the membrane material used. An increase of the operating pressure can make room for a higher water partial pressure in the cathode air and provide more efficient humidification, but the effect is limited because the vapor pressure of the water increases exponentially with temperature. Significantly higher working temperatures thus require a conduction mechanism that is much less dependent on water.

Development of solid-state proton conductors based on mixed metal oxides [6] and solid phosphates [7] is ongoing, but the most successful idea so far has been to introduce a less volatile proton carrier, phosphoric acid, to a polymer material with high thermal stability. Phosphoric acid is used with success in phosphoric acid fuel cells (PAFC) for the same reason and it has the best combination of high protonic conductivity, thermal stability and low vapor pressure at temperatures up to 200 °C.

The polymer most often used for membranes with phosphoric acid is polybenzimidazole (PBI). Membranes of PBI are thermally and chemically stable for operation at 160 °C, and moreover they are capable of imbedding a large amount of phosphoric acid. The proton conductivity at 160 °C is comparable to that of common sulfonic acid-based membranes at 80 °C.

The membrane-acid system is referred to as an *ion solvating membrane* because the acid, water, and ions of both formed by acid–base interaction are dissolved in the membrane phase. The proton conductivity of the membrane–acid system is basically due to the phosphoric acid, but the prevailing understanding is that there is no local phase separation in the PBI-phosphoric acid system. This is in contrast to the sulfonic acid based membranes where hydrophilic and hydrophobic domains exist. The main differences and similarities between the three types of fuel cells discussed are listed in Table 3.1.

3.1.3 The Legacy of the Phosphoric Acid Fuel Cell

Since the proton conductivity of the PBI-phosphoric acid membrane is due to the phosphoric acid absorbed in the membrane, it is sometimes claimed that a fuel cells

Table 3.1 Typical key operational parameters and construction materials of mainstream fuel cells of the three related types

	LT-PEMFC	HT-PEMFC	PAFC
Operating temperature (°C)	60–80	160	200
Water management	Crucial	Little importance	Little importance
CO tolerance (%)	~0.002	~1	1–2
Electrolyte	Perfluorosulfonic acid membrane (humidified)	Ion solvating polymer membrane doped with phosphoric acid	Phosphoric acid in inorganic porous matrix
Anode catalyst (hydrogen)	Platinum on carbon	Platinum on carbon	Platinum on carbon
Cathode catalyst (oxygen)	Platinum on carbon	Platinum on carbon	Platinum on carbon
Electrode structure	Porous carbon	Porous carbon	Porous carbon
Ion conducting phase in electrodes	Ionomer, like the membrane	Phosphoric acid (optional ionomer)	Phosphoric acid (optional ionomer)
Bipolar plates	Carbon composite or coated steel	Carbon composite	Carbon composite
Coolant	Air or water	Air or glycol	Air or glycol

based on this membrane is not a PEMFC, but a PAFC. The most correct interpretation is probably that it is a hybrid between the two [8]. PEMFC in the sense *proton exchange membrane fuel cell* is not a proper match, since PBI, although it can be protonated and deprotonated, is not really a proton exchange material.

The electrolyte of a PAFC is a pure phase of phosphoric acid (with varying water content) contained in a porous inert matrix of silicon carbide. The phosphoric acid in the cell, we discuss here, is not pure phosphoric acid, but a gel-like blend or solution of the acid and the polymer even though the acid content can be as high as 80–95 wt %. The prevailing understanding is that there is no local phase separation like in the PFSA membrane with hydrophilic and hydrophobic domains. Under this view, it is probably more correct that the phosphoric acid-doped PBI membrane is another material, neither a proton exchange membrane nor a phosphoric acid electrolyte. PEMFC in the sense *polymer electrolyte membrane fuel cell* is a better fit, since the material is polymeric, containing what in another context would be seen as a plasticizer. Polymeric materials are still polymeric even if they contain plasticizers. From a practical or a manufacturing point of view, the electrolyte membrane is by all means a polymeric material, soft, flexible, gastight, and easily handled. A strong advantage is that it can be manufactured and handled free-standing in thicknesses of 20–50 μm, in contrast, the more brittle acid containing silicon carbide matrix of 100–200 μm of the PAFC, an electrolyte which, moreover, has to be prepared and supported by one of the electrodes.

Thus, the PBI-phosphoric acid-based fuel cell as a hybrid between the traditional PEMFC and the PAFC. In the following, the different materials and components used in it will be treated in more detail, and it will become apparent that the development of this fuel cell has drawn a great deal on the experience with PAFC.

3.2 Membranes for Elevated Temperature

The concept of combining an ion solvating polymer membrane with an acid for use in HT-PEMFC at temperatures above 100 °C depends on a compatible set of a polymer and an acid with the right properties.

3.2.1 The Acid

The acid must be thermally stable and besides high proton conductivity have an extremely low vapor pressure to minimize evaporation. Phosphoric acid is a very good choice. It has very high conductivity and the vapor pressure of the acid itself (not water) is so low at temperatures below 200 °C that it has only been estimated by extrapolation [9] from values at higher temperatures [10]. The estimated acid vapor pressure at 160 °C is 0.71 μbar [9]. Unfortunately, phosphoric acid brings along two serious drawbacks as well. Phosphate ions adsorb strongly to platinum [11] and this makes the catalytic sites on the platinum surface much less available to oxygen and the oxygen overpotential is much larger in HT-PEMFC than in LT-PEMFC for that reason. Additionally, the oxygen solubility in phosphoric acid is very low, and if the catalyst is covered by a phosphoric acid film, the penetration of oxygen is set back. Both effects are well-known from PAFC and this cell as well as in the HT-PEMFC the full benefit of the higher temperature, as compared with the LT-PEMFC with the same catalyst and a comparable electrolyte resistance is far from realized. Actually, on pure hydrogen, the LT-PEMFC has somewhat higher power density despite the low temperature.

It should be mentioned too that phosphoric acid has a complicated chemistry. Ortho-phosphoric acid (H_3PO_4) can condense to form pyro-phosphoric acid ($H_4P_2O_7$) by liberation of a water molecule. This happens under dry conditions at the working temperature of the HT-PEMFC and the process may continue to trimers ($H_5P_3O_{10}$ or $H_3P_3O_9$) and higher oligomers/polymers [12]. All the condensation processes are governed by equilibria in which the water activity is decisive. As long as the fuel cell produces a current, the water produced shifts in the equilibrium in favor of the ortho-phosphoric acid, but when left idling with a cathodic airflow through, the acid will eventually start condensing and consequently, the conductivity decreases due to the less mobile species. When the cell starts producing a current and consequently also water, the polymerization process is reversed and the acid is hydrolyzed. Humidification of the cathode airflow has the same effect. Other acids to consider for HT-PEMFC are inorganic solid acids such as phosphotungstic acid ($H_3PW_{12}O_{40}$) [13] and phosphosilicates [14] or protic organic salts [15]. Sulfuric acid is tempting, but it is much more volatile and at high concentration it is in equilibrium with water and sulfur trioxide, both of which are vapors at temperature of relevance. Dehydration will thus form a very volatile residue, which escapes easily and the process cannot be reversed by addition of water. Practically, phosphoric acid is exclusively used in HT-PEMFC.

3.2.2 The Polymer Membrane

Many polymeric materials can in principle be used in HT-PEMFC, but the prevailing choice has been heterocyclic mostly aromatic polymers, especially polybenzimidazoles (PBI). This combination was first reported in 1995 by Wainright et al. [16]. The group started by doping a PFSA membrane with phosphoric acid to impart high-temperature conductivity [17], but they were not very successful and later, the poor fuel cell performance with a phosphoric acid doped PFSA membrane was studied further [18]. When an acidic membrane is doped with more acid, water is needed for deprotonation and formation of ions that can support ion conductivity. Phosphoric acid can act as a proton acceptor, but a rather weak one and it is not a strong acid either. Consequently, its self-protolysis is limited and without water, ions are mostly formed by protonation by the strong sulfonic acid of the PFSA polymer, which is limited in number. When applied in a fuel cell, $H_4PO_4^+$ ions carry the current in the absence of water. Water that might be present in the first place will be dragged away from the anode and since water is formed at the cathode only, the anodic side of the electrolyte loses conductivity. The phosphoric acid-doped PFSA membrane has also been applied in electrolyzer cells where water vapor is constantly added, and then the proton conductivity is not lost [19].

The result with acid-doped PBI was a real breakthrough and the way was paved for a new branch of fuel cells, the HT-PEMFC.

3.2.3 Polybenzimidazoles

Polybenzimidazoles constitute a large family of polymers containing the benzimidazole groups as a part of the polymer repeat unit. Poly(2,2′-(*m*-phenylene)-5,5′-bibenzimidazole) (*m*-PBI, Fig. 3.1) is commercially available and has been extensively used for HT-PEMFC. In order to improve chemical stability, mechanical robustness, and ultimately fuel cell durability, a large number of polybenzimidazole structure derivatives have been developed over the years. The more common examples are *p*-PBI and AB-PBI as shown in Fig. 3.1. The only difference between *m*-PBI and *p*-PBI is that the benzene ring is *meta* and *para* coordinated. This looks like a minor difference, but it actually has a strong impact on solubility and stiffness of the polymer. *p*-PBI is less soluble in the solvents uses (DMAc, NMP and formic acid) and the material is stiffer. Numerous other moieties like ether links, sulfones, and fluorinated groups have been included in the polymer to modify the properties. Examples are reviewed in the literature [20, 21].

The un-protonated nitrogen atoms are basic and accept easily a proton each from phosphoric acid. This gives the polymer a high affinity to phosphoric acid and it can be doped to a very high level. The two first phosphoric acid molecules protonate the polymer backbone and when additional acid is introduced, the proton conductivity increases significantly. Acid doping levels (ADL), defined as the number of phosphoric acid molecules per polymer repeat unit, of 6–12 are common in post-doped membranes and by the direct casting technique, ADL as high as 30–40 is possible. Acid doping is explained in Sect. 3.2.4. When the imidazole unit is protonated, it forms a cation, which is balanced charge wise by the formed phosphate ion ($H_2PO_4^-$).

When PBI has proven so successful, it is partly because of the good match with phosphoric acid, and partly because it due to the aromatic structure is very stable, chemically as well as thermally. The common *m*-PBI has superior thermomechanical properties with a glass transition temperature of 420 °C and is therefore used for many high-temperature applications, like heat-resistant clothing for firemen and race drivers [22]. Chemical stability is naturally important too for a material to be applied with hot concentrated phosphoric acid.

PBI possesses both proton donor and proton acceptor sites (the protonated and the un-protonated nitrogen sites) and is, therefore, amphoteric with pK_a around 5 in the fully protonated form and pK_a around 13 in the pristine form [23, 24]. Consequently, PBI can also be doped with potassium hydroxide and form a hydroxide ion conductor. Then, the benzimidazole units are fully deprotonated instead, forming an anion balanced by the potassium ion. Development of alkaline cells on this basis is ongoing [25–27], but beyond the scope of the present treatise.

It is also possible to improve the mechanical stability of the membrane by crosslinking it with different bridges ending in halogens [28–30] or carbon–carbon double bonds [31]. In all these examples, the electrophilic crosslinking agents react with the PBI backbone and attach covalently to the nitrogen sites. The polymer becomes insoluble, swelling is reduced and acid doping may require higher temperature or longer time. Another way of crosslinking PBI is by thermal curing as discovered in the 60s [32, 33]. The technique was later used for fuel cell membranes

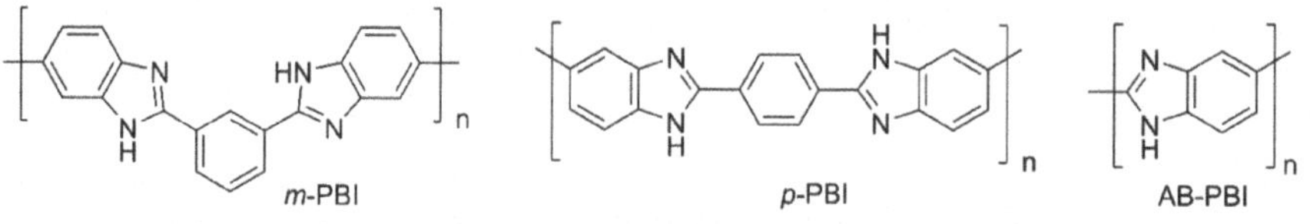

Fig. 3.1 Polymer repeat unit structures of *m*-PBI, *p*-PBI and AB-PBI

where the curing was carried out at 350 °C in inert atmosphere [34]. The exact chemical effect of the curing is not clear, but the cured polymer exhibits properties similar to the covalently crosslinked membranes, namely insolubility and improved mechanical strength. Thermally cured membranes are, therefore, referred to as thermally crosslinked membranes. In addition, crosslinked membranes including the cured membranes tend to appear less sensitive to the Fenton peroxide test. The Fenton test is considered a very tough test for the assessment of oxidative stability and PBI, like many other polymers, is severely affected when subjected. Crosslinked PBI loses mass much slower and remains in one piece for a longer period than its non-linked analog despite the same backbone structure [28–31, 34]. Moreover, thermal crosslinking seems to improve the acid retention. The lowest voltage decay rate over thousands of hours of operation at 200 mA cm^{-2} and 160 °C was 0.5 μV h^{-1} and this was reported recently for a cell with a thermally crosslinked membrane [35].

A related way to improve mechanical strength is ionic crosslinking. Polymer blends of PBI with acidic polymers like polysulfones undergo acid-base interactions forming ionic "polysalts", which improve the strength of blend membranes [20, 36, 37].

3.2.4 Acid Introduction

There are basically two ways to introduce phosphoric acid to the PBI membrane. In the *post-doping process*, the PBI membrane is immersed in concentrated phosphoric acid and left to equilibrate [38]. With 85% phosphoric acid the ADL is typically 10–11 in case of *m*-PBI. With lower acid concentrations, lower doping levels can be realized. Besides providing conductivity, the acid acts as a plasticizer and the desired ADL is a trade-off between conductivity and mechanical robustness. One of the benefits of crosslinking is that a higher ADL can be reached leading to higher conductivity without compromising mechanical strength. Doping can also be achieved at elevated temperatures, but eventually, the polymer dissolves in the acid if it is not crosslinked.

It is possible to obtain extremely high ADLs by the *direct casting process* or the PPA process [39]. The membranes are cast directly from the crude polyphosphoric acid solution used for the polymer synthesis. The polyphosphoric acid is slowly hydrolyzed to ortho-phosphoric acid and the polymer phases out. This is a sol-gel process and the gel structure obtained is crucial for the integrity of the membrane material. Doping levels of 30–40 mol phosphoric acid per polymer repeat unit have been achieved with *p*-PBI [40]. If a similar ADL was aimed for by post-doping the membrane would dissolve completely. An added value of high doping levels is that the membrane contains more acid and that it can afford to lose more acid during long-term operation. This is somehow counterbalanced by the fact that the acid is easier lost from the membrane and the electrodes have to be optimized differently to those for a post-doped membrane to avoid flooding.

3.2.5 Polybenzimidazole Composite Membranes

A large number of PBI-based composites have been developed throughout the years, ultimately aiming at improving fuel cell performance and durability by improving the mechanical stability, acid retention, and conductivity of the membrane or reduce fuel/oxidant crossover. The properties of the composite materials depend on the loading of the inorganic component, and particularly on the particle size and dispersion of the secondary phase. For an overview of PBI composites, refer to Linares et al. [41]. For an older and more general review of inorganic–organic fuel cell membranes, refer to Jones and Rozière [42].

3.2.5.1 Oxide and Phosphate Composites

Composite materials of PBI and various oxides can be readily prepared by addition of nanoparticles to the membrane casting solution. Using this methodology, PBI composites containing 2 wt% TiO_2 have been prepared and extensively studied with respect to fundamental properties and long-term performance in a fuel cell stack [43, 44]. The improved durability observed on the stack level was traced back to the improved acid retention characteristics of the composite membrane.

However, co-casting of composite membranes from a dispersion tend to produce membranes with larger agglomerates of the inorganic phase. This is often undesired, as it results in a poorly defined material. One way to circumvent agglomeration during casting is to work with surface modification strategies that improve the compatibility between the components by introducing new functionalities. For example, composites with a well-dispersed phase of SiO_2 in a PBI matrix have been obtained from nanoparticles that had been surface functionalized with imidazole or sulfonated vinylbenzene groups [45–47]. Another good way to disperse the inorganic component as nanoparticles is to synthesize it in situ. For example, SiO_2 can be formed from the silica precursor tetraethoxysilane (TEOS) [48].

Well-dispersed clay nanocomposites were obtained after extensive optimization by screening an array of organic surface modifiers [49]. Other examples of PBI-clay composites are with montmorillonite and reduced plasticizing effect of the phosphoric acid and of methanol crossover is reported [50, 51].

Zirconium phosphate ($Zr(HPO_4)_2 \cdot nH_2O$, ZrP) is another material with a layered structure in common with the clays. It is of special interest for composites due to its intrinsic proton conductivity. Composites of ZrP and PBI or PBI blends led to slightly increased conductivity [52]. Yamazaki et al. [53, 54] prepared composite membranes of the variant zirconium tricarboxybutylphosphonate ($Zr(O_3PC(CH_2)_3(COOH)_3)_2$) using PBI as the binder. Membranes of doped tin phosphate have also been manufactured with PBI as binder [55].

A special class of solid materials are the heteropolyacids (HPA). This is a special family of acids constituting a metal, a nonmetal, oxygen, and hydrogen. A common structure is the Keggin structure, which contains the Keggin ion $(AM_{12}O_{40})^{n-}$.

Here A is a nonmetal (e.g. P or Si) and M is a metal (e.g. Mo or W). The ion is balanced by protons, hence the acidity. Examples of HPAs used for PBI composites are phosphomolybdic acid ($H_3PMo_{12}O_{40}$, PMA) [56, 57], phosphotungstic acid ($H_3PW_{12}O_{40}$, PWA) [58–60], and silicotungstic acid ($H_4SiW_{12}O_{40}$, SWA) [60–65]. The HPAs interact with the basic sites of PBI. Functionalization of mesoporous silica by PWA showed high proton conductivity and improved stability at temperatures up to 200 °C [66]. The PBI composite membranes with this inorganic filler showed a significant stabilizing effect on the doping phosphoric acid at temperatures up to 200 °C [67]. As revealed, this exceptional durability was due to the formation of a highly conductive and thermally stable phosphosilicate phase in the composite membranes [68].

3.2.5.2 Carbon Composites

Carbon structures in the form of nanotubes and graphene are of particular interest for structural reasons. The extremely high strength make them obvious for composites with a view to improve the mechanical strength and durability. Naturally, care should be taken not to form short circuits with these wide and highly electron conductive materials. Besides strength, graphene layers in the membrane might even reduce gas crossover without limiting the proton transport as much as other barrier layers, since it is claimed that protons can pass through the hexagons of graphene [69].

The aromatic structure of PBI can be expected to ensure good compatibility with sp^2 carbon, i.e., nanotubes and graphene, but suspension in a casting solvent can be difficult and casting might, therefore, be carried out with functionalized carbon materials, i.e., with a partly oxidized surface containing oxygen, hydroxides, carboxylic acids, etc. The editor of this book has published extensively on PBI coated aromatic carbon materials as catalyst supports. See Sect. 3.3.2.

Composites with *carbon nanotubes* (CNT), in most cases multiwalled carbon nanotubes (MWCNT), have been studied by several groups. Membrane fabrication is challenging and surface modification strategies are needed to get a well-dispersed phase. Coleman et al. [70] reviewed in 2006 strengthening of polymers with CNT, but without addressing PBI or fuel cells. Suryani et al. [71] reported significantly improved mechanical strength and also higher ionic conductivity for a composite of PBI and MWCNT as compared to pure PBI. When functionalized, the compatibility with PBI was increased. A marginal fuel cell performance improvement at 150 °C was seen. Composites of PBI and MWNCT has also been used for an alkaline direct methanol fuel cell with improved performance [72]. A slight decrease of methanol crossover was reported too, surprisingly accompanied by increased methanol uptake as the content of nanotubes was increased. As an example of a non-membrane application, Lu et al. [73] functionalized carbon nanofibers (not nanotubes) and synthesized PBI directly on the fibers. Subsequently, the fibers were hot-press sintered to obtain structural materials with high flexural strength for aerospace applications. Improved strength and somewhat improved fuel cell performance can also be achieved by incorporation of phosphonated MWCNT [74].

Increased ionic conductivity as a result of CNT in the membrane structure is not very intuitive, but several groups report that and hence better fuel cell performance. A recurring suggested explanation is that the membrane material is structured as a result of the introduction CNTs leading to formation of domains [74, 75].

Like CNT, *graphene* and *graphene oxide* (GO) have been used for composites with PBI. To achieve good dispersion, GO has been functionalized through nucleophilic reactions between the epoxy group of GO and amine groups in e.g. 3-mercaptopropyl trimethoxysilane [76], 1-aminopropyl-3-methyl-imidazole [77], *p*-aminobenzene sulfonic acid [78] and polydopamine [79].

Graphene in its oxide form, GO, contains hydroxyl, epoxy and carboxyl groups, which may form hydrogen bond channels for proton conduction. The proton conductivity of GO is up to 10^{-2} S cm^{-1} due to the presence of the hydrophilic oxygen-containing groups such as carboxylic acid and epoxy oxygen, that facilitate "hopping" of protons through hydrogen-bonding networks in association with water molecules [80]. Attempt to assemble a fuel cell using a GO-based electrolyte has been made [81]. Chen et al. [82] made in 2010 a comprehensive review of graphene-based materials in electrochemistry. Composites membranes for fuel cells or for ionic conductivity was not treated, but the literature in this field is also more recent.

PBI/GO composites have been reported without fuel cell experiments, but with varying conductivities [83]. PBI has also been synthesized around GO with improved mechanical and thermal properties even with small amounts of graphene [84]. PBI-GO composites have been used for HT-PEMFC, at times with moderate improvement of performance [85]. The primary benefit is probably on the mechanical properties. Yang et al. [86] prepared a composite membrane of triazole functionalized GO and PBI. The tensile strength and elongation at break were improved significantly and fuel cell performance slightly. The triazole functionalized GO dispersed well in the solvent used. Composites of PBI and GO functionalized with ionic liquids has been explored recently by different groups for use in both acidic [87] and alkaline [88] fuel cells. Functionalized GO has also been used for crosslinking quaternized poly(ether ether ketone) membranes [89].

Membranes of a composite of PBI and GO sulfonated by radiation grafting were reported [90]. The sulfonation led to significantly increased strength in the acid doped as well as the un-doped state, compared to an un-sulfonated composite. A somewhat increased proton conductivity was measured too. Superior conductivity of composites of sulfonated GO was previously reported [91].

Instead of reinforcing PBI with graphene, PBI can improve the strength of Nacre-like graphene paper with graphene as a glue [92]. As an alternative to a composite, Deng et al. [93] made a three-layer membrane of polybenzimidazole/graphite oxide/polybenzimidazole and the mechanical strength of the membrane was more than doubled.

3.2.6 Pyridine Containing Aromatic Polyethers

In parallel to the development of PBI membranes, a similar membrane family has been developed at University of Patras and Foundation of Research and Technology-Hellas in Patras. Instead of benzimidazole, pyridine units provide the aromatic nitrogen functionality that makes the membrane work in HT-PEMFC [94, 95]. The behavior of the membrane in a HT-PEMFC is similar to that of the PBIs, but fuel cell tests are often performed at higher temperatures than with PBI. Especially crosslinked membranes are operated at 180–220 °C [96, 97].

Cell development and testing have to some extent been done at the company Advent, which was established based on this pyridine based technology. This can be the reason why studies on durability at the high temperatures are scarce in the literature. However, stable performance for 350 h at temperatures varied between 200 and 220 °C without clear visible degradation [76] and later operation for 500 h at 210 °C with a degradation rate of only 3 $\mu V\ h^{-1}$ [74] were reported by the Greek groups. At this temperature, more severe degradation can be expected with PBI-based cells, even at such a short time. A 2011 prospect from Advent shows a degradation of 9 $\mu V\ h^{-1}$ over 4000 h at 180 °C. Today Advent deals with both the pyridine membranes, called TPS, and PBI membranes based on technology from BASF. In their latest MEA prospect, they state operational temperatures of 120–180 °C for PBI MEAs and 120–200 °C for TPS MEAs, but without indication of durability for TPS MEAs. Examples of polymers with both benzimidazole and pyridine can be found in the literature too [98, 99].

When it comes to fuel cell long-term stability, the crux of the matter is whether acid retention and resistance to condensation are higher in the pyridine system than in the PBI system. If not, acid loss rates or phosphate condensation will most likely be high at temperatures above 200 °C.

3.3 Catalysts

The state-of-art catalyst in HT-PEMFC is platinum or platinum alloys for both anode and cathode and this represents no difference to LT-PEMFC. For both types of cells, increased activity for oxygen reduction is seen when platinum is alloyed with certain other transition metals, like nickel or cobalt, and on the anode side, platinum is alloyed with ruthenium for use in cells converting methanol [100–104] or dimethyl ether [105]. Despite the higher temperature and the presence of phosphoric acid, the catalyst is thus the same and for LT-PEMFC and HT-PEMFC.

The complementing key component of the catalyst is the catalyst support, i.e. the material on which the platinum particles are fixed. Electrocatalysts used in the early years of PAFC were platinum black only. Later, high surface-area carbon supports were introduced in the 1970s as an approach to a higher degree of metal dispersion to reduce the loading and cost of the noble metal. Requirements for the carbon support materials include high electronic conductivity, chemical, and electrochemical stability, sufficiently high surface area as well as suitable pore-size distribution. So far, carbon blacks seems to have the best combination as well as commercial availability and low cost.

However, cost and availability of PGM is a challenge and both the platinum-based catalyst and the carbon-based support do degrade during long-term operation of the HT-PEMFC. On this background, it is desired to develop new catalytic materials as well as catalyst support materials with improved durability and lower cost. This is yet another common trend for LT-EPMFC and HT-PEMFC. In the following paragraphs, the use of alternative carbon-based materials for support as well as for PMG-free catalysts will be treated briefly. Carbon-free catalyst substrates is another active field of research, but so far almost exclusively tested in LT-PEMFC. The major challenges are to obtain sufficient specific surface area, electronic conductivity and chemical stability of such ceramic materials (oxides, nitrides, carbides). Refer to the review by Wang et al. [106].

3.3.1 Carbon Black Supported Noble Metal Catalysts

The high surface area of carbon black is advantageous for a fine dispersion of the 3–5 nm platinum particles, but at the same time, it makes the material more vulnerable to corrosion than materials with a more graphitic structure and a smaller specific surface area. Common carbon black products are Vulcan XC-72 (Cabot Corporation) with a specific surface areas of about 250 $m^2\ g^{-1}$, Black Pearls 2000 (Cabot Corporation) with 1100–1400 $m^2\ g^{-1}$ and Ketjenblack EC-600JD (Akzo Nobel) with 1200–1400 $m^2\ g^{-1}$.

The catalyst nanoparticles can be applied either by direct synthesis on the carbon surface by impregnation and subsequent reduction or they can be synthesized separately and kept suspended before application on the support. With the impregnation method, carbon black is first impregnated with a solution containing a salt or an acid of the catalyst metal (e.g., H_2PtCl_6) by the capillary action with or without surfactants. The metal is then reduced into nanoparticles either in the solution using a reducing agent such as formic acid, formaldehyde, or sodium borohydride, or by gaseous hydrogen or formaldehyde after evaporation of the solvent. Alternatively, the platinum particles are synthesized elsewhere by other methods like the polyol method in glycol or the oil-in-water method in which the particle size is controlled by micelles formed in an oil-water emulsion. Many other techniques are known, and alloy synthesis requires special measures due to different reduction potentials of the precursors.

The higher operational temperatures of HT-PEMFC as well as the presence of phosphoric acid aggravates both carbon corrosion and platinum dissolution. Early studies on dissolution of platinum in concentrated phosphoric acid showed a potential dependence on the process, particularly at cathode potential close to open circuit voltage [107–109]. At even higher potentials, the passivation by platinum oxides may play a role in limiting the metal solubility [110], which showed a peak at around 1.1 V in one study [111]. At a temperature of 160 °C and a cathode potential of 0.7 V, the platinum dissolution rate was estimated to 2×10^{-9} mg Pt cm^{-2} h^{-1}, increasing to 2×10^{-2} mg Pt cm^{-2} h^{-1} at a cathode potential of 1.0 V [112].

Carbon blacks are susceptible to corrosive conditions such as high cathode potentials and low pH in combination with oxygen and water vapor as in HT-PEMFC. Electrochemical corrosion of carbon according to Eq. (3.2) is thermodynamically favored at potentials higher than 0.207 V versus SHE [113]

$$C + 2H_2O \rightarrow CO_2 + 4H^+ + 4e^- \qquad E^0 = 0.207\ \text{V versus SHE} \tag{3.2}$$

The process is proposed to proceed by formation of carbon surface oxide intermediates followed by hydrolysis of the oxides to carbon dioxide [114]. In practice, sluggish kinetics of carbon corrosion makes carbon a feasible candidate as catalyst support, but the corrosion kinetics is promoted by elevated temperature.

Oh et al. [115] quantitatively related the first 30 min of carbon corrosion in a HT-PEMFC to that in a LT-PEMFC at potentials between 1.0 and 1.4 V. The LT-PEMFC was tested at 70 °C with fully humidified oxygen whereas the HT-PEMFC was tested at 150 °C using non-humidified oxygen. The results showed little difference in the carbon corrosion rate for potentials up to 1.1 V. At potentials higher than 1.1 V, however, the carbon corrosion rate for the HT-PEMFC increased dramatically in comparison to that of the LT-PEMFC. This indicates that elevated temperatures have higher influence on the carbon corrosion than what can be offset by a decrease in humidity. For carbon supports with various platinum loadings, Stevens et al. [116] showed significantly enhanced carbon corrosion when temperature increased from 125 to 195 °C. The low water partial pressure often realized in HT-PEMFC was accredited as a moderating effect, but still, at potentials higher than 1.1 V, the carbon corrosion rate was found to be high due to the elevated temperature. Oono et al. [117] investigated the temperature effect on HT-PEMFC degradation in a temperature range from 150 to 190 °C. They attributed the initial performance loss, which was severely intensified with the temperature, to carbon corrosion and platinum particle agglomeration. After prolonged testing, visible thinning of the catalyst layers as well as growth of the platinum catalyst particles was observed [118].

Schmidt et al. [119] focused on the corrosion stability of supported catalysts in PBI-based cells by purging the fuel stream alternately with hydrogen and air. This operation was found to drive the cathode potential cycling between 0.95 and 0.2 V. During the cycling, carbon dioxide was found to correlate well with the catalyst support loss, which resulted in a fuel cell performance drop, particularly seen as an increased mass transport overpotential. For evaluation of catalyst and support, potentials higher than the OCV are typically applied as an accelerated degradation stressor, either by cycling or by potential hold. In doing so, atmospheric humidification seems to be necessary, though it is not a common practice of the fuel cell operation [120]. And the corrosion of carbon supports and sintering of noble metal catalysts are intensified under dynamic tests with thermal, load and shutdown-startup cycling [121].

Several reports of around 18,000 h of operation of HT-PEMFC single cells at 200 mA cm^{-2} with degradation rates of only 5–10 μV h^{-1} can be found in the literature [122, 123]. Recently, a degradation rate of as low as 0.5 μV h^{-1} within a testing period of 9000 h was reported [124]. However operational parameters including stoichiometries of gas flows should be carefully controlled [125].

A suitable strategy to minimize the carbon corrosion is to use carbon materials, which have higher graphite contents and therefore fewer structural defects at which the oxidation initiates [126]. Heat-treatment of carbon blacks at very high temperature is known to impart the graphitic character to the carbon black, though at the expense of a significant decrease in the specific surface area. Cleemann et al. [127] showed that graphitization at 2800 °C of carbon black (before applying the platinum catalyst) improved the stability in a potential cycling test. Note that catalysts on multiwalled carbon nanotubes proved even more stable in the same study. Mader et al. [121] investigated the stability of eight types of carbon for PBI-based fuel cells at a temperature of 180 °C and a potential of 1.0 V versus RHE. A clear correlation of the carbon corrosion rate was observed with the increase of the BET surface area of the materials. At the same time, good dispersion of the noble metal catalysts was somewhat hampered when using synthetic graphite as support.

3.3.2 Carbon Nanotubes and Graphene as Catalyst Support

The stable sp^2 structure of graphite is also present in CNT and graphene. The fibrous shape of carbon nanotubes should be ideal for mechanical stabilization and electronic conduction and at the same time ensures high porosity. Combined with the high thermal and chemical stability of this form of carbon, CNT could be the ideal catalyst support, also for HT-PEMFC. The challenge is that the stable surface is not an easy substrate for a good dispersion of the platinum particles. The usual way to make the CNT more collaborative is to functionalize them by a moderate oxidation. The functional surface groups formed (O, OH, COOH etc.) make the surface less hydrophobic and more receptive. The growth and dispersion of platinum nanoparticles are believed to benefit from these surface functionalities, which serve as effective binding sites for the platinum ions where homogeneous nucleation occurs and well-dispersed nanoparticles are achieved [127]. Platinum clusters synthesized beforehand will also disperse better on the functionalized surface.

The penalty is that functionalization also affects the otherwise superior stability that was the reason for selecting CNT for support. To minimize the degradation of the support material triggered by these surface defects, a unique approach for the growth of platinum nanoparticles on non-oxidized graphitic carbon has been developed by Nakashima and coworkers [128, 129]. By attaching a nanolayer of PBI on the surface of carbon, platinum ions or catalyst clusters are coordinated via the functionalities of the polymers. In this way, homogeneous growth and dispersion of platinum nanoparticles can be realized, with a narrow diameter distribution [130]. The PBI material is selected as it can efficiently be dispersed on CNT through the π–π interaction and provide binding sites for coordination of platinum ions. At the same time, when assembled in fuel cells, the polymer can be protonated by phosphoric acid from the membrane and become proton conductive. The obtained platinum nanoparticles are therefore in good contact with the electronically conducting carbon support and the ionically conducting PBI phase [130]. An eight times higher mass activity [131] and a four times higher durability [132] have been reported.

Even with carbon blacks, impregnation with PBI before platinum loading was found to facilitate the immobilization of the proton conducting path onto the surface of the electrocatalyst when further doped with poly(vinylphosphonic acid) (PVPA) instead of phosphoric acid [133].

PBI-wrapped carbon materials have been applied to other systems than HT-PEMFC. Both PBI-wrapped CNT and carbon blacks were also combined with Nafion, where the acid–base interaction provides high affinity between Nafion and PBI [134, 135]. Increased CO tolerance at room temperature due to the OH groups at OH-PBI was claimed [136]. Moreover, the first case of use for PEM electrolysis at 80 °C was reported in 2017 [137], and finally, the PBI wrapping was used for oxidation of ethanol in alkaline solution without fuel cell tests [138] and for alkaline fuel cell tests with a Tokoyama A201 alkaline membrane [139]. In 2013, Fujigaya and Nakashima [130] published a review about the PBI wrapping technique. In that, numerous kinds of PBI used are listed.

Similar work with *graphene materials* is not as extensive and was initiated later. Fujigaya et al. [140] expanded the wrapping to graphene. Graphite was exfoliated in a PBI solution. Subsequently, platinum was deposited as 2.7 nm nanoparticles with excellent dispersion. In a potential cycling test, the electrochemical surface area was degraded to a lesser degree than the commercial Pt-carbon black catalyst compared with. The work was continued by Li et al. [141] via a graphene oxide route where PBI was synthesized on isophalic acid-functionalized graphene and platinum nanoparticles were deposited. Improved stability with respect to potential cycling was seen as compared to a commercial carbon black base catalyst. Later the group introduced sulfonic acid-functionalized carbon black as a graphene spacer and made fuel cell tests at 80 °C with a Nafion membrane [142, 143]. PBI decorated reduced graphene oxide was used as a support material for a palladium catalyst in an alkaline fuel cell at 90 °C [144]. A study of single-wall carbon nanohorns without PBI wrapping as support for a HT-PEMFC was reported by Boaventura et al. [145].

3.3.3 Non-platinum Group Metal Catalysts

Driven by the vast perspective in replacement of the PGM based catalysts for PEM fuel cells (and the equivalent electrolyzers, although it is probably even more challenging), tremendous effort is made to find active, stable, and inexpensive alternatives. The dominating trend is to study materials containing iron, nitrogen, and carbon (Fe-N-C). The molecular structure of such materials is generally complicated compared to a metallic surface of one or a few elements.

The development began in the 1960s when Jasinski [146] inspired by functional structures in nature discovered that cobalt phthalocyanine was an active oxygen reduction catalyst in alkaline medium. In phthalocyanine, four heterocyclic nitrogen atoms embedded in a carbon structure hold a single metal ion. A similar structure is found in the heme molecule that binds oxygen in our blood, in chlorophyll and in cytochrome c in our respiratory chain. Over the recent years, numerous materials have been synthesized by various heat treatments of precursors containing metal ions (mostly iron), nitrogen and carbon, and high oxygen reduction activities have been demonstrated in acidic environment relevant to PEMFC, as thoroughly reviewed [4, 147–149], with a focus on degradation [150], and treated in more detail elsewhere in this book. It is much more difficult to identify the catalytic site(s) in such materials than in traditional catalysts. However, the general understanding [151] is that the site is associated with a metal ion coordinated by four hetrocyclic nitrogen embedded in carbon, like in phthalocyanine, but that the heterocycles can be pyrrolic (a 5-membered ring) or pyridinic (a 6-membered ring). The low surface density of the sites make it particularly difficult to characterize them but recently Chung et al. [152] showed rather convincing STEM images of a possible site.

The state-of-the-art PEMFC performance based on Fe-N-C catalysts matches that of Pt/C at low current densities, though the needed catalyst layer is around an order of magnitude thicker. This is the reason why good performance is mostly achieved with pure oxygen as oxidant [153, 154]. Pure kinetic current-voltage characteristics, i.e., without negative contributions from internal resistance and mass transport, describes the catalytic activity of the electrode. When plotted logarithmic in current, a linear curve is obtained (a Tafel plot). When this is done for electrodes with Fe-N-C catalysts, mass transport limitations, caused by the thick catalyst layer, kick in at rather low current densities and the curve bends. At times, extrapolation to higher current densities is made to illustrate the potential performance provided that the transport and resistive properties are improved. By extrapolation, this way, the performance of Fe-N-C catalysts reached a volumetric current density of 99 A cm^{-3} at 0.8 V in 2009 and 230 A cm^{-3} in 2011 [155, 156]. A recent breakthrough was with a MOF-based catalyst, with the current density extrapolated to 450 A cm^{-3} at 0.8 V. This exceeded the newly updated DOE target of 300 A cm^{-3} [153].

3.3.4 Non-platinum Group Metal Catalysts for High-Temperature PEMFC

The new PGM-free catalyst can be of particular interest for the further development of HT-PEMFC. It is well-known from the PAFC development as well as from later studies that phosphate ions adsorb strongly on platinum [157–159]. Studies of this with particular relevance to the HT-PEMFC are also available [160–164]. This adsorption is the reason why the oxygen reduction overvoltage in HT-PEMFC (and PAFC) is more severe than in LT-PEMFC. Additionally, the solubility of oxygen is small in phosphoric acid and if the catalyst is covered with an acid film, reduced performance is a consequence of this condition too. Consequently, the improved performance that one might reasonably expect at the higher temperature is not realized. On the contrary, the voltage at any given current density is generally somewhat higher for a LT-PEMFC when operated on pure hydrogen.

This impediment of the kinetics on platinum gives an advantage to other catalyst materials without a strong affinity for phosphate. They don't have to be as active as platinum in general, only as active as platinum inhibited by phosphate.

It is not easily justified to state that all Fe-N-C catalysts have similar properties with respect to anion adsorption, but with that reservation in mind, it is worth noting that phosphate tolerance of two different Fe-N-C catalysts was demonstrated independently by rotating disk electrode experiments [165, 166]. In both studies, a shift of the limiting current was seen for a platinum catalyst as well as for an in-house made Fe-N-C catalysts following increasing additions of phosphoric acid to the perchloric acid electrolyte. This is the expected effect of the lowering of the oxygen solubility in phosphoric acid. However, the kinetic currents were in both studies shifted significantly for the platinum catalyst but not at all for the Fe-N-C catalyst. A complete tolerance to methanol was also observed for the Fe-N-C catalyst in contrast to the platinum catalyst which was severely affected [166]. Other studies confirm that Fe-N-C catalysts are indifferent to carbon monoxide in contrast to platinum [167]. Several other anions, like CN^- [168], SCN^- [167, 169] and Cl^- [167] are known to poison Fe-N-C catalysts.

Recent rotating disk electrode measurements by the authors group show that phosphate even has a positive effect on the oxygen reduction kinetics on a Fe-N-C catalyst when compared to perchloric acid, which is considered the standard electrolyte for rotating disk electrode measurements. In Fig. 3.2 it is shown how the current density differs in perchloric and phosphoric acid for platinum and Fe-N-C, respectively. A better performance of the Fe-N-C catalyst is achieved in phosphoric acid than in perchloric acid, i.e., the opposite of what is seen with platinum. This is a big advantage for the application such non-PGM catalysts in high-temperature PEMFC with concentrated phosphoric acid as the electrolyte.

3.4 Electrodes and Cells

The electrodes in an HT-PEMFC are made of several carbon materials with different properties. In that respect they are very similar to electrodes for LT-PEMFC. The structural part that provides the mechanical stability is made from carbon fibers, which are either woven or non-woven. This layer typically occupy around 90% of the volume. See Fig. 3.3. On the inner side of the fibrous layer is the microporous layer (MPL), which provides a smooth interface to the catalyst layer between that

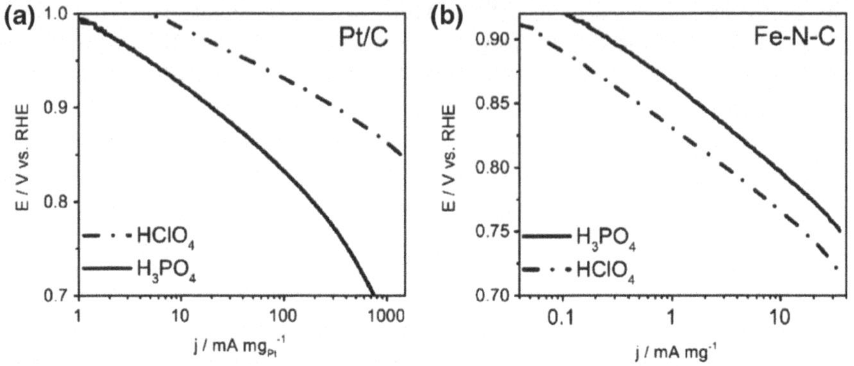

Fig. 3.2 Mass-specific current densities for oxygen reduction on **a** Pt/C and **b** Fe-N-C catalysts in perchloric acidic ($HClO_4$) and phosphoric acid W (H_3PO_4) electrolytes. Conditions: **a** 20 wt% platinum on carbon with a catalyst loading of 100 μg Pt cm^{-2}; **b** Fe-N-C catalysts with a catalyst loading of 600 μg cm^{-2}. Rotation rate: 1600 rpm; Scan rate: 10 mV s^{-1} (Submitted for publication in Applied Catalysis B: Environmental by the author's group)

and the membrane. The MPL is made from carbon nanoparticles similar to the carbon support of the platinum catalysts. Some nanocarbons used in MPLs are carbon blacks with high porosities and high specific surface areas, others are graphitized to improve chemical stability, which lead to a loss of porosity and surface area. The fibrous layer and the microporous layer constitute together the gas diffusion layer (GDL), which is normally a commercial product for which the detailed information about the materials and the manufacturing technique used is proprietary information. Suppliers of GDL materials are Freudenberg and SGL in Germany and Toray Industries and Mitsubishi Chemical in Japan. The carbon layers are typically wet proofed by treatment with polytetrafluoroethylene (PTFE) to avoid flooding by water in LT-PEMFC and by phosphoric acid in PAFC and HT-PEMFC. Since both carbon and PTFE tolerate temperatures above 200 °C most material can probably be used for both LT-PEMFC and HT-PEMFC when it comes to thermal stability. However, stability is one thing; a good balance between hydrophobicity and gas transport properties as well as the right pore size distribution are other decisive parameters. Freudenberg and SGL specifically recommend certain products for HT-PEMFC. Carbon-based GDL materials were originally developed for PAFC so the adaptation to phosphoric acid containing cells is not new.

The hydrophobic properties of the GDL of an HT-PEMFC are not only necessary to mitigate flooding by acid. Sufficient hydrophobicity is also important to limit acid leaching from the catalyst layer and ultimately from the membrane. Acid loss is a serious cause of cell degradation, especially during long-term operation. This is a degradation mechanism the HT-PEMFC shares only with the PAFC. The reason why hydrophobicity can influence the acid leaching is that wetting of the structures of the GDL is a first step for surface diffusion out of the cell. Besides, when distributed over the large surface area of the porous GDL, evaporation can

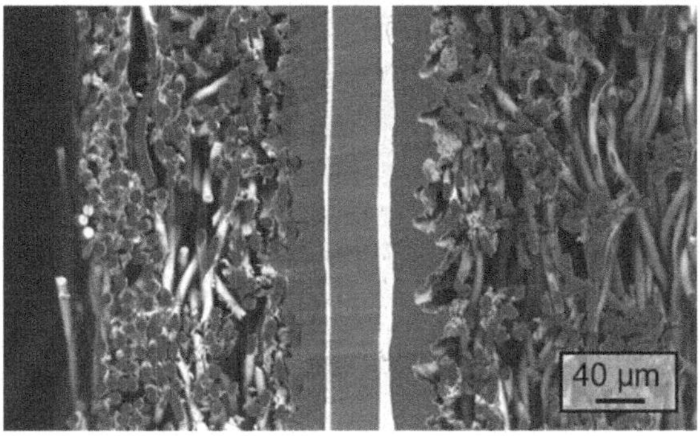

Fig. 3.3 Cross section of an MEA of a HT-PEMFC. The central layer is the polymer membrane. Then follows (left and right) the lighter catalyst layers, the microporous gas diffusion layers and finally the thick fibrous gas diffusion layers. The thickest catalyst layer to the right is for the cathode since oxygen process needs the most catalyst (Image: L. Seerup, DTU Energy)

take place from more sites and the gas phase saturates faster with phosphoric acid which consequently escapes faster by gas phase diffusion and subsequently by convection when it reaches the gas flow in the channel. Kannan et al. [170] have shown how the choice of GDL can affect the acid loss rate and the performance decay rate. Several cells with different combinations of GDLs for the anode and the cathode respectively, were operated at constant current density for ca. 3500 h. The temperature was 180 °C instead of the usual 160 °C to accelerate acid evaporation. During the course of the experiments, the cell voltages were measured continuously and acid was collected form the exits and quantified at intervals. The linear voltage decay rate varied from 11 to 53 $\mu V\ h^{-1}$ depending on the GDLs. The acid collection rates were linear too and generally followed the same trend, although it was concluded that the amount of acid collected was only a small fraction of the acid in the system, and thus not a quantitative measure of the real acid loss from the MEA.

The catalyst layer (CL) can be applied on either the GDL or on the membrane before the whole cell is assembled, typically by hot-pressing. The acid doped membrane can be difficult to coat and it is common to apply the catalyst layer on the GDL. It can be carried out by all the same techniques as for LT-EPMFC, i.e. by screen printing, tape casting, and different spraying techniques. A difference to the traditional LT-PEMFC is that it is not necessary to integrate an ionomer for proton conduction in the CL. In LT-PEMFC, an ionomer in the form of a perfluorosulfonic acid, like that constituting the membrane, is used. When humidified, it becomes proton conducting. In the HT-PEMFC, PBI is often added for the same reason, but phosphoric acid will wet the pores of the CL anyway and it can be questioned if PBI makes a difference to the proton conductivity in the CL at all. Another function of PBI in the CL is to bind the catalyst covered carbon particles and form a mechanically robust structure. However, even this is not a prerequisite. After reducing the amount of binder in the electrodes, Martin et al. [171] took the daring step of making the electrodes completely without a binder. The result was a series of well-performing cells that could be easily handled after hot-pressing. The catalyst used was supported on carbon black of ca. 50 nm (Johnson Matthey) and the agglomerative forces, likely with the aid of phosphoric acid squeezed out of the membrane, were apparently strong enough to secure coherent CLs with thicknesses ranging from 2 to 34 μm. Binderless electrodes were later used for a study targeting reduced platinum loading in HT-PEMFC [172].

The difference between GDLs for HT-PEMFC and LT-PEMFC is limited. Basically, the same requirements are given. Transport of reactants, products, electrons and heat is equivalent and sufficient hydrophobicity is important although it serves to exclude water and phosphoric acid respectively in the two cases. Properties that can make certain materials suited only to LT-PEMFC are chemical and thermal stability in phosphoric acid at the temperature of the HT-PEMFC.

3.5 Utilization of High-Temperature PEMFC

The previous parts on HT-PEMFC have dealt with the cell components and related materials issues only. Similar parts could be added on stacking and system components, the applications and a status for commercialization. These areas are certainly important parts for the understanding of the HT-PEMFC narrative, but in line with the scope of this volume, they will only be touched briefly.

3.5.1 Stacking the Cells

For any application, the cells must be stacked between bipolar plates that distribute the gasses and conduct current and heat through the series of cells. The bipolar plates for PEMFC and PAFC are traditionally made from machined graphite, but for cost reasons and for improvement of toughness, they are now manufactured from carbon composites. Injection or compression molding makes the most sense for large volume production, but until the market and thus the production volumes get larger, the business case might be to machine the bipolar plates from cast composites and obtain the toughness without the expense for the ultimate machinery.

The carbon used is graphite or synthetic graphite since there is no desire for a high surface area. Instead, conductivity and corrosion resistance are key parameters. CNT can be added for increased strength. The statements so far hold for both LT-PEMFC and HT-PEMFC, but when it comes to the binder, the elevated temperature of HT-PEMFC must be considered. Some binders used for LT-PEMFC plates might be soft at the elevated temperature. The binder can be a thermosetting or a thermoplastic polymer. The thermosetting polymer is thermally crosslinked and cannot be re-melted, while the thermoplastic polymer is shaped by melting and solidification only. The thermoplastic polymer is well suited for injection molding. Suitable thermoplastic binders for HT-PEMFC are polyphenylene sulfide (PPS), polyetheretherketone (PEEK) and polyphenylsulfone (PSO). For sealing, fluorinated elastomers like FKM, and Viton are used to cope with the hot phosphoric acid. PTFE is stable, but less adaptive and tends to creep. Kundler and Hickmann [173] provide a review on bipolar plates and gaskets.

Metallic plates were the game changer for thin and tough bipolar plated for automotive LT-PEMFC stacks exceeding 3 kW per liter stack. Their success depended on a suitable coating to prevent a high contact resistance. A similar solution has not been found for the harsh environment of the HT-PEMFC.

3.5.2 System and Applications

The system around the stack, the balance of plant (BoP), resembles that of LT-PEMFC when building for pure hydrogen as fuel. One difference is that the coolant of a liquid cooled stack must have a higher boiling point than water if pressurization shall be avoided. Glycols or heat transfer oils are used. Another difference is that the water management components are omitted. The thermal management is less critical since there is no risk of water condensation during normal operation. Nevertheless, care shall be taken to avoid condensation of water during startup and shut down. Special means for heating the stack during cold-start is mandatory since it cannot operate and self-heat at room temperature.

The most obvious advantage system wise is realized when the HT-PEMFC is integrated with a fuel reformer. Then the reforming system can be simplified because no carbon monoxide removal is necessary downstream of the water gas shift process. The waste heat of the stack can even be utilized for evaporation of water for a steam reformer as well as many liquid fuels [2]. The simplicity is most clearly seen for a methanol-powered system. Methanol can be reformed at temperatures as low and 250–300 °C and no separate water gas shift reactor is needed when reforming is carried out this low temperature. The company Serenergy produces integrated methanol powered systems where the whole BoP including the reformer occupies less space than the fuel cell sack.

The suitability of the HT-PEMFC technology for integration with a reformer makes it a candidate for domestic combined heat and power production. The company Elcore has developed such a unit powered by natural gas. Nevertheless, the most significant market introduction is probably in the telecom industry in Asia where Serenergy provides backup and continuous solutions based on their methanol powered systems.

In the automotive field, HT-PEMFC is mostly absent. Daimler developed a methanol technology for fuel cell vehicles and demonstrated successfully their NECAR 5, which drove across the United States in 2001. The fuel cell was an LT-PEMFC. Later, the automotive companies decided to pursue pure hydrogen solutions instead. Volkswagen developed a HT-PEMFC technology, but gave it up. The very high power density of more than 3 kW per liter stack (pure hydrogen) is impossible to match for HT-PEMFC without metallic bipolar plates and with the somewhat higher oxygen reduction overpotential. Moreover, with the 700 bar hydrogen storage tanks, which is standard today, the driving range for personal vehicles is around 600 km and for need for more compact fuel storage in personal vehicles is not obvious. A remaining argument for methanol is the lack of the hydrogen infrastructure.

HT-PEMFC is not completely absent in mobility. Serenergy has developed a range extender for battery electric vehicles. A Fiat e500 and a Nissan e-NV200 van each had their driving range extended to 800 km by means of a methanol-powered HT-PEMFC range extender. HT-PEMFC stacks were also installed in the German fuel cell powered airplane Antares DLR-H2 and recently the tour boat MS Innogy at Lake Baldeneysee in Germany, which was equipped with 35 kW of methanol-powered HT-PEMFC from Serenergy.

The dominating cell (MEA) manufacturer used to be BASF Fuel cell (formerly PEMEAS). Their technology was based on direct cast membranes. In 2013 the company ceased the cell manufacturing and kept only a minor activity. Present cell manufacturers are Danish Power Systems (post-doped membranes and cells), Advent (pyridine based post-doped membranes and direct cast membranes - BASF technology), Elcore (cells based on post-doped membranes), Dongjin Semichem (cells based on post-doped membranes) and Fumatech (membranes).

The dominating stack and system manufacturer is Serenergy with their HT-PEMFC systems based on methanol. Other players are Elcore with heat and power units powered by natural gas and Siqens with direct methanol systems.

3.6 Conclusions

High-temperature polymer electrolyte membrane fuel cells (HT-PEMFC) is a well-established fuel cells technology. It can be seen as a hybrid between the conventional low-temperature polymer fuel cell and the phosphoric acid fuel cell. It has most components in common with one or both of the other two types of fuel cells. The HT-PEMFC is operated at temperatures around 160 °C and the benefits of this are increased tolerance to fuel impurities like carbon monoxide, less need for water management, efficient cooling and higher value of the waste heat.

The key to the elevated temperature operation is to apply a thermally stable ion solvating polymer membrane, typically of polybenzimidazole, and dope it with phosphoric acid to obtain proton conductivity.

Nanocarbons play important roles in the development of both improved membranes and as catalyst supports. Moreover, the trend in the search for new oxygen reduction catalysts to replace platinum is to study carbonaceous nanostructures containing iron–nitrogen creview papers are publishedomplexes. Candidates must be made from abundant inexpensive materials and prove sufficient activity and stability in the acidic environment of a polymer fuel cell. Platinum is partly poisoned by phosphoric acid, but new catalysts under development are not. This gives them an advantage over platinum in HT-PEMFC. The primary use of HT-PEMFC is in combination with a fuel reformer that converts carbonaceous fuels to hydrogen leaving the system independent of a hydrogen infrastructure.

3.7 Further Reading

HT-PEMFC are treated along with other fuel cells in the scientific literature and at conferences. Besides, a number of specific references and sources for further information have emerged over the years. Specific HT-PEMFC related review papers are published by different groups [121, 174–182]. The edited volume of +500 pages "*High-temperature Polymer Electrolyte Membrane Fuel Cells*" was published by Springer late 2016 [183]. HT-PEMFC is a main topic of the International CARISMA Conference series held biannually since 2008 when it was initiated in France by the European *Coordination Action for Research on Intermediate and high-temperature Specialized Membrane electrode Assemblies* (CARISMA).

References

1. Li Q, He R, Gao J et al (2003) The CO poisoning effect in polymer electrolyte membrane fuel cells operational at temperatures up to 200 °C. J Electrochem Soc 150(12):A1599–A1605
2. Jensen JO, Li Q, Pan C et al (2007) High temperature PEMFC and the possible utilization of the excess heat for fuel processing. Int J Hydrogen Energy 32(10–11):1567–1571
3. Jensen JO, Li Q, He R et al (2005) 100–200 °C polymer fuel cells for use with $NaAlH_4$. J Alloys Compd 404–406:653–656
4. Jaouen F, Proietti E, Lefévre M et al (2011) Recent advances in non-precious metal catalysis for oxygen-reduction reaction in polymer electrolyte fuel cells. Energy Environ Sci 4:114–130
5. Meyer S, Nikiforov AV, Petrushina IM et al (2015) Transition metal carbides (WC, Mo_2C, TaC, NbC) as potential electrocatalysts for the hydrogen evolution reaction (HER) at medium temperatures. Int J Hydrogen Energy 40(7):2905–2911
6. Kochetova N, Animitsa I, Medvedev D et al (2016) Recent activity in the development of proton conducting oxides for high-temperature applications. RSC Adv 6:73222–73268
7. Paschos O, Kunze J, Stimming U, Maglia F (2011) A review on phosphate based, solid state, protonic conductors for intermediate temperature fuel cells. J Phys: Condens Matter 23:234110
8. Jensen JO, Hjuler HA, Aili D et al (2016) Introduction. In: Li Q et al (eds) High temperature polymer electrolyte membrane fuel cells. Springer International Publishing, Switzerland, pp 1–4
9. Jakobsen MTD, Jensen JO, Cleemann LN et al (2016) Durability issues and status of PBI-based fuel cells. In: Li Q et al (eds) High temperature polymer electrolyte membrane fuel cells. Springer International Publishing, Switzerland, pp 487–509
10. Brown EH, Whitt CD (1952) Vapor pressure of phosphoric acid. Ind Eng Chem 44:615–618
11. Park HY, Ahn SH, Kim SK et al (2016) Characterizing coverage of phosphoric acid on carbon-supported platinum nanoparticles using in situ extended X-Ray absorption fine structure spectroscopy and cyclic voltammetry. J Electrochem Soc 163:F210–F215
12. Korte C, Conti F, Wackerl J et al (2016) Phosphoric acid and its interactions with polybenzimidazole-type polymers. In: Li Q et al (eds) High temperature polymer electrolyte membrane fuel cells. Springer International Publishing, Switzerland, pp 169–194
13. Zeng J, Jiang SP (2011) Characterization of high-temperature proton-exchange membranes based on phosphotungstic acid functionalized mesoporous silica nanoconnposites for fuel cells. J Phys Chem C 115:11854–11863

14. Ansari Y, Tucker TG, Huang W et al (2016) A flexible all-inorganic fuel cell membrane with conductivity above Nafion, and durable operation at 150 °C. J Power Sources 303:142–149
15. Luo J, Jensen AH, Brooks N et al (2015) 1,2,4-Triazolium perfluorobutanesulfonate as an archetypal pure protic organic ionic plastic crystal electrolyte for all-solid-state fuel cells. Energy Environ Sci 8(4):1276–1291
16. Wainright JS, Wang JT, Weng D, Savinell RF, Litt M (1995) Acid-doped polybenzimidazoles—a new polymer electrolyte. J Electrochem Soc 142:L121–L123
17. Savinell R, Yeager E, Tryk D et al (1994) A polymer electrolyte for operation at temperatures up to 200 °C. J Electrochem Soc 141:L46–L48
18. Aili D, Savinell RF, Jensen JO et al (2014) The electrochemical behaviour of phosphoric acid doped poly(perfluorosulfonic acid) membranes. ChemElectroChem 1:1471–1475
19. Hansen MK, Aili D, Christensen E et al (2012) PEM steam electrolysis at 130 °C using a phosphoric acid doped short side chain PFSA membrane. Int J Hydrogen Energy 37 (15):10992–11000
20. Kerres J (2016) Applications of acid–base blend concepts to intermediate temperature membranes. In: Li Q et al (eds) High temperature polymer electrolyte membrane fuel cells. Springer International Publishing, Switzerland, pp 59–89
21. Yang J, He R, Aili D (2016) Synthesis of polybenzimidazoles. In: Li Q et al (eds) High temperature polymer electrolyte membrane fuel cells. Springer International Publishing, Switzerland, pp 151–167
22. Chung TS (1997) A critical review of polybenzimidazoles: historical development and future R&D. J Macromol Sci Rev Macromol Chem Phys C37:277–301
23. Leykin AY, Askadskii AA, Vasilev VG et al (2010) Dependence of some properties of phosphoric acid doped PBIs on their chemical structure. J Membr Sci 347:69–74
24. Walba H, Isensee RW (1961) Acidity constants of some arylimidazoles and theier cations. J Org Chem 26:2789–2791
25. Aili D, Hansen MK, Renzaho RF et al (2013) Heterogeneous anion conducting membranes based on linear and crosslinked KOH doped polybenzimidazole for alkaline water electrolysis. J Membr Sci 447:424–432
26. Aili D, Jankova KJ, Li Q et al (2015) The stability of poly(2,2′-(*m*-phenylene)-5,5′-bibenzimidazole) membranes in aqueous potassium hydroxide. J Membr Sci 492:422–429
27. Kraglund MR, Aili D, Jankova K et al (2016) Zero-gap alkaline water electrolysis using ion-solvating polymer electrolyte membranes at reduced KOH concentrations. J Electrochem Soc 163(11):F3125–F3131
28. Li Q, Pan C, Jensen JO (2007) Cross-linked polybenzimidazole membranes for fuel cells. Chem Mater 19:350–352
29. Yang J, Aili D, Li Q et al (2013) Covalently cross-linked sulfone polybenzimidazole membranes by poly (vinylbenzyl chloride) for fuel cell applications. Chemsuschem 6 (2):275–282
30. Yang J, Li Q, Cleemann LN et al (2013) Cross-linked hexafluoropropylidene polybenzimidazole membranes with chloromethyl polysulfone for fuel cell applications. Adv Energy Mater 3:622–630
31. Aili D, Li Q, Christensen et al (2011) Crosslinking of polybenzimidazolemembranes by divinylsulfone post-treatment for high-temperature proton exchange membrane fuel cell applications. Polym Int 60(8):1201–1207
32. Vogel H, Marvel CS (1961) Polybenzimidazoles, new thermally stable polymers. J Polym Sci 50(154):511–539
33. Gillham JK (1963) Polymer structure: cross-linking of a polybenzimidazole. Science 139 (3554):494–495
34. Aili D, Cleemann LN, Li Q et al (2012) Thermal curing of PBI membranes for high temperature PEM fuel cells. J Mater Chem 22:5444–5453
35. Søndergaard T, Cleemann LN, Becker H (2017) Long-term durability of HT-PEM fuel cells based on thermally cross-linked polybenzimidazole. J Power Sources 342:570–578

36. Kerres J, Schönberger F, Chromik A et al (2008) Partially fluorinated arylene polyethers and their ternary blend membranes with PBI and H_3PO_4. Part I. Synthesis and characterisation of polymers and binary blend membranes. Fuel Cells 8(3–4):175–187
37. Li Q, Jensen JO, Pan C et al (2008) Partially fluorinated aarylene polyethers and their ternary blends with PBI and H_3PO_4. Part II. Characterisation and fuel cell tests of the ternary membranes. Fuel Cells 8(3–4):188–199
38. Aili D, Jensen JO, Li Q (2016) Polybenzimidazole membranes by post acid doping. In: Li Q et al (eds) High temperature polymer electrolyte membrane fuel cells. Springer International Publishing, Switzerland, pp 195–215
39. Fishel K, Qian G, Benicewicz BC (2016) PBI membranes via the PPA process. In: Li Q et al (eds) High temperature polymer electrolyte membrane fuel cells. Springer International Publishing, Switzerland, pp 217–238
40. Yu S, Zhang H, Xiao L et al (2009) Synthesis of poly (2,2′-(1,4-phenylene) 5,5′-bibenzimidazole) (para-PBI) and phosphoric acid doped membrane for fuel cells. Fuel Cells 9:318–324
41. Linares JJ, Battirola LC, Lobato J (2016) PBI-based composite membranes. In: Li Q et al (eds) High temperature polymer electrolyte membrane fuel cells. Springer International Publishing, Switzerland, pp 275–295
42. Jones DJ, Rozière J (2008) Advances in the development of inorganic–organic membranes for fuel cell applications. In: Scherer GG (ed) Fuel cells I. Advance polymer science, vol 215. Springer, Berlin, Heidelberg, pp 219–264
43. Lobato J, Cañizares P, Rodrigo MA et al (2011) A novel titanium PBI-based composite membrane for high temperature PEMFCs. J Membr Sci 369:105–111
44. Pinar FJ, Cañizares P, Rodrigo MA et al (2015) Long-term testing of a high-temperature proton exchange membrane fuel cell short stack operated with improved polybenzimidazole-based composite membranes. J Power Sources 274:177–185
45. Namazi H, Ahmadi H (2011) Improving the proton conductivity and water uptake of polybenzimidazole-based proton exchange nanocomposite membranes with TiO_2 and SiO_2 nanoparticles chemically modified surfaces. J Power Sources 196:2573–2583
46. Suryani Liu Y-L (2009) Preparation and properties of nanocomposite membranes of polybenzimidazole/sulfonated silica nanoparticles for proton exchange membranes. J Membr Sci 332:121–128
47. Kurdakova V, Quartarone E, Mustarelli P et al (2010) PBI-based composite membranes for polymer fuel cells. J Power Sources 195:7765–7769
48. Chuang SW, Hsu SLC, Liu YH (2007) Synthesis and properties of fluorine-containing polybenzimidazole/silica nanocomposite membranes for proton exchange membrane fuel cells. J Membr Sci 305(1–2):353–363
49. Plackett D, Siu A, Li QF et al (2011) High-temperature proton exchange membranes based on polybenzimidazole and clay composites for fuel cells. J Membr Sci 383:78–87
50. Hsu SLC, Chang KC (2002) Synthesis and properties of polybenzoxazoleclay nanocomposites. Polymer 43:4097–4101
51. Chuang SW, Hsu SLC, Hsu CL (2007) Synthesis and properties of fluorinecontaining polybenzimidazole/montmorillonite nanocomposite membranes for direct methanol fuel cell applications. J Power Sources 168:172–177
52. He RH, Li QF, Xiao G et al (2003) Proton conductivity of phosphoric acid doped polybenzimidazole and its composites with inorganic proton conductors. J Membr Sci 226:169–184
53. Yamazaki Y, Jang MY, Taniyama T (2004) Proton conductivity of zirconium tricarboxybutylphosphonate/PBI nanocompositemembrane. Sci Technol Adv Mater 5:455–459
54. Jang MY, Yamazaki Y (2005) Preparation and characterization of composite membranes composed of zirconium tricarboxybutylphosphonate and polybenzimidazole for intermediate temperature operation. J Power Sources 139:2–8

55. Heo P, Kajiyama N, Kobayashi K (2008) Proton conduction in $Sn_{0.95}Al_{0.05}P_2O_7$–PBI–PTFE composite membrane. Electrochem Solid State Lett 11:B91–B95
56. Asensio JA, Borros S, Gomez-Romero P (2003) Electrochem Commun 5:967
57. Gómez-Romero P, Asensio JA, Borrós S (2005) Hybrid proton-conducting membranes for polymer electrolyte fuel cells: phosphomolybdic acid doped poly(2,5-benzimidazole)—(ABPBI-$H_3PMo_{12}O_{40}$). Electrochim Acta 50:4715–4720
58. Staiti P, Minutoli M, Hocevar S (2000) Membranes based on phosphotungstic acid and polybenzi-midazole for fuel cell application. J Power Sources 90:231–235
59. He R, Li Q, Xiao G et al (2003) Proton conductivity of phosphoric acid doped polybenzimidazole and its composites with inorganic proton conductors. J Membr Sci 226:169–184
60. Verma A, Scott K (2010) Development of high-temperature PEMFC based on heteropoly-acids and polybenzimidazole. J Solid State Electrochem 14:213–219
61. He R, Li Q, Xiao G et al (2003) Proton conductivity of phosphoric acid doped polybenzimidazole and its composites with inorganic proton conductors. J Membr Sci 226:169–184
62. Staiti P (2001) Proton conductive membranes based on silicotungstic acid/silica and polybenzimidazole. Mater Lett 47:241–246
63. Staiti P (2001) Proton conductive membranes constituted of silicotungstic acid anchored to silica-polybenzimidazole matrices. J New Mater Electrochem Sys 4:181–186
64. Staiti P, Minutoli M (2001) Influence of composition and acid treatment on proton conduction of composite polybenzimidazole membranes. J Power Sources 94:9–13
65. Lee JW, Khan SB, Akhtar K et al (2012) Fabrication of composite membrane based on silicotungstic heteropolyacid doped polybenzimidazole for high temperature PEMFC. Int J Electrochem Sci 7:6276–6288
66. Tang HL, Pan M, Jiang SP (2011) Self assembled 12-tungstophosphoric acid-silica mesoporous nanocomposites as proton exchange membranes for direct alcohol fuel cells. Dalton Trans 40:5220–5227
67. Aili D, Zhang J, Jakobsen MTD et al (2014) Exceptional durability enhancement of PA/PBI based polymer electrolyte membrane fuel cells for high temperature operation at 200 °C. J Mater Chem 4A:4019–4024
68. Zhang J, Aili D, Bradley J et al (2018) In situ formed phosphoric acid/phosphosilicate nanoclusters in the exceptional enhancement of durability of polybenzimidazole membrane fuel cells at elevated high temperatures. J Electrochem Soc 164:F1615–F1625
69. Hu S, Lozada-Hidalgo M, Wang FC et al (2014) Proton transport through one-atom-thick crystals. Nature 516:227–230
70. Coleman JN, Khan U, Blau WJ et al (2006) Small but strong: a review of the mechanical properties of car-bon nanotube–polymer composites. Carbon 1360(44):1624–1652
71. Suryani Chang C-M, Liu Y-L et al (2011) Polybenzimidazole membranes modified with polyelectrolyte-functionalized multiwalled carbon nanotubes for proton exchange membrane fuel cells. J Mater Chem 21:7480–7486
72. Wu J-F, Lo C-F, Li L-Y et al (2014) Thermally stable polybenzimidazole/carbon nano-tube composites for alkaline direct methanol fuel cell applications. J Power Sources 246:39–48
73. Lu Y, Chen J, Cui H et al (2008) Doping of carbon fiber into polybenzimidazole matrix and mechanical properties of structural carbon fiber-doped polybenzimidazole composites. Compos Sci Technol 68:3278–3284
74. Kannan R, Kagalwala HN, Chaudhari HD et al (2011) Improved performance of phosphonated carbon nanotube-polybenzimidazole composite membranes in proton exchange membrane fuel cells. J Mater Chem 21(7223–7231):1384
75. Li N, Zhang F, Wang J et al (2009) Dispersions of carbon nanotubes in sulfonated poly[bis (benzimidazobenzisoquinolinones)] and their proton-conducting composite membranes. Polymer 50:3600–3608

76. Zarrin H, Higgins D, Jun Y et al (2011) Functionalized graphene oxide nanocomposite membrane for low humidity and high temperature proton exchange membrane fuel cells. J Phys Chem C 115:20774–20781
77. Xu C, Liu X, Cheng J et al (2015) A polybenzimidazole/ionic-liquid-graphite-oxide composite membrane for high temperature polymer electrolyte membrane fuel cells. J Power Sources 274:922–927
78. Zhang N, Wang BL, Zhang YR et al (2014) Mechanically reinforced phosphoric acid doped quaternized poly(ether ether ketone) membranes via cross-linking with functionalized graphene oxide. Chem Comm 50:15381–15384
79. He Y, Wang J, Zhang H et al (2014) Polydopamine-modified graphene oxide nanocomposite membrane for proton exchange membrane fuel cell under anhydrous conditions. J Mater Chem A 2(25):9548–9558
80. Karim MR, Hatakeyama K, Matsui T et al (2013) Graphene oxide nanosheet with high proton conductivity. J Am Chem Soc 135:8097–8100
81. Tateishi H, Hatakeyama K, Ogata C et al (2013) J Electrochem Soc 160(11):F1175–F1178
82. Chen D, Tang L, Li J (2010) Graphene-based materials in electrochemistry. Chem Soc Rev 39:3157–3180
83. Xue C, Zou J, Sun Z et al (2014) Graphite oxide/functionalized graphene oxide and polybenzimidazole composite membranes for high temperature proton exchange membrane fuel cells. Int J Hydrogen Energy 39:7931–7939
84. Wang Y, Shi Z, Fang J et al (2011) Direct exfoliation of graphene in methanesulfonic acid and facile synthesis of graphene/polybenzimidazole nanocomposites. J Mater Chem 21:505–512
85. Üregen N, Pehlivanoglu K, Özdemir Y et al (2017) Development of polybenzimidazole/graphene oxide composite membranes for high temperature PEM fuel cells. Int J Hydrogen Energy 42:2636–2647
86. Yang J, Liu C, Gao L et al (2015) Novel composite membranes of triazole modified graphene oxide and polybenzimidazole for high temperature polymer electrolyte membrane fuel cell applications. RSC Adv 5:101049–101054
87. Xu CX, Liu XT, Cheng JG et al (2015) A polybenzimidazole/ionic-liquid-graphite-oxide composite membrane for high temperature polymer electrolyte membrane fuel cells. J Power Sources 274:922–927
88. Wang C, Lin B, Qiao G et al (2016) Polybenzimidazole/ionic liquid functionalized graphene oxide nanocomposite membrane for alkaline anion exchange membrane fuel cells. Mater Lett 173:219–222
89. Zhang N, Wang B, Zhang Y et al (2014) Mechanically reinforced phosphoric acid doped quaternized poly(ether ether ketone) membranes via cross-linking with functionalized graphene oxide. Chem Commun 50:15381–15384
90. Cai Y, Yue Z, Xu S (2017) A novel polybenzimidazole composite modified by sulfonated graphene oxide for high temperature proton exchange membrane fuel cells in anhydrous atmosphere. J Appl Polym Sci 134(25):44986–44993
91. Xu C, Cao Y, Kumar R et al (2011) A polybenzi-midazole/sulfonated graphite oxide composite mem-brane for high temperature polymer electrolyte membrane fuel cells. J Mater Chem 21:11359–11364
92. Wang Y, Yu J, Chen L et al (2013) Nacre-like graphene paper reinforced by polybenzimidazole. RSC Adv 3:20353–20362
93. Deng Y, Wang G, Fei MM et al (2016) A polybenzimidazole/graphite oxide based three layer membrane for intermediate temperature polymer electrolyte membrane fuel cells. RSC Adv 6(76):72224–72229
94. Kallitsis JK, Geormezi M, Neophytides SG (2009) Polymer electrolyte membranes for high-temperature fuel cells based on aromatic polyethers bearing pyridine units. Polym Int 58(11):1226–1233
95. Kallitsis JK, Andreopoulou AK, Daletouet M et al (2016) Pyridine containing aromatic polyether membranes. In: Li Q et al (eds) High temperature polymer electrolyte membrane fuel cells. Springer International Publishing, Switzerland, pp 91–126

96. Papadimitriou KD, Paloukis F, Neophytides SG et al (2011) Cross-linking of side chain unsaturated aromatic polyethers for high temperature polymer electrolyte membrane fuel cell applications. Macromolecules 44:4942–4951
97. Morfopoulou CI, Andreopoulou AK, Daletou MK et al (2013) Cross-linked high temperature polymer electrolytes through oxadiazole bond formation and their applications in HT PEM fuel cells. J Mater Chem A 1:1613–1622
98. Carollo A, Quartarone E, Tomasi C et al (2006) Developments of new proton conducting membranes based on different polybenzimidazole structures for fuel cells applications. J Power Sources 160(1):175–180
99. Kurdakova V, Quartarone E, Mustarelli P et al (2010) PBI-based composite membranes for polymer fuel cells. J Power Sources 195:7765–7769
100. Gubler L, Kramer D, Belack J et al (2007) A polybenzimidazole-based membrane for the direct methanol fuel cell. J Electrochem Soc 154(9):B981–B987
101. Lobato J, Cañizares P, Rodrigo MA et al (2008) Performance of a vapor-fed polybenzimidazole (PBI)-based direct methanol fuel cell. Energy Fuels 22:3335–3345
102. Modestov AD, Tarasevich MR, Pu H (2012) Investigation of methanol electrooxidation on Pt and Pt–Ru in H_3PO_4 using MEA with PBI–H_3PO_4 membrane. J Power Sources 205:207–214
103. Zhao X, Yuan W, Wu Q et al (2015) High-temperature passive direct methanol fuel cells operating with concentrated fuels. J Power Sources 273:517–521
104. Li L-Y, Yu B-C, Shih C-M et al (2015) Polybenzimidazole membranes for direct methanol fuel cell: acid-doped or alkali-doped? J Power Sources 287:386–395
105. Jensen JO, Vassiliev A, Olsen MI et al (2012) Direct dimethyl ether fuelling of a high temperature polymer fuel cell. J Power Sources 211:173–176
106. Wang Y-J, Wilkinson DP, Zhang J (2011) Noncarbon support materials for polymer electrolyte membrane fuel cell electrocatalysts. Chem Rev 111(12):7625–7651
107. Honji A, Mori T, Tamura K et al (1988) Aggloemration of platinum particles supported on carbon in phosphoreic acid. J Electrochem Soc 135:355–359
108. Bindra P, Clouser SJ, Yeager E (1979) Platinum dissolution in concentrated phosphoric acid. J Electrochem Soc 126:1631–1632
109. Aragane J, Murahashi T, Odaka T (1988) Change of Pt distribution in the active components of phosphoric acid fuel cell. J Electrochem Soc 135:844–850
110. Ahluwalia RK, Arisetty S, Wang X et al (2013) Thermodynamics and kinetics of platinum dissolution from carbon-supported electrocatalysts in aqueous media under potentiostatic and potentiodynamic conditions. J Eletrcochem Soc 160:F447–F455
111. Wang XP, Kumar R, Myers DJ (2006) Effect of voltage on platinum dissolution relevance to polymer electrolyte fuel cells. Electrochem Solid State Lett 9:A225–A227
112. Aragane J, Urushibata H, Murahashi T (1996) Effect of operational potential on performance decay rate in a phosphoric acid fuel cell. J Appl Electrochem 26:147–152
113. Kinoshita K (1988) Carbon. Electrochemical and physicochemical properties. Wiley, New York
114. Roen LM, Paik CH, Jarvic TD (2004) Electrocatalytic corrosion of carbon support in PEMFC cathodes. Electrochem Solid State Lett 7:A19–A22
115. Oh H-S, Lee J-H, Kim H (2012) Electrochemical carbon corrosion in high temperature proton exchange membrane fuel cells. Inter J Hydrogen Energy 37:10844–10849
116. Stevens DA, Dahn JR (2005) Thermal degradation of the support in carbon-supported platinum electrocatalysts for PEM fuel cells. Carbon 43:179–188
117. Oono Y, Fukuda T, Sounai A et al (2010) Influence of operating temperature on cell performance and endurance of high temperature proton exchange membrane fuel cells. J Power Sources 195:1007–1014
118. Oono Y, Sounai A, Hori M (2012) Long-term cell degradation mechanism in high-temperature proton exchange membrane fuel cells. J Power Sources 210:366–373
119. Hartnig C, Schmidt TJ (2011) Simulated start-stop as a rapid aging tool for polymer electrolyte fuel cell electrodes. J Power Sources 196:5564–5572

120. Søndergaard T, Cleemann LN, Zhong LJ, Becker H, Steenberg T, Hjuler HA, Seerup L, Li QF, Jensen JO (2017) Catalyst degradation under potential cycling as an accelerated stress test for PBI based high temperature PEM fuel cells—effect of humidification. Electrocatalysis. https://doi.org/10.1007/s12678-017-0427-1
121. Mader J, Xiao L, Schmidt TJ et al (2008) Polybenzimidazole/acid complexes as high-temperature membranes. Adv Polym Sci 216(1):63–124
122. Schmidt TJ, Baurmeister J (2006) Durability and reliability in high-temperature reformed hydrogen PEFCs. ECS Trans 3:861–869
123. Oono Y, Sounai A, Hori M (2013) Prolongation of lifetime of high temperature proton exchange membrane fuel cells. J Power Sources 241:87–93
124. Søndergaard T, Cleemann LN, Becker H et al (2017) Long-term durability of HT-PEM fuel cells based on thermally crosslinked polybenzimidazole. J Power Sources 342:570–578
125. Søndergaard T, Cleemann LN, Becker H et al (2018) Long term durability of PBI based HT-PEM fuel cells under varied flow rates and temperatures. J Electrochem Soc (in press)
126. Landsman DA, Luczak FJ (2003) Catalyst studies and coating technologies. In: Vielstichm W et al (eds) Handbook of fuel cells, vol 3. Wiley, pp 811–831
127. Cleemann LN, Buazar F, Li Q et al (2013) Catalyst degradation in high temperature proton exchange membrane fuel cells based on acid doped polybenzimidazole membranes. Fuel Cells 13:822–831
128. Okamoto M, Fujigaya T, Nakashima N (2009) Design of an assembly of poly(benzimidazole), carbon nanotubes, and Pt nanoparticles for a fuel-cell electrocatalyst with an ideal interfacial Nanostructure. Small 5:735–740
129. Fujigaya T, Okamoto M, Nakashima N (2009) Design of an assembly of pyridine-containing polybenzimidazole, carbon nanotubes and Pt nanoparticles for a fuel cell electrocatalyst with a high electrochemically active surface area. Carbon 47:3227–3232
130. Fujigaya T, Nakashima N (2013) Fuel cell electrocatalyst using polybenzimidazole-modified carbon nanotubes as support materials. Adv Mater 25:1666–1681
131. Hafez IH, Berber MR, Fujigaya T et al (2014) Enhancement of platinum mass activity on the surface of polymer-wrapped carbon nanotube-based fuel cell electrocatalysts. Sci Rep 4 (1):6295
132. Berber MR, Hafez IH, Fujigaya T et al (2014) Durability analysis of polymer-coated pristine carbon nanotube-based fuel cell electrocatalysts under non-humidified conditions. J Mater Chem A 2:19053–19059
133. Berber MR, Fujigaya T, Nakashima N (2014) High-temperature polymer electrolyte fuel cell using poly(vinylphosphonic acid) as an electrolyte shows a remarkable durability. ChemCatChem 6:567–571
134. Fujigaya T, Hirata S, Berber MR et al (2016) Improved durability of electrocatalyst based on coating of carbon black with polybenzimidazole and their application in polymer electrolyte fuel cells. ACS Appl Mater Interfaces 8:14494–14502
135. Yang Z, Li J, Ling Y et al (2017) Bottom-up design of high-performance pt electrocatalysts supported on carbon nanotubes with homogeneous ionomer distribution. ChemCatChem 9:3307–3313
136. Yang Z, Luo F (2017) Pt nanoparticles deposited on dihydroxy-polybenzimidazole wrapped carbon nanotubes shows a remarkable durability in methanol electro-oxidation. Int J Hydrogen Energy 42:507–514
137. Fujigaya T, Shi Y, Yang J et al (2017) A highly efficient and durable carbon nanotube-based anode electrocatalyst for water electrolyzers. J Mater Chem A 5:10584–10590
138. Lin Y, Liu Q, Fan J et al (2016) Highly dispersed palladium nanoparticles on poly(N1, N3-dimethylbenzimidazolium)iodide functionalized multiwalled carbon nanotubes for ethanol oxidation in alkaline solution. RSC Adv 6:102582–102594
139. Matsumoto K, Fujigaya T, Yanagi H et al (2011) Very high performance alkali anion-exchange membrane fuel cells. Adv Func Mater 21(6):1089–1094
140. Fujigaya T, Hirata S, Naotoshi Nakashima (2014) A highly durable fuel cell electrocatalyst based on polybenzimidazole-coated stacked graphene. J Mater Chem A 2(11):3888–3893

141. Li Z-F, Zhang HY, Yang F et al (2014) Pt catalysts supported on polybenzimidazole-grafted graphene for PEMFCs. ECS Trans 64(3):131–136
142. Li Z-F, Xin L, Yang F et al (2015) Hierarchical polybenzimidazole-grafted graphene hybrids as supports for Pt nanoparticle catalysts with excellent PEMFC performance. Nano Energy 16:281–292
143. Xin L, Yang F, Qiu Y et al (2016) Polybenzimidazole (PBI) functionalized nanographene as highly stable catalyst support for polymer electrolyte membrane fuel cells (PEMFCs). J Electrochem Soc 163(10):F1228–F1236
144. Zeng L, Zhao TS, An L et al (2015) A high-performance sandwiched-porous polybenzimidazole membrane with enhanced alkaline retention for anion exchange membrane fuel cells. Energy Environ Sci 8:2768–2774
145. Boaventura M, Brandaõ L, Mendes A (2011) Single-wall nanohorns as electrocatalyst support for high temperature PEM fuel cells. J Electrochem Soc 158(4):B394–B401
146. Jasinski R (1964) A new fuel cell catalyst. Nature 201:1212–1213
147. Chen Z, Higgins D, Yu A et al (2011) A review on non-precious metal electrocatalysts for PEM fuel cells. Energy Environ Sci 4:3167–3192
148. Nie Y, Li L, Wei Z (2015) Recent advancements in Pt and Pt-free catalysts for oxygen reduction reaction. Chem Soc Rev 44:2168–2201
149. Xia ZH, An L, Chen PK et al (2016) Non-Pt nanostructured catalysts for oxygen reduction reaction: synthesis, catalytic activity and its key factors. Adv Energy Mater 6(17):1600458
150. Banham D, Ye S, Pei K et al (2015) A review of the stability and durability of non-precious metal catalysts for the oxygen reduction reaction in proton exchange membrane fuel cells. J Power Sources 285:334–348
151. Kramm UI, Lefèvre M, Larouche N et al (2014) Correlations between mass activity and physicochemical properties of Fe/N/C catalysts for the ORR in PEM fuel cell via Fe-57 mossbauer spectroscopy and other techniques. J Am Chem Soc 136(3):978–985
152. Chung HT, Cullen DA, Higgins D et al (2017) Direct atomic-level insight into the active sites of a high-performance PGM-free ORR catalyst. Science 357:479–484
153. Shui J, Chen C, Grabstanowicz L et al (2015) Highly efficient nonprecious metal catalyst prepared with metal–organic framework in a continuous carbon nanofibrous network. Proc Natl Acad Sci 112:10629–10634
154. Chung HT, Cullen DA, Higgins D et al (2017) Direct atomic-level insight into the active sites of a high-performance PGM-free ORR catalyst. Science 357:479–484
155. Lefevre M, Proietti E, Jaouen F et al (2009) Iron-based catalysts with improved oxygen reduction activity in polymer electrolyte fuel cells. Science 324:71–74
156. Proietti E, Jaouen F, Lefevre M et al (2011) Iron-based cathode catalyst with enhanced power density in polymer electrolyte membrane fuel cells. Nature Com 2(1):416
157. Zelenay P, Scharifker BR, Bockris JO et al (1986) Comparison of the properties of CF_3SO_3H and H_3PO_4 in relation to fuel-cells. J Electrochem Soc 133:2262–2267
158. Zelenay P, Habib MA, Bockris JO'M (1986) Adsorption from solution on platinum: an in situ FTIR and radiotracer study. Langmuir 2:393–405
159. He Q, Yang X, Chen W et al (2010) Influence of phosphate anion adsorption on the kinetics of oxygen electroreduction on low index Pt(hkl) single crystals. Phys Chem Chem Phys 12:12544–12555
160. Zecevic SK, Wainright JS, Litt MH et al (1997) Kinetics of O_2 reduction on a Pt electrode covered with a thin film of solid polymer electrolyte. J Electrochem Soc 144:2973–2982
161. Liu ZY, Wainright JS, Savinell RF (2004) High-temperature polymer electrolytes for PEM fuel cells: study of the oxygen reduction reaction (ORR) at a Pt-polymer electrolyte interface. Chem Eng Sci 59:4833–4838
162. Liu ZY, Wainright JS, Litt MH et al (2006) Study of the oxygen reduction reaction (ORR) at Pt interfaced with phosphoric acid doped polybenzimidazole at elevated temperature and low relative humidity. Electrochim Acta 51:3914–3923
163. Zhang J, Tang Y, Song C et al (2006) Polybenzimidazole-membrane-based PEM fuel cell in the temperature range of 120–200 °C. J Power Sources 172:163–171

164. Neyerlin KC, Singh A, Chu D (2008) Kinetic characterization of a Pt-Ni/C catalyst with a phosphoric acid doped PBI membrane in a proton exchange membrane fuel cell. J Power Sources 176:112–117
165. Li Q, Wu G, Cullen DA et al (2014) Phosphate-tolerant oxygen reduction catalysts. ACS Catal 4:3193–3200
166. Hu Y, Jensen JO, Zhang W et al (2014) Hollow spheres of iron carbide nanoparticles encased in graphitic layers as oxygen reduction catalysts. Angew Chem Int Ed 53(14):3675–3679
167. Wang Q, Zhou ZY, Lai YJ et al (2014) Phenylenediamine-based FeNx/C catalyst with high activity for oxygen reduction in acid medium and its active-site probing. J Am Chem Soc 136:10882–10885
168. Gupta S, Fierro C, Yeager E (1991) The effects of cyanide on the electrochemical properties of transition metal macrocycles for oxygen reduction in alkaline solutions. J Electroanal Chem Interf Electrochem 306:239–250
169. Zhong L, Jensen JO, Cleemann LN et al (2017) Electrochemical probing into the active sites of graphitic-layer encapsulated iron oxygen reduction reaction electrocatalysts. Sci Bull. https://doi.org/10.1016/j.scib.2017.11.017
170. Kannan A, Li Q, Cleemann LN et al (2018) Acid distribution and durability of HT-PEM fuel cells with different electrode supports. Accepted for publication in Fuel cells
171. Martin S, Li Q, Steenberg T et al (2014) Binderless electrodes for high-temperature polymer electrolyte membrane fuel cells. J Power Sources 272:559–566
172. Martin S, Jensen JO, Li Q (2015) Lowering the platinum loading of high temperature polymer electrolyte membrane fuel cells with acid doped polybenzimidazole membranes. J Power Sources 293:51–56
173. Kundler I, Hickmann T (2016) Bipolar plates and gaskets: different materials and processing methods. In: Li Q et al (eds) High temperature polymer electrolyte membrane fuel cells. Springer International Publishing, Switzerland, pp 1–4
174. Li QF, He RH, Jensen JO et al (2003) Approaches and recent development of polymer electrolyte membranes for fuel cells operating above 100 °C. Chem Mater 15(26):4896–4915
175. Chandan A, Hattenberger M, El-kharouf A et al (2013) High temperature (HT) polymer electrolyte membrane fuel cells (PEMFC)—a review. J Power Sources 231:264–278
176. Li QF, He RH, Jensen JO et al (2003) Approaches and recent development of polymer electrolyte membranes for fuel cells operating above 100 °C. Chem Mater 15(26):4896–4915
177. Quartarone E, Angioni S, Mustarelli P (2017) Polymer and composite membranes for proton-conducting, high-temperature fuel cells: a critical review. Materials 10(7):687
178. Haque MA, Sulong AB, Loh KS et al (2017) Acid doped polybenzimidazoles based membrane electrode assembly for high temperature proton exchange membrane fuel cell: a review. Int J Hydrogen Energy 42(14):9156–9179
179. Rosli RE, Sulong AB, Daud WRW et al (2017) A review of high-temperature proton exchange membrane fuel cell (HT-PEMFC) system. Int J Hydrogen Energy 42(14):9293–9314
180. Araya SS, Zhou F, Liso V et al (2016) A comprehensive review of PBI-based high temperature PEM fuel cells. Int J Hydrogen Energy 41(46):21310–21344
181. Zeis R (2015) Materials and characterization techniques for high-temperature polymer electrolyte membrane fuel cells. Beilstein J Nanotechnol 6:68–83
182. Subianto S (2014) Recent advances in polybenzimidazole/phosphoric acid membranes for high-temperature fuel cells. Polym Int 63:1134–1144
183. Li Q, Aili D, Hjuler HA et al (eds) (2016) High temperature polymer electrolyte membrane fuel cells. Springer International Publishing, Switzerland

Chapter 4
The Role of Carbon Blacks as Catalyst Supports and Structural Elements in Polymer Electrolyte Fuel Cells

Masahiro Watanabe and Donald A. Tryk

4.1 Introduction

In polymer electrolyte fuel cells (PEFCs) based on proton exchange membranes (PEM), carbon blacks have been used as supports for platinum and its alloys since the 1970s [1]. In fact, the transition from the use of Pt black to carbon-supported Pt was one of the most significant events in the history of PEFCs. The 1979 patent of Jalan and Bushnell is still used today for the deposition of Pt on carbon blacks [2]. In this review, we intend to examine some of the past and present research trends in carbon-supported catalysts in PEFCs, focusing on work carried out at the University of Yamanashi over a period of around 30 years, from the mid-1980s to the present.

One of the overarching themes in the fuel cell area is that of the microscopic region that has been variously called the "triple-phase boundary", "triple contact", or "three-phase region", which is still not very precisely defined but essentially involves the approximate location where the solid, electronically conducting phase, usually including some form of carbon, the liquid or solid ionically conducting phase, and the gaseous phase, in which reactants such as hydrogen and oxygen and product water are transported (Fig. 4.1). Thus, there are three interfaces: solid–liquid, solid–gas, and liquid–gas. As pointed out by Appleby and Foulkes, the inventor of the fuel, Sir William Grove, recognized the existence of this region, calling it a "notable surface of action" [3].

It was pointed out by Bockris and Cahan that the current flow depends greatly upon the thickness of the electrolyte film through which the gaseous reactants

M. Watanabe (✉) · D. A. Tryk
Fuel Cell Nanomaterials Center, University of Yamanashi, 6-43 Miyamae, Kofu 400-0021, Japan
e-mail: m-watanabe@yamanashi.ac.jp

D. A. Tryk
e-mail: donald@yamanashi.ac.jp

N. Nakashima (ed.), *Nanocarbons for Energy Conversion: Supramolecular Approaches*, Nanostructure Science and Technology,
https://doi.org/10.1007/978-3-319-92917-0_4

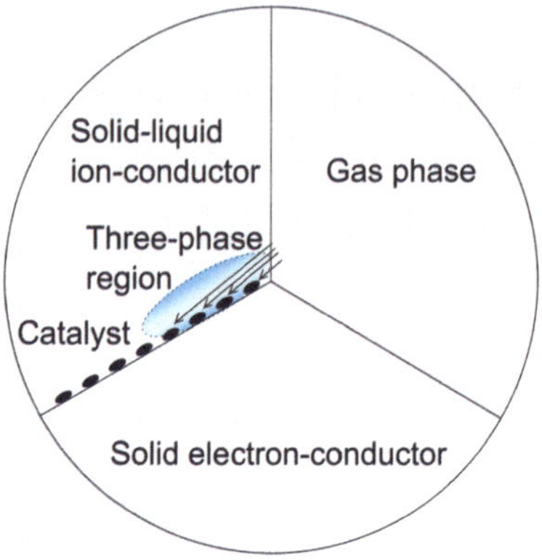

Fig. 4.1 Schematic diagram of the three-phase region. The three-phase region is shown, which does not have distinct boundaries but is the region close to the gas phase, in which most of the current flows, due to the effective mass transport of gas. The flow of electrons, ions, and gas is normal to the surface of the page, while the transport of gas from pores through electrolyte to catalyst is within the plane of the page

(H_2 or O_2) must pass prior to reaching the surface of the electrode [4]. They studied a macroscopic system in which there is a meniscus of liquid electrolyte, with the top edge being extremely thin and supporting the major fraction of the current. This concept is useful not only for the phosphoric acid fuel cell (PAFC) but also the PEFC and other types of fuel cells. In contrast, the solid oxide fuel cell (SOFC) literature makes frequent use of the quantitative length of the triple-phase boundary, since the gases do not diffuse appreciably through the solid oxide ionic conductor.

Each phase has its own function: the electrode provides either a source or sink for electrons, as well as catalytic sites. The ionic phase provides a source or sink for protons, and the gas phase provides a source or sink for reactant and product gases. Thus, for each, the function dictates a particular set of desirable properties, e.g., electronic conductivity of the support, the activity of the catalyst, and so on. For each phase, both the intrinsic function and the durability must be considered.

Fig. 4.2 Schematic diagram of the proposed structure for the high-performance gas diffusion electrode. Reprinted from [5] with permission from Elsevier

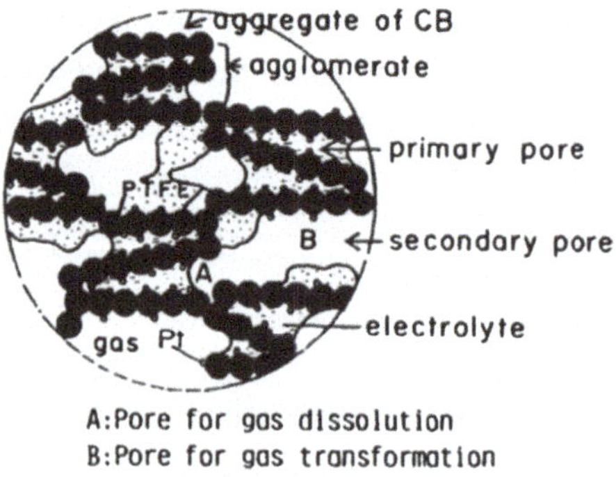

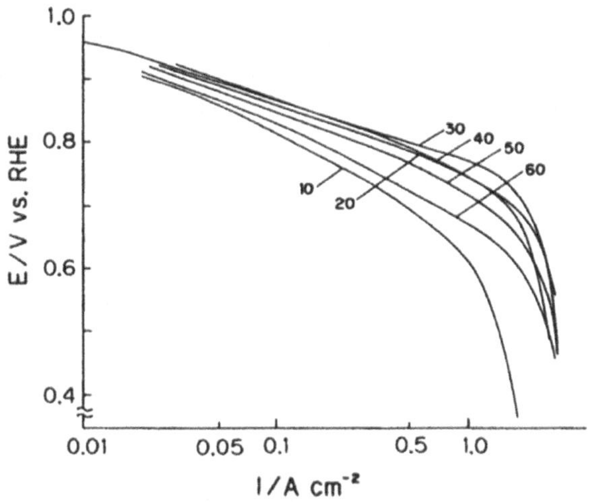

Fig. 4.3 Polarization curves of the electrodes having various PTFE contents for O_2 reduction at 190 °C in 100% H_3PO_4. Platinum was loaded by a colloidal method. The numbers on the curves represent the PTFE content. Reprinted from [5] with permission from Elsevier

We will start in Sect. 4.2 with an overview of the whole system, in order to create perspective and reinforce the three-phase concept. In Sect. 4.3, we focus on the relationship between the Pt catalyst and the support in terms of the performance, and Sect. 4.4, we will discuss the durability of both Pt and support.

4.2 Electrode Structure

Recognizing in the early 1980s, the importance of optimized transport of electrons, ions and gases in the PAFC, the University of Yamanashi (UY) group set about to fabricate gas diffusion electrode (GDE) structures in which there were well-connected networks of gas transport, electron transport, and ion transport. Gas transport channels were provided by the larger, secondary pores, with the aid of PTFE particles that were sintered together during the hot-pressing step (Fig. 4.2) [5]. The ionic transport was provided by the phosphoric acid contained in smaller, primary pores, in which the actual ORR took place.

In Fig. 4.3, we can see the effect of having the proper weight percentage of PTFE on the I-E curves. For low-PTFE percentages, the Tafel slopes were high due to the poor gas mass transport, since the hydrophobic gas channels were not all connected into a continuous network. For high PTFE percentages, the Tafel slopes were also high, due to the lack of continuity of the hydrophilic channels and poor wetting of the Pt catalyst particles. Around 30% PTFE, the Tafel slope was at a minimum and was linear over a wide range of current density, due to the continuity of both hydrophobic and hydrophilic networks. This is an example of the "Goldilocks effect", from the children's story, meaning "not too little, not too much, but just right".

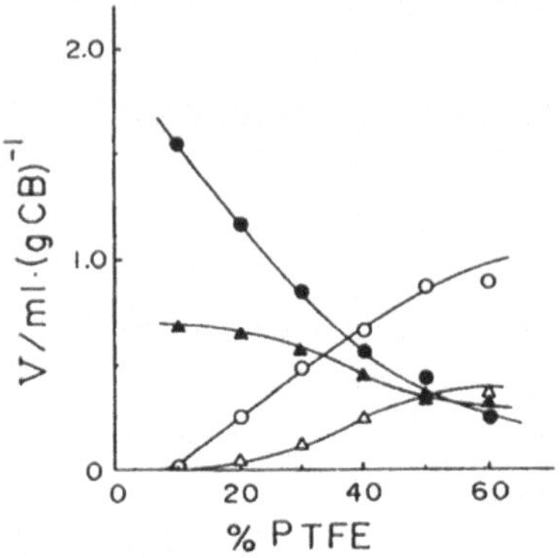

Fig. 4.4 Effect of PFFE content on the degree of occupation of each pore by the electrolyte and gas in the reaction layer of a high-performance gas diffusion electrode. Filled triangles primary pore volumes, filled circles secondary pore volumes filled by electrolyte. Open triangles primary pores, open circles secondary pores, filled by gas. Reprinted from [5] with permission from Elsevier

At 30% PTFE, 36% of the secondary pores were filled with gas, while only 16% of the primary pores were filled with gas (Fig. 4.4). In contrast, 84% of the primary pores were filled with electrolyte, resulting in high utilization of Pt. It was proposed that the electrodes achieved high performance due to the continuous networks of both electrolyte and gas. This was accomplished by proper dispersion of both the catalyzed carbon black and the PTFE particles, the latter then being sintered together with an optimized hot-pressing step.

This work had employed a single type of catalyst powder. As such, some of the Pt catalyst particles were unavoidably trapped in the hydrophobic regions, so that they were not wetted with phosphoric acid, and thus the utilization of Pt (U_{Pt}) defined by S_{CV}/S_d, was a maximum of 75%. The mean surface area of electrolyte-contacted (wetted) Pt particles S_{CV} was determined by applying the electric charge of 210 μC cm^{-2} (Pt) for hydrogen adsorption/reduction in a conventional cyclic voltammogram, whereas the mean surface area without regard for wetting was determined by either of two conventional methods, i.e., with X-ray diffraction (XRD) or transmission electron microscopy (TEM). The work was further refined by preparing two different powders, a hydrophilic catalyst powder and a hydrophobic catalyst-free powder [6]. These two powders were mixed in various proportions, and for lower fractions of the hydrophobic powder, U_{Pt} increased to 100%. The performance was also enhanced greatly as the hydrophobic/hydrophilic powder ratio increased and exhibited a maximum at a ratio of ca. 0.6. The overall optimization process can be described as follows (Fig. 4.5); by usage of hydrophilic catalyst powder, with high U_{Pt}, the current density at high potential i_1 is increased to i_2. However, gas mass transport is lacking, as evidenced by the increase in Tafel slope. The usage of the hydrophobic powder, with the creation of a gas channel network, allows gas to reach all of the wetted catalyst particles. Thus, the Tafel slope decreases to the ideal value, i.e., 90 mV decade^{-1} indicating that all of

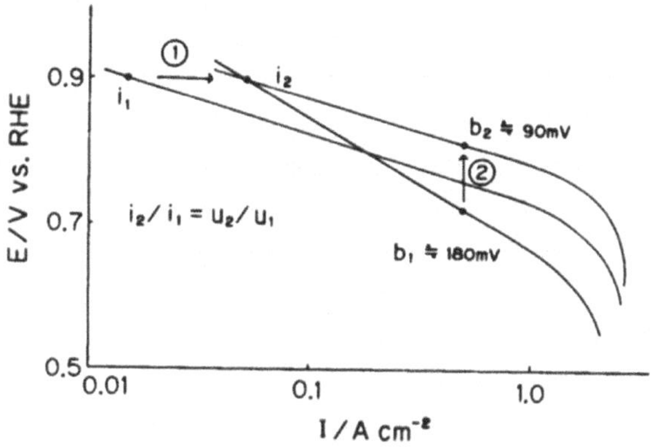

Fig. 4.5 Schematic explanation of the improvement of electrode performance with improving the reaction layer structure (1) on the electrolyte network, and (2) the gas supplying network. Reprinted from [6] with permission from Elsevier

the Pt particles are being utilized without O_2 diffusion limitation. Beyond the optimum addition of the hydrophobic powder, the performance decreases due to the lowered U_{Pt}.

This approach was continued with the use of an improved type of gas-supply powder, in which a new type of gas-supply powder was prepared by coating the carbon black particles with polyethylene and then subjecting them to fluorination (FPE) [7]. This work was further carried out in more detail [8].

These same concepts were then used and further improved in the PEFC system, with a three-layer electrode concept being developed (Fig. 4.6) [8]. This type of structure is still very much used today. The hydrophobic GDL was found to be effective in enhancing the O_2 diffusivity and the catalyst utilization. Due to high O_2 diffusivity, high performance was observed for an electrode with a carbon paper backing layer (CBL) with no FEP treatment at low air utilization. The FEP treatment was found to be suitable for co-generation type fuel cells, which operate at high potential (0.8 V) and high utilization of air (40%).

One of the important differences between the PAFC and PEFC is that, in the former, the hot acid is effective in wetting the catalyst particles mostly located in primary pores, whereas, in the latter, it is more difficult for the ionomer to do so. Song et al. were able to address this problem by preparing a catalyst powder via autoclaving, so that Nafion was able to enter the small, primary pores to better contact the Pt particles residing in these pores (Fig. 4.7) [9]. Thus, significantly higher U_{Pt} values were reached, as high as 90%.

The effect of the autoclave treatment was demonstrated directly via the computed electron tomography (CET) technique (Fig. 4.8) [10]. Silver-stained Nafion was used in order for visualization to be possible in the transmission electron microscope. It became clear that, after the autoclaving treatment, Nafion was able to penetrate the primary pores, due either to a reduction of the effective micelle size of the ionomer or to an increase in the molecular flexibility at high temperature.

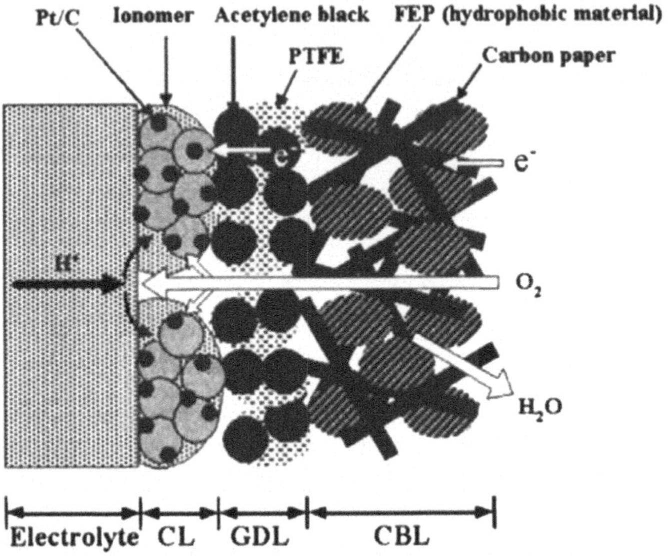

Fig. 4.6 Schematic diagram of the three-layer electrode structure, with catalyst layer (CL), hydrophobic gas diffusion layer (GDL) containing acetylene black and PTFE, and carbon backing layer (CBL), containing carbon paper and hydrophobic FEP. Reprinted from [8] with permission from the Electrochemical Society of Japan

During the assessment of the performance of PEFCs, it was found in our laboratory that there were certain deficiencies in the use of U_{Pt} as an indicator. For example, at higher temperatures, the hydrogen adsorption charge, which is used to measure the actual Pt area in contact with the electrolyte, is less than expected, due to the fact that H adsorption strength decreases with temperature. Also, for Pt alloy catalysts, the H adsorption charge is decreased to an unknown extent due to the weakened H adsorption. Furthermore, even though the ion-conductive network might be able to access all of the catalyst particles at the low current densities used in the cyclic voltammogram (CV), the ionomer film might be too thin for high current density operation.

Thus, it was decided to make use of the Pt effectiveness factor, which is essentially the actual current in a measurement of a porous electrode, with mass transport losses, divided by the current expected with no mass transport losses [11].

The effectiveness factor was originally devised by Thiele in 1939 for use in heterogeneous gas-phase catalysis [12]. Thiele called it the “relative rate of reaction”. Thereafter, its use was taken up in electrochemistry by various workers. Giner and Hunter used the concept in their treatment of the flooded agglomerate model for fuel cells [13], and Mund and Sturm also used it [14].

To measure the intrinsic current density without mass transport is not straightforward. It could be done with a smooth bulk electrode, in principle, but there would be major differences in the surface characteristics. We have found that the channel flow double electrode (CFDE) technique is useful for this purpose, because

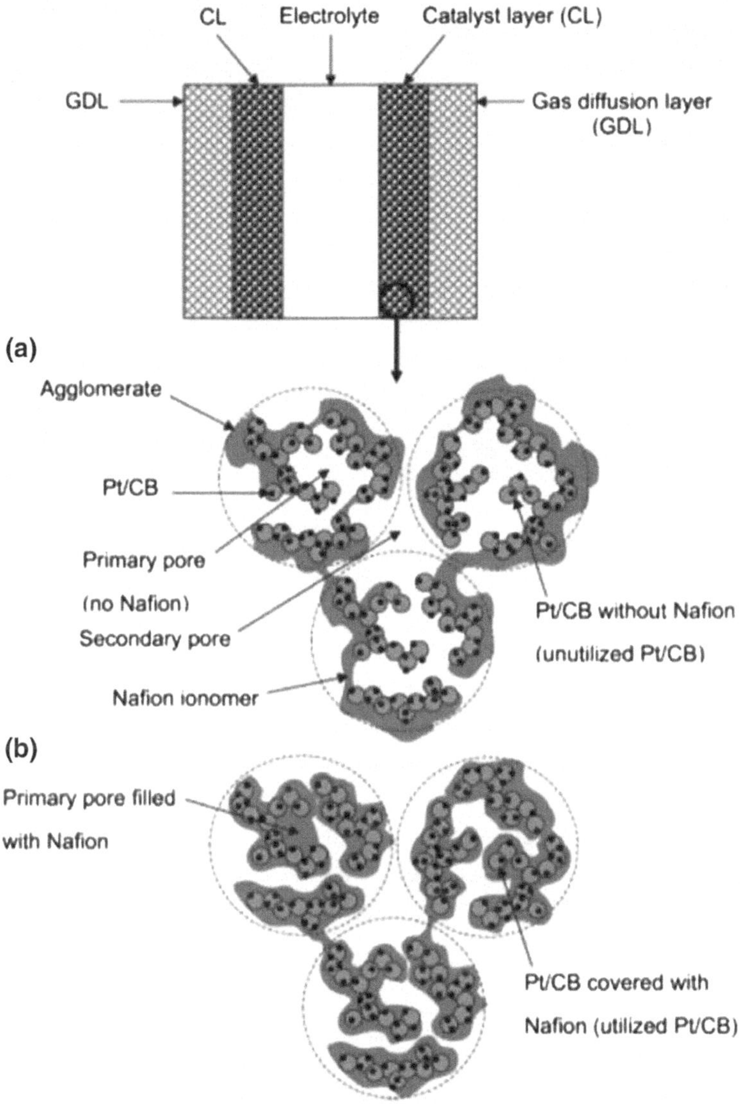

Fig. 4.7 Schematic representation of a catalyst layer structure with **a** low Pt utilization **b** and high Pt utilization. Reprinted with permission from [9]. Copyright 2006 American Chemical Society

it is possible to use a real fuel cell catalyst and to operate in the same temperature range as that in the fuel cell (e.g., 80 °C), and the catalyst wetting should be nearly complete (Fig. 4.9). The well-defined geometry of the experiment enables one to determine the catalyst activity with little effect of mass transport, by use of the same type of approach used with the rotating disk technique, i.e., the Koutecky–Levich relationship:

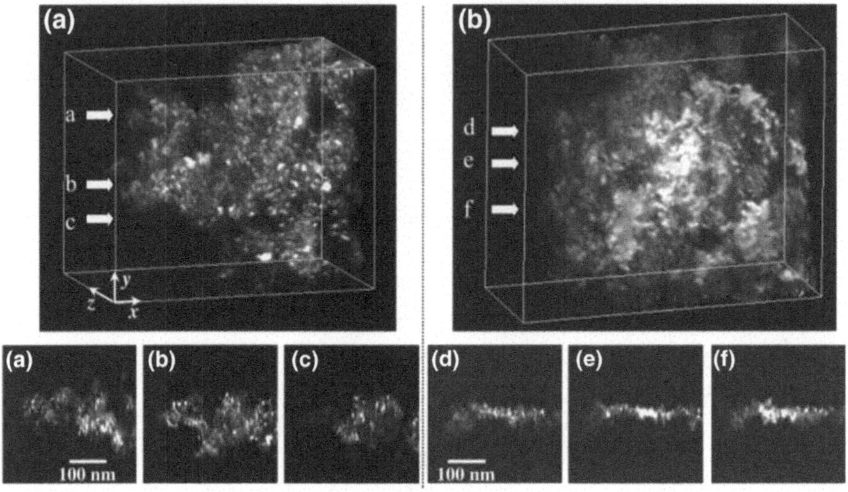

Fig. 4.8 3D reconstructed images of Ag^+-stained Nafion-Pt/CB (N/C = 0.7) **A** with and **B** without autoclave treatment. Images of slices obtained along the y-axis at locations of arrows a–f are shown in panels a–f. Reprinted with permission from [10]. Copyright 2006 American Chemical Society

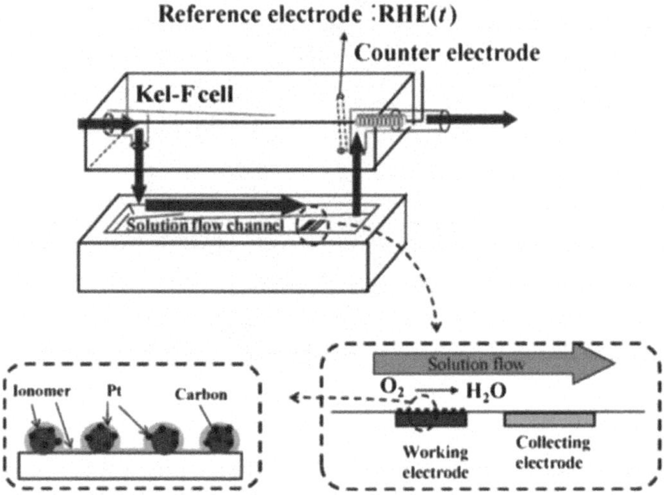

Fig. 4.9 Schematic diagram of the multichannel flow double electrode (CFDE) cell. Reprinted from [11] with permission from Elsevier

$$I^{-1} = I_k^{-1} + I_d^{-1} \tag{4.1}$$

where I is the measured current, I_k is the intrinsic kinetically controlled current, and I_d is the diffusion-controlled current. The I_d is a function of a flow-rate (U_m) of the electrolyte, in which a reactant R is dissolved with a saturated concentration [R] at the reaction temperature, being proportional to $U_m^{1/3}$ and [R]. An intercept value in the extrapolation of the plots of I^{-1} versus $U_m^{-1/3}$ to the infinite U_m value for each [R] or reaction temperature gives the corresponding I_k^{-1} [15].

There are still issues to address, including possible differences in the ORR kinetics comparing (a) liquid acid electrolyte versus ionomer and (b) liquid versus gas phase. Thus, there might still be some uncertainty in the Ef_{Pt} values. Also, there are still some mass transport losses that cannot be fully accounted for, so, at present, the Ef_{Pt} values are reported at a potential sufficiently high to avoid severe mass transport effects (0.85 V).

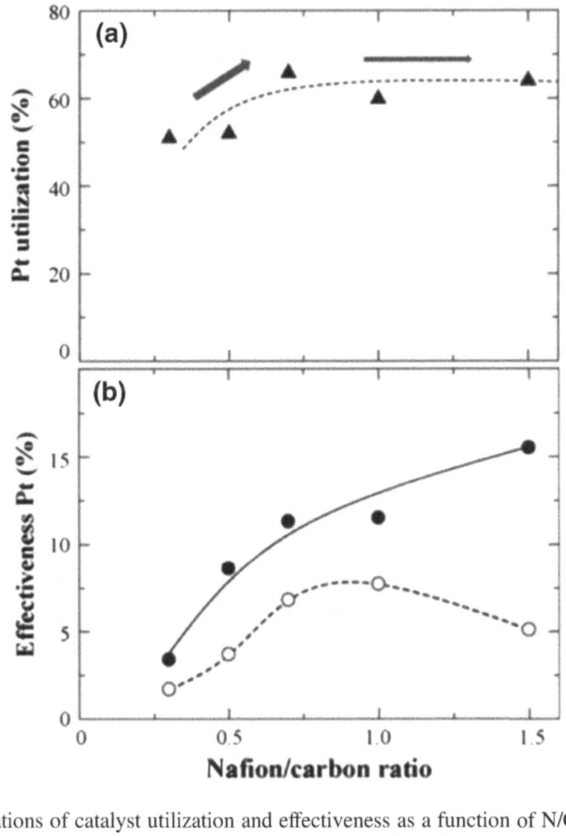

Fig. 4.10 Variations of catalyst utilization and effectiveness as a function of N/C in the CL: **a** Pt utilization (U_{Pt}) determined by CV in humidified N_2 (100% RH) at T_{cell} = 40 °C obtained at a potential scan rate of 20 mV s^{-1}; **b** Ef_{Pt} determined by MA in the MEA and MA_{max} in the CFDE at 0.85 V. The oxidant gases used were O_2 (solid circles) and air (open circles). Reprinted from [11] with permission from Elsevier

There are some interesting results from these measurements. A series of experiments were carried out with membrane-electrode assemblies (MEAs) with various ratios of Nafion to carbon (N/C). In Fig. 4.10, it can be seen that the U_{Pt} values increased and reached a plateau with increasing N/C. However, the effectiveness factors kept on increasing for O_2 feed but reached a peak and then decreased for air feed. However, even at the highest N/C value (1.5), the Ef_{Pt} value under O_2 was only 15%, compared with a U_{Pt} value of 60%. Thus, it appears that there is still much work to do to improve the effectiveness.

Some of the possible problems with low Ef_{Pt} can be traced to the thickness of the Nafion layer, as depicted in Fig. 4.11. The thicker layers can severely impede gas mass transport, while the thinner layers can easily be interrupted, with consequent isolation of catalyst particles.

Further work was carried out with the measurement of Ef_{Pt} values, in which the effects of (a) CL thickness and (b) Pt alloy [16]. The use of thin CLs was found to be useful, with high Ef_{Pt} values of 26.5% for O_2 and 17.4% for air for a CL of ca. 2.5 μm. The use of a Pt-Co alloy also increases the performance. However, it must be noted that the yardstick in that case should also be the same Pt-Co alloy. It was found for the Pt catalysts that the intrinsic ORR catalytic activity extrapolated to 0.9 V versus RHE, 2.26×10^{-4} A cm^{-2}, was in agreement with the aggregate value of 2.0×10^{-4} A cm^{-2} reported by Gasteiger et al. [17] for temperatures between 60 and 80 °C, based on comparisons of values reported by various laboratories for very similar TKK electrocatalysts. However, in our measurements, in which we applied slower potential sweep rates of 0.5 to 5 mV s^{-1}, seeking to achieve a steady-state ORR condition, the value at 80 °C (1.501×10^{-4} A cm^{-2} Pt) was somewhat lower, due to a higher Pt coverage of inert surface oxides, but the value should be quite realistic, corresponding to typical practical operating conditions.

More recently, the UY group has made further progress in optimizing the CL structure, with the overall vision being depicted in Fig. 4.12 for the ideal structure and various strategies for achieving it [18]. These include optimization of the

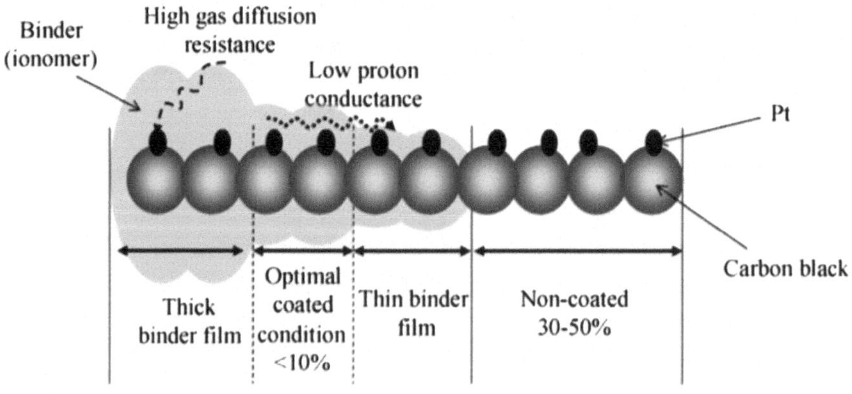

Fig. 4.11 Schematic depiction of the variation of Nafion film thickness with depth within the catalyst layer. Reprinted from [11] with permission from Elsevier

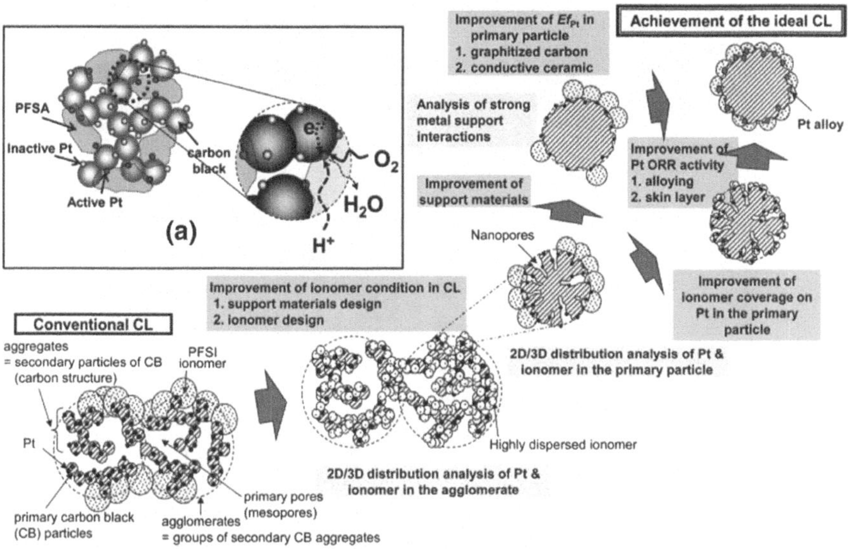

Fig. 4.12 Scenario for the development of catalyst layers (CL) with a high effectiveness of Pt (Ef_{Pt}) in MEAs. Inset (a) is a schematic depiction of functions at the three-phase boundary consisting of Pt/CB, the ionomer and the gas diffusion path for ORR in the CL. Reproduced from Ref. [18] with permission from the PCCP Owner Societies

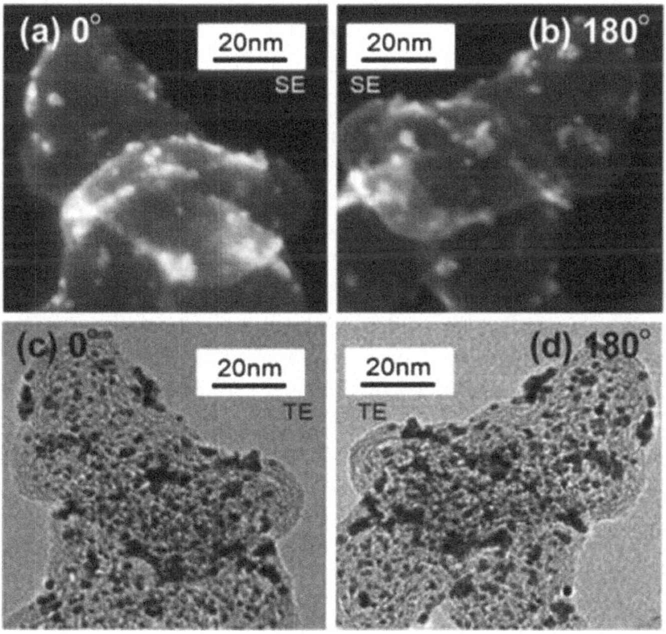

Fig. 4.13 STEM images of the c-Pt/CB cathode catalyst from two observation angles: SEM images at **a** 0° and **b** 180°; TEM images at **c** 0° and **d** 180°. Reproduced from Ref. [18] with permission from the PCCP Owner Societies

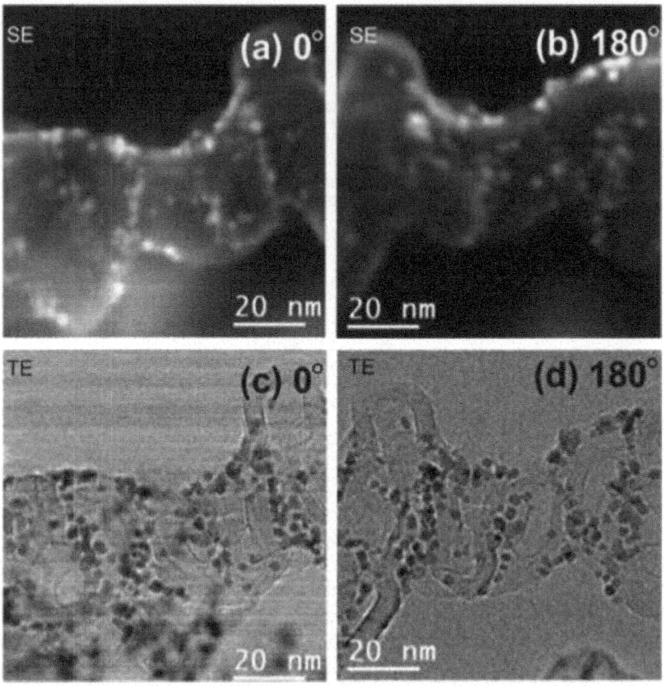

Fig. 4.14 STEM images of n-Pt/GCB cathode catalysts: SEM images at **a** 0° and **b** 180°; TEM images at **c** 0° and **d** 180°. Reproduced from Ref. [18] with permission from the PCCP Owner Societies

support materials, elimination of nanopores, and optimization of ionomer distribution. In this work, a new microscopic technique was developed to characterize the distribution of Pt nanoparticles in the interior of support nanopores versus those on the outer surface of support particles. This technique involves the use of a transmission electron microscope in both the transmission mode and the scanning (STEM) mode (Figs. 4.13 and 4.14) [18]. In the STEM mode, only the particles residing on the outer surface are visible, while, in the TEM mode, all of the particles are visible. In this way, the Pt particles in nanopores can be distinguished. When the sample is rotated 180°, the particles on the outer surface on the back of the particle can be seen. Thus, a new parameter has been proposed, which is the "effective surface area", $S(e)_{Pt}$, which is essentially the cumulative area of the Pt particles that reside on the outer surfaces of the support nanoparticles. The method to calculate this value involves the particle size distribution histograms of the type shown in Fig. 4.15. This area can then be compared with the total Pt surface area S_{Pt}. The remainder is the interior Pt area $S(i)_{Pt}$.

Under various conditions, such as the CFDE method, RDE, and MEA, there were different degrees of Pt wetting for the c-Pt/CB catalyst, which contains a significant percentage of internal surface area, 51% (Fig. 4.16). Under CFDE

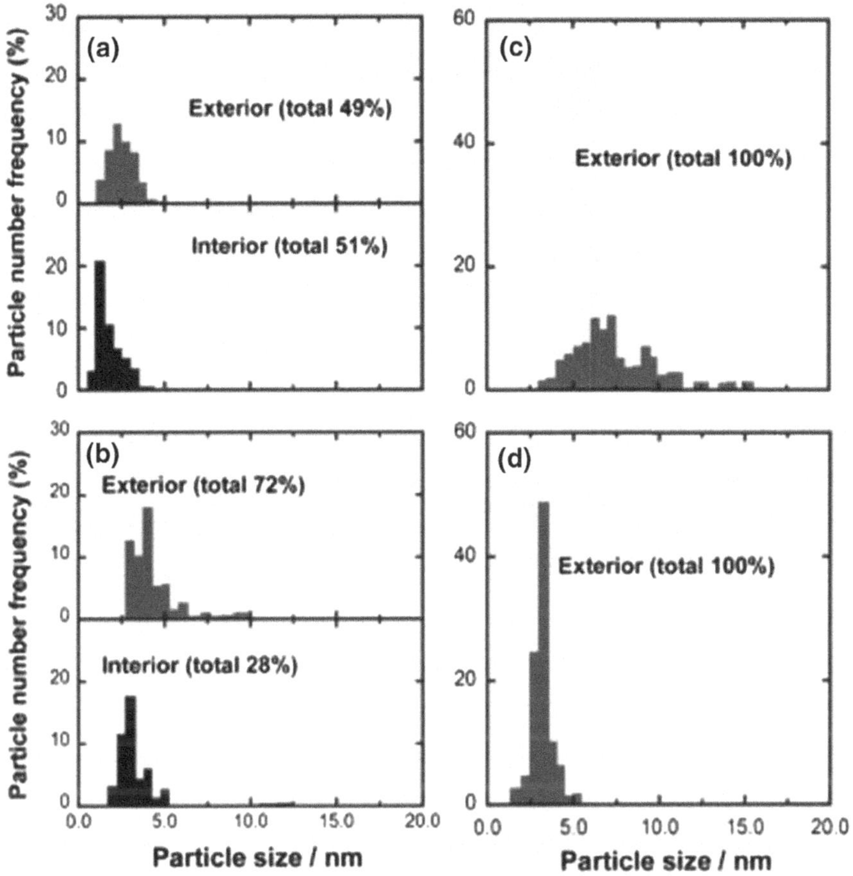

Fig. 4.15 Pt particle size distributions at both the interior and exterior surfaces at different catalysts: **a** c-Pt/CB, **b** c-Pt/GCB, **c** c-Pt/GCB-HT and **d** n-Pt/GCB. Reproduced from Ref. [18] with permission from the PCCP Owner Societies

conditions, all of the Pt surface area can be accessed, but only about 70% can be accessed under MEA conditions. Thus, it is desirable to make use of support materials, such as GCB, with a near absence of nanopores, so that most of the Pt particles are on the exterior surfaces.

In the most recent work, a new method of CL preparation was examined, the electrospray (ES) method [19]. The experimental setup is shown in Fig. 4.17. The advantages of this method are that it can produce an extremely uniform layer, with excellent porosity and essentially no waste of material (Fig. 4.18), in comparison to those prepared by the conventional pulse-swirl-spray method (PSS). The mass activities were found to be higher than those for the PSS method, with more than twice the activity at high humidity, due to the increased porosity and the higher degree of Pt coverage with a uniform ionomer film, resulting in improved Ef_{Pt} (Fig. 4.19).

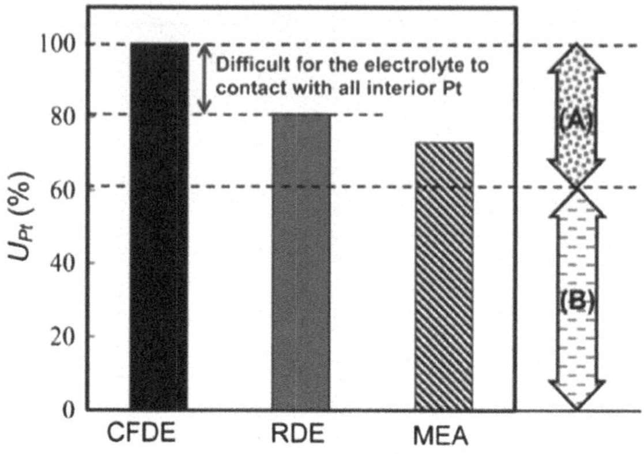

Fig. 4.16 Relationships between Pt utilization and Pt location on CB (c-Pt/CB), evaluated using cyclic voltammograms from the CFDE and RDE in N_2-saturated 0.1 M $HClO_4$ at T_{cell} = 25 °C with sweep rate = 100 mV s^{-1} (CFDE, RDE), or by MEA in a humidified N_2 (100% RH) atmosphere at T_{cell} = 40 °C with sweep rate = 20 mV s^{-1}. Pt surface area, indicating (A) 38% in the CB interior and (B) 62% on the exterior CB surface. Reproduced from Ref. [18] with permission from the PCCP Owner Societies

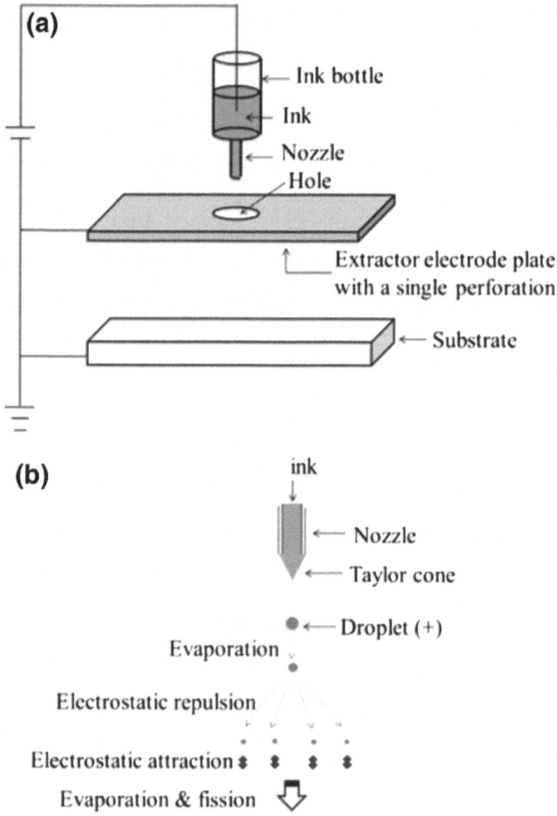

Fig. 4.17 **a** Schematic drawing of electrospray (ES) process. **b** Steps of microparticle and nanoparticle production by ES [19] (CC BY, http://creativecommons.org/licenses/by/4.0/)

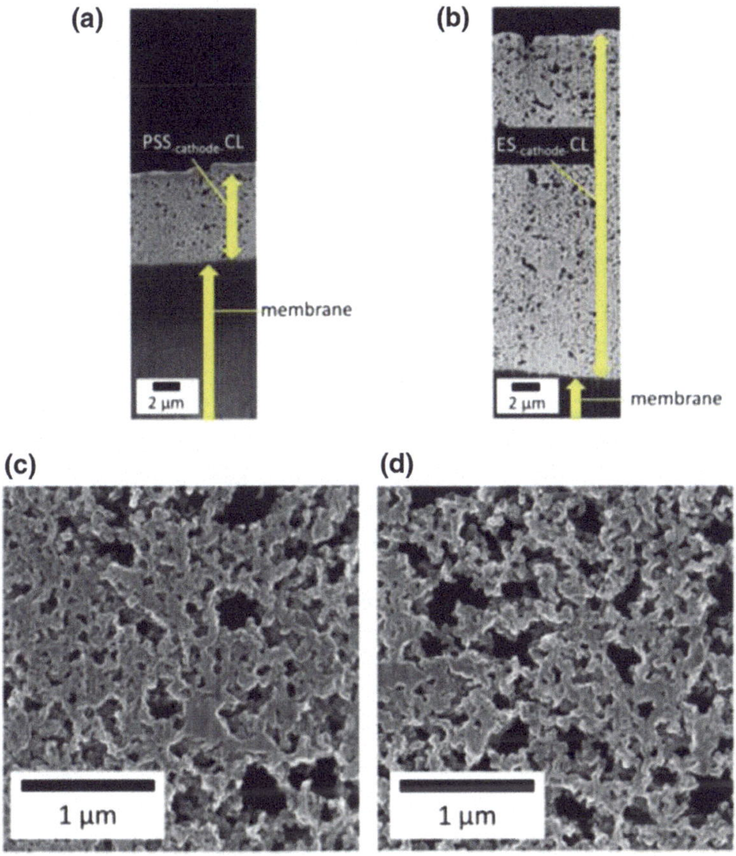

Fig. 4.18 SEM cross-sectional images of cathode CLs prepared by **a** PSS method and **b** ES method, magnified images of the cathode CLs prepared by **c** PSS and **d** ES [19] (CC BY, http://creativecommons.org/licenses/by/4.0/)

Fig. 4.19 Humidity dependence of mass activity on PSS (●) and ES (▲) cathodes at 0.85 V (IR-free), 80 °C, air/H_2, ambient pressure [19] (CC BY, http://creativecommons.org/licenses/by/4.0/)

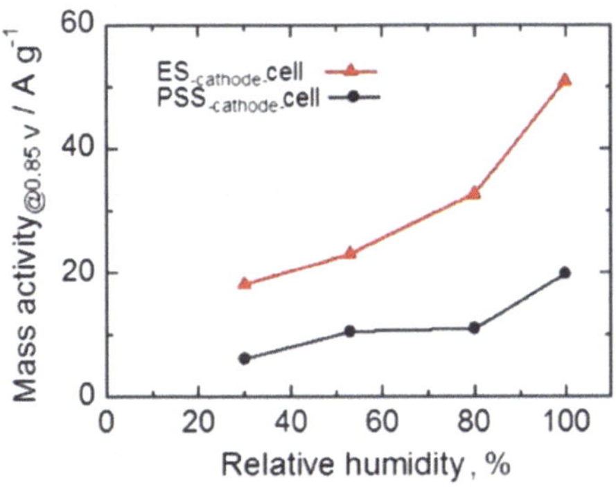

4.3 Catalyst–Support Relationship

An interesting line of investigation was begun in 1988 with the collaboration of Stonehart [20]. The essential concept is that the activity of a catalyst is independent of the particle size, contrary to results of other groups, so that small particles, down to 1.4 nm, can be used in order to increase the mass activity. In order for the activity to be realized, however, it is necessary for the particles to be dispersed adequately on the support surface, at least 15–20 nm between particles, so that radial diffusive mass transport can operate (Fig. 4.20) [20].

For a number of years, it had been considered that the intrinsic catalytic activity for the ORR on Pt nanoparticles decreased for very small particles, for various reasons [21, 22]. Thus, there was thought to be an upper limit on MA_{Pt}. However, the work at UY with Stonehart showed that, in fact, MA_{Pt} could be much higher than hitherto observed, as long as the Pt particles were dispersed on a high area carbon support [23]. The "territory" concept (Fig. 4.21) illustrates the proposed overlap of radial diffusion fields for closely spaced particles. When particles are dispersed adequately, the MA_{Pt} values are shown to increase linearly with Pt surface area (Fig. 4.22).

This concept is important in considering activity and performance but will also be shown to be important in considering durability, as discussed later.

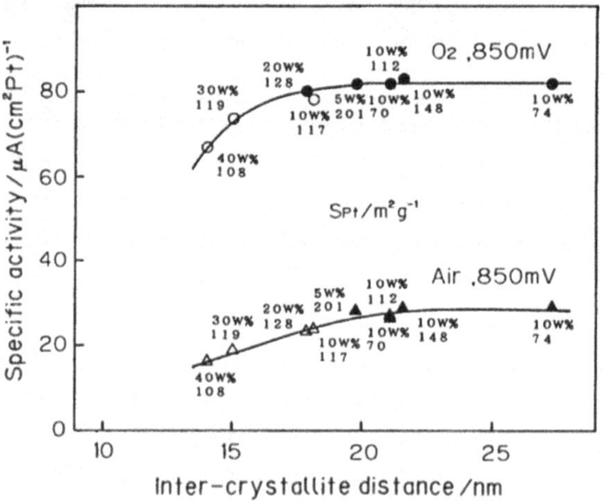

Fig. 4.20 Specific activity of dispersed Pt crystallites for ORR at 60 °C in 1.5 M H_2SO_4 solution as a function of intercrystallite distance of Pt on a series of carbon supports (10 wt% loading) and also at a series of Pt loadings on a CB with specific surface area (S_{CB}: 584 $m^2\ g^{-1}$). The numbers represent the Pt loading (wt%) on CB and the specific Pt surface area (S_{Pt}: $m^2\ g_{Pt}^{-1}$). Reprinted from [20] with permission from the Chemical Society of Japan

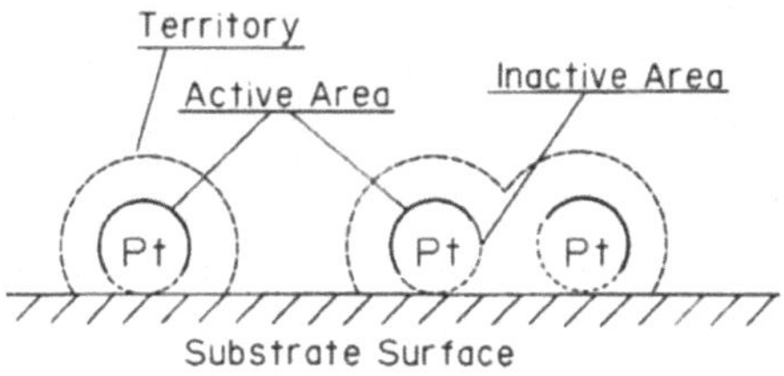

Fig. 4.21 "Territory" concept to explain the possibility of overlap of radial diffusion fields of supported catalyst particles relating to the "lost activity" for small particles, often attributed to a size effect. Reprinted from [23] with permission from Elsevier

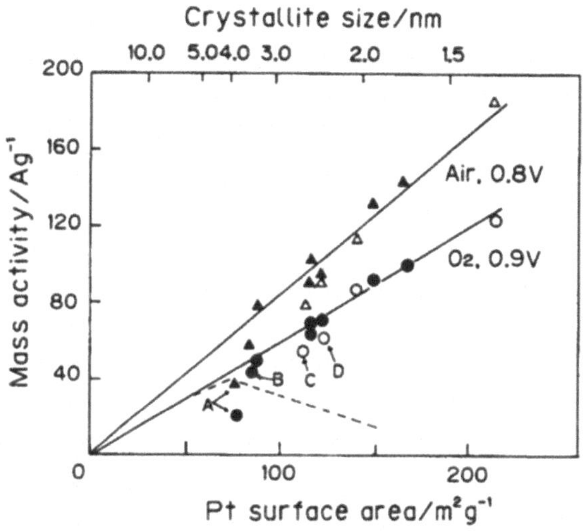

Fig. 4.22 Variation of MA_{Pt} with S_{Pt} for ORR at 0.9 V (in O_2) and 0.85 V (in air) at 190 °C and 100% H_3PO_4. Label A indicates the data for Pt catalysts supported on CB with low S_{CB} (<100 m^2 g^{-1}) and label B indicates those for catalysts with high Pt loading (>30 wt%), which deviate from the proposed functional relationship due to the small interparticle distances [23]. Dashed line reproduced from Ross's work, showing the so-called size effect. Reprinted from [22] with permission from Elsevier

4.4 Catalyst Durability

It is well known that carbon is thermodynamically unstable at the usual operating potentials of the oxygen cathode, even from 0.4 to 0.9 V, particularly during the occasional excursions above 1.0 V versus RHE that can occur under conditions of local starvation of H_2 in the anode at high fuel utilization or startup/shutdown operation, e.g., due to the reaction at the CB support of portions of the cathode that face the fuel-starved portion of the anode:

$$C + 2H_2O \rightarrow CO_2 + 4H^+ + 4e^- \quad (4.2)$$

which has a standard potential of +0.207 V versus standard hydrogen electrode (SHE) at 25 °C [24]. Around 2007, we began a series of studies examining the degradation of Pt/C catalysts under various conditions, particularly those of practical importance for fuel cell-powered electric vehicles (FCEVs) and residential systems [25]. For such systems, which usually undergo daily startup/shutdown cycles, the potentials at the cathode can often exceed 1.5 V versus RHE due to the "reverse current" effect, which proceeds whenever H_2 is present on one part of the fuel anode and air on another part [26]. Essentially, this occurs due to the HOR forcing the potential of the anode to be low, which in turn leads to the ORR occurring on the air-containing part of the electrode, which then forces an anodic reaction to occur at the cathode, in order to supply protons for the cathodic reaction occurring at the anode. The two candidate reactions are the oxygen evolution reaction and carbon oxidation, the easier one being the latter.

In an initial study, the effect of 0.9 V was examined for operation at 90 °C, with the surface area of Pt decreasing dramatically. There are several possible processes involved, including (a) carbon corrosion, with detachment and agglomeration of Pt, (b) Pt corrosion and redeposition on other Pt particles (Ostwald ripening), and (c) Pt corrosion and redeposition in the membrane to form a Pt "band" (Fig. 4.23). In this work, a method was devised to distinguish between carbon corrosion and Pt corrosion. The former was obtained by means of CV in both presence and absence of CO, to measure the double layer charging current, while the latter was obtained from the H adsorption charge in the CV.

The carbon corrosion rate can be decreased via the use of graphitized carbon black, as reported in the 1970s [27]. The reason that graphitized carbon is more corrosion resistant is that the graphene layers are more chemically inert within the layers than at the edges. Graphitized carbon blacks can often form continuous sheets of graphene layers, without any edges, as shown later.

However, one problem with GCB is that the deposition of Pt particles can result in their accumulation at defects on the surface. At UY, a deposition method that had been developed by IBM researchers [28] was modified for use with carbon blacks; the modified method was termed the "nanocapsule" method, since it employed a reverse micelle approach. This method has turned out to be particularly suitable for Pt deposition on GCB, since it results in an excellent dispersion on the surface, as well as in a highly uniform particle size.

These two aspects, i.e., high dispersion and uniform particle size, were found to be instrumental in providing superior corrosion resistance for Pt/GCB catalysts. The reasons for this effect are related to the "territories" concept already discussed above (Fig. 4.21). Although not fully explained, we propose that the dispersion allows the actual current density on each particle to be as low as possible, which then leads to a minimal rate of hydrogen peroxide production. The reason that higher current densities and/or lower potentials lead to higher peroxide production rates is that, at low potentials, hydrogen adsorbs atomically at (111) surfaces and

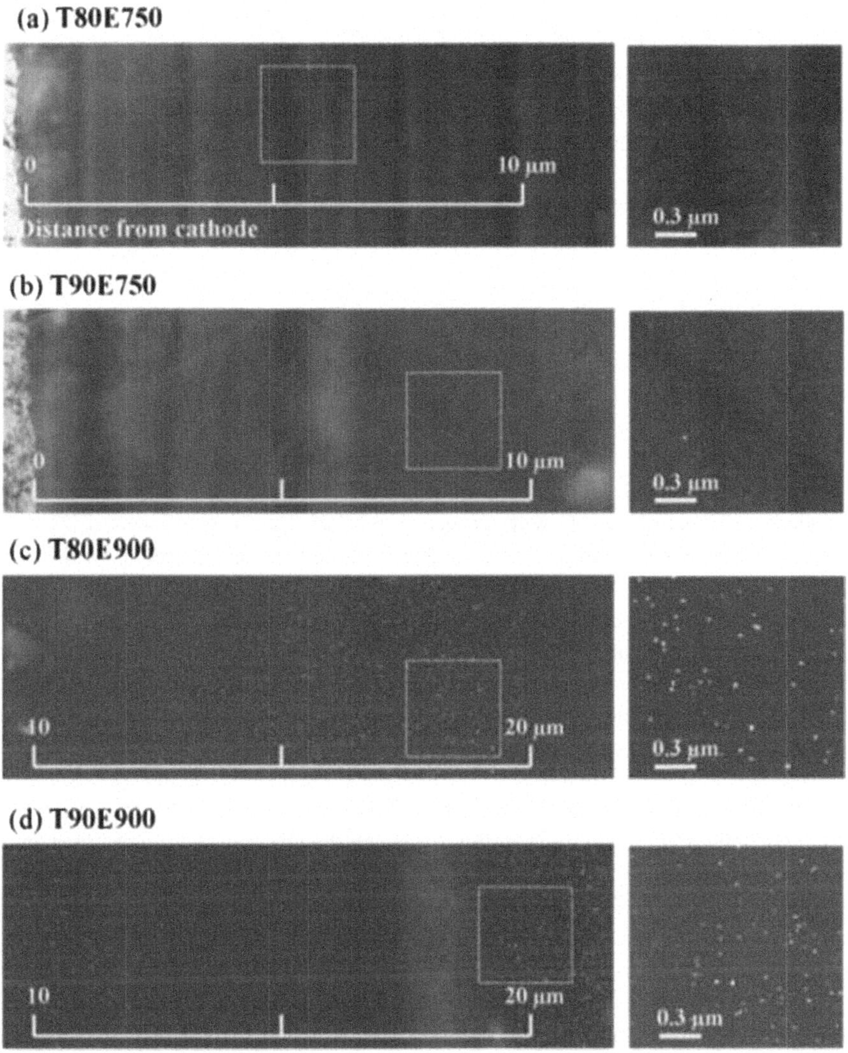

Fig. 4.23 STEM images of the cross section of the electrolyte membrane: **a** T_{cell} = 80 °C, E = 0.75 V for 460 h; **b** T_{cell} = 90 °C, E = 0.75 V for 480 h; **c** T_{cell} = 80 °C, E = 0.90 V for 450 h; **d** T_{cell} = 90 °C, E = 0.90 V for 400 h. Distance from the cathode interface is shown. The right panel is high-magnification image of the white square part in the left panel. Reprinted from [25] with permission from Elsevier

blocks the adsorption of oxygen, which is necessary for its dissociation [29]. Hydrogen peroxide is thought to be highly damaging to not only carbon supports but to the membrane. It can react directly or indirectly, after being converted to hydroxyl radicals via the Fenton reaction.

The uniform particle size is also important in impeding the Ostwald ripening process, in which smaller particles, tend to dissolve at a higher rate than larger ones.

This process accelerates, because, as the particle size decreases, the tendency to dissolve increases rapidly. The reason that smaller particles are more vulnerable to dissolution is said to be that they have higher surface energy, but this is a very generic concept, which is also used for water droplets. A more physically appropriate model for Pt particles is that each of the surface atoms has a certain coordination number (CN), which would be 9 for a perfect (111) surface on an fcc bulk solid. If there are defects, however, these atoms will have CN < 9 and will thus be easier to remove [30]. The end result of Ostwald ripening is also the severe loss of surface area, with higher ORR current densities at each particle and lower potentials, resulting in increased peroxide production rates.

A Pt/GCB catalysts prepared via the nanocapsule method (n-Pt/GCB) was found to degrade much more slowly with potential cycling than commercial Pt/CB (c-Pt/CB) and Pt/GCB (c-Pt/GCB), the latter having a much less uniform dispersion (Figs. 4.24 and 4.25) [31].

An additional factor involved in the degradation process is that the carbon corrosion process tends to produce quinone groups on the surface, which, when reduced, can react efficiently with oxygen to produce hydrogen peroxide [32]. The presence of quinone groups can be seen as a broad CV peak centered around 0.6 V versus RHE, which was particularly evident for the non-graphitized CB catalyst (Fig. 4.26c). This process also tends to accelerate, with peroxide attacking the surface, leading to changes that enhance peroxide production.

The startup/shutdown, processes have a deleterious effect, particularly during the shutdown, when air flows through the anode gas distribution system, replacing hydrogen. With the use of a novel visualization system for oxygen partial pressure devised at UY, it was shown that this process was much slower than expected based on simple replacement (Fig. 4.27) [33]. This is due to the fact that the oxygen is being consumed via reduction as it flows through the channel. After a gas-exchange cycling accelerated degradation test involving 500 cycles, the replacement process was faster, due to the fact that less oxygen was consumed, but this difference was not large, because the anode does not undergo such significant degradation in comparison with the cathode.

Subsequent work has been carried out, as discussed later. Also, the UY group is now working on the direct visualization of the carbon corrosion process via measurement of pCO_2.

The benefits of the nanocapsule preparation were also examined further by Lee et al. using a standard potential cycling protocol involving potential steps between 0.9 and 1.3 V with single cell tests at 65 °C [34]. Again, it was found that the n-Pt/GCB catalyst was far more stable than c-Pt/CB or c-Pt/GCB.

The startup/shutdown processes and their effects on carbon-supported Pt catalysts were further studied with a combination of TEM and Raman spectroscopy [35]. The samples were tested by means of a standard potential step protocol (0.9 V ↔ 1.3 V vs. RHE), holding 30 s at each potential and taking 1 min for one cycle, 3000 potential cycles for commercial Pt/CB and 10,000 cycles for commercial Pt/GCB, Pt/GCB-HT (heat treated for 30 min at 900 °C under a reducing atmosphere (H_2-N_2), and nanocapsule Pt/GCB) in membrane-electrode assemblies at

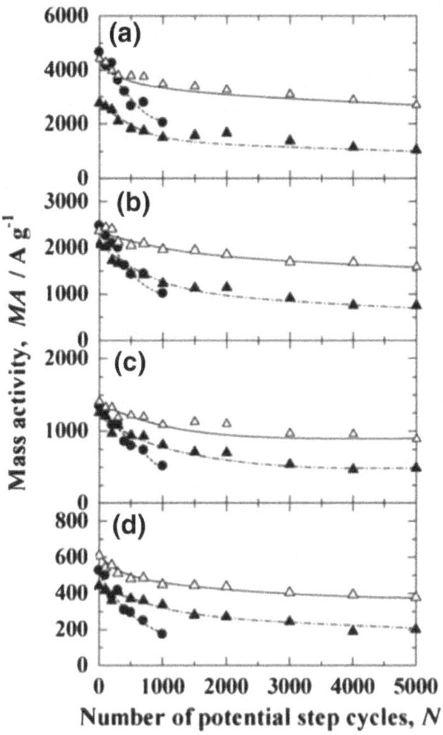

Fig. 4.24 Plots of MA at Nafion-coated (open triangles) n-Pt/GC, (solid triangles) c-Pt/GC, and (solid circles) c-Pt/CB electrodes as a function of N. The potential of the working electrode was **a** 0.70 V, **b** 0.76 V, **c** 0.80 V, and **d** 0.85 V versus RHE. Reproduced from Ref. [31] with permission from the PCCP Owner Societies

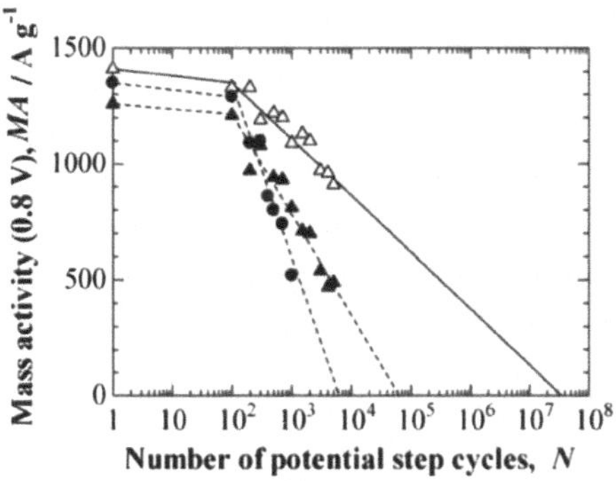

Fig. 4.25 Plots of $MA_{0.8V}$ at Nafion-coated (open triangles) n-Pt/GC, (solid triangles) c-Pt/GC, and (solid circles) c-Pt/CB electrodes at 0.80 V as a function of log [N]. Reproduced from Ref. [31] with permission from the PCCP Owner Societies

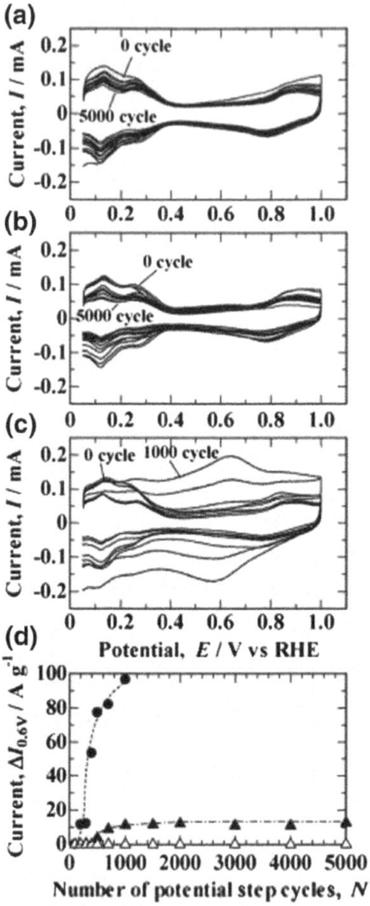

Fig. 4.26 Cyclic voltammograms to determine the ECSA at Nafion-coated **a** n-Pt/GC, **b** c-Pt/GC, and **c** c-Pt/CB electrodes in 0.1 M $HClO_4$ solution deaerated with N_2 at 25 °C. **d** Plots of $\Delta I_{0.6V}$ at these electrodes as a function of N for the durability test. Reproduced from Ref. [31] with permission from the PCCP Owner Societies

65 °C with 100% RH H_2 (anode) and N_2 (cathode). Pt/CB was found to undergo rapid degradation, as measured by the decrease of electrochemically active surface area (ECA) and ORR performance, as well as TEM observation (Figs. 4.28 and 4.29). Raman spectra exhibited clear changes before versus after testing, and a highly detailed analysis was carried out based on a curve-fitting analysis involving five distinct peaks (Fig. 4.30). Interestingly, only small changes were found before vs. after testing for Pt/GC, due to the fact that the outer material was simply converted to Raman-inactive species, including CO_2. The degree of degradation of the graphitized carbon black support was evaluated from changes in the band area ratio of the G band (ca. 1575 cm^{-1}) to the D1 band (ca. 1325 cm^{-1}). The changes in

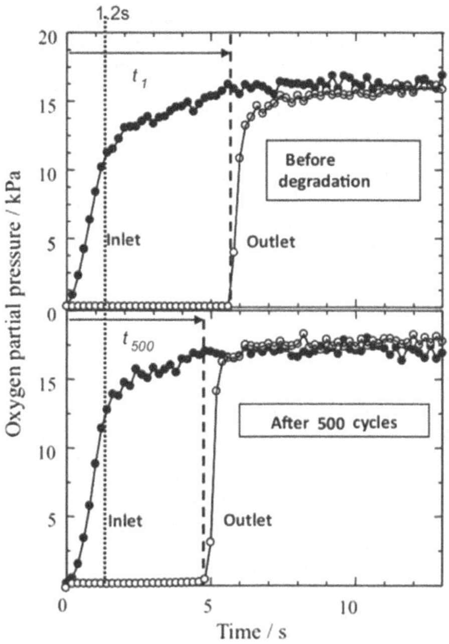

Fig. 4.27 pO_2 at the inlet and the outlet of the flow channel during the gas exchange from H_2 to air, analyzed from the data in the visualization. Reprinted from [33] with permission from Elsevier

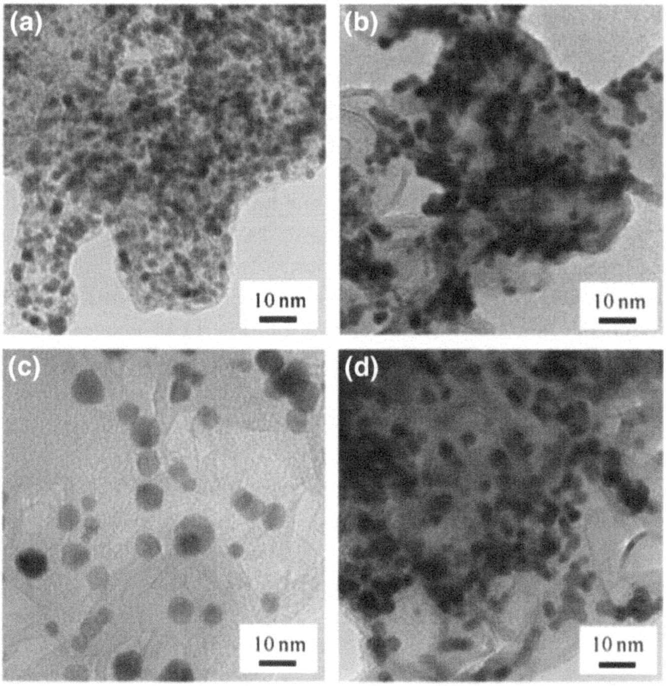

Fig. 4.28 TEM images of uncycled catalysts **a** c-Pt/CB, **b** c-Pt/GCB, **c** Pt/GCB-HT, and **d** n-Pt/GCB. Reprinted from [35] with permission from Elsevier

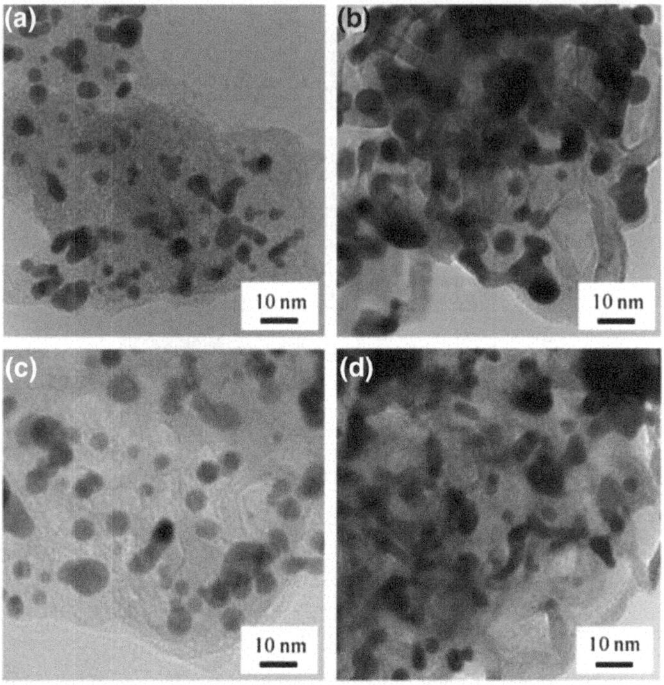

Fig. 4.29 TEM images of durability-test cycled catalysts: **a** c-Pt/CB, **b** c-Pt/GCB, **c** Pt/GCB-HT, and **d** n-Pt/GCB. Reprinted from [35] with permission from Elsevier

the Raman spectra tended to support the earlier findings that the dispersion of Pt on the GCB surface is important in decreasing degradation. For the n-Pt/GCB catalyst, it was found that, in contrast to c-Pt/GCB, the Pt particles were dispersed on the basal plane as well as at defects, which led to decreased attack. The heat treatment of the Pt/GCB-HT sample stabilized the graphitic shell, even though there were changes in Pt particle size. Figure 4.31 shows a schematic diagram of the various types of degradation for each type of support.

These results were further supported by long-term endurance testing with step cycles between 0.9 and 1.3 V [36]. High-resolution TEM images show the large differences in durability (Fig. 4.32). The number of cycles at which the ORR mass activity at 0.8 V $MA_{0.8V}$ decreased to 1/2 of the initial value, $N_{1/2}$, was 19 times longer for n-Pt/ GCB than that for c-Pt/GCB and 156 times longer than that for c-Pt/CB.

Further examination of the durability of non-graphitized Pt/CB catalysts with two different test protocols [37]: (1) the previously mentioned square-wave cycling (2007 FCCJ protocol, potential E = 0.9–1.3 V vs. RHE, holding 30 s at each E value, 1 min per cycle) and a newer, triangular wave cycling (2011 FCCJ protocol, E = 1.0–1.5 V vs. RHE, at a scan rate of 0.5 V s^{-1}, 2 s per cycle). The triangular

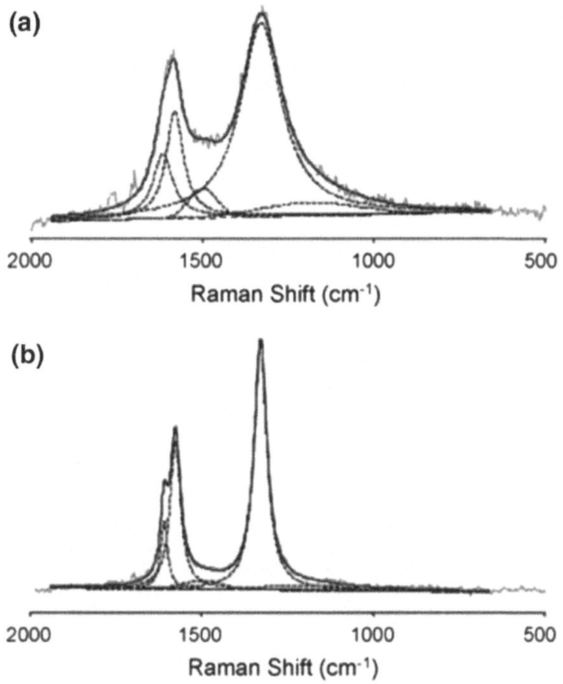

Fig. 4.30 Curve fits based on component bands for Raman spectra of **a** c-Pt/CB and **b** c-Pt/GCB catalysts obtained before potential cycle testing. Reprinted from [35] with permission from Elsevier

wave protocol is designed to more closely mimic the potential changes that occur during gas-exchange cycles during startup and shutdown.

It is somewhat problematic to directly compare the effects of these two protocols, given the great difference in timescale. In Fig. 4.33, we can compare the effect of MA of the two protocols in terms of cycle number, in which case, the square-wave protocol is more destructive, or in terms of time, in which case the triangular wave protocol is more destructive. The main factor involved with this difference is the dynamic nature of the triangular sweep, which does not allow a stable, protective oxide layer to build up [30].

The effect of sweep rate was also examined and found not to be strong (Fig. 4.34). Even at slower sweep rates, the degradation was not affected greatly. On the other hand, the effect of the lower potential used in the square-wave protocol was found to be quite important. When the lower potential was raised from its standard value of 0.9–1.0 V, the degradation rate was lessened significantly, so that it was only slightly worse than that for the triangular wave protocol. When it was decreased to 0.8 V, the degradation was increased significantly. The reasons for this have not been fully elucidated but could include the following: (a) the carbon

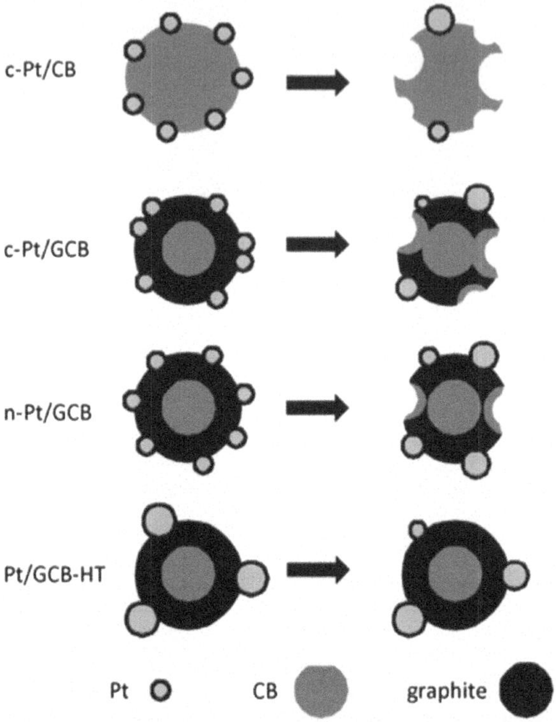

Fig. 4.31 Schematic illustration of the degradation and corrosion mechanism models of the carbon supports on c-Pt/CB, c-Pt/GCB, n-Pt/GCB, and Pt/GCB-HT during the start–stop potential cycle test. Reprinted from [35] with permission from Elsevier

corrosion is enhanced by the presence of fully reduced Pt, and/or (b) during the reduction of PtO_2, Pt^{2+} is formed according to the reaction:

$$PtO_2 + 4H^+ + 2e^- \longrightarrow Pt^{2+} + 2H_2O \qquad E^0 = 0.837\ \text{V vs. SHE} \tag{4.3}$$

which can attack the carbon surface as a rather strong oxidant, according to the reaction [38]:

$$Pt^{2+} + 2e^- \longrightarrow Pt \qquad E^0 = 1.188\ \text{V vs. SHE} \tag{4.4}$$

From the previous results, it is clear that two factors are important in causing degradation of Pt/C catalysts: (a) a wide potential range and (b) potential sweep, although the latter was seen as less important. Thus, the very low potential limit used in CV measurements of ECSA would be expected to be quite a severe condition. Nevertheless, such CV measurements are routinely made during accelerated degradation testing (ADT). The question naturally arises: what is the effect of CV during the course of an ADT? Park et al. addressed this question and found that

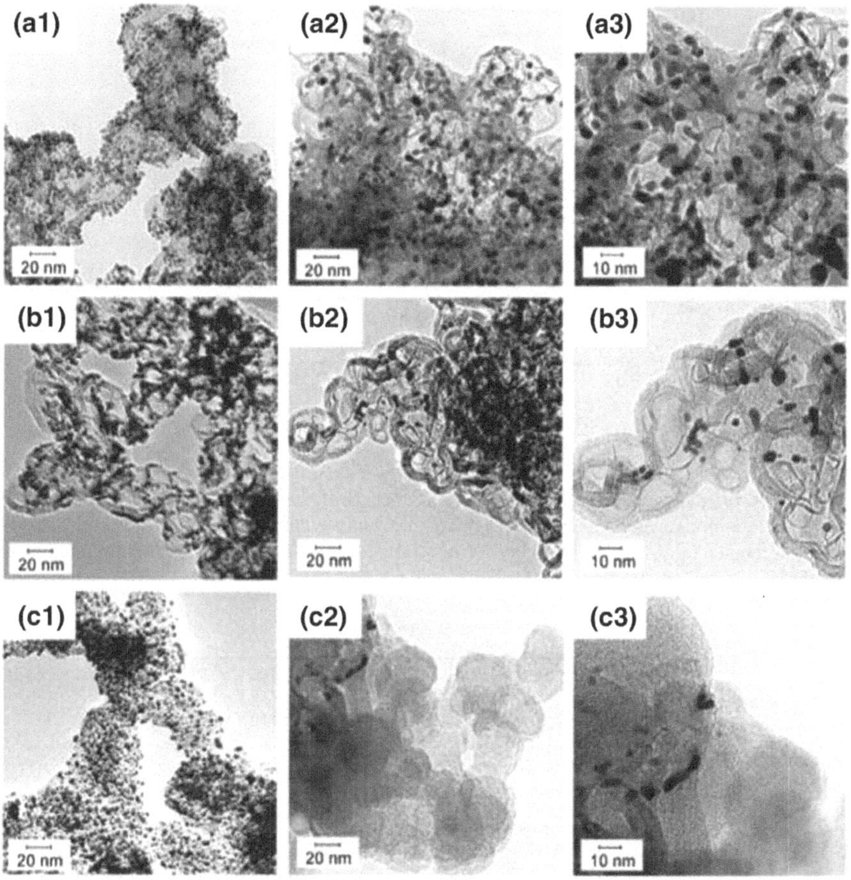

Fig. 4.32 TEM images **a1**, **b1**, and **c1**: pristine powders, **a2**, **b2**, and **c2**: after durability testing (N = 3000), and **a3**, **b3**, and **c3**: high-magnification images after the durability test) of **a** n-Pt/GCB, **b** c-Pt/GCB, and **c** c-Pt/CB. Reprinted from [36] with permission from Elsevier

indeed there is a significant effect (Figs. 4.35 and 4.36) [39]. If the CV is only measured at the end of the ADT, the loss in ECSA is much less than if CVs are run periodically during the ADT. Thus, it is likely that there are many ADT results that overestimate the degradation.

The procedures were as follows. Initial tests of the I-E (current–potential) curve and CV were first conducted. Then, the corrosion testing of carbon support was conducted by triangular wave cycling, in which the cell potentials were swept between 1.0 and 1.5 V at a scan rate of 0.5 V s^{-1}. After the potential cycle tests, the changes of electrochemically active surface area (ECSA) at the cathode were examined by CV tests in the third step. In this study, (1) the consecutive cycling mode included interim CV measurements after a certain number of triangular wave cycles (N = 100, 200, 300, 500, 1000, 3000, 5000, 7000, and 10,000 cycles).

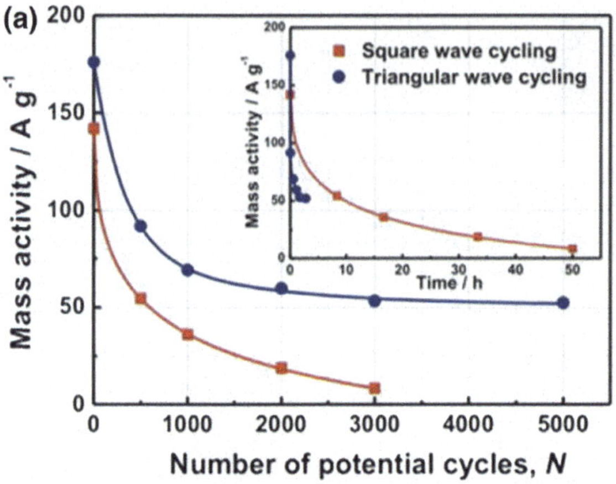

Fig. 4.33 Changes of mass activity during the durability test as a function of the number of square-wave or triangular-wave cycles, and as a function of time, shown as an inset in (**a**). The operating conditions were: anode gas—H_2 (70% utilization), cathode gas—O_2 (40% utilization), operation temperature—80 °C, humidity—100% RH. Reprinted from [37] with permission from Elsevier

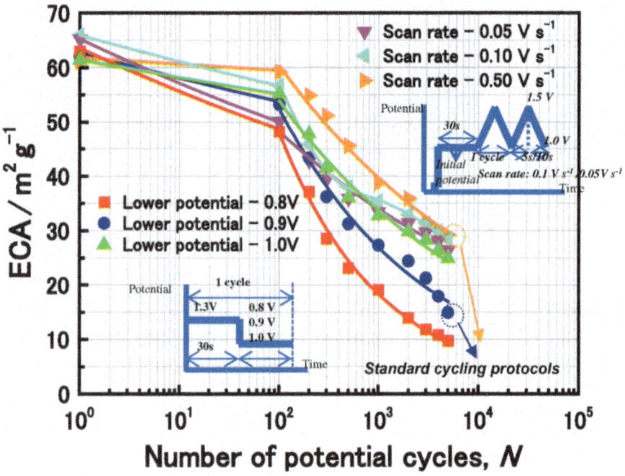

Fig. 4.34 Effect of scan rate (0.50, 0.10, and 0.05 V s^{-1}) for the triangular wave cycling and lower potential limit (0.8, 0.9, and 1.0 V) for the square-wave cycling. Reprinted from [37] with permission from Elsevier

In (2) the individual cycling mode, CV was measured only one time at the end of each potential cycling test (N = 500, 1000, 3000, 5000, 7000, and 10,000 cycles) using six individual MEAs, one for each number of cycles. N denotes the number of the potential cycles (1 cycle = 1.0 V ↔ 1.5 V). In the last step, the degradation of the MEA performance was evaluated by comparing the initial and final I-E curves.

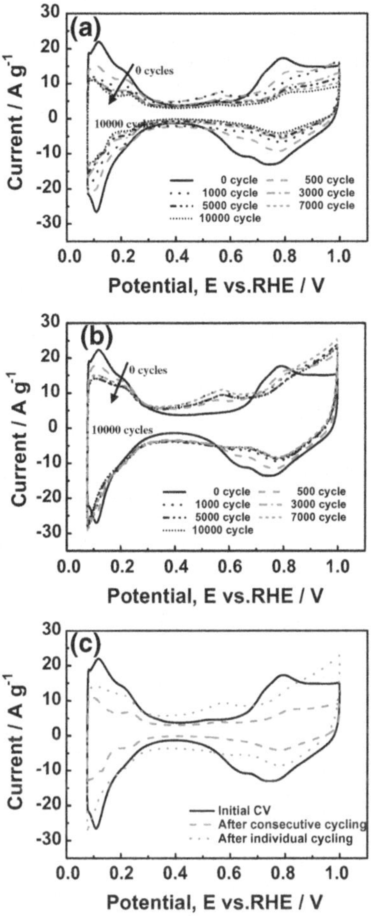

Fig. 4.35 Changes of CVs of Pt/CB catalysts by the two types of cycling tests: **a** consecutive cycling for various cycle numbers, **b** individual cycling for various cycle numbers and **c** consecutive cycling compared with individual cycling after 10,000 cycles; CV sweep rate, 20 mV s^{-1}. Reprinted from [39] with permission from Elsevier

The idea that CV can be destructive has been known in the fuel cell industry for a long time but has not been well documented. After these effects become more widely known, it is likely that its use will become more limited.

The potential cycling method for ADT is convenient, but it is necessary to check the extent of degradation under actual gas-exchange conditions, since there can possibly be important differences. With actual gas exchange, it was found that degradation occurred not only in the outlet region but also in the inlet region of the cathode [40]. Two reference electrodes were used, one each in the inlet and outlet regions. That in the outlet region showed the expected rise in potential due to the reverse current effect (Fig. 4.37). That in the inlet region showed a momentary

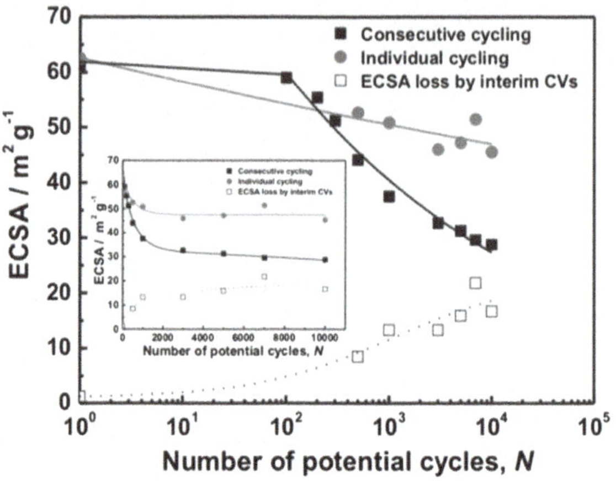

Fig. 4.36 Changes of ECSA values of Pt/CB catalyst resulting from the consecutive cycling and individual cycling. Reprinted from [39] with permission from Elsevier

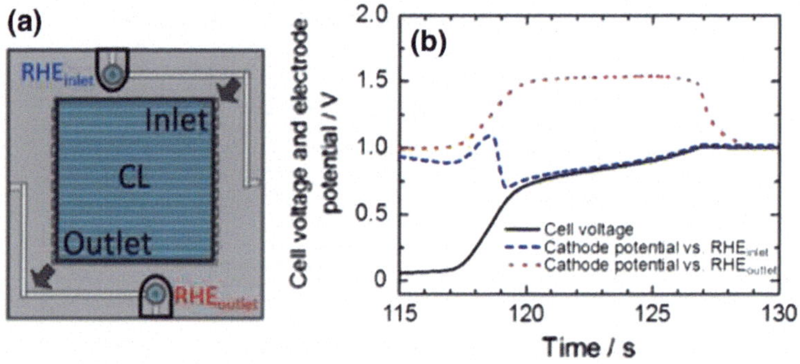

Fig. 4.37 **a** Schematic of the anode separator with RHE at the inlet region, RHE_{inlet} and outlet region, RHE_{outlet} and **b** changes of the cell and cathode potential during the gas-exchange cycling [40] (CC BY, http://creativecommons.org/licenses/by/4.0/)

increase in potential and then a quick decrease. The increase is most likely due to the reverse current effect, since the presence of H_2 quickly causes the potential to decrease and ORR to commence, but then the effect of the H_2 pressure reaching 1.0 atm at the inlet would let the cell reestablish its normal potential, even as the outlet region would still contain a significant O_2 pressure.

This study made use of several characterization techniques to assess the degradation at the two locations, including TEM, Raman, and the glancing incidence X-ray diffraction (GIXD) technique. The latter was used for the first time here to examine the degree and location of Pt particle growth in a PEFC. The advantages of this technique are that it is nondestructive, and it is rapid.

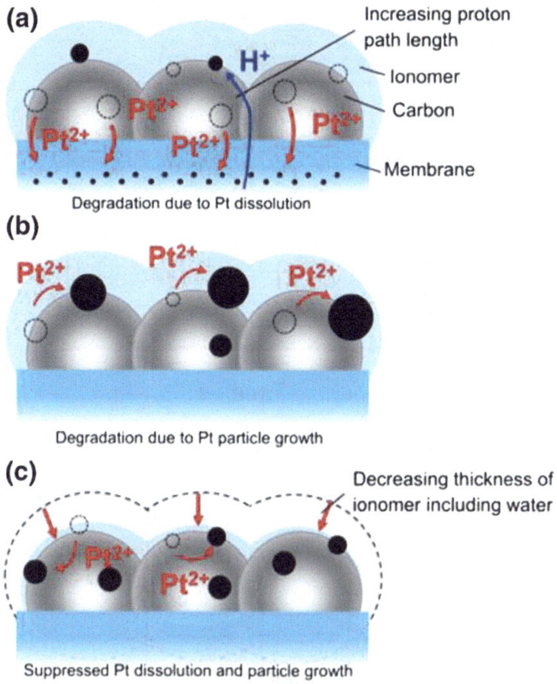

Fig. 4.38 Schematic of Pt (black) degradation on the GCB (gray) support in the cathode for **a** load cycles with a longer open circuit voltage (OCV) holding time (A3, ca. 0.63 V @ 0.5 A cm^{-2} for 3 s; OCV for 60 s, 100% RH), **b** that with a low current density range (A5, ca. 0.73 V @ 0.1 A cm^{-2} for 3 s; OCV for 60 s, 100% RH), **c** that at low RH (A6, ca. 0.63 V @ 0.2 A cm^{-2} for 3 s; OCV for 60 s, 50% RH) [41] (CC BY, http://creativecommons.org/licenses/by/4.0/)

Further work was carried out on the effects of various types of operating conditions on the degradation process, focusing on the dissolution and reprecipitation of Pt nanoparticles [41]. Several different load cycling protocols were tested, with varying time periods of an open circuit, load, and humidity. As expected from previous studies, the Pt degradation is aggravated by long periods at high potential (open circuit) coupled with low potential, so that large amounts of Pt oxides are reduced. Such conditions are made even worse if the humidity is high, because the excess water helps to transport Pt cations out of the CL and into the membrane. Three different scenarios are shown in Fig. 4.38. Panel (a) depicts the effect of long open circuit conditions, (b) depicts the effect of low current density, with Pt particle coarsening but no Pt band formation, and (c) low humidity, with depressed degradation.

Clearly, there are trade-offs with low humidity operation. Even though the Pt degradation rate is decreased, as seen above, there can be adverse effects. For example, if aromatic hydrocarbon ionomers are used as the CL binder, these can adsorb on the Pt surface, decreasing the ORR activity and/or increasing the production of hydrogen peroxide.

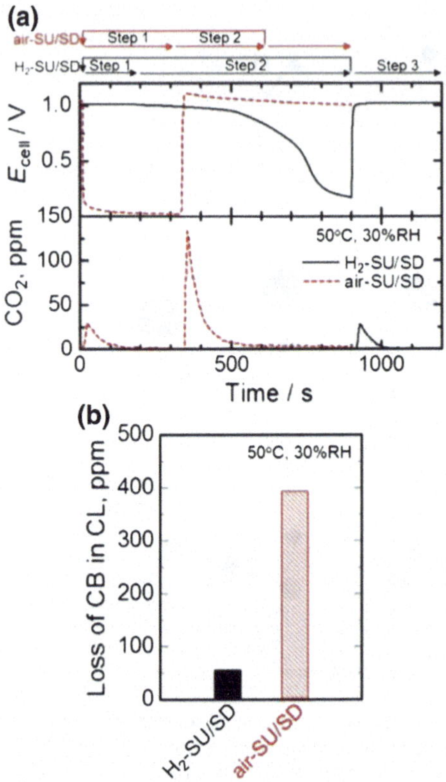

Fig. 4.39 **a** Changes of the E_{cell} and CO_2 concentration and **b** ratio of the oxidized CB to the total CB in the CL during both the air-startup/shutdown (SU/SD) and H_2-SU/SD processes [45] (CC BY, http://creativecommons.org/licenses/by/4.0/)

As previously mentioned, the ORR activity of smaller Pt particles per unit surface area was found to be nearly the same as that of larger ones, so the MA is proportionately larger. Then, a natural question to ask is, what is particle size effect on durability? This question was addressed recently, making use of the advantages of the nanocapsule method in producing samples with very narrow particle size distribution and high dispersion [42, 43]. These characteristics turned out to be quite important in achieving high durability, most likely because the degradative effects of the Pt on the carbon are decreased and spread out. Those effects can include the catalytic effect on carbon corrosion due to metallic Pt, that due to Pt^{2+}, and the production of hydrogen peroxide. It will be necessary to elucidate these possible pathways in more detail in the future.

In order to alleviate the degradation associated with startup and shutdown, Miyata and Wake devised the "hydrogen passivation" process, in which, just at shutdown, the air chamber is isolated by closing it off with valves, and hydrogen from the anode is allowed to permeate through the cell and consume the remaining

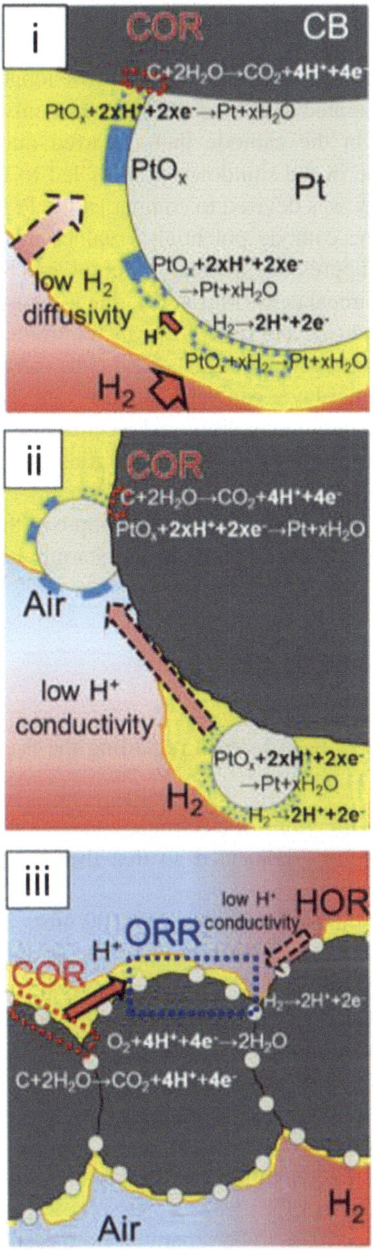

Fig. 4.40 Oxidation mechanisms of the carbon support due to local cells during step 2 (H_2 permeating to the cathode from the anode) both with (i) a single Pt particle and (ii) two or more Pt particles, and (iii) during step 3 (artificial air resupplied) during the H_2-SU/SD process, with respect to the operation shown in Fig. 4.39 [45] (CC BY, http://creativecommons.org/licenses/by/4.0/)

air [44]. This procedure was found to decrease the degradation significantly, but some carbon corrosion was still experienced (Fig. 4.39) [45]. This remaining amount was still considered to be detrimental for the longevity of the cell, so the mechanisms were investigated via a series of experiments. It was found that the reduction of Pt oxides in the cathode that occurred during the permeation of hydrogen from the anode in the shutdown process led to increased carbon corrosion. An effective strategy was devised to counteract the Pt reduction, which was to potentiostatically hold the cathode potential at values up to 1.0 V, at which the corrosion was largely suppressed. Several different mechanisms were proposed based on local electrochemical cells (Fig. 4.40): due to hydrogen permeation during the shutdown step, (i) and (ii), and during the air reintroduction (startup) step (iii). Generally, whenever inhomogeneities exist, local cells can also exist. For example, the thickness of the ionomer layer can vary and generate such cells (Fig. 4.40a, b). It is also possible that the previously discussed mechanisms for Pt-catalyzed carbon corrosion can operate during the hydrogen permeation (shutdown) step, either due to the presence of reduced (metallic) Pt surfaces or to Pt^{2+}. Overall, this work contains a number of important insights that can improve the hydrogen passivation strategy and help to alleviate degradation during startup and shutdown.

4.5 Conclusions

The main lessons that have been learned regarding the use of carbon materials in PEFCs include the following:

1. It is necessary for the functions of electronic conduction, proton conduction and gas transport be carefully delineated so that there are effective networks for each.
2. At the primary carbon particle level (20–100 nm), it is necessary to have excellent wetting, i.e., contact with the ionomer, so that all of the Pt particles are electrochemically active.
3. In secondary pores (>0.1 μm), it is necessary to have hydrophobic characteristics for effective gas transport.
4. Graphitized carbon blacks are attractive as supports due to their lack of nanopores, in which Pt particles would otherwise be trapped and isolated from contact with the ionomer.
5. The electrospray technique might be an effective means of improving the structure and uniformity of CLs.
6. It is necessary for Pt particles to be separated from each other by at least 15 nm so that O_2 will be able to access particle effectively.
7. The nanocapsule method is well suited for the production of catalysts with such a high state of dispersion even on low surface area graphitized carbon supports.

8. The major conditions that lead to degradation of both the support carbon and Pt catalyst particles include long periods of time at high potentials (>1.0 V) coupled with periods at potentials sufficiently low (< ca. 0.6 V) to reduce the Pt oxide.
9. Potential cycling (ramping) is more damaging than potential steps (constant potential).
10. Pt degradation is also aggravated by operation at high humidity due to the washing out of Pt cations with generated water.
11. Interim CV can be very damaging to both carbon and platinum, and thus its unnecessary use should be limited.
12. Startup and shutdown-induced degradation can be greatly decreased by use of the hydrogen passivation strategy, but further improvements can also be made, e.g., potential holding of the cathode to avoid Pt oxide reduction.
13. Nonhomogeneous ionomer layers on Pt/CB should be avoided in order to avoid the occurrence of local cells.
14. Other strategies should also be investigated to avoid the simultaneous presence of hydrogen and oxygen at the anode.
15. The further optimization of carbon materials for use in PEFCs is necessary in order to maximize both performance and durability.
16. Although not discussed here, alternatives to carbon are also needed and are being pursued in our laboratory.

References

1. Kunz HR, Gruver GA (1975) The catalytic activity of platinum supported on carbon for electrochemical oxygen reduction in phosphoric acid. J Electrochem Soc 122(10):1279–1287. https://doi.org/10.1149/1.2134000
2. Jalan VM, Bushnell CL (1979) Method for producing highly dispersed catalytic platinum. USA Patent 4136059
3. Appleby AJ, Foulkes FR (1989) Fuel cell handbook. Van Nostrand Reinhold, New York
4. Bockris JOM, Cahan BD (1969) Effect of a finite-contact-angle meniscus on kinetics in porous electrode systems. J Chem Phys 50(3):1307–1324. https://doi.org/10.1063/1.1671193
5. Watanabe M, Tomikawa M, Motoo S (1985) Preparation of a high performance gas diffusion electrode. J Electroanal Chem 182(1):193–196. https://doi.org/10.1016/0368-1874(85)85453-8
6. Watanabe M, Makita K, Usami H, Motoo S (1986) New preparation method of a high performance gas diffusion electrode working at 100% utilization of catalyst clusters and analysis of the reaction layer. J Electroanal Chem 197(1–2):195–208. https://doi.org/10.1016/0022-0728(86)80149-8
7. Watanabe M, Shimura C, Tsurumi K, Hara N (1991) A new wet-proof technique of gas-diffusion electrodes. Chem Lett 20(7):1113–1116. https://doi.org/10.1246/cl.1991.1113
8. Song JM, Uchida H, Watanabe M (2005) Effect of wet-proofing treatment of carbon backing layer in gas diffusion electrodes on the PEFC performance. Electrochem 73(3):189–193
9. Song JM, Suzuki S, Uchida H, Watanabe M (2006) Preparation of high catalyst utilization electrodes for polymer electrolyte fuel cells. Langmuir 22(14):6422–6428. https://doi.org/10.1021/la060671w

10. Uchida H, Song JM, Suzuki S, Nakazawa E, Baba N, Watanabe M (2006) Electron tomography of nafion ionomer coated on Pt/carbon black in high utilization electrode for PEFCs. J Phys Chem B 110(27):13319–13321. https://doi.org/10.1021/jp062678s
11. Lee M, Uchida M, Yano H, Tryk DA, Uchida H, Watanabe M (2010) New evaluation method for the effectiveness of platinum/carbon electrocatalysts under operating conditions. Electrochim Acta 55(28):8504–8512. https://doi.org/10.1016/j.electacta.2010.07.071
12. Thiele EW (1939) Relation between catalytic activity and size of particle. Ind Eng Chem 31 (7):916–920. https://doi.org/10.1021/ie50355a027
13. Giner J, Hunter C (1969) The mechanism of operation of the teflon-bonded bas diffusion electrode: a mathematical model. J Electrochem Soc 116(8):1124–1130. https://doi.org/10.1149/1.2412232
14. Mund K, Fv Sturm (1975) Degree of utilization and specific effective surface area of electrocatalysts in porous electrodes. Electrochim Acta 20(6–7):463–467. https://doi.org/10.1016/0013-4686(75)90035-3
15. Wakabayashi N, Takeichi M, Itagaki M, Uchida H, Watanabe M (2005) Temperature-dependence of oxygen reduction activity at a platinum electrode in an acidic electrolyte solution investigated with a channel flow double electrode. J Electroanal Chem 574(2):339–346. https://doi.org/10.1016/j.jelechem.2004.08.013
16. Lee M, Uchida M, Tryk DA, Uchida H, Watanabe M (2011) The effectiveness of platinum/carbon electrocatalysts: dependence on catalyst layer thickness and Pt alloy catalytic effects. Electrochim Acta 56(13):4783–4790. https://doi.org/10.1016/j.electacta.2011.03.072
17. Gasteiger HA, Kocha SS, Sompalli B, Wagner FT (2005) Activity benchmarks and requirements for Pt, Pt-alloy, and non-Pt oxygen reduction catalysts for PEMFCs. Appl Catal B: Environ 56(1–2):9–35. https://doi.org/10.1016/j.apcatb.2004.06.021
18. Uchida M, Park YC, Kakinuma K, Yano H, Tryk DA, Kamino T, Uchida H, Watanabe M (2013) Effect of the state of distribution of supported Pt nanoparticles on effective Pt utilization in polymer electrolyte fuel cells. Phys Chem Chem Phys 15(27):11236–11247. https://doi.org/10.1039/C3cp51801a
19. Takahashi K, Kakinuma K, Uchida M (2016) Improvement of cell performance in low-Pt-loading PEFC cathode catalyst layers prepared by the electrospray method. J Electrochem Soc 163(10):F1182–F1188. https://doi.org/10.1149/2.0611610jes
20. Watanabe M, Saegusa S, Stonehart P (1988) Electro-catalytic activity on supported platinum crystallites for oxygen reduction in sulphuric acid. Chem Lett 17(9):1487–1490. https://doi.org/10.1246/cl.1988.1487
21. Bregoli LJ (1978) The influence of platinum crystallite size on the electrochemical reduction of oxygen in phosphoric acid. Electrochim Acta 23(6):489–492. https://doi.org/10.1016/0013-4686(78)85025-7
22. Ross PN (1986) In: Rao UV (ed) Precious Metals 1986. International Precious Metals Institute, Allentown, PA, pp 355–363
23. Watanabe M, Sei H, Stonehart P (1989) The influence of platinum crystallite size on the electroreduction of oxygen. J Electroanal Chem 261:375–387. https://doi.org/10.1016/0022-0728(89)85006-5
24. VanMuylder J, Pourbaix M (1966) Carbon. In: Pourbaix M (ed) Atlas of electrochemical equilibria in aqueous solutions. Pergamon Press, Oxford, pp 449–457
25. Yoda T, Uchida H, Watanabe M (2007) Effects of operating potential and temperature on degradation of electrocatalyst layer for PEFCs. Electrochim Acta 52(19):5997–6005. https://doi.org/10.1016/j.electacta.2007.03.049
26. Reiser CA, Bregoli L, Patterson TW, Yi JS, Yang JD, Perry ML, Jarvi TD (2005) A reverse-current decay mechanism for fuel cells. Electrochem Solid-State Lett 8(6):A273–A276. https://doi.org/10.1149/1.1896466
27. Bett JA, Kinoshita K, Stonehart P (1974) Crystallite growth of platinum dispersed on graphitized carbon black. J Catal 35(2):307–316. https://doi.org/10.1016/0021-9517(74)90209-7

28. Sun S, Murray CB, Weller D, Folks L, Moser A (2000) Monodisperse FePt nanoparticles and ferromagnetic FePt nanocrystal superlattices. Science 287(5460):1989–1992. https://doi.org/10.1126/science.287.5460.1989
29. Markovic NM, Adzic RR, Cahan BD, Yeager EB (1994) Structural effects in electrocatalysis: oxygen reduction on platinum low index single-crystal surfaces in perchloric acid solutions. J Electroanal Chem 377(1–2):249–259. https://doi.org/10.1016/0022-0728(94)03467-2
30. Sugawara Y, Okayasu T, Yadav AP, Nishikata A, Tsuru T (2012) Dissolution mechanism of platinum in sulfuric acid solution. J Electrochem Soc 159(11):F779–F786. https://doi.org/10.1149/2.017212jes
31. Yano H, Akiyama T, Bele P, Uchida H, Watanabe M (2010) Durability of Pt/graphitized carbon catalysts for the oxygen reduction reaction prepared by the nanocapsule method. Phys Chem Chem Phys 12(15):3806–3814. https://doi.org/10.1039/B923460H
32. Hossain MS, Tryk D, Yeager E (1989) The electrochemistry of graphite and modified graphite surfaces: the reduction of O_2. Electrochim Acta 34:1733–1737. https://doi.org/10.1016/0013-4686(89)85057-1
33. Ishigami Y, Takada K, Yano H, Inukai J, Uchida M, Nagumo Y, Hyakutake T, Nishide H, Watanabe M (2011) Corrosion of carbon supports at cathode during hydrogen/air replacement at anode studied by visualization of oxygen partial pressures in a PEFC—Start-up/shut-down simulation. J Power Sources 196(6):3003–3008. https://doi.org/10.1016/j.jpowsour.2010.11.092
34. Lee M, Uchida M, Okaya K, Uchida H, Watanabe M (2011) Durability of Pt/graphitized carbon catalyst prepared by the nanocapsule method for the start/stop operating condition of polymer electrolyte fuel cells. Electrochem Commun 79(5):381–387. https://doi.org/10.5796/electrochemistry.79.381
35. Hara M, Lee M, Liu C-H, Chen B-H, Yamashita Y, Uchida M, Uchida H, Watanabe M (2012) Electrochemical and Raman spectroscopic evaluation of Pt/graphitized carbon black catalyst durability for the start/stop operating condition of polymer electrolyte fuel cells. Electrochim Acta 70(Suppl C):171–181. https://doi.org/10.1016/j.electacta.2012.03.043
36. Yano H, Akiyama T, Watanabe M, Uchida H (2013) High durability of Pt/graphitized carbon catalysts for polymer electrolyte fuel cells prepared by the nanocapsule method. J Electroanal Chem 688(Suppl C):137–142. https://doi.org/10.1016/j.jelechem.2012.09.028
37. Park YC, Kakinuma K, Uchida M, Tryk DA, Kamino T, Uchida H, Watanabe M (2013) Investigation of the corrosion of carbon supports in polymer electrolyte fuel cells using simulated start-up/shutdown cycling. Electrochim Acta 91:195–207. https://doi.org/10.1016/J.Electacta.2012.12.082
38. Muylder JV, Zoubov ND, Pourbaix M (1966) Platinum. In: Pourbaix M (ed) Atlas of electrochemical equilibria in aqueous solutions. Pergamon Press, Oxford, pp 378–383
39. Park Y-C, Kakinuma K, Uchida M, Uchida H, Watanabe M (2014) Deleterious effects of interim cyclic voltammetry on Pt/carbon black catalyst degradation during start-up/shutdown cycling evaluation. Electrochim Acta 123(Suppl C):84–92. https://doi.org/10.1016/j.electacta.2013.12.120
40. Yamashita Y, Itami S, Takano J, Kodama M, Kakinuma K, Hara M, Watanabe M, Uchida M (2016) Durability of Pt catalysts supported on graphitized carbon-black during gas-exchange start-up operation similar to that used for fuel cell vehicles. J Electrochem Soc 163(7):F64–F650. https://doi.org/10.1149/2.0101704jes
41. Takei C, Kakinuma K, Kawashima K,Tashiro K, Watanabe M, Uchida M (2016) Load cycle durability of a graphitized carbon black-supported platinum catalyst in polymer electrolyte fuel cell cathodes. J Power Sources 324:729–737. https://doi.org/10.1016/j.jpowsour.2016.05.117
42. Yano H, Watanabe M, Iiyama A, Uchida H (2016) Particle-size effect of Pt cathode catalysts on durability in fuel cells. Nano Energy 29:323–333. https://doi.org/10.1016/j.nanoen.2016.02.016

43. Watanabe M, Yano H, Uchida H, Tryk DA (2017) Achievement of distinctively high durability at nanosized Pt catalysts supported on carbon black for fuel cell cathodes. J Electroanal Chem. https://doi.org/10.1016/j.jelechem.2017.11.017
44. Miyate K, Wake C (2014) Japan Patent 5647079
45. Yamashita Y, Itami S, Takano J, Kakinuma K, Uchida H, Watanabe M, Iiyama A, Uchida M (2017) Degradation mechanisms of carbon supports under hydrogen passivation startup and shutdown process for PEFCs. J Electrochem Soc 164(4):F181–F187. https://doi.org/10.1149/2.0101704jes

Chapter 5
Understanding the Stability of Nanoscale Catalysts in PEM Fuel Cells by Identical Location TEM

Somaye Rasouli and Paulo J. Ferreira

5.1 Introduction

A thorough knowledge of the atomic structure and composition of electrocatalyst nanoparticles is paramount to the development of advanced materials for proton exchange membrane fuel cells (PEMFC), one of the most promising energy conversion devices for automotive and stationary applications. The efficiency of the fuel cells is largely restricted by two main challenges: (1) the slow oxygen reduction reaction (ORR) kinetics catalyzed by platinum (Pt) based nanocatalysts in the cathode and (2) the instability of nanocatalyst particles during fuel cell operation. Pt nanoparticles (NP) are currently used as the catalyst to promote the kinetics of hydrogen oxidation and ORR in the anode and cathode of the fuel cell, respectively. However, Pt-based alloys have being investigated to replace Pt on the cathode as a way to improve the efficiency of the fuel cell and reduce cost [1–5]. Although the enhancement in the ORR activity of Pt-alloys is well established, the durability of the catalysts remains the main issue for their commercialization.

Due to their large surface area-to-volume ratio, Pt and Pt-alloy NPs have a strong tendency to grow in size over short time scales, which can lead to significant changes in their desirable properties with subsequent changes in the behavior and performance of PEMFC. In particular, the growth of the Pt-based NPs catalysts is responsible for a reduction in their electrochemically active surface area, which

S. Rasouli · P. J. Ferreira (✉)
Materials Science and Engineering Program, University of Texas at Austin, Austin TX 78712, USA
e-mail: ferreira@mail.utexas.edu

S. Rasouli
e-mail: rasouli@utexas.edu

N. Nakashima (ed.), *Nanocarbons for Energy Conversion: Supramolecular Approaches*, Nanostructure Science and Technology,
https://doi.org/10.1007/978-3-319-92917-0_5

leads to an undesired catalyst deactivation and subsequent reduction in cell performance after several cycles [6–8]. Thus, it is critical to understand the fundamentals behind catalyst instability, to be able to develop more fruitful strategies towards preventing catalyst activity loss.

Ferreira et al. proposed four mechanisms describing the loss of electrochemical active surface area of Pt NPs in the cathode of the PEMFC during cycling operation, namely (1) dissolution and re-deposition of Pt on large particles on the nanometer scale or Ostwald ripening, (2) particle migration on the carbon support and coalescence, (3) particle detachment from carbon support, and (4) particle dissolution and re-precipitation on the ionomer phase or/and membrane by chemical reduction of soluble Pt with permeated hydrogen from the anode side [7]. Many studies, including post mortem transmission electron microscopy (TEM) studies have been performed to address the mechanisms of catalyst degradation [6, 7, 9–14]. In these studies, a batch of randomly selected NPs were characterized before cycling and then compared to another batch of randomly selected NPs in the cathode of the membrane electrode assembly (MEA) after voltage cycling.

Although several issues have been addressed in the aforementioned studies, it is still unclear what the dominant degradation mechanism is, particularly because direct evidence of each mechanism is not available and thus a fundamental understanding of particle behavior in real time, at different stages of fuel cell cycling, is still challenging. Furthermore, when using Pt-alloy NPs, such as Pt-Ni or Pt-Co, we expect that new factors, such as heterogeneous dissolution of the elements during voltage cycling and enthalpy of mixing of the elements, might affect the mechanisms of loss of specific activity and the active surface area of the nanoparticles during fuel cell operation.

As a consequence, various key questions remain unanswered, namely: (1) Is modified Ostwald ripening or coalescence the main mechanism for particle growth within the cathode? (2) How does voltage cycling affect particle movement on the carbon support and consequently the coalescence of particles? (3) How does carbon degradation during voltage cycling affect the coalescence of particles? (4) How does alloying influence the degradation mechanism?

In order to assess the degradation of fuel cell components including the catalyst NPs, several accelerated stress test (AST) protocols have been previously developed [14–17]. These ASTs can be performed either in situ in PEMFCs (single cell) or ex situ in a classical three-electrode electrochemical cell using liquid electrolytes (half-cell) [13, 15–18]. Although the durability test in PEMFCs provides an invaluable insight into the degradation of the fuel cells, it is difficult to separate the contribution of other components such as membrane and gas diffusion layer. ASTs in three-electrode electrochemical cells have the advantage of being less expensive and allow the ability to provide controlled information on the degradation of the catalytic layer excluding any contribution from other parts of the PEMFC. In fact, several studies have shown a good agreement between the ECA loss mechanisms and half-cell durability tests [9, 19, 20].

To understand the details about each degradation mechanism, and analyze the behavior of Pt and Pt alloy NPs at different stages of fuel cell cycling, the best way would be performing in situ voltage cycling of the MEAs inside the TEM.

However, operating a fuel cell inside the microscope is extremely challenging, considering the simultaneous presence of hydrogen and oxygen (or nitrogen) on the anode and cathode side of the MEA inside the microscope. Thus, the closest way is to perform potential cycling in a half-cell in the presence of liquid electrolyte, instead of a single cell with solid electrolyte. Identical location S/TEM (IL S/TEM) and electrochemical in situ liquid cell TEM are the two approaches which provide direct evidence for the catalysts degradation mechanisms.

5.2 Identical Location TEM

Identical location TEM is an affordable and useful approach to study morphological and structural changes of electron microscopy compatible samples [21–23]. Although IL TEM does not provide true in situ information and suffers from time-consuming sample transfer, the approach has the advantage that several locations of the samples can be analyzed before and after the electrochemical cycling. Thus, this method gives a thorough information of the phenomena that occur; during which both electrochemical analysis and electron microscopy can be used without constraints.

This method is based on the ex situ electrocatalyst accelerated stress test in a classical three-electrode electrochemical cell using liquid electrolytes (half-cell). For this method, the TEM grid itself is used as a working electrode for the accelerated stress tests to replicate on a TEM grid the effect of potential cycling on the cathode side of the fuel cell outside of the microscope. In order to find the same location before and after the stress test, specific features on the grid and a series of TEM images at different magnification are utilized to provide the map of the area of interest. TEM grid is usually made of gold, because it has to be stable in the electrolyte at the voltage range during potential cycling. The identical location TEM procedure is schematically shown in Fig. 5.1. It consists of the following steps: (1) Deposition of the catalyst on the TEM grid, (2) Preliminary TEM observation to find the area of interest (R.O.I) on the TEM grid, (3) Potential cycling of the TEM grid in the electrochemical cell and (4) Re-characterize the R.O.I.

5.3 Capabilities of IL TEM

Compared to ex situ TEM studies, where a batch of randomly selected NPs are characterized before cycling and then compared to another batch of randomly selected NPs after voltage cycling, the identical location approach opens up new opportunities for the visualization of degradation processes of fuel cell nanocatalysts on the atomic/nanoscale. This method provides the possibility of tracking the behavior of a single nanoparticle during different stages of potential cycling. The progress that has been made in understanding the instability mechanisms of fuel cell catalysts by developing IL TEM is reviewed below.

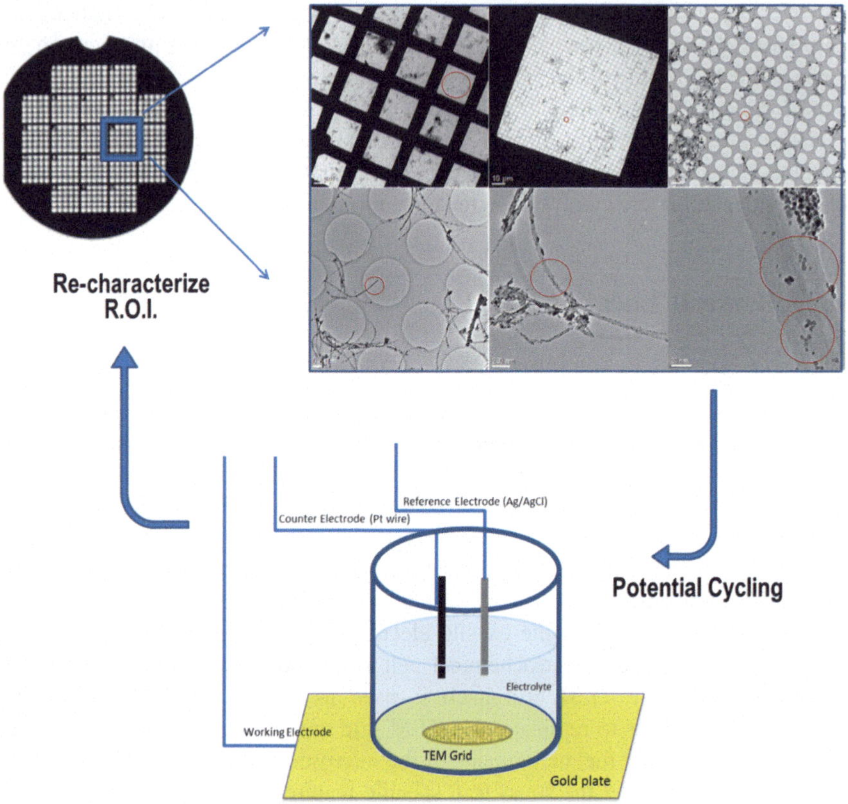

Fig. 5.1 Schematic of identical location procedure

5.3.1 *Pt NPs Supported on Carbon Nanotubes*

The degradation mechanisms of Pt NPs supported on carbon nanotubes (CNT) was demonstrated at different stages of potential cycling by Rasouli et al., through IL TEM [24]. They showed that, in the first 1000 cycles, the main mechanism responsible for the loss of electrochemical active surface area of Pt/CNT electrocatalysts is particle motion, followed by coalescence. This is clearly shown in Fig. 5.2a where two well-separated NPs on the surface of a CNT (yellow square 1), move until they contact each other and finally coalesce (Fig. 5.2b). Changes in orientation of individual NPs were also observed after voltage cycling. Figures 5.2a and 5.2b (yellow square 2) depict this process where an individual particle goes through rotation (FFTs show in insets) after cycling.

Severe structural deformation of the carbon nanotubes during voltage cycling is another source of degradation (yellow square 3 in Figs. 5.2a and 5.2b). The wave-like structure of the carbon formed after voltage cycling is the result of the

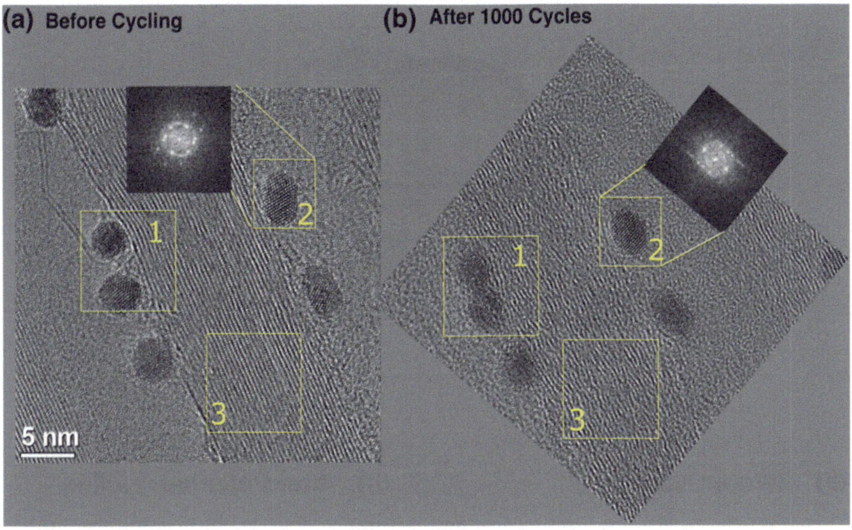

Fig. 5.2 Aberration corrected TEM images of the Pt nanoparticles on carbon nanotubes, **a** before, and **b** after 1000 potential cycles

appearance of the defect sites on the carbon nanotubes, which convert the flat hexagon structure into heptagon and pentagon carbon rings [25].

In order to understand if there is any correlation between carbon degradation and particle movement on the carbon support, carbon degradation was accelerated under the electron beam, while the behavior of the nanoparticles was observed. It was shown that the particles start to move as soon as carbon atoms at the interface of the carbon/particle interface are removed. During voltage cycling, carbon corrosion occurs through a series of reactions at the carbon/ionomer, ionomer/particle, and carbon/particle interfaces [26]. These reactions result in the formation of carbon dioxide and removal of carbon atoms at the particle/carbon interface. When carbon atoms at the carbon/particle interface are removed from the surface of carbon, leaving a carbon defect site, the particle tends to make new bonds with the next carbon atoms to possibly decrease its free surface energy (Fig. 5.3). Upon particle migration, the NPs are likely to contact each other and subsequently coalesce into a single particle, driven by a reduction in surface energy. Rasouli et al. showed the process of coalescence in real time through phase contrast TEM images of two adjacent particles. According to their observation, atoms move towards the neck between the two adjacent particles through surface diffusion. As diffusion towards the neck continues, the dihedral angle is reduced leaving a dislocation at the grain boundary. Finally, the dislocation is eliminated during the late stages of coalescence leading to a defect-free single particle [24].

In the postmortem TEM study of Pt alloy NPs on the MEA of a fuel cell [27], cluster bridging was proposed as a new mechanism for surface area loss of particles

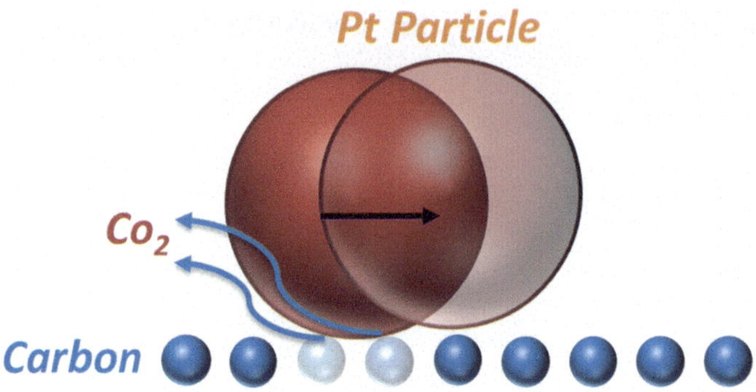

Fig. 5.3 Schematic of the proposed particle movement mechanism on the surface of carbon

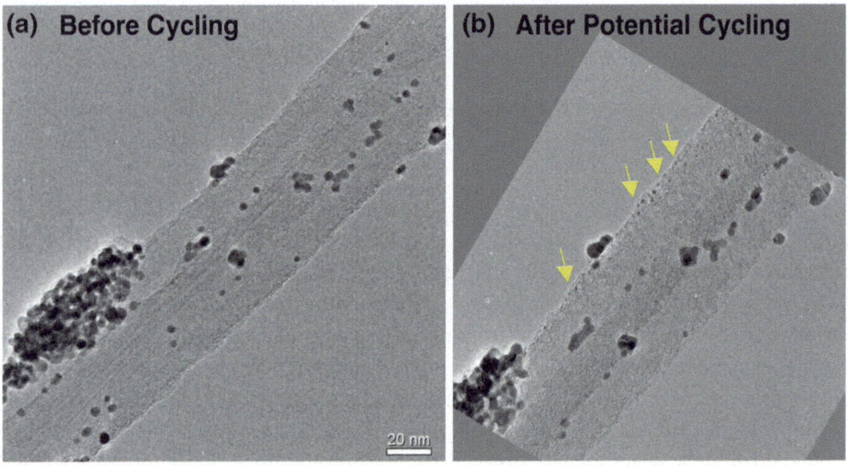

Fig. 5.4 Aberration-corrected TEM images of Pt clusters and single atoms on the surface of CNT

during voltage cycling. IL TEM, was able to confirm this mechanism through direct observations. Figure 5.4 shows that many single atoms and clusters appear on the surface of the CNT after 2000 cycles, which plays an important role in the surface area loss of the nanoparticles through re-deposition of single atoms and/or cluster bridging.

5.3.2 Pt NPs Supported on Amorphous Carbon

Similarly to Pt NPs on carbon nanotubes, particle migration and coalescence and particle dissolution are shown to be responsible for the degradation of Pt NPs supported on amorphous carbon (CB). As it is shown in the 3-D image of Pt/CB,

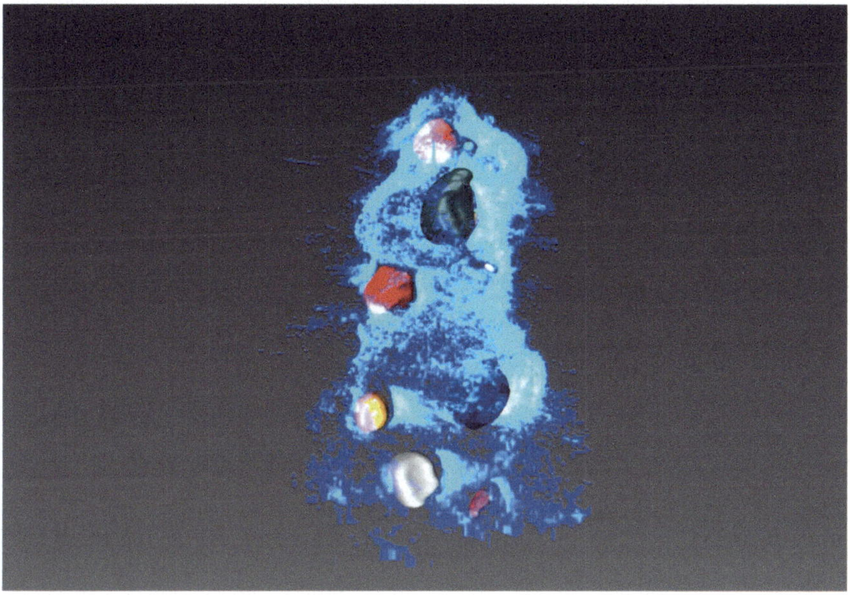

Fig. 5.5 3D reconstructed image of Pt NPs supported on amorphous carbon

reconstructed from a total of 60 TEM images over a tilt range of −60 to +60 degrees, with a 2° tilting (Fig. 5.5), the carbon substrate is not flat and has a convoluted 3D structure. As the nanoparticles are actually within the 3-D carbon network, the movement of NPs in the direction perpendicular to the 2D images cannot be identified. TEM images of 3.2 nm Pt nanocatalysts supported on CB before and after potential cycling shows similar behavior to the NPs on carbon nanotubes. These particles are mobile on the surface of amorphous carbon. Almost all of the particles change their relative position after potential cycling. They either move toward or drift apart from each other. In some cases, the particles move toward each other until they make contact and coalesce. On the other hand some particles seem to de-sinter after potential cycling. This type of behavior has been shown previously to occur in MgO particles [28]. However, we should point out that these nanoparticles are sitting on a carbon substrate that is not completely flat, as seen from the 3-D reconstructed image. Thus, it is possible that the nanoparticles that seem to de-sinter or move apart may actually not be in contact, but instead may be in different planes.

In contrast to the NPs on carbon nanotubes, the appearance of atomic clusters is very difficult to observe on the surface of amorphous carbon. The reason why no atomic clusters appear on the surface of amorphous carbon, in contrast to carbon nanotubes, can be due to the lack of ionomer phase on the surface of amorphous carbon. The CNTs in the previous experiment (Fig. 5.4) were wrapped in ionomer during synthesis, while no ionomer was used in the synthesis of the Pt NPs on amorphous carbon. The ionomer phase serves as a medium for the atomic clusters to move on the surface of carbon. Thus, the lack of ionomer on amorphous carbon

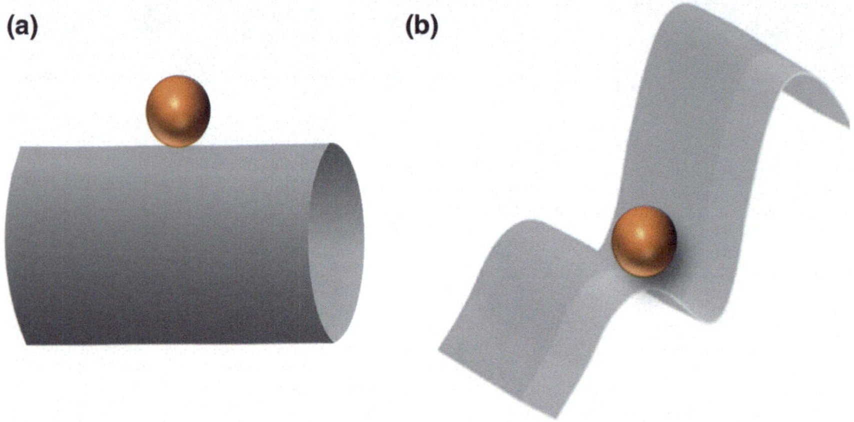

Fig. 5.6 Schematic of a particle in contact with **a** carbon nanotube, and **b** amorphous carbon

results in the release of Pt ions in the liquid electrolyte. This is likely why no atomic clusters can be observed on the surface of amorphous carbon, and no significant particle growth due to Ostwald ripening can be seen.

Another important observation is the fact that particles seem to be more mobile on the surface of CNTs than amorphous carbon. This can be due to the difference in morphology between CNT and CB. Amorphous carbon exhibits a convoluted 3D structure (Fig 5.6), in contrast to carbon nanotubes with a flat surface. As a result, the particles on the surface of carbon black are likely to have larger fraction of surface area in contact with carbon than particles supported on CNTs. Thus, more vacancies are needed at the carbon/particle interface for particles to start their motion.

5.3.3 Pt-Alloys Supported on Amorphous Carbon

The behavior of Pt-alloy nanocatalysts such as Pt-Ni and Pt-Co supported on amorphous carbon has also been investigated during the various stages of fuel cell potential cycling, using identical location aberration-corrected STEM [22, 24]. In a study on Pt-Ni NPs [24], the characterization of the initial powder showed that most of the particles exhibit a disordered structure where Pt and Ni form a face centered cubic solid solution. This structure shows a heterogeneous composition distribution, where Pt-rich and Ni-rich regions are observed within the nanoparticles. It was also shown that Pt seems to segregate to the surface of the nanoparticles. The Pt concentration and the thickness of the Pt-rich shell are higher in smaller particles compared to larger ones. The reason for this behavior starts with the leaching process of Ni, the less noble element, which occurs preferentially at the outermost layers of the NP, leading to the formation of a Pt-rich surface. In smaller NPs, the fraction of surface atoms with respect to bulk atoms increases, and thus smaller NPs are more

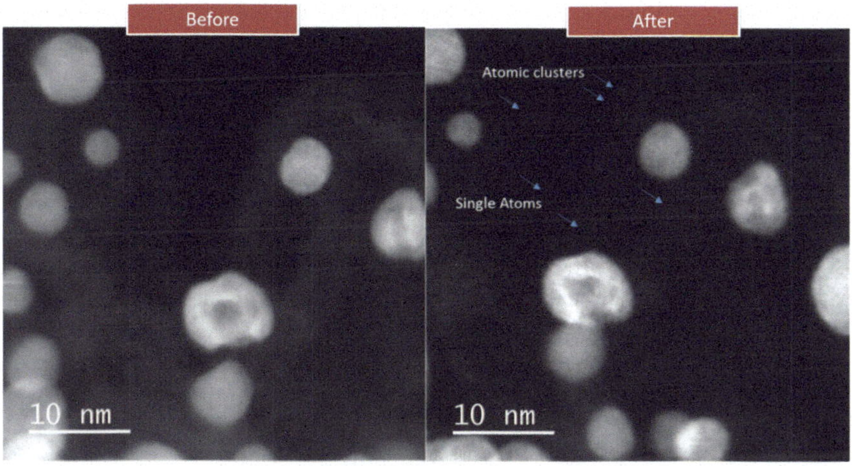

Fig. 5.7 Appearance of single atoms and atomic clusters on the carbon support after voltage cycling

prone to Ni leaching due to the higher density of low coordination surface sites, such as steps and kinks. This results in the formation of a thicker Pt-rich layer and higher Pt/Ni ratio in these particles [10, 29–31].

During the initial voltage cycling, carbon corrosion occurs, which leads to particle motion; however, no particle coalescence could be observed. After 2000 cycles, single atoms and atomic clusters appear on the carbon support (Fig. 5.7).

The experiments done by Rasouli et al. also showed that the atomic re-deposition consists of Pt and not Ni. We expect steps and kinks to be more prone to dissolution, in contrast to low energy surfaces [32]. Due to the low Ni/Ni^{2+} redox potential, the re-deposition of Ni on particles is not thermodynamically favorable [33]. As discussed before, smaller particles have lower amounts of Ni and thicker Pt-rich surface layers. This Pt-rich shell can act as a heterogeneous nucleation site for the re-deposition of Pt. In fact, it has been shown in the literature that the presence of a Pt-rich shell induces a lattice strain on the surface layer, which leads to a downshift in the d-band center of Pt as a result of the ligand effect from the transition metal in the core [34, 35]. According to Schlapka et al., this ligand effect would considerably be reduced if the Pt-rich shell thickness is beyond four Pt monolayers (approximately 1 nm) [36]. In this case, the ligand effect and consequently the compressive strain on the surface can be neglected, assuming that the presence of the transition metal four monolayers beneath the surface has a negligible impact on the electronic structure. In this regard, we can assume that smaller particles, poor in Ni and exhibiting thicker Pt-rich shells, have surfaces similar to that of pure Pt, particularly with the same lattice parameter. Therefore, the re-deposition of Pt on the surface of these smaller particles will cause minimal lattice mismatch between the newly deposited material and the original particle. On the other hand, due to the shift in the d-band center of Pt in particles with thin Pt-rich shells, these particles have lower binding energy for Pt single atoms, compared to pure Pt particles or those with thick Pt shells. As a result, dissolved Pt atoms present on the carbon support are more prone to re-deposit on smaller particles with higher amount of Pt and thicker Pt-rich surface layers. In summary, it seems that the re-deposition of single atoms on these particles is thermodynamically preferred due to the lower lattice mismatch and higher Pt-Pt binding energy on the surface.

Particle migration and particle dissolution are also observed for Pt_3Co NPs supported on amorphous carbon during potential cycling. As shown in Fig. 5.8, particle dissolution and particle movement happens simultaneously during potential cycling. Almost all of the particles change their location on the surface of carbon

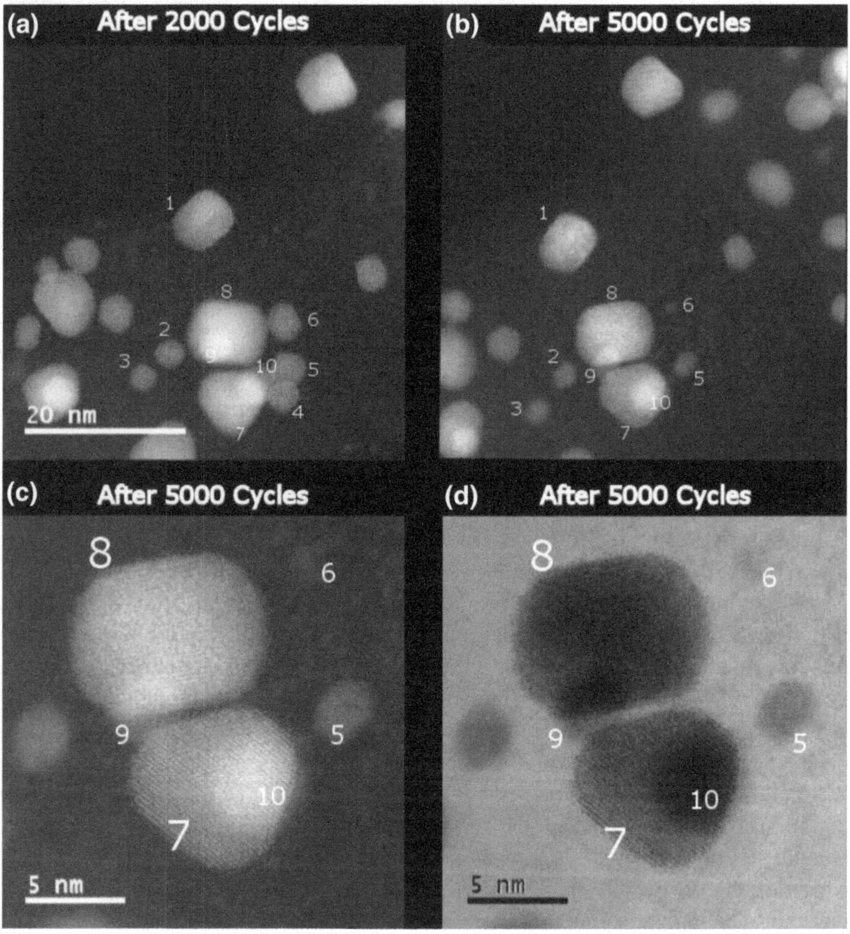

Fig. 5.8 HAADF STEM images of the Pt3Co NPs **a** after 2000 cycles, and **b** after 5000 cycles, **c** and **d** high magnification dark field and bright field images of the particles labeled as 5–10

support. For instance, particle 1 located at the edge of the carbon support in Fig. 5.8a, moves away from the edge in Fig. 5.8b. Similarly, particles 2 and 3 change their relative position. As it was shown in the previous sections, this can be due to a change in the shape of carbon support due voltage, in addition to the particle movement on the surface of carbon as a result of carbon corrosion. Since carbon has a 3D convoluted shape, the details of the changes in the carbon support are not clear from the 2-D projected STEM images. However, the fact that some of the particles are in focus and some are out of focus reveals that the particle are located at different heights relative to the eccentric height.

Modification in the composition distribution is another source of degradation for Pt alloy nanocatalysts. Figure 5.9 shows HAADF STEM images along with EDS mapping of the nanoparticles inside the red boxes labeled 1 and 2 before and after potential cycling. In these images Pt and Co are illustrated in red and green, respectively. Co leaching outside of the particles happen for all of the particles. However, it occurs more severely for smaller particles than the larger ones. The small particles shown in Figs. 5.9c and 5.9d are almost depleted in Co while larger particles only reduce their Co content. In smaller NPs, the fraction of surface atoms with respect to bulk atoms increases, and thus smaller NPs are more prone to Co leaching due to the higher density of low coordination surface sites.

5.3.4 Experimental Requirement and Challenges for IL TEM

In order for IL TEM investigations to be reliable and successful, it is required to ensure that nothing will affect the investigated material between the electrochemical cycling and the electron microscopy. This can be confirmed by comparing the images of the catalyst before and after the transfer to an electrochemical cell with either no potential cycling or cycling in a potential window where no significant catalyst degradation is expected. Additionally, the TEM grid used for the experiment should be chemically stable in the environment and condition of the accelerated stress test (AST) as it is a corrosive environment. Typically, gold finder TEM grids are used for IL TEM to avoid interfering effects from its degradation in acidic environment. Standard gold finder grids have been shown to be stable in AST conditions with potential cycles up to 1.4 VRHE at room temperature [37]. Another important consideration is the stability of the carbon coating of the TEM grid. Schlogl et al. showed that the maximum potential for the experiment is limited at temperature higher than room temperature. For example, at 75 °C, upper potential limit should be lower than 1.3 V as no increased carbon coating rupture was observed in the TEM grids, although some signs of degradation of the coating might already be seen [37]. By increasing the potential and temperature gold dissolution might occur, leaving artifacts on the investigated catalyst materials.

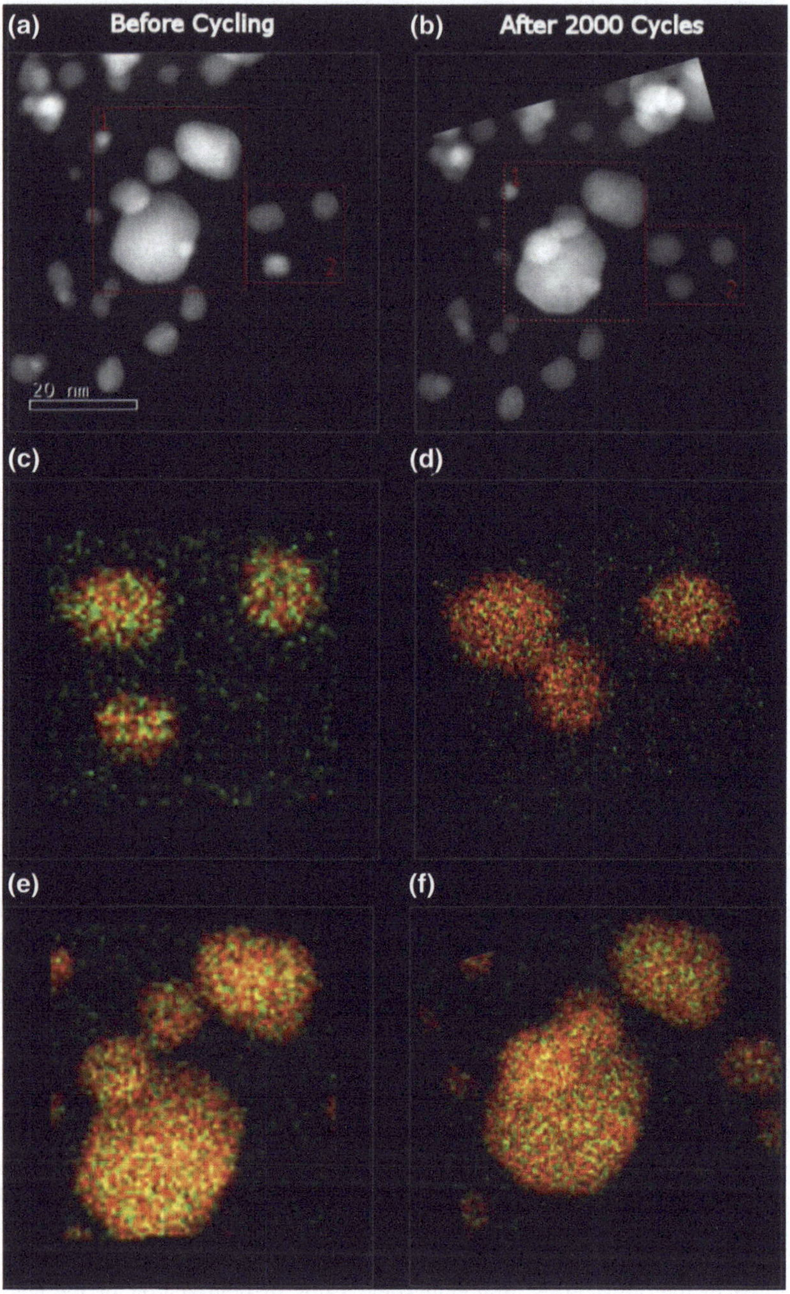

Fig. 5.9 HAADF STEM images of Pt_3Co NPs **a** before and **b** after 2000 cycles, **c** and **d** EDS Mapping of the NPs in box 2, **e** and **f** EDS mapping of the NPs in box 1

Thus, the stability window of the grid need to be considered for performing reliable IL TEM measurements.

Another important consideration is effect of electron high-energy electrons on the investigated material that might interfere with the electrochemical degradation. In general, there are two main types of primary damage that occur in samples following exposure to high-energy electrons: radiolysis and knock-on damage [38]. Both knock-on damage and radiolysis are examples of inelastic scattering in which a portion of the incident energy of an imaging electron is lost due to a collision event. Knock-on damage occurs when an incident electron directly interacts with the nucleus and transfers enough energy to eject the atom from the specimen. Thus, knock-on damage is manifested as the displacement of atoms within a sample and results in the formation of point defects in a periodic lattice. Radiolysis damage arises due to sample ionization by incident electrons interacting with valence electrons and breaking chemical bonds [39]. In order to avoid the effect of electrons on the materials investigated through IL TEM, the electron dose in the TEM should be controlled. The comparison between the images before and after exposure to electron beam can be used to confirm the minimal interference of the electrons in the degradation of the catalysts.

5.4 Conclusions

To understand the mechanisms of catalysts nanoparticles in PEM fuel cells, it is required to determine the behavior of the nanoparticles during different stages of fuel cell cycling. The first challenge is to find a way to ensure that the exact same nanoparticles during the various stages of cycling are observed and analyzed.

To accomplish this, an experimental setup that replicates on a TEM grid the effect of voltage cycling on the cathode of an MEA is required. Using this approach, which is called identical location TEM, it is possible to track the behavior of a single nanoparticle at different stages of voltage cycling on the nano-atomic scale. Through these direct observations, it is demonstrated that due to carbon corrosion, single point defects appear at the carbon/nanoparticle interface, which in turn result in particle migration and consequently coalescence. In addition, the commonly held view of the mechanism of particle dissolution and deposition has been also revisited by IL TEM. Thus, during the later stages of cycling, when the concentration of dissoluble Pt reaches a critical amount, single atoms and atomic clusters appear on the carbon support, which consequently move toward other particles and re-deposit on their surface. This dissolution happens preferentially at the corners and steps of the nanoparticle, while re-deposition occurs on low energy planes. It has also been revealed that re-deposition is not necessarily an isotropic process as atomic clusters can deposit between two or more particles, possibly bridging them.

Furthermore, the atomic surface evolution and phase segregation of alloyed nanoparticles under the effect of voltage has been investigated through IL TEM using advanced spectroscopy techniques. While it is generally accepted in the literature that larger particles grow at the expense of smaller ones, it has been shown that in case of alloys, deposition of Pt occurs on the surface of smaller particles rather than larger ones. This is due to the thicker Pt-rich surfaces on the smaller particles, since the Pt-rich surface act as nucleation sites for re-precipitation of Pt.

References

1. Holby EF, Sheng W, Shao-Horn Y, Morgan D (2009) Pt nanoparticle stability in PEM fuel cells: influence of particle size distribution and crossover hydrogen. Energy Environ Sci 2:865. https://doi.org/10.1039/b821622n
2. Jalan V, Taylor EJ (1983) Importance of interatomic spacing in catalytic reduction of oxygen in phosphoric acid. J Electrochem Soc 130:2299–12302
3. Landsman DA, Luczak FJ (2003) Handbook of fuel cells: fundamentals, technology and applications
4. Gasteiger HA, Kocha SS, Sompalli B, Wagner FT (2005) Activity benchmarks and requirements for Pt, Pt-alloy, and non-Pt oxygen reduction catalysts for PEMFCs. Appl Catal 56:9–35
5. Stamenkovic VR, Mun BS, Arenz M, Mayrhofer KJJ, Lucas CA, Wang G, Ross PN, Markovic NM (2007) Trends in electrocatalysis on extended and nanoscale Pt-bimetallic alloy surfaces. Nat Mater 6:241–247
6. Ferreira PJ, la O' GJ, Shao-Horn Y, Morgan D, Makharia R, Kocha S, Gasteiger HA (2005) Instability of Pt/C electrocatalysts in proton exchange membrane fuel cells. J Electrochem Soc 152:A2256. https://doi.org/10.1149/1.2050347
7. Shao-Horn Y, Sheng WC, Chen S, Ferreira PJ, Holby EF, Morgan D (2007) Instability of supported platinum nanoparticles in low-temperature fuel cells. Top Catal 46:285–305. https://doi.org/10.1007/s11244-007-9000-0
8. Ferreira PJ, Shao-Horn Y (2007) Formation mechanism of Pt single-crystal nanoparticles in proton exchange membrane fuel cells. Electrochem Solid-State Lett 10:B60–B63
9. Gilbert JA, Kropf AJ, Kariuki NN, DeCrane S, Wang X, Rasouli S, Yu K, Ferreira PJ, Morgan D, Myers DJ (2015) In-operando anomalous small-angle X-ray scattering investigation of Pt 3 Co catalyst degradation in aqueous and fuel cell environments. J Electrochem Soc 162:F1487–F1497. https://doi.org/10.1149/2.0531514jes
10. Chen S, Gasteiger HA, Hayakawa K, Tada T, Shao-Horn Y (2010) Platinum-alloy cathode catalyst degradation in proton exchange membrane fuel cells: nanometer-scale compositional and morphological changes. J Electrochem Soc 157:A82. https://doi.org/10.1149/1.3258275
11. Gummalla M, Ball S, Condit D, Rasouli S, Yu K, Ferreira P, Myers D, Yang Z (2015) Effect of particle size and operating conditions on Pt_3Co PEMFC cathode catalyst durability. Catalysts 5:926–948. https://doi.org/10.3390/catal5020926
12. Dubau L, Maillard F, Chatenet M, Guetaz L, André J, Rossinot E (2010) Durability of Pt_3Co/C cathodes in a 16 Cell PEMFC stack: macro/microstructural changes and degradation mechanisms. J Electrochem Soc 157:B1887. https://doi.org/10.1149/1.3485104
13. Borup R, Meyers J, Pivovar B, Kim YS, Mukundan R, Garland N, Myers D, Wilson M, Garzon F, Wood D, Zelenay P, More K, Stroh K, Zawodzinski T, Boncella J, McGrath JE, Inaba M, Miyatake K, Hori M, Ota K, Ogumi Z, Miyata S, Nishikata A, Siroma Z, Uchimoto Y, Yasuda K, Kimijima KI, Iwashita N (2007) Scientific aspects of polymer electrolyte fuel cell durability and degradation. Chem Rev 107:3904–3951. https://doi.org/10.1021/cr0501821
14. Hayre R, Colella W, Cha S, Prinz F (2009) Fuel cell fundamentals. Wiley, New York

15. Yu X, Ye S (2007) Recent advances in activity and durability enhancement of Pt/C catalytic cathode in PEMFC: Part II: Degradation mechanism and durability enhancement of carbon supported platinum catalyst. J Power Sources 172:145
16. Shao Y, Kou R, Wang J, Viswanathan VVKJH, Liu J, Wang Y, Lin Y (2008) The influence of the electrochemical stressing (potential step and potential-static holding) on the degradation of polymer electrolyte membrane fuel cell electrocatalysts. J Power Sources 280:280
17. Kocha S (2012) Electrochemical degradation: electrocatalyst and support durability. In: Mench MM, Kumbur EC, Veziroglu TN (eds) Polymer electrolyte fuel cell degradation. Academic Press, Oxford, p 89
18. Yuan X-Z, Li H, Zhang S, Martin J, Wang H (2011) A review of polymer electrolyte membrane fuel cell durability test protocols. J Power Sources 196:9107
19. Ricea CA, Urchaga P, Pistono AO, McFerrin BW, McComb BT, Hu J (2015) Platinum dissolution in fuel cell electrodes: enhanced degradation from surface area assessment in automotive accelerated stress tests. J Electrochem Soc 162:F1175
20. Marcua Alina, Totha Gabor, Patrick Pietraszb JW (2014) Cathode catalysts degradation mechanism from liquid electrolyte to membrane electrode assembly. Comptes Rendus Chim 17:752
21. Zorko M, Jozinović B, Bele M, Hodnik N, Gaberšček M (2014) SEM method for direct visual tracking of nanoscale morphological changes of platinum based electrocatalysts on fixed locations upon electrochemical or thermal treatments. Ultramicroscopy 140:44–50. https://doi.org/10.1016/j.ultramic.2014.02.006
22. Janbroers S, Louwen JN, Zandbergen HW, Kooyman PJ (2009) Insights into the nature of iron-based Fischer-Tropsch catalysts from quasi in situ TEM-EELS and XRD. J Catal 268:235–242. https://doi.org/10.1016/j.jcat.2009.09.021
23. Fiordaliso EM, Sharafutdinov I, Carvalho HWP, Grunwaldt JD, Hansen TW, Chorkendorff I, Wagner JB, Damsgaard CD (2015) Intermetallic GaPd2 nanoparticles on SiO_2 for low-pressure CO_2 hydrogenation to methanol: catalytic performance and in situ characterization. ACS Catal 5:5827–5836. https://doi.org/10.1021/acscatal.5b01271
24. Rasouli S (2017) Degradation mechanisms of Pt and Pt Alloy nanocatalysts in proton exchange membrane fuel cells. University of Texas at Austin
25. Evora MC, Klosterman D, Lafdi K, Li L, Silva LGA (2012) Mechanism of electron beam radiation damage on carbon nanofiber surface. In: UV EB technical conference proceedings
26. Pandy A, Yang Z, Gummalla M, Atrazhev VV, Kuzminyh NY, Sultanov VI, Burlatsky S (2013) A carbon corrosion model to evaluate the effect of steady state and transient operation of a polymer electrolyte membrane fuel cell. J Electrochem Soc 160:F972–F979. https://doi.org/10.1149/2.036309jes
27. Rasouli S, Ortiz Godoy RA, Yang Z, Gummalla M, Ball SC, Myers D, Ferreira PJ (2017) Surface area loss mechanisms of Pt_3Co nanocatalysts in proton exchange membrane fuel cells. J Power Sources 343:571–579. https://doi.org/10.1016/j.jpowsour.2017.01.058
28. Rankin J, Boatner LA (1994) Unstable neck formation during initial-stage sintering. J Am Ceram Soc 77:1987–1990. https://doi.org/10.1111/j.1151-2916.1994.tb07088.x
29. Chen S, Sheng W, Yabuuchi N, Ferreira PJ, Allard LF, Shao-Horn Y (2008) Origin of oxygen reduction reaction activity on Pt_3Co nanoparticles: atomically resolved chemical compositions and structures. J Phys Chem C 113:1109–1125
30. Watanabe M, Tsurumi K, Mizukami T, Nakamura T, Stonehart P (1994) Activity and stability of ordered and disordered Co/Pt alloys for phosphoric acid fuel cells. J Electrochem Soc 141:2659
31. Snyder RC, Doherty MF (2007) Faceted crystal shape evolution during dissolution or growth. AIChE J 53:1337–1348. https://doi.org/10.1002/aic.11132
32. Wise H, Oudar J (1990) Materials and concepts in surface reactivity and catalysis. Dover Publications, New York
33. Bard AJ, Faulkner LR (2002) Electrochemical methods: fundamentals and applications. Wiley New York;, (2001) Russ J Electrochem 38:1505–1506. https://doi.org/10.1023/a:1021637209564

34. Hammer B, Norskov JK (2000) Theoretical surface science and catalysis-calculations and concepts. Adv Catal 45:71–129
35. Adzic RR, Zhang J, Sasaki K, Vukmirovic MB, Shao M, Wang JX, Nilekar AU, Mavrikakis M, Valerio JA, Uribe F (2007) Platinum monolayer fuel cell electrocatalysts. Top Catal 46:249–262
36. Schlapka A, Lischka M, Gross A, Kasberger U, Jakob P (2003) Surface strain versus substrate interaction in heteroepitaxial metal layers: Pt on Ru(0001). Phys Rev Lett 91:016101/1-016101/4. https://doi.org/10.1103/physrevlett.91.016101
37. Schlögl K, Mayrhofer KJJ, Hanzlik M, Arenz M (2011) Identical-location TEM investigations of Pt/C electrocatalyst degradation at elevated temperatures. J Electroanal Chem 662:355–360. https://doi.org/10.1016/j.jelechem.2011.09.003
38. Ugurlu O, Haus J, Gunawan AA, Thomas MG, Maheshwari S, Tsapatsis M, Mkhoyan KA (2011) Radiolysis to knock-on damage transition in zeolites under electron beam irradiation. Phys Rev B - Condens Matter Mater Phys 83. https://doi.org/10.1103/physrevb.83.113408
39. Brown R by PD (1999) Transmission electron microscopy—a textbook for materials science, by David B. Williams and C. Barry Carter. Microsc Microanal 5:452–453. https://doi.org/10.1017/s1431927699990529

Chapter 6
Synthesis of Nanoporous Carbon and Their Application to Fuel Cell and Capacitor

Koki Urita, Isamu Moriguchi and Naotoshi Nakashima

6.1 Synthesis and Structure of Nanoporous Carbon

6.1.1 Introduction

Porous carbons are important materials for adsorbent, catalysts, electrode materials, and so on. Activated carbons (ACs), which commonly possess broad pore size distributions (PSDs) from micropore (<2 nm) to mesopore (2–50 nm) or macropore (>50 nm) ranges, were widely used so far for their applications, and the porosity and PSD of ACs were controlled by activation processes using alkaline, water, or CO_2 at high temperature to some extent. Recently, precise pore size control of nanoporous carbons has been actively studied so as to meet the requirements of electrodes for fuel cells, capacitors, and Li-ion batteries from the viewpoints of not only increasing surface area (reaction interface) but also the formation of effective mass transport paths and utilization of specific nanospace effect. Some effective synthetic processes of nanoporous carbons are introduced according to the classification of pore size range.

K. Urita · I. Moriguchi (✉)
Graduate School of Engineering, Nagasaki University, Bunkyo,
Nagasaki 852-8521, Japan
e-mail: mrgch@nagasaki-u.ac.jp

K. Urita
e-mail: urita@nagasaki-u.ac.jp

N. Nakashima
International Institute for Carbon-Neutral Energy Research (WPI-I2CNER),
Kyushu University, Motooka, Nishi-ku, Fukuoka 819-0395, Japan
e-mail: nakashima.naotoshi.614@m.kyushu-u.ac.jp

N. Nakashima (ed.), *Nanocarbons for Energy Conversion: Supramolecular Approaches*, Nanostructure Science and Technology,
https://doi.org/10.1007/978-3-319-92917-0_6

6.1.2 Synthesis of Microporous Carbons

Microporous carbons were successfully synthesized by a high-temperature chlorination of carbides such as SiC and TiC, whereby metal and metalloids are removed as chlorides. The carbide-derived carbons (CDCs) have a narrow PSD with a mean value that is tunable with better than 0.05 nm accuracy in the range of ~0.5 to ~3 nm [1] and a specific surface area up to 2000 $m^2\ g^{-1}$ [2]. The ease and accuracy of pore tunability in CDC allowed experimental determination of the optimal pore size for hydrogen storage [3] and electric double-layer capacitor [4]. However, the chlorination process is not always convenient because it needs safety equipment. As an easier process, an ordered microporous carbon with a Y-type zeolite-replicated structure was synthesized by an infiltration of carbon source such as furfuryl alcohol into the micropores of zeolite and then a carbonization and a removal of the template zeolite by HF etching [5]. Microporous carbon materials with narrow PSDs and high specific surface areas were also synthesized from an organic–inorganic hybrid polymer with covalently bonded phenolic oligomer and siloxane moieties, which play a role of carbon and SiO_2 sources, respectively (Fig. 6.1). The hybrid polymer was converted into a SiO_2/carbon nanocomposite by sol–gel reaction under acidic or basic condition and a heating up to 800–1000 °C in an Ar atmosphere, and then microporous carbons were obtained by removal of the residual SiO_2 part by HF etching [6]. Continuous micropore channels and micropores below 1 nm were successfully formed in the carbons as shown in Fig. 6.2.

6.1.3 Synthesis of Mesoporous and Macroporous Carbons

A lot of studies have focused on template processes for the synthesis of mesoporous and macroporous carbons, which can be classified into the following categories: (1) carbonization of polymer blends with silica gel [7, 8], (2) the use of inorganic porous materials as templates, in which a carbon source is infiltrated into pores and is carbonized, then the porous template is removed by dissolution [9–14], and (3) colloidal crystal-assisted template synthesis [15–18]. Disordered mesoporous carbons with the high surface area are obtained for (1). In contrast, ordered

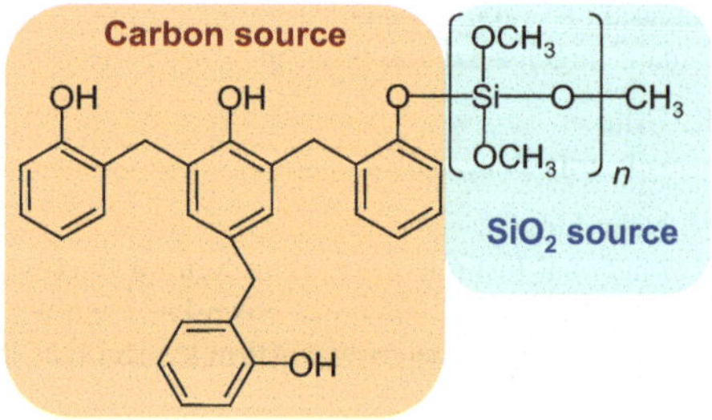

Fig. 6.1 Chemical structure of organic–inorganic hybrid polymer

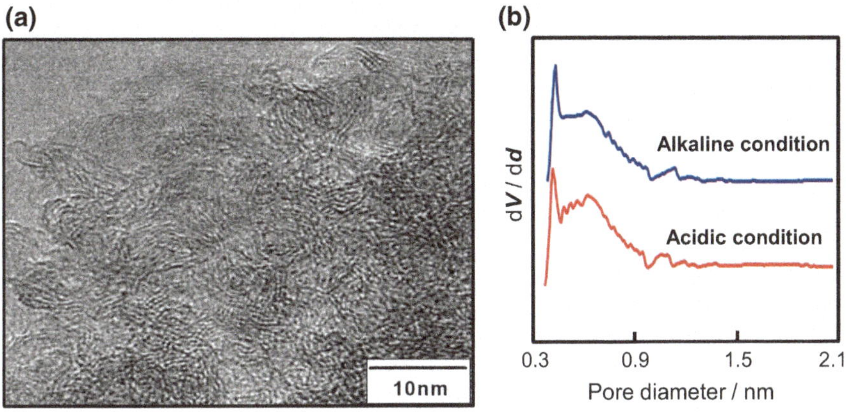

Fig. 6.2 **a** TEM image and **b** micropore size distributions of microporous carbons obtained from the organic–inorganic hybrid polymer

microporous, mesoporous, and macroporuos carbons with narrow PSDs can be synthesized by (2) and (3) methodologies. A mesoporous carbon with a narrow pore size distribution and an ordered porous structure (CMK-1) was successfully synthesized by a repeating infiltration of sucrose in pores of MCM-48 mesopous silica, followed by treatments with H_2SO_4, carbonization by heating in inert gas, and then removal of silica by HF etching [9]. A colloidal crystal template process was successfully applied to the synthesis of ordered porous carbons with narrow PSDs in the pore size range from mesopore to macropore [15, 16]. A carbon source such as a mixture solution of phenol and formaldehyde was infiltrated into the interstitial space of a silica colloidal crystal, and then was in situ polymerized and carbonized. Ordered porous carbons could be obtained finally by dissolving the silica template with *aq*. HF or *aq*. NaOH. Formation of skeletal carbon frameworks consisting of spherical void arrays, the void size of which is comparable to the diameter of template silica particles, can be observed as shown in Fig. 6.3. It was also confirmed by N_2 adsorption–desorption isotherm measurements that the colloidal crystal-derived porous carbons possessed mesopores or macropores originating from silica template and micropores in the mesopore or macropore wall, which were inherently produced for hard carbon sources (Fig. 6.4). The mesopore and/or macropore surface area of colloidal crystal-templated porous carbons increases with decreasing the silica particle size. For example, porous carbons obtained at 1000 °C heating from a phenol–formaldehyde carbon source by using 120 and 45 nm silica particles had total surface areas of 662 and 1162 $m^2\ g^{-1}$ and macropore or mesopore surface areas of 262 and 797 $m^2\ g^{-1}$ while keeping the almost same micropore surface area around 400 $m^2\ g^{-1}$.

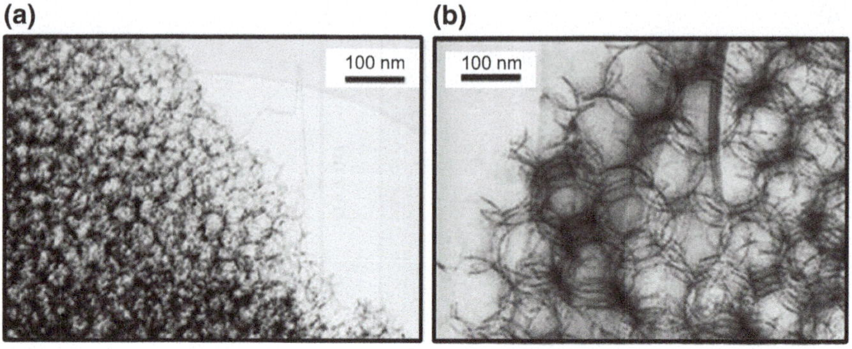

Fig. 6.3 TEM images of colloidal crystal-derived nanoporous carbons obtained by using the average SiO_2-particle size of **a** 45 nm, and **b** 120 nm

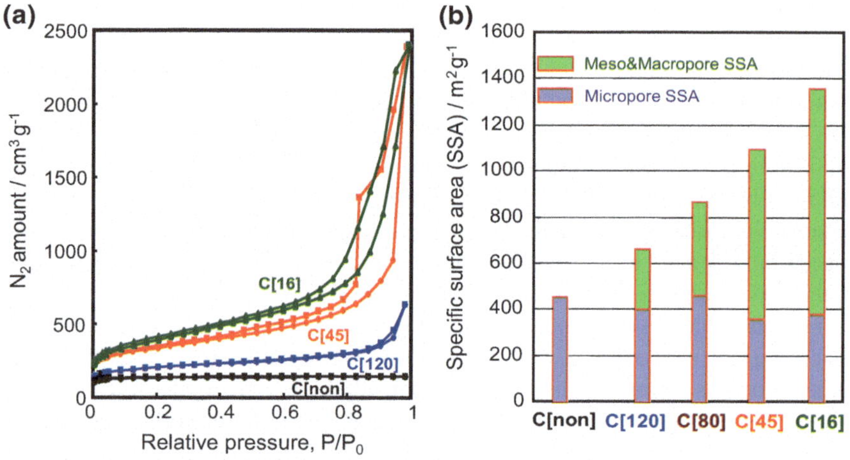

Fig. 6.4 **a** N_2 adsorption–desorption isotherms of colloidal crystal-derived nanoporous carbons, C[*X*], where *X* indicates the average size (nm) of SiO_2 particle used, and **b** their microporous, mesoporous, and microporous and total specific surface areas. C[non] is the carbon synthesized from a phenol-formaldehyde carbon source without the template

6.2 Application to Fuel Cell Catalyst

6.2.1 *Introduction*

As described above, the colloidal crystal-derived nanoporous carbon (NanoPC) has a porous structure with a high surface area ($\sim$1,000 m^2 g^{-1}) [6, 15, 16]. Such nanocarbons have a high potential to be a supporting carbon of polymer electrolyte fuel cells (PEFCs) [19–21]; such structure facilitates the mass transport of reactants to the active sites and products of the fuel cell reaction, and thus it is expected to

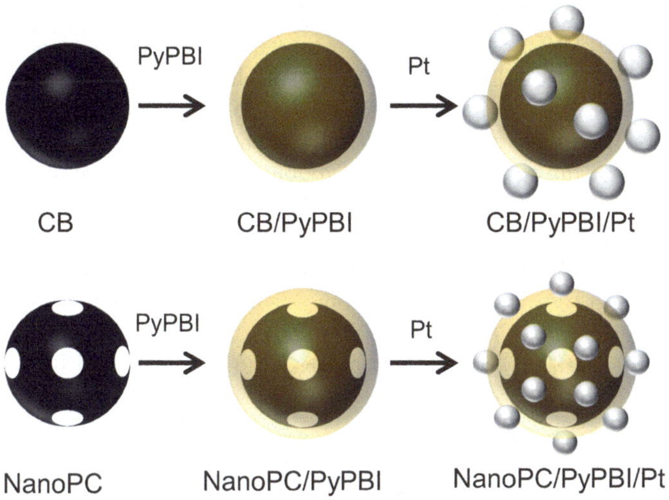

Fig. 6.5 Schematic images of CB/PyPBI/Pt and NanoPC/PyPBI/Pt electrocatalysts. Reproduction from [34] with permission of The American Chemical Society

avoid diffusion limitations, which is important to improve the fuel cell performance [22–25]. Arenz et al. reported that Pt loaded on Ketjen black (KB) showed higher durability than that of Pt loaded on Vulcan, namely the KB shows ~3 times higher specific surface area than that of Vulcan [26, 27]. Pt-NPs loaded on a porous carbon are reported to be more stable than those deposited on carbon black (CB) [28], and the Pt-NPs deposited on a polymer-wrapped NanoPC are durable against carbon corrosion [29, 30]. While such electrocatalysts use Nafion ionomers, thus use of the material in high-temperature PFECs is limited due to the lack of proton conduction in the catalyst layer at temperatures over 100 °C [31–33].

As describe in Chap. 1 of this book, polybenzimidazole (PBI) and pyridine-based PBI (PyPBI) (for chemical structures, see Fig. 1.4) provide the anchor sites for the Pt-NPs, and enhanced the durability of fuel cell catalysts fabricated using PBI-Wrapped CNTs as a carbon support. Such method was applied to NanoPC. Schematic drawing of a PBI-wrapped NanoPC before Pt loading to provide NanoPC/PyPBI/Pt electrocatalyst is presented in Fig. 6.5 [34]. The Pt-NPs were well dispersed on the PyPBI-wrapped NanoPC with the diameter range of 2.2 ± 0.2 nm.

6.2.2 PEFC Performance

The PEFC performance was tested [34]. The electrochemical surface area (ECSA) value of NanoPC/PyPBI/Pt was 63.7 $m^2\ g^{-1}$ of Pt (Fig. 6.6a), which was close of higher than those of CB/Pt (60.6 $m^2\ g^{-1}$ of Pt) and CB/PyPBI/Pt (46.8 $m^2\ g^{-1}$ of Pt). The durability test of the catalyst was carried out based on the protocol of the

Fuel Cell Commercialization Conference of Japan (FCCJ) (potential cycling from 1 to 1.5 V vs. RHE) (Fig. 6.6b) [35], which reflects carbon corrosion ($C + 2H_2O \rightarrow CO_2 + 4H^+ + 4e^-$, 0.207 V vs. RHE) [36–38]. The hydroquinone–quinone (HQ/Q) redox peaks are observed for the electrocatalyst at 0.5 V versus RHE, implying carbon corrosion [39]. With increasing potential cyclings, the current of the hydrogen absorption/desorption and Pt oxidation/reduction peaks decreased due to the detachment of the Pt-NPs from the bare CB and CB/PyPBI leading to 46% and 32% loss in the ECSA, respectively. It is evident that after wrapping with PyPBI, the polymer protected the CB from serious corrosion. Thus, the CB/PyPBI/Pt showed a higher durability than that of the conventional CB/Pt. Notably, for the NanoPC/PyPBI/Pt (Fig. 6.6c, d), the ECSA remained the initial value even after 10,000 cyclings, indicating that the NanoPc/PyPBI/Pt possesses very high durability.

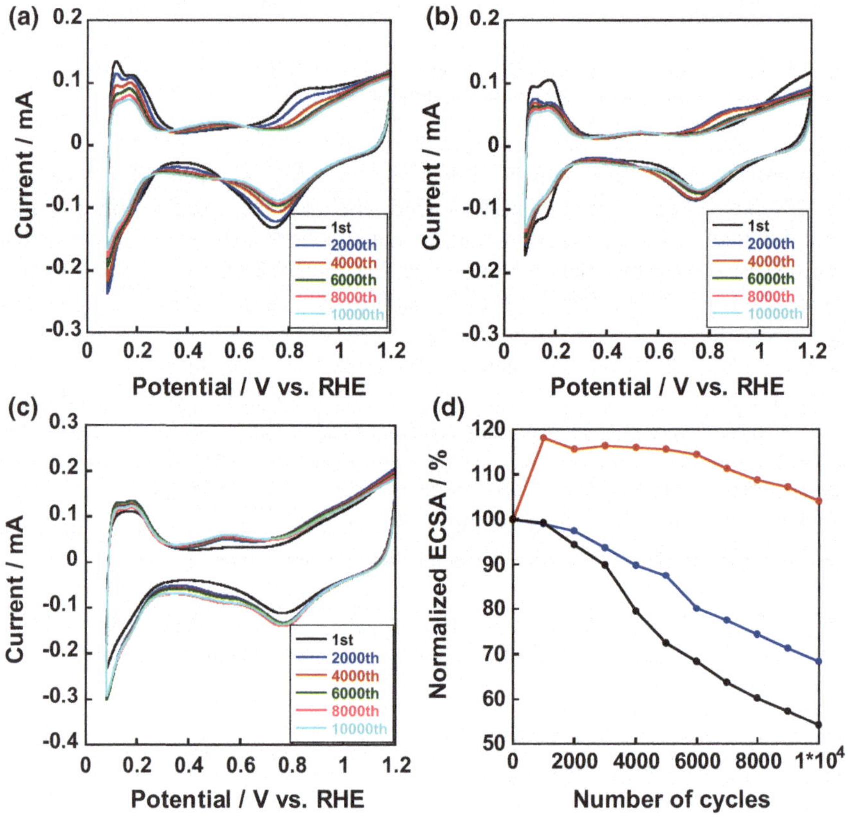

Fig. 6.6 CV curves of the CB/Pt (**a**), CB/PyPBI/Pt (**b**), and NanoPC/PyPBI/Pt (**c**) after 2000, 4000, 6000, 8000, and 10,000 potential cycles test. **d** Normalized ECSAs of CB/Pt (black line), CB/PyPBI/Pt (blue line), and NanoPC/PyPBI/Pt (red line) as a function of the number of potential cycles in the range of 1.0–1.5 V versus RHE. Reproduction from [34] with permission of The American Chemical Society

What is a mechanism for such high durability? The morphologies of the three electrocatalysts after durability test observed by TEM, in which the formation of Pt-NP aggregates with a diameter of ∼5.0 nm was observed for CB/Pt and PyPBI/Pt catalysts due to carbon corrosion, which causes the ECSA loss [40, 41]. After the durability test, the diameter of the Pt-NPs in the NanoPC/PyPBI/Pt also increased to ∼3.1 nm, while as shown in Fig. 6.7a, b, and some smaller Pt-NPs with a diameter of ∼1.8 nm were observed. Such smaller Pt-NPs are important to maintain the ECSA even after the durability test, while the micropores became larger owing to the carbon corrosion. Such behavior is schematically drawn in Fig. 6.7c. Noto et al. reported that the H_2O_2 pretreatment removed the micropores of the support and made the deeply buried electrocatalyst available after the pretreatment [42, 43].

The gas diffusion electrodes (GDEs) of three different electrocatalysts were fabricated according to our previous reports [44–48], and their fuel cell performances were examined. Figure 6.8 shows the polarization curves of the electrocatalysts operated at 120 °C using dry hydrogen and air under atmospheric pressure for the anode and cathode. Before applying the current to the single cell, the open-circuit voltages (OCV) were 0.90, 0.92, and 0.91 V for the CB/Pt, CB/PyPBI/Pt, and NanoPC/PyPBI/Pt, respectively. The potential of the CB/PyPBI/Pt gradually decreased with the increasing the loaded current compared to that of the CB/Pt.

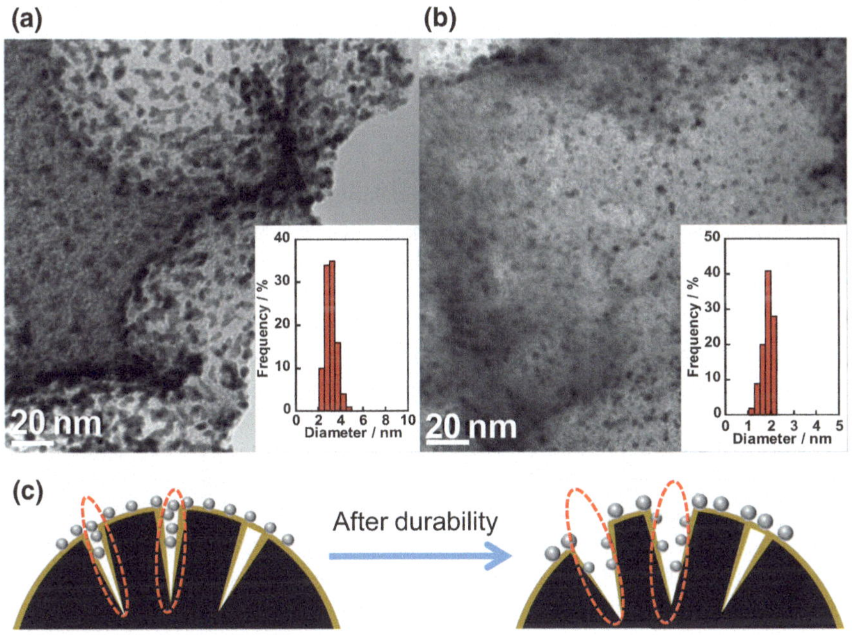

Fig. 6.7 TEM images of the NanoPC/PyPBI/Pt after durability test: **a** Larger Pt-NPs area, **b** Smaller Pt-NPs area. Inset: Histograms of particle size distribution for 100 particles in the TEM images. **c** Schematic illustration of the NanoPC/PyPBI/Pt after the durability test. Reproduction from [34] with permission of The American Chemical Society

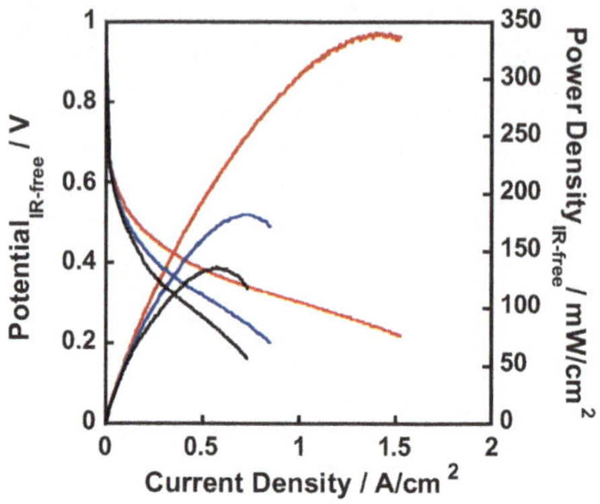

Fig. 6.8 I-V and power density polarization curves of different MEAs based on CB/Pt (black line), CB/PyPBI/Pt (blue line), and NanoPC/PyPBI/Pt (red line) at 120 °C under non-humidified atmosphere. Reproduction from [34] with permission of The American Chemical Society

Compared to the power density of the conventional CB/Pt (115 mW cm^{-2}), for CB/PyPBI/Pt, it was improved to 183 mW cm^{-2}, which is due to higher Pt utilization efficiency. Notably, the power density of NanoPC/PyPBI/Pt showed the power density of 342 mW cm^{-2}, which was almost 2 times higher than that of the CB/PyPBI/Pt. Such high performance would be derived from the nanoporous structure of the NanoPC that protects aggregation of Pt-NPs and facilitates the diffusion of the reactant.

6.2.3 *CO Tolerance*

The fuel cells, especially the direct methanol fuel cells (DMFCs), have received much attention due to their higher energy density (5.04 kWh L^{-1}), simple storage, and transportation compared to the hydrogen fuel cells (0.53 kWh L^{-1}) [49–56]. The direct conversion of methanol has a voltage similar ($CH_3OH + 3/2O_2 \rightarrow CO_2 + 2H_2O + 3e^-$, $E = 1.19$ V) to that of hydrogen ($E = 1.23$ V). The DMFCs, however, still suffer from three problems on the anode side, namely, (i) the carbon monoxide (CO) poisoning of the platinum nanoparticles (Pt-NPs), which is generated from the uncompleted methanol oxidation reaction (MOR) [57–62]; (ii) the sluggish MOR compared to the hydrogen oxidation reaction (HOR) [63–65]; and (iii) the low durability of the electrocatalyst in terms of Pt stability and carbon corrosion, which degrades the fuel cell performance [66, 67].

The development of a new method to address the CO poisoning problem in the DMFCs is important. Recently, Nakashima et al. reported that poly(vinylphosphonic acid) (PVPA) plays an important role in enhanced CO tolerance of the electrocatalyst due to the formation of the $Pt(OH)_{ads}$ that consumes the $Pt(CO)_{ads}$ [68, 69], and the durability was improved because of the polymer wrapping which protected the electrocatalyst from carbon corrosion that stabilized the Pt-NPs.

We here summarize the MOR and durability of an electrocatalyst composed of PyPBI-wrapped NanoPC, PVPA (denoted as NanoPC/PyPBI/Pt/PVPA, Fig. 6.9)

[70] evaluated at 60 °C under high methanol concentrations (4 M and 8 M), which is important for practical operations of DMFCs.

The amounts of the PVPA evaluated from the TGA analysis of the NanoPC/PyPBI/Pt and NanoPC/PyPBI/Pt/PVPA were 49.9 wt% and 42.8 wt%, respectively, in which the higher PVPA in the composite was attributed to the higher specific surface area [69]. The diameter of the Pt-NP calculated from the TEM image was 2.3 ± 0.2 nm which was almost comparable to the Pt size of the NanoPC/PyPBI/Pt [68, 69].

Figure 6.10a, b shows CO stripping voltammograms, in which a sharp peak due to the oxidation of the CO species absorbed on the Pt surfaces is observed at ~0.9 V versus RHE. The CO oxidation peak of the electrocatalyst (NanoPC/PyPBI/Pt/PVPA) appears at 877 mV, indicating a negative shift compared to that of the commercial CB/Pt (950 mV). This onset potential of the CO oxidation on the electrocatalyst was lower than that of the commercial CB/Pt [71], suggesting that the NanoPC/PyPBI/Pt/PVPA has a higher CO tolerance. This result would be due to the PVPA layer on the electrocatalyst that would weaken the binding energy between the Pt-NPs and CO species, which accelerates the formation of the Pt$(OH)_{ads}$ that consumed the $Pt(CO)_{ads}$ [68].

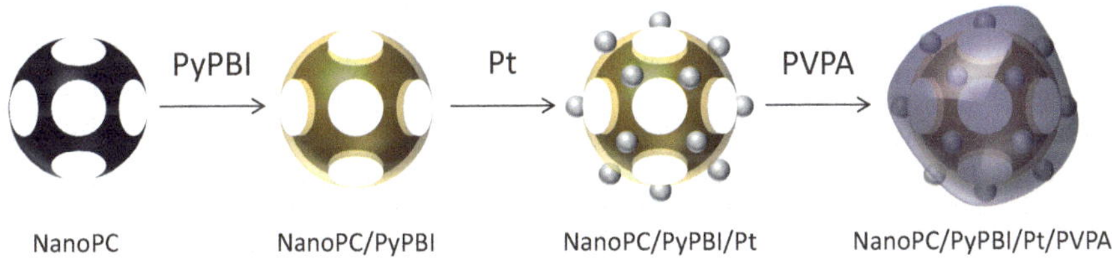

Fig. 6.9 Schematic drawing for the fabrication of NanoPC/PyPBI/Pt/PVPA. Reproduction from [70] with permission of The American Chemical Society

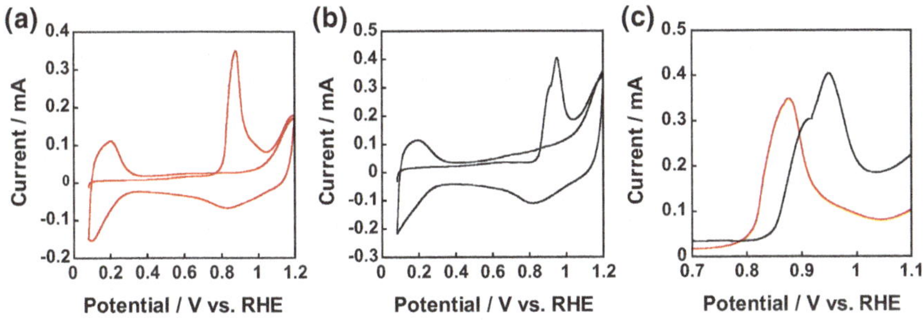

Fig. 6.10 CO stripping voltammograms of the NanoPC/PyPBI/Pt/PVPA (**a**) and CB/Pt (**b**) before durability test at the scan rate of 50 mV/s at 60 °C. **c** Magnification of the CO stripping profile in the range of 0.7–1.1 V versus RHE. Reproduction from [70] with permission of The American Chemical Society

The long-term durability was evaluated in an N_2-saturated 0.1 M $HClO_4$ electrolyte at 60 °C by potential cycling in the range of 1.0–1.5 V versus RHE in order to accelerate the carbon corrosion whose protocol was proposed by FCCJ [35].

MOR is the anodic reaction in DMFCs and the CO species generated during the MOR poison the Pt-NP, which leads to the covered Pt-NPs becoming inactive. Thus, the CO poisoning is still a serious problem in DMFCs. The CO tolerance can be evaluated from the ratios of the I_f and I_b in the MOR curves [72, 73], where I_f and I_b are the anodic and reverse anodic peaks, respectively, as used in the following equation [74–76].

Anodic peak (I_f) [77]:

$$Pt(CH_3OH)_{ads} + H_2O \rightarrow CO_2 + 6H_2O + 6e^- \quad (6.1)$$

$$Pt(CH_3OH)_{ads} \rightarrow Pt(CO)_{ads} + 4H^+ + 4e^- \quad (6.2)$$

$$Pt + H_2O \rightarrow Pt(OH)_{ads} + H^+ + e^- \quad (6.3)$$

Reverse anodic peak (I_b):

$$Pt(OH)_{ads} + Pt(CO)_{ads} \rightarrow Pt + CO_2 + H^+ + e^- \quad (6.4)$$

Higher I_f/I_b ratios imply a higher CO tolerance. Figure 6.11a, b shows the MOR of the two electrocatalysts of NanoPC/PyPBI/Pt/PVPA and Pt/CB measured in the presence of 4 M and 8 M methanol before and after the durability test at 60 °C, in which highly concentrated methanol (4 M or 8 M) was fed to the anode to address the sluggish MOR [78]. The I_f/I_b ratio is summarized in Table 6.1 [70].

From the table, it is evident that the durability of NanoPC/PyPBI/Pt/PVPA in 4 M and 8 M methanol exhibited an ~3 times higher CO tolerance than that of the commercial CB/Pt. Such a high CO tolerance would be derived from the PVPA coating which accelerated the water adsorption to form the $Pt(OH)_{ads}$ according to Eq. (6.3), and weakened the binding energy between the Pt-NPs and CO species. The I_f/I_b values of the NanoPC/PyPBI/Pt/PVPA decreased by 37.6% and 32.6% for the 4 M methanol and 8 M methanol after the durability test, respectively. The CO tolerance of the NanoPC/PyPBI/Pt/PVPA was still ~3 times higher than that of the CB/Pt even after the durability test. Also, the PVPA layer on the NanoPC/PyPBI/Pt/PVPA was stable for long because of the strong multipoint base–acid interaction between the PyPBI and PVPA. After the durability test, the mass activities of both NanoPC/PyPBI/Pt/PVPA and CB/Pt decreased by ~35%, while the mass activity of NanoPC/PyPBI/Pt/PVPA was still ~2 times higher than that of CB/Pt.

Figure 6.12 shows the I-V curve and power density of NanoPC/PyPBI/Pt/PVPA measured under 70 °C with 100%RH. The maximum power density of 132 mW/cm^2 was ~1.6 times higher than that of the previously reported commercial CB/Pt (81 mW/cm^2) [62], indicating that NanoPC/PyPBI/Pt/PVPA is an efficient DMFC catalyst.

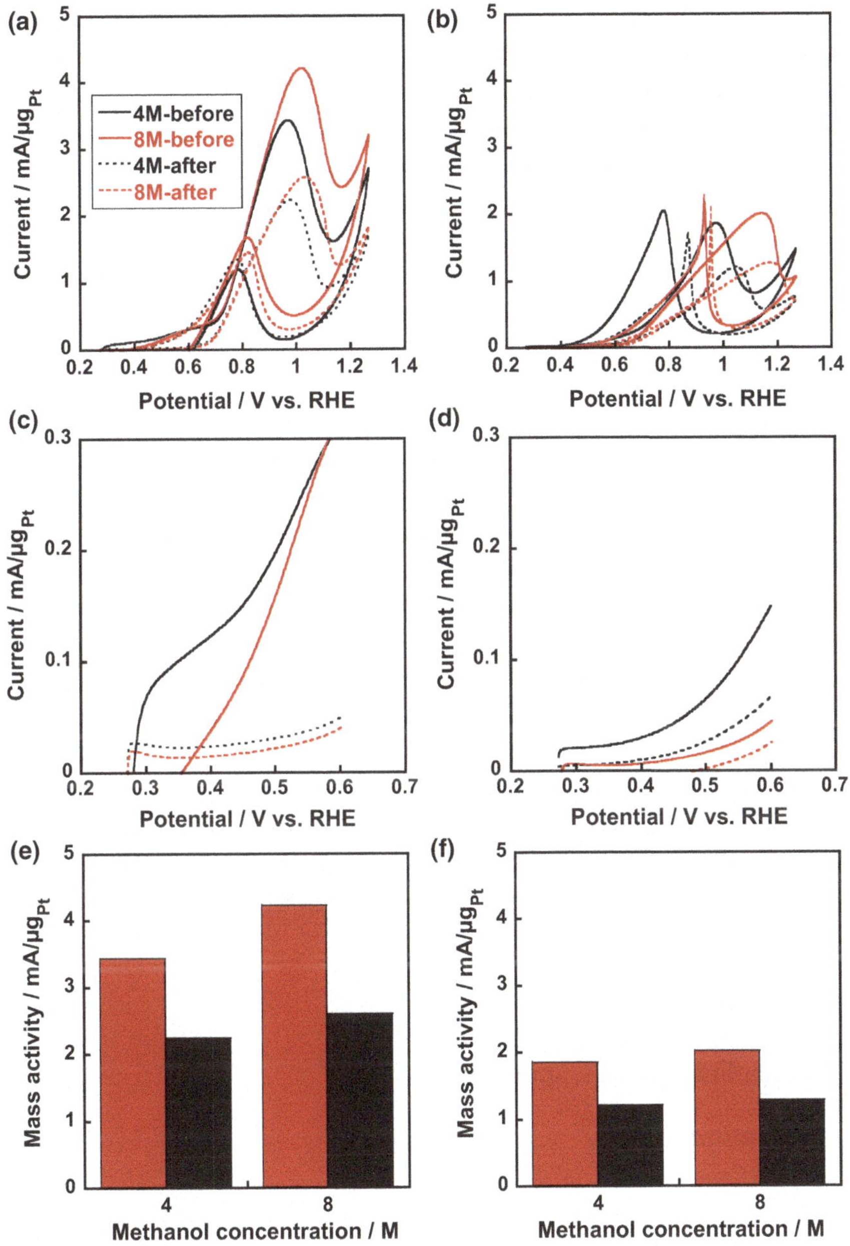

Fig. 6.11 MOR curves measured in an N_2-saturated 0.1M $HClO_4$ with 4 M (red line) and 8 M (black line) methanol at 60 °C of the NanoPC/PyPBI/Pt/PVPA (**a**) and CB/Pt (**b**) before (solid line) and after (dotted line) the durability test. Magnified MOR curves in the range of 0.27–0.6 V versus RHE for the NanoPC/PyPBI/Pt/PVPA (**c**) and CB/Pt (**d**). Mass activities of the NanoPC/PyPBI/Pt/PVPA (**e**), and CB/Pt (**f**) as a function of the methanol concentration before (red column) and after (black column) the durability test. Reproduction from [70] with permission of The American Chemical Society

Table 6.1 Comparisons of the NanoPC/PyPBI/Pt/PVPA and CB/Pt for CO tolerance (I_f/I_b ratio) before and after the durability test. Reproduction from [70] with permission of The American Chemical Society

	NanoPC/PyPBI/Pt/PVPA (I_f/I_b ratio)	CB/Pt (I_f/I_b ratio)
4 M (before durability)	2.86	0.91
8 M (before durability)	2.52	0.87
4 M (after durability)	1.78	0.69
8 M (after durability)	1.72	0.60

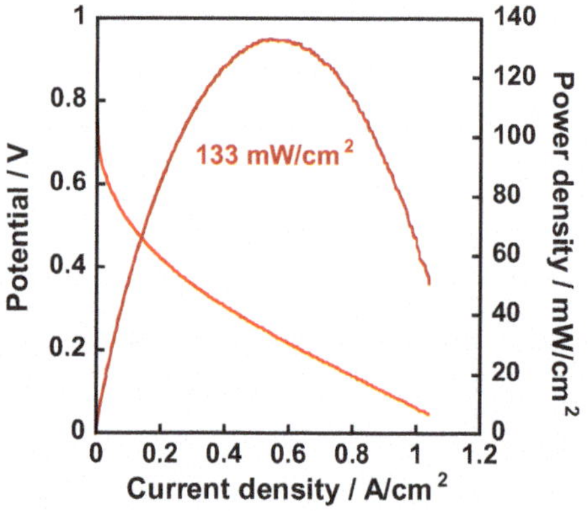

Fig. 6.12 Polarization I-V and power density curve of the MEA fabricated from NanoPC/PyPBI/Pt/PVPA under 70 °C with 8 M methanol (9 mL/min) and 100%RH humidified air (200 mL/min) for the anode and cathode, respectively. Reproduction from [70] with permission of The American Chemical Society

In conclusion, NanoPC/PyPBI/Pt/PVA is a high CO tolerant, high MOR, durable, and efficient DMFC catalyst that works under high methanol concentration (4 M, 8 M) and high temperature (60 °C) that simulates operating conditions from industry. Development of such catalyst is of importance for the design and fabrication of a DMFC anodic electrocatalyst with a high performance for use in real world.

6.3 Application to Capacitor

6.3.1 Introduction

Electric double-layer capacitors (EDLCs) have much higher power density, lower internal resistance, applicable in wider temperature range, and longer cycle life than other energy storage devices such as Li-ion batteries. When a voltage is applied to electrodes and thus electrolyte ions are placed in an electric field between the electrodes, the ions move on the electrodes with opposite charges. The ion transfer is a trigger for large current between the electrodes. The transferred ions afterward form concentrated charge layer on electrodes, calling electric double-layer

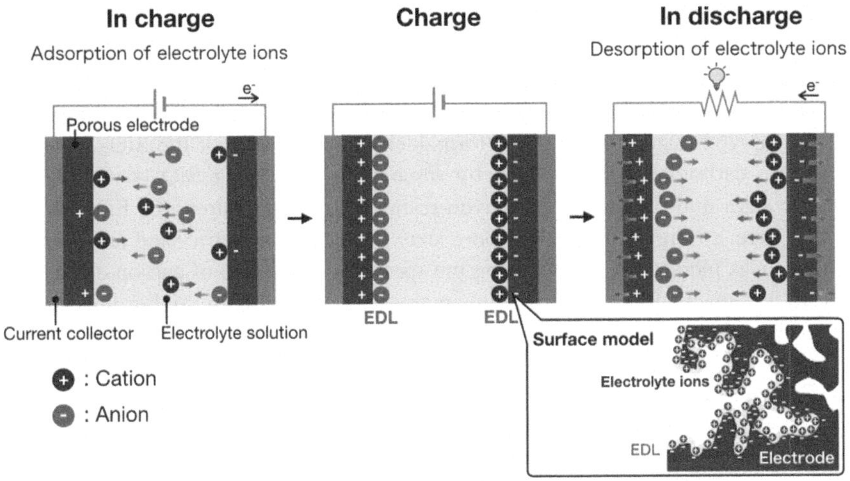

Fig. 6.13 Schematic image of energy storage in EDLCs

(EDL) [79]. Accordingly, the energy storage in EDLCs is based on the forming EDL of electrolyte ions on electrode surfaces (Fig. 6.13). The EDL capacitance (C) is theoretically expressed by the following equation:

$$C = \varepsilon_0 \varepsilon_r \frac{S}{d}$$

where ε_0, ε_r, S, and d are the permittivity of vacuum, relative permittivity of dielectric material (electrolyte, e.g., acetonitrile: 37.5, propylene carbonate: 64.9 at 293 K), the surface area of electrode and the thickness of EDL at the electrode/electrolyte interface, respectively: the EDL capacitance is proportional to the electrode surface area as well as to the reciprocal of thickness of the EDL. The nanoporous carbons with a high surface area are therefore desired materials as an electrode for EDLCs and have been extensively studied [80–93]. Such strenuous studies on various types of porous carbon materials have recently made progress on the improvement of EDLC performance and decreasing cost of EDLCs [94].

6.3.2 *EDLC Property of Mesoporous Carbons*

Until around the beginning of the 2000s, mesoporous carbons with average pore size from 2 to 50 nm have vigorously studied as EDLC electrodes due to relatively high EDLC properties [15, 16, 82, 83, 95–97]. Since mesopore size is large enough for electrolyte ions to access into the pores, the pore walls in mesoporous carbons effectively contribute to form EDL and provide high EDL capacitance. Figure 6.14 shows the relationship between a surface area and a gravimetric/area-specific capacitance on porous carbons in an aqueous H_2SO_4 electrolyte. The numeral in brackets is the templated-SiO_2 particle size corresponding to an average pore size of the porous carbons. The pore size of porous carbons is easily tuned up by a colloidal crystal template process as described in Sect. 6.1 [15, 16]. The nonporous carbon is synthesized from phenol/formaldehyde resin without the SiO_2 particles. Since the carbons are synthesized by same procedure using SiO_2 with different particle sizes and phenol/formaldehyde resin as a carbon source, the EDL capacitances, here, strongly attribute the pore size. Both of gravimetric- and area-specific capacitances increase with increasing the specific surface area of carbon electrodes. The EDL capacitances of mesoporous carbons are additionally higher than that of nonporous and macroporous carbons. The result suggests that mesoporous carbons are more useful as EDLC electrodes than macroporous carbons. Figure 6.15 shows a rate performance of mesoporous carbons. The capacitive property at high rate is also important to yield high power density. The high capacity above 100 F g^{-1} is maintained for the mesoporous carbon with average pore size of 16 nm (C[16]) even up to the high rate of 100 mV s^{-1} in an aqueous electrolyte. By contrast, for one with smaller average pore size of 2.6 nm (C[2.6]) [98], the capacitance is steeply decreased with increasing the scan rate. Although C[2.6] has larger BET surface area of 1540 m^2 g^{-1} than that of C [16], which is 1402 m^2 g^{-1}, the EDL capacitance is low: the porous structure constructed only by relatively small mesopores should not be suitable for the effective access of electrolyte ions. Furthermore, in different porous structure types of carbons, a linear relationship between the BET surface area and the gravimetric capacitance is not found. For designing mesoporous carbon, electrodes need to consider not only for the total surface area but also the porous structures (e.g., pore size distribution, pore shape) based on the ion transport in electrodes.

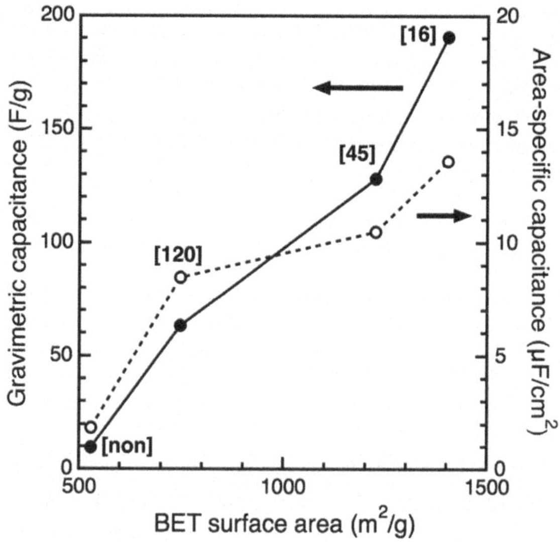

Fig. 6.14 Gravimetric (open circle) and area-specific (closed circle) capacitances in an aqueous H_2SO_4 against the BET surface area of nonporous, mesoporous, and macroporous carbon electrodes. The numeral in brackets indicates average pore size

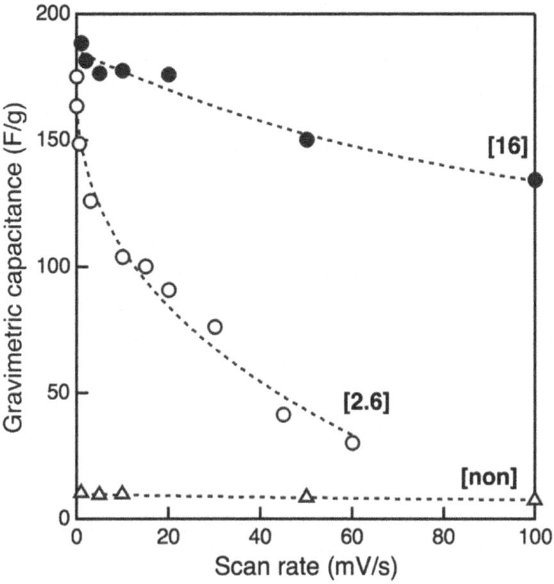

Fig. 6.15 Gravimetric capacitance in an aqueous H_2SO_4 against scan rate of the nonporous carbon (open triangle) and mesoporous carbons with pore size of 2.6 nm (open circle) and 16 nm (closed circle) size

6.3.3 EDLC Property of Microporous Carbons

Microporous carbons with higher specific surface area have been considered not to provide higher EDL capacitance than mesoporous carbons due to the smaller pore size than solvated electrolyte ions [99]. Since electrolyte ions form solvated state in organic solutions, they hardly access into micropores and thus will not significantly contribute to the EDL capacitance. In 2006, an advanced study suggested that micropores should work effectively to improve a specific capacitance in an organic electrolyte solution (tetraethylammonium tetrafluoroborate (TEA-TFB) in acetonitrile (ACN)) [4]. They initially implied that partial desolvation of electrolyte ions are occurred in micropores and the EDL capacitance is improved by decreasing EDL thickness and effective use of micropore surfaces. Although the TEA^+-7ACN solvate ion size is 1.3 nm [100], the carbide-derived carbons (CDCs) with average pore size of $\sim$0.75 nm provide quite high EDL capacitances ($\sim$140 F g^{-1} and $\sim$11 μF cm^{-2}). From viewpoint of ion size, due to the bare TEA ion size of 0.68 nm, the desolvated TEA ions access into such the narrow pores and can form EDL. Comparable results on desolvation in micropores were also demonstrated by theoretical studies and in situ measurements [101–109]. Remarking that such desolvation has been also occurred even on Li-ions in propylene carbonate (PC) solution whose theoretical desolvation energy is 218 kJ/mol [110] and the interaction of Li-PC is stronger than TEA-PC [111]. The solvation number of Li-ions in micropores, mesopores, and macropores and/or on external surface of porous carbons can be determined from Raman spectra and ^{7}Li NMR spectra. Raman spectroscopy is a useful technique to determine average solvation number in low concentration (<2 M) of electrolyte solutions [112–114]. When the result of ^{7}Li NMR spectra is combined with the solvation number determined from Raman spectra, we can see solvation numbers from the chemical shifts in NMR spectra. The results indicated that the solvation number is dramatically decreased in micropores [6]. Figure 6.16 points out that micro/mesoporous carbons occurring desolvation show high area-specific capacitances [6]. On the microporous carbon with mainly ultramicropores (<0.5 nm) showing low capacitance, desolvation is not occurred due to a quite small pore that even solvent molecules cannot enter the pores: EDL does not form on the ultramicropores and such pore wall cannot efficiently contribute to provide EDL capacitance. A microporous carbon also shows lower EDL capacitance than the mesoporous carbons although the desolvation occurs in the micropores. The result implies that all of microporous carbons are not applicable to EDLC electrode which provides high EDL capacitance. Figure 6.17a–d shows representative pore shapes of porous carbons. The pore shape is classified roughly into two groups, a slit-shaped pore and a worm-like (WL)-shape pore which is here constructed by connected pores of each cage-shaped pore. The gravimetric capacitance against total specific surface area of microporous and micro/mesoporous carbons are shown in Fig. 6.17e. The specific surface area is determined by applying the grand canonical Monte Carlo (GCMC) method [115] to N_2 adsorption isotherms at 77 K. From the viewpoint of using molecular simulated adsorption isotherm to determine the porosities of samples, when we know pore shapes (e.g., slit, cylinder, and cage), the GCMC method provides more accurate information on porous structures (e.g., specific surface area, pore volume, and pore size distribution) . The BET method is commonly used to determine the surface area of electrodes, but the method based on multilayer adsorption does not accurately

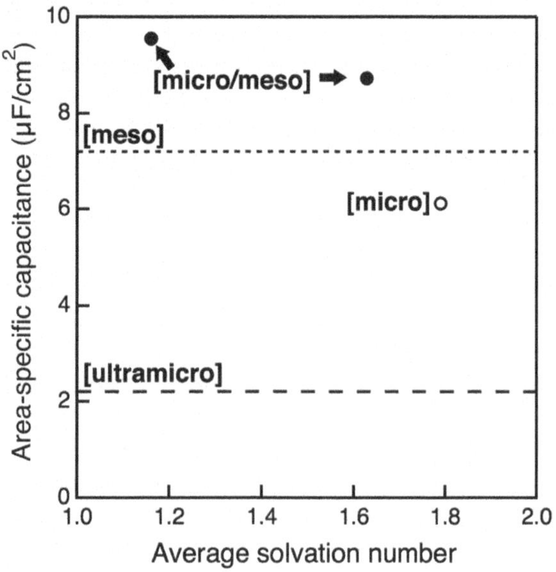

Fig. 6.16 Area-specific capacitance against average solvation number in micropores of porous carbons (closed circle: micro/mesoporous carbons, open circle: microporous carbon, short-dashed line: mesoporous carbon with a pore size of 17 nm, long-dashed line: microporous carbon mainly with ultramicropores)

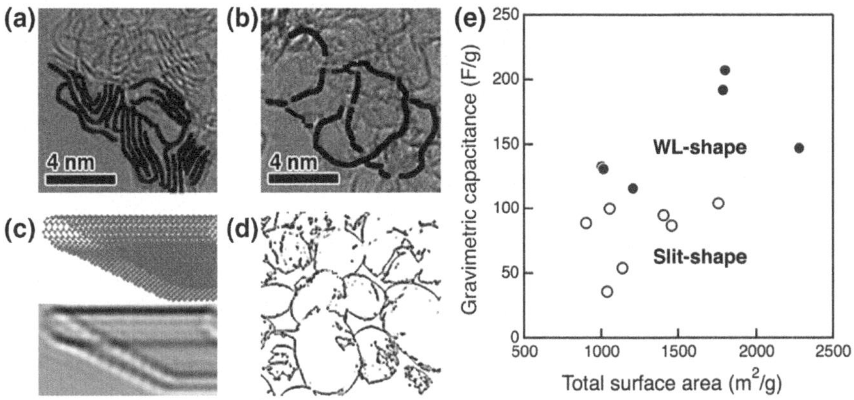

Fig. 6.17 TEM image (**a**, **b**) and schematic image (**c**, **d**) of slit-shaped (**a**, **c**) and WL-shaped (**b**, **d**) porous carbons. In TEM images, carbon walls constructing pores are partially highlighted in black. **e** Gravimetric capacitance against total surface area determined by GCMC method of slit-shaped (open circle) and WL-shaped (closed circle) carbons

represent the specific surface area of microporous carbons [116]. Figure 6.17e clearly indicates that the gravimetric capacitance of WL-shaped carbons is generally higher than that of slit-shaped carbons. However, in each pore shape, higher surface area is not entirely to provide higher EDL capacitance as mentioned in mesoporous carbons. We know here that EDL capacitance depends on the pore shape but not only on the surface area. Another important factor in designing porous carbon electrodes to provide high EDL capacitances is the ratio of surface area of the pores with the solvated ion size against the desolvated ion size [117]. Figure 6.18 shows gravimetric capacitance against the ratio of the value of specific surface area (R). Both of the gravimetric capacitance in slit-shaped and WL-shaped carbons increase with increasing the surface ratio until $R \sim 0.5$, and then the capacitance of WL-shaped lowers down to the value of slit-shaped carbon. Such low gravimetric capacitance at the low R implies that most of the spaces where desolvated ions can access do not contribute to the EDL formation regardless of the pore shape. The availability of spaces for desolvated ions must depend strongly on amount of the spaces, where solvated ions can access, connecting the pores for desolvated ions. The desolvation should be accelerated by the strong confinement effect from slightly larger spaces than solvate ion size, and the desolvated ions smoothly form EDL on the smaller pores. The results presented in this section indicate that a pore size distribution considering solvated and desolvated ion sizes of electrolytes is quite important as well as the pore shape and the surface area for an optimization of electrode structures.

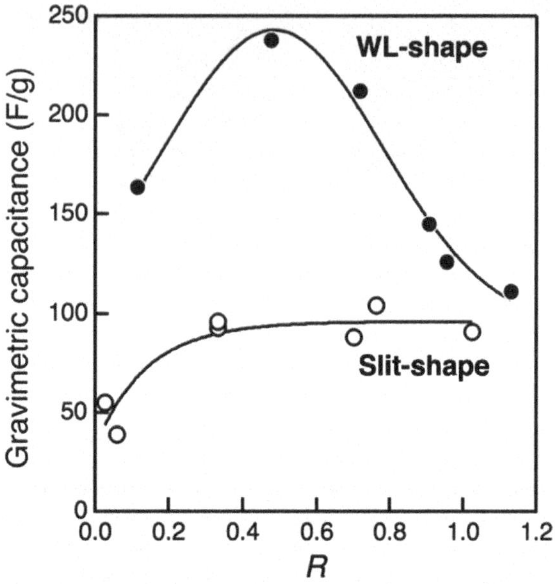

Fig. 6.18 Gravimetric capacitance against the rate of specific surface area (*R*) of slit-shaped (open circle) and WL-shaped (closed circle) porous carbons. *R* is given by the surface area of pores, where solvated ions can access, divided by the surface area of pores, where desolvated ions can access: *R* < 1 implies that the pore where ions can access by desolvation is a dominant pore

References

1. Gogotsi Y, Nikitin A, Ye H, Zhou W, Fischer JE, Yi B, Foley HC, Barsoum MW (2003) Nanoporous carbide-derived carbon with tunable pore size. Nature 2:591–594
2. Dash RK, Yushi G, Gogotsi Y (2005) Synthesis, structure and porosity analysis of microporous and mesoporous carbon derived from zirconium carbide. Microporous Mesoporous Mater 86:50–57
3. Gogotsi Y, Dash RK, Yushin G, Yildirim T, Laudislo G, Fischer JE (2005) Tailoring of nanoscale porosity in carbide-derived carbons for hydrogen storage. J Am Chem Soc 127:16006–16007
4. Chmiola J, Yushin G, Gogotsi Y, Portet C, Simon P, Taberna PL (2006) Anomalous increase in carbon capacitance at pore sizes less than 1 nanometer. Science 313:1760–1763
5. Ma Z, Kyotani T, Liu Z, Terasaki O, Tomita A (2001) Very high surface area microporous carbon with a three-dimensional nano-array structure: synthesis and its molecular structure. Chem Mater 13:4413–4415
6. Urita K, Ide N, Isobe K, Furukawa H, Moriguchi I (2014) Enhanced electric double-layer capacitance by desolvation of lithium ions in confined nanospaces of microporous carbon. ACS Nano 8:3614–3619
7. Kawashima D, Aihara T, Kobayashi Y, Kyotani T, Tomita A (2000) Preparation of mesoporous carbon from organic polymer/silica nanocomposite. Chem Mater 12:3397–3401
8. Han S, Sohn K, Hyeon T (2000) Fabrication of new nanoporous carbons through silica templates and their application to the adsorption of bulky dyes. Chem Mater 12:3337–3341
9. Ryoo R, Joo SH, Jun S (1999) Synthesis of highly ordered carbon molecular sieves via template-mediated structural transformation. J Phys Chem B 103:7743–7746
10. Lu A-H, Schmidt W, Spliethoff B, Schüth F (2003) Synthesis of ordered mesoporous carbon with bimodal pore system and high pore volume. Adv Mater 15:1602–1606
11. Che S, Lund K, Tatsumi T, Iijim S, Joo SH, Ryoo R, Terasaki O (2003) Direct observation of 3D mesoporous structure by scanning electron microscopy (SEM): SBA-15 silica and CMK-5 carbon. Angew Chem Int Ed Engl 42:2182–2185

12. Ryoo R, Joo SH, Kruk M, Jaroniec M (2001) Ordered mesoporous carbons. Adv Mater 13:677–681
13. Che S, Garcia-Bennett AE, Liu X, Hodgkins RP, Wright PA, Zhao D, Terasaki O (2003) Synthesis of large-pore la3d mesoporous silica and its tubelike carbon replica. Angew Chem Int Ed Engl 42:3930–3934
14. Zhang W-H, Liang C, Sun H, Shen Z, Guan Y, Ying P, Li C (2002) Synthesis of ordered mesoporous carbons composed of nanotubes via catalytic chemical vapor deposition. Adv Mater 14:1776–1778
15. Moriguchi I, Nakahara F, Furukawa H, Yamada H, Kudo T (2004) Colloidal crystal-templated porous carbon as a high performance electrical double-layer capacitor material. Solid State Lett 7:A221–A223
16. Moriguchi I, Nakahara F, Yamada H, Kudo T (2005) Electrical double-layer capacitive properties of colloidal crystal-templated nanoporous carbons. Stud Surf Sci Catal 156:589–594
17. Tu J-S, Kang S, Yoon SB, Chai G (2002) Fabrication of ordered uniform porous carbon networks and their application to a catalyst supporter. J Am Chem Soc 124:9382–9383
18. Lei Z, Zhang U, Wang H, Ke Y, Li J, Li F, Xing J (2001) Fabrication well-ordered macroporous active carbon with a microporous framework. J Mater Chem 11:1975–1977
19. Bosco JP, Sasaki K, Sadakane M, Ueda W, Chen JG (2010) Synthesis and characterization of three-dimensionally ordered macroporous (3DOM) tungsten carbide: application to direct methanol fuel cells. Chem Mater 22:966–973
20. Joo SH, Choi SJ, Oh I, Kwak J, Liu Z, Terasaki O, Ryoo R (2001) Ordered nanoporous arrays of carbon supporting high dispersions of platinum nanoparticles. Nature 412:169–172
21. Chang H, Joo SH, Pak C (2007) Synthesis and characterization of mesoporous carbon for fuel cell applications. J Mater Chem 17:3078–3088
22. Hayashi A, Kimijima KI, Miyamoto J, Yagi I (2009) Oxygen transfer and storage processes inside the mesopores of platinum-deposited mesoporous carbon catalyst thin-layer electrode. J Phys Chem C 113:12149–12153
23. Minamida Y, Zhao X, Noda Z, Hayashi A, Sasaki K (2013) Characterization of MEAs fabricated by a carbon support with the nano-channel structure. ECS Trans 58:1105–1111
24. Fang B, Kim JH, Kim MS, Yu JS (2013) Hierarchical nanostructured carbons with meso-macroporosity: design, characterization, and applications. Acc Chem Res 46:1397–1406
25. Lai X, Halpert JE, Wang D (2012) Recent advances in micro-/nano-structured hollow spheres for energy applications: from simple to complex systems. Energy Environ Sci 5:5604–5618
26. Speder J, Zana A, Spanos I, Kirkensgaard JJK, Mortensen K, Arenz M (2013) On the influence of the Pt to carbon ratio on the degradation of high surface area carbon supported PEM fuel cell electrocatalysts. Electrochem Commun 34:153–156
27. Speder J, Zana A, Spanos I, Kirkensgaard JJK, Mortensen K, Hanzlik M, Arenz M (2014) Comparative degradation study of carbon supported proton exchange membrane fuel cell electrocatalysts—the influence of the platinum to carbon ratio on the degradation rate. J Power Sources 261:14–22
28. Balgis R, Sago S, Anilkumar GM, Ogi T, Okuyama K (2013) Self-organized macroporous carbon structure derived from phenolic resin via spray pyrolysis for high-performance electrocatalyst. ACS Appl Mater Interfaces 5:11944–11950
29. Tintula KK, Jalajakshi A, Sahu AK, Pitchumani S, Sridhar P, Shukla AK (2013) Durability of Pt/C and Pt/MC-PEDOT catalysts under simulated start-stop cycles in polymer electrolyte fuel cells. Fuel Cells 13:158–166
30. Tintula KK, Sahu AK, Shahid A, Pitchumani S, Sridhar P, Shukla AK (2010) Mesoporous carbon and poly(3,4-ethylenedioxythiophene) composite as catalyst support for polymer electrolyte fuel cells. J Electrochem Soc 157:B1679–B1685
31. Holdcroft S (2013) Fuel cell catalyst layers: a polymer science perspective. Chem Mater 26:381–393

32. Aili D, Hansen MK, Pan C, Li Q, Christensen E, Jensen JO, Bjerrum NJ (2011) Phosphoric acid doped membranes based on Nafion®, PBI and their blends—membrane preparation, characterization and steam electrolysis testing. Int J Hydrogen Energy 36:6985–6993
33. Mauritz KA, Moore RB (2004) State of understanding of Nafion. Chem Rev 104:4535–4585
34. Yang Z, Moriguchi I, Nakashima N (2015) Durable Pt electrocatalyst supported on a 3D nanoporous carbon shows high performance in a high-temperature polymer electrolyte fuel cell. ACS Appl Mater Interfaces 7:9800–9806
35. Ohma A, Shinohara K, Iiyama A, Yoshida T, Daimaru A (2011) Membrane and catalyst performance targets for automotive fuel cells by FCCJ membrane, catalyst. MEA WG. ECS Trans. 41:775–784
36. Huang S-Y, Ganesan P, Park S, Popov BN (2009) Development of a titanium dioxide-supported platinum catalyst with ultrahigh stability for polymer electrolyte membrane fuel cell applications. J Am Chem Soc 131:13898–13899
37. Maass S, Finsterwalder F, Frank G, Hartmann R, Merten C (2008) Carbon support oxidation in PEM fuel cell cathodes. J Power Sources 176:444–451
38. Huang S-Y, Ganesan P, Popov BN (2011) Titania supported platinum catalyst with high electrocatalytic activity and stability for polymer electrolyte membrane fuel cell. Appl Catal B 102:71–77
39. Gharibi H, Amani M, Pahlavanzadeh H, Kazemeini M (2013) Investigation of carbon monoxide tolerance of platinum nanoparticles in the presence of optimum ratio of doped polyaniline with para toluene sulfonic acid and their utilization in a real passive direct methanol fuel cell. Electrochim Acta 97:216–225
40. Zana A, Speder J, Roefzaad M, Altmann L, Bäumer M, Arenz M (2013) Probing degradation by IL-TEM: the influence of stress test conditions on the degradation mechanism. J Electrochem Soc 160:F608–F615
41. Zana A, Speder J, Reeler NEA, Vosch T, Arenz M (2013) Investigating the corrosion of high surface area carbons during start/stop fuel cell conditions: a Raman study. Electrochim Acta 114:455–461
42. Di Noto V, Negro E, Polizzi S, Vezzù K, Toniolo L, Cavinato G (2014) Synthesis, studies and fuel cell performance of "core-shell" electrocatalysts for oxygen reduction reaction based on a PtNix carbon nitride "shell" and a pyrolyzed polyketone nanoball "core". Int J Hydrogen Energy 39:2812–2827
43. Negro E, Polizzi S, Vezzù K, Toniolo L, Cavinato G, Di Noto V (2014) Interplay between morphology and electrochemical performance of "core-shell" electrocatalysts for oxygen reduction reaction based on a PtNix carbon nitride "shell" and a pyrolyzed polyketone nanoball "core". Int J Hydrogen Energy 39:2828–2841
44. Berber MR, Hafez IH, Fujigaya T, Nakashima N (2014) Durability analysis of polymer-coated pristine carbon nanotube-based fuel cell electrocatalysts under non-humidified conditions. J Mater Chem A 2:19053–19059
45. Matsumoto K, Fujigaya T, Sasaki K, Nakashima N (2011) Bottom-up design of carbon nanotube-based electrocatalysts and their application in high temperature operating polymer electrolyte fuel cells. J Mater Chem 21:1187–1190
46. Berber MR, Fujigaya T, Sasaki K, Nakashima N (2013) Remarkably durable high temperature polymer electrolyte fuel cell based on poly(vinylphosphonic acid)-doped polybenzimidazole. Sci Rep 3:1–7
47. Berber MR, Fujigaya T, Nakashima N (2014) High-temperature polymer electrolyte fuel cell using poly(vinylphosphonic acid) as an electrolyte shows a remarkable durability. ChemCatChem 6:567–571
48. Yang Z, Hafez IH, Berber MR, Nakashima N (2015) An enhanced anode based on polymer-coated carbon black for use as a direct methanol fuel cell electrocatalyst. ChemCatChem 7:808–813
49. Zhao X, Yin M, Ma L, Liang L, Liu C, Liao J, Lu T, Xing W (2011) Recent advances in catalysts for direct methanol fuel cells. Energy Environ Sci 4:2736–2753

50. Koenigsmann C, Wong SS (2011) One-dimensional noble m electrocatalysts: a promising structural paradigm for direct methanol fuel cells. Energy Environ Sci 4:1161–1176
51. Ahmad H, Kamarudin SK, Hasran UA, Daud WRW (2011) A novel hybrid Nafion-PBI-ZP membrane for direct methanol fuel cells. Int J Hydrogen Energy 36:14668–14677
52. Li X, Chen W-X, Zhao J, Xing W, Xu Z-D (2005) Microwave polyol synthesis of Pt/CNTs catalysts: effects of pH on particle size and electrocatalytic activity for methanol electrooxidization. Carbon 43:2168–2174
53. Heinzel A, Barragán VM (1999) Review of the state-of-the-art of the methanol crossover in direct methanol fuel cells. J Power Sources 84:70–74
54. Paoletti C, Cemmi A, Giorgi L, Giorgi R, Pilloni L, Serra E, Pasquali M (2008) Electro-deposition on carbon black and carbon nanotubes of Pt nanostructured catalysts for methanol oxidation. J Power Sources 183:84–91
55. Bosco JP, Sasaki K, Sadakane M, Ueda W, Chen JG (2010) Synthesis and characterization of three-dimensionally ordered macroporous (3DOM) tungsten carbide: application to direct methanol fuel cells. Chem Mater 22:966–973
56. Bahrami H, Faghri A (2013) Review and advances of direct methanol fuel cells: Part II: Modeling and numerical simulation. J Power Sources 230:303–320
57. Alia SM, Zhang G, Kisailus D, Li D, Gu S, Jensen K, Yan Y (2010) Porous platinum nanotubes for oxygen reduction and methanol oxidation reactions. Adv Funct Mater 20:3742–3746
58. Ding LX, Wang AL, Li GR, Liu ZQ, Zhao WX, Su CY, Tong YX (2012) Porous Pt-Ni-P composite nanotube arrays: highly electroactive and durable catalysts for methanol electrooxidation. J Am Chem Soc 134:5730–5733
59. Lv H, Peng T, Wu P, Pan M, Mu S (2012) nano-boron carbide supported platinum catalysts with much enhanced methanol oxidation activity and co tolerance. J Mater Chem 22:9155–9160
60. Li W, Liang C, Zhou W, Qiu J, Zhou Z, Sun G, Xin Q (2003) Preparation and characterization of multiwalled carbon nanotube-supported platinum for cathode catalysts of direct methanol fuel cells. J Phys Chem B 107:6292–6299
61. Su F, Zeng J, Bao X, Yu Y, Lee JY, Zhao XS (2005) Preparation and characterization of highly ordered graphitic mesoporous carbon as a Pt catalyst support for direct methanol fuel cells. Chem Mater 17:3960–3967
62. Xue X, Ge J, Tian T, Liu C, Xing W, Lu T (2007) Enhancement of the electrooxidation of ethanol on Pt-Sn-P/C catalysts prepared by chemical deposition process. J Power Sources 172:560–569
63. Huang H, Wang X (2014) Recent progress on carbon-based support materials for electrocatalysts of direct methanol fuel cells. J Mater Chem A 2:6266–6291
64. Zhou Z, Wang S, Zhou W, Wang G, Jiang L, Li W, Song S, Liu J, Sun G, Xin Q (2003) Novel synthesis of highly active Pt/C cathode electrocatalyst for direct methanol fuel cell. Chem Commun 9:394–395
65. Wang CH, Du HY, Tsai YT, Chen CP, Huang CJ, Chen LC, Chen KH, Shih HC (2007) High performance of low electrocatalysts loading on CNT directly grown on carbon cloth for DMFC. J Power Sources 171:55–62
66. Sharma S, Pollet BG (2012) Support materials for PEMFC and DMFC rlectrocatalysts—a review. J Power Sources 208:96–119
67. Liu H, Song C, Zhang L, Zhang J, Wang H, Wilkinson DP (2006) A review of anode catalysis in the direct methanol fuel cell. J Power Sources 155:95–110
68. Yang Z, Berber MR, Nakashima N (2014) A polymer-coated carbon black-based fuel cell electrocatalyst with high CO-tolerance and durability in direct methanol oxidation. J Mater Chem A 2:18875–18880
69. Yang Z, Kim C, Hirata S, Fujigaya T, Nakashima N (2015) Facile enhancement in CO-tolerance of a polymer-coated Pt electrocatalyst supported on carbon black: comparison between Vulcan and Ketjenblack. ACS Appl Mater Interfaces 7:15885–15891

70. Yang Z, Moriguchi I, Nakashima N (2016) A highly-durable CO-torelant poly(vinylphosphoric acid)-coated electrocatalysy supported on a nanoporous carbon. ACS Appl Mater Interfaces 8:9030–9036
71. Yang M, Guarecuco R, Disalvo FJ (2013) Mesoporous chromium nitride as high performance catalyst support for methanol electrooxidation. Chem Mater 25:1783–1787
72. Zhao Y, Fan L, Zhong H, Li Y, Yang S (2007) Platinum nanoparticle clusters immobilized on multiwalled carbon nanotubes: electrodeposition and enhanced electrocatalytic activity for methanol oxidation. Adv Funct Mater 17:1537–1541
73. Sun S, Zhang G, Gauquelin N, Chen N, Zhou J, Yang S, Chen W, Meng X, Geng D, Banis MN, Li R, Ye S, Knights S, Botton G A, Sham T-K, Sun X (2013) Single-atom catalysis using Pt/graphene achieved through atomic layer deposition. Sci Rep 3:1775
74. Ding LX, Wang AL, Li GR, Liu ZQ, Zhao WX, Su CY, Tong YX (2012) Porous Pt-Ni-P Composite nanotube arrays: highly electroactive and durable catalysts for methanol electrooxidation. J Am Chem Soc 134:5730–5733
75. Liao S, Holmes K-A, Tsaprailis H, Birss VI (2006) High performance PtRuIr catalysts supported on carbon nanotubes for the anodic oxidation of methanol. J Am Chem Soc 128:3504–3505
76. Yu LH, Kuo CH, Yeh CT (2007) Poly(vinylpyrrolidone)-modified graphite carbon nanofibers as promising supports for PtRu catalysts in direct methanol fuel cells. J Am Chem Soc 129:9999–10010
77. Samant PV, Rangel CM, Romero MH, Fernandes JB, Figueiredo JL (2005) Carbon supports for methanol oxidation catalyst. J Power Sources 151:79–84
78. Gao MR, Gao Q, Jiang J, Cui CH, Yao WT, Yu SH (2011) A methanol-tolerant Pt/$CoSe_2$ nanobelt cathode catalyst for direct methanol fuel cells. Angew Chem Int Ed 50:4905–4908
79. Conway B E (1999) Electrochemical supercapacitors, 1st ed. Kluwer Academic/Plenum Publishers, New York, pp 105–124
80. Moriguchi I (2014) Nanostructure-controlled materials for electrochemical charging-discharging. Chem Lett 43:740–745
81. Li W, Reichenauer G, Fricke J (2002) Carbon aerogels derived from cresol-resorcinol-formaldehyde for supercapacitors. Carbon 37:2049–2055
82. Zhou H, Shu S, Hibino M, Honma I (2003) electrochemical capacitance of self-ordered mesoporous carbon. J Power Sources 122:219–223
83. Jang JH, Han S, Hyeon T, Oh SM (2003) electrochemical capacitor performance of hydrous ruthenium oxide/mesoporous carbon composite electrodes. J Power Sources 123:79–85
84. Qu D, Shi H (1998) Studies of activated carbons used in double-layer capacitors. J Power Sources 74:99–107
85. Emmenegger Ch, Mauron Ph, Sudan P, Wenger P, Hermann V, Gallay R, Züttel A (2003) Investigation of electrochemical double-layer (ECDL) capacitors electrodes based on carbon nanotubes and activated carbon materials. J Power Source 124:321–329
86. Gamby J, Taberna PL, Simon P, Fauvarque JF, Chesneau M (2001) Studies and characterisations of various activated carbons used for carbon/carbon supercapacitors. J Power Sources 101:109–116
87. García-Gómez A, Moreno-Fernández G, Lobato B, Centeno TA (2015) Constant capacitance in nanopores of carbon monoliths. Phys Chem Chem Phys 17:15687–15790
88. Jäckel N, Rodner M, Schreiber A, Jeongwook J, Zeiger M, Aslan M, Weingarth D, Presser V (2016) Anomalous or Regular capacitance? the influence of pore size dispersity on double-layer formation. J Power Sources 326:660–671
89. Jäckel N, Simon P, Gogotsi Y, Presser V (2016) Increase in capacitance by subnanometer pores in carbon. ACS Energy Lett 1:1262–1265
90. Lozano-Castelló D, Lillo-Ródenas MA, Cazorla-Amorós D, Linares-Solano A (2001) Preparation of activated carbons from Spanish anthracite: I. Activation by KOH. Carbon 39:741–749

91. Portet C, Yang Z, Korenblit Y, Gogotsi Y, Mokaya R, Yushin G (2009) Electrical double-layer capacitance of zeolite-templated carbon in organic electrolyte. J Electrochem Soc 156:A1–A6
92. Wei L, Sevilla M, Fuertes AB, Mokaya R, Yushin G (2011) Hydrothermal carbonizatoin of abundant renewable natural organic chemicals for high-performance supercapacitor electrodes. Adv Energy Mater 1:356–361
93. Wei L, Yushin G (2012) Nanostructured activated carbons from natural precursors for electrical double layer capacitors. Nano Energy 1:552–565
94. Gu W, Yushi G (2014) Review of nanostructured carbon materials for electrochemical capacitor applications: advantages and limitations of activated carbon, carbide-derived carbon, zeolite-templated carbon, carbon aerogels, carbon nanotubes, onion-like carbon, and graphene. WIREs Energy Environ 3:424–473
95. Lee J, Yoon S, Oh SM, Shin C-H, Hyeon T (2000) Development of a new mesoporous carbon using an HMS aluminosilicate template. Adv Mater 12:359–362
96. Shiraishi S, Kurihara H, Tsubota H, Oyam A, Soneda Y, Yamada Y (2001) Electric double layer capacitance of highly porous carbon derived from lithium metal and PTFE. Solid-State Lett 4:A5–A8
97. Li W, Reichenauer G, Fricke J (2002) Carbon aerogels derived from cresol-resorcinol-formaldehyde for supercapacitors. Carbon 40:2955–2959
98. Fuertes AB, Pico F, Rojo JM (2004) Influence of pore structure on electric double-layer capacitance of template mesoporous carbons. J Power Sources 133:329–336
99. Fuertes AB, Lota G, Centeno TA, Frackowiak E (2005) Templated mesoporous carbons for supercapacitor application. Electrochim Acta 50:2799–2805
100. Kim Y-J, Masutzawa Y, Ozaki S, Endo M, Dresselhaus MS (2004) PVDC-based carbon material by chemical activation and its application to nonaqueous EDLC. J Electrochem Soc 151:E199–E205
101. Wang H, Köster TK-J, Trease NM, Ségalini J, Taberna PL, Simon P, Gogotsi Y, Grey CP (2011) Real-time NMR studies of electrochemical double-layer capacitors. J Am Chem Soc 133:19270–19273
102. Merlet C, Péan C, Rotenberg B, Madden PA, Daffos B, Taberna PL, Simon P, Salanne M (2013) Highly confined ions store charge more efficiently in supercapacitors. Nat Commun 4:2701
103. Ohba T, Kaneko K (2013) Competition of desolvation and stabilization of organic electrolytes in extremely narrow nanopores. J Phys Chem C 117:17092–17098
104. Tsai WY, Taberna PL, Simon P (2014) Electrochemical quartz crystal microbalance (EQCM) study of ion dynamics in nanoporous carbons. J Am Chem Soc 136:8722–8728
105. Forse AC, Merlet C, Griffin JM, Grey CP (2016) New perspectives on the charging mechanisms of supercapacitors. J Am Chem Soc 138:5731–5744
106. Burt R, Breitsprecher K, Daffos B, Taberna PL, Simon P, Birkett G, Zhao XS, Holm C, Salanne M (2016) Capacitance of nanoporous carbon-based supercapacitors is a trade-off between the concentration and the separability of the ions. J Phys Chem Lett 7:4015–4021
107. Salanne M, Rotenberg B, Naoi K, Kaneko K, Taberna PL, Grey CP, Dunn B, Simon P (2016) Efficient storage mechanisms for building better supercapacitors. Nat Energy 1:16070
108. Force AC, Griffin JM, Merlet C, Carretero-Gonzalez J, Raji ARO, Trease NM, Grey CP (2017) Direct observation of ion dynamics in supercapacitor eelectrodes using in-situ diffusion NMR spectroscopy. Nat Energy 2:16216
109. Tanak A, Iiyama T, Ohba T, Ozeki S, Urita K, Fujimori T, Kanoh H, Kaneko K (2010) Effect of a quaternary ammonium salt on propylene carbonate structure in slit-shape carbon nanopores. J Am Chem Soc 132:2112–2113
110. Okoshi M, Yamada Y, Yamada S, Nakai H (2013) Theoretical analysis on de-solvation of lithium, sodium, and magnesium cations to organic electrolyte solvents. J Electrochem Soc 160:A2160–A2165

111. Hu L, Guo D, Feng G, Li H, Zahi T (2016) Asymmetric behavior of positive and negative electrodes in carbon/carbon supercapacitors and its underlying mechanism. J Phys Chem C 120:24675–24681
112. Hyodo S, Okabayashi K (1989) Raman intensity study of local structure in non-aqueous electrolyte solutions-I. cation-solvent interaction in $LiClO_4$/ethylene carbonate. Electrochim Acta 34:1551–1556
113. Morita M, Asai Y, Yoshimoto N, Ishikawa M (1998) A Raman spectroscopic study of organic electrolyte solutions based on binary solvent systems of ethylene carbonate with low viscosity solvents which dissolve different lithium salts. J Chem Soc Faraday Trans 94:3451–3456
114. Yamada Y, Koyama Y, Abe T, Ogumi Z (2009) Correlation between charge-discharge behavior of graphite and solvation structure of the lithium ion in propylene carbonate-containing electrolytes. J Phys Chem C 113:8948–8953
115. Miyahara MT, Numaguchi R, Hiratsuka T, Nakai K, Tanaka H (2014) Fluids in nanospaces: molecular simulation studies to find out key mechanisms for engineering. Adsorption 20:213–223
116. Brunauer S, Emmett PH, Teller D (1938) Adsorption of gases in multimolecular layers. J Am Chem Soc 60:309–319
117. Urita K, Urita C, Fujirta K, Horio K, Yoshida M, Moriguchi I (2017) The ideal porous structure of EDLC carbon electrodes with extremely high capacitance. Nanoscale 9:15643–15649

Chapter 7
Theoretical Approach for Nanocarbon-Based Energy Catalyst Design

Gergely Juhasz and Aleksandar Staykov

7.1 Electronic and Optical Properties of Covalent Modification of Single-Wall Carbon Nanotubes

7.1.1 Introduction

The structure of a single-wall carbon nanotube (SWNT) can be conveniently described as a graphene sheet rolled up into a cylinder with a diameter typically 0.5–50 nm and often over several μm in length. The unit cell of the tube can be defined using a chirality vector, **C**: this vector connects to two carbon atoms on the original graphene lattice, which becomes equivalent when the tube is rolled up. **C** can be expressed with a pair of integers (n, m) to represent all the possible nanotube structures: $\mathbf{C} = n\mathbf{a}_1 + m\mathbf{a}_2$, where $\mathbf{a}_1$ and $\mathbf{a}_2$ are the unit cell base vectors of the original graphene sheet and defined as $n \geq m$. The nanotubes with $n = m$ and $m = 0$ are called armchair and zigzag tubes, respectively, referring to geometry of edges on the two end of the tubes. The tubes with $n \neq m$ and $m > 0$ are chiral.

Chirality (n, m) is a convenient way to indicate the geometry (unit cell, radius) and properties of a nanotube. Tubes with $\mathrm{mod}(n - m, 3) = 0$ are metallic conductors, while $\mathrm{mod}(\mathrm{n} - \mathrm{m}) = 1$ or 2 are semiconductors with the exception of very small diameter tubes with strong curvature effects. While the unit cell of armchair and zigzag carbon SWNTs is of reasonable size, chiral tubes can have large unit cells even when the diameter of the tube is small. This makes the simulations of

G. Juhasz (✉)
School of Science, Tokyo Institute of Technology, Meguro-ku, Tokyo, Japan
e-mail: juhasz@chem.titech.ac.jp

A. Staykov
International Institute for Carbon-Neutral Energy Research (WPI-I2CNER), Kyushu University, Motooka, Nishi-ku, Fukuoka 819-0395, Japan
e-mail: alex@i2cner.kyushu-u.ac.jp

N. Nakashima (ed.), *Nanocarbons for Energy Conversion: Supramolecular Approaches*, Nanostructure Science and Technology,
https://doi.org/10.1007/978-3-319-92917-0_7

carbon nanotubes potentially costly, moreover, chemical modifications of the tube wall further increase the computational costs.

There are numerous sophisticated methods evolved in the past century to simulate the electronic structure of molecules and materials based on Schrödinger equation. In practice, the success of these calculations largely relies on the fine balance between the accuracy of the method and the cost of the calculation. As we have seen above, this cost can be large for carbon nanotubes, which permits the use of only few methods.

7.1.2 Calculation Methods for Simulation of Carbon Nanotube

Recently, DFT emerged as one of the most widely used methods for atomistic simulations. DFT has its root in the seminal work of Hohenberg, Kohn, and Sham [1, 2], and has huge impact on chemistry and materials science [3–6]. It is based on the approximation of the total energy of the ground state from the electron density (ρ). It is generally solved using the Kohn–Sham (KS) equation

$$\left[-\frac{1}{2}\nabla^2 + v(r) + \int \frac{\rho(r)}{|r - r'|} dr' + v_{XC}(r)\right] \varphi_i(r) = \varepsilon_i \varphi_i(r), \tag{7.1}$$

where the terms in the parenthesis correspond to the kinetic energy of the noninteracting KS system, the external potential of the nuclei–electron attraction, the Coulomb potential at **r** due to the electronic charge, and the exchange potential. The *i*th KS orbital is φ_i and ε_i is the corresponding KS orbital energy.

The KS equation is itself exact, however, we have to choose between several different approximations for the exchange potential, as its exact form is not known. One common choice for exchange-correlation is using the Generalized Gradient Approximation (GGA) (e.g., PBE or BLYP). Hybrid functionals (e.g., B3LYP), which includes a mixture of Hartree–Fock exchange with DFT exchange-correlation, are also very popular. A more recent development is the application of range separated density functionals, such as CAM-B3LYP to eliminate the long-range self-repulsion.

The KS equation can be solved on localized or periodic basis set. For nonperiodic structures a natural choice is the Slater-type orbitals (STOs), centered around the nuclei. These basis sets are inspired by the exact orbitals for hydrogen atom. On the other hand, it is much more computationally efficient and more common to use Gaussian-type orbitals (GTOs).

$$\Phi_{z,n,l,m}(r, \Theta, \phi) = N Y_{l,m}(\Theta, \phi) r^{2n-2-l} \exp(-Zr^2) \tag{7.2}$$

To compensate for the different shape of GTOs, generally, the linear combination of a number of GTOs are used to describe better the behavior of the electron in different regions. These localized basis sets are convenient for performing calculations on nonperiodic systems, for example, in Chemistry. Systems with periodic boundary conditions, it is much more convenient to use plane wave basis sets

$$\begin{aligned} \phi_k(\vec{r}) &= e^{-i\vec{k}\vec{r}} \\ \vec{k}\vec{r} &= 2\pi n \end{aligned} \tag{7.3}$$

for n being any integer, and **r**, **k** is restricted to the first Brillouin zone. The accuracy of the method is controlled by the highest possible value of n. One advantage of plane wave methods is that the accuracy is controlled by this single parameter while creating balanced STO or GTO basis sets or scaling their accuracy is a much more challenging task.

In spite of the success of DFT, it is still an expensive method when systems with thousands of atoms should be simulated. One problem is that every step requires the evaluation of a large number of integrals to evaluate the matrix elements. This can be solved with a drastic simplification of the KS equation, and turn to semi-empirical methods, where matrix elements are not calculated as integrals every time but approximated from well parametrized simple algebraic formulas.

Density Functional Tight-Binding (DFTB) is a low-order Taylor-expansion of the KS equation. The energy of the system is expressed as the sum of one-electron energies (ε) and a repulsion potential (v) which depends only on interatomic distances

$$E_{DFTB} = \sum_{i}^{N} \varepsilon_i + \frac{1}{2}\sum_{j \neq k}^{M} v^{j,k}(|R_j - R_k|) \tag{7.4}$$

Both the atomic terms and pair-potentials are interpolated from tabulated values (Slater–Koster files), and have a cut off distance. The speed of the method comes from the fact that no integrals are calculated, the matrix elements are all calculated with simple interpolation and the resultant matrix is sparse due to the cutoff. The most common forms of DFTB are the first-order (DFTB1, nonSC-DFTB) [7, 8], second-order (DFTB2 or SC-DFTB) [9], and third-order (DFTB3) expansions [10]. In simple systems of, e.g., pure elements nonSC-DFTB can be sufficiently accurate, however, systems with charges and polarization effect SC-DFTB improves the total energies and forces. The third-order expansion, DFTB3, substantially improves the description of charged systems and the accuracy of proton affinity calculations.

The accuracy and transferability of the method are also determined by the parametrization. There is no unique choice for parameters used for the repulsion: universal transferability cannot be expected. While DFTB is generally called semi-empirical, strictly speaking, no empirical parameter is needed for the calculations. On the other hand, including a wider range of experimental and theoretical results to the parametrization generally improves the transferability of the parameter set.

To test the accuracy of DFTB in reproducing electronic properties of CNTs, we chose spectroelectrochemical properties of semiconducting SWNTs as a reference for comparison. We were interested in if we can reproduce the correct trends as well as quantitative reproduction of the redox potentials.

There are several reports on the band gap of semiconducting SWNTs experimentally determined using scanning tunneling spectroscopy [11, 12], and spectroelectrochemistry [13, 14]. Semiconducting nanotubes have well-defined E_{11} peaks/E_{22} peaks in the near-IR region, characteristic to their chirality (*n,m*). While these energies overlap on regular absorption or emission spectra, a two-dimensional simultaneous scan of these energies provides a well-resolved map, where each peak represents a E_{11}/E_{22} pair, which belongs to a given (m,n) chirality that allows us to monitor the electrochemistry of mixture of nanotubes, as only neutral SWNTs show photoluminescent signals.

Spectroelectrochemical measurements monitor the intensity of the E_{11} luminescence peak versus the electrode potential and provide the redox potentials of a wide range of tubes with a diameter in the $d = 0.6–1.1$ nm. Tubes with $d > 1.5$ nm diameter are difficult to dissolve in suitable polymer matrix, therefore, such experimental data is always limited to relatively smaller tubes.

The calculations were performed using DFTB as implemented in DFTB+ [15] on periodic models of SWNTs with regular convergence criteria and mio SK set. The size of the unit cell was optimized in *z* direction, and the size of *x* and *y* were chosen so at least 10 nm separation between the individual nanotubes to avoid any interaction. The *k*-points for the integration of the Brillouin zone were chosen using Monkhorst–Pack sampling $(1 \times 1 \times 4)$ [16].

The measured reduction and oxidation potentials follow an approximately linear trend when plotted as a function of inverse diameter (1/*d*) of the nanotubes (see in Fig. 7.1) [17]. The reduction potentials are also showing a clear chirality family behavior. These characteristic trends have been well reproduced by the DFTB calculations. To compare experimental data with the calculations, we used the correspondence between the reduction potential of the tube with the lower edge of conduction band and that of the oxidation potential with the top edge of valence band. The DFTB calculations reproduced both the linear trend as well as chirality family behavior, however, with different accuracy. The linear trends of the band edges have an error, up to approx. 0.2 eV and the slope of the linear trend is also clearly off. One possible source of inaccuracy is that all the simulations were all carried out in vacuum, while is there is a potential shift induced solvent/polymer matrix effects. On the other hand, the calculations well reproduced that the family behavior for both the oxidation potential (Eox) values and the more pronounced family in the reduction potentials (Ered), too.

DFTB can give an estimation for the band structure and the electrochemical band gap, however, excitonic effects cannot be described at tight-binding level, therefore, such electrochemical values are more appropriate to test the accuracy of the method than optical transition energies. Let us emphasize that these calculations

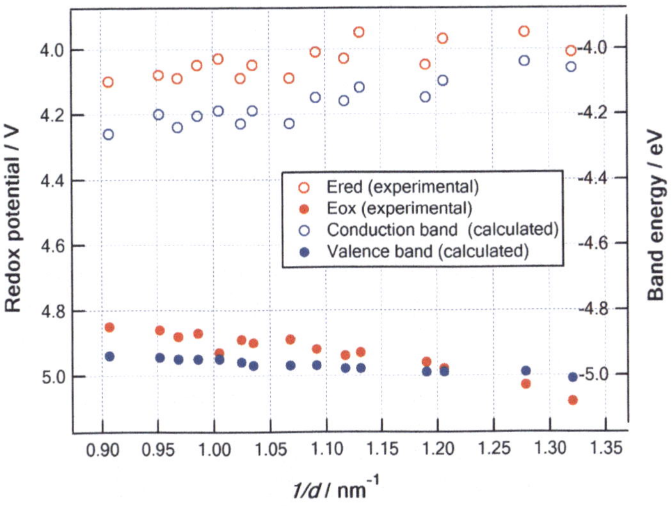

Fig. 7.1 Comparison of the experimental reduction and oxidation potential values (left axis, red marks) of the SWNTs and the computed lowest conduction band energies and the highest valence band energies (right axis, blue marks)

were carried out with the rather generic *mio* SK parameter set. As we previously mentioned, there is no unique "best" parameter set for DFTB and it can be freely parametrized for the problem. Also, experimental data like electrochemical properties can be included in the parametrization process, therefore it is possible to create specialized parameter sets for a given problem.

7.1.3 Simulation of Excitonic States Using TD-DFT Method

Electronic and optical properties of semiconducting single-walled carbon nanotubes are largely influenced by their one-dimensional nature and the confinements of electrons around their circumference. The small dielectric constant of SWNTs leads to large electron-hole binding energies and the formation of strongly bound excitons.

Due to the cylindrical symmetry of the tubes, both optically active (bright) and optically forbidden (dark) excitons can be found around the band gap. The relative position of dark and bright excitons strongly affect the optical properties of the SWNTs, however, the calculation of these energies are challenging. Calculations based on the time-dependent density functional theory (TD-DFT) are maybe the most promising technique that is able to tube models with several hundreds of atoms. However, excitonic effects can only be simulated with hybrid functionals containing significant Hartree–Fock (HF) exchange [18]. In Fig. 7.2, from Ref transition density matrices of a (9,1), tube with the lowest excitonic transition are compared. For the TD-DFT calculations different funtionals have been used: PBE, B3LYP (20% HF contribution) and the costume made BHandHLYP, which is similar to B3LYP, but with 50% HF exchange. The diagonal direction shows the excitonic wavefunction along the tube. Off-diagonal direction is characteristic to the coherence size of the excitons. Pure functionals like PBE cannot properly reproduce

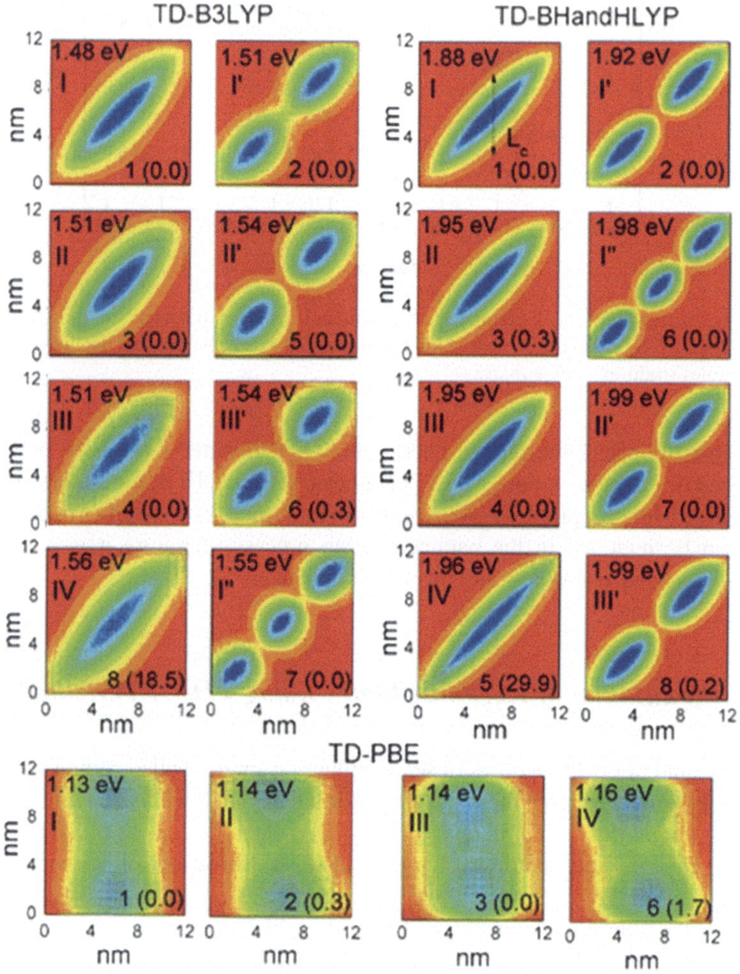

Fig. 7.2 Excitonic transition density matrices of a (9,1) tube calculated using TD-DFT (B3LYP, BHandHLYP with 50% HF exchange and PBE). The first eight excitons are presented by two-dimensional contour plots as a function of the electron (vertical axis, nanometers) and hole (horizontal axis, nanometers) coordinates along the tube axis. The amplitudes scale from red (0) to violet (1) through the natural rainbow color sequence. Excitons are labeled according to their order number with respect to the ground state (the right corner of each plot). The transition energy (in eV) and the oscillator strength (in parenthesis) are also shown for each excitonic state. Reproduced from Ref. [18] with permission from the PCCP Owner Societies

excitonic effects, so all the excitations are delocalized. Hybrid functionals with increasing HF contribution show more and more localized excitons, with smaller coherence size.

When we compare the calculated transition energies to the experimental ones (Fig. 7.3, [18]), we can see that the calculation errors can be significant. One source of the error is the small basis set that was applied. Since the nanotube models were finite size (about 10 nm), errors coming from confinement effects may affect the results.

In pristine tubes, the low energy dark states lying below the lowest bright states are significantly limiting the fluorescence quantum yield (QY) of the SWNTS. By breaking the symmetry of the tube with chemical doping, and introducing emissive

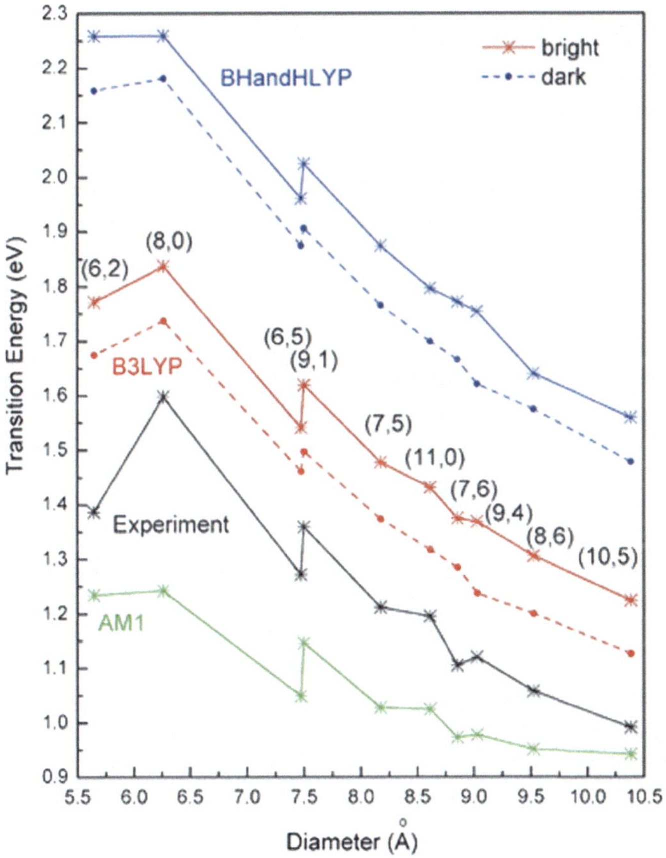

Fig. 7.3 Calculated bright and dark excitonic energies compared to the experimental (bright) values. Circles and dashed lines correspond to energies of the first dark excitons and stars and solid lines correspond to energies of the first bright excitons in different tubes calculated using TD-BHandHLYP, TD-B3LYP, and TD-HF-AM1 methods. Reproduced from Ref. [18] with permission from the PCCP Owner Societies

states below the dark states, this QY can be drastically improved. Recently, there are several examples that were reported when low-concentration oxygen dopant [19, 20] or diazonium sp^3 dopant [21] result in the appearance of new, intense fluorescence peaks.

More detailed studies on the oxygen-doped systems using spectroelectrochemical measurements [22] and photoluminescence measurements on individual tubes [23] revealed that there are several different doping positions with different strength to bind excitons and different fluorescence peak. Successful simulation of the doping sites and their emission frequencies [23] demonstrates that computational tools can be essential in interpreting and designing optical properties of doped nanotubes.

7.2 Surface Curvature of Single-Wall Carbon Nanotubes and Its Relation to Metal Nanoparticle Immobilization on the Nanotube Surfaces

Single-wall carbon nanotubes (SWCNTs) are seen as one-dimensional polymers of single layer sp^2-hybridized carbon atoms having graphene-like wrapped structure. Since the original discovery by Iijima [24], SWCNTs are expected to become key materials in the fields of nanoelectronics [25], nanomechanics [26], and energy-related chemistry. Recently, they have major application in the polymer electrolyte fuel cells (PEFC) where they successfully replace other carbon product like amorphous carbon and carbon black as electron transporting polymers and active surfaces for metal nanoparticle catalyst deposition [27, 28]. The π-conjugates delocalized electron system of SWCNTs provides high electrical conductivity with metallic or semiconducting properties. The electronic properties are linked to the SWCNT structure. The SWCNT can be imaginary built be wrapping a two-dimensional graphene sheet in tubular shape. The wrapping depends on the tube vectors (n,m). In case of $n = m$, the resulting tube has armchair periphery, e.g., na imaginary cut perpendicular to the tube and unwrapping would result in graphene nanoribbon with armchair edge, and the tube is characterized with metallic electronic properties. In case of $m = 0$, the tube has zigzag periphery and the tube is characterized with semiconducting properties. The tubes with $n \neq m \neq 0$ the tube has chiral periphery and metallic electronic properties, if $n - m$ is multiple of 3 and semiconducting properties in all other cases.

One of the obstacles for use of SWCNTs in nanoelectronics is that usually they are synthesized as a mixture of armchair, zigzag, and chiral tubes with heterogeneous electronic properties. Separation of the tubes is challenging due to their identical mass, solubility, and reactivity. Recently, Papadimitrakopoulos and coauthors have proposed an efficient method for separation of chiral nanotubes using derivatives of the flavin molecule [29]. Other studies have proposed separation of metallic nanotubes using porphyrin derivatives [30]. Recently, experimental and theoretical efforts have been employed in the understanding the separation mechanism. It is based on the difference of surface adsorption strength of given molecule on tubes with different chirality. The separation using flavin derivatives is based on its supermolecular helical structure which can align with chiral tubes while porphyrin derivatives selectively chemisorb on non-chiral tubes' surfaces.

Beside organic molecules, metal nanoparticle immobilization on SWCNTs' surfaces is actively researched in the field of PEFC and precious metal catalysis. In PEFC, platinum nanoparticles are deposited on SWCNTs. Due to the high cost, it is crucial to prevent the platinum nanoparticles from aggregation, and thus prevent the high surface area of the catalyst. Additionally, platinum nanoparticles smaller than 5 nm show supreme catalytic activity compared to larger particles [31]. To avoid aggregation the SWCNTs' surfaces are wrapped with conducting polymers characterized with strong affinity to bind to the platinum surface. In addition, single platinum atom deposition on the SWCNT surfaces has been achieved by deposition on the surfaces SWCNTs [32].

In a recent work related to aerobic oxidation of alkanes on gold nanoparticles, Tsukuda and coauthors have used (6,0) SWCNT with diameter of approximately 0.5 nm [33]. In that work, they have shown that it is possible to deposit atomically controlled gold clusters consisting of 25 gold atoms. They have shown that the effect of SWCNT support was similar to the hydroxyapatite support. This report had two important consequences for the mechanism of the aerobic oxidation reaction: the supporting surface did not participate in the catalytic process; and similar to hydroxyapatite (6,0) SWCNT could delay the merging of gold nanoparticles [34]. This result suggests for stronger gold nanoparticle–SWCNT interaction compared to the interaction with graphene surface. While the main difference between graphene and SWCNT is in the electronic properties, zero-band semiconductor for graphene and metallic or semiconductor for SWCNT, surface curvature may also play role in the binding interaction between SWCNT and metal nanoparticles.

While SWCNT are often considered to be one-dimensional structures, in reality, they are two-dimensional graphene surfaces wrapped in the three-dimensional space. Their three-dimensional structure is negligible compared to the length of the tube, however, it might be crucial for their interaction with metal nanoparticles. To understand better the relation between surface curvature and binding properties we will look at the structure of flat graphene and curved graphene. Pristine graphene is infinite two-dimensional plane of hexagons consisting of sp^2 hybridized carbon atoms. The three sp^2 hybrid orbitals build the σ-skeleton of the hexagons while the remaining $2p_z$ AOs overlap to create the delocalized π-conjugated bond, which is responsible for many of the remarkable properties of graphene. To study the electronic properties of graphene, it is often sufficient to take into account only the $2p_z$ AOs, as they are responsible for the bands below and above the Fermi level. That is why, the tight-binding approximation (Hückel molecular orbital theory) showed remarkable accurate prediction results for graphene and graphene nanoribbons [35]. However, its applications to SWCNTs yield in less impressive results. The reason behind is in the surface curvature.

When graphene sheet is curved, the spatial distortion between carbon atoms leads to reduced overlapping between the $2p_z$ AOs as it is shown in Scheme 7.1. Stronger curvature would significantly reduce the π-overlap and eventually lead to dangling bonds on the SWCNTs surfaces. Thus, the strongly curved carbon surface can be considered as a superposition between the three-dimensional sp^2 hybridized carbon atom and the three-dimensional sp^3 hybridized carbon atom. Tubes with smaller diameter would show more sp^3 like character, while tubes with larger diameter would show more sp^2 like character. While sp^2 hybridized carbon atom builds three σ-bonds, the sp^3 hybridized carbon builds four σ-bonds. Thus, tubes with smaller diameter would show bigger tendency to bind metal nanoparticles compared to tubes with larger diameter [36].

In order to properly investigate the binding interaction between SWCNTs and metal nanoparticles one have to estimate properly the degree of sp^2/sp^3 hybridization of the carbon atoms. Thus, accurate estimation of the electronic properties is required and usually the applied methods should be selected carefully. In this work, periodic, plane wave density functional theory calculations were performed using Vienna Ab initio Software Package (VASP) [37, 38] Perdew–Burke–Ernzerhof parametrization of the generalized gradient approximation (GGA-PBE) was employed using projector augmented wave (PAW) pseudopotentials [39]. The results were verified with the optB86b-vdW functional, which includes the long-range van der Waals interactions. Both functionals showed similar trends for optimized geometries and binding energies. An electron cutoff energy of 400 eV was employed for all geometry optimizations and density of states (DOS) calculations. An electron cutoff energy of 300 eV was employed for the first-principle molecular dynamics simulations. Owing to the relatively large supercells, 1 × 1 × 1 gamma k-point sampling was used. Density of states and site-projected density of states analysis for the CNTs were performed with 1 × 1 × 10 k-point sampling. Spin polarization was

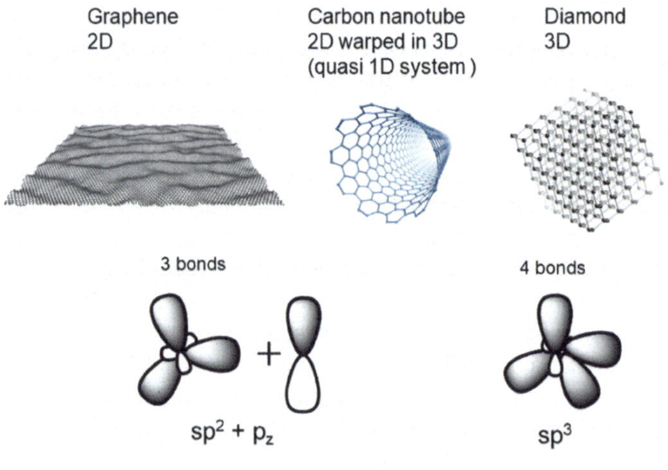

Scheme 7.1 Orbital hybridization of carbon

not employed during the calculations. Absolute values of the Fermi energies were determined as relative to the vacuum potential. Throughout this study, the graphical visualization package VESTA was used to analyze and visualize the computed DFT electron density distribution [40].

Binding energies of the metal nanoparticles to the SWCNTs surfaces were estimated with the following relation:

$$E_{bind} = E_{all} - E_{nt} - E_{np}, \quad (7.5)$$

where E_{bind} denotes the binding energy, E_{all} is the energy of the nanoparticle–nanotube complex, E_{np} is the energy of the nanoparticle, and E_{nt} the energy of the nanotube. In addition to the electron density distribution, electron density differences were compared, which were obtained with the following relation:

$$p_{bind} = p_{all} - p_{nt} - p_{np} \quad (7.6)$$

Here, p_{diff} represents the electron density difference, p_{all} denotes the electron density of the nanoparticle–nanotube complex, p_{np} is the electron density of the nanoparticle, and p_{nt} is the electron density of the nanotube. Bader population analysis was performed to allocate electron density to the ions. First-principle molecular dynamics calculations were performed for particles adsorbed on flat graphene surface and on curved SWCNT surfaces. The dynamics simulations were performed for 1000 steps of 0.5 fs and 500 °C.

We have calculated the density of states of CNT(6,0); CNT(9,0); CNT(12,0), and CNT(18,0) with diameters of 0.47, 0.71, 0.94, and 1.41 nm, respectively. The DOS plots and PDOS plots are shown in Fig. 7.4. The DOS of the large diameter CNT(18,0) resembles closely the DOS of graphene. However, with decrease in diameter, i.e., increase of curvature, a peak emerges below the Fermi level. The peak is denoted with an arrow shown in Fig. 7.4. The intensity of the peak increases for CNT(9,0) and becomes dominant for CNT(6,0). To elucidate the nature of the peak, we look at the PDOS plots. In case of CNT(18,0), the peaks above and below the Fermi level are resulting from the $2p_z$ AOs on the carbon atoms. This result clearly indicates sp^2 hybridization. However, with the increase of surface curvature peaks of $2p_x$ AOs and $2p_y$ AOs contribute to the peak below the Fermi level which is clear indication for partial sp^3 character. Thus, the curvature indeed controls the electronic properties of the carbon atoms, however, the effect is significant only for small diameter SWCNTs and disappears for tubes with diameter above 1 nm.

Calculations were performed for 13 atoms palladium nanoparticle, Pd13 adsorbed on the tube surfaces. The particle was characterized with cuboctahedra symmetry. Binding energy between the particle and the tube was estimated using Eq. 7.5. For comparison, the binding energy with the flat graphene surface was also calculated. The binding energy of Pd13 to CNT(6,0) is −2.84 eV which points to significant chemisorption. With an increase in the tube's diameter, for example,

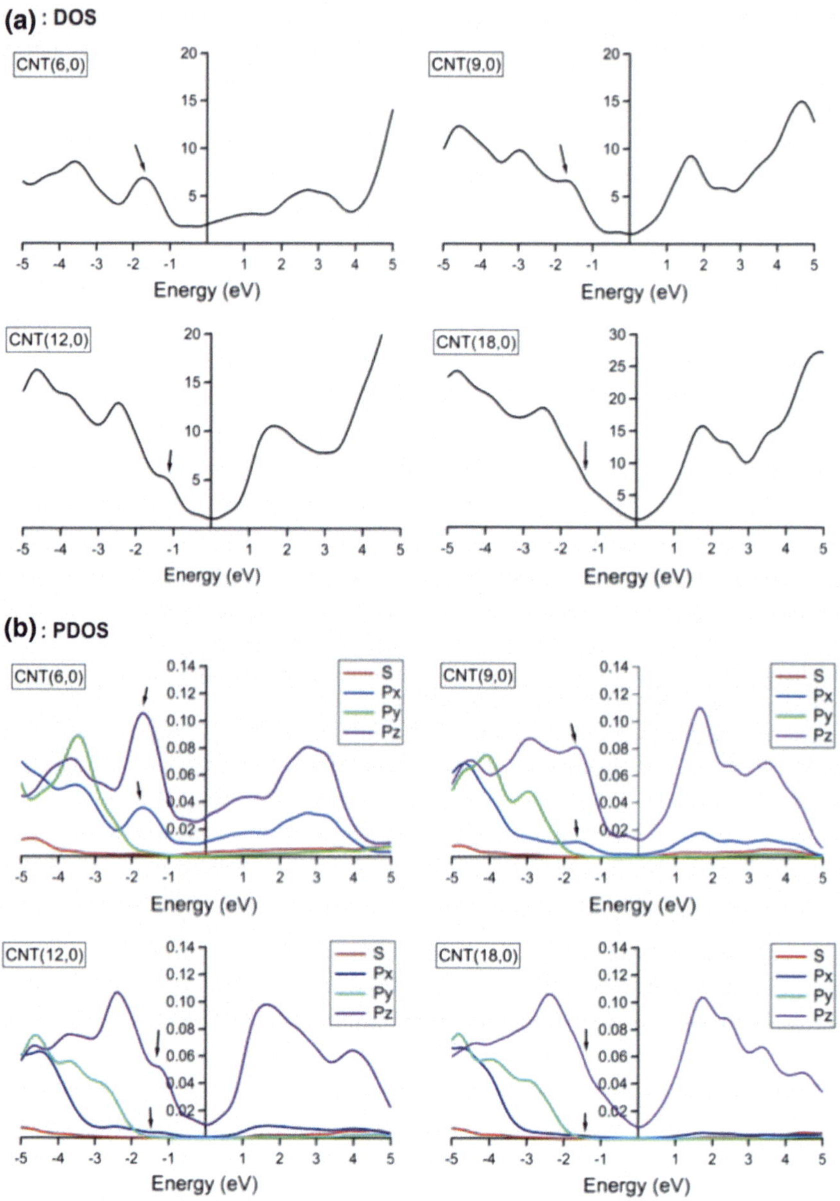

Fig. 7.4 Density of states and site-projected density of states spectra of zigzag CNTs with metallic properties. The spectra are plotted as a function of the electron energy in the interval −5 to 5 eV. The energies are relative to the Fermi energy (EF = 0). The coordinates' directions are defined locally. **a** DOS spectra; **b** PDOS spectra. Reproduced from Ref. [36] with permission

CNT(9,0), the binding energy is reduced to −2.24 eV. A further increase in the tube's diameter results in binding energies of −1.71 eV and −1.59 eV for CNT (12,0) and CNT(18,0), respectively. The binding energy of Pd13 to the flat graphene surface is −0.85 eV. The results are presented in Fig. 7.5. The dependence between the binding energy and the tube diameter, e.g., the tube curvature, is plotted in Fig. 7.5. The plot indicates that the curvature effect is strong only for tubes with diameter below 1 nm. Larger tubes will show small dependence of particle-tube binding energies on the tube diameter. Thus, the curvature effect might remain unnoticed by experimentalists as often mixture of tubes is used. To understand the nature of the tube–nanoparticle interaction we have performed electron density difference analysis. The results indicate that electron density from the d_z^2 AOs of the Pd atoms has been transferred to the region between the tube surface and the particle. Such interaction resembles weak covalent type bonding. With the decrease of the tube surface curvature the electron transfer is reduced.

Finally, we have performed first-principle molecular dynamics to elucidate the probability for nanoparticle aggregation on curved and flat surfaces. In the computational setup, two Pd13 nanoparticles were placed at 0.5 nm distance on the CNT(12,0) surface. Computational annealing was performed to 500 °C for 0.5 ps. As a result, the particles remained at their original positions. Similar computational annealing was performed for two Pd13 nanoparticles placed at distance of 0.5 nm on the flat graphene surface. As a result of the annealing on the flat surface both particles merge for less than 0.5 ps. The reason for the high mobility of the particles on the flat surface is due to the weak binding interaction between carbon and palladium. In case of the curved carbon surface, the curvature increasing the

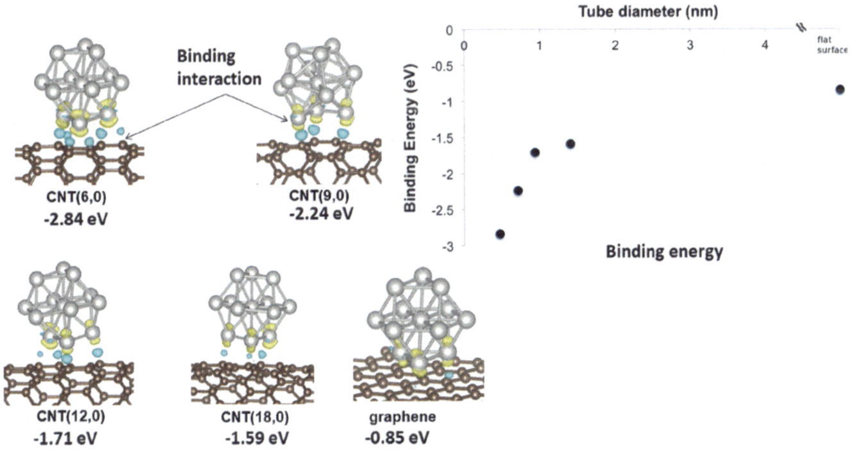

Fig. 7.5 Pd13 adsorbed on CNT (6,0); CNT (6,0); CNT (6,0); CNT (6,0) and graphene. Binding energy values calculated with Eq. 7.5 are given and are plotted versus tube diameter. Electron density difference calculated with Eq. 7.6. Blue color denotes positive density, yellow color denotes negative density

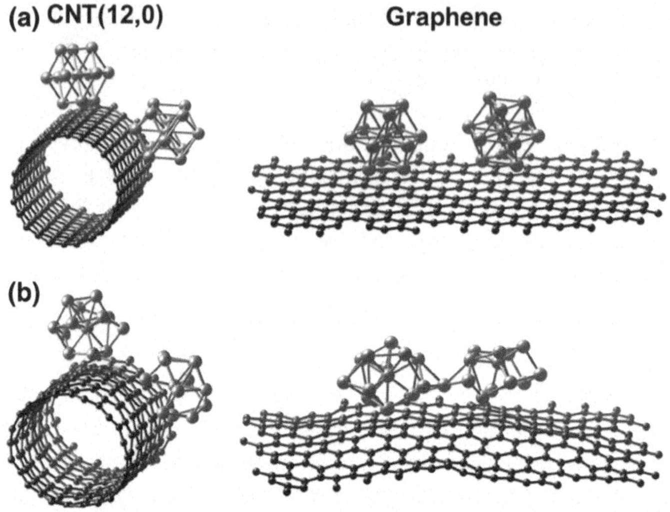

Fig. 7.6 First-principle molecular mechanics. **a** Starting geometry. **b** Final geometry

binding interaction and as a result the particles are immobile. The results of the first-principle molecular dynamics calculations are summarized in Fig. 7.6.

In conclusion, with an increase in surface curvature in sp^2 hybridized 2D carbon materials, the electronic properties of carbon atoms are changing. As a result, the carbon atoms show some sp^3 hybridized character with an increase in the curvature. The superposition between sp^2/sp^3 hybridized character strongly depends on the curvature, thus in case of SWCNTs it depends on the nanotube diameter. Usually, effects resulting from the tube surface curvature are noticeable for tubes below 1 nm of diameter. The sp^2 hybridized builds three covalent bonds while the sp^3 hybridized carbon tends to build four covalent bonds. Thus, with the increase in surface curvature one increases the affinity of the tube surface to bind substrates. In particular, this is noticeable with metal nanoparticles. We have shown successfully that palladium nanoparticles can be immobilized on curved SWCNT surface with diameter below 1 nm. In this case, upon high-temperature annealing the articles remain immobile and prevent their fine size. In case of flat surface, the particles possess significant mobility and in short time aggregated to form larger particles. That result might be crucial for catalysis where sized dependent catalytic activity is often important for the catalyst performance.

References

1. Hohenberg P, Kohn W (1964) Inhomogeneous electron gas. Phys Rev B 136:B864–B871
2. Kohn W, Sham LJ (1965) Self-consistent equations including exchange and correlation effects. Phys Rev 140:A1133–A1138

3. Peverati R, Truhlar DG (2014) Quest for a universal density functional: the accuracy of density functionals across a broad spectrum of databases in chemistry and physics. Philos Trans R Soc A 372:20120476
4. Perdew JP, Burke K (1996) Comparison shopping for a gradient-corrected density functional. Int J Quantum Chem 57:309–319
5. Fiolhais C, Nogueira F (2003) Marques MAL a primer in density functional theory. Springer, New York
6. Engel E, Dreizler RM (2011) Density functional theory: an advanced course. Springer, New York
7. Porezag D, Frauenheim Th, Köhler T, Seifert G, Kaschner R (1995) Construction of tightbinding-like potentials on the basis of density-functional theory: application to carbon. Phys Rev B 51:12947–12957
8. Seifert G, Porezag D, Frauenheim T (1996) Calculations of molecules, clusters, and solids with a simplified LCAO-DFT-LDA scheme. Int J Quant Chem 58:185–192
9. Elstner M, Porezag D, Jungnickel G, Elsner J, Haugk M, Frauenheim T, Suhai S, Seifert G (1998) Self-consistent-charge density-functional tight-binding method for simulations of complex materials properties. Phys Rev B 58:7260–7268
10. Yang Y, Yu H, York D, Cui Q, Elstner M (2007) Extension of the self-consistent-charge density-functional tight-binding method: third-order expansion of the density functional theory total energy and introduction of a modified effective coulomb interaction. J Phys Chem A 111:10861–10873
11. Odom TW, Huang J-L, Kim P, Lieber CM (1998) Atomic structure and electronic properties of single-walled carbon nanotubes. Nature 391:62–64
12. Wilder JWG, Venema LC, Rinzler AG, Smalley RE, Dekker C (1998) Electronic structure of atomically resolved carbon nanotubes. Nature 391:59–62
13. Okazaki K, Nakato Y, Murakoshi K (2003) Absolute potential of the Fermi level of isolated single-walled carbon nanotubes. Phys Rev B 68:035434
14. Tanaka Y, HiranaY Niidome Y, Kato K, Saito S, Nakashima N (2009) Experimentally determined redox potentials of individual (n, m) single-walled carbon nanotubes. Angew Chem Int Ed 48:7655–7659
15. Aradi B, Hourahine B, Frauenheim Th (2007) DFTB+, a sparse matrix-based implementation of the DFTB method. J Phys Chem A 111(26):5678–5684
16. Monkhorst HJ, Pack JD (1976) Special points for Brillouin-zone integrations. Phys Rev B 13:5188–5192
17. Hirana Y, Juhasz G, Miyauchi Y, Mouri SMatsuda K, Nakashima N (2013) Empirical prediction of electronic potentials of single-walled carbon nanotubes with a specific chirality (n,m). Sci Rep 3:2959
18. Kilina S, Badaeva E, Piryatinski A, Tretiak S, Saxena A, Bishop AR (2009) Bright and dark excitons in semiconductor carbon nanotubes: insights from electronic structure calculations. Phys Chem Chem Phys 11:4113–4123
19. Ghosh S, Bachilo SM, Simonette RA, Beckingham KM, Weisman RB (2010) Oxygen doping modifies near-infrared band gaps in fluorescent single-walled carbon nanotubes. Science 330:1656–1659
20. Miyauchi Y, Iwamura M, Mouri S, Kawazoe T, Ohtsu M, Matsuda K (2013) Brightening of excitons in carbon nanotubes on dimensionality modification. Nat Photonics 7:715–719
21. Piao YM, Meany B, Powell LR, Valley N, Kwon H, Schatz GC, Wang YH (2013) Brightening of carbon nanotube photoluminescence through the incorporation of sp^3 defects. Nat Chem 5:840–845
22. Shiraishi T, Gergely Juhász G, Shiraki T, Akizuki N, Miyauchi Y, Matsuda K, Nakashima N (2016) Determination of precise redox properties of oxygen-doped single-walled carbon nanotubes based on in situ photoluminescence electrochemistry. J Phys Chem C 120 (29):15632–15639

23. Ma X, Adamska L, Yamaguchi H, Yalcin SE, Tretiak S, Doorn SK, Htoon H (2014) Electronic structure and chemical nature of oxygen dopant states in carbon nanotubes. ACS Nano 8(10):10782–10789
24. Iijima S (1991) Helical microtubules of graphitic carbon. Nature 354:56
25. Won Bong C, Eunju B, Donghun K, Soodoo C, Byung-ho C, Ju-hye K, Eungmin L, Wanjun P (2004) Aligned carbon nanotubes for nanoelectronics. Nanotechnology 15:S512
26. Coleman JN, Khan U, Blau WJ, Gun'ko YK (2006) Small but strong: a review of the mechanical properties of carbon nanotube-polymer composites. Carbon 44:1624–1652
27. Hafez IH, Berber MR, Fujigaya T, Nakashima N (2014) Enhancement of platinum mass activity on the surface of polymer-wrapped carbon nanotube-based fuel cell electrocatalysts. Sci Rep 4:6295
28. Fujigaya T, Nakashima N (2013) Fuel cell electrocatalyst using polybenzimidazole-modified carbon nanotubes as support materials. Adv Mater 25:1666–1681
29. Ju S-Y, Abanulo DC, Badalucco CA, Gascón JA, Papadimitrakopoulos F (2012) Handedness enantioselection of carbon nanotubes using helical assemblies of flavin mononucleotide. J Am Cheml Soc 134:13196–13199
30. Satake A, Miyajima Y, Kobuke Y (2005) Porphyrin–carbon nanotube composites formed by noncovalent polymer wrapping. Chem Mater 17:716–724
31. Bai L, Wang X, Chen Q, Ye Y, Zheng H, Guo J, Yin Y, Gao C (2016) Explaining the size dependence in platinum-nanoparticle-catalyzed hydrogenation reactions. Ang Chem Int Ed 55:15656–15661
32. Liu J et al (2017) High performance platinum single atom electrocatalyst for oxygen reduction reaction. Nat Comm 8:15938
33. Xie S, Tsunoyama H, Kurashige W, Negishi Y, Tsukuda T (2012) Enhancement in aerobic alcohol oxidation catalysis of Au25 clusters by single Pd atom doping. ACS Catalysis 2:1519–1523
34. Liu Y, Tsunoyama H, Akita T, Xie S, Tsukuda T (2011) Aerobic oxidation of cyclohexane catalyzed by size-controlled Au clusters on hydroxyapatite: size effect in the sub-2 Nm regime. ACS Catalysis 1:2–6
35. Staykov A, Dietz LGF, Nikolai Tyutyulkov N (2003) Is 2-D graphite an ultimate large hydrocarbon? Iv. Structure and energy spectra of large polycyclic aromatic hydrocarbons with different symmetry. Z Naturforsch B 58:965–970
36. Staykov A, Ooishi Y, Ishihara T (2014) Immobilizing metal nanoparticles on single wall nanotubes. Effect of surface curvature. J Phys Chem C 118:8907–8916
37. Kresse G, Furthmüller J (1996) Efficiency of ab-initio total energy calculations for metals and semiconductors using a plane-wave basis set. Comput Mater Sci 6:15–50
38. Kresse G, Furthmüller J (1996) Efficient iterative schemes for ab initio total-energy calculations using a plane-wave basis set. Phys Rev B 54:11169–11186
39. Blöchl PE (1994) Projector augmented-wave method. Phys Rev B 50:17953–17979
40. Momma K, Izumi F (2011) Vesta 3 for three-dimensional visualization of crystal, volumetric and morphology data. J Appl Crystallogr 44:1272–1276

Chapter 8
Doped and Decorated Carbon Foams for Energy Applications

Stephen M. Lyth

8.1 Carbon in Fuel Cells

Carbon is a unique element existing in a wide range of different forms, each with an array of differing properties. Elemental carbon is a major constituent in common gasses such as methane and carbon dioxide; in hydrocarbons such as alcohols and fossil fuels; and in polymers/plastics. Carbon also exists in purer solid forms as graphite, diamond, carbon fibre, activated carbons, graphene, fullerenes, charcoal and soot. As such, carbon permeates our everyday life.

Carbon is also a key material for a wide range of energy-related applications, and it plays a particularly important role in electrochemistry. For example, in polymer electrolyte membrane fuel cells (PEMFCs), carbon is a crucial element in almost every component. Graphite is used for the bipolar plates due to its high conductivity, resistance to corrosion and malleability [1]. The electrolyte membrane is a sulfonated fluoropolymer, with a carbon backbone [2]. Carbon fibre is used as the gas diffusion layer (GDL) due to high strength, electrical conductivity and high porosity [3]. A layer of carbon black is used as the microporous layer (MPL), due to high electrical conductivity and porosity [4]. Finally, in the electrochemical heart of the PEMFC, carbon black is used as a platinum nanoparticle support in the anode and

S. M. Lyth (✉)
Kyushu University Platform of Inter/Transdisciplinary Energy Research (Q-PIT),
Kyushu University, 744 Motooka, Nishi-ku, Fukuoka 819-0395, Japan
e-mail: lyth@i2cner.kyushu-u.ac.jp

S. M. Lyth
International Institute for Carbon-Neutral Energy Research (WPI-I2CNER),
Kyushu University, 744 Motooka, Nishi-ku, Fukuoka 819-0395, Japan

S. M. Lyth
Department of Mechanical Engineering, Energy2050, University of Sheffield,
The Arts Tower, Sheffield S10 2TN, UK

N. Nakashima (ed.), *Nanocarbons for Energy Conversion: Supramolecular Approaches*, Nanostructure Science and Technology,
https://doi.org/10.1007/978-3-319-92917-0_8

cathode electrocatalyst layers. At this location, the fundamental processes of electrochemical conversion of hydrogen and oxygen to electricity and water take place.

Carbon black currently dominates the field as a catalyst support material for electrochemistry and PEMFCs. It is generally composed of aggregates of carbon nanoparticles ~50–100 nm in diameter. Despite the expensive and energy-intensive synthesis process, carbon black is relatively cheap due to the huge scale of production, for use in a variety of industrial applications. Some of the most common applications include polymer or rubber reinforcement (e.g., for tyres), inks/black dyes and light-weight composite materials, especially for the automobile and aerospace industries. Carbon black is also favoured as a catalyst support in electrochemistry due to its low cost (<150 USD/tonne) [5], large surface area (e.g., ~300 m^2/g), suitable porosity, good electronic conductivity (e.g., 558 S/m) [6] and electrochemical stability. Therefore, any new material designed to compete with carbon black must have similarly large surface area, porosity, conductivity and low cost.

Carbon black is generally produced by spray pyrolysis of heavy petroleum products such as coal tar. As such, it is primarily produced from fossil fuels, and thus cannot be considered as a sustainable material. Indeed, the carbon black industry is already facing shortages as more stringent environmental protection regulations are rolled out across the globe [7]. In particular, there is pressure on carbon black producers to limit sulphur oxide and nitrogen oxide emissions. Therefore, alternative sustainable carbon materials should be found.

One candidate is carbon nanotubes (CNTs), which are becoming increasingly popular as an catalyst support in electrochemical devices [8–10]. CNTs have extremely high conductivity and can form a strong, porous network [11]. They are already being used in some industrial PEMFC applications, and have real potential for the future, as outlined in some of the other chapters of this book. However, CNTs are much more expensive than carbon black at present and are not produced at anywhere near the same scale. An additional problem is the weak interaction between metal nanoparticles and the CNT surface, leading to detachment and thus reduced durability under PEMFC operating conditions [12]. Graphene has also been explored extensively as an alternative electrocatalyst support material, with promising results [13–15]. In particular, the conductivity of graphene is very high, due to the highly graphitic structure. However, restacking of the individual graphene layers can lead to much lower surface area than required for catalytic applications. In addition, the cost of graphene powders is also still relatively high. Thus, there is a potential gap in the market for alternative nanostructured carbon materials in catalysis and electrochemical applications.

8.2 Carbon Foams in General

Carbon foams offer a possible alternative to carbon blacks and other carbon nanomaterials for use in energy-related applications. A foam is a material in which pockets of gas are trapped in a liquid or solid matrix. Foams can be classed into two broad types: closed-cell (in which the gas is isolated in discreet pockets) and open-cell (in which the voids are interconnected). For example, soap bubbles are a closed-cell foam system where the gas is trapped, while sponges are an open-cell foam system in which water can percolate through the entire structure. The terminology employed in the discussion of carbon foams is given in Fig. 8.1. In electrochemistry, open-cell carbon foams have the potential to be hugely useful. They have a large surface area available where chemical reactions can occur. Reactants and products can easily diffuse through the open pore network. They have a continuous electronic pathway over large length scales. They can also be produced fairly cheaply, by several different methods.

A common method to produce low-cost carbon foams is to use in situ foaming during precursor decomposition to create porosity in a single step. This is known as 'self-blowing' or 'gas templating'. For example, coal, coal tar pitch and petroleum pitch are commonly used as a precursor to create carbon foams using this method [17], while alternatives such as cyanate ester [18] or tannin–formaldehyde polycondensates can be used [19]. One method involves simply heating the precursor, and volatiles evolved during heating serve as bubble agents to create hollow cells in the viscous precursor material, followed by curing and carbonization at high temperature. Another similar method involves controlled heating of the precursors to above their melting point in an inert atmosphere, under high pressure, followed by a fast release of pressure. The decrease in pressure results in the formation of gas-filled cells, and curing/carbonization steps follow. The viscosity and volume increases of the precursors affect the ultimate foam cell structure. These types of carbon foams are extremely cheap and easy to make. However, there is limited control of the chemical structure and microporosity of the final product, and they are generally derived from fossil fuels.

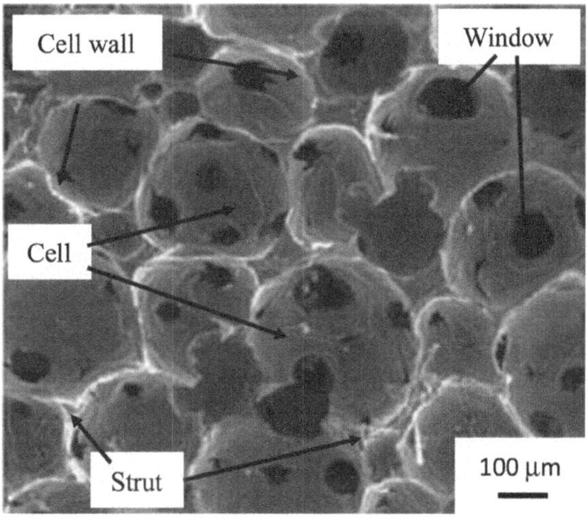

Fig. 8.1 Terminology employed to characterize carbon foams. Reproduced with permission from Elsevier [16]

Another method to prepare carbon foams is simply by pyrolysis of polymer foams. In this case, a polymer foam (such as a polyurethane foam [20] or a phenolic-based foam [21]) is simply carbonized at high temperature in an inert atmosphere. This results in a carbonized foam that retains most of the microstructure of the precursor. This has the advantage of being cheap and straightforward. However, control of the final microstructure structure and chemical structure is difficult and relies mainly on the properties of the precursor polymer, and the heating conditions applied.

Carbon 'foams' (or more strictly, mesoporous carbons) can also be formed via soft templating [22, 23]. For example, self-assembly of resorcinol and formaldehyde with Pluronic F127 surfactant, followed by carbonization, results in a mesoporous carbon material. In this case, the pore size can be tuned by changing the molar ratios of the precursors and the polymerization time [24]. This is a very simple and controllable method to produce porous carbon 'foams'. However, the very small pore size in these mesoporous carbons can lead to problems with mass diffusion for electrochemical applications.

One more method to form carbons foams with an open-porous structure is to utilize a sacrificial metal oxide template (e.g., silica, or alumina) [25, 26]. For example, a polymer precursor is mixed with metal oxide particles, and this mixture is pyrolysed to convert the polymer into graphitic carbon. Subsequently, the metal oxide template is removed by washing the powder in concentrated acid or alkaline solution to dissolve the inorganic materials. This leaves behind pores in the carbon that have the same size and shape as the original template, which can be selected depending on the desired microstructure of the final product. This method gives a large degree of control over the pore size and distribution, since templates of arbitrary size and shape can be selected. Indeed, this sacrificial templating method has been used with great success to synthesize Pt-free Fe-N-C electrocatalysts [26]. However, the use of sacrificial templates adds several processing steps compared with the simple spray pyrolysis technique of carbon black. In addition, the sacrificial template is usually literally sacrificed, which could be seen as wasteful. As such, alternative methods to create porous carbon foams with fewer processing steps are desirable.

8.3 Carbon Foams Derived from Sodium Alkoxides

Herein, we describe a class of unique graphene-like carbon foams produced by foaming, or self-blowing of different sodium alkoxide precursors. The nanostructured carbons formed using this process are presented as a possible alternative to carbon black and related nanomaterials for different energy-related applications. Using alkoxides as precursors lends an extra degree of control to the carbon foam product, as the chemical structure of the precursor can be straightforwardly modified. For example, the chain length or molecular weight can be varied to modify the viscosity, or heteroatoms (such as nitrogen, sulphur or fluorine) can be routinely introduced to dope the final product. The sodium oxides produced during decomposition produce a 'self-activating' effect, resulting in a high degree of microporosity and large surface area. In addition, alkoxides can be produced from bioethanol, and therefore the use of fossil fuels can be avoided.

In 2009, Choucair et al. published a key paper entitled 'Gram-scale production of graphene based on solvothermal synthesis and sonication' in Nature Nanotechnology [27]. In this work, sodium metal and ethanol were reacted at 220 °C for 72 h in a polytetrafluoroethylene (PTFE) reactor. The intermediate solid powder formed by this process was speculated to have a clathrate-like structure, with solid sodium ethoxide encapsulating ethanol-filled cells. This intermediate material was then flash pyrolysed, decomposing to form a foam-like carbonaceous powdery product, which they referred to as graphene (Fig. 8.2). The porous foam-like structure was speculatively attributed to the clathrate-like cells. However, since sodium ethoxide is soluble in ethanol, it seems unlikely that this type of clathrate structure would be mechanically stable. For the clathrate-like structure survive, the flash pyrolysis step to form carbon foam with such well-defined cells also seems unlikely.

Whilst this 'graphene' had almost atomically thin carbon walls, as measured by atomic force microscopy (AFM), there were indicators suggesting that this material should not strictly be referred to as graphene. First of all, elemental analysis revealed an oxygen content of >13 wt%. This is a similar oxygen content to that observed in graphene oxide rather than graphene [28]. Second, no characteristic 2D peak (a common indicator of graphene-like crystal structure) was observed in Raman spectroscopy (Fig. 8.2). This suggests that there is little or no long-range crystallinity in this carbon foam material, and that the structure is closer to that of an amorphous carbon. Third, the conductivity of pellets made from this carbon foam was just 0.05 S/m, which is many orders of magnitude lower than that of graphene or conventional graphite powders [29]. Despite the speculative formation mechanism and semantics issues, the graphene-like material presented in this work was extremely interesting and new. The surface area was large at >600 m^2/g, as measured via nitrogen adsorption using Brunauer–Emmett–Teller (BET) theory. In addition, this carbon foam was produced using very few steps, at gram scale. As such, it was an ideal material to further develop in various energy-related applications.

In 2010, the same group investigated the superconducting properties of magnesium diboride (MgB_2) decorated onto their carbon foams [30]. These composite materials were prepared by ball milling carbon foam with crystalline boron powder in toluene, followed by vacuum drying and pressing into pellets. The pellets were then placed in an iron tube with magnesium powder, and heated to 850 °C for 10 h, under flowing argon gas to form MgB_2. By supporting the MgB_2 on the carbon foam, they obtained an enhancement in the critical current density (J_c) by an order of magnitude at 5 K, under a magnetic field of 8 T. This work shows promise for large-scale MgB_2 superconducting wire applications.

In 2012, the same group investigated how the preparation method affected the microstructure of these carbon foams, via nitrogen adsorption [31]. Here, as well as

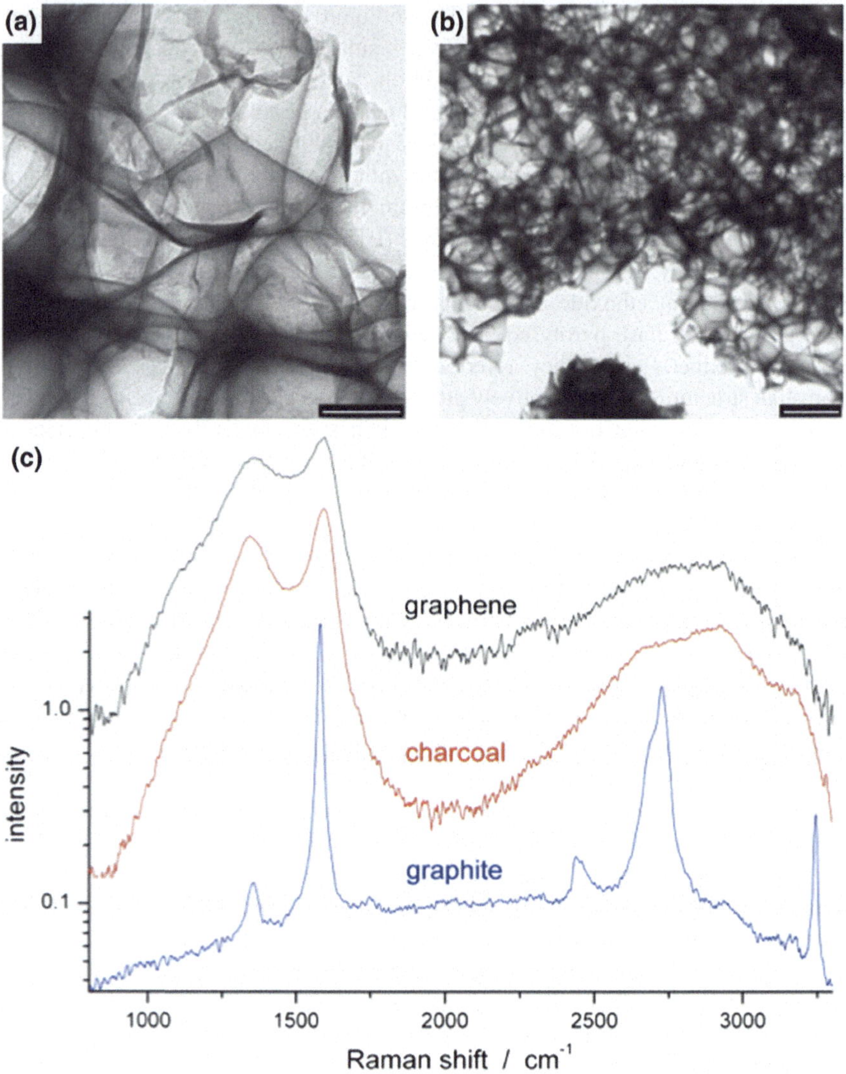

Fig. 8.2 Characterization of the carbon foam produced by Choucair et al. **a**, **b** Transmission electron microscopy (TEM) images reveal the porous foam-like nature of the material. **c** Raman spectroscopy shows the highly amorphous nature of the carbon foam, similar to that of charcoal. Reproduced with permission from Nature Publishing Group [27]

using ethanol as a carbon feedstock as in their original paper, they also tried methanol. A 1:1 molar ratio of alcohol and sodium metal was reacted in a PTFE-lined reactor, at 220 °C for 72 h (presumably forming sodium methoxide from methanol). This intermediate product was flash pyrolysed in air, and then washed/purified using several different protocols. The different effects of stirring, sonication and solvent-type (i.e. deionized water or acidified ethanol) were investigated. The substitution of ethanol with methanol resulted in slightly different microstructures, with ethanol in this case forming a more open sheet-like structure, and methanol forming a foam with pores ~ 500 nm in diameter (Fig. 8.3). The BET surface areas achieved were relatively low, ranging from 11.3 m^2/g for the ethanol-derived carbon to 484 m^2/g for the methanol-derived carbon. The thickness of the cell walls of the methanol-derived carbon foams was measured to be around 1 nm by AFM.

In 2014, the same group replaced ethanol with propanol as the carbon feedstock, reacting it with sodium in the same way as previously described [32]. In this case, TEM revealed that the carbon powder formed comprises graphene-like platelets rather than a foam-like structure. This may be attributed to the use of a propanol precursor in place of ethanol or methanol, affecting the viscosity of the melting alkoxide, or due to the sonication step used in washing, destroying the foam-like

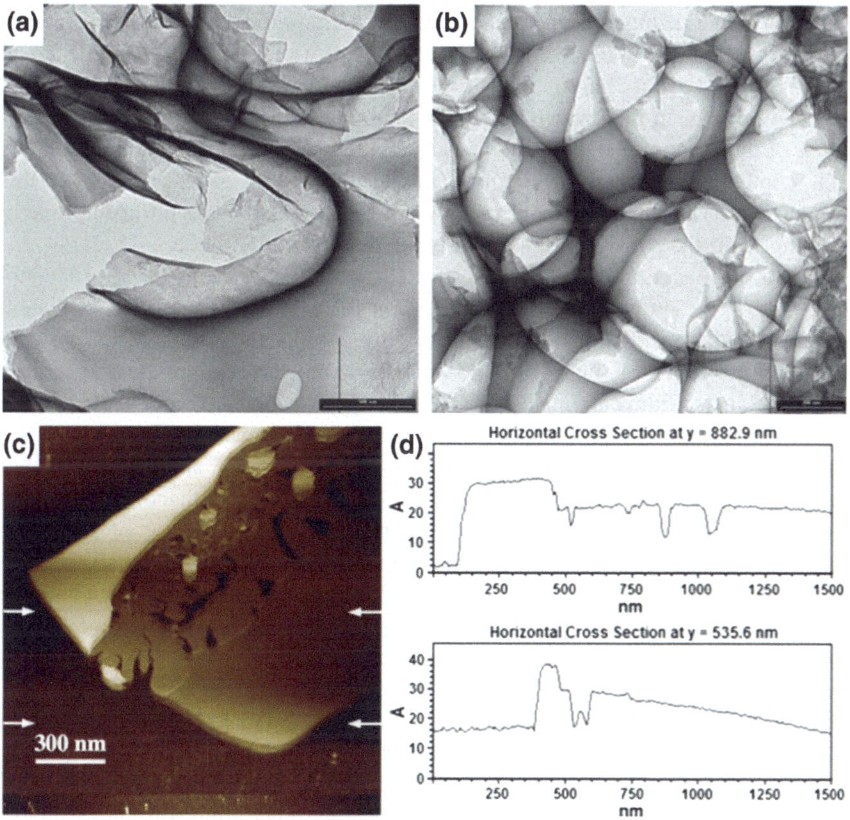

Fig. 8.3 TEM images of **a** ethanol-derived, and **b** methanol-derived carbon foams. **c** AFM image and **d** line profiles of the methanol-derived carbon foam sample. Reproduced with permission from Elsevier [31]

microstructure. They measured the spin lifetimes of electrons in this carbon powder to be as long as 65 ns, which is in the range required for spintronics applications. In addition, this group investigated their ethoxide-derived carbon foam using muon spectroscopy to probe muon–hydrogen nuclear dipolar interactions [33].

In 2015, a different group, Speyer et al., produced a detailed study on the same class of material, again produced by solvothermal reaction between ethanol and sodium [34]. Here, they clarified that the 'intermediate material' was simply sodium ethoxide, and investigated the effect of pyrolysis temperature, heating rate and time on the properties of the final 'graphenic powder'. In this work, some regions of more crystalline multilayer graphitic carbon were observed by TEM, X-ray diffraction (XRD) and Raman spectroscopy, especially after pyrolysis at 850 °C. They observed very large BET surface area, almost 2500 m^2/g, which slightly decreased with increasing pyrolysis temperature, and with increasing pyrolysis time. They also discussed the formation of sodium carbonate as a contaminant phase. One possible issue with the more crystalline graphitic materials produced in this work is the Inconel crucibles used for pyrolysis (a nickel–chromium–iron alloy). In general, much higher temperatures are required for extensive graphitization of carbon (e.g. >2600 °C), unless a catalyst is used [35]. Ni, Cr and Fe are all well known to have a catalytic effect for the formation of crystalline carbon at high temperature (e.g. in the growth of carbon nanotubes) [36]. For example, graphitization of the carbon in steel generally occurs at much lower temperatures [37]. As such, contamination from the pyrolysis vessel must be taken into account in this case, and this graphitization can probably not be attributed as an intrinsic property resulting from the pyrolysis of sodium ethoxide alone at 850 °C.

In 2017, the same group also investigated the influence of the type of alcohol feedstock on the properties of the final carbon foam [38]. They reacted methanol, ethanol or butan-1-ol with sodium, followed by pyrolysis at 850 °C under flowing nitrogen for 4 h, with a final washing step. They confirmed the presence of sodium hydroxide and sodium carbonate as solid by-products which are mostly removed by washing (although some encapsulated contaminants remain). All the materials displayed some crystallinity in XRD and Raman spectroscopy, and the thickness of the carbon layers was determined to be 10, 11 and 12 for methanol, ethanol and butan-1-ol precursors, respectively. This points towards the formation of 'few-layer graphene'. Some influence of the carbon precursor chain length on the crystallinity of the final product was also found, with methanol resulting in carbon foam with the lowest degree of disorder (Fig. 8.4). However, again, this crystallinity may result from the catalytic action of the Inconel crucibles used during pyrolysis rather than being a fundamental property of the combustion product of sodium alkoxides. Meanwhile, the BET surface area increased as the chain length increased from 1430, to 2316, to 2372 m^2/g. For methanol-derived carbon foams, the microporosity was less developed and hysteresis in the nitrogen adsorption isotherms was large, whereas for butan-1-ol, the hysteresis was minimal, indicating fewer mesopores and more micropores.

Summarizing the above fundamental works, Choucair et al. produced carbon foams by reacting alcohol with sodium at elevated temperature in a sealed PTFE

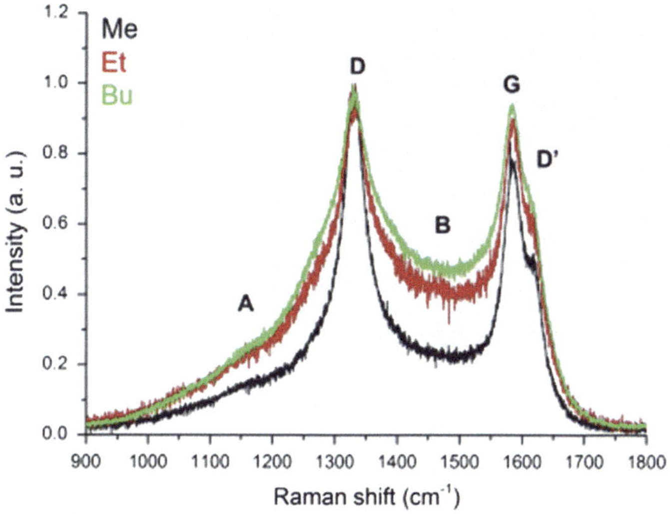

Fig. 8.4 Normalized Raman spectra (λ = 632.8 nm) of carbon foam samples derived from methanol, ethanol and butan-1-ol precursors. Reproduced with permission from Elsevier [38]

reaction vessel. They utilized methanol, ethanol and propanol as different carbon feedstocks, sometimes resulting in a foam-like structure, and sometimes in a flatter, platelet-type structure. They observed surface areas of up to ~600 m^2/g (BET) and showed that this material could be applied in spintronics, or superconductivity applications. Speyer et al. focussed more on the formation mechanisms of carbon foams, utilizing methanol, ethanol and butan-1-ol as different carbon feedstocks, and investigated the effect of pyrolysis temperature on the final carbon foam product. They obtained higher surface areas of up to ~2500 m^2/g. The above studies explore some of the more fundamental aspects of carbon foams derived from sodium alkoxides. Herein, we look at some of the potential applications of this new and interesting graphene-like carbon foam.

8.4 Carbon Foams for Hydrogen Storage

In 2014, Lyth et al. reported the hydrogen uptake properties of carbon foams [39]. In this case, sodium ethoxide was produced by the reaction between sodium and ethanol at 220 °C in a sealed PTFE reactor. The resulting powder was then burned in air to form the carbonaceous product. This product was then pyrolysed at 1000 °C under nitrogen gas flow, in aluminium oxide boats. Although the micron-scale cells had almost atomically thin walls, these were confirmed to be largely amorphous by Raman spectroscopy. The BET surface area was measured to be 1270 m^2/g (Fig. 8.5), and the pore size distribution revealed a high proportion of

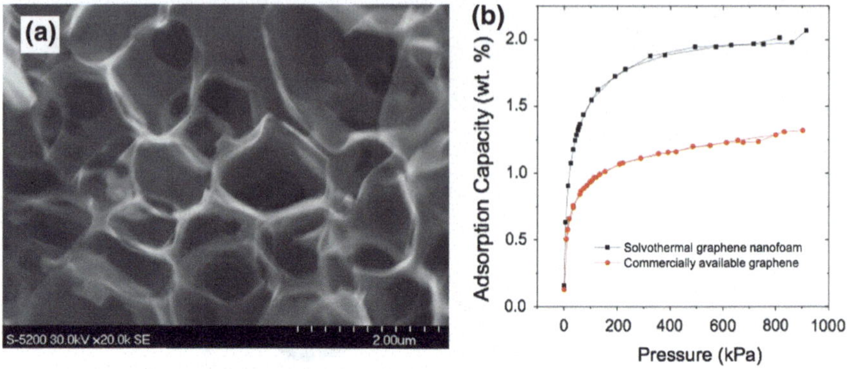

Fig. 8.5 **a** SEM image and **b** hydrogen uptake (at 77 K) of sodium ethoxide-derived carbon foams. Reproduced with permission from Elsevier [39]

pores in the 3–4 nm range. Due to the large surface area, the carbon foam was explored as a hydrogen storage material. The hydrogen uptake was measured at 77 K and 1 MPa to be 2.1 wt%. This was significantly higher than commercially available graphene powders, mainly due simply to the larger surface area.

In 2015, Choucair et al. published similar results on the hydrogen uptake of their original carbon foam materials, at 77 K (Fig. 8.6) [40]. As well as looking at the pure material, they also composited their carbon foams with magnetite, silver, titania, nickel and silicon, by solid-state mechanical mixing. In this case, the carbon foam supports had slightly low surface areas of <500 m^2/g. The maximum pressure used to measure hydrogen uptake was 1 atm (0.1 MPa), which is lower than generally utilized in the hydrogen storage literature, making it difficult to make direct comparisons. However, the maximum hydrogen uptake of the pure carbon foam measured here was 0.5 wt% at 77 K and 0.1 MPa. This is lower than the

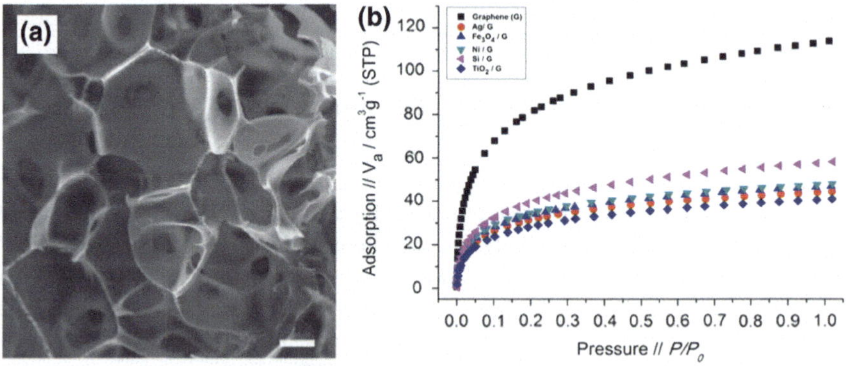

Fig. 8.6 **a** SEM image and **b** hydrogen uptake (at 77 K) of sodium ethoxide-derived carbon foams. Reproduced with permission from Elsevier [40]

values of around 1.5 wt% obtained in [39] at the same pressure (Fig. 8.5). This lower hydrogen uptake is directly attributable simply to the lower surface area. Compositing their carbon foam with the various coatings resulted in a significant decrease in surface area and hydrogen uptake in all cases, partly due to the addition of materials with higher molecular weight, and partly due to added processing steps, which can destroy the porosity in such materials.

8.5 Carbon Foam Supercapacitors

In 2015, Cui et al. studied the formation mechanism of carbon foams (called graphene frameworks in this case) derived from pyrolysis of sodium alkoxides [41]. Here, they attributed the relatively high crystallinity of their materials to the catalytic action of sodium metal on the carbon formation. However, this group also performed the solvothermal reaction in a sealed stainless steel reactor at 200 °C. As previously discussed, this makes it highly likely that iron contamination from the reaction vessel contributes to the catalytic graphitization of the carbon foam at relatively low temperature (1000 °C in this case). This group obtained carbon foams using both propanol and acetic acid as carbon feedstocks, adding to the list of precursor materials previously employed. The surface areas were measured to be around 750, 540 and 180 m^2/g for methanol-, ethanol- and n-propanol-derived carbon foams, respectively. This group measured the electrochemical supercapacitor properties of their carbon foams, obtaining a specific capacitance of 310.7 F/g at a current density of 0.2 A/g (Fig. 8.7). The excellent performance as a supercapacitor was attributed to the large specific surface area; the presence of some oxygen-containing functional groups, enhancing the wettability between the carbon and electrolyte; and the three-dimensional structure, reducing the diffusion length of electrons and ions.

8.6 Sulphur-Decorated Carbon Foams for Lithium-Ion Batteries

In 2010, Choucair et al. used their carbon foam for electrochemical applications by simply mixing sulphur nanoparticles with their carbon foam, and studying the performance in lithium-ion batteries [42]. In 2011, they improved on this work by mixing their carbon foam with elemental sulphur, heating this mixture first to 200 °C to infiltrate melted sulphur into the carbon, and then heating at 300 °C to coat the carbon foam with vaporized sulphur (Fig. 8.8) [43]. Thermogravimetric analysis showed that the loading of sulphur was around 22 wt%. The sulphur-decorated carbon foam was then mixed with carbon black and polyvinylidene fluoride (PVDF), pasted on onto aluminium foil substrates, dried

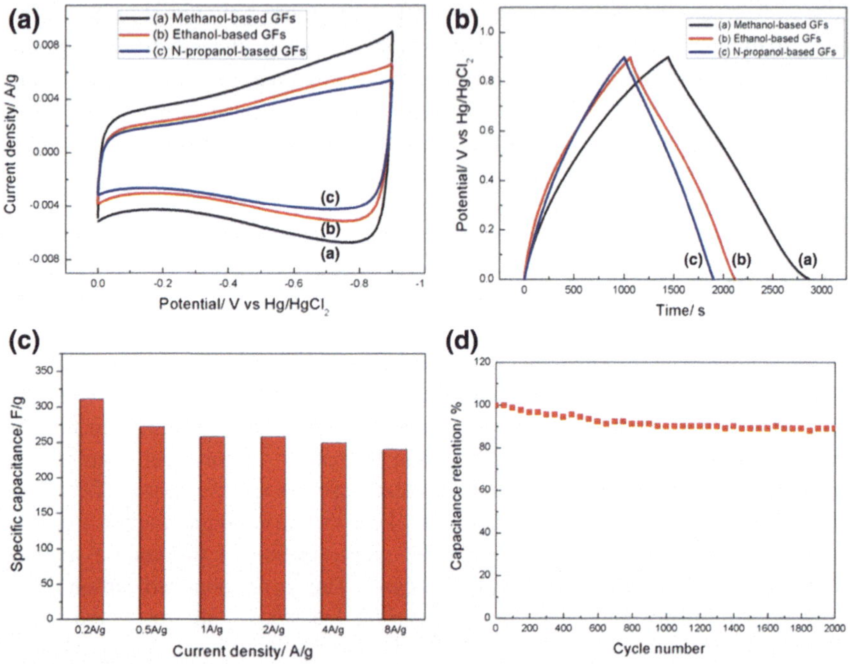

Fig. 8.7 Capacitive performance of carbon foam electrodes in 6 M KOH aqueous solution: **a** CV curves (30 mV/s); **b** Galvanostatic charge /discharge curves at 0.2 A/g; **c** capacitance of a methanol-derived carbon foam electrode; **d** cycling stability of the methanol-derived carbon foam electrode at 2 A/g. Reproduced with permission from ACS Publications [41]

and pressed. This composite electrode was then incorporated into a rechargeable lithium battery cell, in which sulphur is converted first to lithium polysulfide and then lithium sulphide. They found that the capacity was improved to 1611 mA h g^{-1}, compared with 1100 mA h g^{-1} for the reference electrode. The cycling stability of the sulphur-decorated carbon foam was also improved compared to a carbon black/sulphur electrode without carbon foam inclusion. This was first attributed to enhanced conductivity of the electrode due to the inclusion of the carbon foam, and then to enhanced reaction kinetics, as evidenced by electrochemical impedance spectroscopy measurements (EIS). This work effectively showed that this carbon foam material is highly suited to energy applications, and especially to electrochemistry.

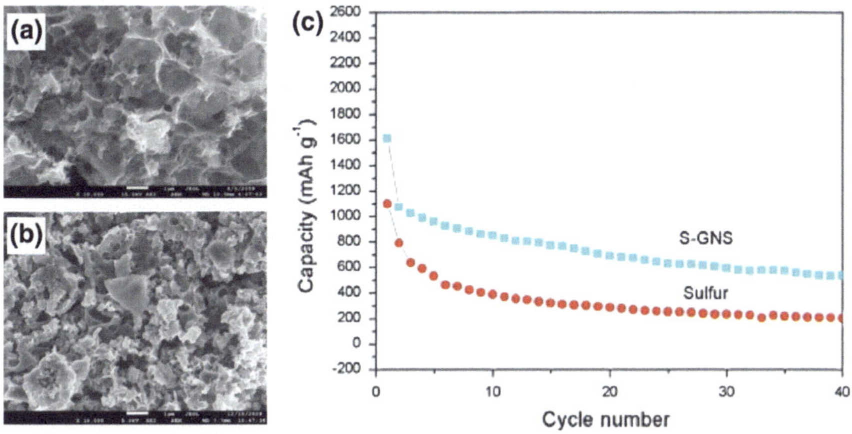

Fig. 8.8 SEM images of **a** carbon foam, and **b** sulphur-decorated carbon foam (S-GNS). **c** Capacity versus cycle number for the composite electrodes. Reproduced with permission from Elsevier [43]

8.7 Functionalized Carbon Foams as Electrochemical Sensors

Carbon foam was recently used as an electrochemical sensor by Choucair et al., in 2017. Carbon foam derived from ethanol was functionalized with carboxylic acid groups in order to facilitate the adsorption and crosslinking of enzyme cytochrome C [44]. They confirmed direct electrical communication between the redox centre of the cytochrome C and the functionalized carbon foam, and utilized this composite electrode as a hydrogen peroxide (H_2O_2) biosensor in aqueous phosphate buffer. A rapid amperometric response (5 s) was recorded upon exposure to hydrogen

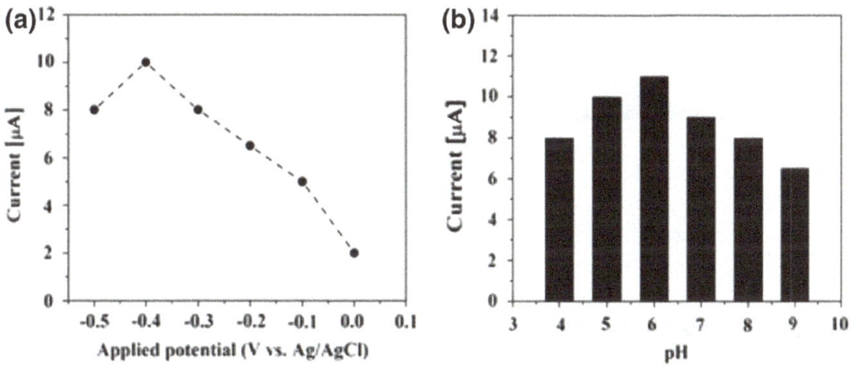

Fig. 8.9 **a** Dependence of the current response of the enzyme-decorated carbon foam biosensor at different applied potentials. **b** Effect of pH on the catalytic current of the biosensor at −0.4 V. Reproduced with permission from Springer-Verlag Berlin Heidelberg [44]

peroxide. The linear range of response to concentration ranged from 5×10^{-7} to 2×10^{-4} M, with a detection limit of 2×10^{-7} M (Fig. 8.9). This work displayed yet further evidence of the suitability of this class of carbon foam as a material for electrochemistry.

8.8 Platinum-Decorated Carbon Foams for PEMFCs

In 2014, Liu et al. decorated carbon foams with platinum nanoparticles to investigate their application as catalyst supports in PEMFCs [45]. Carbon foam was synthesized simply by burning commercially available sodium ethoxide in air, followed by pyrolysis in flowing nitrogen at 800 or 1000 °C. This showed that the solvothermal reaction between sodium and ethanol is not a required step—any sodium ethoxide is a suitable precursor. The surface area was >1500 m^2/g, and the structure was confirmed by Raman spectroscopy to be highly amorphous. These carbon foams were decorated with platinum nanoparticles, by thermal decomposition of platinum bis(acetylacetonate)platinum(II). The resulting nanoparticles were more uniformly distributed and smaller (2.3 nm) than those decorated on carbon black or graphene powders supports (Fig. 8.10). This was attributed to the large number of defects in this amorphous material, providing ample sites for nanoparticle nucleation.

In cyclic voltammograms, the electrochemical surface area (ECSA) of the Pt-decorated carbon foam was ~100 m^2/g, compared with ~70 m^2/g for the carbon black support. In linear sweep voltammetry (LSV) measurements, the mass activity for the oxygen reduction reaction (ORR) was higher (176 A/g) than that of

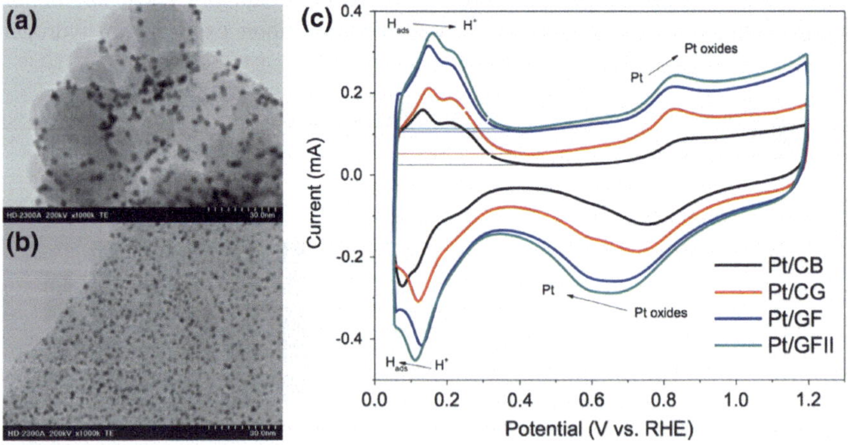

Fig. 8.10 SEM images of **a** Pt-decorated carbon black, and **b** Pt-decorated carbon foam. **c** Cyclic voltammograms of Pt-decorated carbon black, graphene, and carbon foam pyrolysed at 800 or 1000 °C [45]

Pt-decorated carbon black (138 A/g) or graphene powder (169 A/g). These improvements were attributed to the improved platinum nanoparticle size and distribution. The stability of the Pt nanoparticles during potential cycling was also much improved, with little change in average particle size even after 60,000 potential cycles, due to strong binding between Pt and defects in the carbon foam support.

8.9 Silver-Decorated Carbon Foams for Electrochemical CO_2 Conversion

Conversion of CO_2 to useful fuels or plastics is a potentially a way to reduce CO_2 emissions. For example, silver is a common catalyst for the electrochemical conversion of CO_2 to carbon monoxide. In 2017, Ma et al. decorated sodium ethoxide-derived carbon foams with silver nanoparticles, via the citrate-protecting method [46]. This carbon foam had a large surface area of $\sim$900 m^2/g (much higher than the graphene or carbon black reference samples). Raman spectroscopy revealed little or no crystallinity in the carbon foam, with a signal similar to that of carbon black. The Ag nanoparticle distribution was much improved on the carbon foam supports compared with graphene or carbon black (Fig. 8.11). In addition, the

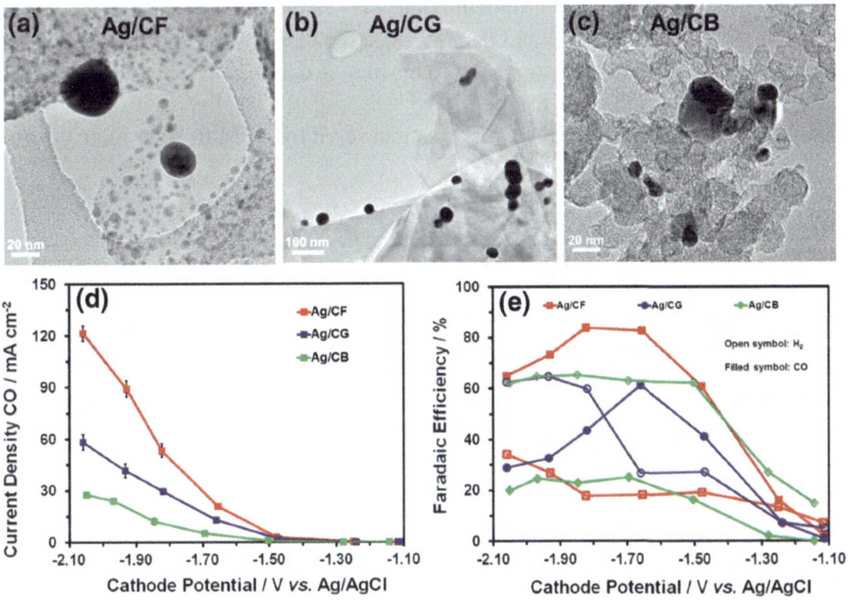

Fig. 8.11 SEM images of silver-decorated **a** carbon foam (CF), **b** graphene (CG) and **c** carbon black (CB). **d** Partial electrochemical current density for CO formation, and **e** Faradaic efficiency of electrochemical CO_2 conversion cells. Reproduced with permission from John Wiley and Sons [46]

Ag nanoparticle size on carbon foam was around 5 nm, compared with >20 nm on both graphene and carbon black. This is similar to the case of platinum-decorated carbon foam [45] and is again attributed to the amorphous nature of the material, providing defects in the carbon lattice for facile nanoparticle nucleation.

The Ag-decorated carbon foams were incorporated into an electrochemical CO_2 conversion flow cell, showing by far the highest current density for the formation of CO (120 mA/cm^2), compared to graphene (58 mA/cm^2) or carbon black (30 mA/cm^2). The maximum Faradaic efficiency for the carbon foam support was 84% (at −1.82 V vs. Ag/AgCl), which was also significantly higher than for graphene or carbon black (around 65% in both cases). This improved performance was attributed simply to the enhanced nanoparticle distribution in the case of the carbon foam support.

8.10 Nitrogen-Doped Carbon Foams as Metal-Free Electrocatalysts for PEMFCs

In the above studies, relatively pure carbon foams are produced by the thermal decomposition of sodium ethoxide. However, by switching to alternative alkoxide materials, it is possible to dope the carbon with different heteroatoms. In 2012, Lyth et al. reacted sodium with ethanolamine to form a nitrogen-containing metal alkoxide (probably sodium 2-aminoethoxide) [47]. This was then flash pyrolysed in air at 600 °C, washed in water and then pyrolysed under flowing nitrogen at various temperatures. The same macroporous foam-like structure was obtained as in the above works, although here the pore size was slightly smaller at just ~200 nm (Fig. 8.12). The thickness of the walls was measured by AFM to be around 1.7 nm.

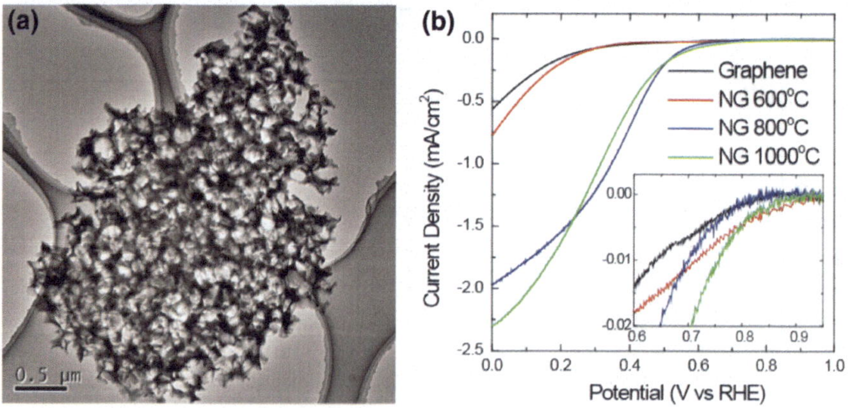

Fig. 8.12 **a** TEM image of nitrogen-doped carbon foam derived from ethanolamine. **b** Linear sweep voltammograms (LSVs) of this nitrogen-doped carbon foam pyrolysed at different temperatures. Reproduced with permission from the Surface Science Society of Japan [47]

The surface area was rather low (between 60 and 170 m^2/g), and the nitrogen content varied from 15.1, to 8.8, to 4.8 wt% after pyrolysis at 600, 800 and 1000 °C, respectively. This showed that high nitrogen content can be achieved using this technique, putting it in the range of an amorphous carbon nitride. X-ray photoelectron spectroscopy (XPS) revealed that the nitrogen was largely in pyridinic and tertiary (graphite-like) atomic configurations.

These materials were tested as 'metal-free' electrocatalysts for the oxygen reduction reaction (ORR). Non-platinum-group metal (non-PGM) catalysts for fuel cells are a major area of research in the quest to decrease the cost of hydrogen-related technologies. The most successful materials for this application to date are generally Fe-N-C catalysts—carbon-based powders doped with nitrogen and iron. These catalysts are now starting to meet US DOE targets for automotive applications [48–50]. However, there is an ongoing debate about the nature of the active site for the ORR in such materials [51]. The three main candidates for the active sites are (1) atomically dispersed Fe in porphyrin-like configuration [52]; (2) Fe nanoparticles encapsulated in graphitic carbon [53] or (3) metal-free nitrogen-doped carbon [54]. One particular problem with the third theory is that iron (or other transition metal) contamination is often found in carbon black [55] or carbon nanotube-supported 'metal-free' catalysts [56]. Since the nitrogen-doped carbon foams produced from sodium alkoxides are potentially free from iron contamination and have very large surface area, they are ideal to test the activity of truly 'metal-free' electrocatalysts. The maximum onset potential for the ORR in these nitrogen-doped carbon foams was 0.87 V vs RHE, and the maximum current density was -2.3 mA/cm^2 (at 0 V vs. RHE and 1600 rpm). The highest electron transfer number recorded was $n = 2.9$, reflecting a mix of 2- and 4-electron transfer. These were relatively good results for a 'metal-free' electrocatalyst at the time. However, due to the stainless steel reaction vessels used in this work, metal contamination could not be ruled out.

In 2016, the same group followed up on this work. Here, Liu et al. reacted sodium with a mixture of anhydrous ethanol and diethanolamine to form the alkoxide precursor [57, 58]. Significant steps were taken to ensure that iron contamination was not present, including the use of a PTFE reaction vessel, alumina boats and even avoiding the use of stainless steel spatulas. The precursor was ignited and burned in air, washed, dried and pyrolysed at 1000 °C under nitrogen, then at 800 °C under hydrogen gas flow (to reduce the oxygen content). A second sample was then also graphitized at 1400 °C to improve the conductivity. The same micron-scale cells were observed by SEM (Fig. 8.13), and the surface area in both cases was $\sim$700 m^2/g. Inductively coupled plasma-atomic emission spectrometry (ICP-AES) was used as a more sensitive technique than XPS to confirm that the material was metal-free, to below the detection limit if the machine.

The catalyst heat-treated at 1000 °C had an onset potential for the ORR of 0.8 V vs RHE, a mass activity 0.46 A/g (at 0.6 V vs. RHE) and a maximum current density of 2.9 mA/cm^2 (at 1600 rpm and 0.2 V vs. RHE). Graphitization at 1400 °C resulted in an increase in onset potential to 0.85 V vs RHE, a mass activity of 2.79 A/g (at 0.6 V vs. RHE) and a maximum current density of 3.5 mA/cm^2

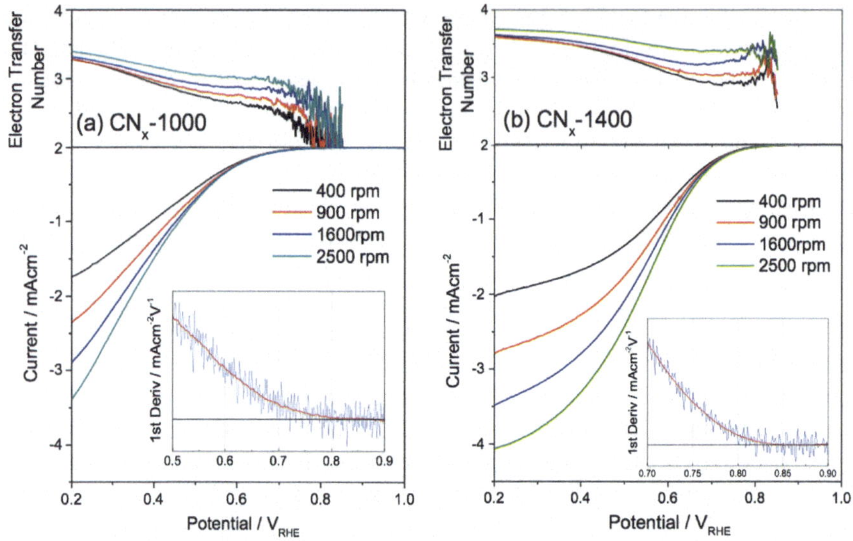

Fig. 8.13 **a** TEM images of nitrogen-doped carbon foams derived from diethanolamine. LSVs (with ring current, top) in acid solution of these catalysts pyrolysed at **b** 1000 °C and **c** 1400 °C [58]

(at 1600 rpm). This indicated that the increase in conductivity associated with graphitization was important to increase electrochemical activity. Despite the serious efforts to remove possible sources of iron contamination, the activity of these catalysts was much improved compared with their previous work [47].

In the above studies, the ORR activity was eventually attributed to tertiary nitrogen sites in the carbon matrix, and the reasonable current density was achieved by increasing the conductivity by pyrolysis at higher temperature. However, it is clear that the activity of these 'metal-free' electrocatalysts is lower than that of most Fe-N-C electrocatalysts. Therefore, it was essentially concluded that whilst some 4-electron transfer is probably possible in the absence of iron, most of the active sites in Fe-N-C electrocatalysts probably much faster Fe-containing moieties.

The same samples were also measured in alkaline solution (0.1 M KOH) [57]. It is well known that metal-free electrocatalysts perform better in alkaline conditions compared to acidic conditions, and that the reaction mechanisms are probably quite different. In this case, the ORR activity of the nitrogen-doped carbon foam was much higher, with an onset potential of 0.95 V vs RHE, a half-wave potential of 0.8 V vs RHE and a maximum current density of −4 mA/cm^2 (Fig. 8.14). However, again, these materials could not compete with Fe-containing electrocatalysts. Despite the slightly lower activity compared with Fe-N-C electrocatalysts, the durability was extraordinary. After subjecting the samples to 60,000 potential cycles, almost no change in the cyclic voltammograms was observed, and only slight changes were seen in the mass diffusion limited current of LSVs. In comparison, Pt/C underwent large changes under the same conditions, and the Fe-N-C

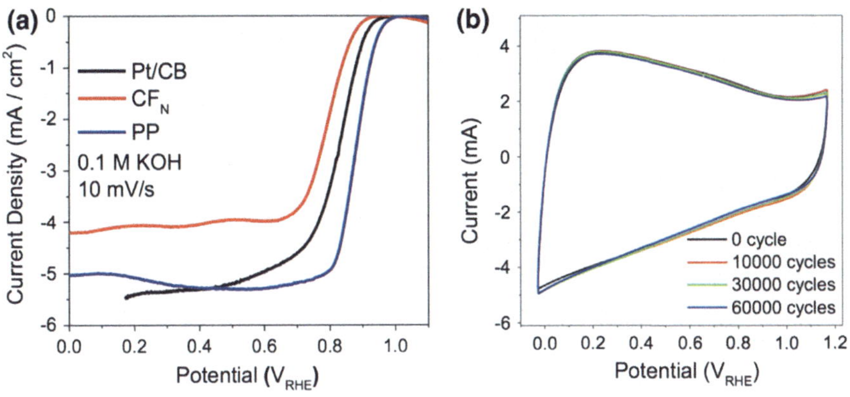

Fig. 8.14 Electrochemical performance in alkaline solution. **a** Linear sweep voltammograms (LSVs) comparing metal-free nitrogen-doped carbon foams (CF_N, red), with platinum (black), and a commercially available Fe-N-C catalyst (PP, blue). **b** Cyclic voltammograms of CF_N showing no change in response over 60,000 potential cycles [57]

reference sample also displayed some variation in activity. It was concluded even in alkaline conditions, Fe-containing active sites are required for the ORR to proceed efficiently, although some slower metal-free active sites may exist.

8.11 Fe-N-C Carbon Foams for PEMFCs

Whilst the study of 'metal-free' nitrogen-doped carbon ORR electrocatalysts is academically interesting, their activity is not sufficient for applications. Fe-N-C catalysts that include iron, nitrogen and carbon generally have much higher activity. Therefore, Mufundirwa et al. fabricated sodium alkoxide-derived nitrogen-doped carbon foams and then decorated these with iron [59]. In this case, a new method for the synthesis of the alkoxide precursor was used, in order to avoid the traditional solvothermal method. This significantly improved the safety of the reaction, by avoiding the use of high pressure and high temperature, and slowing down the reaction between sodium metal and the alcohol. Triethanolamine and anhydrous ethanol were diluted in tetrahydrofuran (THF) in sealed glassware, under flowing nitrogen gas. This solution was cooled in an ice bath and magnetically stirred, and then sodium lumps were added. After 48 h, the reaction was complete, the THF was evaporated under reduced pressure, and the solid product pyrolysed at 600 °C under flowing nitrogen. After washing, a second pyrolysis step was performed at 1000 °C under nitrogen. The resulting sample was dispersed in water, mixed with iron acetate (FeAc), and dried. Finally, the second pyrolysis step was performed at 900 °C in a mixed gas stream of ammonia and nitrogen to decompose the FeAc. These iron-decorated nitrogen-doped carbon (Fe-N-C) foams had a surface area in the region of 1600 m^2/g and were decorated with many carbon-encapsulated iron

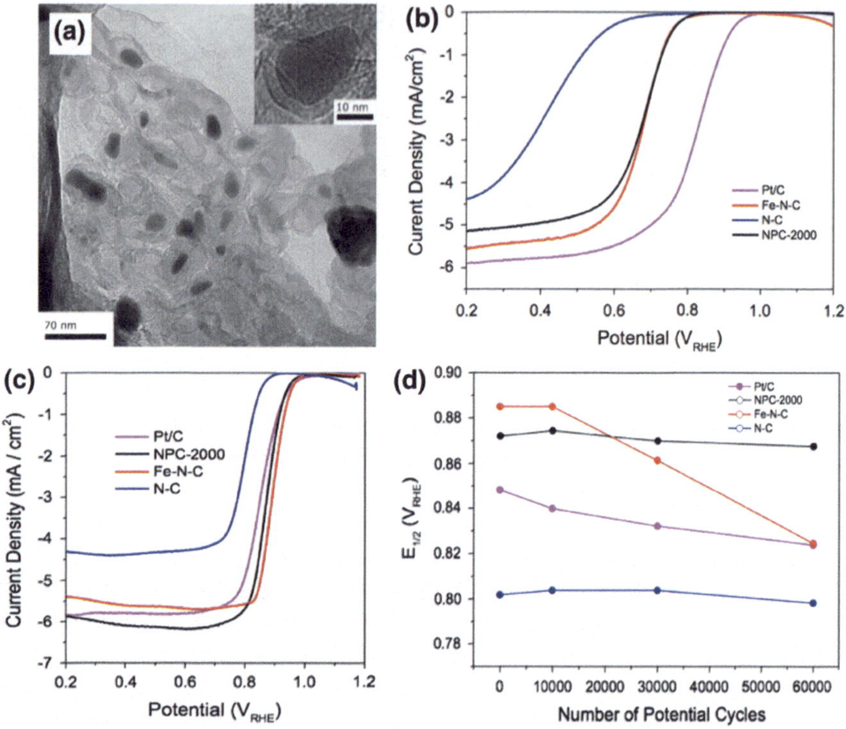

Fig. 8.15 **a** TEM image of Fe nanoparticles in Fe-N-C foams. LSVs in **b** acid solution and **c** alkaline solution. **d** Cycling durability in 0.1 M KOH. Reproduced with permission from Elsevier [59]

nanoparticles, as observed by TEM (Fig. 8.15). The nitrogen content was 2.2 wt% and the iron content was ~3 wt%.

The ORR activity of these Fe-N-C carbon foam electrocatalysts far outperformed the previous metal-free incarnations. In acid solution, the onset potential was 0.88 V vs RHE, the half-wave potential was 0.68 V vs RHE and the mass activity (at 0.8 V vs. RHE) was 6.32 A/g. The activity was slightly better than that of commercially available Fe-N-C electrocatalysts—in particular, the mass diffusion limited current density was −5.56 mA/cm^2 for the Fe-N-C foam, compared with −5.11 mA/cm^2 for the commercially available Pt-free catalyst.

In alkaline solution, these Fe-N-C foams outperformed not only the commercially available Pt-free catalysts but also platinum. In KOH, Fe-N-C had an onset potential of ~1.0 V vs RHE, an impressive half-wave potential of 0.89 V vs RHE, a high mass diffusion limited current density of −5.7 mA cm^{-2} and a mass activity of 9 A/g. The cycling durability was measured, and the initial activity was very high for the first 10,000 potential cycles. However, after this, the performance degraded more quickly than the commercially available Fe-N-C catalyst.

As such, Fe-N-C foam is a highly promising material with the potential to compete with the best non-precious electrocatalysts in the literature and in industry. However, the stability should be improved for long-term applications.

In summary, the performance of metal-free and Fe-N-C foam electrocatalysts derived from nitrogen-doped carbon foams has steadily improved over the past 5 years (Table 8.1). This has been due to improved synthesis and processing techniques that have resulted in higher surface area, optimized porosity, improved conductivity and a larger number of active sites for the ORR. These carbon foam catalysts can now compete with other world-class Fe-N-C catalysts produced by other methods. This is all due to the unique controllable nature of sodium alkoxide-based carbon foams.

8.12 Pt-Decorated Nitrogen-Doped Carbon Foams for PEMFCs

Earlier, we saw how Liu et al. utilized carbon foam as a Pt support in PEMFC electrocatalysts. They followed up this work in 2014, decorating platinum on nitrogen-doped carbon foam [60]. Diethanolamine and ethanol were reacted with sodium. The alkoxide product was burned in air, crushed, sonicated, washed and dried. The samples were pyrolysed at 800 °C under nitrogen, washed and finally pyrolysed in hydrogen at 800 °C. The BET surface area was 705 m^2/g, and the nitrogen content was $\sim$3–6%. Pt decoration (20 wt%) was via thermal decomposition of bis(acetylacetonate)platinum(II). The Pt nanoparticle distribution and homogeneity were improved in the nitrogen-doped carbon foam, attributed to the enhanced nucleation of nanoparticles at nitrogen sites (Fig. 8.16). The initial electrochemical performance was slightly worse than un-doped carbon foams, due to the lower surface area and reduced electronic conductivity. However, the durability was significantly enhanced, with improved ECSA retention during potential cycling, attributed to increased binding energy between Pt and nitrogen-doped carbon.

In similar work by a different group, in 2017, Oh et al. utilized carbon foam as a support material for mesoporous nitrogen-doped carbon (essentially carbon nitride) and platinum [61]. Carbon foam was produced by rapid decomposition of sodium ethoxide at 800 °C for 1 min, followed by crushing, washing and drying. Carbon nitride was coated onto the carbon foam by impregnation with 1-ethyl-3-methylimidazolium dicyanamide, followed by a second pyrolysis at 600 °C for 1 min under ambient atmosphere. In this work, the high surface area and microporosity of alkoxide-derived carbon foams was attributed to self-activation of the carbon by the NaOH, Na_2O and Na_2O_2 formed during pyrolysis. The macroporous structure is clearly attributed to H_2O, CO and CO_2 gas evolution (i.e. foaming). The thickness of the carbon foam walls was 11.4 nm, and the coating thickness was 4–6 nm. The final surface area was 833 $m^2\ g^{-1}$, and Raman spectroscopy

Table 8.1 Comparison of different Pt-free ORR electrocatalysts based on carbon foams

Reference, year	Precursors	Pyrolysis temp. (°C)	Surface area (m^2/g)	Electrolyte	Onset potential (V vs. RHE)	Current density (@0.2 V vs. RHE) (mA/cm^2)	Half-wave potential (V vs. RHE)
[47], 2012	Ethanolamine	600	170	$HClO_4$	0.80	−2.0	0.35
[47], 2012	Ethanolamine	1000	700	$HClO_4$	0.80	−2.3	0.30
[58], 2014	Ethanol + Diethanolamine	1000	700	$HClO_4$	0.80	−2.9	0.45
[58], 2014	Ethanol + Diethanolamine	1400	700	$HClO_4$	0.85	−3.5	0.55
[57], 2014	Ethanol + Diethanolamine	1400	700	KOH	0.95	−4.0	0.80
[59], 2017	Ethanol + Triethanolamine	900	1200	$HClO_4$	0.88	−5.6	0.68
[59], 2017	Ethanol + Triethanolamine + Iron Acetate	900	1600	KOH	1.0	−5.7	0.89

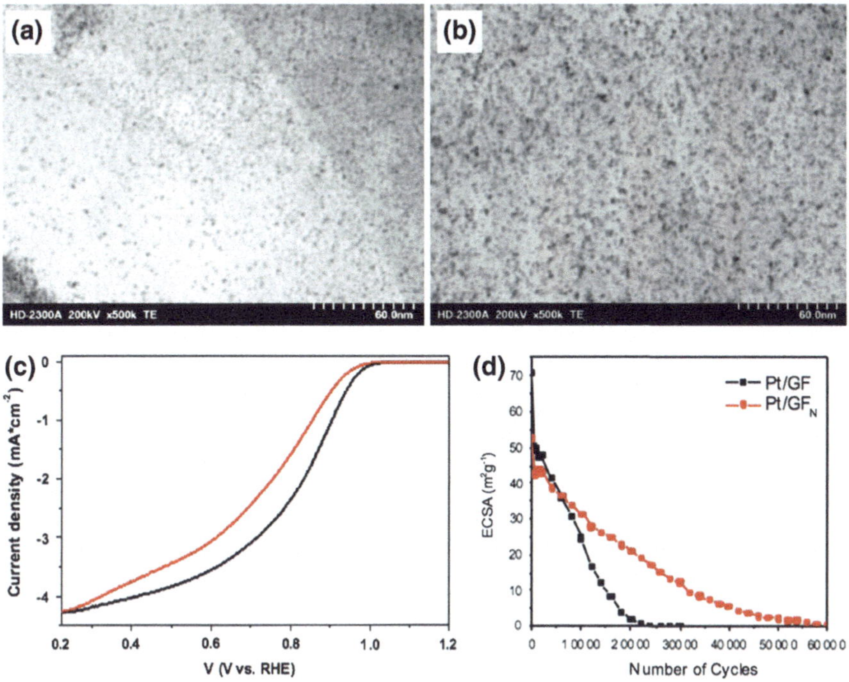

Fig. 8.16 TEM images of **a** Pt-decorated carbon foam, and **b** Pt-decorated nitrogen-doped carbon foam. **c** LSVs comparing the ORR activities. **d** Change in ECSA with potential cycling. Reproduced with permission from John Wiley and Sons [60]

revealed a highly defective material. Electrochemical impedance spectroscopy (EIS) revealed that carbon nitride coating resulted in increased electronic conductivity. However, the mechanism for this increase is unclear, since the conductivity of carbon nitride is much lower than that of pure carbon [10]. EIS is probably not an ideal technique to measure the conductivity of electrocatalyst powders.

This carbon nitride-coated carbon foam was decorated with Pt nanoparticles via electrochemical reduction of hexachloroplatinic acid. Uncoated, Pt-decorated carbon foam had reasonable initial ECSA of 60 m^2/g, which quickly decreased to <5 m^2/g after just 1500 potential cycles. In contrast, the coated composite carbon foam support displayed very low initial ECSA of <5 m^2/g, which increased to 50 m^2/g after 1000 potential cycles (Fig. 8.17). This 'activation' mechanism was attributed to differences in wetting and diffusion in the coated sample. However, such changes in the ECSA of electrocatalysts are commonly observed in the first few 1000 potential cycles, and therefore rigorous cycling tests are generally conducted over 10,000 s of cycles to draw meaningful conclusions [62]. Fuel cell membrane electrode assemblies (MEAs) were fabricated with impressive performance. The polarization curves revealed significantly enhanced mass diffusion in

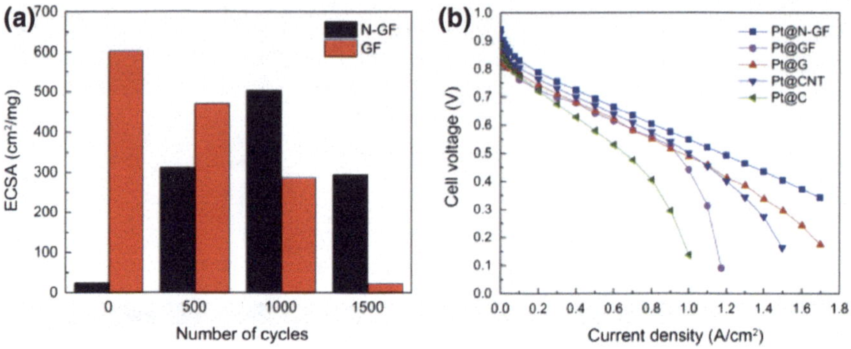

Fig. 8.17 **a** Electrochemical surface area (ECSA) of Pt-decorated carbon foam (red) and Pt-decorated carbon nitride-coated carbon foam. **b** Polarization curves of MEAs using Pt-decorated carbon nitride-coated carbon foams (blue), and others. Reproduced with permission from Elsevier [61]

the coated carbon foam samples, compared with carbon black, graphene or carbon nanotube supports. The power density was not reported but is estimated here to be ~ 600 mW/cm^2.

8.13 Fluorinated Carbon Foams

Finally (for now), carbon foams derived from sodium alkoxide have been doped with fluorine. In 2015, Lyth et al. reacted sodium metal with a long-chain fluorinated alcohol ($C_6F_{13}CH_2CH_2OH$) in a sealed PTFE pressure reactor, to form a fluorinated sodium alkoxide [63]. Decomposition of this resulted in the production of a fluorinated carbon foam. The microstructure was significantly different from that of the carbon foams or nitrogen-doped carbon foams mentioned above. In this case, instead of an open-cell carbon foam with micron-scale voids, clusters of closed-cell hollow nanoparticles were formed (Fig. 8.18). This difference in microstructure was attributed to the higher viscosity of the fluorinated alkoxide as it melted and foamed. Several crystalline graphitic phases were formed in the material, as observed by X-ray diffraction (XRD). The fluorine content was quite high at 17 at.%, and mainly covalently bonded to carbon. The fluorinated carbon powder was printed onto various substrates, and the water contact angle was measured to be 168°. This indicated that the material was superhydrophobic. These nanostructured, superhydrophobic carbon foams could have important applications in water management in electrochemical devices such as fuel cells and electrolysers.

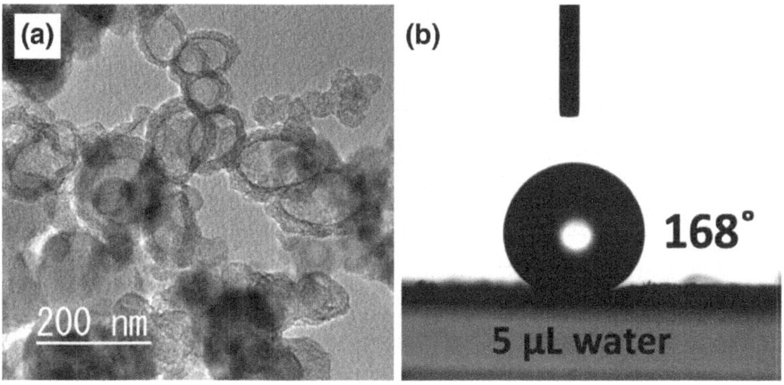

Fig. 8.18 **a** TEM image of fluorinated carbon foam. **b** Water contact angle measurements (168°). Reproduced with permission from The Royal Society of Chemistry [63]

8.14 Summary and Conclusions

Carbon foams derived from the decomposition of sodium alkoxides are interesting, low cost and highly versatile nanomaterials whose chemical structure can be straightforwardly tailored. The surface areas of these carbon foams have reached up to 2500 m^2/g, due to the very thin graphene-like carbon walls, and high degree of microporosity achieved through 'self-activation'. As well as being interesting from a purely materials science point-of-view, these novel materials have found use in a wide variety of potential energy-related applications. Pure carbon foams have been used for hydrogen storage and supercapacitors due to their large surface area. Sulphur-decorated carbon foams have been used as electrode materials in lithium-ion batteries. Functionalized carbon foams have been utilized as electrochemical sensors. Decorating carbon foams with platinum nanoparticles results in electrocatalysts with enhanced mass activity for the oxygen reduction reaction (ORR). Decorating carbon foams with silver nanoparticles results in enhanced activity for the carbon dioxide reduction reaction to carbon monoxide. Nitrogen-doped carbon foams have been used as metal-free electrocatalysts for the ORR, with reasonable activity and high durability, especially in alkaline solution. Adding iron to nitrogen-doped carbon foams results in highly active Fe-N-C foam electrocatalysts for the ORR that outperform other state-of-the-art Pt-free electrocatalysts in both acid and alkaline solutions. Fluorinated carbon foams are superhydrophobic, and have potential uses in water management in electrochemical energy conversion devices. With further improvements in materials design and new applications in the pipeline, these highly versatile carbon foams have the potential to improve the efficiency and viability of many energy-based applications, at a lower cost.

Acknowledgements The author gratefully acknowledges the support of the Kyushu University Platform of Inter/Transdisciplinary Energy Research (Q-PIT); and the International Institute for Carbon Neutral Energy Research (WPI-I2CNER), Kyushu University, Japan.

References

1. Kakati BK, Ghosh A, Verma A (2013) Efficient composite bipolar plate reinforced with carbon fiber and graphene for proton exchange membrane fuel cell. Int J Hydrogen Energy 38:9362–9369. https://doi.org/10.1016/j.ijhydene.2012.11.075
2. Bayer T, Pham H-CC, Sasaki K et al (2016) Spray deposition of Nafion membranes: electrode-supported fuel cells. J Power Sources 327:319–326. https://doi.org/10.1016/j.jpowsour.2016.07.059
3. Li G, Kaneko K, Ozeki S et al (1995) Water rejective nature of fluorinated microporous carbon fibers. Langmuir 11:716–717. https://doi.org/10.1021/la00003a008
4. Ross PN, Sokol H. The corrosion of carbon black anodes in alkaline electrolyte I. Acetylene black and the effect of cobalt catalyzation. US Nav Ammunit Depot PA Barnes E Kirton, Anal Calorim. J Therm Anal 8:99–133
5. Sarawan S, Wongwuttanasatian T (2013) A feasibility study of using carbon black as a substitute to coal in cement industry. Energy Sustain Dev 17:257–260. https://doi.org/10.1016/J.ESD.2012.11.006
6. Marinho B, Ghislandi M, Tkalya E et al (2012) Electrical conductivity of compacts of graphene, multi-wall carbon nanotubes, carbon black, and graphite powder. Powder Technol 221:351–358. https://doi.org/10.1016/j.powtec.2012.01.024
7. Grahame TJ, Klemm R, Schlesinger RB (2014) Public health and components of particulate matter: the changing assessment of black carbon. J Air Waste Manag Assoc 64:620–660. https://doi.org/10.1080/10962247.2014.912692
8. Fujigaya T, Nakashima N (2013) Fuel cell electrocatalyst using polybenzimidazole-modified carbon nanotubes as support materials. Adv Mater 25:1666–1681. https://doi.org/10.1002/adma.201204461
9. Fujigaya T, Uchinoumi T, Kaneko K, Nakashima N (2011) Design and synthesis of nitrogen-containing calcined polymer/carbon nanotube hybrids that act as a platinum-free oxygen reduction fuel cell catalyst. Chem Commun (Camb) 47:6843–6845. https://doi.org/10.1039/c1cc11303h
10. Lyth SM, Nabae Y, Islam NM et al (2012) Oxygen reduction activity of carbon nitride supported on carbon nanotubes. J Nanosci Nanotechnol 12:4887–4891. https://doi.org/10.1166/jnn.2012.4947
11. Lyth SM, Silva SR (2007) Field emission from multiwall carbon nanotubes on paper substrates. Appl Phys Lett 90:173124
12. Daio T, Staykov A, Guo L et al (2015) Lattice strain mapping of platinum nanoparticles on carbon and SnO_2 supports. Sci Rep 5:13126. https://doi.org/10.1038/srep13126
13. Chen D, Tang L, Li J (2010) Graphene-based materials in electrochemistry. Chem Soc Rev 39:3157. https://doi.org/10.1039/b923596e
14. Brownson DAC, Banks CE (2010) Graphene electrochemistry: an overview of potential applications. Analyst 135:2768. https://doi.org/10.1039/c0an00590h
15. Bayer T, Bishop SR, Perry NH et al (2016) Tunable mixed ionic/electronic conductivity and permittivity of graphene oxide paper for electrochemical energy conversion. ACS Appl Mater Interfaces 8:11466–11475. https://doi.org/10.1021/acsami.6b01670
16. Inagaki M, Qiu J, Guo Q (2015) Carbon foam: preparation and application. Carbon N Y 87:128–152. https://doi.org/10.1016/J.CARBON.2015.02.021
17. Chen C, Kennel EB, Stiller AH et al (2006) Carbon foam derived from various precursors. Carbon N Y 44:1535–1543. https://doi.org/10.1016/J.CARBON.2005.12.021
18. Lin Q, Luo B, Qu L et al (2013) Direct preparation of carbon foam by pyrolysis of cyanate ester resin at ambient pressure. J Anal Appl Pyrolysis 104:714–717. https://doi.org/10.1016/J.JAAP.2013.05.007
19. Tondi G, Fierro V, Pizzi A, Celzard A (2009) Tannin-based carbon foams. Carbon N Y 47:1480–1492. https://doi.org/10.1016/J.CARBON.2009.01.041

20. Inagaki M, Morishita T, Kuno A et al (2004) Carbon foams prepared from polyimide using urethane foam template. Carbon N Y 42:497–502. https://doi.org/10.1016/J.CARBON.2003.12.080
21. Lei S, Guo Q, Shi J, Liu L (2010) Preparation of phenolic-based carbon foam with controllable pore structure and high compressive strength. Carbon N Y 48:2644–2646. https://doi.org/10.1016/J.CARBON.2010.03.017
22. Hayashi A, Kimijima K, Miyamoto J, Yagi I (2009) Oxygen transfer and storage processes inside the mesopores of platinum-deposited mesoporous carbon catalyst thin-layer electrode. J Phys Chem C 113:12149–12153. https://doi.org/10.1021/jp901298r
23. Hayashi A, Notsu H, Kimijima K et al (2008) Preparation of Pt/mesoporous carbon (MC) electrode catalyst and its reactivity toward oxygen reduction. Electrochim Acta 53:6117–6125. https://doi.org/10.1016/J.ELECTACTA.2008.01.110
24. Suryavanshi UB, Ijima T, Hayashi A et al (2011) Simple methods for tuning the pore diameter of mesoporous carbon. Chem Commun 47:10758. https://doi.org/10.1039/c1cc13471j
25. Serov A, Workman MJ, Artyushkova K et al (2016) Highly stable precious metal-free cathode catalyst for fuel cell application. J Power Sources 327. https://doi.org/10.1016/j.jpowsour.2016.07.087
26. Gokhale R, Chen Y, Serov A et al (2017) Novel dual templating approach for preparation of highly active Fe-N-C electrocatalyst for oxygen reduction. Electrochim Acta 224:49–55. https://doi.org/10.1016/j.electacta.2016.12.052
27. Choucair M, Thordarson P, Stride Ja (2009) Gram-scale production of graphene based on solvothermal synthesis and sonication. Nat Nanotechnol 4:30–33. https://doi.org/10.1038/nnano.2008.365
28. Daio T, Bayer T, Ikuta T et al (2015) In-situ ESEM and EELS observation of water uptake and ice formation in multilayer graphene oxide. Sci Rep 5:11807. https://doi.org/10.1038/srep11807
29. Marinho B, Ghislandi M, Tkalya E et al (2012) Electrical conductivity of compacts of graphene, multi-wall carbon nanotubes, carbon black, and graphite powder. Powder Technol 221:351–358. https://doi.org/10.1016/J.POWTEC.2012.01.024
30. Xu X, Dou SX, Wang XL et al (2010) Graphene doping to enhance the flux pinning and supercurrent carrying ability of a magnesium diboride superconductor. Supercond Sci Technol 23:85003. https://doi.org/10.1088/0953-2048/23/8/085003
31. Choucair M, Tse NMK, Hill MR, Stride JA (2012) Adsorption and desorption characteristics of 3-dimensional networks of fused graphene. Surf Sci 606:34–39. https://doi.org/10.1016/J.SUSC.2011.08.016
32. Náfrádi B, Choucair M, Forró L (2014) Spin lifetime of itinerant electrons in chemically synthesized graphene multi-layers. Carbon N Y 74:346–351. https://doi.org/10.1016/J.CARBON.2014.03.046
33. Riccò M, Pontiroli D, Mazzani M et al (2011) Muons probe strong hydrogen interactions with defective graphene. Nano Lett 11:4919–4922. https://doi.org/10.1021/nl202866q
34. Speyer L, Fontana S, Cahen S et al (2015) Multi-scale characterization of graphenic materials synthesized by a solvothermal-based process: influence of the thermal treatment. Solid State Sci 50:42–51. https://doi.org/10.1016/j.solidstatesciences.2015.10.009
35. Sevilla M, Fuertes AB (2006) Catalytic graphitization of templated mesoporous carbons. Carbon N Y 44:468–474. https://doi.org/10.1016/j.carbon.2005.08.019
36. Watts PCP, Lyth SM, Henley SJ, Silva SRP (2008) Secondary nanotube growth on aligned carbon nanofibre arrays for superior field emission. J Nanosci Nanotechnol 8:2147–2150
37. Rosen A, Taub A (1962) The kinetics of graphitization in steel at subcritical temperatures. Acta Metall 10:501–509. https://doi.org/10.1016/0001-6160(62)90193-1
38. Speyer L, Fontana S, Ploneis S, Hérold C (2017) Influence of the precursor alcohol on the adsorptive properties of graphene foams elaborated by a solvothermal-based process. Microporous Mesoporous Mater 243:254–262. https://doi.org/10.1016/j.micromeso.2017.02.035

39. Lyth SM, Shao H, Liu J, Sasaki K, Akiba E (2014) Hydrogen adsorption on graphene foam synthesized by combustion of sodium ethoxide. Int J Hydrogen Energy 39:376
40. Choucair M, Mauron P (2015) Versatile preparation of graphene-based nanocomposites and their hydrogen adsorption. Int J Hydrogen Energy 40:6158–6164. https://doi.org/10.1016/J.IJHYDENE.2015.03.065
41. Cui H, Zheng J, Yang P et al (2015) Understanding the formation mechanism of graphene frameworks synthesized by solvothermal and rapid pyrolytic processes based on an alcohol-sodium hydroxide system. ACS Appl Mater Interfaces 7:11230–11238. https://doi.org/10.1021/acsami.5b01201
42. Chou S-L, Wang J-Z, Choucair M et al (2010) Enhanced reversible lithium storage in a nanosize silicon/graphene composite. Electrochem Commun 12:303–306. https://doi.org/10.1016/J.ELECOM.2009.12.024
43. Wang J-Z, Lu L, Choucair M et al (2011) Sulfur-graphene composite for rechargeable lithium batteries. J Power Sources 196:7030–7034. https://doi.org/10.1016/J.JPOWSOUR.2010.09.106
44. Kafi AKM, Yusoff MM, Choucair M, Crossley MJ (2017) A conductive crosslinked graphene/cytochrome c networks for the electrochemical and biosensing study. J Solid State Electrochem 21:2761–2767. https://doi.org/10.1007/s10008-017-3598-z
45. Liu J, Takeshi D, Sasaki K, Lyth SM (2014) Defective graphene foam: a platinum catalyst support for PEMFCs. J Electrochem Soc 161:F838–F844. https://doi.org/10.1149/2.0231409jes
46. Ma S, Liu J, Sasaki K et al (2017) Carbon foam decorated with silver nanoparticles for electrochemical CO_2 conversion. Energy Technol. https://doi.org/10.1002/ente.201600576
47. Lyth SMSM, Nabae Y, Islam NMNM et al (2012) Solvothermal synthesis of nitrogen-containing graphene for electrochemical oxygen reduction in acid media
48. Workman MJ, Serov A, Tsui L et al (2017) Fe–N–C catalyst graphitic layer structure and fuel cell performance. ACS Energy Lett 2:1489–1493. https://doi.org/10.1021/acsenergylett.7b00391
49. Chen Y, Gokhale R, Serov A et al (2017) Novel highly active and selective Fe-N-C oxygen reduction electrocatalysts derived from in-situ polymerization pyrolysis. Nano Energy. https://doi.org/10.1016/j.nanoen.2017.05.059
50. Strickland K, Miner E, Jia Q et al (2015) Highly active oxygen reduction non-platinum group metal electrocatalyst without direct metal-nitrogen coordination. Nat Commun 6:7343. https://doi.org/10.1038/ncomms8343
51. Dodelet J-P (2013) The controversial role of the metal in Fe- or Co-based electrocatalysts for the oxygen reduction reaction in acid medium. In: Shao M, Higgins D, Chen Z (eds) Electrocatalysis in fuel cells. Springer, London, London, pp 271–338
52. Zitolo A, Goellner V, Armel V et al (2015) Identification of catalytic sites for oxygen reduction in iron- and nitrogen-doped graphene materials. Nat Mater 14:937–942. https://doi.org/10.1038/nmat4367
53. Varnell JA, Tse ECM, Schulz CE et al (2016) Identification of carbon-encapsulated iron nanoparticles as active species in non-precious metal oxygen reduction catalysts. Nat Commun 7:12582. https://doi.org/10.1038/ncomms12582
54. Dai L, Xue Y, Qu L et al (2015) Metal-free catalysts for oxygen reduction reaction. Chem Rev. https://doi.org/10.1021/cr5003563
55. Lepri FG, Borges DLG, Araujo RGO et al (2010) Determination of heavy metals in activated charcoals and carbon black for Lyocell fiber production using direct solid sampling high-resolution continuum source graphite furnace atomic absorption and inductively coupled plasma optical emission spectrometry. Talanta 81:980–987
56. Qu L, Liu Y, Baek J-BJB, Dai L (2010) Nitrogen-doped graphene as efficient metal-free electrocatalyst for oxygen reduction in fuel cells. ACS Nano 4:1321–1326. https://doi.org/10.1021/nn901850u

57. Liu J, Cunning BV, Daio T et al (2016) Nitrogen-doped carbon foam as a highly durable metal-free electrocatalyst for the oxygen reduction reaction in alkaline solution. Electrochim Acta. https://doi.org/10.1016/j.electacta.2016.10.090
58. Liu J, Takeshi D, Orejon D et al (2014) Defective nitrogen-doped graphene foam: a metal-free, non-precious electrocatalyst for the oxygen reduction reaction in acid. J Electrochem Soc 161:F544–F550. https://doi.org/10.1149/2.095404jes
59. Mufundirwa A, Harrington GF, Smid B et al (2017) Durability of template-free Fe-N-C foams for electrochemical oxygen reduction in alkaline solution. J Power Sources. https://doi.org/10.1016/j.jpowsour.2017.07.025
60. Liu J, Takeshi D, Sasaki K, Lyth SM (2014) Platinum-decorated nitrogen-doped graphene foam electrocatalysts. Fuel Cells 14:728–734. https://doi.org/10.1002/fuce.201300258
61. Oh E-J, Hempelmann R, Nica V et al (2017) Coating procedure for chemical and morphological functionalization of multilayer-graphene foams. Carbon N Y 121:170–180. https://doi.org/10.1016/J.CARBON.2017.05.058
62. Ohma A, Shinohara K, Iiyama A et al (2011) Membrane and catalyst performance targets for automotive fuel cells by FCCJ membrane, catalyst, MEA WG. In: ECS transactions. The Electrochemical Society, pp 775–784
63. Lyth SM, Ma W, Liu J et al (2015) Solvothermal synthesis of superhydrophobic hollow carbon nanoparticles from a fluorinated alcohol. Nanoscale 7:16087–16093. https://doi.org/10.1039/c5nr03484a

Chapter 9
Hydrogen-Evolving CNT-Photocatalysts for Effective Use of Solar Energy

Yutaka Takaguchi, Tomoyuki Tajima and Hideaki Miyake

9.1 Introduction

Overall water splitting using photocatalysts has attracted considerable interest due to its potential for CO_2-free production of hydrogen from water by utilizing abundant solar light energy [1–7]. To improve the productivity of hydrogen of the overall water splitting, expanding active wavelengths of the photocatalyst from UV light to visible or near-IR (NIR) lights is required, because maximum solar light conversion efficiencies depend on the active wavelength region in the solar spectrum, e.g., ca. 2% (300 nm < λ < 400 nm), 16% (300 nm < λ < 600 nm), and 32% (300 nm < λ < 800 nm) [8]. In this context, Z-scheme photocatalysis systems, i.e., two-step photoexcitation processes using an O_2-evolving photocatalyst and a H_2-evolving photocatalyst combined with a redox shuttle have been developed [9–12]. However, the active wavelengths of the photocatalytic water splitting are not at satisfactory level. For instance, Kudo and coworkers described an efficient Z-scheme photocatalytic system based on $BiVO_4$ as an O_2-evolving photocatalyst, Ru/$SrTiO_3$:Rh as a H_2-evolving photocatalyst, and $[Co(bpy)_3]^{2+/3+}$ as a redox couple exhibited overall water-splitting activity under the illumination by the light with wavelength less than 520 nm [13]. In this chapter, a novel category of H_2-evolving photocatalysts based on semiconducting SWCNTs (s-SWCNTs) for the

Y. Takaguchi (✉) · T. Tajima
Graduate School of Environmental and Life Science, Okayama University,
Okayama 700-8530, Japan
e-mail: yutaka@cc.okayama-u.ac.jp

T. Tajima
e-mail: tajimat@cc.okayama-u.ac.jp

H. Miyake
Graduate School of Sciences and Technology for Innovation, Yamaguchi University,
Ube 755-8611, Japan
e-mail: h-miyake@yamaguchi-u.ac.jp

N. Nakashima (ed.), *Nanocarbons for Energy Conversion: Supramolecular Approaches*, Nanostructure Science and Technology,
https://doi.org/10.1007/978-3-319-92917-0_9

photocatalytic overall water splitting is described. These CNT-photocatalysts shows H_2-evolving activity under visible and even NIR illumination since s-SWCNTs act as light absorbers. Although several examples of H_2-evolution reaction (HER) using a photocatalytic system containing SWCNTs as a conducting material have been reported [14–16], HER triggered by the photoexcitation of s-SWCNTs has never been proved until our report in 2017 [17–19].

s-SWCNTs have very unique properties that span organic and inorganic semiconducting materials [20–29]. Importantly, s-SWCNTs are quite efficient light absorbers from the viewpoint of solar energy harvesting. Because the absorption bands of s-SWCNTs are depending on their chiral indices (*n*,*m*) [30, 31], a hypothetical 150 nm thick film comprised of ten different particular s-SWCNT species would absorb 86% of solar spectrum below 1.1 eV [32]. Tune and Shapter reported the potential power conversion efficiency of a hypothetical tandem device fabricated from the combination of four single-species, (6,4), (9,1), (7,3), and (7,5)tubes, could be estimated to as high as 28% [33]. These calculations highlight the potential of solar energy harvesting through the highly absorptive s-SWCNT excitonic transitions. In this context, Blackburn reviewed various photovoltaic systems, except for photocatalysts, based on s-SWCNT for the effective use of solar energy [34–36].

H_2-evolving CNT-photocatalysts based on s-SWCNTs as a solar light absorber are quite rare due to several obstacles: (1) Large exciton dissociation energy of s-SWCNT, (2) Bundle formation and energy migration between s-SWCNTs, (3) Single helicity is necessary for high performance, and (4) Low dispersibility in water. To solve these problems, we propose using a coaxial heterojunction nanowire structure based on s-SWCNTs as a core and fullerodendron as a shell material (Fig. 9.1). From the viewpoint of photovoltaic (PV) device application of s-SWCNTs, s-SWCNT/C_{60} heterojunctions are necessary for the high internal quantum efficiencies (IQEs) of the solar cells [37–40], because the exciton dissociation at the donor–acceptor (SWCNT-C_{60}) interface, which occurs due to the band offset energy between the conduction band of SWCNT and the LUMO of C_{60}, derives the generation of the mobile carriers, holes and electrons, despite the large binding energy (>100 meV) of the excitons produced by the optical absorption of

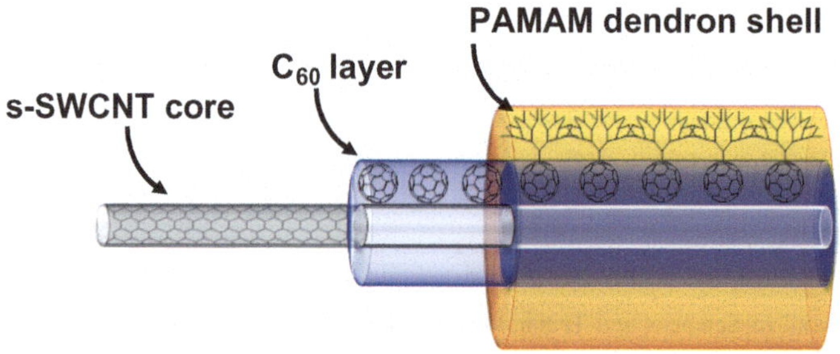

Fig. 9.1 The structure of a typical CNT-photocatalyst having a coaxial SWCNT/C_{60} heterojunction

the s-SWCNTs [41–43]. Therefore, s-SWCNT/C_{60} heterojunction of the coaxial nanowire could generate the electron that be consumed by HER at the co-catalyst. Furthermore, the coaxial nanowire structure serves the isolation of the individual s-SWCNTs very stable. Hence, energy migration and/or carrier trapping because of the interaction between s-SWCNTs with different helicities will never occur. In other words, time-consuming purification processes of s-SWCNTs to obtain a highly pure single helicity of SWCNT is unnecessary. The shell of the coaxial nanowire acts as not only a protecting layer of the SWCNT/C_{60} heterojunction but also a dispersant layer in water.

9.2 Coaxial Heterojunction Nanowires: CNT-Photocatalysts

The coaxial SWCNT/C_{60} heterojunction with s-SWCNT-core was fabricated by a physical modification of s-SWCNTs using an amphiphilic fullerodendron [44]. Schematic illustration of the formation of SWCNT/fullerodendron supramolecular nanocomposites via self-organization of the fullerodendron on the lateral surface of the SWCNT is shown in Fig. 9.2. In a typical experiment, SWCNTs (1.0 mg) was added to a water solution (10 mL) of fullerodendron (25.5 mg, 0.01 μmol), which was then sonicated in a bath-type ultrasonicator at 17–25 °C for 3 h. After the suspension was centrifuged (3000 g, 30 min), the black supernatant dispersion containing SWCNT/fullerodendron was collected and purified by dialysis for 3 days to remove any excess of fullerodendron molecules. AFM and TEM observations of SWCNT/fullerodendron nanocomposites reveal nanofibers with diameters of 2–5 nm consisting with their coaxial nanowire structures [45]. The isolation

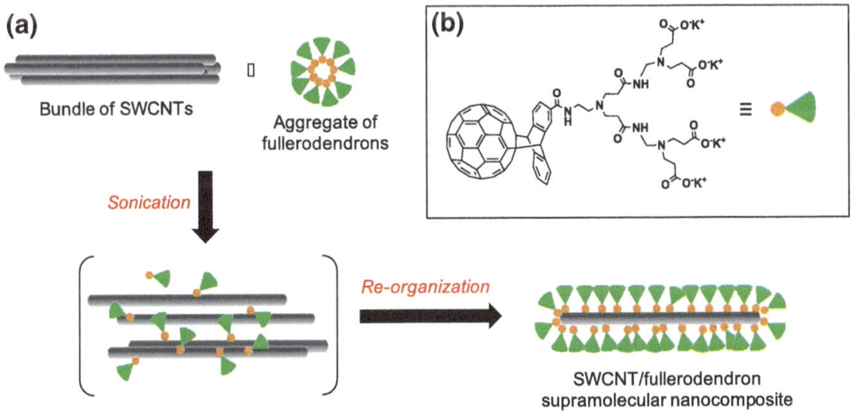

Fig. 9.2 **a** Schematic illustration of the formation of a SWCNT/fullerodendron supramolecular nanocomposite that was used for a CNT-photocatalyst. **b** The structure of fullerodendron

of the individual s-SWCNT is evidenced by 2D photoluminescence spectra of SWCNT/fullerodendron nanocomposites in D_2O [46].

Photophysical properties of a coaxial heterojunction made by a physical modification of SWCNTs were first reported by Nakashima and colleagues [47]. They observed the fluorescence quenching of pyrene derivatives that adsorbed on the SWCNTs [48]. Photoinduced electron transfer (PET) process of a physically modified SWCNT with a coaxial SWCNT/anthracene heterojunction was then investigated by means of a fluorescence decay profile of the anthracene moiety and nanosecond transient absorption spectra (Fig. 9.3a) [49]. Upon 355-nm laser light irradiation, the PET from an amphiphilic anthryl dendron to the SWCNT-core afforded the radical ion pair ($SWCNT^{\bullet -}$/$anthracene^{\bullet +}$), of which lifetime (τ) was calculated to be 1000 ns. This result indicated that the s-SWCNT-core might hamper the charge recombination due to its high charge carrier mobility (>100 000 cm^2/Vs) [50]. Furthermore, PET of a coaxial heterojunction between SWCNT and C_{60} has been reported by the use of SWCNT/fullerodendron supramolecular nanocomposites (Fig. 9.3b) [51, 52]. On addition of methylviologene dication (MV^{2+}) as an electron pooling reagent, the radical cation $MV^{\bullet +}$ was accumulated by visible light illumination of the SWCNT/fullerodendron supramolecular nanocomposites. Nanosecond transient absorption spectra shows the increase of $MV^{\bullet +}$ (620 nm) was observed as the decrease of $C_{60}^{\bullet -}$ (1020 nm) (Fig. 9.3c). The fluorescence lifetimes of the fullerene moiety and efficient accumulation of $MV^{\bullet +}$ suggest that the photosensitized charge separation occurs between C_{60} and SWCNTs, followed by electron-mediating and hole-shifting

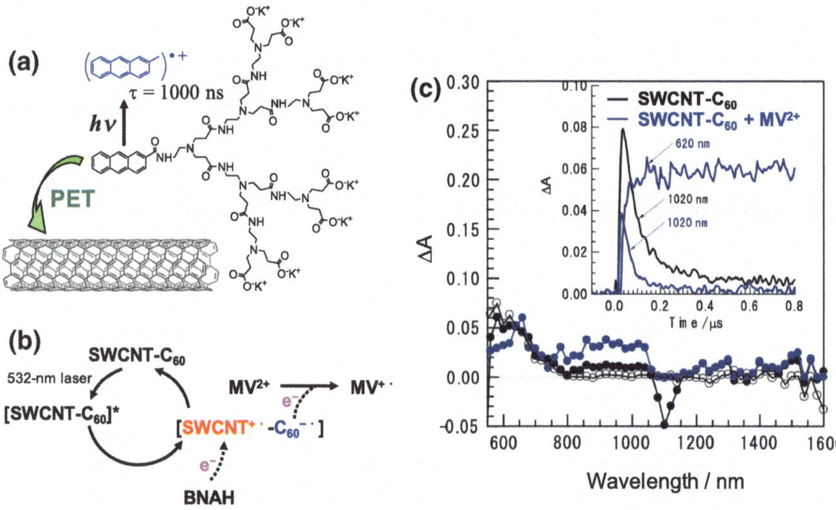

Fig. 9.3 **a** PET from anthryl dendron to SWCNT leading to the generation of anthracene cation radical. **b** PET from SWCNT-C_{60} heterojunction to MV^{2+}. **c** Nanosecond transient absorption spectra of SWCNT/C_{60} in D_2O observed by 532-nm laser irradiation. Inset: Absorption time profile

processes, resulting in the persistent $MV^{\bullet+}$. This result indicated that SWCNT/C_{60} coaxial heterojunctions could act as photosensitizers of photoredox reaction, such as HER from water.

9.3 Shell Modification of the CNT-Photocatalysts

Since the accumulation of $MV^{\bullet+}$ can be used for HER from water in the presence of Pt nanoparticles [53], we explored the HER using the SWCNT/fullerodendron nanocomposites as photosensitizers (Fig. 9.4). Upon photoirradiation ($\lambda > 422 \pm 5$ nm), continuous and steady HER (5.3 μmol/h) was observed using a water solution (150 mL) of SWCNT/fullerodendron nanocomposites (SWCNT content was 0.48 mg), MV^{2+} (2.4 mM), benzyldihydronicotinamide (BNAH; 1.2 mM), and PVA-Pt (740 μM). The apparent quantum yield (AQY) of this photosensitized H_2 evolution was estimated to be 0.28 under 450-nm light illumination. Furthermore, the AQY of HER can be improved by shell modification of the coaxial heterojunction nanowire. Since there are many functional groups on the surface of SWCNT/fullerodendron nanocomposites, such as carboxy and amine groups of the poly(amidoamine) (PAMAM) dendron moieties, a chemical modification of the coaxial nanowire is quite easy. After introduction of the SiO_2 shell into the SWCNT/fullerodendron by sol-gel processing using tetraethoxy silane (Si $(OEt)_4$), the AQY of HER improved to 0.31 from 0.28 (AQY of SWCNT/fullerodendron) [54]. Moreover, the electron extracting TiO_x layer, which is incorporated into the shell of SWCNT/fullerodendron using a sol-gel condensation of $Ti(O_iPr)_4$, is quite effective to boost the HER from water, of which AQY is estimated to be 0.47 under 450-nm light illumination [55]. It is notable that the thicknesses of metal oxide layers of SWCNT/fullerodendron/SiO_2 and SWCNT/fullerodendron/TiO_x are easily controlled to thin (10–30 nm) owing to the scaffold

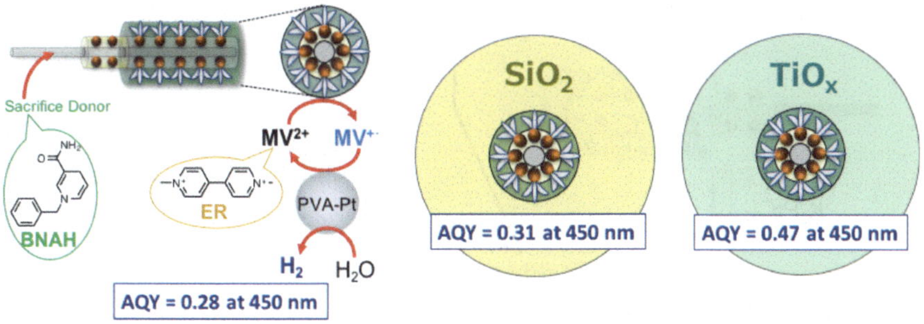

Fig. 9.4 AQYs of HER using CNT-photocatalysts, SWCNT/fullerodendron, SWCNT/fullerodendron/SiO_2, and SWCNT/fullerodenron/TiO_x

effect of the dendrimer moieties of the coaxial nanowires. As the result, the light absorption of the SWCNT-core, that is important for HER activity, is not interrupted by the scattering of the metal oxide shell.

9.4 Co-catalyst Loading on the CNT-Photocatalysts

The PAMAM dendron moieties of CNT-photocatalysts can be used for the coordination site of the metals as well as the scaffold of the sol-gel processing [56–59]. Since Crooks and colleagues reported the fabrication and electrocatalytic activity of versatile metal nanoparticles stabilized by PAMAM dendrimers via the formation of a metal complex with PAMAM dendrimer [60, 61], the same strategy could employ for the construction of an interconnecting system, where the photosensitizer (SWCNT/fullerodendron) and co-catalyst (metal complex or metal nanoparticle) are directly connected through the dendron moiety without the help of an electron relay molecule, such as MV^{2+}. In contrast to the three-component system consisting of the photosensitizer, the electron relay molecule, and the co-catalyst, the interconnecting system can be used for the H_2-evolving photocatalyst as a component of a Z-scheme photocatalytic system that is two-step photoexcitation process based on a H_2-evolving photocatalyst and O_2-evolving photocatalyst combined with a redox shuttle.

Figure 9.5 shows the schematic illustration of the fabrication of the interconnecting system, i.e., CNT-photocatalysts having Pt(II) complex on the shell. Complexation between K_2PtCl_4 and SWCNT/fullerodendron supramolecular

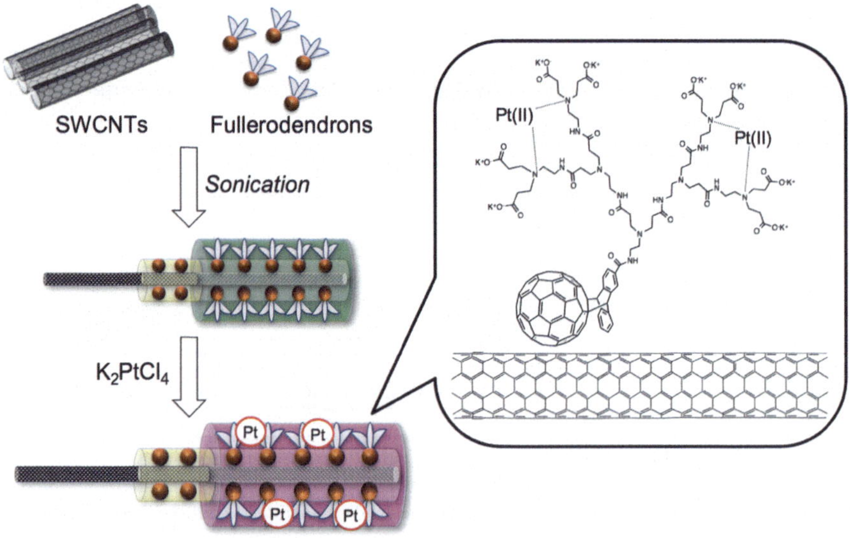

Fig. 9.5 Complexation of the shell of the fullerodendron with Pt(II) to afford the interconnecting system, SWCNT/fullerodendron/Pt(II)

nanocomposites afforded SWCNT/fullerodendron/Pt(II) coaxial nanowires exhibiting H_2 evolving activity, of which AQY was 0.16 upon irradiation of monochromatic light at 450 nm [62]. Although this AQY (0.16) of the SWCNT/fullerodendron/Pt(II) is a bit lower than the AQY of the SWCNT/fullerodendron (0.28) because of the absorption of Pt(II) complexes at the dendron moieties of the CNT-photocatalyst, the H_2-evoluving rate of the SWCNT/fullerodendron/Pt(II) (8.6 μmol/h) is higher than that of SWCNT/fullerodendron (5.3 μmol/h). This result indicates that the direct incorporation of a co-catalyst into the shell of CNT-photocatalyst is useful to construct the interconnecting system for HER that could be applied for the overall water splitting. The high-hole mobility on SWCNTs might contribute the interruption of the charge recombination between the co-catalyst and SWCNT-core.

9.5 Controlling Active Wavelengths by Changing the SWCNT-Core

From the viewpoint of the overall water splitting using a Z-scheme photocatalytic system (two-step photoexcitation system), the wavelength of 1000 nm might be a challenging target of the active wavelength of photocatalytic HER. Because the wavelength of 1000 nm, which equals to 1.23 eV, is the minimal energy to fulfill the redox potentials of the H^+/H_2 and O_2/H_2O pairs. In other words, in order to use the wavelength longer than 1000 nm for HER, the construction of a Z-scheme photocatalytic system is requisite. In this context, the CNT-photocatalyst having (8,3)tube at the core is of interest, since the absorption bands of (8,3)tube are at 680 nm and 1000 nm (Fig. 9.6a). (8,3)SWCNT/fullerodendron and (8,3)SWCNT/fullerodendron/Pt(II) complex were fabricated by the use of commercially available (6,5)-enriched SWCNTs, which contain (6,5), (7,5), and (8,3)tubes [17]. The AQYs

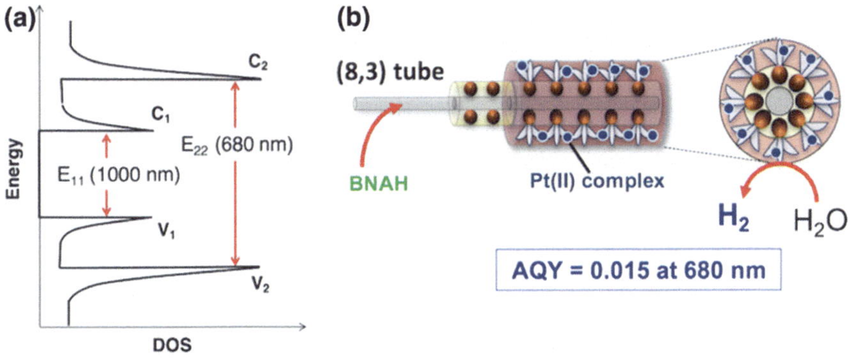

Fig. 9.6 **a** DOS of (8,3)tube at the core of the SWCNT/fullerodendron nanocomposite. **b** Schematic illustration of HER using an interconnecting system, (8,3)tube/fullerodendron/Pt(II)

of HER using (8,3)SWCNT/fullerodendron were estimated to be 0.17 and 0.073 under 680-nm and 1000-nm light illumination, respectively. Importantly, this is the first example of HER triggered by the photoexcitation of SWCNTs, since absorption of the light by C_{60} moieties of the CNT-photocatalyst could occur in previous cases [54, 55, 62].

Compared to good AQYs of HER using (8,3)SWCNT/fullerodendron, (8,3) SWCNT/fullerodendron/Pt(II) showed a relatively low AQY, which was 0.015 (λ_{irr} = 680 nm) (Fig. 9.6b). Since efficient PET from SWCNT-core to Pt(II) complexes was evidenced by photoluminescence spectra exhibiting the quenching of the E_{11} emission of (8,3)SWCNT after the Pt(II) loading on the shell of the (8,3) SWCNT/fullerodendron nanocomposite, one-tenth efficiency of (8,3)SWCNT/ fullerodendron/Pt(II) compared to (8,3)SWCNT/fullerodendron might attributed to the difference of co-catalyst activity between Pt(II) complexes and Pt(0) nanoparticles. It is notable that these AQYs under NIR-light illuminations are quite high even by using (8,3)SWCNT/fullerodendron/Pt(II), so CNT-photocatalysts are potentially useful for the photocatalytic overall water-splitting system by the use of the solar energy (Fig. 9.7). But, unfortunately, the photocatalytic activity of the (8,3)SWCNT/fullerodendron/Pt(II) coaxial nanowire is not sufficient to construct a Z-scheme photocatalytic system. Recently, the author's group has found a highly efficient metal co-catalyst for CNT-photocatalysts that is capable to act as a H_2-evolving photocatalyst in the Z-scheme photocatalytic system: (CNT-photocatalysts)-($BiVO_4$)-($[Co(bpy)_4]^{2+/3+}$).

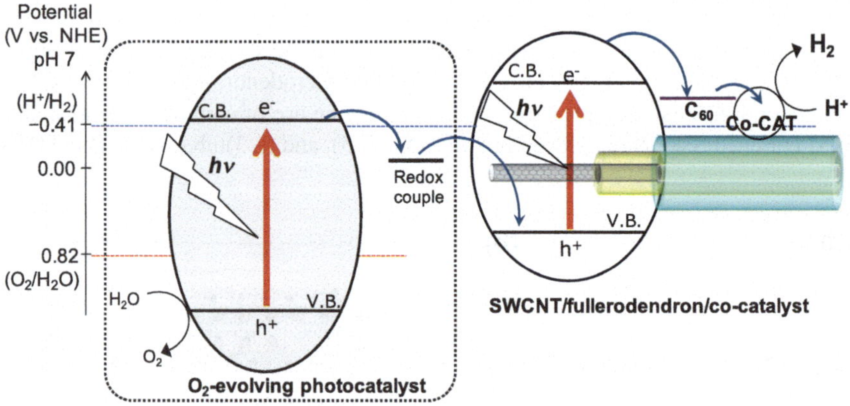

Fig. 9.7 Possible Z-scheme water splitting system based on the CNT-photocatalyst and an O_2-evolving photocatalyst

9.6 Stepping over the Flavel's Limitation of the SWCNT-Core

Although CNT-photocatalysts show high HER activities upon photoirradiation by visible and NIR lights, they cannot use all areas of the solar spectrum because of the diameter limit set by Flavel and colleagues [63]. They reported that the diameter limit for SWCNTs in SWCNT/C_{60} solar cells is 0.95 nm ((8,6)tube) in terms of the s-SWCNTs' ability to generate mobile charge carriers upon photoirradiation of the s-SWCNT/C_{60} heterojunction. In order to overstep the Flavel's limitation, a novel heterojunction generating mobile carriers to produce H_2 is required. In this context, the authors proposed two strategies by the use of (1) spontaneous exciton dissociation in water-dispersible CNT-photocatalysts, (2) three-component heterojunction based on dye-encapsulated s-SWCNTs and C_{60}, wherein the s-SWCNT acts as an intermediate layer.

At this point, CNT-photocatalysts exhibit HER activity upon the photoexcitation of SWCNT-core via the PET process from s-SWCNT to C_{60} after the exciton dissociation and the mobile carrier generation at the s-SWCNT/C_{60} coaxial heterojunction. However, it remains unclear whether the C_{60} layer is necessary for the dissociation of excitons. As the exciton binding energy strongly depends on the surrounding dielectric constant (ε) of the environment [41–43], the exciton dissociation could be facilitated in aqueous ($\varepsilon_{H2O} = 80$) dispersion systems, relative to the case of photovoltaic film devices, wherein s-SWCNTs are surrounded by the acceptor C_{60} ($\varepsilon_{C60} = 4.4$) [63, 64]. To obtain a water-dispersible SWCNT/dendrimer nanocomposite that does not contain the s-SWCNT/C_{60} heterojunction, the poly(amidoamine) dendrimer having a 1,10-bis(decyloxy)decane core and carboxylate (COO^-) terminals was used for the construction a new CNT-photocatalyst (Fig. 9.8) [18]. Surprisingly, 3.7 μmol/h of H_2 was generated upon photoirradiation (λ > 422 nm) using SWCNT/dendrimer nanocomposites as photosensitizers, BNAH, MV^{2+}, and Pt-PVP. For a comparison between pristine SWCNT and the SWCNT/C_{60} heterojunctions, we evaluated the AQYs of HERs by using SWCNT/dendrimer and SWCNT/fullerodendron supramolecular

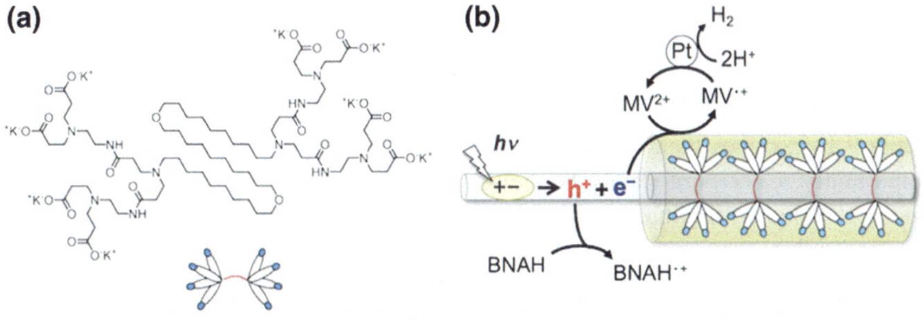

Fig. 9.8 **a** Molecular structure of the dendritic dispersant. **b** HER via spontaneous exciton dissociation using SWCNT/dendrimer supramolecular nanocomposite

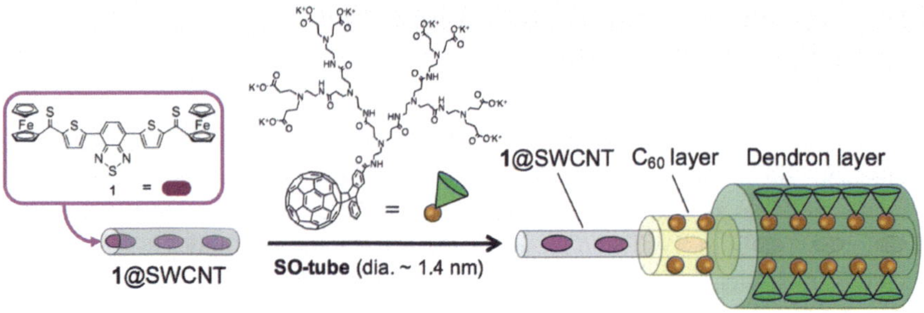

Fig. 9.9 Fabrication of the **1**@SWCNT/fullerodendron nanocomposite as H_2-evolving CNT-photocatalyst

nanocomposites. Under exposure to light of 650 nm wavelength, the SWCNT/C_{60} heterojunctions exhibited a AQY of 0.11 (SWCNT/fullerodendron), which is, as expected, higher than that of SWCNT/dendrimer (0.03). However, it is important that SWCNT/C_{60} heterojunctions are not essential for the photosensitizing property of s-SWCNT for HER.

Another approach for stepping over the Flavel's limit was encapsulation of dye molecules into the SWCNT-core of CNT-photocatalyst to absorb the light that cannot use by the SWCNT/C_{60} coaxial heterojunctions. In order to prove this concept, organic dye **1** [65], which is expected to act as an absorber of visible light and as a donor for C_{60}, was encapsulated within s-SWCNTs to produce the novel dye-encapsulated s-SWCNT (**1**@SWCNT), whose diameter ($\sim$1.4 nm) exceeds the limit set by Flavel's rule [63]. Then CNT-photocatalyst consisting of **1**@SWCNT was fabricated by physical modification of **1**@SWCNT with a fullerodendron (Fig. 9.9) [66]. Owing to the three-component heterojunction, dye **1**/SWCNT/C_{60}, the photoinduced electron transfer from dye **1** to C_{60} proceeded smoothly, resulting in the photosensitized evolution of H_2 from H_2O in the presence of BNAH, MV^{2+} and PVP-Pt, which was confirmed by the action spectra. AQYs of this HER were 0.075 and 0.013 under 550-nm and 650-nm light illumination, respectively.

9.7 Conclusion

Table 9.1 describes the AQYs of photocatalytic HERs using CNT-photocatalysts. Importantly, CNT-photocatalysts shows high efficiency of HER from water even under NIR-light illumination. The active wavelengths of the CNT-photocatalysts are controllable by changing the helicity of s-SWCNTs or encapsulated dye molecules. In marked contrast to CNT solar cells, strict purification of s-SWCNTs to obtain single (*n*,*m*) species is not necessary, because each CNT-photocatalysts have an individual s-SWCNT at the isolated core. From the viewpoint of the

Table 9.1 AQYs of HER using CNT-photocatalysts

CNT-photocatalyst	SWCNT	Wavelength (nm)	AQY (%)	Ref.
SWCNT/fullerodendron	HiPco	450	28	[54]
		650	11	[17]
	SO tube	450	11	[66]
		510	5.4	[66]
		550	0.43	[66]
		650	0.018	[66]
	(6,5)tube	570	3.7	[17]
	(7,5)tube	650	7.0	[17]
	(8,3)tube	680	17	[17]
		1000	7.3	[19]
SWCNT/fullerodendron/SiO_2	HiPco	450	31	[54]
SWCNT/fullerodendron/TiO_x	HiPco	450	47	[55]
SWCNT/fullerodendron/Pt(II)	HiPco	450	16	[62]
	(6,5)tube	570	0.35	[17]
	(7,5)tube	650	0.17	[17]
	(8,3)tube	680	1.5	[17]
SWCNT/dendrimer	HiPco	650	3.0	[18]
dye@SWCNT/fullerodendrimer	SO tube	450	15	[66]
		510	8.6	[66]
		550	1.2	[66]
		650	1.3	[66]

avoidance of time consuming purification process of s-SWCNT to fabricate the CNT solar cells, solar energy conversion using CNT-photocatalyst has the big advantage. Moreover, CNT-photocatalysts are quite promising for the H_2-evolving photocatalyst of Z-scheme photocatalytic systems, because they could combine with an O_2-evolving photocatalyst (e.g., $BiVO_4$) and a redox couple (e.g., $[Co(bpy)_3]^{2+/3+}$). In order to improve the solar energy conversion efficiency, further investigation on the PET process on the CNT-photocatalyst to discover the best combination of heterojunction materials, co-catalysts, redox couples, and O_2 evolving photocatalysts, is now undergoing.

References

1. Chen S, Takata T, Domen K (2017) Particulate photocatalysts for overall water splitting. Nat Rev Mater 2:17050
2. Tee SY, Win KY, Koh L-D, Liu S, Teng CP, Han M-Y (2017) Recent progress in energy-driven water splitting. Adv Sci 4:1600337
3. Kaneko H, Minegishi T, Domen K (2017) Recent progress in the surface modification of photoelectrides toward efficient and stable overall water splitting. Chem Eur J 23:1–11

4. Hisatomi T, Domen K (2017) Progress in the demonstration and understanding of water splitting using particulate photocatalysts. Curr Opin Electrochem 2:148–154
5. Takata T, Pan C, Domen K (2016) Design and development of oxynitride photocatalysts for overall water splitting under visible light irradiation. ChemElectroChem 3:31–37
6. Maeda K, Domen K (2016) Development of novel photocatalyst and cocatalyst materials for water splitting under visible light. Bull Chem Soc Jpn 89:627–648
7. Maeda K (2011) Photocatalytic water splitting using semiconductor particles: history and recent developments. J Photochem Photobiol C: Photochem Rev 12:237–268
8. Abe R (2011) Development of a new system for photocatalytic water splitting into H_2 and O_2 under visible light irradiation. Bull Chem Soc Jpn 84:1000–1030
9. Maeda K (2013) Z-scheme water splitting using two different semiconductor photocatalysts. ACS Catal 3:1486–1503
10. Hisatomi T, Kubota J, Domen K (2014) Recent advances in semiconductors for photocatalytic and photoelectrochemical water splitting. Chem Soc Rev 43:7520–7535
11. Yuan Y-P, Ruan L-W, Barber J, Loo SCJ, Xue C (2014) Hetero-nanostructured suspended photocatalysts for solar-to-fuel conversion. Energy Environ Sci 7:3934–3951
12. Wang Q, Hisatomi T, Suzuki Y, Pan Z, Seo J, Katayama M, Minegishi T, Nishiyama H, Takata T, Seki K, Kudo A, Yamada T, Domen K (2017) Particulate photocatalyst sheets based on carbon conductor layer for efficient Z-scheme pure-water splitting at ambient pressure. J Am Chem Soc 139:1675–1683
13. Sasaki Y, Kato H, Kudo A (2013) $[Co(bpy)_3]^{3+/2+}$ and $[Co(phen)_3]^{3+/2+}$ electron mediators for overall water splitting under sunlight irradiation using Z-scheme photocatalyst system. J Am Chem Soc 135:5441–5449
14. Arai T, Nobukuni S, Sandanayaka ASD, Ito O (2009) Zinc porphyrins covalently bound to the side walls of single-walled carbon nanotubes via flexible bonds: photoinduced electron transfer in polar solvent. J Phys Chem C 113:14493–14499
15. Ahmmad B, Kusumoto Y, Somekawa S, Ikeda M (2008) Carbon nanotubes synergistically enhance photocatalytic activity of TiO_2. Catal Commun 9:1410–1413
16. Chai B, Peng T, Zhang X, Mao J, Li K, Zhang X (2013) Synthesis of C_{60}-decotated SWCNTs (C_{60}-*d*-CNTs) and its TiO_2-based nanocomposite with enhanced photocatalytic activity for hydrogen production. Dalton Trans 42:3402–3409
17. Murakami N, Tango Y, Miyake H, Tajima T, Nishina Y, Kurashige W, Negishi Y, Takaguchi Y (2017) SWCNT photocatalyst for hydrogen production from water upon photoexcitation of (8,3) SWCNT at 680-nm light. Sci Rep 7:43445
18. Ishimoto K, Tajima T, Miyake H, Yamagami M, Kurashige W, Negishi Y, Takaguchi Y (2018) Photo-induced H_2 evolution from water via the dissociation of excitons in water-dispersible single-walled carbon nanotube sensitizer. Chem Commun 54:393–396
19. Kurniawan K, Murakami N, Tango Y, Izawa T, Nishikawa K, Watanabe K, Miyake H, Tajima T, Takaguchi Y (2017) H_2-evolving SWCNT photocatalyst for effective use of solar energy. Proc Nat Res Soc 1:01004
20. Kong J, Franklin NR, Zhou C, Chapline MG, Peng S, Cho K, Dai H (2000) Nanotube molecular wires as chemical sensors. Science 287:622–625
21. Snow ES, Perkins FK, Robinson JA (2006) Chemical vaper detection using single-walled carbon nanotubes. Chem Soc Rev 35:790–798
22. Kauffman DR, Star A (2008) Carbon nanotube gas and vapor sensors. Angew Chem Int Ed 47:6550–6570
23. Schnorr JM, Swager TM (2011) Emerging applications of carbon nanotubes. Chem Mater 23:646–657
24. Fennell JF Jr, Liu SF, Azzarelli JM, Weis JG, Rochat S, Mirica KA, Ravnsbæk JB, Swager TM (2016) Nanowire chemical/biological sensors: status and a roadmap for the future. Angew Chem Int Ed 55:1266–1281
25. Tans SJ, Verschueren ARM, Dekker C (1998) Room-temperature transistor based on a single carbon nanotube. Nature 393:49–52

26. Collins PG, Arnold MS, Avouris P (2001) Engineering carbon nanotubes and nanotube circuits using electrical breakdown. Science 292:706–709
27. Javey A, Guo J, Wang Q, Lundstrom M, Dai H (2003) Ballistic carbon nanotube field-effect transistors. Nature 424:654–657
28. Hu R, Chu L, Zhang J, Li X, Huang W (2017) Carbon materials for enhancing charge transport in the advancements of perovskite solar cells. J Power Sources 361:259–275
29. Kumar S, Nehra M, Kedia D, Dilbaghi N, Tankeshwar K, Kim K-H (2018) Carbon nanotubes: a potential material for energy conversion and storage. Prog Energy Combust Sci 64:219–253
30. Bachilo SM, Strano MS, Kittell C, Hauge RH, Smalley RE, Weisman RB (2002) Structure-assigned optical spectra of single-walled carbon nanotubes. Science 298:2361–2366
31. Nanot S, Hároz EH, Kim J-H, Hauge RH, Kono J (2012) Optoelectronic properties of single-walled carbon nanotubes. Adv Mater 24:4977–4994
32. Bindl DJ, Arnold MS (2013) Efficient exciton relaxation and charge generation in nearly monochiral (7,5) carbon nanotube/C_{60} thin-film photovoltaics. J Phys Chem C 117:2390–2395
33. Tune DD, Shapter JG (2013) The potential sunlight harvesting efficiency of carbon nanotube solar cells. Energy Environ Sci 6:2572–2577
34. Blackburn JL (2017) Semiconducting single-walled carbon nanotubes in solar energy harvesting. ACS Energy Lett 2:1598–1613
35. Ihly R, Mistry KS, Ferguson AJ, Clikeman TT, Larson BW, Reid O, Boltalina OV, Strauss SH, Rumbles G, Blackburn JL (2016) Tuning the driving force for exciton dissociation in single-walled carbon nanotube heterojunctions. Nat Chem 8:603–609
36. Dowgiallo A-M, Mistry KS, Johnson JC, Reid OG, Blackburn JL (2016) Probing exciton diffusion and dissociation in single-walled carbon nanotube-C_{60} heterojunctions. Phys Chem Lett 7:1794–1799
37. Bindl DJ, Wu M-Y, Prehn FC, Arnold MS (2011) Efficiently harvesting excitons from electronic type-controlled semiconducting carbon nanotube films. Nano Lett 11:455–460
38. Jain RM, Howden R, Tvrdy K, Shimizu S, Hilmer AJ, McNicholas TP, Gleason KK, Strano MS (2012) Polymer-free near-infrared photovoltaics with single chirality (6,5) semiconducting carbon nanotube active layers. Adv Mater 24:4436–4439
39. Shea MJ, Arnold MS (2013) 1% solar cells derived from ultrathin carbon nanotube photoabsorbing films. Appl Phys Lett 102:243101
40. Ye Y, Bindl DJ, Jacobberger RM, Wu M-Y, Roy SS, Arnold MS (2014) Semiconducting carbon nanotube aerogel bulk heterojunction solar cells. Small 10:3299–3306
41. Park J, Reid OG, Blackburn JL, Rumbles G (2015) Photoinduced spontaneous free-carrier generation in semiconducting single-walled carbon nanotubes. Nat Commun 6:8809
42. Yuma B, Berciaud S, Besbas J, Shaver J, Santos S, Ghosh S, Weisman RB, Conget L, Gallart M, Ziegler M, Hönerlage B, Lounis B, Gilliot P (2013) Biexciton, single carrier, and trion generation dynamics in single-walled carbon nanotubes. Phys Rev B: Condens Matter Mater Phys 87:205412
43. Santos SM, Yuma B, Berciaud S, Shaver J, Gallart M, Gilliot P, Conget L, Lounis B (2011) All-optical trion generation in single-walled carbon nanotubes. Phys Rev Lett 107:187401
44. Takaguchi Y, Sako Y, Yanagimoto Y, Tsuboi S, Motoyoshiya J, Aoyama H, Wakahara T, Akasaka T (2003) Facile and reversible synthesis of an acidic water-soluble poly (amidoamine) fullerodendrimer. Tetrahedron Lett 44:5777–5780
45. Takaguchi Y, Tamura M, Sako Y, Yanagimoto Y, Tsuboi S, Uchida T, Shimamura K, Kimura S, Wakahara T, Maeda Y, Akasaka T (2005) Fullerodendron-assisted dispersion of single-walled carbon nanotubes via noncovalent functionalization. Chem Lett 34:1608–1609
46. Suzuki H, Iizumi Y, Tange M, Joung S-K, Furube A, Wada T, Tajima T, Takaguchi Y, Okazaki T (2014) Spectroscopic characterization of nanohybrids consisting of single-walled carbon nanotubes and fullerodendron. Fuller Nanotub Carbon Nanostruct 22:75–87

47. Nakashima N, Tomonari Y, Murakami H (2002) Water-soluble single-walled carbon nanotubes via noncovalent sidewall-functionalization with a pyrene-carrying ammonium ion. Chem Lett 31:638–639
48. Tomonari Y, Murakami H, Nakashima N (2006) Solubilization of single-walled carbon nanotubes by using polycyclic aromatic ammonium amphiphiles in water-strategy for the design of high-performance solubilizers. Chem Eur J 12:4027–4034
49. Sandanayaka ASD, Takaguchi Y, Uchida T, Sako Y, Morimoto Y, Araki Y, Ito O (2006) Light-induced electron transfer on the single wall carbon nanotube surrounded in anthracene dendron in aqueous solution. Chem Lett 35:1188–1189
50. Dürkop T, Getty SA, Cobas E, Fuhrer MS (2004) Extraordinary mobility in semiconducting carbon nanotubes. Nano Lett 4:35–39
51. Sandanayaka ASD, Takaguchi Y, Sako Y, Tamura M, Ito O (2010) Photoinduced electron transfer of single walled carbon nanotubes surrounded by fullerodendrimers in aqueous media. Adv Sci Lett 3:353–357
52. Sandanayaka ASD, Zhang H, Takaguchi Y, Sako Y, Tamura M, Araki Y, Ito O (2005) Photoinduced charge separation and charge recombination of fullerene bearing dendritic poly (amidoamine) with carboxylates at the terminal in aqueous media. Chem Commun 5160–5162
53. Kiwi J, Grätzel M (1979) Hydrogen evolution from water induced by visible light mediated by redox catalysis. Nature 281:657–658
54. Tajima T, Sakata W, Wada T, Tsutsui A, Nishimoto S, Miyake M, Takaguchi Y (2011) Photosensitized hydrogen evolution from water using a single-walled carbon nanotube/fullerodendron/SiO_2 coaxial nanohybrid. Avd Mater 23:5750–5754
55. Kurniawan K, Tajima T, Kubo Y, Miyake H, Kurashige W, Negishi Y, Takaguchi Y (2017) Incorporating a TiO_x shell in single-walled carbon nanotube/fullerodendron coaxial nanowires: increasing the photocatalytic evolution of H_2 from water under irradiation with visible light. RSC Adv 7:31767–31770
56. Kadib AE, Katia N, Bousmina M, Majoral JP (2012) Dendrimer-silica hybrid mesoporous materials. New J Chem 36:241–255
57. Knecht MR, Sewell SL, Wright DW (2005) Size control of dendrimer-templated silica. Langmuir 21:2058–2061
58. Knecht MR, Wright DW (2004) Amine-terminated dendrimers as biomimetic templates for silica nanosphere formation. Langmuir 20:4728–4732
59. Larsen G, Lotero E (2000) Amine dendrimers as templates for amorphous silicas. J Phys Chem B 104:4840–4843
60. Crooks RM, Zhao M, Chechick V, Yeung LK (2001) Dendrimer-encapsulated metal nanoparticles: synthesis, characterization, and applications to catalysis. Acc Chem Res 34:181–190
61. Weir MG, Myers VS, Frenkel AI, Crooks RM (2010) In situ X-ray absorption analysis of $\sim$1.8 nm dendrimer-encapsulated Pt nanoparticles during electrochemical CO oxidation. ChemPhysChem 11:2942–2950
62. Sasada Y, Tajima T, Wada T, Uchida T, Nishi M, Ohkubo T, Takaguchi Y (2013) Photosensitized hydrogen evolution from water using single-walled carbon nanotube/fullerodendron/Pt(II) coaxial nanohybrids. New J Chem 37:4214–4219
63. Pfohl M, Graf K, Mertens A, Tune DD, Puerckhauer T, Alam A, Wei L, Chen Y, Zaumseil J, Colsmann A, Krupke R, Flavel BS (2016) Probing the diameter limit of single walled carbon nanotubes in SWCNT: fullerene solar cells. Adv Energy Mater 1600890
64. Perebeinos V, Tersoff J, Avouris P (2004) Scaling of excitons in carbon nanotubes. Phys Rev Lett 92:257402
65. Miyake H, Tajima T, Takaguchi Y (2017) Synthesis and light-absorption characteristics of thiophene derivatives bearing ferrocenylthiocarbonyl groups. Chem Lett 46:48–50
66. Murakami N, Miyake H, Tajima T, Nishikawa K, Hirayama R, Takaguchi Y (2018) Enhanced photosensitized hydrogen production by encapsulation of ferrocenyl dyes into single-walled carbon nanotubes. J Am Chem Soc (in press). https://doi.org/10.1021/jacs.7b12845

Chapter 10
Carbon-Based Electrodes and Catalysts for the Electroreduction of Carbon Dioxide (CO_2) to Value-Added Chemicals

Sumit Verma, Uzoma O. Nwabara and Paul J. A. Kenis

10.1 Introduction

Starting with the industrial revolution in the mid-eighteenth century, primarily naturally occurring fossil fuel resources (coal, oil, and natural gas) have been used to meet the energy demands of the growing global population. While fossil fuels offer an incredibly dense source of raw energy, its consumption (burning) results in large-scale anthropogenic carbon dioxide (CO_2) emissions. As a result, the atmospheric CO_2 levels have risen from $\sim$320 ppm in 1960 to more than 400 ppm in 2016 [1]. CO_2 being a greenhouse gas has been shown to negatively affect the energy balance on earth, thereby correlating to deleterious climate change effects such as increased temperature anomalies (global warming), rising sea levels, and ocean acidification [2, 3]. Thus, developing technological solutions that can mitigate, reduce, or utilize excess anthropogenic CO_2 emissions while maintaining and/or improving the standard of living for the growing world population, remains one of the grand challenges of the twenty-first century [4].

In the early 2000s, Socolow et al. proposed the stabilization wedges approach to help design a framework for stabilizing and eventually reducing the excess anthropogenic CO_2 emissions [5]. The strategy suggested that owing to the enormous scale of excess CO_2 emissions (on the order of 4 GtC year^{-1}) [6], a variety of solutions (referred to as stabilization wedges) need to be implemented together. The solutions include carbon intensity reduction methods such as increasing the energy efficiency of vehicles, buildings, and fossil fuel powered power plants, increasing

S. Verma · U. O. Nwabara · P. J. A. Kenis (✉)
Department of Chemical and Biomolecular Engineering, University of Illinois at Urbana-Champaign, 600 South Mathews Avenue, Urbana, IL 61801, USA
e-mail: kenis@illinois.edu

S. Verma · U. O. Nwabara · P. J. A. Kenis
International Institute for Carbon Neutral Energy Research (WPI-I2CNER), Kyushu University, 744 Moto-oka, Nishi-ku, Fukuoka 819-0395, Japan

N. Nakashima (ed.), *Nanocarbons for Energy Conversion: Supramolecular Approaches*, Nanostructure Science and Technology,
https://doi.org/10.1007/978-3-319-92917-0_10

the penetration of low carbon or renewable energy sources such as wind, solar, nuclear, and/or biofuels into the electricity grid and the transportation sector, as well as environmental conservation techniques such as reduced deforestation, reforestation, afforestation, and conservation tillage. Another emerging area of research is the utilization of CO_2 as a renewable resource to produce value-added products such as plastics, fire extinguishers, fire suppressants, refrigerants, carbon chemicals, etc., or utilize excess CO_2 to drive industrial scale processes such as enhanced fuel recovery (Fig. 10.1) [7, 8]. Of the different CO_2 utilization approaches, of particular interest is the electroreduction of CO_2 to value-added carbon chemical feedstocks such as formate/formic acid ($HCOO^-$/HCOOH), carbon monoxide (CO), methanol (CH_3OH), ethylene (C_2H_4), ethanol (C_2H_5OH), etc. (Fig. 10.2) [9–13]. When driven by intermittent renewable wind or solar electricity, CO_2 electroreduction has the potential to be utilized as a platform process to store excess electricity (when supply exceeds demand) or to manufacture industrially relevant carbon chemicals that are currently manufactured on the large scale using fossil fuel-based feedstocks and/or require fossil fuel based energy to drive the process.

In this chapter, we focus on the electroreduction of CO_2 to carbon chemicals such as $HCOO^-$/HCOOH, CO, CH_3OH, C_2H_4, and C_2H_5OH. We begin by providing an overview of the thermodynamics and reaction pathways for the electroreduction of CO_2 followed by a description of the figure of merits used to characterize electrochemical performance and a concise review of recent progress made in the field with regards to the study and design of improved electrochemical reactors, electrodes, catalysts, and electrolytes. Next, we turn our attention to an emerging class of CO_2 electroreduction electrodes and catalysts *based on carbon materials*, forming the core of this chapter. Carbon materials have been receiving

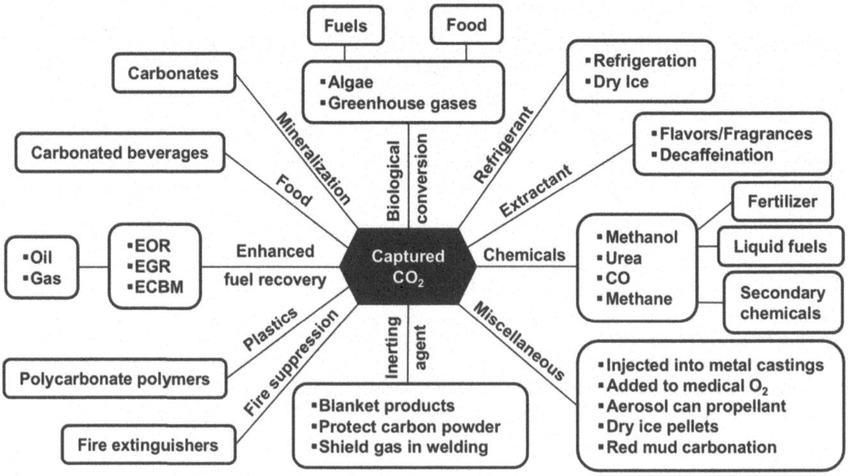

Fig. 10.1 An overview of the many ways in which excess anthropogenic CO_2 can be utilized. Adapted from Ref. [8] with permission from the U.S. Department of Energy's National Energy Technology Laboratory

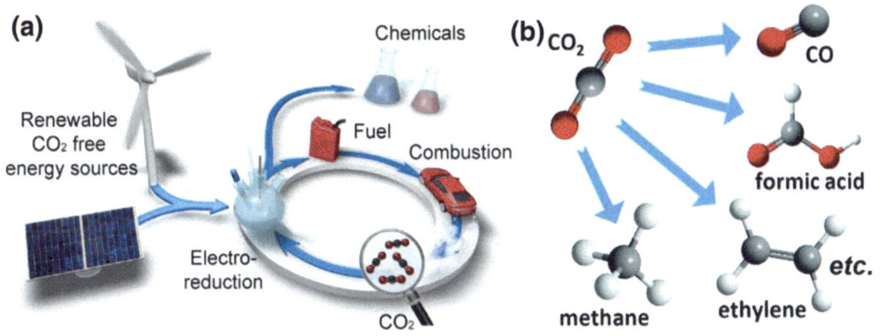

Fig. 10.2 **a** Schematic representation of the renewable electricity driven electroreduction of CO_2. Reproduced with permission from Ref. [13]. Copyright 2014, American Chemical Society. **b** Some of the value-added carbon chemicals that can be produced via the electroreduction of CO_2. Reproduced with permission from Ref. [12]. Copyright 2010, American Chemical Society

significant research attention in the past decade due to their ability to replace the conventionally used transition metal catalysts without sacrificing performance [14, 15]. This review concludes with a summary of the remaining challenges in the field and a proposal for strategies that can be utilized moving forward.

10.2 Electroreduction of CO_2 to Value Added Carbon Chemicals

10.2.1 Thermodynamics and Reaction Mechanism

A typical CO_2 electrolysis process consists of a CO_2 electroreduction reaction taking place at the cathode (negative) side of an electrolyzer coupled to an oxygen (O_2) evolution reaction taking place at the anode (positive) side of an electrolyzer. Depending on the catalyst, electrolyte, or reaction condition being used, the electroreduction of CO_2 can produce more than 16 different carbon products [16–18]. Table 10.1 provides a list of select CO_2 electroreduction products along with the standard reduction potentials (E^0), standard cell potentials (E^0_{cell}), and number of electrons required (z) [19]. A quick analysis of the E^0 and E^0_{cell} values for different CO_2 electroreduction products suggests that the energy requirements are very similar to that required for the electrolysis of water (H_2O) to hydrogen (H_2). Hence, a common concern with many CO_2 electrolysis processes (especially when utilizing aqueous electrolytes) is the parasitic H_2 evolution reaction that tends to compete with the electroreduction of CO_2 for catalytic sites. Thus, the design of catalysts, electrolytes, and electrochemical reactors that can suppress the H_2 evolution reaction and as a result enhance the CO_2 electroreduction reaction remains an active area of research [20–23].

Table 10.1 Standard reduction potential (E^0), standard cell potentials (E^0_{cell}) , and number of electrons exchanged (z) for different products of CO_2 electroreduction. E^0 and E^0_{cell} values are reported in aqueous media under standard conditions of 1 atm and 25 °C [19]

Product	M (g mol^{-1})	Cathode reaction	z	E^0 (V vs. RHE)	Anode reaction	E^0 (V vs. RHE)	E^0_{cell} (V)
H_2	2	$2H_2O + 2e^- \rightarrow H_2 + 2OH^-$	2	0.00	$4OH^- \rightarrow$ $O_2 + 2H_2O + 4e^-$	1.23	–1.23
HCOOH	46.02	$CO_2 + 2H_2O + 2e^- \rightarrow HCOOH + 2OH^-$	2	–0.25			–1.48
CO	28.01	$CO_2 + H_2O + 2e^- \rightarrow CO + 2OH^-$	2	–0.11			–1.34
CH_3OH	32.04	$CO_2 + 5H_2O + 6e^- \rightarrow CH_3OH + 6OH^-$	6	0.02			–1.21
CH_4	16.04	$CO_2 + 6H_2O + 8e^- \rightarrow CH_4 + 8OH^-$	8	0.17			–1.06
C_2H_4	28.05	$2CO_2 + 8H_2O + 12e^- \rightarrow C_2H_4 + 12OH^-$	12	0.06			–1.17
C_2H_5OH	46.07	$2CO_2 + 9H_2O + 12e^- \rightarrow$ $C_2H_5OH + 12OH^-$	12	0.08			–1.15

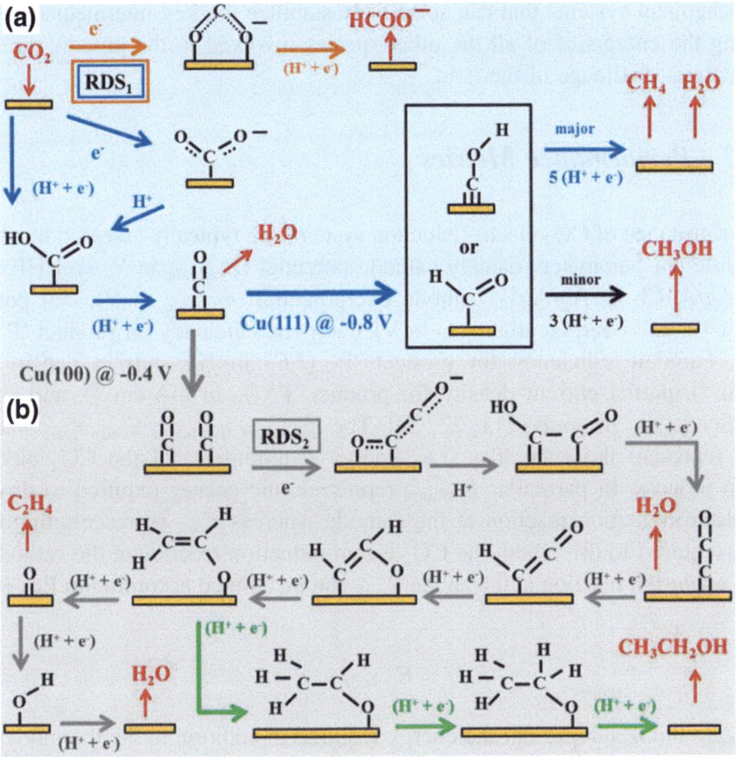

Fig. 10.3 Schematic illustration of the reaction pathways for the electroreduction of CO_2 to **a** CO, CH_4, CH_3OH, $HCOO^-$, and **b** C_2H_4 and C_2H_5OH. Potentials are reported versus the reversible hydrogen electrode (RHE). RDS = rate-determining step. Reproduced with permission from Ref. [26]. Copyright 2015, American Chemical Society

While the theoretical thermodynamic analysis of CO_2 electroreduction suggests that the reaction requires fairly low amounts of energy to proceed, typically a very high overpotential (i.e., energy in addition to the thermodynamic minimum) is required [22, 23]. For the simple case of the electroreduction of CO_2 to $HCOO^-$/HCOOH or CO (i.e., a 2-step reaction), the formation of the rate determining $CO_2^{\bullet-}$ radical anion ($E^0 = -1.9$ V vs. SHE) drives up the energy requirement for this process (Fig. 10.3) [24–26]. As a result, most of the catalysts and electrolytes designed and/or studied for the electroreduction of CO_2 (especially for $HCOO^-$/HCOOH and CO production) aim at lowering the energy requirement for the formation of the $CO_2^{\bullet-}$ radical anion [11, 20–23, 27]. The production of CO_2 electroreduction products such as CH_3OH, CH_4, C_2H_4, and C_2H_5OH that require the transfer of more than 2 electrons is even more complex. This is because many of the products involve the formation of multiple rate determining intermediates and/or share common intermediates (Fig. 10.3) [13, 26, 28]. Developing catalysts and/or electrochemical systems that can selectively stabilize the key intermediate without affecting the energetics of all the other species involved in the process remains a key catalysis challenge in the field.

10.2.2 Performance Metrics

The performance of CO_2 electroreduction systems are typically assessed in terms of eight different parameters namely cathode potential ($E_{cathode}$ in V vs. RHE, SHE, SCE, Ag/AgCl, or Ag/Ag^+), cathode overpotential ($\eta_{cathode}$ in V), cell potential (E_{cell} in V), cell overpotential (η_{cell} in V), energetic efficiency for product 'P' (EE_P in %), Faradaic efficiency for product 'P' (FE_P in %), current density (j in mA cm^{-2}), partial current density for product 'P' (j_P in mA cm^{-2}), and catalyst durability (t_{catdur} in hours) [11, 22, 29]. The $E_{cathode}$, $\eta_{cathode}$, E_{cell}, η_{cell}, and EE_P values represent the energetics (i.e., energy requirement) of the CO_2 electroreduction process. In particular $E_{cathode}$ represents the energy required to drive the CO_2 electroreduction reaction at the cathode whereas E_{cell} represents the overall energy required to drive both the CO_2 electroreduction reaction at the cathode and the O_2 evolution reaction at the anode. E_{cell} can be defined according to Eq. 10.1 as follows:

$$E_{cell} = E_{cathode} - E_{anode} \tag{10.1}$$

$\eta_{cathode}$ and η_{cell} represent the energy required in addition to the thermodynamic minimum cathode reduction potential ($E^0_{cathode}$) and the cell reduction potential (E^0_{cell}), respectively. Assuming that the CO_2 electroreduction experiments were performed under standard conditions of 1 atm, 25 °C, and unit activity of all species involved, $\eta_{cathode}$ and η_{cell} can be defined according to Eqs. 10.2 and 10.3 as follows:

$$\eta_{cathode} = E_{cathode} - E^0_{cathode} \tag{10.2}$$

$$\eta_{cell} = E_{cell} - E^0_{cell} \tag{10.3}$$

EE_P or the energetic efficiency for product 'P' represents the second law of thermodynamics efficiency for the electroreduction of CO_2 to the product 'P' and is defined according to Eq. 10.4 as follows (assuming that the CO_2 electroreduction experiments were performed under standard conditions of 1 atm, 25 °C, and unit activity of all species involved):

$$EE_P = \frac{E^0_{cell} * FE_P}{E_{cell}} \tag{10.4}$$

Note that for CO_2 electroreduction experiments performed under nonstandard conditions (pressure, temperature, and activity of involved species different than

1 atm, 25 °C, or unity, respectively), the equilibrium potential estimated using the Nernst equation should be used as the thermodynamic minimum for Eqs. 10.2–10.4 instead of E^0_{cathode} and E^0_{cell} [30]. FE_{P} or the Faradaic efficiency for product 'P' (also referred to as current efficiency) represents the selectivity of the CO_2 electroreduction process towards the product 'P' and is defined per Eq. 10.5 as follows:

$$FE_{\mathrm{P}} = \frac{znF}{Q} * 100, \tag{10.5}$$

where z is the number of electrons exchanged to form the product 'P' (see Table 10.1 for a list of z values for the different products of CO_2 electroreduction), n represents the moles of the product 'P', F is the Faraday's constant (96,485 C mol^{-1}), and Q is the amount of charge passed. The current density (j) represents the overall activity or the electrochemical reaction rate. Usually, the current density (j) values are obtained by normalizing the total current with the geometric surface area of the electrode being studied. In rare instances, the electrochemically active surface area (ECSA) is used to normalize the current instead of the geometric surface area. For all the subsequent descriptions of j in this chapter, we will be using the j values normalized by the geometric surface area as that represents the most common method of reporting current density data in the literature. j_{P} or partial current density for product 'P' represents the activity or the electrochemical reaction rate for the formation of product 'P' and can be calculated according the Eq. 10.6 as follows.

$$j_{\mathrm{P}} = j * FE_{\mathrm{P}} \tag{10.6}$$

Finally, t_{catdur} represents the durability of the catalyst under investigation or in other words for how long can the catalyst can be used without the need for any catalyst regeneration step.

10.2.3 Recent Progress

The early pioneering work in the area of CO_2 electroreduction was performed by Yoshio Hori and coworkers at Chiba University in Japan in the 1980s and 1990s where they screened different transitions metals to classify them according to their activity and selectivity toward different products of CO_2 electroreduction [17, 18]. Since then, significant research efforts have focused on the study, design, and development of electrochemical reactors, electrodes, catalysts, and electrolytes that can improve the activity and selectivity of CO_2 electroreduction toward a target product. This section provides a brief overview of the promising efforts being undertaken in this area.

10.2.3.1 Electrochemical Reactors and Electrodes

Typical lab scale CO_2 electroreduction studies are performed using either an analytical H-cell or a 3-electrode cell and/or employ a CO_2 saturated aqueous electrolyte as the source of CO_2 feed [16, 31, 32]. However, the solubility of CO_2 in aqueous solutions is low ($\sim$35 mM under ambient conditions) [33]. Hence, studies using dissolved CO_2 as the reactant feed are typically limited by the mass transport of CO_2 to the electrode surface and result in low activity (current densities much less than 50 mA cm^{-2}). An alternative to using dissolved CO_2 as the reactant feed would be the use of a gas diffusion layer (GDL) electrode based flow electrolyzer as demonstrated by us and others [34–45]. Figure 10.4 shows the schematic illustration of a single as well as a dual electrolyte channel flow electrolyzer setup regularly used in our lab [30]. The setup consists of a catalyst coated GDL cathode and a catalyst coated GDL anode separated by a flowing electrolyte stream (for a single electrolyte channel flow electrolyzer) and two flowing electrolyte streams separated by an anion exchange membrane (for a dual electrolyte channel flow electrolyzer). The GDL electrode forms the most important component of this setup as the porous structure of the GDL enables a continuous supply of CO_2 to the electrode–electrolyte interface (i.e., surface at which the CO_2 electroreduction reaction takes place), thus avoiding mass transport limitations associated with the low solubility of CO_2 in aqueous solutions, resulting in high activity (current densities typically in the 100–500 mA cm^{-2} range). GDLs are usually comprised of carbon-based materials and will be discussed in detail in Sect. 10.3.1 of this chapter. The use of a

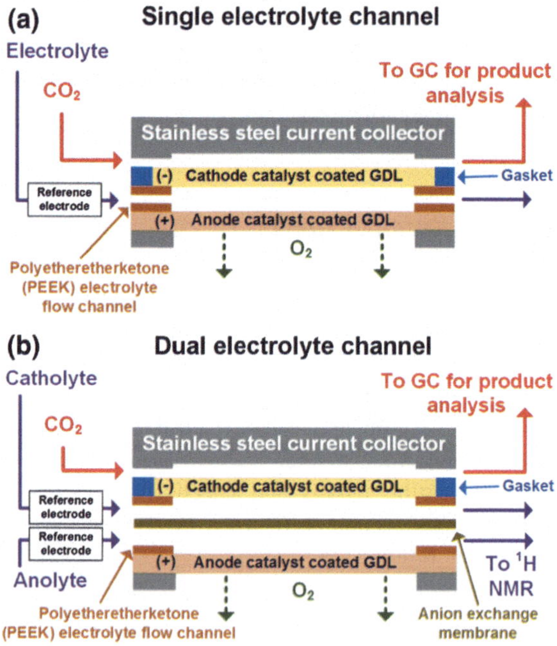

Fig. 10.4 Schematic illustration of a gas diffusion layer (GDL) electrode based **a** single electrolyte channel and **b** dual electrolyte channel flow electrolyzer. Reproduced with permission from Ref. [30]. Copyright 2017, American Chemical Society

flowing electrolyte further enables the study of electrolytes with alkaline pH without having to worry about significant carbonate formation and/or precipitation [46]. Furthermore, the results obtained with a GDL electrode based flow electrolyzers can in principle be scaled to industrial scale stacks by imagining combining multiple of these flow electrolyzers to form a stack. The reader is referred to an excellent in-depth review by Endrődi and coworkers on the different CO_2 electrolyzer designs utilizing a continuous CO_2 stream as the reactant feed [47].

10.2.3.2 Catalysts

Hori and coworkers in their early seminal work classified transition metal catalysts for the electroreduction of CO_2 into four different groups on the basis of their selectivity toward different products [17, 18]. The first group of metals, bismuth (Bi), cadmium (Cd), tin (Sn), indium (In), thallium (Tl), mercury (Hg), and lead (Pb) showed selectivity toward the electroreduction of CO_2 to $HCOO^-$. The second group consisted of metals like gallium (Ga), palladium (Pd), zinc (Zn), silver (Ag), and gold (Au) that were selective toward the electroreduction of CO_2 to CO. The third group consisted of copper (Cu), the only catalyst that showed selectivity

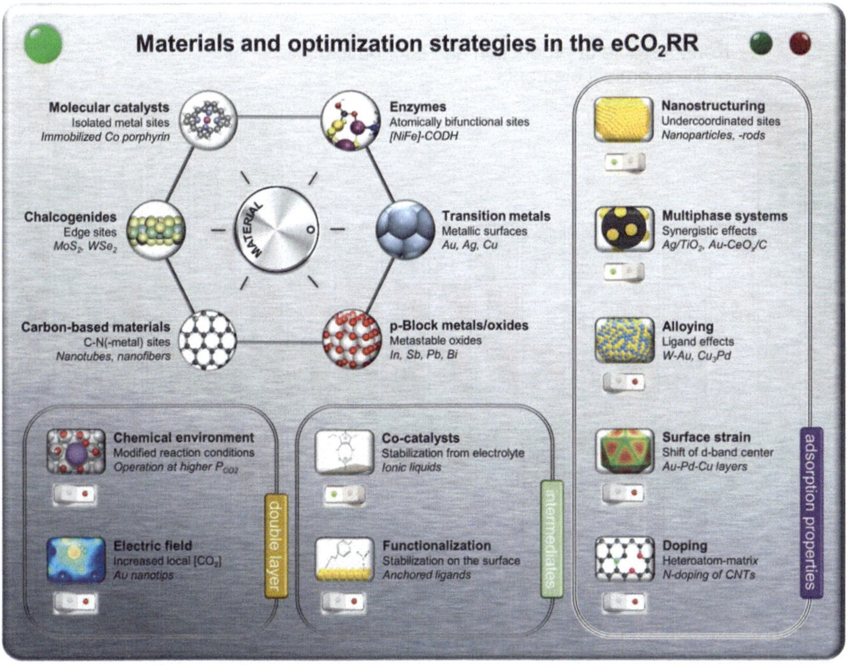

Fig. 10.5 Schematic illustration of the different classes of catalytic materials developed in the literature for the electroreduction of CO_2 along with the different design strategies being employed. Reproduced from Ref. [23] under the creative commons attribution (CC-BY) license

toward production of a variety of hydrocarbons and oxygenates such as CH_4, C_2H_4, C_2H_5OH, and propanol. The fourth group consisted of transition metals such as titanium (Ti), platinum (Pt), iron (Fe), and nickel (Ni) that did not show any CO_2 electroreduction activity and thus resulted in parasitic H_2 evolution. Since this seminal work by Hori over 25 years ago, a variety of catalyst design strategies such as nanostructuring, alloying, surface strain engineering, doping, and the use of catalyst supports have been employed to either improve the intrinsic activity of catalytic sites or increase the number of catalytic sites (Fig. 10.5). See some excellent recent reviews by Larrazabal et al. [23], Seh et al. [48], Qiao et al. [27], and Khezri et al. [49], for an in-depth analysis of the different catalytic materials designed for the electroreduction of CO_2 based on these strategies.

Many of the transition metal catalysts (Ag, Au, etc.) that exhibit high activity and selectivity for the electroreduction of CO_2, unfortunately, are expensive as well as rare and in most cases require an energy-intensive mining process to procure them (Fig. 10.6) [50]. The availability of such materials can thus severely impede the deployment of the CO_2 electroreduction process on a large scale. To circumvent such limitations, an emerging area of research has been to utilize *precious metal-free carbon-based catalysts or support materials for the electroreduction of CO_2*. In addition to providing cost benefits, carbon-based materials also possess attractive material properties such as high surface area, excellent conductivity,

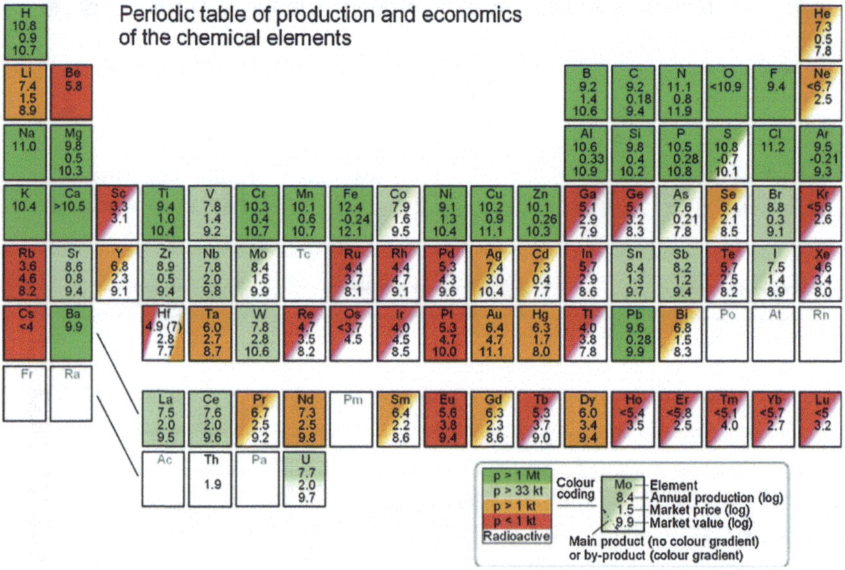

Fig. 10.6 Overview of the production level, production type (main product from an ore or a byproduct), and market price for different elements in the periodic table. The three numbers under each element represent its annual production (kg), price (\$ kg^{-1}) and market value (\$ $year^{-1}$), respectively, and are reported on the logarithmic scale (base 10). Reproduced with permission from Ref. [50]. Copyright 2012, Royal Society of Chemistry

tunable porosity, composition, and structure that can, in turn, be used to develop catalysts with high activity and selectivity toward different CO_2 electroreduction products [14, 15]. Such material properties have attracted significant research attention in the past few years and will be discussed in detail in Sect. 10.3.

10.2.3.3 Electrolytes

In addition to the design and development of catalysts for the electroreduction of CO_2, electrolyte engineering has also been investigated as an additional tool to develop active and selective electrochemical systems while reducing the energy requirements for the same. For example, in our prior work in collaboration with Dr. Richard Masel we showed that the use of an aqueous solution of 1-ethyl-3-methylimidazolium tetrafluoroborate (EMIM BF_4; an ionic liquid) as the electrolyte significantly reduced the overpotential requirement for the electroreduction of CO_2 to CO by stabilizing the rate determining $CO_2^{\bullet-}$ intermediate [51]. The concept of using EMIM BF_4 as the electrolyte was later followed up Masel et al. with the development of anion exchange membranes that incorporate an imidazolium group into styrene backbone [52, 53]. Stable CO_2 electroreduction performance of up to 6 months have been successfully demonstrated with these anion exchange membranes. In some other recent work, we have shown that the size of the electrolyte cation being used improves selectivity for the electroreduction of CO_2 to CO by suppressing H_2 evolution [38], and the use of high concentrations of alkaline electrolytes lowers the overpotentials for CO_2 electroreduction as well as improves activity in comparison to neutral electrolytes [34–36, 54]. The reader is advised to look at an excellent review by Sharma and Zhou that provides a detailed summary of the effect of electrolytes on the electroreduction of CO_2 [55].

10.3 Carbon Electrodes and Catalysts for the Electroreduction of CO_2

Early pioneering work in the area of carbon-based electrocatalysts was performed by Gong et al. who developed nitrogen-doped carbon nanotube arrays that showed superior activity for the oxygen reduction reaction compared to the activity exhibited by the commercially available precious metal based Pt/C catalyst [56]. Since then, many researchers have focused on gaining a fundamental mechanistic understanding of how carbon-based electrocatalysts work and how that understanding can be used to tailor the material properties of carbon, including allotropes such as graphene, diamond, carbon nanotubes, carbon nanofoams, etc. [57]. Two of the most promising material design strategies being utilized include: (i) the modulation of the electronic properties of carbon by doping it with heteroatoms such as nitrogen (N), boron (B), and sulfur (S); and (ii) the utilization of carbon-based materials as supports for catalytically active precious metal nanoparticles. The use of carbon-based catalyst support materials can help incorporate desirable properties of carbon such as high surface area, excellent electronic conductivity, tunable porosity, tunable hydrophobicity, and stability under acidic, basic, or high-temperature conditions into the catalytic system. This, in turn, enables the fine-tuning of the properties (geometric as well as mass activity, selectivity, and durability) of the metal catalyst under investigation. Unsurprisingly, significant research efforts in the past decade or so have focused on the investigation of such material design strategies for the CO_2 electroreduction reaction as will be discussed later in this section. In addition to the carbon-based electrocatalysts, active CO_2 electroreduction systems also employ graphitic carbon-based gas diffusion layer (GDL) electrodes to avoid mass transport limitations associated with the low solubility of CO_2 in aqueous solutions. This section will provide an in-depth review of first the different carbon-based GDL electrodes and second the different catalysts that have been reported in the CO_2 electroreduction literature to date.

10.3.1 Carbon-Based Gas Diffusion Layer (GDL) Electrodes

As mentioned in Sect. 10.2.3.1, a gas diffusion layer (GDL) electrode based flow electrolyzer separates the gaseous and liquid flows, thereby circumventing mass transport issues related to the low solubility of CO_2 in aqueous electrolytes. Conventionally, the gas diffusion layer consists solely of a carbon fiber substrate (CFS) made of Teflon-treated carbon cloth or paper. The role of the GDL is to: (i) separate the liquid (electrolyte) and the gas (CO_2) phase, (ii) support the catalyst layer, (iii) allow CO_2 to diffuse through to the catalyst-electrolyte interface where the reaction occurs, and (iv) prevent the electrolyte from entering into the gas stream (flooding). However, catalyst deposition on a bare CFS results in irregularly

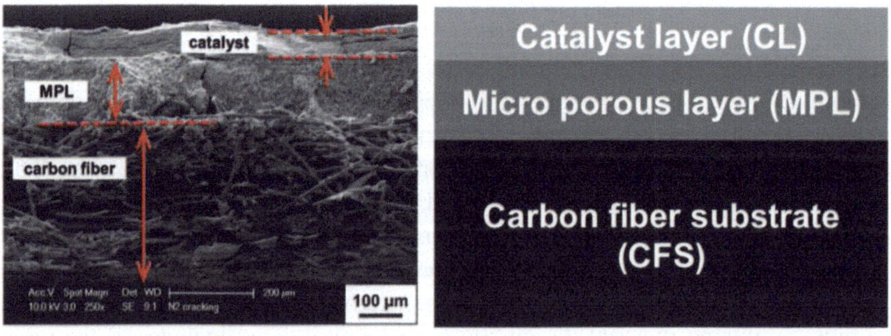

Fig. 10.7 Scanning electron microscopy (SEM) image (*left*) of a gas diffusion layer (GDL) coated with Ag nanoparticles with a corresponding cartoon (*right*) depicting the different layers

distributed particles and ultimately a poorly defined catalyst layer [58]. A nonuniform catalyst layer also exposes the CFS during the electrochemical reaction, which enhances the parasitic H_2 evolution reaction. Therefore, in modern practices, the gas diffusion layer also has a microporous layer (MPL) , which sits on top of the CFS (Fig. 10.7). The presence of a MPL reduces contact resistance between the CFS and the catalysts layer by forming a strong, flat interfacial layer, allowing for a better distribution of the catalyst layer. Typically, the MPL is comprised of a carbon powder held together by a wet-proofing binder such as polytetrafluorethylene (PTFE), both of which contribute to the MPL's hydrophobicity.

The thickness and wet-proofing of both the CFS and the MPL also can be tuned to optimize mass transport, durability, and electrical conductivity. Recently, our group investigated these properties of the GDL for the electroreduction of CO_2 to CO on a Ag nanoparticle catalyst [58]. The optimal levels of wet-proofing for the MPL and CFS were found to be 20 wt% PTFE and 10 wt% PTFE, respectively, after which the catalytic activity (current density) dropped (Fig. 10.8a–c). Very high carbon content in the MPL and the CFS without sufficient wet-proofing leads

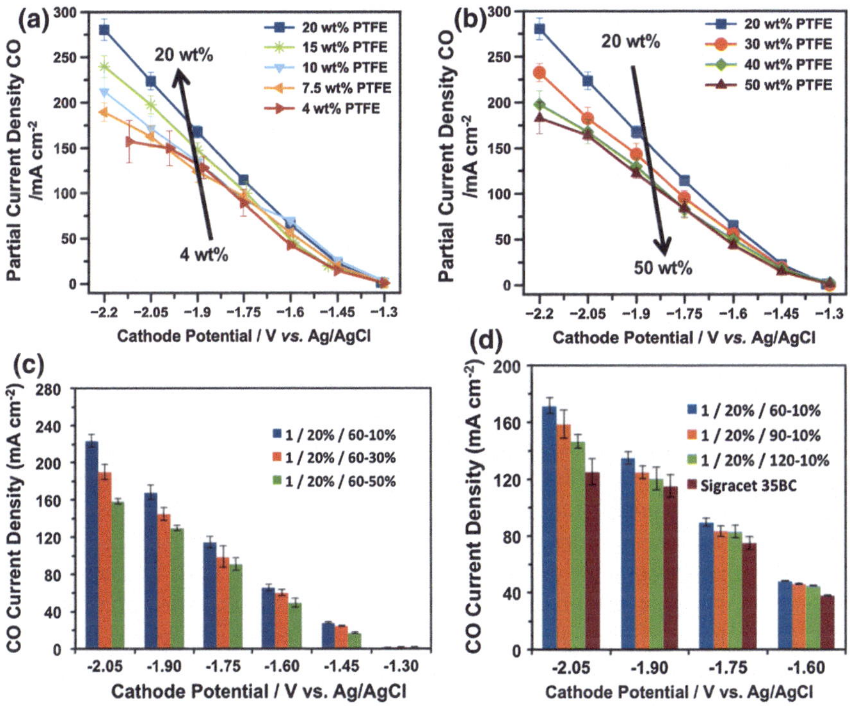

Fig. 10.8 Plots depicting the variation in the partial current density for CO as a function of the cathode potential. **a**, **b** Effect of the varying the polytetrafluoroethylene (PTFE) content in the microporous layer (MPL). **c** Effect of varying wet-proofing levels and **d** thickness of the carbon fiber substrate (CFS). Reproduced with permission from Ref. [58]. Copyright 2016, Elsevier

to electrolyte flooding; on the other hand, too much binder in the MPL and the CFS can result in mass transport issues and in some cases, a decrease in conductivity (charge transfer). In addition, the study showed that the optimal thickness of the CFS should be ~190 μm (Toray-60) as shown in Fig. 10.8d. A higher CFS thickness leads to lower activity because the diffusion length increases, making it more difficult for CO_2 from the gas stream to travel through the GDL to the catalyst layer.

The use of binder-free gas diffusion electrodes was explored by Wang and coworkers [59]. Conventionally used binders such as Nafion or PTFE are insulating materials, and therefore decrease the electrical conductivity of gas diffusion electrodes. Moreover, the use of carbon powder, such as in the MPL, result in low mechanical strength, thus lowering the stability and durability of carbon-based gas diffusion electrodes. In this work, the authors proposed a hierarchically structured porous N-doped carbon membranes (HNCMs) supported on carbon nanotubes (CNTs) as binder-free electrodes for electroreduction of CO_2, producing $HCOO^-$ with a high Faradaic efficiency of 81% at –0.8 V versus RHE. The hierarchal porous architecture of the membrane allows the HNCMs to function as a gas diffusion electrode with enhanced three-phase boundary layer area (catalyst, electrolyte, and CO_2) and the charge transport within it. Additionally, the presence of the CNTs seems to favor pyridinic nitrogen doping, which helps increase the activity and selectivity by stabilizing the reaction intermediates.

10.3.2 Carbon Nanofiber and Porous Carbon-Based CO_2 Electroreduction Catalysts

The earliest example of a metal-free carbon-based electrocatalyst in the CO_2 electroreduction literature was reported by Kumar et al. as a carbon nanofiber (CNF) catalyst with a high density of nitrogen atoms [60]. The CNF catalyst was synthesized by the pyrolysis of electrospun nanofibers of polyacrylonitrile and exhibited a high surface area due to the presence of nano-corrugated fractal-like structures (Fig. 10.9a, b). In the presence of EMIM BF_4 as the electrolyte (beneficial effects described earlier in Sect. 10.2.3.3), the CNF catalyst showed high selectivity for CO production (FE_{CO} ~ 98%) and a fourfold improvement in the current density when compared to the state of the art 5-nm Ag nanoparticles (Fig. 10.9c) under an applied potential of –0.573 V versus SHE. Using a combination of X-ray photoelectron spectroscopy (XPS) and long-term chronoamperometric measurements, the authors concluded that the selective formation of CO on the CNF catalyst can be attributed to the positive charges on the carbon atoms created due to the neighboring negatively charged nitrogen atoms within the CNF lattice (Fig. 10.9d).

Drawing inspiration from the CNF catalyst, Varela et al. reported metal containing nitrogen-doped porous carbon-black catalysts (referred to as "M-N-C"

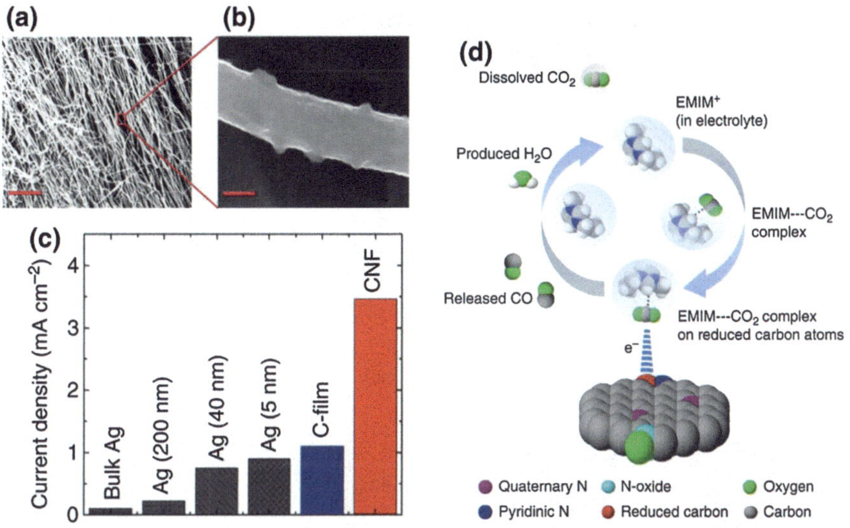

Fig. 10.9 Scanning electron microscopy (SEM) image of the carbon nanofiber (CNF) catalyst displaying **a** the entangled fibers (scale bar = 5 μm) and **b** the individual fiber (scale bar = 200 nm). **c** A comparison of the current density for the electroreduction of CO_2 with CNF as the catalyst in comparison to control samples of Ag and C-film. **d** Schematic representation of the proposed CO_2 electroreduction mechanism on the CNF catalyst. Reproduced with permission from Ref. [60]. Copyright 2013, Nature Publishing Group

catalysts) for the electroreduction of CO_2 to CO and hydrocarbons [61]. The metals explored in this work included Fe, Mn, and FeMn with all three-metal containing nitrogen-doped porous carbon-black catalysts (Fe-N-C, Mn-N-C, and FeMn-N-C) showing an improvement in the CO_2 electroreduction activity when compared to a Au foil (Fig. 10.10a). The selectivity for CO production was found to be insensitive to the identity of the metal incorporated into the nitrogen-doped porous carbon catalyst indicating the nitrogen moiety to be more important for the observed catalytic activity (Fig. 10.10b). In a related effort, Li et al. reported a dual sulfur, nitrogen-doped polymer derived carbon (CPSN) catalyst and a sulfur-doped polymer derived carbon (CPS) catalyst for the electroreduction of CO_2 [62]. The CPSN catalyst showed a higher selectivity towards CO and CH_4 formation (FE_{CO} ~ 11.3% and FE_{CH_4} ~ 0.18%) in comparison to the CPS catalyst (FE_{CO} ~ 2% and FE_{CH_4} ~ 0.1%). However, the low overall selectivity of the CPSN and CPS catalyst toward CO_2 electroreduction products severely limit the practical applicability of such systems. To move beyond the design of porous carbon-based catalysts for CO formation, Song et al. reported a new class of nitrogen-doped ordered cylindrical mesoporous carbon (c-NC) catalysts that enabled the selective electroreduction of CO_2 to C_2H_5OH ($FE_{C_2H_5OH}$ as high as 77%) at a low cathode potential of –0.56 V versus RHE (Fig. 10.10c, d) [63]. Using a set of carefully designed control experiments with inverse mesoporous nitrogen-doped carbon (i-NC) and theoretical density functional theory (DFT)

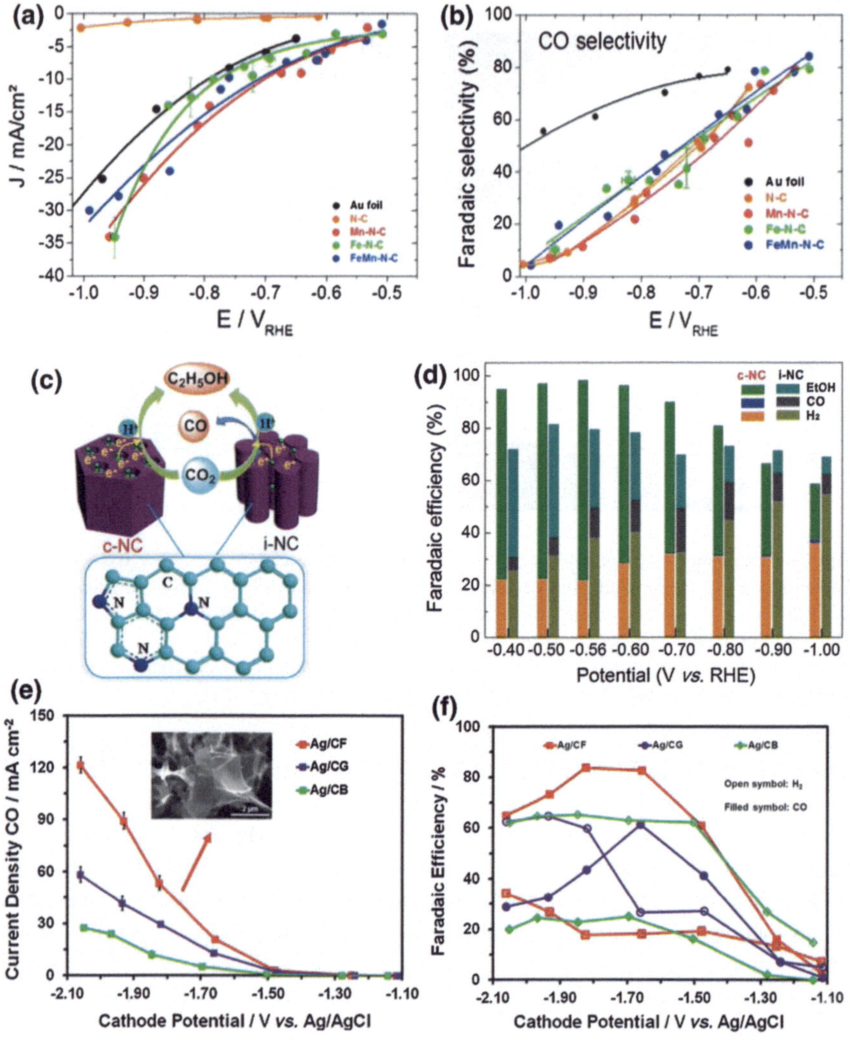

Fig. 10.10 **a** Total current density (*j*) and **b** Faradaic efficiency for the electroreduction of CO_2 to CO using metal (Fe, Mn, and FeMn) containing nitrogen-doped porous carbon black catalyst. Reproduced with permission from Ref. [61]. Copyright 2015, Wiley-VCH Verlag GmbH & Co. **c** Schematic illustration of the cylindrical mesoporous nitrogen-doped carbon (c-NC) and the inverse mesoporous nitrogen-doped carbon (i-NC) along with their **d** Faradaic efficiency for different CO_2 electroreduction products. Reproduced with permission from Ref. [63]. Copyright 2017, Wiley-VCH Verlag GmbH & Co. A comparison of the **e** partial current density for CO production and **f** the Faradaic efficiency for CO and H_2 production when using Ag nanoparticles supported on carbon foam (CF), graphene nanoplatelets (CG), and carbon black (CB). Reproduced with permission from Ref. [43]. Copyright 2017, Wiley-VCH Verlag GmbH & Co

calculations, the authors showed that the high density of pyridinic and pyrrolic nitrogen sites deposited on the inner surface of the c-NCs cylindrical channels facilitated the dimerization of the rate determining CO* intermediate (see Sect. 10.2.1 a review of the commonly accepted mechanism for the electroreduction of CO_2) to form C_2H_5OH. In a more recent effort, our group has also explored the use of high surface area microporous carbon foams (CF) as support materials to deposit Ag nanoparticles (Fig. 10.10e) [43]. In comparison to Ag nanoparticles deposited on commercially available graphene nanoplatelets (CG) and carbon black (CB) supports, Ag/CF shows a 2 and fourfold improvement in the activity for the electroreduction of CO_2 to CO, respectively (Fig. 10.10f). The improved electrochemical performance when using Ag/CF was attributed to large surface areas, porosity, and improved Ag nanoparticle size distribution.

10.3.3 Carbon Nanotube-Based CO_2 Electroreduction Catalysts

Owing to their high electrical conductivity and the ability to introduce interesting curvature effects that can affect reaction kinetics and product selectivity [64], carbon nanotubes (CNTs) have been explored in the CO_2 electroreduction literature as both metal-free catalysts and support materials for metal nanoparticle catalysts. In one of the earliest examples of CNT based catalysis, Zhang et al. reported nitrogen-doped carbon nanotubes with an overlayer of polyethylenimine (PEI) deposited on a glassy carbon electrode (PEI-NCNT/GC) for the selective electroreduction of CO_2 to $HCOO^-$ [65]. The PEI-NCNT/GC electrodes were prepared by a simple method of drop casting a suspension of CNTs in dimethylformamide on a glassy carbon electrode followed by an exposure to ammonia plasma to dope the CNTs with nitrogen, and the subsequent addition of a polyethylenimine overlayer via dip coating (Fig. 10.11a). Compared to control

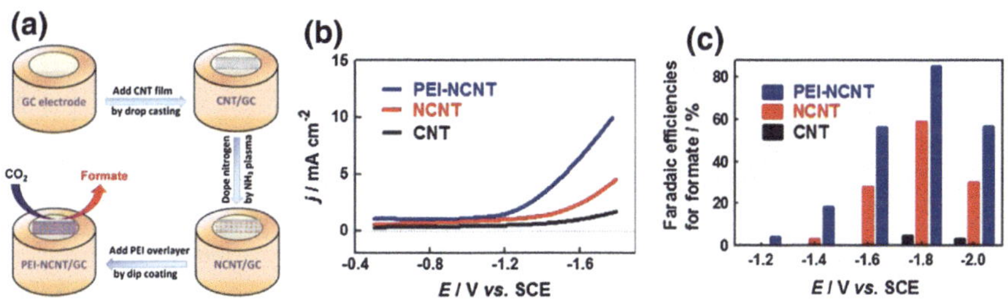

Fig. 10.11 **a** Schematic illustration of the fabrication of nitrogen-doped carbon nanotubes with an overlayer of polyethylenimine (PEI) deposited on a glassy carbon electrode (PEI-NCNT/GC). **b** Linear sweep voltammetry scans and **c** Faradaic efficiency for formate as a function of the cathode potential for the PEI-NCNT/GC, NCNT/GC, and CNT/GC electrodes. Reproduced with permission from Ref. [65]. Copyright 2014, American Chemical Society

samples of pristine CNTs and nitrogen-doped CNT (NCNT), the PEI-NCNT/GC electrode showed higher activity and selectivity towards $HCOO^-$ production with FE_{HCOO^-} as high as 84% being obtained at a cathode potential of −1.8 V versus SCE (Fig. 10.11b, c). The improved performance when using the PEI-NCNT/GC electrode was hypothesized to be a result of the CO_2 molecule first binding to the basic nitrogen sites followed by the reductive formation of the rate-limiting $CO_2^{\bullet -}$ intermediate which was then stabilized by the PEI overlayer by a H-bond interaction.

Metal-free nitrogen-doped CNTs were further explored for the CO_2 electroreduction reaction by Wu and coworkers [66]. The nitrogen-doped CNTs (NCNTs) were prepared by a liquid chemical vapor deposition method and resulted in the formation of bamboo-like nanostructures due to the incorporation of nitrogen into the CNTs (Fig. 10.12a). In terms of the CO_2 electroreduction performance, the NCNTs were found to be stable (minimal drop in performance over 10 h) and selective for CO production with a FE_{CO} as high as 80% obtained at a potential of −0.78 V (Fig. 10.12b, c). In contrast, pristine CNTs showed negligible selectivity and poor activity for the CO_2 electroreduction reaction. Using theoretical DFT calculations, the authors suggested that the rate determining step for the electroreduction of CO_2 to CO on NCNTs involves the formation of the adsorbed COOH* intermediate which becomes downhill in free energy on a pyridinic nitrogen site. Meanwhile, the pyridinic nitrogen sites bind weakly to the adsorbed CO* intermediate which enables an easy desorption of CO* from the surface, resulting in high CO selectivity. This hypothesis was further verified by the same group in subsequent work [67], where they synthesized NCNTs with different nitrogen content and defect density. The experimental results as well as DFT calculations suggested that the presence of easily accessible pyridinic nitrogen with a lone pair of electrons facilitated selective CO production by binding with the CO_2 molecule.

In addition to the nitrogen-doped CNTs, CNT based composite materials have also been explored in the CO_2 electroreduction literature. For example, Lu et al.

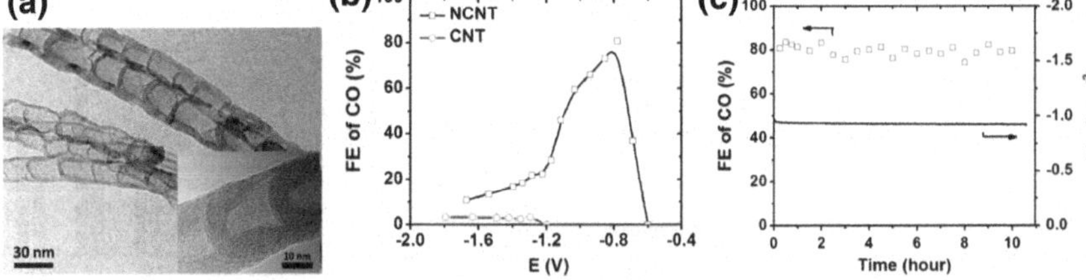

Fig. 10.12 **a** TEM image of the nitrogen-doped CNTs (NCNTs) with a single NCNT shown in inset. **b** Faradaic efficiency for CO (FE_{CO}) as a function of the applied potential for the electroreduction of CO_2 on NCNTs as well as CNTs, and **c** FE_{CO} as well as current (j) as a function of time for the electroreduction of CO_2 at a constant potential of −0.8 V on the NCNT catalyst. Reproduced with permission from Ref. [66]. Copyright 2015, American Chemical Society

reported a covalent carbon–nitrogen bond containing graphitic carbon nitride/carbon nanotube composite catalyst (g-C_3N_4/MWCNT) with a layered structure (Fig. 10.13a), for the electroreduction of CO_2 to CO [68]. In terms of the electrochemical performance, the g-C_3N_4/MWCNT composite catalyst exhibited a maximum FE_{CO} of 60% at a cathode potential of −0.75 V (Fig. 10.13b), with the authors hypothesizing the carbon–nitrogen covalent sites within the g-C_3N_4/MWCNT composite to be responsible for the high CO selectivity. In a separate work, our group reported a pyrolyzed carbon nitride and carbon nanotube composite (CN/MWCNT) catalyst for the electroreduction of CO_2 [69]. The CN/MWCNT catalyst showed a high selectivity for the production of CO (FE_{CO} ~ 98%) and also a 3.5-fold improvement in activity compared to some of the best-known Ag nanoparticle catalysts at intermediate cathode potentials of −1.46 V versus Ag/AgCl (Fig. 10.13c, d), highlighting the promise of using

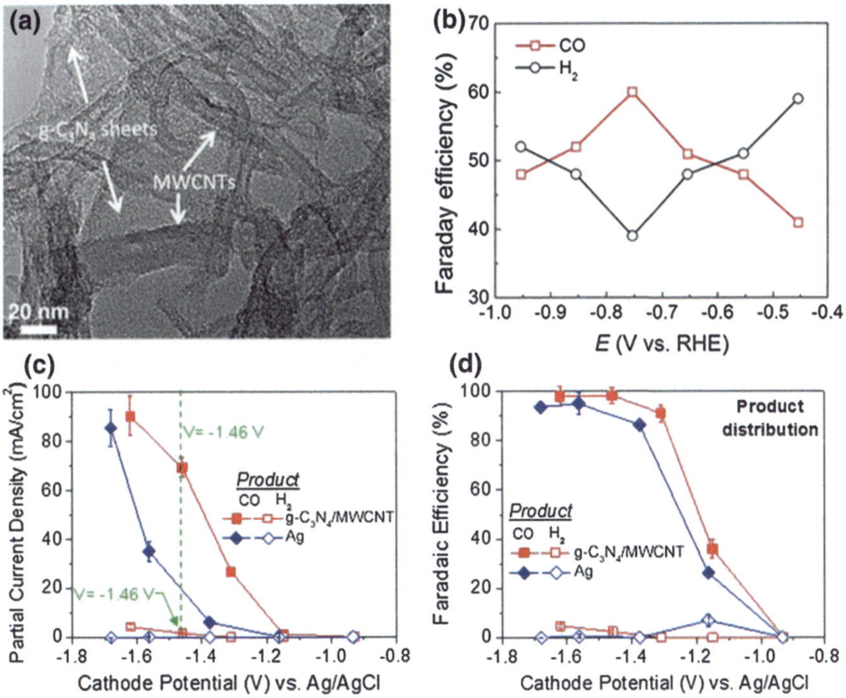

Fig. 10.13 **a** Transmission electron microscopy (TEM) image showing the layered structure of the graphitic carbon nitride/carbon nanotube composite catalyst (g-C_3N_4/MWCNT). **b** Faradaic efficiency as a function of cathode potential for the electroreduction of CO_2 on the g-C_3N_4/MWCNT composite. Reproduced with permission from Ref. [68]. Copyright 2016, Wiley-VCH Verlag GmbH & Co. **c** Partial current density and **d** Faradaic efficiency as a function of the cathode potential for the electroreduction of CO_2 on a pyrolyzed g-C_3N_4/MWCNT composite. Reproduced with permission from Ref. [69]. Copyright 2017, Wiley-VCH Verlag GmbH & Co

carbon-based materials as CO_2 electroreduction catalysts without having to sacrifice activity and selectivity.

The use of CNT supported metal nanoparticle catalysts is another active area of research in the CO_2 electroreduction literature and has been pursued to a good extent by us and others. For example, we have explored the incorporation of multiwalled carbon nanotubes (MWCNT) into the Ag nanoparticle catalyst layer to form three kinds of structures on a gas diffusion layer (GDL) as shown in Fig. 10.14a [70]. The structures include: (i) Ag catalyst layers deposited directly on a GDL (ES1); (ii) Ag catalyst layers on top of MWCNT layers deposited on a GDL (ES2); and (iii) a homogenous mixture of Ag and MWCNT deposited on top of GDL (ES3). Of the different structures, ES3 performed the best with j_{CO} as high as 338 mA cm^{-2} being obtained at a cathode potential of −0.77 V versus RHE (Fig. 10.14b). This represents a twofold enhancement in activity when compared to the conventionally used ES1 structure which was attributed to a lower charge transfer resistance when using the ES3 structure. More recently, we have also reported Au nanoparticles supported on polybenzimidazole wrapped multiwall carbon nanotubes (MWNT/PyPBI/Au) as catalysts for the electroreduction of CO_2 to CO [71]. While Au nanoparticles are well-known to be active and selective for the electroreduction of CO_2 to CO, their stability is often an issue. Utilizing carbon nanotubes wrapped with a pyridine-containing polybenzimidazole polymer provides nucleation sites that can be used for the in situ growth of monodisperse Au nanoparticles (Fig. 10.14c), resulting in improved stability and high electrochemically active surface area. As expected, the MWNT/PyPBI/Au catalyst outperformed other carbon black supported catalysts (CB/PyPBI/Au and CB/Au), unsupported Au nanoparticles, and control samples of MWNT/PyPBI and CB/PyPBI with j_{CO} as high as 160 mA cm^{-2} being obtained at a cathode potential of −1.78 V versus Ag/AgCl and a low Au loading of just 0.17 mg cm^{-2} (Fig. 10.14d). In other work, Hossain et al. reported the electrochemical characterization of 3–60 nm Cu nanoparticles supported on CNTs (Cu/CNT) as catalysts for the electroreduction of CO_2 [72]. Cu/CNT catalyst with a Cu loading of 20 wt% Cu resulted in the highest current density. CH_3OH with a Faradaic efficiency of 38.4% at a cathode potential of −1.7 V versus SCE was reported as the only CO_2 electroreduction product.

10.3.4 Graphene Based CO_2 Electroreduction Catalysts

The integration of graphene materials into electrocatalysts has been presented as another approach to move away from precious transition metals. Atomically thin graphene is typically synthesized by reducing graphene oxide (GO) under high temperatures, which partially restores the semi-metallicity of graphene and increases its conductivity [73]. Moreover, the synthesis becomes more specific depending on the desired graphene morphology and dimensions, as both these properties can enable the tuning of the density of the defects. Several research

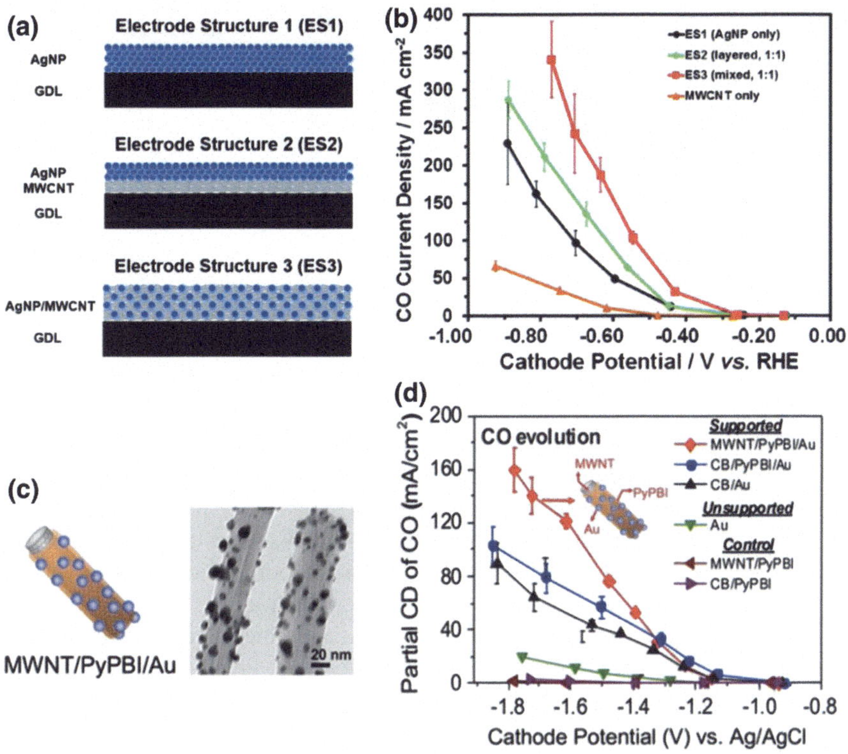

Fig. 10.14 **a** Schematic illustration of the three different Ag nanoparticle-MWCNT composite architectures along with the **b** corresponding partial current density for CO obtained as a function of the cathode potential, when utilizing these architectures as catalysts for the electroreduction of CO_2. Reproduced with permission from Ref. [70]. Copyright 2016, Royal Society of Chemistry. **c** Schematic illustration as well as TEM image for Au nanoparticles supported on polybenzimidazole wrapped multiwall carbon nanotubes (MWNT/PyPBI/Au). **d** Partial current density as a function of cathode potential for the electroreduction of CO_2 on the MWNT/PyPBI/Au catalyst. Reproduced with permission from Ref. [71]. Copyright 2017, Wiley-VCH Verlag GmbH & Co

groups have looked into the utilization of graphene sheets, graphene quantum dots, etc. for the electroreduction of CO_2. In particular, different forms of graphene have been used as supports for traditional metal catalysts to improve their stability, durability, and product selectivity. For example, Alves et al. synthesized copper supported on reduced graphene oxide (Cu/rGO) catalysts as a way to solve instability issues with copper film. Performing electroanalysis in a traditional three-electrode cell, Cu/rGO showed a fourfold improvement in current density (-0.97 mA cm^{-2}) over copper films (-0.24 mA cm^{-2}) at the same potential of -1.54 V versus NHE [73]. However, CO, carbonate, and oxalate were observed as the only CO_2 electroreduction products. In another work, graphene was used to solve stability issues of metal catalysts while increasing activity. Lei et al. created ultrathin metallic Sn quantum sheets confined by graphene as an alternative to bulk

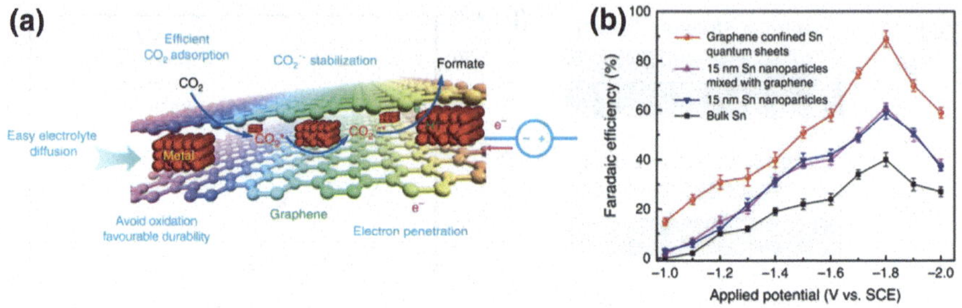

Fig. 10.15 **a** Schematic illustration and **b** Faradaic efficiency as a function of the cathode potential for the electroreduction of CO_2 to $HCOO^-$ on graphene confined Sn quantum sheets. Reproduced from Ref. [74] under the creative commons attribution (CC-BY) license

Sn and Sn nanoparticles (Fig. 10.15a), for the electroreduction of CO_2 to $HCOO^-$ [74]. The confined Sn exhibits a 9-times larger CO_2 absorption capacity compared to bulk Sn, thus improving activity. Selectivity as high as 89% for the production of $HCOO^-$ was realized at an applied potential of −1.8 V versus SCE (Fig. 10.15b). This Sn confined by graphene exhibits a lower Sn-Sn coordination number, verified by X-ray absorption fine structure spectroscopy, which allows for the better stabilization of the rate determining $CO_2^{\bullet-}$ intermediate which is crucial for $HCOO^-$ production. The sandwich-like structure of this catalyst greatly increases the stability and durability of Sn, which was revealed in a 50-h durability test.

Other works have also looked into heteroatom-doped graphene supports to modulate the surface energy of the supported copper. Song et al. supported Cu on nitrogen-doped carbon nanospike (CNS) films as a more selective catalyst for

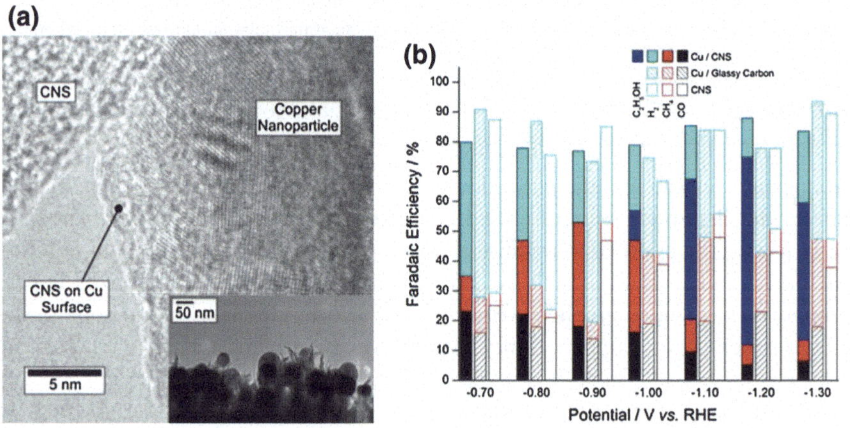

Fig. 10.16 **a** HR-TEM of electrodeposited Cu nanoparticles on nitrogen-doped carbon nanospikes (CNS). **b** Faradaic efficiency for the electroreduction of CO_2 on Cu/CNS, Cu/Glassy carbon, and plain CNS as a function of the cathode potential. Reproduced with permission from Ref. [75]. Copyright 2016, Wiley-VCH Verlag GmbH & Co

C_2H_5OH formation than Cu supported on glassy carbon, reaching a Faradaic efficiency of 63% for C_2H_5OH at −1.2 V versus RHE (Fig. 10.16) [75]. The CNS films are made possible by the high nitrogen defect content in the carbon leading to disorder, which discourages stacking and introduces the curvature of the CNS films, as shown using high resolution transmission electron microscopy (HR-TEM) (Fig. 10.16a). This specific morphology enhances interactions between Cu and the CNS support while possibly strengthening the bond between the CNS film and the C_2 intermediates required for C_2H_5OH production over C_2H_4. Similarly, Li et al. compared Cu nanoparticles supported on pyridinic nitrogen-rich graphene (p-NG) to monodisperse Cu nanoparticles supported on conventional carbon and pristine GO for selectively producing C_2H_4. The Faradaic efficiency for C_2H_4 reached 19% at −0.9 V versus RHE when using Cu/p-NG compared to just 1% and 5% for carbon and GO supports, respectively [76]. Specifically, p-NG exhibits stronger Lewis basicity induced by the presence of pyridinic nitrogen, causing H_2 and protons to become concentrated around the supported Cu and facilitate the CO_2 adsorption, the suggested rate-determining step for the reaction.

With promising results from graphene support materials, many efforts have turned towards metal-free graphene catalysts. However, plain graphene almost exclusively promotes the parasitic H_2 evolution reaction during CO_2 electroreduction. Therefore, the use of heteroatom doping is important to change the properties of graphene. Early on, Sreekanth et al. used boron-doped graphene (BG) to reduce CO_2 to $HCOO^-$/HCOOH, achieving a FE_{HCOO^-} of 66% at −1.4 V versus SCE, which is a threefold improvement over conventionally used bismuth [77]. Through DFT calculations, this work also highlighted the symmetrically distributed electron delocalization of pristine graphene, which boron doping helps to interrupt by providing asymmetric charge and spin density. In addition, boron or carbon can form the important *COOH intermediate in the BG, but carbon in pristine graphene cannot. Also aiming to produce $HCOO^-$, Wang et al. synthesized nitrogen-doped graphene sheet catalysts [78]. As mentioned before, the presence of nitrogen distorts the ordered graphene, producing folded or wrinkled thin carbon sheets. The N-doped graphene sheets resulted in FE_{HCOO^-} of 73% at −0.84 V versus RHE and exhibited high durability of up to 12-h. Similarly, Sun et al. developed various nitrogen-doped carbon (graphene-like) materials on carbon paper (NGM/CP) electrodes by changing the nitrogen-containing base during synthesis [79]. The different bases led to a wide range of surfaces and total nitrogen content in the NGMs, resulting in diverse morphologies. A high CH_4 selectivity of 93.5% was achieved at −1.4 V versus SHE for the electroreduction of CO_2 on the NGM catalyst with the highest amount of total and surface nitrogen, specifically pyridinic and pyridonic/pyrrolic N. The authors postulated that the positively-charged carbon atoms in the electron-withdrawing graphene π system negatively polarize the nitrogen atoms, thereby improving the binding energy for CO_2 and the reaction intermediates.

Looking beyond the utilization of graphene sheets for the electroreduction of CO_2, Wu et al. introduced nitrogen-doped three-dimensional (3D) graphene foam (NG) catalysts as a metal-free alternative to Ag or Au for CO production [80]. The

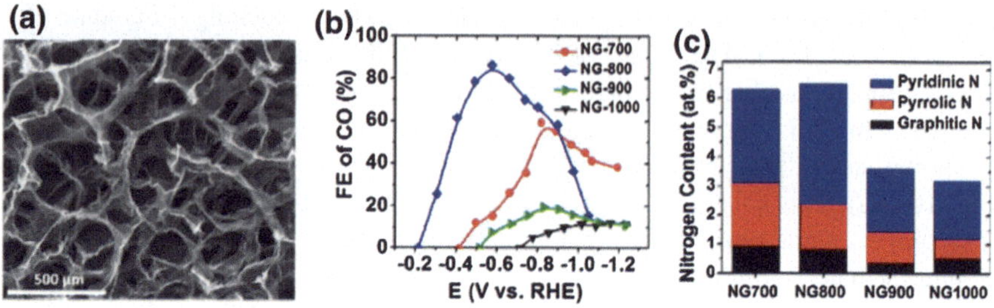

Fig. 10.17 **a** Representative SEM image of the three-dimensional (3D) graphene foam (NG) catalyst showing the open frame structure. **b** Faradaic efficiency for CO production as a function of the cathode potential and **c** N content distribution, for the different NG catalysts prepared by varying the doping temperature (i.e. NG700 implies 700 °C doping temperature). Reproduced with permission from Ref. [80]. Copyright 2016, American Chemical Society

benefit of using graphene foams is its 3D hierarchical structure, shown in Fig. 10.17a, that allows the easy penetration of the electrolyte, therefore increasing the electrochemically active surface area. The NG catalyst performed comparably to Ag and Au at low overpotentials with a FE_{CO} of about 85% being obtained at −0.58 V versus RHE corresponding to an overpotential of −0.47 V. In addition, this work explored the effect of doping temperature on dopant (N) concentration and found that different temperatures led to different concentrations of different types of nitrogen. The NG with the highest concentration of pyridinic nitrogen achieved the highest FE_{CO}, agreeing with the mechanistic conclusions made earlier in the literature (Fig. 10.17b, c).

Recently, in collaboration with Dr. Ajayan's group, we investigated another metal-free 3D graphene structure as an alternative to Cu for hydrocarbon and oxygenate formation: nitrogen-doped graphene quantum dots (NGQDs) with a high density of nitrogen-doped defects at the edge sites (Fig. 10.18a) [81]. The electroreduction of CO_2 on the NGQD catalysts resulted in the selective production of C_2 products such as C_2H_4 (maximum $FE_{C_2H_4}$ ~ 31% at −0.75 V vs. RHE) and C_2H_5OH (maximum $FE_{C_2H_5OH}$ ~ 16% at −0.78 V vs. RHE), that are typically known to be produced only on Cu (Fig. 10.18b). In follow up work, DFT calculations were performed to understand the mechanism of CO_2 electroreduction on NGQDs, and thus explain the formation of C_2 products as well as the preferential formation of CH_4 over CH_3OH [82]. Looking into the thermodynamic free energy diagrams and the localized density of states (Fig. 10.18c), NGQDs were found to lower the free energy of the formation of the *COOH intermediate. Furthermore, the data also suggested the presence of a high density of exposed edges on NGQDs that favored the formation of the pyridinic nitrogen sites, which are presumably the active sites required for carbon–carbon coupling to form C_2 products. For the preferential formation of CH_4 over CH_3OH, water-mediated proton shuttling was shown to be the key as it lowered the energy barrier for the formation of the *CH_2 intermediate.

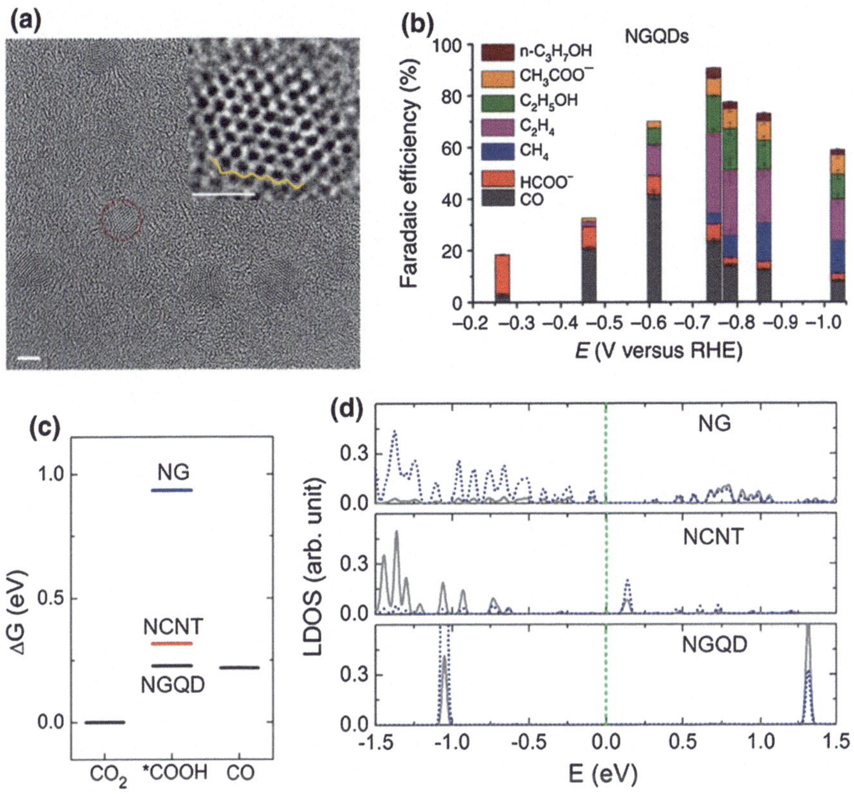

Fig. 10.18 **a** High resolution TEM image of the nitrogen-doped graphene quantum dots (NGQDs) with a single NGQD shown in inset. **b** Faradaic efficiency as a function of the cathode potential for the electroreduction of CO_2 on NGQDs. Reproduced from Ref. [81] under the creative commons attribution (CC-BY) license. **c** Free energy diagram for the electroreduction of CO_2 and localized density of states (LDOS) for different nitrogen-doped carbon materials. NG and NCNT represent nitrogen-doped graphene and carbon nanotubes, respectively. Reproduced with permission from Ref. [82]. Copyright 2017, American Chemical Society

10.3.5 Nanodiamond Based CO_2 Electroreduction Catalysts

Heteroatom (nitrogen, boron, etc.)-doped nanodiamond materials form another interesting class of catalysts for the electroreduction of CO_2. Nanodiamonds exhibit desirable mechanical and electrical properties like stability under a wide potential window, chemical inertness, high mechanical durability, and can be obtained in large volumes at a relatively low cost [83]. Nakata et al. performed the first investigation of such nanodiamond based electrodes by characterizing the catalytic performance of boron-doped diamond (BDD) for the electroreduction of CO_2 using three different electrolytes, i.e., methanol, seawater, and aqueous NaCl [84].

The BDD electrodes exhibited high selectivity for formaldehyde (HCHO) with a maximum Faradaic efficiency of 74% being obtained at a cathode potential of −1.7 V versus Ag/Ag+ (Fig. 10.19a). In a follow up effort by the same group, the authors showed that by changing the electrolyte to an aqueous ammonia solution, the selectivity of the BDD electrodes could be tuned to produce CH_3OH as the main CO_2 electroreduction product with a Faradaic efficiency of 24.3% at a cathode potential of −1.3 V versus Ag/AgCl [85]. Continuing with the trend of investigating nanodiamond materials, Liu et al. proposed a N-doped nanodiamond/Si rod array catalyst (NDD/Si RA) for the selective electroreduction of CO_2 to acetate (CH_3COO^-) [86]. The Faradaic efficiency for acetate production was between 77.3 to 77.6% for an applied cathode potential between −0.8 and −1.0 V versus RHE (Fig. 10.19b). Based on a combination of the electrokinetic data and in situ infrared spectroscopy, the authors predicted the pathway for the formation of CH_3COO^-) on NDD/Si RA to be $CO_2 \rightarrow CO_2^{\bullet-} \rightarrow (COO)_2^{\bullet} \rightarrow CH_3COO^-$. Inspired by the impactful results of doping nanodiamonds with boron and nitrogen separately to tune the selectivity of CO_2 electroreduction, Liu et al. reported a boron and nitrogen co-doped nanodiamond (BND) electrode for the selective electroreduction of CO_2 to C_2H_5OH [87]. Three different kinds of BND electrodes for this study were prepared by depositing a film of BND on a Si substrate via hot filament chemical vapor deposition using a gas mixture of CH_4 (2.5%)/B_2H_6 (12.5%)/N_2 (2.5% for BND1, 5.0% for BND2, and 10.0% for BND3)/H_2. A maximum $FE_{C_2H_5OH}$ of 93.2% was achieved when using the BND3 catalyst at a cathode potential of −1.0 V versus RHE (Fig. 10.19c), thus overcoming the limitation of producing C_2H_5OH at high selectivity via the electroreduction of CO_2. Using a combination of electrochemical experiments and theoretical DFT calculations, the authors suggested the selective electroreduction of CO_2 to C_2H_5OH to be a result of synergistic effects due to the co-doping of boron and nitrogen.

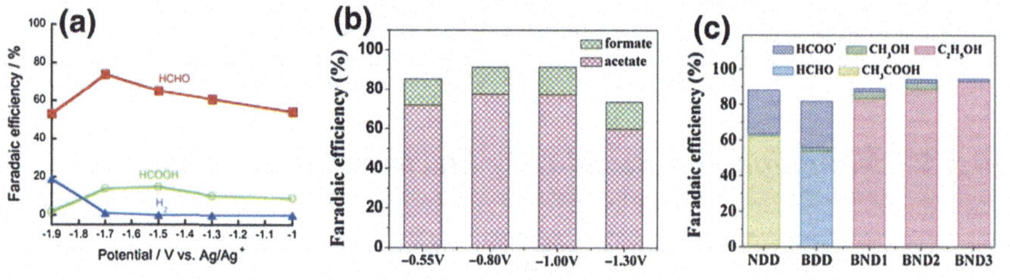

Fig. 10.19 Faradaic efficiency for the electroreduction of CO_2 as a function of the cathode potential when using **a** boron-doped diamond (reproduced with permission from Ref. [84]. Copyright 2014, Wiley-VCH Verlag GmbH & Co.); **b** N-doped nanodiamond/Si rod array (reproduced with permission from Ref. [86]. Copyright 2015, American Chemical Society); and **c** as a function of the different nitrogen-doped nanodiamond (NDD), boron doped nanodiamond (BDD) as well as boron and nitrogen co-doped nanodiamond (BND1, BND2, BND3) catalyst at a constant cathode potential of 1.0 V *vs.* RHE. Reproduced with permission from Ref. [87]. Copyright 2017, Wiley-VCH Verlag GmbH & Co

10.4 Summary and Outlook

In conclusion, the electroreduction of CO_2 to value-added chemicals could offer interesting opportunities for recycling and/or reducing excess wasteful CO_2. Carbon-based catalysts and electrodes with tunable properties form an emerging class of materials that have recently attracted significant interest in the CO_2 electroreduction research community. In addition to providing cost benefits when compared to the traditional transition metal-based catalysts (such as Au, Ag, etc.) for the electroreduction of CO_2, many of the carbon-based materials have matched and even exceeded the performance levels obtained with state of the art metal-based catalysts. A star example of carbon-based materials for the electroreduction of CO_2 is the boron and nitrogen co-doped nanodiamond (BND) catalyst that can facilitate the production of C_2H_5OH at selectivity greater than 90%, thus surpassing a long-standing issue related to the selective production of multicarbon hydrocarbons and oxygenates in the field.

While the efforts highlighted in this chapter definitely show the promise of carbon-based materials for the electroreduction of CO_2, moving forward many different research strategies will have to be utilized to take the performance of carbon-based materials to the next level. For example, identifying active sites that are responsible for the high activity and selectivity of heteroatom-doped carbon materials will be crucial for the future development of this field. Combinations of in situ or in operando spectroscopy with DFT calculations will be needed to understand the mechanism at the molecular level. Once the active sites have been identified, better material synthesis tools will have to be developed to design catalysts that exhibit a high density of the active sites. Another interesting method to improve the performance of carbon-based materials and catalysts could be the integration of electrolyte effects to enhance the CO_2 electroreduction performance. Some of our previous work has shown that using either an alkaline or ionic liquid-based electrolyte can significantly enhance the activity and selectivity of well-known catalytic materials. Electrolyte optimization studies could thus be a very simple tool for improving performance. Another key research question that needs to be further investigated pertains to the stability and durability of different catalysts and electrodes for the electroreduction of CO_2. Electrochemical experiments that evaluate the stability and/or durability of CO_2 electroreduction catalysts and electrodes at time scales exceeding 1000s of hours and current density exceeding 100 mA cm^{-2} i.e., performance levels required for industrial implementation of the process, are almost nonexistent. Hence, better experimental tools as well as robust durability testing protocols will have to be developed to gain a better understanding of how catalysts and electrodes degrade over extended periods of time.

References

1. Trends in atmospheric carbon dioxide. National Oceanic and Atmospheric Administration, Earth System Research Laboratory. https://www.esrl.noaa.gov/gmd/ccgg/trends/weekly.html. Accessed 24 Nov 2017
2. Hansen J, Kharecha P, Sato M, Masson-Delmotte V, Ackerman F, Beerling DJ, Hearty PJ, Hoegh-Guldberg O, Hsu SL, Parmesan C, Rockstrom J, Rohling EJ, Sachs J, Smith P, Steffen K, Van Susteren L, von Schuckmann K, Zachos JC (2013) Assessing "Dangerous climate change": required reduction of carbon emissions to protect young people, future generations and nature. Plos One 8(12):e81648. https://doi.org/10.1371/journal.pone.0081648
3. Feldman DR, Collins WD, Gero PJ, Torn MS, Mlawer EJ, Shippert TR (2015) Observational determination of surface radiative forcing by CO_2 from 2000 to 2010. Nature 519(7543):339–343. https://doi.org/10.1038/nature14240
4. Chu S, Majumdar A (2012) Opportunities and challenges for a sustainable energy future. Nature 488(7411):294–303. https://doi.org/10.1038/nature11475
5. Pacala S, Socolow R (2004) Stabilization wedges: solving the climate problem for the next 50 years with current technologies. Science 305(5686):968–972. https://doi.org/10.1126/science.1100103
6. Intergovernmental panel on climate change 2013: the physical science basis
7. Aresta M, Dibenedetto A, Angelini A (2014) Catalysis for the valorization of exhaust carbon: from CO_2 to chemicals, materials, and fuels. Technological use of CO_2. Chem Rev 114 (3):1709–1742. https://doi.org/10.1021/cr4002758
8. CO_2 utilization focus area. National Energy Technology Laboratory, U.S. Department of Energy. Accessed 24 Nov 2017
9. Appel AM, Bercaw JE, Bocarsly AB, Dobbek H, DuBois DL, Dupuis M, Ferry JG, Fujita E, Hille R, Kenis PJA, Kerfeld CA, Morris RH, Peden CHF, Portis AR, Ragsdale SW, Rauchfuss TB, Reek JNH, Seefeldt LC, Thauer RK, Waldrop GL (2013) Frontiers, opportunities, and challenges in biochemical and chemical catalysis of CO_2 fixation. Chem Rev 113(8):6621–6658. https://doi.org/10.1021/cr300463y
10. Herron JA, Kim J, Upadhye AA, Huber GW, Maravelias CT (2015) A general framework for the assessment of solar fuel technologies. Energy Environ Sci 8(1):126–157. https://doi.org/10.1039/c4ee01958j
11. Jhong HRM, Ma S, Kenis PJA (2013) Electrochemical conversion of CO_2 to useful chemicals: current status, remaining challenges, and future opportunities. Curr Opin Chem Eng 2(2):191–199. https://doi.org/10.1016/j.coche.2013.03.005
12. Whipple DT, Kenis PJA (2010) Prospects of CO_2 utilization via direct heterogeneous electrochemical reduction. J Phys Chem Lett 1(24):3451–3458. https://doi.org/10.1021/jz1012627
13. Kuhl KP, Hatsukade T, Cave ER, Abram DN, Kibsgaard J, Jaramillo TF (2014) Electrocatalytic conversion of carbon dioxide to methane and methanol on transition metal surfaces. J Am Chem Soc 136(40):14107–14113. https://doi.org/10.1021/ja505791r
14. Vasileff A, Zheng Y, Qiao SZ (2017) Carbon solving carbon's problems: recent progress of nanostructured carbon-based catalysts for the electrochemical reduction of CO_2. Adv Energy Mater 7(21). https://doi.org/10.1002/aenm.201700759
15. Duan XC, Xu JT, Wei ZX, Ma JM, Guo SJ, Wang SY, Liu HK, Dou SX (2017) Metal-free carbon materials for CO_2 electrochemical reduction. Adv Mater 29(41). https://doi.org/10.1002/adma.201701784
16. Kuhl KP, Cave ER, Abram DN, Jaramillo TF (2012) New insights into the electrochemical reduction of carbon dioxide on metallic copper surfaces. Energy Environ Sci 5(5):7050–7059. https://doi.org/10.1039/c2ee21234j
17. Hori Y (2008) Electrochemical CO_2 reduction on metal electrodes. Modern aspects of electrochemistry, no 42. Springer, New York, pp 89–189

18. Hori Y, Wakebe H, Tsukamoto T, Koga O (1994) Electrocatalytic process of CO selectivity in electrochemical reduction of CO_2 at metal-electrodes in aqueous-media. Electrochim Acta 39(11–12):1833–1839. https://doi.org/10.1016/0013-4686(94)85172-7
19. Bard AJ, Parsons R, Jordan J (1985) Standard potentials in aqueous solutions. CRC Press, Boca Raton
20. Lu Q, Jiao F (2016) Electrochemical CO_2 reduction: electrocatalyst, reaction mechanism, and process engineering. Nano Energy 29:439–456. https://doi.org/10.1016/j.nanoen.2016.04.009
21. Kumar B, Brian JP, Atla V, Kumari S, Bertram KA, White RT, Spurgeon JM (2016) New trends in the development of heterogeneous catalysts for electrochemical CO_2 reduction. Catal Today 270:19–30. https://doi.org/10.1016/j.cattod.2016.02.006
22. Martin AJ, Larrazabal GO, Perez-Ramirez J (2015) Towards sustainable fuels and chemicals through the electrochemical reduction of CO_2: lessons from water electrolysis. Green Chem 17(12):5114–5130. https://doi.org/10.1039/c5gc01893e
23. Larrazabal GO, Martin AJ, Perez-Ramirez J (2017) Building blocks for high performance in electrocatalytic CO_2 reduction: materials, optimization strategies, and device engineering. J Phys Chem Lett 8(16):3933–3944. https://doi.org/10.1021/acs.jpclett.7b01380
24. Schwarz HA, Dodson RW (1989) Reduction potentials of CO_2^- and the alcohol radicals. J Phys Chem 93(1):409–414. https://doi.org/10.1021/j100338a079
25. Chandrasekaran K, Bockris LM (1987) In-situ spectroscopic investigation of adsorbed intermediate radicals in electrochemical reactions: CO_2^- on platinum. Surf Sci 185(3):495–514. https://doi.org/10.1016/S0039-6028(87)80173-5
26. Kortlever R, Shen J, Schouten KJP, Calle-Vallejo F, Koper MTM (2015) Catalysts and reaction pathways for the electrochemical reduction of carbon dioxide. J Phys Chem Lett. 6 (20):4073–4082. https://doi.org/10.1021/acs.jpclett.5b01559
27. Qiao JL, Liu YY, Hong F, Zhang JJ (2014) A review of catalysts for the electroreduction of carbon dioxide to produce low-carbon fuels. Chem Soc Rev 43(2):631–675. https://doi.org/10.1039/c3cs60323g
28. Peterson AA, Abild-Pedersen F, Studt F, Rossmeisl J, Norskov JK (2010) How copper catalyzes the electroreduction of carbon dioxide into hydrocarbon fuels. Energy Environ Sci 3 (9):1311–1315. https://doi.org/10.1039/c0ee00071j
29. Verma S, Kim B, Jhong HRM, Ma S, Kenis PJA (2016) a gross-margin model for defining technoeconomic benchmarks in the electroreduction of CO_2. Chemsuschem 9(15):1972–1979. https://doi.org/10.1002/cssc.201600394
30. Verma S, Hamasaki Y, Kim C, Huang W, Lu S, Jhong HRM, Gewirth AA, Fujigaya T, Nakashima N, Kenis PJA (2018) Insights into the low overpotential electroreduction of CO_2 to CO on a supported gold catalyst in an alkaline flow electrolyzer. ACS Energy Lett. 3 (1):193–198. https://doi.org/10.1021/acsenergylett.7b01096
31. Chen YH, Li CW, Kanan MW (2012) Aqueous co_2 reduction at very low overpotential on oxide-derived Au nanoparticles. J Am Chem Soc 134(49):19969–19972. https://doi.org/10.1021/ja309317u
32. Lu Q, Rosen J, Zhou Y, Hutchings GS, Kimmel YC, Chen JGG, Jiao F (2014) A selective and efficient electrocatalyst for carbon dioxide reduction. Nat Commun 5. https://doi.org/10.1038/ncomms4242
33. Weiss RF (1974) Carbon dioxide in water and seawater: the solubility of a non-ideal gas. Mar Chem 2:203–215. https://doi.org/10.1016/0304-4203(74)90015-2
34. Verma S, Lu X, Ma S, Masel RI, Kenis PJA (2016) The effect of electrolyte composition on the electroreduction of CO_2 to CO on Ag based gas diffusion electrodes. Phys Chem Chem Phys 18(10):7075–7084. https://doi.org/10.1039/c5cp05665a
35. Kim B, Ma S, Jhong HRM, Kenis PJA (2015) Influence of dilute feed and pH on electrochemical reduction of CO_2 to CO on Ag in a continuous flow electrolyzer. Electrochim Acta 166:271–276. https://doi.org/10.1016/j.electacta.2015.03.064
36. Ma S, Luo R, Moniri S, Lan Y, Kenis PJA (2014) Efficient electrochemical flow system with improved anode for the conversion of CO_2 to CO. J Electrochem Soc 161(10):F1124–F1131. https://doi.org/10.1149/2.1201410jes

37. Ma S, Lan YC, Perez GMJ, Moniri S, Kenis PJA (2014) Silver supported on titania as an active catalyst for electrochemical carbon dioxide reduction. Chemsuschem 7(3):866–874. https://doi.org/10.1002/cssc.201300934
38. Thorson MR, Siil KI, Kenis PJA (2013) Effect of cations on the electrochemical conversion of CO_2 to CO. J Electrochem Soc 160(1):F69–F74. https://doi.org/10.1149/2.052301jes
39. Salehi-Khojin A, Jhong HRM, Rosen BA, Zhu W, Ma SC, Kenis PJA, Masel RI (2013) Nanoparticle silver catalysts that show enhanced activity for carbon dioxide electrolysis. J Phys Chem C 117(4):1627–1632. https://doi.org/10.1021/jp310509z
40. Jhong HRM, Brushett FR, Kenis PJA (2013) The effects of catalyst layer deposition methodology on electrode performance. Adv Energy Mater 3(5):589–599. https://doi.org/10.1002/aenm.201200759
41. Tornow CE, Thorson MR, Ma S, Gewirth AA, Kenis PJA (2012) Nitrogen-based catalysts for the electrochemical reduction of CO_2 to CO. J Am Chem Soc 134(48):19520–19523. https://doi.org/10.1021/ja308217w
42. Whipple DT, Finke EC, Kenis PJA (2010) Microfluidic reactor for the electrochemical reduction of carbon dioxide: the effect of pH. Electrochem Solid-State Lett 13(9):B109. https://doi.org/10.1149/1.3456590
43. Ma S, Liu JF, Sasaki K, Lyth SM, Kenis PJA (2017) Carbon foam decorated with silver nanoparticles for electrochemical CO_2 conversion. Energy Technol 5(6):861–863. https://doi.org/10.1002/ente.201600576
44. Delacourt C, Ridgway PL, Kerr JB, Newman J (2008) Design of an electrochemical cell making syngas (CO + H_2) from CO_2 and H_2O reduction at room temperature. J Electrochem Soc 155(1):B42–B49. https://doi.org/10.1149/1.2801871
45. Dufek EJ, Lister TE, McIlwain ME (2011) Bench-scale electrochemical system for generation of CO and syngas. J Appl Electrochem 41(6):623–631. https://doi.org/10.1007/s10800-011-0271-6
46. Naughton MS, Brushett FR, Kenis PJA (2011) Carbonate resilience of flowing electrolyte-based alkaline fuel cells. J Power Sources 196(4):1762–1768. https://doi.org/10.1016/j.jpowsour.2010.09.114
47. Endrődi B, Bencsik G, Darvas F, Jones R, Rajeshwar K, Janáky C (2017) Continuous-flow electroreduction of carbon dioxide. Prog Energy Combust Sci 62(133–154). https://doi.org/10.1016/j.pecs.2017.05.005
48. Seh ZW, Kibsgaard J, Dickens CF, Chorkendorff IB, Norskov JK, Jaramillo TF (2017) Combining theory and experiment in electrocatalysis: insights into materials design. Science 355(6321). https://doi.org/10.1126/science.aad4998
49. Khezri B, Fisher AC, Pumera M (2017) CO_2 reduction: the quest for electrocatalytic materials. J Mater Chem A 5(18):8230–8246. https://doi.org/10.1039/c6ta09875d
50. Vesborg PCK, Jaramillo TF (2012) Addressing the terawatt challenge: scalability in the supply of chemical elements for renewable energy. RSC Adv 2(21):7933–7947. https://doi.org/10.1039/c2ra20839c
51. Rosen BA, Salehi-Khojin A, Thorson MR, Zhu W, Whipple DT, Kenis PJA, Masel RI (2011) Ionic liquid-mediated selective conversion of CO_2 to CO at low overpotentials. Science 334 (6056):643–644. https://doi.org/10.1126/science.1209786
52. Kutz RB, Chen QM, Yang HZ, Sajjad SD, Liu ZC, Masel IR (2017) Sustainion imidazolium-functionalized polymers for carbon dioxide electrolysis. Energy Technol 5 (6):929–936. https://doi.org/10.1002/ente.201600636
53. Yang HZ, Kaczur JJ, Sajjad SD, Masel RI (2017) Electrochemical conversion of CO_2 to formic acid utilizing sustainion (TM) membranes. J CO_2 Util 20:208–217. https://doi.org/10.1016/j.jcou.2017.04.011
54. Ma S, Sadakiyo M, Luo R, Heima M, Yamauchi M, Kenis PJA (2016) One-step electrosynthesis of ethylene and ethanol from CO_2 in an alkaline electrolyzer. J Power Sources 301:219–228. https://doi.org/10.1016/j.jpowsour.2015.09.124

55. Sharma PP, Zhou XD (2017) Electrocatalytic conversion of carbon dioxide to fuels: a review on the interaction between CO_2 and the liquid electrolyte. Wires Energy Environ 6(4). https://doi.org/10.1002/wene.239
56. Gong KP, Du F, Xia ZH, Durstock M, Dai LM (2009) Nitrogen-doped carbon nanotube arrays with high electrocatalytic activity for oxygen reduction. Science 323(5915):760–764. https://doi.org/10.1126/science.1168049
57. Liu X, Dai LM (2016) Carbon-based metal-free catalysts. Nat Rev Mater 1(11). https://doi.org/10.1038/natrevmats.2016.64
58. Kim B, Hillman F, Ariyoshi M, Fujikawa S, Kenis PJA (2016) Effects of composition of the micro porous layer and the substrate on performance in the electrochemical reduction of CO_2 to CO. J Power Sources 312:192–198. https://doi.org/10.1016/j.jpowsour.2016.02.043
59. Wang H, Jia J, Song PF, Wang Q, Li DB, Min SX, Qian CX, Wang L, Li YF, Ma C, Wu T, Yuan JY, Antonietti M, Ozin GA (2017) Efficient electrocatalytic reduction of CO_2 by nitrogen-doped nanoporous carbon/carbon nanotube membranes: a step towards the electrochemical CO_2 refinery. Angew Chem Int Edit 56(27):7847–7852. https://doi.org/10.1002/anie.201703720
60. Kumar B, Asadi M, Pisasale D, Sinha-Ray S, Rosen BA, Haasch R, Abiade J, Yarin AL, Salehi-Khojin A (2013) Renewable and metal-free carbon nanofibre catalysts for carbon dioxide reduction. Nat Commun 4. https://doi.org/10.1038/ncomms3819
61. Varela AS, Sahraie NR, Steinberg J, Ju W, Oh HS, Strasser P (2015) Metal-doped nitrogenated carbon as an efficient catalyst for direct CO_2 electroreduction to CO and hydrocarbons. Angew Chem Int Edit 54(37):10758–10762. https://doi.org/10.1002/anie.201502099
62. Li WL, Seredych M, Rodriguez-Castellon E, Bandosz TJ (2016) Metal-free nanoporous carbon as a catalyst for electrochemical reduction of CO_2 to CO and CH_4. Chemsuschem 9 (6):606–616. https://doi.org/10.1002/cssc.201501575
63. Song YF, Chen W, Zhao CC, Li SG, Wei W, Sun YH (2017) Metal-free nitrogen-doped mesoporous carbon for electroreduction of CO_2 to ethanol. Angew Chem Int Edit 56 (36):10840–10844. https://doi.org/10.1002/anie.201706777
64. Chai GL, Guo ZX (2016) Highly effective sites and selectivity of nitrogen-doped graphene/CNT catalysts for CO_2 electrochemical reduction. Chem Sci 7(2):1268–1275. https://doi.org/10.1039/c5sc03695j
65. Zhang S, Kang P, Ubnoske S, Brennaman MK, Song N, House RL, Glass JT, Meyer TJ (2014) Polyethylenimine-enhanced electrocatalytic reduction of CO_2 to formate at nitrogen-doped carbon nanomaterials. J Am Chem Soc 136(22):7845–7848. https://doi.org/10.1021/ja5031529
66. Wu JJ, Yadav RM, Liu MJ, Sharma PP, Tiwary CS, Ma LL, Zou XL, Zhou XD, Yakobson BI, Lou J, Ajayan PM (2015) Achieving highly efficient, selective, and stable CO_2 reduction on nitrogen-doped carbon nanotubes. ACS Nano 9(5):5364–5371. https://doi.org/10.1021/acsnano.5b01079
67. Sharma PP, Wu JJ, Yadav RM, Liu MJ, Wright CJ, Tiwary CS, Yakobson BI, Lou J, Ajayan PM, Zhou XD (2015) Nitrogen-doped carbon nanotube arrays for high-efficiency electrochemical reduction of CO_2: on the understanding of defects, defect density, and selectivity. Angew Chem Int Edit 54(46):13701–13705. https://doi.org/10.1002/anie.201506062
68. Lu XY, Tan TH, Ng YH, Amal R (2016) Highly selective and stable reduction of CO_2 to CO by a graphitic carbon nitride/carbon nanotube composite electrocatalyst. Chem-Eur J 22 (34):11991–11996. https://doi.org/10.1002/chem.201601674
69. Jhong HRM, Tornow CE, Smid B, Gewirth AA, Lyth SM, Kenis PJA (2017) A nitrogen-doped carbon catalyst for electrochemical CO_2 conversion to CO with high selectivity and current density. Chemsuschem 10(6):1094–1099. https://doi.org/10.1002/cssc.201600843

70. Ma S, Luo R, Gold JI, Yu AZ, Kim B, Kenis PJA (2016) Carbon nanotube containing Ag catalyst layers for efficient and selective reduction of carbon dioxide. J Mater Chem A 4 (22):8573–8578. https://doi.org/10.1039/c6ta00427j
71. Jhong HRM, Tornow CE, Kim C, Verma S, Oberst JL, Anderson PS, Gewirth AA, Fujigaya T, Nakashima N, Kenis PJA (2017) Gold nanoparticles on polymer-wrapped carbon nanotubes: an efficient and selective catalyst for the electroreduction of CO_2. ChemPhysChem. https://doi.org/10.1002/cphc.201700815
72. Hossain SS, Rahman SU, Ahmed S (2014) electrochemical reduction of carbon dioxide over CNT-supported nanoscale copper electrocatalysts. J Nanomater. https://doi.org/10.1155/2014/374318
73. Alves DCB, Silva R, Voiry D, Asefa T, Chhowalla M (2015) Copper nanoparticles stabilized by reduced graphene oxide for CO_2 reduction reaction. Mater Renew Sustain 4(1). https://doi.org/10.1007/s40243-015-0042-0
74. Lei FC, Liu W, Sun YF, Xu JQ, Liu KT, Liang L, Yao T, Pan BC, Wei SQ, Xie Y (2016) Metallic tin quantum sheets confined in graphene toward high-efficiency carbon dioxide electroreduction. Nat Commun 7. https://doi.org/10.1038/ncomms12697
75. Song Y, Peng R, Hensley DK, Bonnesen PV, Liang LB, Wu ZL, Meyer HM, Chi MF, Ma C, Sumpter BG, Rondinone AJ (2016) High-selectivity electrochemical conversion of CO_2 to ethanol using a copper nanoparticle/N-doped graphene electrode. ChemistrySelect 1 (19):6055–6061. https://doi.org/10.1002/slct.201601169
76. Li Q, Zhu WL, Fu JJ, Zhang HY, Wu G, Sun SH (2016) Controlled assembly of Cu nanoparticles on pyridinic-N rich graphene for electrochemical reduction of CO_2 to ethylene. Nano Energy 24:1–9. https://doi.org/10.1016/j.nanoen.2016.03.024
77. Sreekanth N, Nazrulla MA, Vineesh TV, Sailaja K, Phani KL (2015) Metal-free boron-doped graphene for selective electroreduction of carbon dioxide to formic acid/formate. Chem Commun 51(89):16061–16064. https://doi.org/10.1039/c5cc06051f
78. Wang HX, Chen YB, Hou XL, Ma CY, Tan TW (2016) Nitrogen-doped graphenes as efficient electrocatalysts for the selective reduction of carbon dioxide to formate in aqueous solution. Green Chem 18(11):3250–3256. https://doi.org/10.1039/c6gc00410e
79. Sun XF, Kang XC, Zhu QG, Ma J, Yang GY, Liu ZM, Han BX (2016) Very highly efficient reduction of CO_2 to CH_4 using metal-free N-doped carbon electrodes. Chem Sci 7(4):2883–2887. https://doi.org/10.1039/c5sc04158a
80. Wu JJ, Liu MJ, Sharma PP, Yadav RM, Ma LL, Yang YC, Zou XL, Zhou XD, Vajtai R, Yakobson BI, Lou J, Ajayan PM (2016) Incorporation of nitrogen defects for efficient reduction of CO_2 via two-electron pathway on three-dimensional graphene foam. Nano Lett 16(1):466–470. https://doi.org/10.1021/acs.nanolett.5b04123
81. Wu JJ, Ma S, Sun J, Gold JI, Tiwary C, Kim B, Zhu LY, Chopra N, Odeh IN, Vajtai R, Yu AZ, Luo R, Lou J, Ding GQ, Kenis PJA, Ajayan PM (2016) A metal-free electrocatalyst for carbon dioxide reduction to multi-carbon hydrocarbons and oxygenates. Nat Commun 7. https://doi.org/10.1038/ncomms13869
82. Zou XL, Liu MJ, Wu JJ, Ajayan PM, Li J, Liu BL, Yakobson BI (2017) How nitrogen-doped graphene quantum dots catalyze electroreduction of CO_2 to hydrocarbons and oxygenates. ACS Catal 7(9):6245–6250. https://doi.org/10.1021/acscatal.7b01839
83. Mochalin VN, Shenderova O, Ho D, Gogotsi Y (2012) The properties and applications of nanodiamonds. Nat Nanotechnol 7(1):11–23. https://doi.org/10.1038/Nnano.2011.209
84. Nakata K, Ozaki T, Terashima C, Fujishima A, Einaga Y (2014) High-yield electrochemical production of formaldehyde from CO_2 and seawater. Angew Chem Int Edit 53(3):871–874. https://doi.org/10.1002/anie.201308657
85. Jiwanti PK, Natsui K, Nakata K, Einaga Y (2016) Selective production of methanol by the electrochemical reduction of CO_2 on boron-doped diamond electrodes in aqueous ammonia solution. RSC Adv 6(104):102214–102217. https://doi.org/10.1039/c6ra20466j

86. Liu Y, Chen S, Quan X, Yu H (2015) Efficient electrochemical reduction of carbon dioxide to acetate on nitrogen-doped nanodiamond. J Am Chem Soc 137(36):11631–11636. https://doi.org/10.1021/jacs.5b02975
87. Liu Y, Zhang Y, Cheng K, Quan X, Fan X, Su Y, Chen S, Zhao H, Zhang Y, Yu H, Hoffmann MR (2017) Selective electrochemical reduction of carbon dioxide to ethanol on a boron- and nitrogen-Co-doped nanodiamond. Angew Chem Int Edit 56:15607–15611. https://doi.org/10.1002/anie.201706311

Chapter 11
Recent Progress in Non-precious Metal Fuel Cell Catalysts

Yuta Nabae and Akimitsu Ishihara

11.1 Fe/N/C and N/C Carbon-Based Cathode Catalysts

11.1.1 Introduction

Polymer electrolyte fuel cells (PEFCs) have received a great deal of attention for their utility in applications such as transportation, portable devices, and combined heat and power systems due to their high energy efficiency and scalability. Figure 11.1 contains a schematic diagram of a typical PEFC. The fuel, H_2, is oxidized using an anode catalyst and the resultant proton and electron are transported through a proton exchange membrane (PEM) and an external circuit, respectively, at which point O_2 from air is reduced by the proton and electron on a cathode catalyst. PEFCs with alkaline membranes have also been proposed but because they suffer from poor durability, a proton exchange membrane functionalized with sulfonic acid is used in all commercialized PEFCs. Because such PEMs are strongly acidic, the range of potential materials for the anode and cathode catalysts is quite limited from a durability standpoint; therefore, Pt nanoparticles loaded on a carbon support (Pt/C) have been the first choice for both anode and cathode catalysts. The precious metal content in these catalysts must be greatly reduced to globalize PEFCs, but because the oxygen reduction reaction (ORR) is a slow electrochemical reaction, the cathode catalysts in modern commercial PEFCs require a relatively large amount of Pt. Thus, it is extremely important to develop

Y. Nabae
Department of Materials Science and Engineering, Tokyo Institute of Technology,
2-12-1 S8-26, Ookayama, Meguro-ku, Tokyo 152-8552, Japan
e-mail: nabae.y.aa@m.titech.ac.jp

A. Ishihara (✉)
Institute of Advanced Sciences, Yokohama National University,
79-5 Tokiwadai, Hodogaya-ku, Yokohama, Kanagawa 240-8501, Japan
e-mail: ishihara-akimitsu-nh@ynu.ac.jp

N. Nakashima (ed.), *Nanocarbons for Energy Conversion: Supramolecular Approaches*, Nanostructure Science and Technology,
https://doi.org/10.1007/978-3-319-92917-0_11

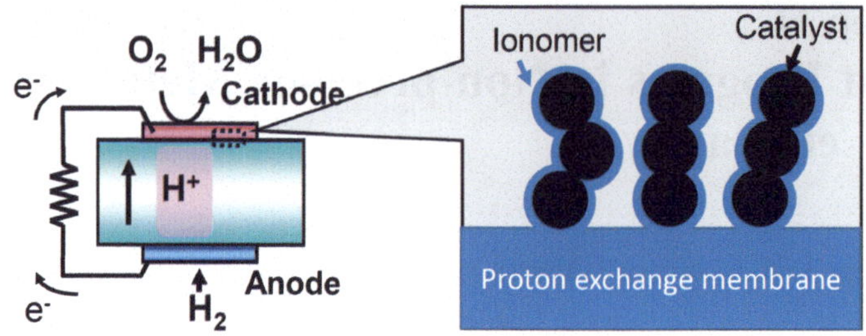

Fig. 11.1 Schematic of a PEFC with proton exchange membrane

non-precious metal (NPM) cathode catalysts. Because commercial PEFCs utilize PEM for the electrolyte, this book chapter focuses on the ORR over NPM catalysts in acidic media.

The history of the development of NPM ORR catalysts begins with the discovery of the catalytic activity of Co phthalocyanine for ORR by Jasinski and coworkers [1]. Chemists were inspired by the similarity of such macrocyclic compounds to natural enzymes and many macrocyclic compounds were studied as ORR catalysts [2]; however, their catalytic activities and durabilities were not sufficient to warrant their consideration for commercial fuel cell catalysts. In this context, since Jahnke reported that heat-treatment can improve the ORR catalytic activity and durability of macrocyclic compounds [3], numerous attempts have been made to develop NPM cathode catalysts by pyrolyzing precursors containing transition metals (mainly Fe or Co) and nitrogen and carbon sources [4, 5]. In brief, when Fe, N, and C-containing precursors are pyrolyzed at 600–1000 °C, ORR catalysis occurs to some degree, however, the nature of the catalytic activity depends on the precursors and pyrolysis protocol.

11.1.2 ORR Reaction Scheme and Active Sites on Fe/N/C and N/C Catalysts

Figure 11.2a shows a typical reaction model for ORRs [6]. In the ORR, a four-electron reduction pathway produces H_2O and releases a relatively large free energy (1.23 V vs. RHE), which is suitable for fuel cell applications, while the two-electron reduction pathway produces H_2O_2 at a lower potential (0.7 V). Some of the H_2O_2 produced could be further reduced to H_2O, which results in a quasi-four-electron reduction [7]. Typical analyses of the Fe/N/C catalysts, including rotating disk electrode (RDE) voltammetry with Koutecky–Levich (KL) and rotating ring–disk electrode (RRDE) voltammetry, tend to exhibit relatively high electron numbers (close to four) [8–10]. However, these analyses cannot resolve the quasi-four-electron reduction, and the experimental data does not support an actual four-electron pathway during NPM catalysis. To address these

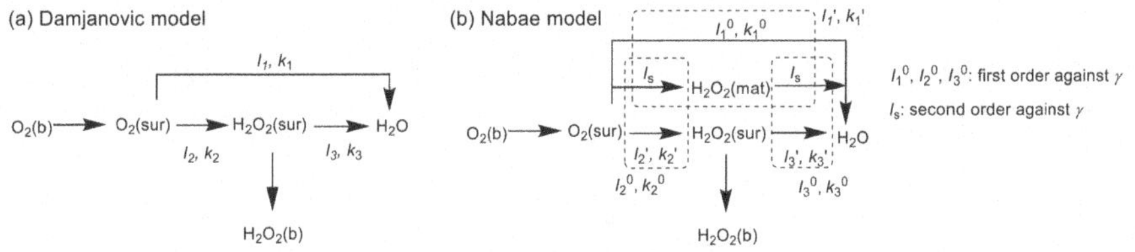

Fig. 11.2 **a** Damjanovic model, which consider only H_2O_2 on the disk electrode surface, H_2O_2(surf), and **b** Nabae model, which consider H_2O_2 to be further reduced to H_2O in the catalyst layer matrix, H_2O_2(mat), for the analysis of RRDE voltammograms of the ORR. I_s is the current via the series reaction in the catalyst layer matrix. γ is the catalyst loading density [11]

concerns, Nabae and coworkers have recently developed a new model (Fig. 11.2b) and method for RRDE analysis [11]. This new method avoids overestimating the contribution of the four-electron reduction and revealed that the contribution of the four-electron (I_1) pathway using a Fe/N/C catalyst was only 51% while the rest of the current originates from either of the two-electron pathways (I_2 and I_3).

The structure of the catalytically active sites in this class of catalysts is still hotly debated. Figure 11.3 shows the most commonly proposed active sites for catalytic ORRs. One particularly convincing model involves a catalytic center based on metal-N coordination, which has been proposed based on experimental evidence such as ToF-SIMS, HAADF-STEM, and EXAFS [12–14]. In the meantime, metal-free N-doped carbon species with pyridinic nitrogen and graphitic nitrogen have been also proposed as responsible for the observed ORR activity based on experimental [15–17] and theoretical studies [18, 19]. However, typical Fe-free N-doped carbon catalysts (N/C catalysts) have a lower onset potential accompanied by higher selectivity to H_2O_2 [20, 21]; therefore, it is likely that N/C catalysts in acidic media generally catalyze the two-electron pathway to form H_2O_2, although this might contribute to the quasi-four-electron reduction pathway by assisting the first two-electron pathway.

To obtain highly active Fe/N/C catalysts, sufficient nitrogen content is essential, and the precursor and pyrolysis protocols must be carefully designed to carbonize the precursor while preventing the loss of nitrogen species. It should also be noted that Fe species are important during the pyrolysis of precursors because they catalyze the growth of the carbon network [22, 23].

11.1.3 Polyimide Nanoparticles as Precursors for Fe/N/C Catalysts

This section focuses on polyimide nanoparticles as a precursor for Fe/N/C cathode catalysts. As shown in Fig. 11.1, the catalyst materials in PEFCs must be in contact with the ionomer to promote the proton conductivity of the catalyst layer. In addition, the catalyst powder should be fine to increase the interfacial area between

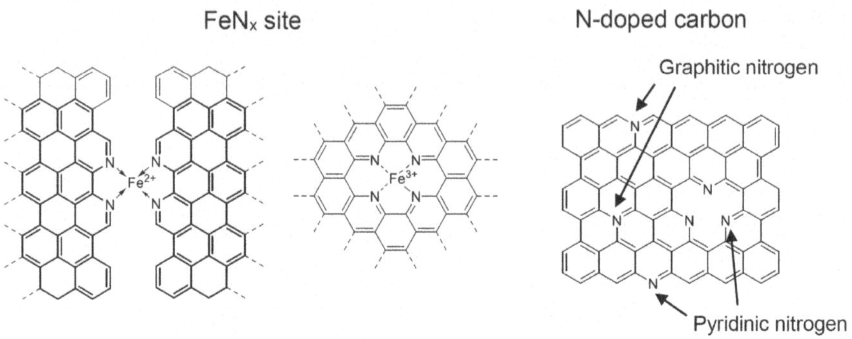

Fig. 11.3 Proposed catalytically active sites for ORR in acidic media

the catalyst and ionomer. Infusible polyimides are suitable for this purpose because their morphology can be controlled at the polymerization stage and retained after high-temperature pyrolysis because of its high thermal stability. Figure 11.4 shows SEM images of polyimide nanoparticles prepared using the precipitation polymerization of quite common polyimide precursors, pyromellitic acid dianhydride (PMDA) and 4,4′-oxidianiline (ODA) [24]. Precipitation polymerization is a homogeneous process that begins as a homogenous system in the continuous phase where the monomer is completely soluble, but the polymer is insoluble, and therefore precipitates upon formation. As shown in Fig. 11.4, the particle size can be controlled by changing the temperature and monomer concentration during polymerization. The smallest particle size obtained using the combination of PMDA and ODA was 100 nm. As shown in Fig. 11.5, a fine morphology was retained after carbonization to obtain the Fe/N/C catalyst. The chemical

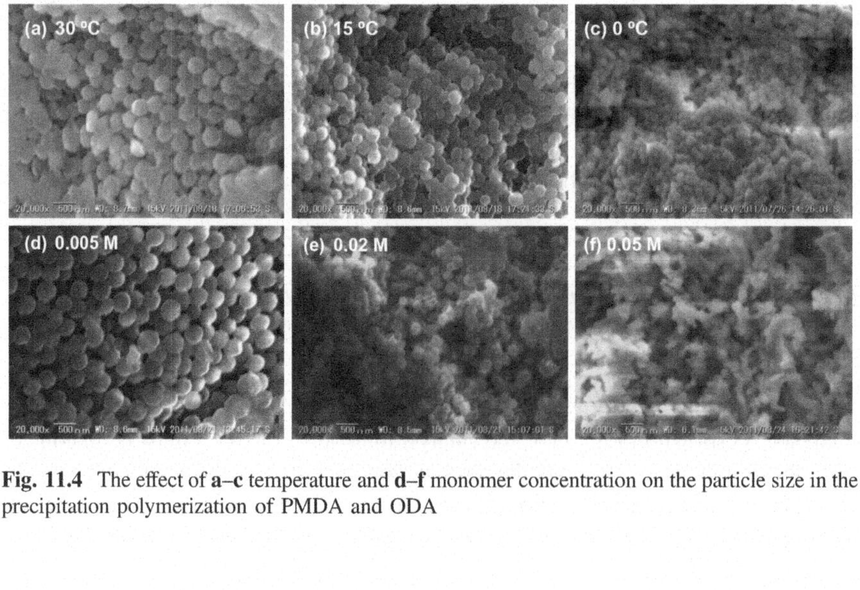

Fig. 11.4 The effect of **a–c** temperature and **d–f** monomer concentration on the particle size in the precipitation polymerization of PMDA and ODA

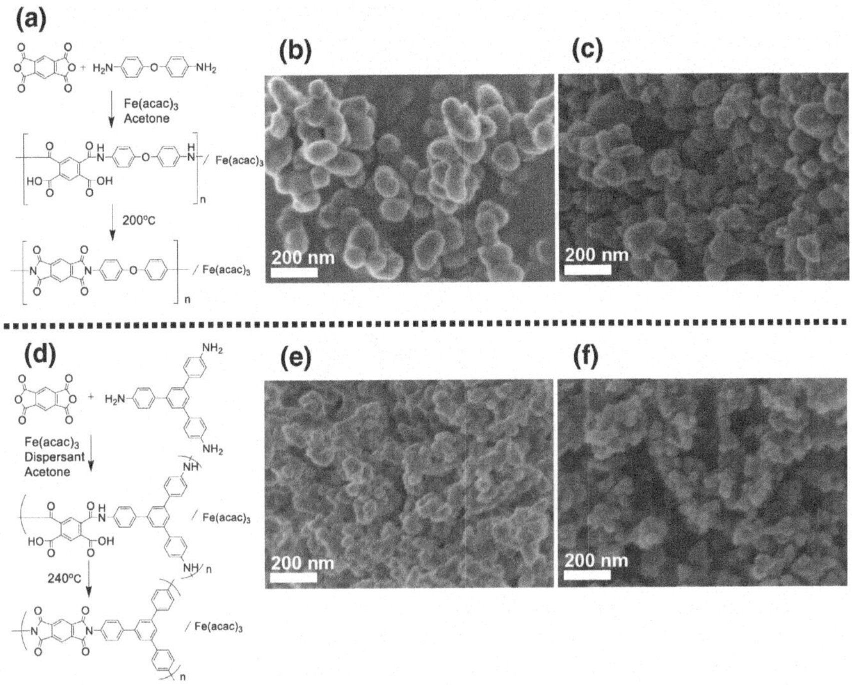

Fig. 11.5 **a** Synthetic route of the polyimide from PMDA and ODA and its FE-SEM images **b** before and **c** after carbonization. **d**, **e** Those of the polyimide synthesized from PMDA and TAPB in the presence of dispersant [25]

composition of the resultant carbon was determined using a CHN elemental analyzer and an electron probe microanalyzer: C 84 wt%, H 1.2, N 2.6 wt%, and Fe 1.1 wt%. The Brunauer, Emmett, and Teller (BET) surface area was determined to be 1200 $m^2\ g^{-1}$.

Polyimide nanoparticles of an even smaller size (60 nm) were synthesized by adding a dispersant, *N,N*-dimethyldodecylamine, during polymerization. However, the compact particle size was not retained after the carbonization in polyimides prepared from PMDA and ODA. This was probably because the low molecular weight of the polyimide led to decreased thermal stability, resulting in the fusion of polyimide nanoparticles. To address this concern, ODA was replaced with a tri-amine monomer, 1,3,5-tris(4-aminophenyl)benzene (TAPB), to afford higher thermal stability via cross-linking [25]. Figure 11.5d–f shows the synthetic route and SEM images of the polyimide and resultant carbon from PMDA and TAPB. The 60 nm particle size was successfully retained, even following carbonization. The chemical composition of the resultant carbon was determined to be C 91 wt%, H trace, N 3.0 wt%, and Fe 1.5 wt%. The BET surface area was 1217 $m^2\ g^{-1}$.

A more detailed analysis of the Fe and N species was conducted using powder X-ray diffraction (XRD), X-ray photoemission spectroscopy (XPS), and X-ray

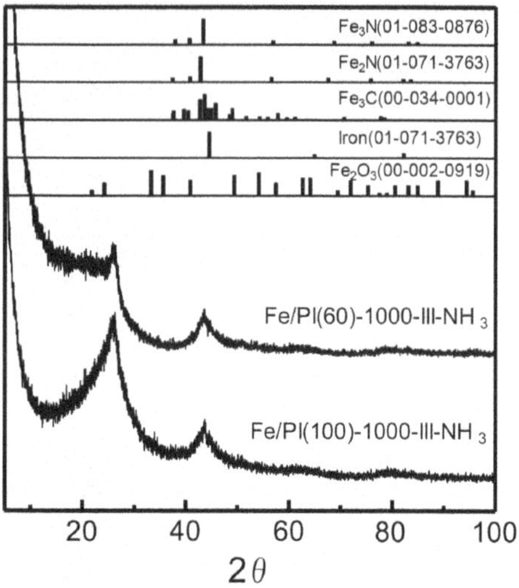

Fig. 11.6 XRD patters of Fe/N/C catalysts prepared from polyimide nanoparticles [25]

adsorption near edge structure (XANES) spectroscopy. The XRD patterns (Fig. 11.6) suggest that both samples contain considerable amounts of iron carbide. Linear combination fitting analyses of the XANES spectra are shown in Fig. 11.7b, d. The majority of Fe species can be assigned to clustered Fe species such as metallic iron, carbide, and oxide. In addition, mononuclear Fe species, such as FePc-like (Fe with D_{4h} symmetry) and $(FePc)_2O$-like (Fe with C_{4v} symmetry), are probably formed in small quantities. The structures of N species were analyzed by XPS spectra, which are shown in Fig. 11.7. The N 1*s* spectra were deconvoluted into four peaks: pyridinic (398.4–398.5 eV), pyrrolic (400.0–400.3 eV), graphitic (401.2 eV), and oxidized (402.9 eV) nitrogen species [26]. The two largest peaks were assigned to pyridinic and graphitic nitrogen species. These nitrogen species have been proposed as catalytically active sites for the ORR [19, 27], although they may be active in two-electron reduction rather than four-electron reduction in acidic media [20]. The peak at 398.4–398.5 eV may obscure the signal from any FeN_x type structures, if present, which has been reported at 398.7 eV [28]. These XPS and XANES data suggest the coexistence of Fe-free nitrogen species (pyridinic and graphitic) and FeN_x species.

11.1.4 Fuel Cell Performance of Fe/N/C Catalysts Prepared from Spherical Polyimide Nanoparticles

The abovementioned Fe/N/C catalysts were tested under practical fuel cell conditions. FE-SEM cross-sectional images of the membrane electrode assembly (MEA) are shown in Fig. 11.8. Four mg of Fe/N/C catalyst was applied onto the cathode side with Nafion binder, which resulted in a cathode layer with a thickness

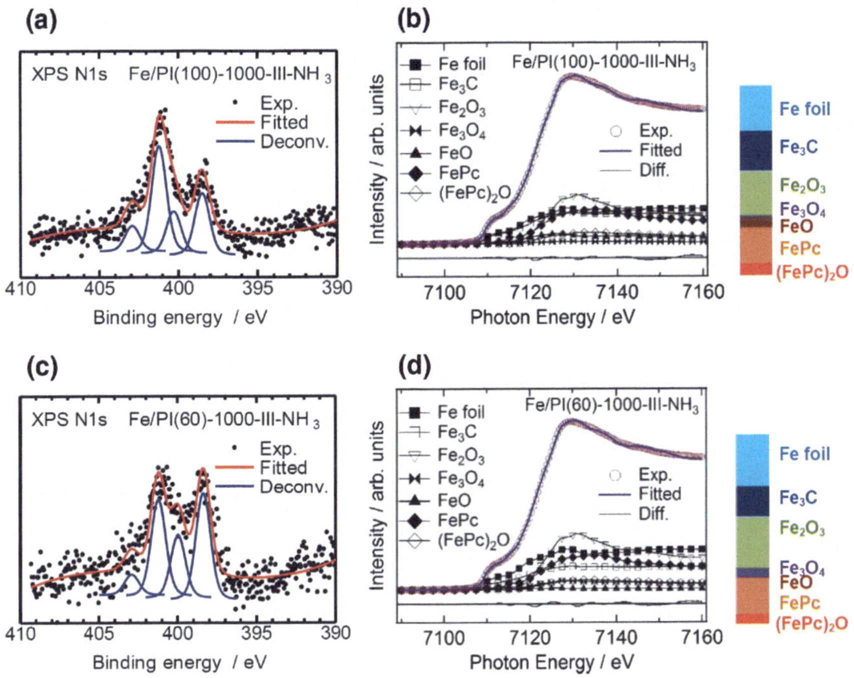

Fig. 11.7 **a** N 1s XPS spectrum with deconvoluted curves and **b** Fe K-edge XANES spectrum with reference spectra weighted by their contribution to the Fe/N/C catalyst (100 nm). **c** N 1s XPS and **d** Fe K-edge XANES spectra for the smaller Fe/N/C catalyst (60 nm). The bar graphs on the right provide visual representations of the relative compositions [25]

of approximately 60 μm. A quite porous and uniform catalyst layer is clearly observed in the enlarged images. This ideal morphology could contribute to the successful mass transport diffusion in the catalyst layer.

Figure 11.9a shows *I-V* performance curves for the MEA prepared using Fe/N/C cathode catalysts. The MEA with the 100 nm cathode catalyst had open circuit voltages of 0.96 and 0.90 V under pure O_2 and air, respectively, and the current density reached 1 A cm^{-2} at 0.57 V (O_2) and 0.32 V (air). The MEA with the 60 nm cathode showed similar open circuit voltages of 0.94 V (O_2) and 0.90 V (air) but higher voltages of 0.62 V (O_2) and 0.46 V (air) at a current density of 1 A cm^{-2}. The performance of the polyimide-derived Fe/N/C catalyst under an air atmosphere is better than that of any of the state of the art technologies summarized in Table 11.1. Figure 11.9b contains Tafel plots for the MEAs with both of the prepared catalysts. The plots suggest that the most important difference between the performances of these two catalysts is derived from their success in mass transport diffusion at high current densities, which is probably enhanced by the smaller particle size rather than the kinetics which are important at low current densities.

The durability of the MEAs was investigated by operating the cells over long periods. Figure 11.10 shows the changes in the cell voltage during operation at 0.2

Fig. 11.8 FE-SEM cross-section images of the MEA with the polyimide-derived Fe/N/C cathode catalysts

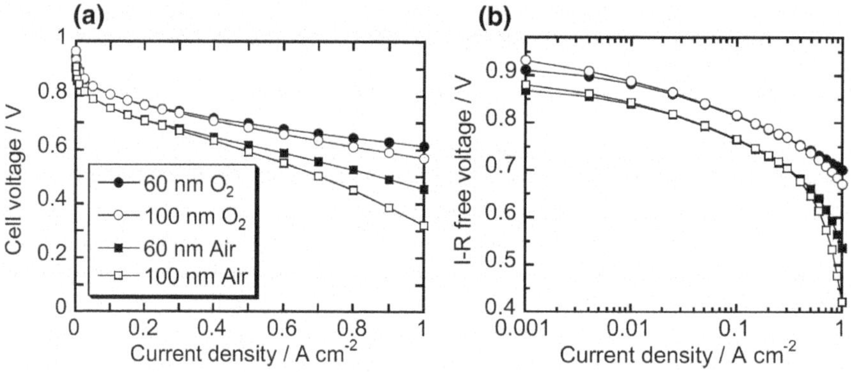

Fig. 11.9 **a** *I*-*V* performance curves and **b** Tafel plots of the *I*-*V* curves under air for the Fe/N/C cathode catalysts from polyimide nanoparitcles. Anode: PtRu/C catalyst with 0.4 mg-PtRu cm^{-2} loading, humidified H_2 at 80 °C, 1 bar backpressure. Cathode: 4 mg cm^{-2} catalyst loading, pure or balanced O_2 (humidified) at 80 °C, 1 bar backpressure. Electrolyte: Nafion NR211. *T*: = 80 °C [25]

A cm^{-2} for 600 h. Although the cell voltage certainly decreased, the cells were successfully operated for 600 h. While most of the reported NPM catalysts degrade significantly within 100 h under practical fuel cell conditions, the durability demonstrated in Fig. 11.10 is quite promising.

Table 11.1 State of the arts of H_2-air fuel cell performance with NPM cathode catalysts

Group	Performance	Temp. and back pressure	Ref.
Y. Nabae (Japan)	0.46 V@1 A cm^{-2} 0.5 V@0.8 A cm^{-2}	80 °C and 1 bar	[25]
J.P. Dodelet (Canada)	0.5 V@0.7 A cm^{-2}	80 °C and 2 bar	[29]
S. Mukerjee (US)	0.44 V@1 A cm^{-2}	80 °C and 2 bar	[30]
P. Zelenay (US)	0.4 A@1 A cm^{-2}	80 °C and 1 bar	[31]

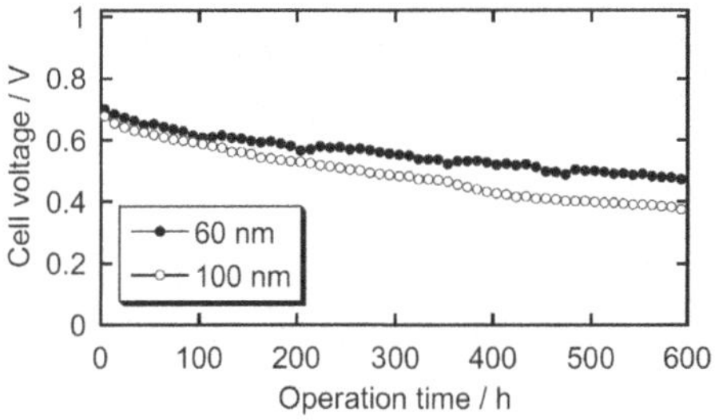

Fig. 11.10 Cell voltage stability curves at 0.2 A cm^{-2} with air as the cathode gas. The conditions were the same as those detailed in Fig. 11.9 [25]

11.1.5 Summary

Fine nanocarbons containing Fe/N/C and N/C catalytic centers can be prepared in a multistep pyrolysis of polyimide nanoparticles. The H_2-air fuel cell performance with Fe/N/C catalysts from 60 nm polyimide particles is especially remarkable. This is probably because the fine morphology of the catalyst layer enhances mass transport diffusion, which is especially important for the cathode gas. Although further study is required to improve the catalytic performance and clarify the features of the active sites, the contributions of nanocarbons will be central to the successful commercialization of NPM catalysts for fuel cells.

11.2 Group 4 and 5 Oxide-Based Cathodes with Nanocarbons

11.2.1 Introduction

In spite of extensive studies to find new non-platinum cathode catalysts for PEFCs, the performance of existing catalysts is still insufficient in regards to the electrocatalytic activity of the ORR and long-term stability. In particular, high stability is required for the cathode of PEFCs because the cathode catalysts are exposed to an acidic and oxidative atmosphere, i.e., a strong corrosive environment. From this perspective, we focused on group 4 and 5 transition metal oxide-based materials, known as valve metals, which are expected to be stable under corrosive electrochemical environments. Further, the adsorption state of the intermediate species of the ORR governs the activity. The adsorption state may be controlled by the modification of the surface state of these oxides. However, since perfect group 4 and 5 metal oxides are almost insulators, it is difficult to apply the insulators to electrocatalysts because of the insufficient electrical conductivity. Thus, we attempted at applying carbon materials to insulating oxide-based materials to produce catalytically active sites for the ORR and to form the electron conduction path.

11.2.2 Emergence of ORR Activity of Insulating Oxide-Based Catalysts by Nanocarbons

To produce active sites for the ORR and to form the electron conduction path, we have applied nanocarbons to insulating oxide-based catalysts. First, we focused on group 4 and 5 metal carbonitrides, such as Ta_2CN or Zr_2CN, as precursors [32, 33]. Ta_2CN powders were heat-treated using a rotary kiln furnace at 1000 °C for several hours under nitrogen gas containing 2% hydrogen and 0.5% oxygen to prepare the catalyst powders. Excess hydrogen caused a very low partial pressure of oxygen ($p_{O2} = 2.7 \times 10^{-15}$ atm) owing to the water formation reaction. Because the flowing gas contained oxygen, the oxidation of carbonitride proceeded gradually during the heat-treatment. We describe the oxidized tantalum carbonitride powders as Ta-CNOs.

We defined a parameter, DOO, as an indication of the degree of oxidation using XRD intensities for the specific reflections. A typical powder X-ray diffraction pattern of Ta-CNO has both Ta_2CN and Ta_2O_5 peaks. The DOO for Ta-CNOs, which contains Ta_2CN and Ta_2O_5 phases is defined using the integrated intensity of 1 1 1 reflection of c-Ta_2CN and 1 11 0 reflection of o-Ta_2O_5.

$$\mathrm{DOO} = \frac{I_{\mathrm{Ta_2O_5}}}{I_{\mathrm{Ta_2O_5}} + I_{\mathrm{Ta_2CN}}}$$

Regarding the evaluation of the ORR activity, Ketjenblack EC 300 J powders (7 wt%) were added to maintain the electrical contact between the substrate electrode and the Ta-CNO powders. In Fig. 11.11a, b, we plotted the onset potential for the ORR, E_{ORR}, and the oxygen reduction reaction current, i_{ORR} at 0.6 V versus RHE, for the Ta-CNOs with respect to the DOO. E_{ORR} and i_{ORR} for commercial Ta_2O_5 are also shown as reference. The starting material, Ta_2CN (DOO = 0), and commercial Ta_2O_5 (DOO = 1) showed no remarkable ORR activity. The Ta-CNOs, however, exhibit clear catalytic activity for the ORR. E_{ORR} abruptly increased to 0.9 V with the increase in the DOO up to 0.2 and was constant in a wide DOO range from 0.25 to 0.98. This result indicates that the active surface formed immediately by the slight oxidation of Ta_2CN, and the quality of active sites did not change even if the oxidation proceeded. Nevertheless, the i_{ORR} at 0.6 V first increased at the DOO of 0.2. This suggests that the density of effective active sites increased at the DOO of 0.2. In the DOO from 0.2 to 0.4, the i_{ORR} at 0.6 V did not increase. However, i_{ORR} at 0.6 V increased the second time at the DOO of 0.4.

To analyze the local structure and electronic structure of the near surface-oxide regions in the Ta-CNOs, we conducted X-ray absorption spectroscopy measurements in the conversion-electron-yield (CEY) mode [34]. Figure 11.12a shows the X-ray absorption near edge structure (XANES) spectra at Ta L_3 absorption edge (white line) for the Ta-CNO catalysts, taken in the CEY mode. As Ta L_3 absorption arises from a dipolar transition from 1 s to unoccupied 5d states, the absorption (white-line) intensity, that is, the peak area from approximately −5 to 10 eV from E_0 (absorption edge) increases when the tantalum 5d band vacancy increases. We observed that the white-line intensity increases as the DOO increases. As DOO increases, the XANES spectra display discontinuous behavior near the DOO of 0.15. Such behaviors, which can be observed clearly, show the dependences of the peak position and white-line intensity of the XANES spectra (Fig. 11.12b, c). The peak position slightly increases up to DOO = 0.2, and then remains at almost the same level up to 0.98. This dependence was similar to that of the E_{ORR} on the DOO (Fig. 11.11a). These results indicated that the surface was immediately oxidized and the surface state did not change up to a DOO of 0.98. The white-line intensities, however, gradually increases up to DOO = 0.4, and then decreases from 0.4 to 0.7. Subsequently, the intensity remains at almost the same level up to 0.96 and finally, it increases again at DOO = 1.0. This indicates that the number of 5d band vacancies gradually decreases from 0.4 to 0.98. Compared with the DOO variations

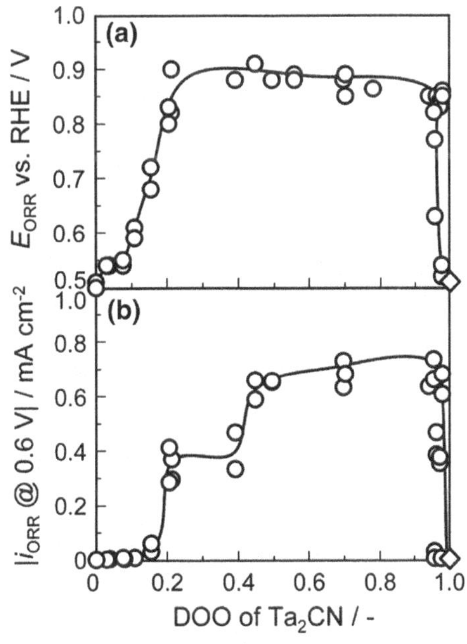

Fig. 11.11 Onset potential of ORR for Ta-CNOs (○) and commercial Ta_2O_5 (◇) (**a**) and oxygen reduction current i_{ORR} at 0.6 V for Ta-CNOs (○) and commercial Ta_2O_5 (◇) (**b**) against DOO [32]

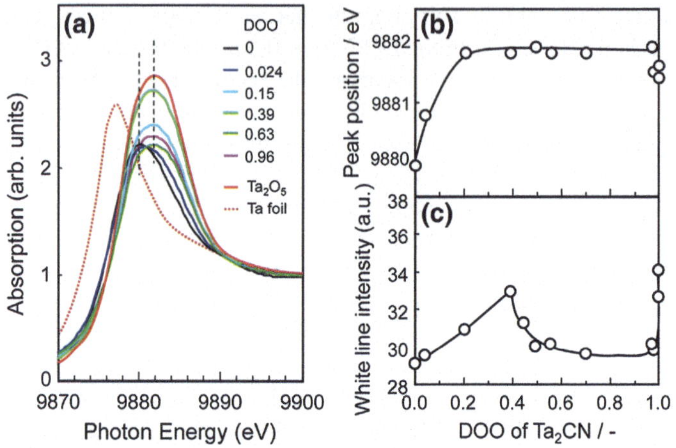

Fig. 11.12 **a** XANES spectra at Ta L_3 absorption edge (white line) for the Ta-CNO catalysts, taken in the CEY mode. Dependence of peak position (**b**) and white-line intensity (**c**) of XANES spectra at Ta L_3 absorption edge for Ta-CNO catalysts taken in the CEY mode [32]

of the ORR current (Fig. 11.11b), we can conclude that the density of 5d vacancies that should be related to the density of oxygen-vacancy defects correlates with the ORR current.

Formation of the oxygen vacancies seems to relate to the deposition of carbon during the oxidation of carbonitride. Figure 11.13a shows the Raman spectra of the Ta-CNOs with the DOO of 0, 0.029, 0.21, 0.39, 0.49, 0.71, 0.98, and Ta_2O_5. The starting material with the DOO of 0, Ta_2CN, had a strong peak at around 100–180 cm^{-1}. As the DOO increases, the broad peaks at around 500–700 cm^{-1} that correspond to Ta_2O_5 can be observed. Above 0.49, the Ta_2CN peaks disappear. In addition, we observed two other large peaks that are ascribed to defective graphite (the G band attributed to the graphitic structure at around 1580 cm^{-1} and the D band attributed to the defect structure at around 1330 cm^{-1}). These behaviors are a direct evidence of carbon deposition during the partial oxidation process. As we did not observe diffraction peaks that originated from the well-crystallized carbon in XRD measurements, the deposited carbon is rather amorphous. Such deposited carbons should be oxidized during the heat-treatment process of carbonitrides, producing a reductive atmosphere that is suitable for introducing oxygen vacancies on the surface of the tantalum oxide particles. The nature of such deposited carbons is almost independent of the DOO, because the ratio of the G band to D band was unchanged from DOO = 0.21–0.98. Somewhat surprisingly, however, a large amount of graphitic carbon remains on the surface even at a DOO of 0.98, i.e., well-oxidized regions, implying an additional role of the deposited carbon. Figure 11.13b shows the TEM image of the Ta-CNO surface with a DOO of 0.98. It was clearly observed that the deposited carbon with a thickness of $\sim$1 nm completely covered the oxide surface.

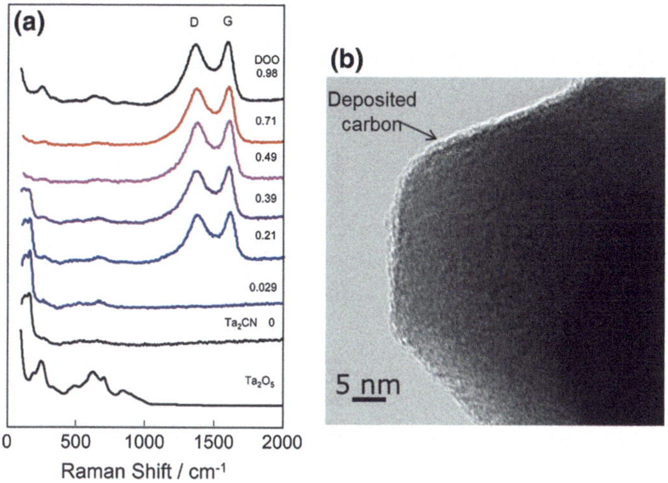

Fig. 11.13 **a** Raman spectra of Ta-CNOs with DOO of 0, 0.029, 0.21, 0.39, 0.49, 0.71, 0.98, and Ta_2O_5. **b** TEM images of the surface of Ta-CNO with DOO of 0.98 [32]

To estimate the content of the deposited carbon, elemental analysis for O, N, H, and C in the catalysts was performed. First, we estimated the O/Ta ratio in oxides. The average composition of the bulk Ta-CNO with a DOO of 0.98 was $Ta_2O_{4.77}$. This result indicates that catalytically active Ta-CNO catalysts indeed have oxygen vacancies. We assumed that all remaining nitrogen must exist in the form of Ta_2CN. Ta_2CN was oxidized as the following equation:

$$\begin{aligned} & 2\mathrm{Ta_2CN} + \frac{2x + 2 - y}{2}\mathrm{O_2} \\ & = 2\mathrm{Ta_2O}_x + (2 - y)\mathrm{CO} + y\mathrm{C} + \mathrm{N_2} \end{aligned}$$

Therefore, the amount of deposited carbon was calculated by subtracting the carbon content involved in Ta_2CN from the amount of the detected carbon.

Figure 11.14a, b shows the DOO dependence on the content of the related compounds and the DOO variation in the C content during oxidation. As oxidation starts, Ta_2CN gradually oxidized to tantalum oxides and solid carbon simultaneously deposits on the surface of the oxides. As DOO increases, the deposited carbon would be oxidized to CO and/or CO_2 (one half of C in Ta_2CN is deposited and the rest of C is oxidized at DOO = 0.2, for example). The amount of deposited carbon increases up to DOO = 0.4. Above 0.4, the deposited carbons are cautiously oxidized to CO and/or CO_2. The produced CO reductive atmosphere should introduce oxygen vacancies on the oxide surface. As shown in Fig. 11.12b, the white-line intensity of Ta-L_3 XANES initially increases, but above DOO = 0.4, it decreases. This behavior indicates that the tantalum oxide formed at the initial stage of oxidation below 0.2 have few oxygen vacancies, while for tantalum oxide formed

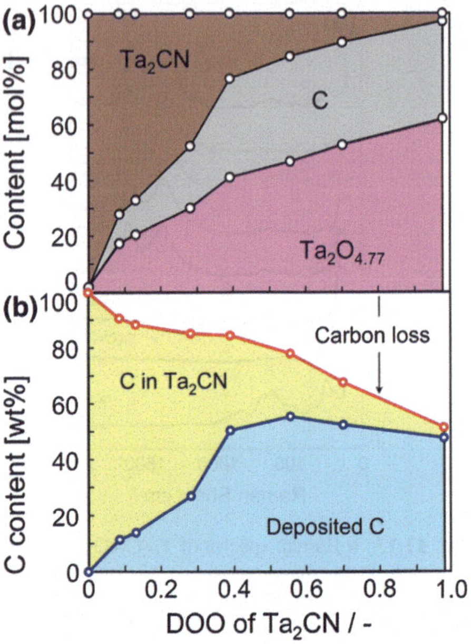

Fig. 11.14 Dependence of estimated content on compounds such as Ta_2CN, deposited C, and $Ta_2O_{4.77}$ on DOO (**a**) and dependence of C content on DOO (**b**) [32]

above DOO = 0.4, the oxide has enough vacancies to produce the ORR activity. Namely, below DOO = 0.2, presumably, because the amount of deposited carbon is small thus the reductive CO density is small. Therefore, the oxygen vacancies are not introduced as much. However, above 0.2, because many carbons are oxidized, the local reduction atmosphere is enough to reduce the oxides. Another role of the deposited carbon is to enhance electrical conduction on the catalyst surface. As shown in Fig. 11.12b, since the white-line intensity is almost unchanged at the DOO range from 0.4 to 0.96, the number of ORR active sites (oxygen vacancies) is almost constant in this region. However, i_{ORR} gradually increases from 0.4 to 0.98. This indicates that the deposited carbon plays a role in supporting the local electron conduction near the ORR active sites on the oxide surface. Figure 11.15 shows the image of the Ta-CNO on a glassy carbon substrate. As a result, we found that carbon was deposited on the oxide surface during the oxidation of tantalum carbonitrides. The deposited carbon seems to play two important roles in the formation of oxygen vacancies (ORR active sites): providing a reductive atmosphere, and producing electron conduction paths on the rather insulating oxide surface.

11.2.3 Preparation of Oxide-Based Nanoparticles with ORR Activity by Nanocarbons

We synthesized oxide-based ORR catalysts as described above by pyrolyzing metal carbonitrides of several hundred nanometers in size under low oxygen partial

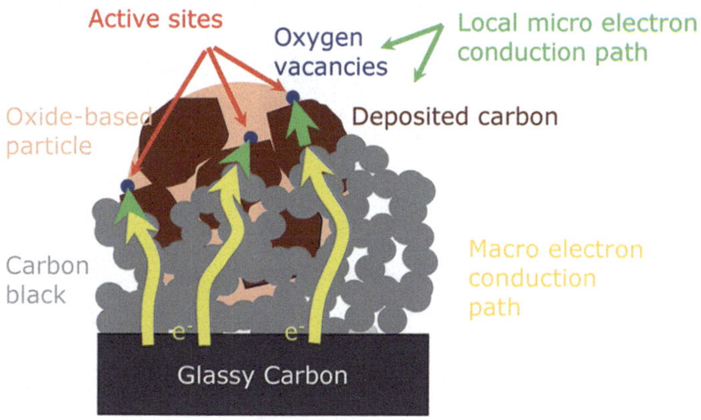

Fig. 11.15 Schematic image of Ta-CNO on glassy carbon substrate [32]

pressures. Although the onset potential for the ORR was relatively high, their ORR currents remained insufficient because carbonitrides were synthesized above 1000 °C to produce large particle sizes, that is, low surface areas of a few $m^2\ g^{-1}$. We, therefore, attempted at developing a new synthesis route for an oxide catalyst containing one of the group 4 and 5 metals. We focused on oxy-tantalum phthalocyanine (TaOPc) as a new precursor instead of carbonitrides [35–37]. The new catalyst was synthesized via the decomposition of oxy-tantalum phthalocyanine (TaOPc) at desired temperatures for 3 h under 4%H_2 and 96%N_2 gas to control the size of tantalum oxide (TaO_x) particles to a few nanometers and to connect TaO_x to a support that is multiwalled carbon nanotubes (MWCNTs). Figure 11.16 shows the TEM image of the TaO_x/MWCNT after pyrolysis at 1173 K. The inset shows an enlarged image of the central area. Although some aggregation of TaO_x into particles larger than 10 nm had occurred, most of them were nanoparticles and were successfully supported on MWCNTs. As more clearly shown in the inset, the light gray amorphous carbon species derived from phthalocyanine connected the black TaO_x particles with the MWCNT.

Figure 11.17 shows the mass activity of the ORR versus the potential curves of two catalysts synthesized from different precursors, TaOPc, and metal-free phthalocyanine (HPc). The TaO_x/MWCNT catalyst synthesized at 1173 K clearly showed a high onset potential of ~0.9 V for the ORR, whereas the onset potential of the HPc-derived catalyst remained at ~0.7 V. The mass activities at 0.8 V were plotted as a function of pyrolyzing temperatures in the inset. The mass activities of the HPc-derived catalysts were negligible compared to those of TaO_x/MWCNT catalysts at all pyrolyzing temperatures, demonstrating that the activity of the latter originates from the activity of TaO_x. Neither the TaOPc-derived amorphous carbon that provided the electron conduction path between the surface of TaO_x particles and the MWCNT supports nor the MWCNT supports themselves contributed to the ORR activity. In this case, the deposited carbons derived from the phthalocyanine

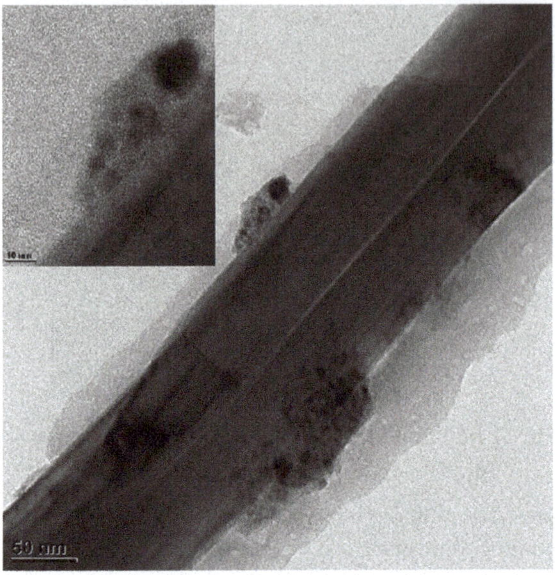

Fig. 11.16 TEM image of TaO_x/MWCNT after pyrolysis at 1173 K. The inset shows an enlarged image of the central area [35]

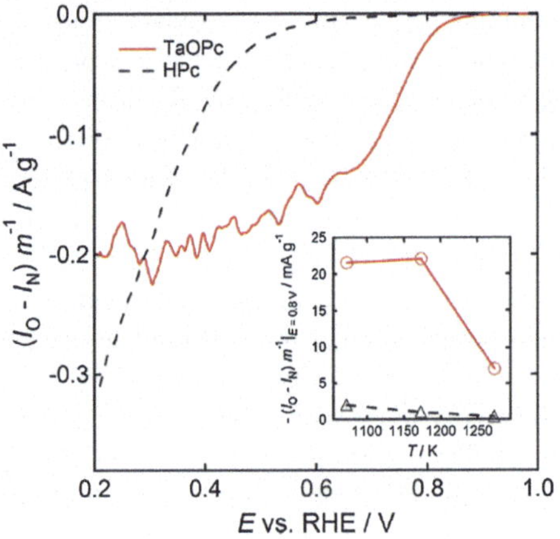

Fig. 11.17 Mass activity of ORR versus potential curves of TaO_x/MWCNT and HPc-dispersed MWCNT after pyrolysis at 1173 K using their precursors, TaOPc and HPc, respectively. The inset shows mass activity at 0.8 V versus T curves of these two catalyst types [35]

structure successfully provided electron conduction paths to TaO_x insulating particles.

Decomposition of phthalocyanine produced carbons that were almost completely covered with the oxide surface. Although the deposited carbon formed electron conduction paths to the oxide surface, the covered carbon hindered ORR from occurring. Thus, we tried to control the amount of the deposited carbon to optimize the balance between the exposure of the oxide surface and the formation of electron conduction paths. The deposited carbon was gradually oxidized by heat-treatment at 900 °C for several hours under 2% hydrogen and 0.05% oxygen, where equilibrium O_2 partial is 1.3×10^{-19} atm. We prepared nano-sized titanium oxide-based cathodes using oxy-titanium tetra-pyrazino-porphyrazine (TiOTPPz) as a precursor dispersed on multiwalled carbon nanotubes (MWCNTs) [38, 39]. To investigate the contribution of the carbon materials to the catalytic reactions, the samples without oxides were prepared by removing the oxides with hydrofluoric acid.

Figure 11.18a–c shows the oxygen reduction activities of the catalysts that were heat-treated at 900 °C for a different time. The samples that were heat-treated for more than 2 h exhibited catalytic activities associated with the presence of titanium oxides and gradually decreases with further heat-treatment. The titanium oxide particles were deposited on a carbon matrix (Fig. 11.18d), formed by the pyrolysis of the organic ligand of the titanium complex and were deposited on carbon nanotubes used as current collectors for the electrochemical tests. Compared with the samples without the oxides by removing titanium oxide with HF, the activities up to 1 h are attributable to the carbon materials, whereas the difference in the catalyst activities between the samples with and without oxides shown in Fig. 11.18 is attributable to the activities of the oxide particles.

We analyzed their atomic structures using the X-ray pair distribution functions (PDFs), which can show the structures of all of the materials, even for nanocrystals and amorphous materials, where broad features in the XRD patterns are shown (Fig. 11.19). Figure 11.19a shows that a structural transformation occurs from 1 h to 3 h. This is consistent with the trend of the catalytic activity. The structure of the heat-treated sample from only increasing the temperature up to 900 °C (heat-treatment time = 0 h) was found to be a lepidocrocite-type titanium oxide. The sample annealed for 1 h shows the XRD and PDF having the features of samples annealed at both 0.5 and 3 h, clearly indicating the commencement of the structural transformation, which may be triggered by the removal of nitrogen dopants and the deposition of carbon. Through the transformation, the PDF peak reflecting the Ti–Ti distance in edge-shared linkage ($\sim$3 Å) decreased, whereas the PDF peak reflecting that in the corner-shared linkage ($\sim$3.8 Å) increased. This is reasonable because the lepidocrocite structure, wherein TiO_6 octahedra form a two-dimensional structure through edge-sharing connectivity, is rich in edge-shared linkage of TiO_6 octahedra (or denser) than the typical TiO_2 crystal structures (anatase, brookite, and rutile; Fig. 11.19b). To understand the structure of the sample annealed for 3 h, which showed the highest ORR activity, we tested a number of possible structure models of titanium oxides (e.g., rutile, anatase, and brookite) with/without the defects analyzed by curve fitting the PDFs. We finally revealed that large rutile particles and particles having brookite-like octahedral linkage were formed. The formation of these structures from lepidocrocite-type titanium oxide is reasonable in light of the similar connectivity. Further annealing saw particles growing on the external surface of the carbon matrix, which are well-crystalline rutile (47 wt% at 5 h and 46.7 wt% at 10 h). The PDF analysis revealed that the content of the brookite-like domains decreased from 3 to 10 h, whereas another phase assignable to a TiO-like cubic structure was grown from 5 h. Therefore, we concluded that the abrupt increase in the ORR activity corresponded to a structural transformation from a reduced lepidocrocite-type layered titanium oxide to a disordered structure consisting of domains of brookite-like TiO_6

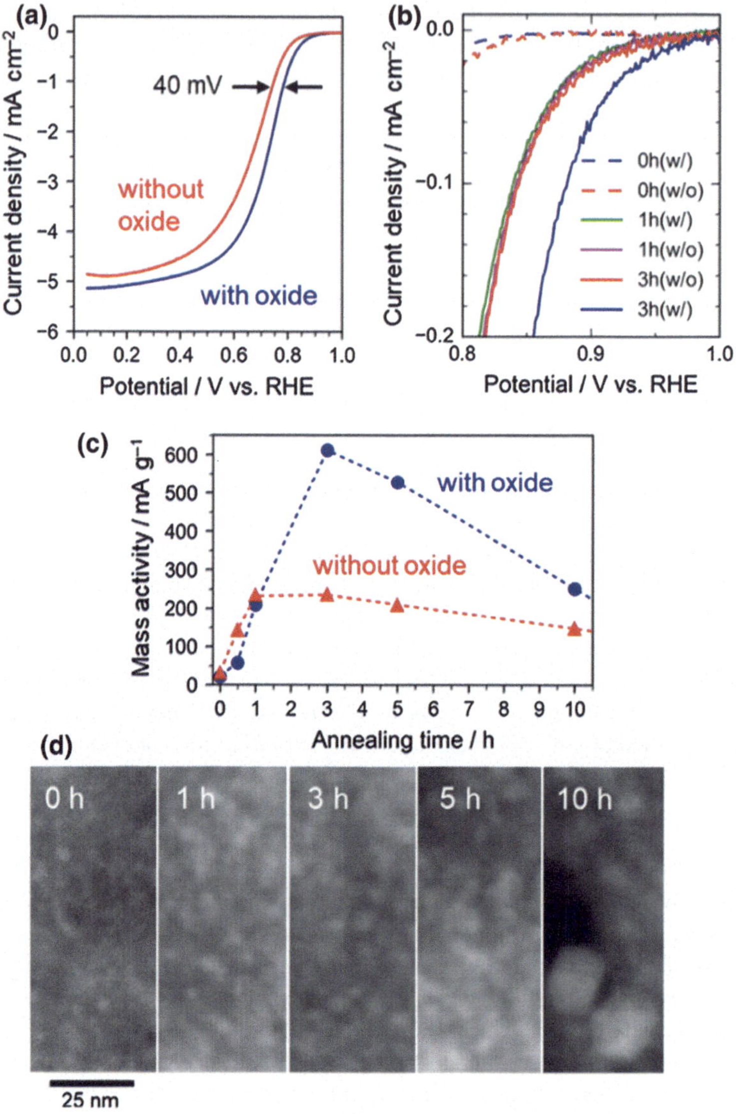

Fig. 11.18 Electrochemical oxygen reduction activities of the catalysts heat-treated for a different time. The samples with (blue) and without (red) titanium oxides are compared. **a** Linear sweep voltammograms of the samples heat-treated for 3 h. The current densities are based on the geometric areas. **b** Comparison of kinetically controlled current densities. **c** Trend of mass activity of the samples heat-treated for a different time. The activities are normalized by the carbon contents and obtained at 0.80 V versus RHE. **d** High angle annular dark field scanning transmission electron microscopy images of the samples heat-treated for 0, 1, 3, 5, and 10 h (at the same magnification) [8]

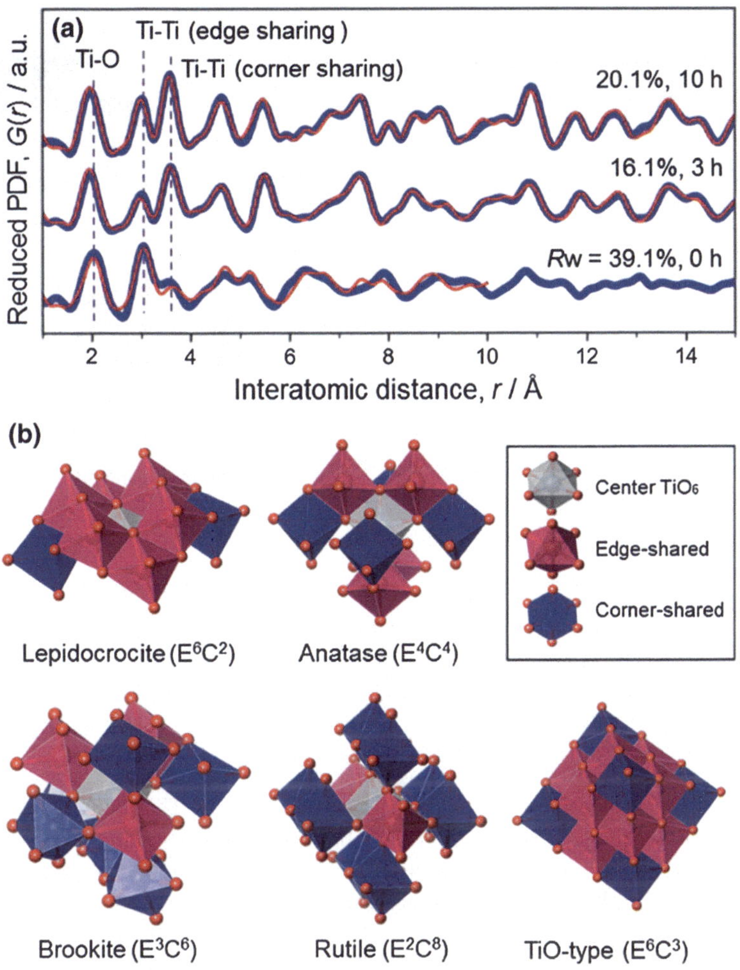

Fig. 11.19 X-ray PDF structure analysis. **a** Pair distribution functions (Q_{max} = 17.5 Å − 1). The experimental data (blue) were fitted with simulated curves (red) based on structure models: (i) lepidocrocite-type TiO_2 (0 h), (ii) rutile and brookite (3 h), and (iii) rutile, brookite, and TiO-type structure (10 h). The fitting range was 1–10 Å for 0 and 1–20 Å for others. **b** Structure models and polyhedral connectivity. E_mC_n represents the polyhedral connectivity: numbers of edge-sharing (E_m) and corner sharing (C_n) [39]

octahedral linkages (Fig. 11.20). Such a structure transformation might be caused by the deposition of carbon and the removal of nitrogen. We believe that such an octahedral network existing in the brookite-like domains contains active sites for the ORR. The further treatment transformed these brookite-like domains into another phase having more edge-sharing sites like the TiO-type cubic structure. This finding would position noncrystalline, disordered structures as a possible origin of the catalytic activity.

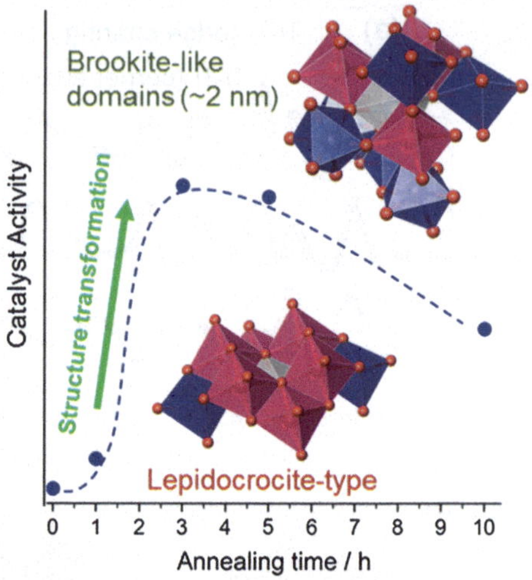

Fig. 11.20 Schematic image of the relationship between ORR activity and crystalline structure of titanium oxide-based cathodes [39]

11.2.4 Fuel Cell Performance of Zirconium Oxynitride Cathode Prepared from Oxy-Phthalocyanine with MWCNT Support

We demonstrated single-cell performance of zirconium oxynitride (ZrO_xN_y) cathodes [40]. We already revealed that pyrolysis in an NH_3 atmosphere was useful to enhance the ORR activity of the oxide-based catalysts [41]. Thermally decomposed oxy-zirconium phthalocyanine (ZrOPc) in the presence of the multiwalled carbon nanotube (MWCNT) supports was used to connect ZrO_xN_y nanoparticles to the supports above, followed by pyrolysis in an atmosphere of gaseous NH_3.

The transmission electron microscopy (TEM) image of ZrO_xN_y/MWCNTs after pyrolysis in an atmosphere of NH_3 at 973 K (Fig. 11.21a) shows that the black nanometer-sized ZrO_xN_y particles are well dispersed in light gray MWCNTs. The deposited amorphous carbon species derived from phthalocyanine connected the nano-sized ZrO_xN_y particles with the MWCNT to form efficient electron conduction paths. In addition, the carbon deposition hindered the size increase in the oxide particles. Figure 11.21b reveals that the nanoparticles feature a single tetragonal ZrO_2 phase. The Zr 3d signal of the 973 K sample was split into Zr $3d_{5/2}$ and $3d_{3/2}$ peaks by spin-orbit coupling, appearing as a doublet at 182.6 and 184.9 eV assigned to Zr^{4+} in ZrO_2. As shown in Fig. 11.21c, compared to the signals observed before NH_3 pyrolysis (dashed line), these peaks were slightly shifted to lower binding energies, suggesting that the surface was partially reduced by NH_3 pyrolysis at 973 K. Both samples showed noisy N 1s spectra, with peaks above 398 eV attributed to zirconium oxynitride and nitrogen-doped carbon species. Because of the high degree of overlap, these peaks could not be deconvoluted;

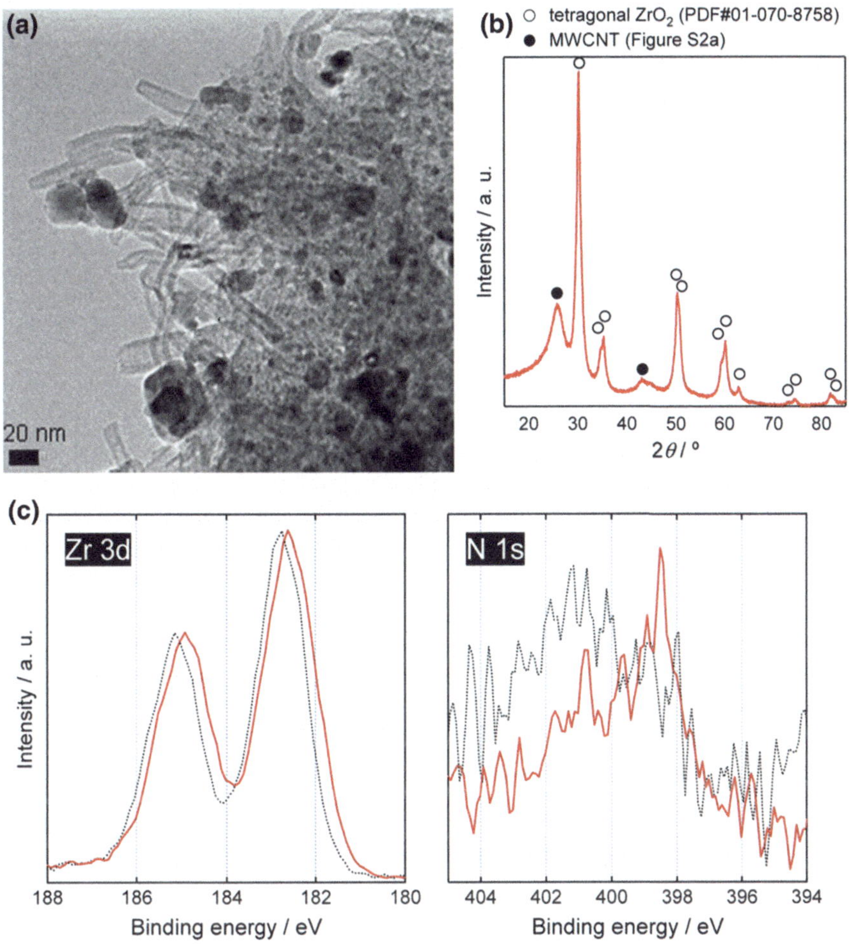

Fig. 11.21 **a** TEM image, **b** XRD pattern, and **c** Zr 3d and N 1s spectra of the ZrO_xN_y/MWCNT catalyst after NH_3 pyrolysis at 973 K. For comparison, Zr 3d and N 1s spectra of ZrO_xN_y/MWCNTs before NH_3 pyrolysis are shown by dashed lines in (**c**) [40]

however, it can be concluded that NH_3 pyrolysis at 973 K did not significantly enhance the level of ZrO_2 surface nitrogen doping.

The activity of ZrO_xN_y/MWCNTs was evaluated for a single-cell cathode, with the voltage versus current density (V-j) curve shown in Fig. 11.22. The performance displayed by the present ZrO_xN_y/MWCNT composite is the highest ever reported for an oxide-based catalyst, with j = 10 mA cm^{-2} at V = 0.9 V without IR correction. The corresponding open circuit voltage and the maximum power density are 0.96 V and 0.63 W cm^{-2} at V = 0.45 V, respectively.

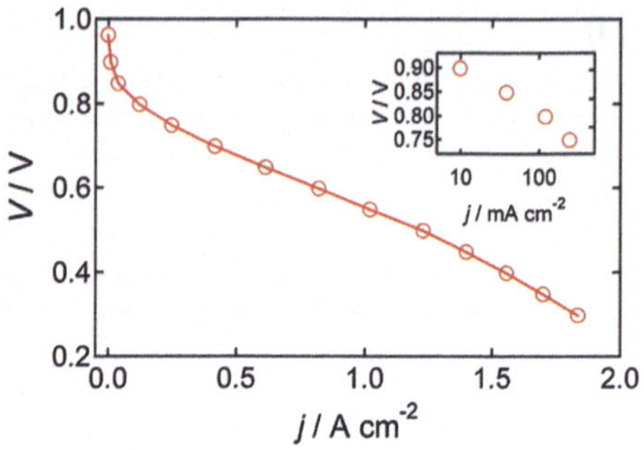

Fig. 11.22 Cell voltage versus current density (V-j) curve for a membrane electrode assembly fabricated using a ZrO_xN_y/MWCNT (10 mg cm^{-2}) cathode and a Pt/C anode (0.6 mg cm^{-2}, corresponding to a Pt loading of 0.3 mg cm^{-2}) [40]

11.2.5 Summary

Group 4 and 5 metal oxide-based compounds have been used to apply to PEFC cathodes by combining these oxides with nanocarbons. Because group 4 and 5 metal oxide-based compounds are almost insulators, the formation of sufficient electron conduction paths is required for continuous ORR. The deposited carbon derived from precursors by pyrolysis plays an important role in increasing the electrical conductivity of the catalyst. In addition, the deposited carbon produced the active sites for the ORR on the oxide surface owing to the formation of strong reductive atmosphere near the surface. Although further study is required to improve the catalytic performance and clarify the mechanism of the formation of the active sites, the contributions of nanocarbons will be essential to exhibit their abilities as high-performance ORR catalysts.

Acknowledgements The majority of the authors' research described in this chapter was financially supported by the New Energy and Industrial Technology Development Organization (NEDO).

References

1. Jasinski R (1964) A new fuel cell cathode catalyst. Nature 201:1212–1213
2. Masa J, Ozoemena KI, Schuhmann W, Zagal JH (2013) Fundamental studies on the electrocatalytic properties of metal macrocyclics and other complexes for the electroreduction of O_2. Springer, London, pp 157–212
3. Jahnke H, Schönborn M, Zimmermann G (1976) Organic dyestuffs as catalysts for fuel cells. Top Curr Chem 61:133–181
4. Dodelet J-P (2013) The controversial role of the metal in Fe- or Co-based electrocatalysts for the oxygen reduction reaction in acid medium. Springer, London, pp 271–338
5. Elbaz L, Wu G, Zelenay P (2013) Heat-treated non-precious-metal-based catalysts for oxygen reduction. Springer, London, pp 213–246
6. Damjanovic A, Genshaw MA, Bockris JO (1966) Distinction between intermediates produced in main and side electrodic reactions. J Chem Phys 45:4057
7. Olson TS, Pylypenko S, Fulghum JE, Atanassov P (2010) Bifunctional oxygen reduction reaction mechanism on non-platinum catalysts derived from pyrolyzed porphyrins. J Electrochem Soc 157:B54

8. Nallathambi V, Lee J-W, Kumaraguru SP et al (2008) Development of high performance carbon composite catalyst for oxygen reduction reaction in PEM Proton Exchange Membrane fuel cells. J Power Sources 183:34–42
9. Serov A, Artyushkova K, Atanassov P (2014) Fe-N-C oxygen reduction fuel cell catalyst derived from carbendazim: synthesis, structure, and reactivity. Adv Energy Mater 4:1301735
10. Iwazaki T, Obinata R, Sugimoto W, Takasu Y (2009) High oxygen-reduction activity of silk-derived activated carbon. Electrochem Commun 11:376–378
11. Muthukrishnan A, Nabae Y (2016) Estimation of the inherent kinetic parameters for oxygen reduction over a Pt-free cathode catalyst by resolving the quasi-four-electron reduction. J Phys Chem C 120:22515–22525
12. Lefvre M, Dodelet JP, Bertrand P (2002) Molecular oxygen reduction in PEM fuel cells: evidence for the simultaneous presence of two active sites in Fe-based catalysts molecular oxygen reduction in PEM fuel cells: evidence for the simultaneous presence of two active sites in Fe-based catalysts. J Phys Chem B 106:8705–8713
13. Chung HT, Cullen DA, Higgins D et al (2017) Direct atomic-level insight into the active sites of a high-performance PGM-free ORR catalyst. Science 357:479–484
14. Wu J, Nabae Y, Muthukrishnan A, Ohsaka T (2016) Electrochemical deposition and dissolution of Fe species for N-doped carbon to understand the degradation mechanism of Pt-free oxygen reduction catalysts. Electrochim Acta 214:307–312
15. Guo D, Shibuya R, Akiba C et al (2016) Active sites of nitrogen-doped carbon materials for oxygen reduction reaction clarified using model catalysts. Science 351:361–365
16. Wang X, Lee JS, Zhu Q et al (2010) Ammonia-treated ordered mesoporous carbons as catalytic materials for oxygen reduction reaction. Chem Mater 22:2178–2180
17. Yu D, Zhang Q, Dai L (2010) Highly-efficient metal-free growth of nitrogen-doped single-walled carbon nanotubes on plasma-etched substrates for oxygen reduction. J Am Chem Soc 132:15127–15129
18. Sidik RA, Anderson AB, Subramanian NP et al (2006) O_2 reduction on graphite and nitrogen-doped graphite: experiment and theory. J Phys Chem B 110:1787–1793
19. Ikeda T, Boero M, Huang SF et al (2008) Carbon alloy catalysts: active sites for oxygen reduction reaction. J Phys Chem C 112:14706–14709
20. Park J, Nabae Y, Hayakawa T, Kakimoto M (2014) Highly selective two-electron oxygen reduction catalyzed by mesoporous nitrogen-doped carbon. ACS Catal 4:3749–3754
21. Muthukrishnan A, Nabae Y, Okajima T, Ohsaka T. A kinetic approach to investigate the mechanistic pathways of oxygen reduction reaction on Fe-containing N-doped carbon catalysts. ACS Catal 5:5194–5202
22. Nabae Y, Sonoda M, Yamauchi C et al (2014) Highly durable Pt-free fuel cell catalysts prepared by multi-step pyrolysis of Fe phthalocyanine and phenolic resin. Catal Sci Technol 4:1400
23. Matter PH, Wang E, Millet J-MM, Ozkan US (2007) Characterization of the iron phase in CNx-based oxygen reduction reaction catalysts. J Phys Chem C 111:1444–1450
24. Nabae Y, Kuang Y, Chokai M et al (2014) High performance Pt-free cathode catalysts for polymer electrolyte membrane fuel cells prepared from widely available chemicals. J Mater Chem A 2:11561–11564
25. Nabae Y, Nagata S, Hayakawa T et al (2016) Pt-free carbon-based fuel cell catalyst prepared from spherical polyimide for enhanced oxygen diffusion. Sci Rep 6:23276
26. Raymundo-Piñero E, Cazorla-Amorós D, Linares-Solano a et al (2002) Structural characterization of N-containing activated carbon fibers prepared from a low softening point petroleum pitch and a melamine resin. Carbon N Y 40:597–608
27. Ikeda T, Hou Z, Chai GL, Terakura K (2014) Possible oxygen reduction reactions for graphene edges from first principles. J Phys Chem C 118:17616–17625
28. Gottfried JM, Bai Y, Buchner F et al (2008) Direct metalation of a phthalocyanine monolayer on Ag (111) with coadsorbed iron atoms direct metalation of a phthalocyanine monolayer on Ag (111) with coadsorbed iron atoms. Society 6087–6092

29. Proietti E, Jaouen F, Lefevre M et al (2011) Iron-based cathode catalyst with enhanced power density in polymer electrolyte membrane fuel cells. Nat Commun 2:1–6
30. Li J, Ghoshal S, Liang W et al (2016) Structural and mechanistic basis for the high activity of Fe–N–C catalysts toward oxygen reduction. Energy Environ Sci 9:2418–2432
31. Chung HT, Cullen DA, Higgins D et al (2017) Direct atomic-level insight into the active sites of a high-performance PGM-free ORR catalyst. Science 80(357):479–484
32. Ishihara A, Tamura M, Ohgi Y, Matsumoto M, Matsuzawa K, Mitsushima S, Imai H, Ota K (2013) Emergence of oxygen reduction activity in partially oxidized tantalum carbonitrides: roles of deposited carbon for oxygen-reduction-reaction-site creation and surface electron conduction. J Phys Chem C 117:18837–18844
33. Ohgi Y, Ishihara A, Matsuzawa K, Mitsushima S, Matsumoto M, Imai H, Ota K (2013) Factors for improvements of catalytic activity for zirconium oxide-based oxygen-reduction electrocatalysts. J Electrochem Soc 160:F162–F167
34. Imai H, Matsumoto M, Miyazaki T, Fujieda S, Ishihara A, Tanura M, Ota K (2010) Structural defects working as active oxygen-reduction sites in partially oxidized Ta-carbonitride core-shell particles probed by using surface-sensitive conversion-electron-yield x-ray absorption spectroscopy. Appl Phys Lett 96:191905
35. Ishihara A, Chisaka M, Ohgi Y, Matsuzawa K, Mistushima S, Ota K (2015) Synthesis of nano-TaO_x oxygen reduction reaction catalysts on multi-walled carbon nanotubes connected via a decomposition of oxy-tantalum phthalocyanine. Phys Chem Chem Phys 17:7643–7647
36. Uehara N, Ishihara A, Matsumoto M, Imai H, Kohno Y, Matsuzawa K, Mitsushima S, Ota K (2015) Tantalum oxide-based electrocatalysts made from oxy-tantalum phthalocyanines as non-platinum cathodes for polymer electrolyte fuel cells. Electrochim Acta 179:146–153
37. Uehara N, Ishihara A, Nagai T, Matsumoto M, Imai H, Kohno Y, Matsuzawa Mitsushima S, Ota K (2015) Kinetic study of oxygen reduction reaction on tantalum oxide-based electrocatalysts produced from oxy-tantalum phthalocyanines in acidic media. Electrochim Acta 182:789–794
38. Hayashi T, Ishihara A, Nagai T, Arao M, Imai H, Kohno Y, Matsuzawa K, Mistushima S, Ota K (2016) Temperature dependence of oxygen reduction mechanism on a titanium oxide-based catalyst made from oxy-titanium tetra-pyrazino-porphyrazine using carbon nano-tubes as support in acidic solution. Electrochim Acta 209:1–6
39. Tominaka S, Ishihara A, Nagai T, Ota K (2017) Noncrystalline titanium oxide catalysts for electrochemical oxygen reduction reactions. ACS Omega 2:5209–5214
40. Chisaka M, Ishihara A, Morioka H, Nagai T, Yin S, Ohgi Y, Matsuzawa K, Mitsushima S, Ota K (2017) Zirconium oxynitride-catalyzed oxygen reduction reaction at polymer electrolyte fuel cell cathodes. ACS Omega 2:678–684
41. Chisaka M, Ishihara A, Uehara N, Matsumoto M, Imai H, Ota K (2015) Nano-TaO_xN_y particles synthesized from oxytantalum phthalocyanine: how to prepare precursors to enhance the oxygen reduction reaction activity after ammonia pyrolysis? J Mater Chem A 3:16414–16418

Chapter 12
Carbon Nanotube-Based Non-Pt Fuel Cell Catalysts

Jun Yang and Naotoshi Nakashima

12.1 Introduction

Polymer electrolyte fuel cells (PEFCs) have attracted extensive interests from academic and industrial communities due to their high energy conversion efficiency, low pollutant emission, and potential for automotive applications. Since 2009, residential fuel cell cogeneration systems have been commercially available in Japan. In 2014, Toyota proposed the first commercial fuel cell vehicle "Mirai", in which PEFC systems have been successfully applied as the main power source. However, until now the high cost is still one of the most important factors that limit the wide application of PEFCs. The development of low-cost and reliable electrocatalysts is highly important for realizing eco-friendly society because the electrocatalysts cost is usually over 40% in fuel cell systems.

Electrode reactions in a hydrogen-fueled PEFC include oxygen reduction reaction on cathode (ORR) and hydrogen oxidation on anode. The electrocatalytic activity of cathode materials for the ORR significantly affects the cell performance because of the sluggish kinetics. Platinum-based electrocatalysts are the most widely used in current PEFC prototypes which significantly raise the cost. Hence, a great effort has been made to develop non-Pt electrocatalysts for ORR.

In most cases, non-Pt electrocatalysts need to be loaded on an electrically-conductive support to gain sufficient electron-conducting paths and

J. Yang (✉)
Ningbo Institute of Materials Technology and Engineering, Chinese Academy of Sciences, No. 1219, Zhongguan West Road, Zhenhai District, Ningbo 315201, Zhejiang, People's Republic of China
e-mail: yangjun@nimte.ac.cn

N. Nakashima
International Institute for Carbon-Neutral Energy Research (WPI-I2CNER), Kyushu University, 744 Motooka, Fukuoka 819-0395, Japan
e-mail: nakashima.naotoshi.614@m.kyushu-u.ac.jp

N. Nakashima (ed.), *Nanocarbons for Energy Conversion: Supramolecular Approaches*, Nanostructure Science and Technology,
https://doi.org/10.1007/978-3-319-92917-0_12

large specific surface area, which are important for high electrochemical performance. Carbon nanotubes (CNTs) are a promising support material for the non-Pt electrocatalysts due to their large specific surface area, high electrical conductivity, and excellent resistance to corrosion during long-term operation of fuel cells. Furthermore, CNTs can easily form self-standing network with high porosity, allowing a fast mass transport through the electrode. The electrocatalytic properties of doped carbon nanotubes for fuel cells have also attracted many attentions in recent years. Since the cathode dominates the overall performance of a single cell due to the sluggish kinetics of the ORR, only CNT-based non-Pt electrocatalysts for ORR or bifunctional catalysts are discussed in this chapter.

12.2 Pristine Carbon Nanotubes as the Supports for M-N_4 Catalysts

Jasinski reported the catalytic activity of transition metal-N4 chelates for the ORR for the first time in 1964 [1]. Since then, a variety of metal porphyrins and phthalocyanines have been explored as the ORR catalysts. These non-Pt catalysts possess high catalytic activity and negligible H_2O_2 production even when polarized at a high overpotential. Moreover, they are inert to methanol oxidation, which allow them to be used as the cathode for direct methanol fuel cells (see Chap. 2). However, most of these organic electrocatalysts are in lack of electrical conductivity. Thus, it is necessary to combine them with conductive supports, such as nanocarbons.

Pristine CNTs have highly crystallized graphitic surfaces. As a result, polycyclic aromatic molecules such as metal phthalacyanines and porphyrins as well as their derivatives can be easily absorbed on their surfaces via non-covalent π-π interactions, providing CNT-based hybrid electrocatalysts for the ORR. Morozan et al. deposited iron(II) phthalocyanine (FePc), cobalt(II) tetra-tert-butylphthalocyanine (Co(Bu)Pc), cobalt(II) 2,3,7,8,12,13,17,18-octaethylporphine (CoEtP), and cobalt (II) 5,10,15,20-tetrakis(4-tert-butylphenyl)-porphyrine (Co(BuPh)P) on a variety of carbon nanotubes, including single-wall nanotubes (SWCNTs), double-wall nanotubes (DWCNTs) and multiwall nanotubes (MWNTs) [2]. The composite catalysts were simply prepared by ultrasonicating the catalysts dispersed in tetrahydrofuran (THF). The samples were evaluated as the ORR electrocatalysts in alkaline media. As shown in Fig. 12.1, the catalysts supported on MWNTs exhibited superior ORR performance to those supported on SWCNTs and DWNTs. A possible reason is that MWNTs possess higher electrical conductivity than SWCNTs and DWCNTs [3]. Moreover, it was confirmed that FePc has a higher electrocatalytic activity for the ORR than the other catalysts (see Fig. 12.2).

It is also reported that FePc can be easily anchored on the surfaces of the CNTs by sonicating or even shaking in organic solvents at room temperature based on π-π interactions [4, 5]. The raw FePc powder has a stacking microstructure as shown in

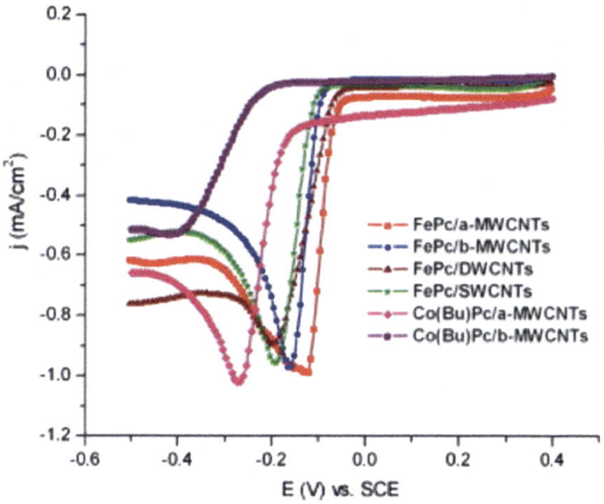

Fig. 12.1 Linear sweep voltammograms of FePc/a-MWCNTs (red), FePc/b-MWCNTs (blue), FePc/DWCNTs (triangle), FePc/SWCNTs (green), Co(Bu)Pc/a-MWCNTs (pink), Co(Bu)Pc/b-MWCNTs (purple) in O_2-saturated 0.1 M NaOH solution. Scan rate: 20 mV s^{-1} For FePc/a, see Fig. 12.3; the term "MWCNT" in the figure and caption refers to multiwalled carbon nanotubes, the same as "MWNT". Reproduction from [2] with permission of Elsevier

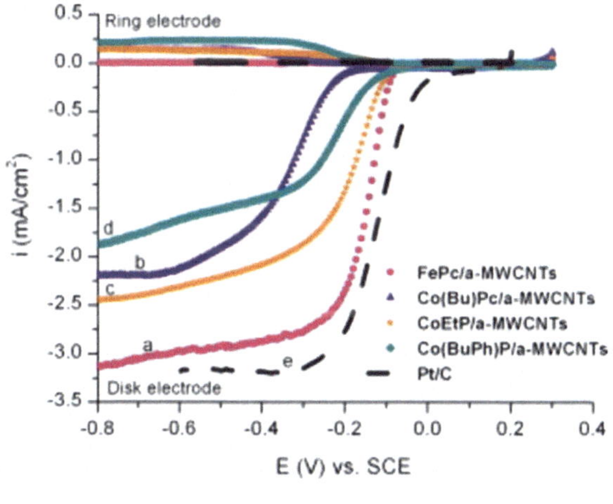

Fig. 12.2 RRDE measurements of oxygen reduction (negative current) and hydrogen peroxide oxidation (positive current) on FePc/a-MWNTs (pink curves), Co(Bu)Pc/a-MWNTs (purple curves), CoEtP/a-MWNTs (orange curves), Co(BuPh)P/a-MWNTs (green curves), Pt/C (black curves) electrodes in O_2-saturated 0.1 M NaOH solution. The ring electrode was polarized at 0.223 V versus SCE. Rotation rate: 500 rpm. Scan rate: 5 mV s^{-1}; the term "MWCNT" in the figure and caption refers to multiwalled carbon nanotubes, the same as "MWNT". Reproduction from [2] with permission of Elsevier

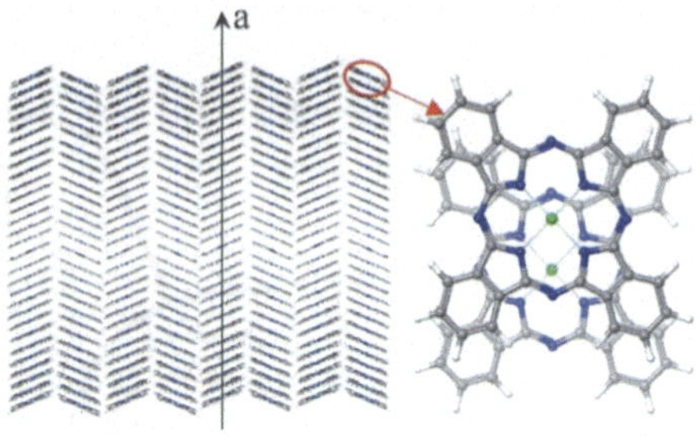

Fig. 12.3 Schematic illustration of a-form FePc (denoted as FePc/a). Reproduction from [5] with permission of Royal Society of Chemistry

Fig. 12.3. When added to organic solvents, such as THF and EtOH, the extremely low solubility of FePc makes a very small proportion of FePc delaminate to a single molecule, which could adsorb on the surfaces of the CNTs [5].

Yang et al. investigated the effects of the parameters in preparation of FePc/CNT hybrids in details, including mixing methods, solvent effect, and the molar ratios of the FePc and CNTs. By carefully adjusting the parameters of preparation procedure, the obtained ORR performance was found to be higher than that of a commercial Pt/C catalyst (Fig. 12.4).

Furthermore, the microstructure of FePc/MWNCNT was simulated by the density function theory (DFT) calculation (Fig. 12.5). When deposited on the surfaces of the MWNT, the flat FePc molecules are bent like an umbrella. Moreover, the charge density surrounding the centered Fe ions increased and the O–O bond of OOH adsorbed on the Fe ion is elongated. These results indicate that the combination of MWNTs and FePc does not only improve the electrical conductivity but also promote the cleavage of the O–O bond on FePc [5].

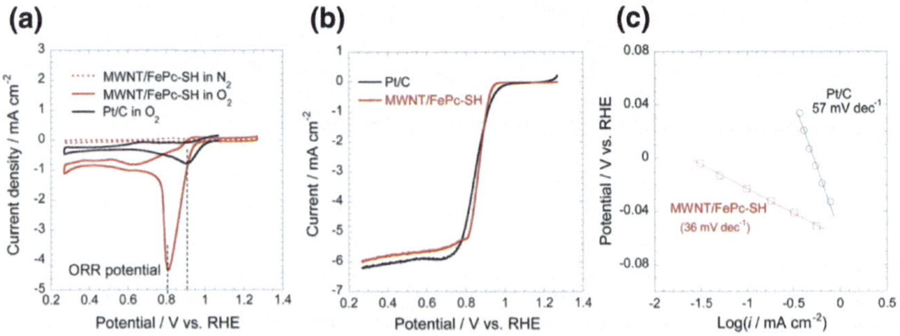

Fig. 12.4 **a** CV profiles of MWNT/FePc-SH measured at 10 mV s^{-1} in O_2- (red-solid line) and N_2-saturated (red-dot line) 0.1 M KOH, and the CV profile of Pt/C measured at 10 mV s^{-1} in O_2-0.1 M KOH (black-solid line) **b** LSV profiles measured at 10 mV s^{-1} and 1600 rpm in O_2-saturated 0.1 M KOH, and **c** Tafel plots. Reproduction from [5] with permission of Royal Society of Chemistry

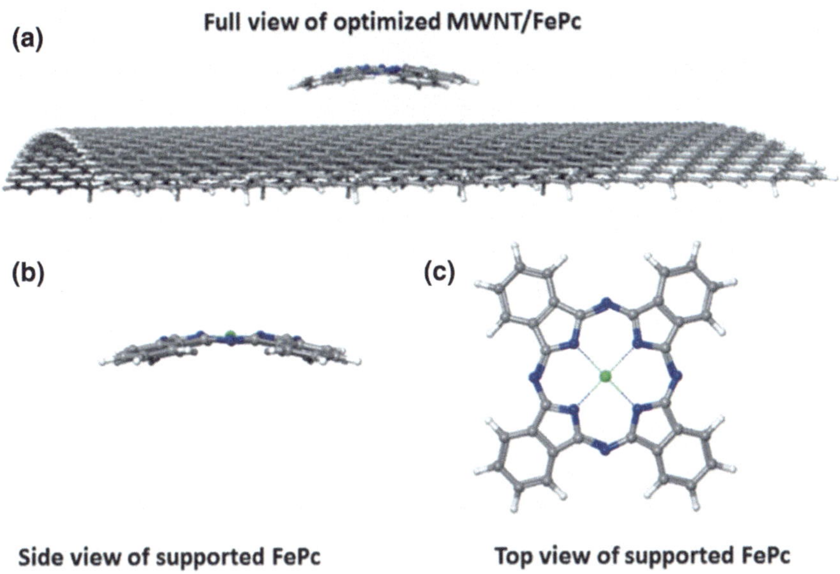

Fig. 12.5 **a** Overall, **b** side, and **c** top images of the optimized FePc/MWNT structure generated by the DFT calculations. Reproduction from [5] with permission of Royal Society of Chemistry

12.3 Functionalized Carbon Nanotubes as the Supports for M-N_4 Catalysts

It is well accepted that the centered metal ions of M-N_4 molecules act as the active site for oxygen reduction reaction [6]. Oxygen molecules being adsorbed on the metal ions converted to the OOH, and then to the OH. Thus, the electronic state of the centered metal ions significantly affects the electrocatalytic activity of the M-N_4 catalysts.

Cao et al. deliberately provided the axial ligand for the centered Fe ions in the FePc molecules [7]. They used pyridine-functionalized single-walled carbon nanotubes (Py-CNTs) as the support. Then, FePc was combined to the Py-CNTs by refluxing the two materials in THF under argon atmosphere to provide FePc-Py-CNTs (Fig. 12.6a, b). In this case, the Fe ions contained a five-coordinated structure, like the cytochrome c oxidase in the respiratory chain does. Such a well-designed nanostructure resulted in an excellent ORR performance and durability of the FePc-Py-CNTs. As is shown in Fig. 12.7a, the half-wave potential of the FePc-Py-CNTs is even higher than that of Pt/C. Moreover, it did not decrease even after 1000 cycles of cyclic voltammetry, while that of Pt/C decreased significantly (Fig. 12.7d). According to the DFT calculations, the five-ligand structure of the FePc on Py-CNTs increased the charge surrounding the centered Fe ions. As a result, the O–O bond in OOH adsorbed on the Fe ions is elongated, leading to the promoted ORR activity [7].

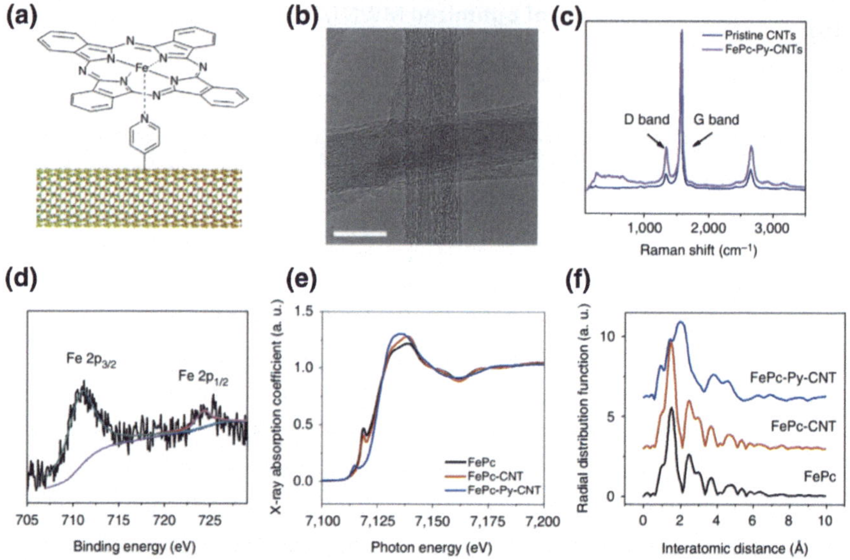

Fig. 12.6 **a** Schematic diagram of the structure of FePc-Py-CNTs composite. **b** HR-TEM image of FePc-Py-CNTs composite. Scale bar, 10 nm. **c** Raman spectra of pristine CNTs and FePc-Py-CNTs composite. D-band and G-band peaks are marked with arrows. **d** XPS Fe 2p spectra of FePc-Py-CNTs composite. **e** X-ray absorption near edge structure (XANES) spectra of FePc, FePc-CNTs and FePc-Py-CNTs. **f** Fourier-transformed module of EXAFS signals for FePc, FePc-CNTs and FePc-Py-CNTs. Reproduction from [7] with permission of Nature Publishing Group

12.4 Oxidized Carbon Nanotubes as the Supports for Transition Metal Oxides and Phosphides

As aforementioned, well-formed CNTs have highly crystallized surfaces, resulting in the lack of binding sites for the precursors of ORR catalysts. In order to in situ synthesize catalysts on the CNTs, it is necessary to functionalize the outer surfaces of CNTs.

The creation of defects on the outer surfaces of CNTs by oxidation is a convenient and common way to functionalize CNTs. CNTs can be effectively oxidized by wet chemical methods [8–10], photo-oxidation [11], oxygen plasma [12], or gas phase treatment [13]. After the oxidation, the CNTs are mainly decorated with the carboxyl as well as hydroxyl groups. These oxygen-containing groups facilitate the exfoliation of CNT bundles, increase the solubility in polar media, and provide abundant binding sites for the precursors of the catalysts [14]. Afterward, the precursors of catalysts can be anchored on the surfaces of the CNTs, and form transition metal oxide, phosphide, or sulfide catalysts by subsequent treatments.

Following such a strategy, Liang et al. prepared a highly active ORR electrocatalyst composed of cobalt oxide (CoO) nanocrystals and CNTs [15]. The TEM

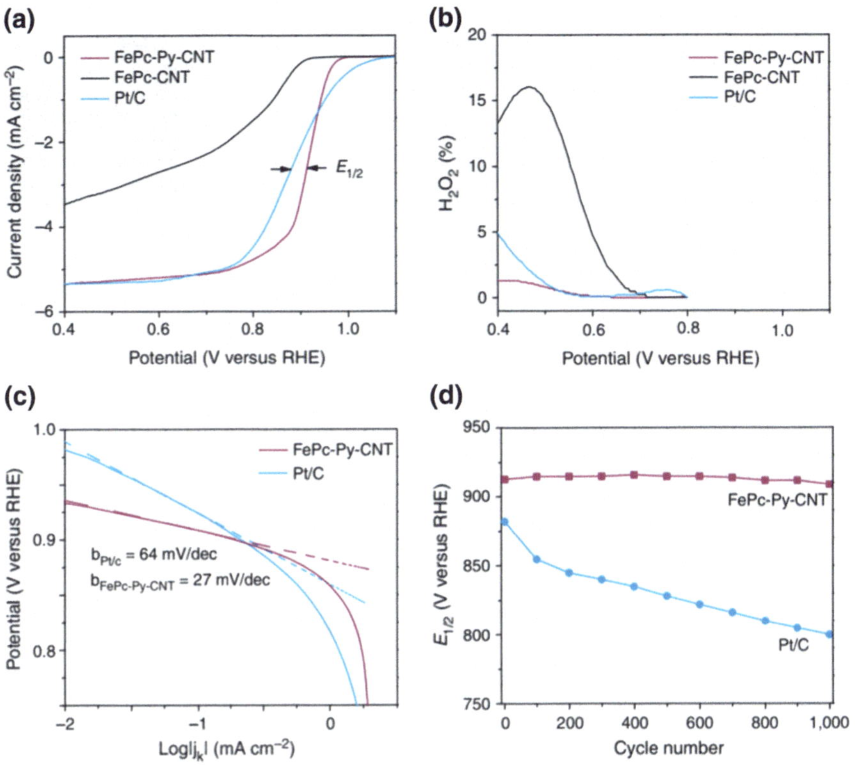

Fig. 12.7 **a** Linear scanning voltammograms of FePc-CNTs, FePc-Py-CNTs and commercial Pt/C catalyst. **b** H_2O_2 yield measured with various catalyst of FePc-CNTs, FePc-Py-CNTs and commercial Pt/C catalyst. **c** Tafel plots of FePc-Py-CNTs and commercial Pt/C catalyst. **d** Half-wave potential as a function of cycle number of FePc-Py-CNTs and commercial Pt/C catalyst during durability test. Electrolyte: 0.1 M KOH, scan rate: 10 mV s^{-1}. Reproduction from [7] with permission of Nature Publishing Group

images of the sample are shown in Fig. 12.8. The CoO nanocrystals ranging from 5 to 10 nm were homogeneously deposited on mildly oxidized MWNTs. Thanks to the well-controlled nanostructure, the hybrid electrocatalyst exhibited an attractive ORR performance close to Pt/C [15].

Hou et al. utilized the Co_3O_4/CNTs as the precursor of CoP/CNTs [16]. The Co_3O_4 nanoparticles were deposited on mildly-oxidized CNTs by a facile solvothermal synthesis method [15]. Then, the Co_3O_4/CNTs and NaH_2PO_4 were placed on an upstream and downstream of a quartz tube, respectively. After an annealing treatment at 300 °C in Ar for 2 h, the Co_3O_4 nanoparticles were transformed to CoP.

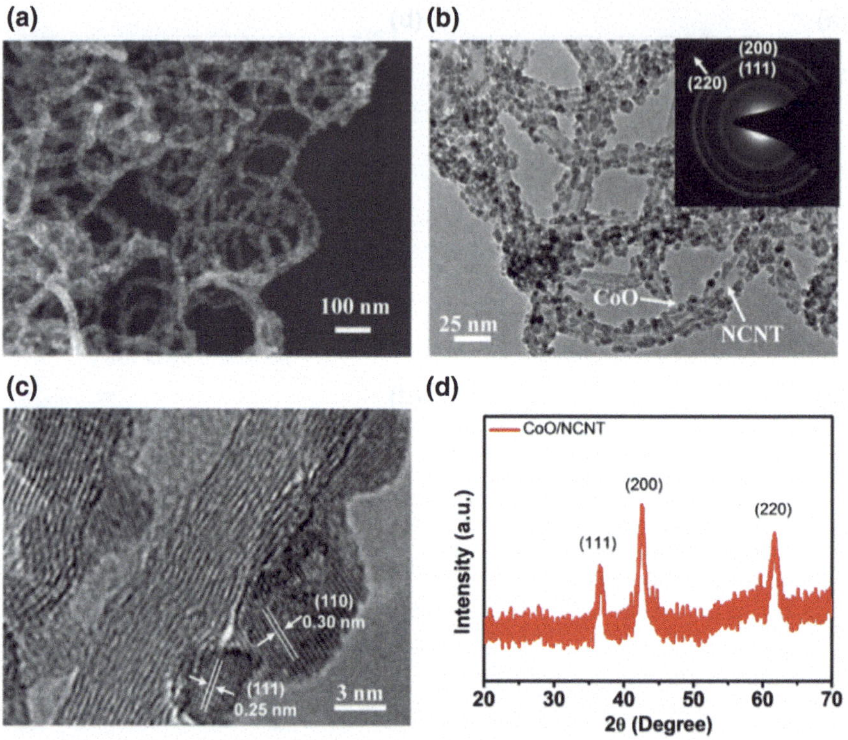

Fig. 12.8 **a** SEM, **b**, **c** TEM images, and **d** XRD profile of the CoO/NCNT hybrid catalyst. Reproduction from [15] with permission of American Chemical Society

As can be seen in Fig. 12.9, the CoP nanoparticles with an average diameter ~ 1.5–2.0 nm were homogeneously anchored on the CNTs. Such a well-controlled nanostructure of the CoP/CNTs is a promising electrocatalyst for both ORR and OER in alkaline media [16].

12.5 Polymer-Functionalized Carbon Nanotubes as the Supports for Transition Metal Oxide and Phosphate Catalysts

In addition to oxidation, wrapping the CNTs with a conducting polymer is also a convenient way to functionalize the inert outer surfaces of CNTs. One of the most widely used conducting polymers is polyaniline (PANI) and its derivatives that possess a high electrical conductivity and environmental stability due to their π-conjugated structures [17]. Moreover, PANIs have numerous amino groups as anchor sites for metal ions. Therefore, PANI-wrapped MWNTs are frequently reported as a promising composite support for precious metal nanoparticles [18–22].

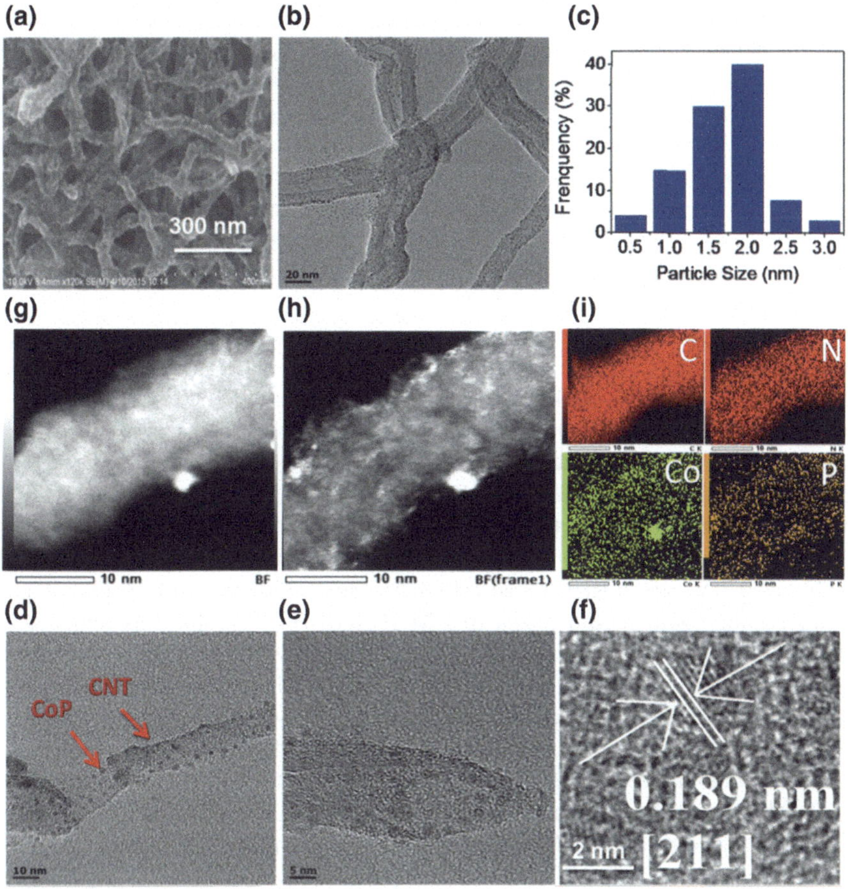

Fig. 12.9 **a** SEM, **b** and **d** TEM, **c** distribution of the particle size, and **e** and **f** HR-TEM images. **g** Cs-corrected HAADF-STEM image, **h** bright field image, and **i** corresponding EDX elemental mapping of Co, P, C, and N of the CoP-CNT composite. Reproduction from [16] with permission of American Chemical Society

The typical preparation procedure of PANI/MWNTs is as follows. MWNTs are oxidized by HNO_3 to form binding sites on their surfaces. Then, the oxidized CNTs are dissolved in an organic solvent, such as CH_2Cl_2 together with excess aniline. Subsequently, PANI is formed by an efficient interfacial polymerization and wraps the whole surfaces of the CNTs [22].

In recent years, a great effort have been made to deposit non-precious metal and metal oxides on PANI-wrapped CNTs. Hu et al. deposited iron and cobalt nanoparticles on such PANI/MWNT support for ORR [23, 24]. Aniline monomer, CNTs, and citric acid pretreated metal chlorides are mixed and ultrasonicated in an organic solvent to which $NaBH_4$ is added, then ultrasonicated in an ice bath. During this process, the formation of PANI and iron nanoparticles occurred simultaneously. It is proposed that a two-step redox process occurs for the coordination of metal and aniline. First, metal cations oxidize the polymer unit (aniline), and then reduced to metallic nuclei, which subsequently bind to imine nitrogen in aniline be oxidized to cation in situ, leading to the formation of the radical cation segments [25].

The typical resultant catalyst, Fe-PANI/CNT(+CA), is shown in Fig. 12.10. The CNTs were covered with a PANI film (see Fig. 12.10a, b), and uniform Fe nanoparticles with 2–4-nm diameters was coated on the PANI/CNTs homogeneously (see Fig. 12.10c–e). The crystal phase of the nanoparticles was confirmed to be iron. However, the X-ray photoelectron spectroscopy (XPS) revealed that the Fe atoms at the surfaces were bivalent and trivalent, i.e., a core-shell structure of the Fe nanoparticles was formed.

It was demonstrated that the pretreatment of $FeCl_3$ by citric acid (CA) was necessary to form a uniform distribution of Fe nanoparticles. Otherwise, the Fe nanoparticles would be agglomerated as shown in Fig. 12.10e. However, in the presence of CA, only leucoemeraldine base was formed in the PANI structure, which lacks electronic conductivity. Therefore, the Tafel slope of the Fe-PANI/CNT(+CA) was $\sim$60 and 142.8–172.9 mV dec^{-1} at the low and high overpotential regions, respectively, indicating that further improvement of the performance is necessary. However, it is worth noting that the current density of the Fe-PANI/CNT(+CA) at 0.7 V versus RHE remained 94.2% after 1350 cycles of cyclic voltammetry for the ORR in acid media (0.5 M H_2SO_4) (Fig. 12.11). This remarkable durability was attributed to its specific structure; namely, the Fe nanoparticles were mounted in the porous structure of PANI. Yin et al. prepared PANI-Co-CNT catalysts using a similar procedure described above [24]. The core-shell structure was also observed for the cobalt nanoparticles coated on the PANI-CNTs, and excellent durability for the ORR in acid media was obtained.

Besides iron and cobalt nanoparticles, spinel oxide such as $CoFe_2O_4$ can also be deposit on PANI-wrapped CNTs. Liu et al. wrapped oxidized CNTs with PANI by the in situ polymerization of aniline using $(NH_4)_2S_2O_8$ and $FeSO_4 \cdot 7H_2O$ as the oxidant [26]. Then, the formed uniform $CoFe_2O_4$ nanoparticles were deposited on the PANI-MWNTs by a reduction method using hydrazine as the reductant. The catalyst exhibited a good electrochemical performance and remarkable durability for oxygen evolution reaction.

Nakashima et al. have developed a facile method to functionalize CNTs using polybenzimidazole (PBI) and pyridine-based polybenzimidazole (PyPBI) [27–33]. PBI and PyPBI can be easily dissolved in organic solvents, such as dimethyl acetamide (DMAc). By ultrasonicating PyPBI or PBI and CNTs together in DMAc at room temperature for several hours, CNTs are wrapped by forming a very thin layer of PyPBI or PBI, which is aided CNT dispersion in the organic solvent, as well as provides abundant binding sites for many different types of metal cations, such as platinum and palladium. Comparing with the functionalization of the CNTs by in situ polymerization of aniline, the advantage of this method is that it is not necessary to oxidize CNTs to provide binding sites for metal ions. As a result, highly crystallized surfaces are protected from severe damage upon high electric potential application. As a result, metal and metal oxide nanoparticles can be readily anchored on the pristine CNT supports. Following this strategy, Yang et al.

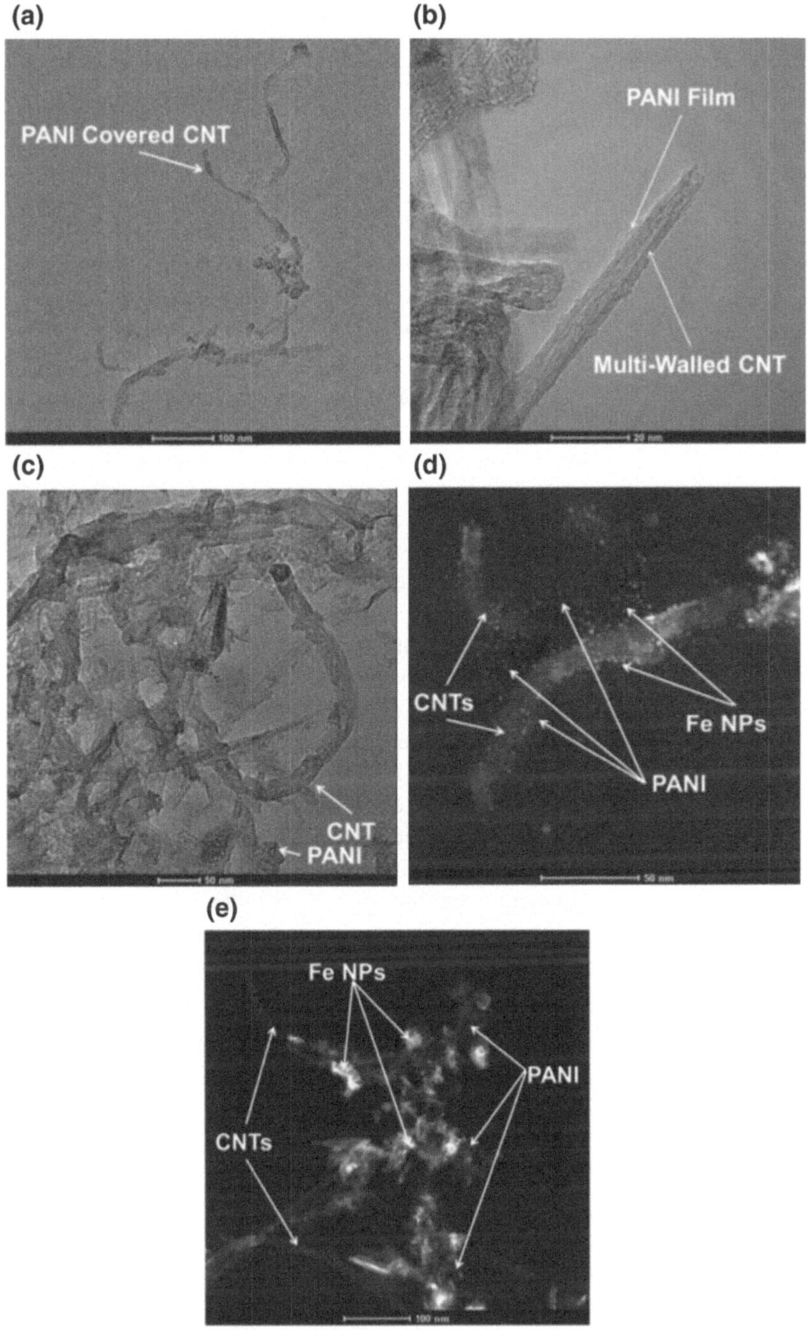

Fig. 12.10 **a** TEM of Fe-PANI/CNT(+CA); **b** HRTEM of Fe-PANI/CNT(+CA); **c** TEM of Fe-PANI/CNT(−CA); **d** STEM of Fe-PANI/CNT(+CA). Reproduction from [23] with permission of Elsevier

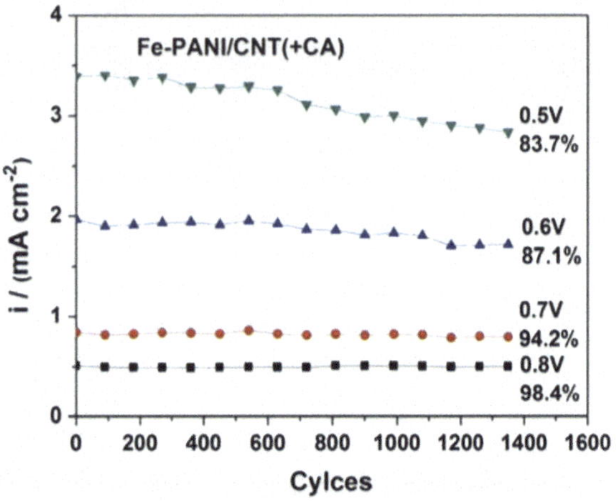

Fig. 12.11 The ORR stabilities after 1350 cycles for Fe-PANI/CNT(+CA) in O_2-saturated 0.5 M H_2SO_4. Rotation rate: 1600 rpm. Reproduction from [23] with permission of Elsevier

successfully decorated pristine CNTs with $Ni(OH)_2$, Co_3O_4, and $Ni_xCo_{3-x}O_4$ nanoparticles via a facile solvothermal method as shown in Fig. 12.12 [34].

As is shown in Fig. 12.13a, MWNT/PyPBI is composed of the pristine CNTs and a layer of PyPBI thinner than ~1 nm. After the solvothermal synthesis, uniform $Ni_xCo_{3-x}O_4$ nanoparticles with a diameter of ~5 nm were homogeneously coated on the MWNT/PyPBI. The authors have demonstrated that the presence of Ni is of significant importance to the formation of well-dispersed uniform nanoparticles of $Ni_xCo_{3-x}O_4$. The reason is that during the solvothermal synthesis, plate-like $Ni(OH)_2$ would be formed and cover the surfaces of the CNTs well (see Fig. 12.13f–h). In this case, the cobalt oxide can be anchored tightly in the Ni(OH) $_2$ and form spinel-type complex oxide ($Ni_xCo_{3-x}O_4$). In the absence of Ni, Co_3O_4, nanoparticles would agglomerate on the surfaces of the CNTs, as shown in Fig. 12.13d, e. Owing to the good electrical conductivity of the CNTs, the high electrochemical activity of the spinel oxide, and the well-controlled nanostructures

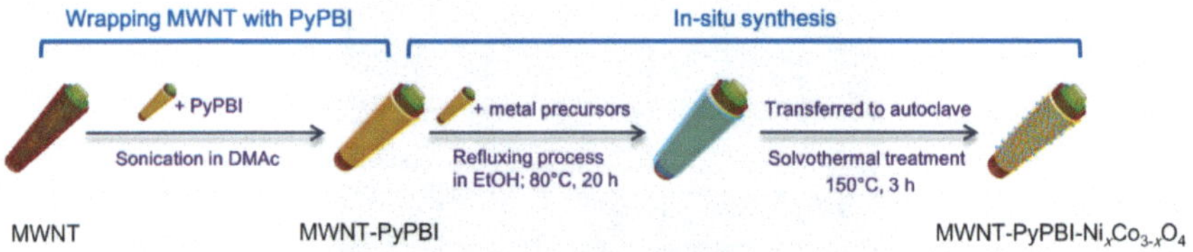

Fig. 12.12 Preparation procedure of MWNT-PyPBI-$Ni_xCo_{3-x}O_4$. MWNTs were wrapped with PyPBI, and then $Ni_xCo_{3-x}O_4$ was in situ synthesized by a facile solvothermal treatment. Reproduction from [34] with permission of Nature Publishing Group

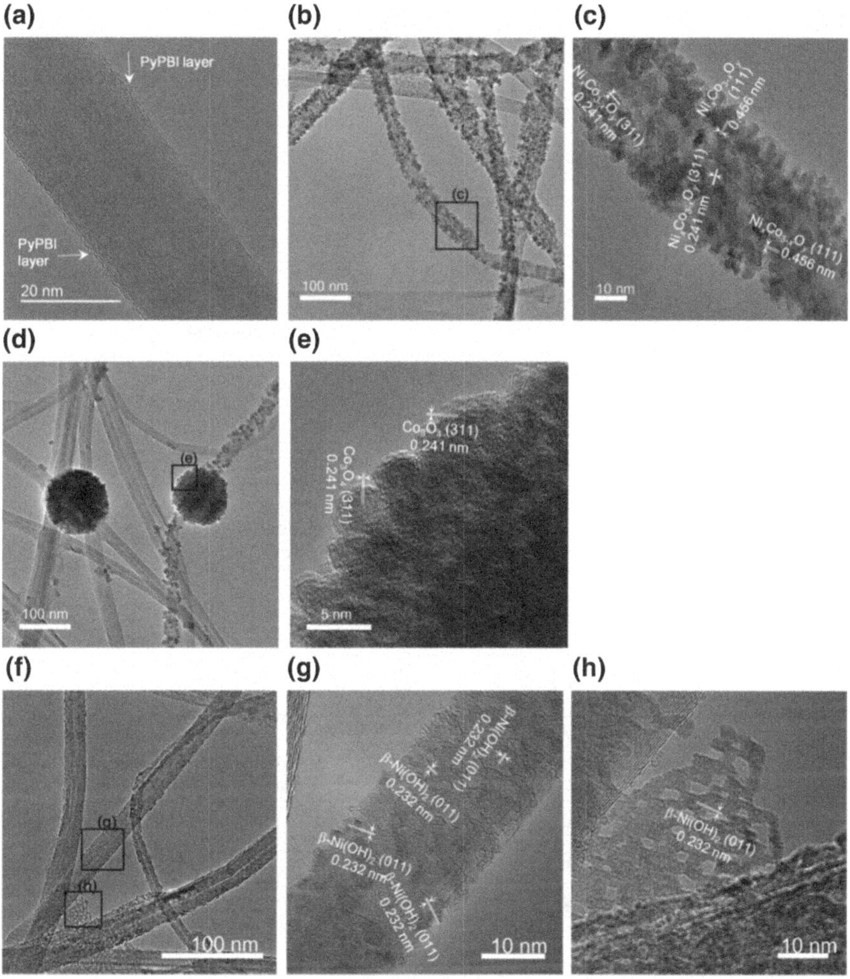

Fig. 12.13 TEM images of **a** MWNT-PyPBI, **b**, **c** MWNT-PyPBI-$Ni_xCo_{3-x}O_4$, **d**, **e** MWNT-PyPBI-Co_3O_4, and **f**, **g**, **h** MWNT-PyPBI-$Ni(OH)_2$; the images of the black rectangle-selected area in (**b**, **d** and **f**) are enlarged in (**c**, **e**, **g** and **h**). Reproduction from [34] with permission of Nature Publishing Group

of the hybrid catalysts, the MNWT-PyPBI-$Ni_xCo_{3-x}O_4$ exhibited a significantly enhanced efficiency and durability for both ORR and OER (Fig. 12.14).

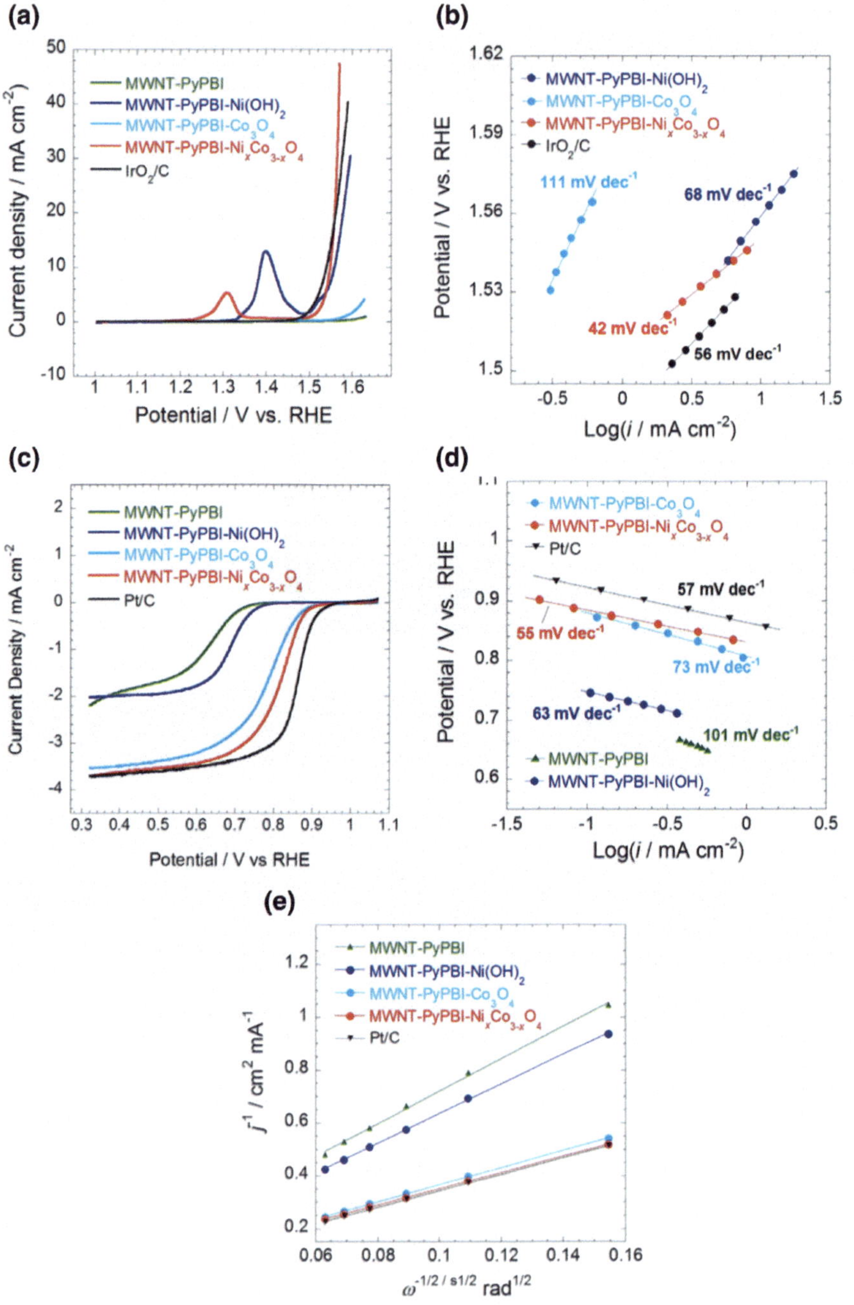
(a)
MWNT-PyPBI
MWNT-PyPBI-Ni(OH)2
MWNT-PyPBI-Co3O4
MWNT-PyPBI-NixCo3-xO4
IrO2/C
Current density / mA cm-2
Potential / V vs. RHE
(b)
MWNT-PyPBI-Ni(OH)2
MWNT-PyPBI-Co3O4
MWNT-PyPBI-NixCo3-xO4
IrO2/C
111 mV dec-1
68 mV dec-1
42 mV dec-1
56 mV dec-1
Potential / V vs. RHE
Log(i / mA cm-2)
(c)
MWNT-PyPBI
MWNT-PyPBI-Ni(OH)2
MWNT-PyPBI-Co3O4
MWNT-PyPBI-NixCo3-xO4
Pt/C
Current Density / mA cm-2
Potential / V vs RHE
(d)
MWNT-PyPBI-Co3O4
MWNT-PyPBI-NixCo3-xO4
Pt/C
57 mV dec-1
55 mV dec-1
73 mV dec-1
63 mV dec-1
101 mV dec-1
MWNT-PyPBI
MWNT-PyPBI-Ni(OH)2
Potential / V vs. RHE
Log(i / mA cm-2)
(e)
MWNT-PyPBI
MWNT-PyPBI-Ni(OH)2
MWNT-PyPBI-Co3O4
MWNT-PyPBI-NixCo3-xO4
Pt/C
j-1 / cm2 mA-1
ω-1/2 / s1/2 rad1/2

◂**Fig. 12.14** **a** OER polarization curves, **c** ORR polarization curves, and **e** K-L plots of MWNT-PyPBI, MWNT-PyPBI-Ni$(OH)_2$, MWNT-PyPBI-Co_3O_4, and MWNT-PyPBINixCo$_{3-x}O_4$; the ORR and K-L curves of Pt/C and the OER curve of IrO2/carbon black are also presented for comparison; the corresponding Tafel plots are shown in (**b**, **d**); electrolyte: O_2-saturated 1 M KOH solution; scan rate: 10 mV s^{-1}; rotating rate: 1600 rpm; catalyst loading for non-platinum catalysts: 0.3 mg cm^{-2}; catalyst loading for Pt/C: 14.2 mg Pt cm^{-2}; catalyst loading for IrO_2/C: 10.2 mg Pt cm^{-2}; temperature: 25 °C. Reproduction from [34] with permission of Nature Publishing Group

12.6 N-Doped CNT-Based Electrocatalysts for ORR

It is convenient to dope nanocarbons (CNTs and graphene) with N atoms during and the resultant products exhibit promising electrocatalytic activities for ORR. Dai et al. prepared vertically aligned N-doped carbon nanotubes (VA-NCNT) by pyrolysis of iron phthalocyanine [35]. The as-prepared N-doped CNT array was found to be an efficient ORR catalyst by Gong et al. [36].

Besides, the CNTs are also used as the support materials for N-doped carbons, such as amorphous carbon coating and graphene. Such hybrid ORR elecatalysts have attracted much attention in recent years [37–40].

12.7 Concluding Remarks

In summary, efficient and low-cost electrocatalysts for fuel cells can be obtained by hybridizing carbon nanotubes and various kinds of catalysts. Carbon nanotubes do not only provide electrical conductivity and large surface area but also affect the electron conditions of the supported catalyst, resulting in the promotion of the overall electrocatalytic activities.

References

1. Jasinski R (1964) A new fuel cell cathode catalyst. Nature 201:1212
2. Morozan A, Campidelli S, Filoramo A, Jousselme B, Palacin S (2011) Catalytic activity of cobalt and iron phthalocyanines or porphyrins supported on different carbon nanotubes towards oxygen reduction reaction. Carbon 49:4839–4847
3. Roche S, Akkermans E, Chauvet O, Hekking F, Issi J (2006) Understanding carbon nanotubes from basics to applications. Springer Lect Notes Phys 677
4. Dong G, Huang M, Guan L (2012) Iron phthalocyanine coated on single-walled carbon nanotubes composite for the oxygen reduction reaction in alkaline media. Phys Chem Chem Phys 14:2557–2559
5. Yang J, Toshimitsu F, Yang Z, Fujigaya T, Nakashima N (2017) Pristine carbon nanotube/iron phthalocyanine hybrids with a well-defined nanostructure show excellent efficiency and durability for the oxygen reduction reaction. J Mater Chem A 5:1184–1191

6. Shi Z, Zhang J (2007) Density functional theory study of transitional metal macrocyclic complexes' dioxygen-binding abilities and their catalytic activities toward oxygen reduction reaction. J Phys Chem C 111:7084–7090
7. Cao R, Thapa R, Kim H, Xu X, Gyu K (2013) Promotion of oxygen reduction by a bio-inspired tethered iron phthalocyanine carbon nanotube-based catalyst. Nat Commun 4:2076
8. Chen J, Hamon MA, Hu H, Chen Y, Rao A (1998) Solution properties of single-walled carbon nanotubes. Science 282:95–98
9. Rosca ID, Watari F, Uo M, Akasaka T (2005) Oxidation of multiwalled carbon nanotubes by nitric acid. Carbon 43:3124–3131
10. Zhang J, Zou H, Qing Q, Yang Y, Li Q (2003) Effect of chemical oxidation on the structure of single-walled carbon nanotubes. J Phys Chem B 107:3712–3718
11. Grujicic M, Cao G, Rao A, Tritt T, Nayak S (2003) UV-light enhanced oxidation of carbon nanotubes. Appl Surf Sci 214:289–303
12. Felten A, Bittencourt C, Pireaux J (2006) Gold clusters on oxygen plasma functionalized carbon nanotubes: XPS and TEM studies. Nanotechnology 17:1954
13. Tsang S, Harris P, Green M (1993) Thinning and opening of carbon nanotubes by oxidation using carbon dioxide. Nature 362:520–522
14. Datsyuk V, Kalyva M, Papagelis K, Parthenios J, Tasis D (2008) Chemical oxidation of multiwalled carbon nanotubes. Carbon 46:833–840
15. Liang Y, Wang H, Diao P, Chang W, Hong G (2012) Oxygen reduction electrocatalyst based on strongly coupled cobalt oxide nanocrystals and carbon nanotubes. J Am Chem Soc 134:15849–15857
16. Hou C-C, Cao S, Fu W-F, Chen Y (2015) Ultrafine CoP nanoparticles supported on carbon nanotubes as highly active electrocatalyst for both oxygen and hydrogen evolution in basic media. ACS Appl Mater Interfaces 7:28412–28419
17. Wan M (2008) A template-free method towards conducting polymer nanostructures. Adv Mater 20:2926–2932
18. Lee H-Y, Vogel W, Chu PP-J (2011) Nanostructure and surface composition of Pt and Ru binary catalysts on polyaniline-functionalized carbon nanotubes. Langmuir 27:14654–14661
19. Wu G, Li L, Li J-H, Xu B-Q (2005) Polyaniline-carbon composite films as supports of Pt and PtRu particles for methanol electrooxidation. Carbon 43:2579–2587
20. Konyushenko EN, Stejskal J, Trchová M, Hradil J, Kovářová J (2006) Multi-wall carbon nanotubes coated with polyaniline. Polymer 47:5715–5723
21. He D, Zeng C, Xu C, Cheng N, Li H (2011) Polyaniline-functionalized carbon nanotube supported platinum catalysts. Langmuir 27:5582–5588
22. Shi L, Liang R-P, Qiu J-D (2012) Controllable deposition of platinum nanoparticles on polyaniline-functionalized carbon nanotubes. J Mater Chem 22:17196–17203
23. Hu T-H, Yin Z-S, Guo J-W, Wang C (2014) Synthesis of Fe nanoparticles on polyaniline covered carbon nanotubes for oxygen reduction reaction. J Power Sources 272:661–671
24. Yin Z-s Hu, T-h Wang J-l, Wang C, Z-x Liu, J-w Guo (2014) Preparation of highly active and stable polyaniline-cobalt-carbon nanotube electrocatalyst for oxygen reduction reaction in polymer electrolyte membrane fuel cell. Electrochim Acta 119:144–154
25. Tang L, Wu T, Kan J (2009) Synthesis and properties of polyaniline–cobalt coordination polymer. Synth Met 159:1644–1648
26. Liu Y, Li J, Li F, Li W, Yang H, Zhang X, Liu Y, Ma J (2016) J Mater Chem A 4:4472–4478
27. Fujigaya T, Shi Y, Yang J, Li H, Ito K, Nakashima N (2017) A highly efficient and durable carbon nanotube-based anode electrocatalyst for water electrolyzers. J Mater Chem A 5:10584–10590
28. Fujigaya T, Nakashima N (2013) Fuel cell electrocatalyst using polybenzimidazole-modified carbon nanotubes as support materials. Adv Mater 25:1666–1681
29. Yang Z, Nakashima N (2015) A simple preparation of very high methanol tolerant cathode electrocatalyst for direct methanol fuel cell based on polymer-coated carbon nanotube/platinum. Sci Rep 5:12236

30. Yang Z, Nakashima N (2015) Poly (vinylpyrrolidone)–wrapped carbon nanotube-based fuel cell electrocatalyst shows high durability and performance under non-humidified operation. J Mater Chem A 3:23316–23322
31. Yang Z, Moriguchi I, Nakashima N (2016) A highly-durable CO-tolerant poly (vinylphosphonic acid)-coated electrocatalyst supported on a nanoporous carbon. ACS Appl Mater Interfaces 8:9030–9036
32. Berber MR, Fujigaya T, Nakashima N (2014) High-temperature polymer electrolyte fuel cell using poly(vinylphosphonic acid) as an electrolyte shows a remarkable durability. ChemCatChem 6:567–571
33. Berber MR, Hafez IH, Fujigaya T, Nakashima N (2014) Durability analysis of polymer-coated pristine carbon nanotube-based fuel cell electrocatalysts under non-humidified conditions. J Mater Chem A 2:19053–19059
34. Yang J, Fujigaya T, Nakashima N (2017) Decorating unoxidized-carbon nanotubes with homogeneous Ni-Co spinel nanocrystals show superior performance for oxygen evolution/reduction reactions. Sci Rep 7:45384
35. Dai L, Patil A, Gong X, Guo Z, Liu L (2003) Aligned nanotubes. ChemPhysChem 4: 1150–1169
36. Gong K, Du F, Xia Z, Durstock M, Dai L (2009) Nitrogen-doped carbon nanotube arrays with high electrocatalytic activity for oxygen reduction. Science 323:760
37. Cai X, Xia B, Franklin J, Li B, Wang X, Wang Z, Chen L, Lin J, Lai L, Shen Z (2017) Free-standing vertically-aligned nitrogen-doped carbon nanotube arrays/graphene as air-breathing electrodes for rechargeable zinc–air batteries. J Mater Chem A 5:2488–2495
38. Oh E, Hempelmann R, Nica V, Radev I, Natter H (2017) New catalyst supports prepared by surface modification of graphene- and carbon nanotube structures with nitrogen containing carbon coatings. J Power Sources 341:240–249
39. Li Z, Shao M, Yang Q, Tang Y, Wei M, Evans D, Duan X (2017) Directed synthesis of carbon nanotube arrays based on layered double hydroxides toward highly-efficient bifunctional oxygen electrocatalysis. Nano Energy 37:98–107
40. Yang L, Lv Y, Cao D (2018) Co, N-codoped nanotube/graphene 1D/2D heterostructure for efficient oxygen reduction and hydrogen evolution reactions. J Mater Chem A 6:3926–3932

Chapter 13
Polymer Electrolyte Membranes: Design for Fuel Cells in Acidic Media

Kenji Miyatake

13.1 Introduction

Polymer electrolyte fuel cells (PEFCs) have been commercialized in electric vehicles and residential power generation in recent years. The initial success of the commercialization of PEFCs owes, at least partly, to the development of highly proton conductive and durable, thin polymer electrolyte membranes (PEMs). Most PEFCs utilize perfluorinated sulfonic acid (PFSA) ionomer membranes. Nevertheless, there still is a great demand for alternative membranes to state-of-the-art PFSA ionomer membranes. While the PFSA ionomer membranes are highly proton conductive and chemically and physically stable, high gas permeability, high cost, and environmental incompatibility of the fluorinated materials are potential drawbacks for the wide-spread dissemination of PEFCs in a variety of fields. In the past couple of decades, a significant number of proton conductive polymeric materials have been proposed as alternative membranes [1–3].

Aromatic hydrocarbon polymers, acid-doped polymers, ionic liquids, and their composites fall into this category. Some of these emerging materials are claimed to exhibit high proton conductivity, very low gas permeability, and reasonable stability. However, none of them could compete with PFSA ionomer membranes under a wide range temperature and humidity in practical fuel cell operating conditions. The most critical issues associated with the alternative membranes are insufficient durability and significant dependence of the proton conductivity upon humidity. In addition, interfaces between these proton conductive materials and the catalyst layers have not been well-explored. In this chapter, some of the recent approaches in aromatic polymer-based PEMs are reviewed, in particular, focusing on the proton conductivity, mechanical stability, and interfacial problems.

K. Miyatake (✉)
Clean Energy Research Center, University of Yamanashi, Kofu, Japan
e-mail: miyatake@yamanashi.ac.jp

N. Nakashima (ed.), *Nanocarbons for Energy Conversion: Supramolecular Approaches*, Nanostructure Science and Technology,
https://doi.org/10.1007/978-3-319-92917-0_13

13.2 Design of Aromatic Ionomer Membranes for Improved Conductivity and Stability

Aromatic ionomers with acidic functions have been extensively investigated as possible alternatives to PFSA ionomer membranes. Fluorine-free aromatic (co) polymers are attractive for their high thermal stability, gas impermeability, environmental compatibility, and cost-effectiveness. Such polymers include sulfonated poly(arylene ether)s [4–7], polyimides [8], polyphenylenes [9–12], and others [13–15]. It has been demonstrated by several research groups that block copolymer based PEMs composed of sequenced sulfonated and unsulfonated aromatic components are much more proton conductive than the random copolymers of the same components, since the block copolymer membranes tend to form developed phase separated morphology based on the hydrophilic/hydrophobic differences, which eventually results in interconnected ionic channels as proton transport pathways [16].

A typical example is a block copolymer, SPK (Fig. 13.1a), composed of highly sulfonated benzophenone groups as hydrophilic components [17]. Because of densely sulfonated hydrophilic component, SPK membranes are highly proton conductive, comparable to PFSA ionomer (e.g., Nafion) membranes over a wide range of humidity. In order to achieve comparably high performance in PFSA ionomer membranes, SPK requires relatively high concentration of sulfonic acid groups or ion exchange capacity (IEC), which causes large water absorption and swelling under fully hydrated conditions and may cause mechanical failure [18].

Many attractive properties of PEMs are related with their hydrophilic components since acid groups and the hydrated protons play a major role in proton conduction. For better performing aromatic PEMs, sulfophenylene group was used as a simple and effective hydrophilic component. The basic idea is that sulfophenylene units without carrying heteroatom linkages (e.g., ether, ketone, sulfone groups) are expected to contribute to good mechanical and chemical stabilities of the resulting PEMs. The sulfophenylene group has high IEC value as the

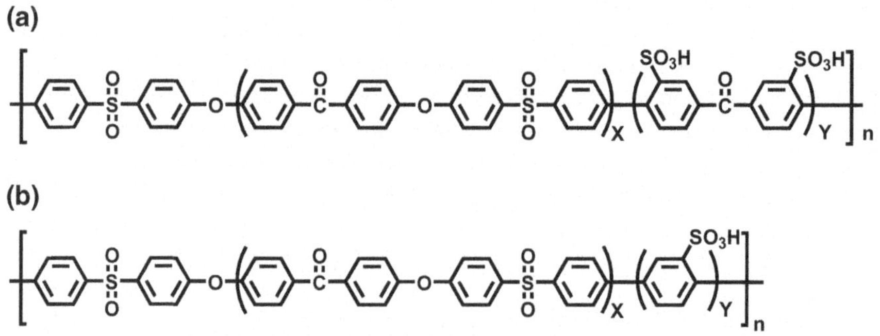

Fig. 13.1 Chemical structures of sulfonated aromatic block copolymers, **a** SPK and **b** SPP

hydrophilic segment to contribute to high proton conductivity under low humidity conditions. An SPP copolymer (Fig. 13.1b) containing sulfophenylene hydrophilic unit and the same hydrophobic component as that of SPK copolymer has been developed [19].

TEM images of SPK and SPP membrane stained with lead (Pb^{2+}) ions are shown in Fig. 13.2. SPK and SPP membranes exhibited similar phase-separated morphology composed of hydrophilic (dark) and hydrophobic (bright) domains. The similarity in the morphology is reasonable since these two polymers share the same hydrophobic component and the slightly different hydrophilic component. The hydrophobic domains were rather predominant and their sizes were approximately 10 nm in width. The hydrophilic domain sizes were approximately 5 nm in width, which seemed large for their chemical structures. The results suggest inter- and/or intermolecular aggregations of each component.

In Fig. 13.3 are plotted water uptake and proton conductivity of SPK and SPP membranes at 80 °C as a function of relative humidity (RH). Data for PFSA ionomer (Nafion) membrane taken under the same conditions are also included. It is reasonable that both water uptake and proton conductivity increased as increasing the humidity for all membranes. Despite its slightly higher IEC value, SPP membrane exhibited similar (or lower at some RHs) water uptake compared to that of SPK membrane. Absence of polar ketone groups in the hydrophilic component, which are suggested to absorb water via hydrogen bonding, might contribute to the lower water uptake. Compared to Nafion, both aromatic copolymer membranes absorbed much larger amount of water due to their higher IEC values.

The proton conductivity was somewhat more complicated and seemed more related with the chemical structure of the hydrophilic components. SPK and SPP membranes showed comparable proton conductivity to Nafion membrane at a wide range of humidity. At 60% RH and higher, SPP membrane exhibited higher proton conductivity than SPK membrane similar to the water uptake. For example, SPK

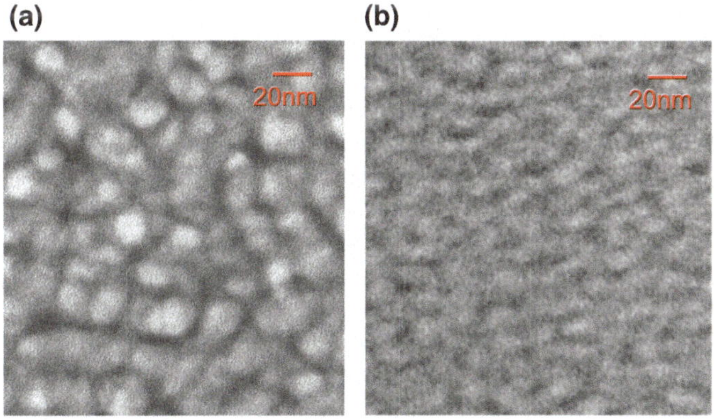

Fig. 13.2 TEM images of **a** SPK (IEC = 2.53 meq/g) and **b** SPP (IEC = 2.67 meq/g) membranes stained with lead ions

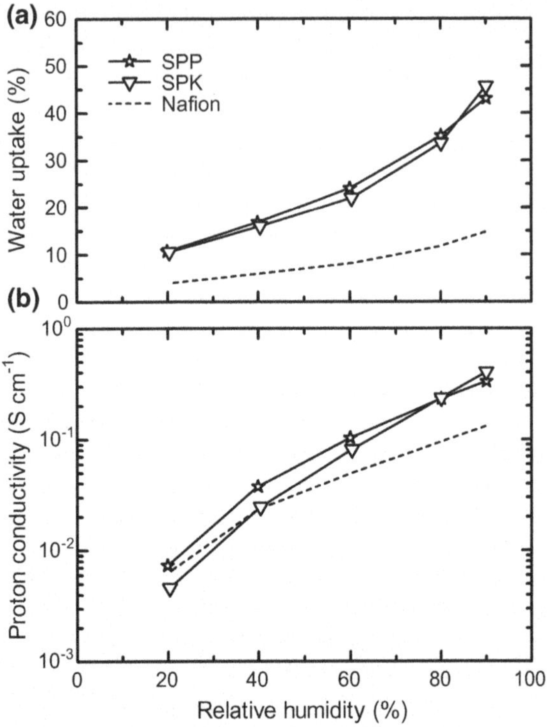

Fig. 13.3 **a** Water uptake and **b** proton conductivity of SPP (IEC = 2.53 meq/g), SPK (IEC = 2.67 meq/g), and Nafion (IEC = 0.91 meq/g) membranes at 80 °C and various humidities

membrane (54% water uptake) showed higher proton conductivity (543 mS/cm) than that (360 mS/cm) of SPP membrane (46% water uptake). At lower RH, the difference in the conductivity became larger than that of the water uptake. SPP membrane showed higher proton conductivity (7.3 mS/cm) than that (4.6 mS/cm) of SPK membrane at 20% RH. It is considered that higher local concentration of the sulfonic acid groups (6.40 meq/g) in the hydrophilic component of SPP than that (5.88 meq/g) of SPK is responsible for higher proton conductivity. SPK and SPP membranes exhibited high proton conductivity at higher temperatures. The conductivity of SPK and SPP membranes at 120 °C and 50% RH was ca. 60 mS/cm, was comparable to that at 80 °C and 50% RH. Although the water uptake of these membranes decreased with increasing temperature, the proton conductivity showed nearly the same values. The results indicate that the lower water content at higher temperature was probably countervailed by higher mobility of the hydronium ions.

Figure 13.4 shows storage modulus (E'), loss modulus (E''), and tan δ (E'/E'') of SPK, SPP, and Nafion membranes at 80 °C as a function of RH (the same measurement conditions as for the water uptake and proton conductivity in Fig. 13.3). SPP and SPK membranes showed higher E' values than those of Nafion membrane. There were significant differences in the mechanical properties between SPK and SPP membranes. SPP membrane showed excellent viscoelastic properties with slight dependence on the humidity, a slight decrease in E' with increasing the humidity and no transitions in E'' and tan δ curves in a wide range of the humidity. E' value was higher than 1 GPa even at high humidity. In contrast, SPK membrane showed a drop in E' and a transition (possible glass transition) in E'' and tan δ curves at ca. 60% RH. Lack of ketone groups in the hydrophilic component was likely to responsible for better mechanical properties of SPP membrane. The high mechanical stability of SPP membrane under humidified conditions was also observed by tensile tests, where SPP membrane possessed a high initial Young's

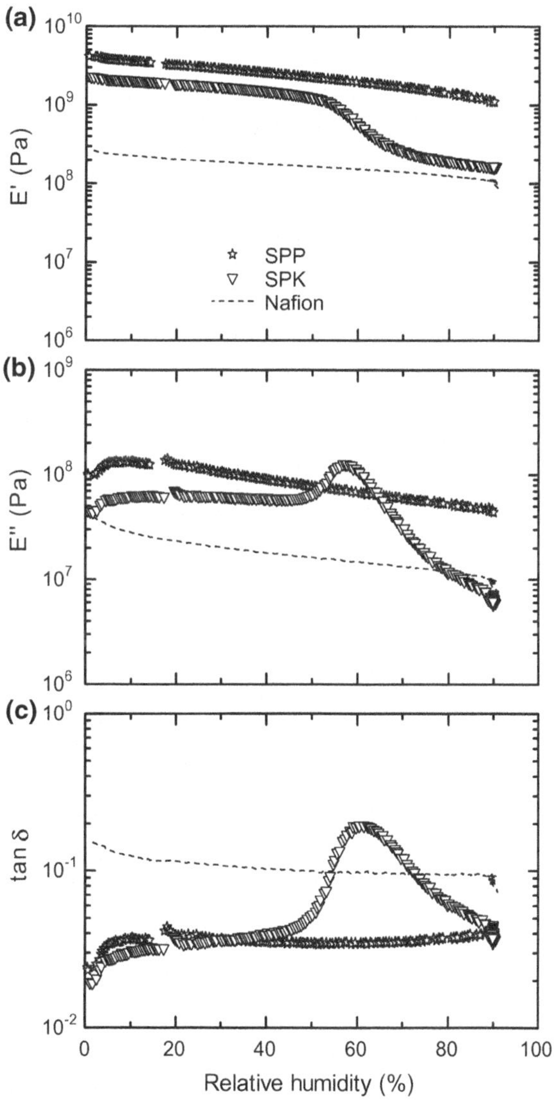

Fig. 13.4 DMA analyses of SPK, SPP, and Nafion membranes at 80 °C and various humidities, **a** storage modulus (E'), **b** loss modulus (E''), and **c** tan $\delta = E'/E''$

modulus (>1 GPa), maximum stress at breakpoint (>30 MPa), and reasonable strain (>66%). SPK membrane showed a much lower Young's modulus (<0.5 GPa).

Fuel cell performance was evaluated for SPK and SPP membranes at 80 °C, 100% RH, and 30% RH. As shown in Fig. 13.5, open circuit voltage (OCV) was 0.972 V for SPP and 0.970 V for SPK at 100% RH, suggesting low gas permeability of the membranes. SPK and SPP membranes showed comparable fuel cell performance and ohmic resistance at 100% RH. At 30% RH, SPP membrane showed much better fuel cell performance than that of SPK membrane. The higher fuel cell performance with SPP membrane was due to its lower ohmic resistance reflecting the higher proton conductivity at low RH as shown in Fig. 13.3.

13.3 Design of Aromatic Ionomer Membranes for Improved Interface

As discussed above, well-designed hydrocarbon ionomer membranes exhibit comparable fuel cell performance to that with Nafion membrane. However, recent studies revealed that the cathode catalytic performance was not as good with hydrocarbon ionomer membranes as that with Nafion membrane even when the same cathode catalyst layers (same Pt loading and same Nafion-based binder) were used. It is assumed that the interfacial incompatibility between the hydrocarbon

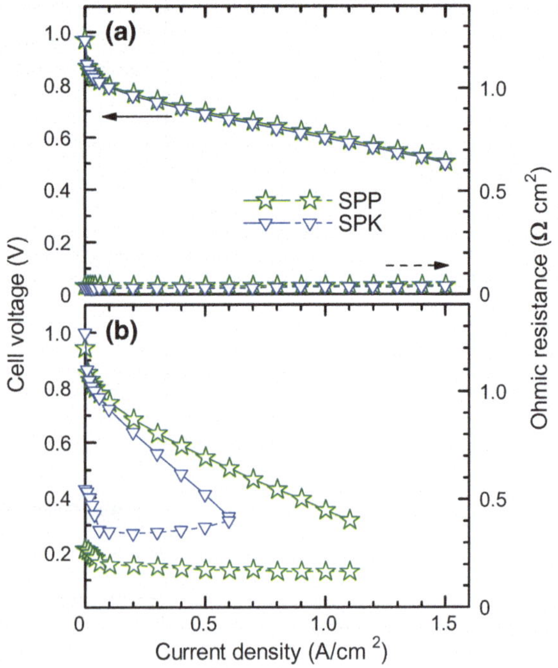

Fig. 13.5 Fuel cell performance of SPK and SPP membranes at 80 °C, **a** 100% RH and **b** 30% RH (hydrogen and air were supplied to the anode and cathode, respectively)

ionomer membrane and the catalyst layer causes lower cathode performance. As a typical example, IR-corrected (IR free) polarization curves and ohmic resistances of fuel cells with SPK and Nafion membranes are shown in Fig. 13.6. Under fully humidified conditions (100% RH), SPK membrane cell exhibited only slightly lower fuel cell performance to that of Nafion membrane cell. The ohmic resistance was in good accordance with that calculated from the proton conductivity (Fig. 13.3) and the thickness of the membranes. The difference in the performance between SPK and Nafion membrane cells became greater at lower humidity (30% RH). There are some intrinsic properties of SPK membrane at low humidity responsible for the results: (1) the proton conductivity of SPK membrane was slightly lower than that of Nafion membrane, (2) SPK membrane tended to dehydrate more than Nafion membrane (although SPK absorbed more water).

Then, to analyze quantitatively the incompatibility of SPK membrane and the cathode catalyst layer, mass activities at 0.85 V which are regarded as a measure of the effectiveness of Pt catalysts were calculated and plotted as a function of the humidity in Fig. 13.7. The mass activity of Nafion membrane cell was 108 A/g at 100% RH and ca. two times larger than that of SPK membrane cell. The results indicate that the catalytic performance could be significantly affected by the membranes even when the same Nafion-based electrodes were being used. From detailed morphological analyses of the membranes, it was found that the proton transport network at the interface of the membrane and the catalyst layer was

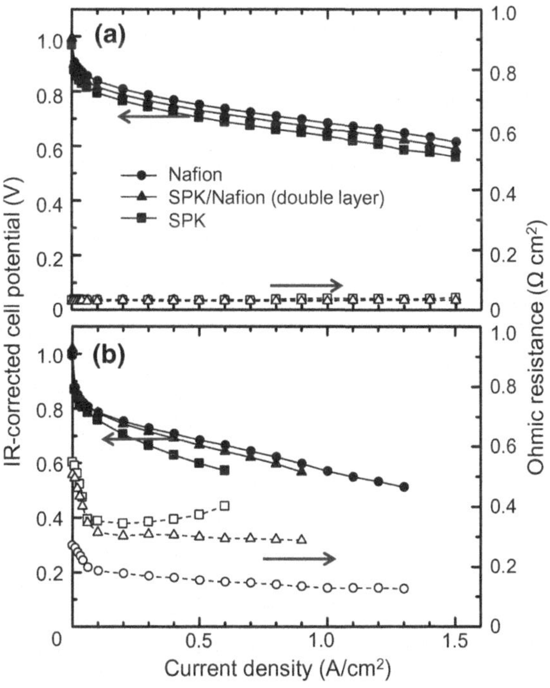

Fig. 13.6 IR-corrected polarization curves and ohmic resistances of fuel cells with SPK, SPK/Nafion (double layer), and Nafion membranes at 80 °C, **a** 100% RH and **b** 30% RH

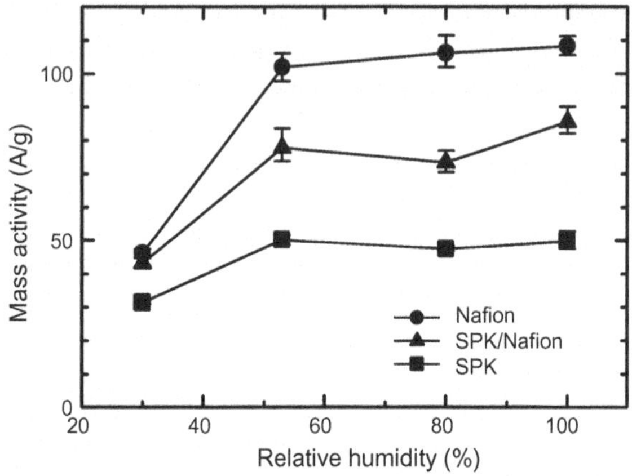

Fig. 13.7 Mass activity at 0.85 V of fuel cells with SPK, SPK/Nafion (double layer), and Nafion membranes at 80 °C at various humidities

responsible. Since SPK membrane contained a large-scale hydrophilic/hydrophobic phase-separated morphology (several tens of nanometer) as suggested by TEM and SAXS analyses, there existed some portion of the Pt nanoparticle catalysts (2–3 nm in diameter) covered with Nafion binder that would not have close contact with the proton conductive hydrophilic domains of the membrane but attached onto non-proton conductive hydrophobic domains as depicted in Fig. 13.8a. Pt particles on such hydrophobic domains did not have enough supply of protons from the SPK membrane, and thus are not very active for the electrocatalytic reduction of oxygen under dynamic (fuel cell operating) conditions.

In order to mitigate the incompatibility issue, effect of Nafion interlayer on the cathode performance was investigated [20]. The *so-called* double layer membrane was composed of a thin (ca. 0.55 μm) flat Nafion layer on a thick (24 μm) SPK membrane, of which phase-separated morphologies were very similar to those of the original components as confirmed by TEM image (Fig. 13.8c). The double layer membrane exhibited comparable water uptake and proton conductivity and better fuel cell performance to those of SPK membrane. Because of its tendency to dehydrate and its large phase-separated morphology, SPK membrane exhibited larger ohmic resistance and lower fuel cell performance compared to those of Nafion membrane at low humidity. In the SPK/Nafion double layer membrane, effect of Nafion interlayer on the fuel cell performance was significant. The interlayer lowered the ohmic resistance of SPK membrane due to mitigation of the membrane from drying excessively and improved the cathode catalytic performance. The mass activity at 0.85 V of the double layer membrane cell was much higher than that of SPK membrane cell, comparable to that of Nafion membrane cell at 30% RH. The very thin Nafion interlayer successfully compensated for the shortcomings of aromatic ionomer membrane.

Preparation of double layer membrane, however, could be often cumbersome and may not be appropriate for practical applications. It would be more preferable if a single membrane functions well and achieves good interfacial contact with the catalyst layer. For the purpose, partially fluorinated hydrocarbon ionomer,

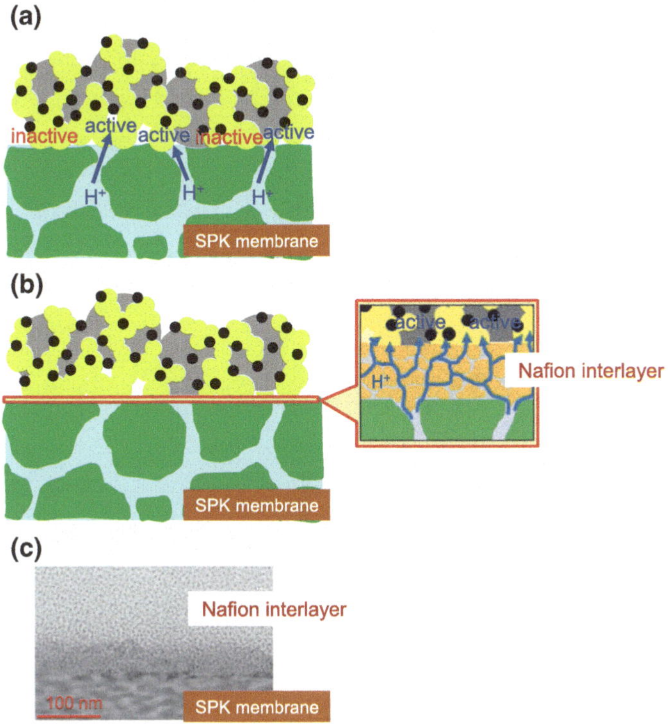

Fig. 13.8 Schematic model of the interface of **a** SPK membrane and the catalyst layer and **b** SPK/Nafion (double layer) membrane and the catalyst layer, and **c** cross-sectional TEM image of the double layer membrane

sulfonated poly(phenyene perfluoroalkylene) (SPAF) composed of linear perfluoroalkylene chains as the hydrophobic component and sulfonated oligophenylene groups as the hydrophilic component was designed [21]. Despite its partially fluorinated main chain structure similar to that of Nafion, SPAF could be more easily prepared by typical polycondensation reaction (Scheme 13.1). The copolymerization of bis(3-chlorophenyl)-1,6-perfluorohexane and dichlorobenzenes proceeded well via Ni(0) promoted reaction to provide high molecular weight precursor copolymers (PAF). The degree of sulfonation and IEC of SPAF copolymers were controllable by changing the sulfonation reaction conditions.

Figure 13.9 shows the cross-sectional TEM image of SPAF membrane. SPAF membrane exhibited characteristic phase-separated morphologies similar to those of Nafion membrane. The hydrophilic domains were spherical and smaller (ca. 1–2 nm in diameter) than that (ca. 3–6 nm in diameter) of Nafion membrane and highly distributed throughout the sight of view. Compared to SPK and SPP membranes (Fig. 13.2), the domain sizes of SPAF membrane were much smaller. The rigid and compact sulfonated phenylene units and their random distribution along the polymer main chains would be responsible. Despite the random

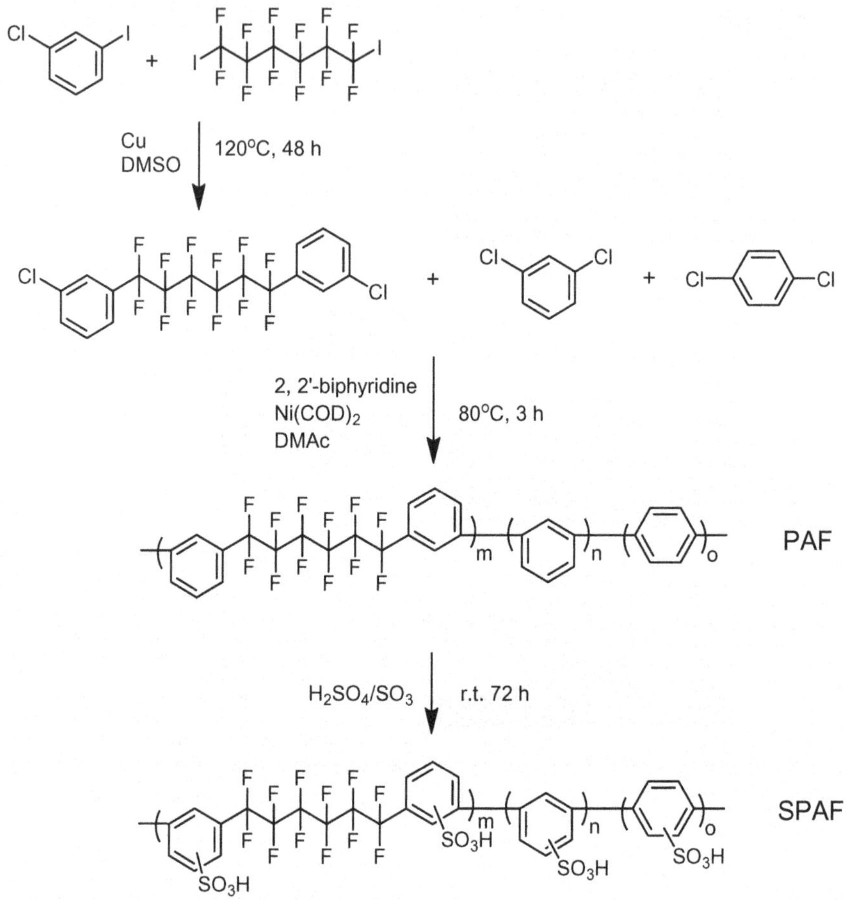

Scheme 13.1 Synthesis of SPAF copolymer

copolymer structure, the hydrophilic/hydrophobic interfaces of the SPAF membrane were more distinct and clearer than those of SPK and SPP membranes, presumably because of the larger hydrophilic/hydrophobic differences of the components.

Figure 13.10 shows IR-corrected polarization curves and ohmic resistance of a fuel cell with SPAF membrane at 100% and 30% RH, respectively. At 100% RH, the fuel cell exhibited low ohmic resistance (ca. 0.04 Ω cm^2) because of the high proton conductivity (>0.1 S/cm) of SPAF membrane. At 30% RH, the cell exhibited higher ohmic resistance (ca. 0.33 Ω cm^2 at current densities above 0.1 A/cm^2), reflecting its low proton conductivity. It was noticeable that the ohmic resistance was lower than that expected from the proton conductivity and the thickness of SPAF membrane (0.59 Ω cm^2 at 30% RH). The lower ohmic resistance than that calculated from the proton conductivity of the membrane indicated that the product

Fig. 13.9 TEM image of SPAF (IEC = 1.59 meq/g) membrane stained with lead ions

water in the cathode catalyst layer was effectively utilized by the membrane via back-diffusion. The IR-corrected I-V performance of the SPAF membrane cell was better than those of Nafion, SPK, and SPK/Nafion double layer membranes shown in Fig. 13.6. SPAF membrane with smaller ionic domains was able to promote interfacial proton transport, resulting in improved cathode performance. It is reasonably assumed that such interfacial compatibility of SPAF membrane and the catalyst layers also contributes to the back-diffusion of water from the cathode to the membrane.

The mass activity of Pt catalysts at 0.85 V was estimated from the polarization curves and plotted as a function of humidity in Fig. 13.11. SPAF membrane cell

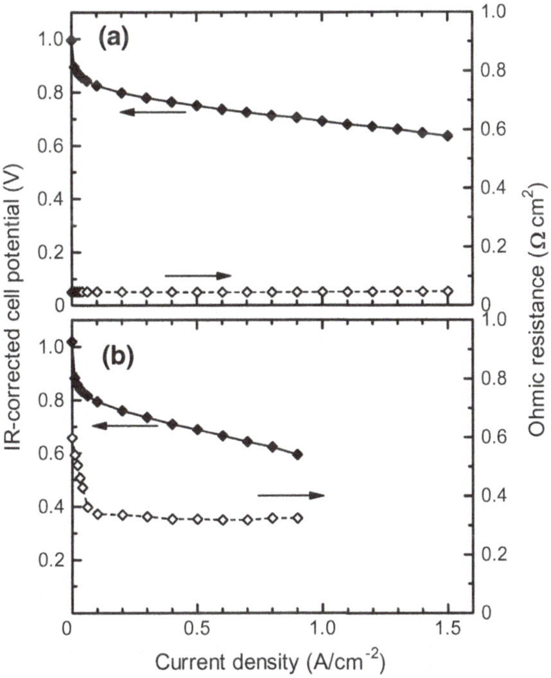

Fig. 13.10 IR-corrected polarization curves and ohmic resistances of a fuel cell with SPAF membrane at 80 °C, **a** 100% RH and **b** 30% RH

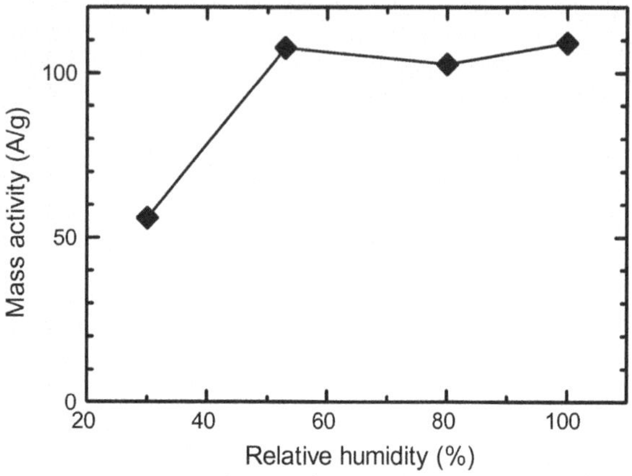

Fig. 13.11 Mass activity at 0.85 V of a fuel cell with SPAF membrane at 80 °C at various humidities

achieved high mass activity values; higher than 100 A/g at >50% RH. Mass activity of the SPAF membrane cell was ca. two times greater than that of the SPK membrane cell at a wide range of the humidity. The results indicate that SPAF membrane with a small-scale, phase-separated morphology is more suitable for practical fuel cell operating (dynamic) conditions than typical hydrocarbon ionomer membranes.

13.4 Conclusions

Considerable effort has been consumed for developing alternative proton conductive membranes for fuel cells that contain no or less fluorine atoms, to replace state-of-the-art perfluorinated ionomer membranes. In this chapter are shown some of the recent examples of aromatic polymer-based ionomer membranes to improve the proton conductivity and interfacial issues under practically operating conditions. While the emerging membranes have to be subjected to further tests under a wide range of conditions to check other properties such as chemical and mechanical durability and gas impermeability, the results described herein emphasize the importance and possibilities of suitable molecular design for even better-performing membranes.

References

1. Hickner MA, Ghassemi H, Kim YS, Einsla BR, McGrath JE (2004) Alternative polymer systems for proton exchange membranes (PEMs). Chem Rev 104:4587–4612
2. Park CH, Lee CH, Guiver MD, Lee YM (2011) Sulfonated hydrocarbon membranes for medium-temperature and low-humidity proton exchange membrane fuel cells (PEMFCs). Prog Polym Sci 36:1443–1498

3. Kreuer KD (2014) Ion conducting membranes for fuel cells and other electrochemical devices. Chem Mater 26:361–380
4. Miyatake K, Chikashige Y, Higuchi E, Watanabe M (2007) Tuned polymer electrolyte membranes based on aromatic polyethers for fuel cell applications. J Am Chem Soc 129:3879–3887
5. Tian S, Meng Y, Hay AS (2009) Membranes from poly(aryl ether)-based ionomers containing randomly distributed nanoclusters of 6 or 12 sulfonic acid groups. Macromolecules 42:1153–1160
6. Bae B, Yoda T, Miyatake K, Uchida H, Watanabe M (2010) Proton-conductive aromatic ionomers containing highly sulfonated blocks for high-temperature-operable fuel cells. Angew Chem Int Ed 49:317–320
7. Li N, Wang C, Lee SY, Park CH, Lee YM, Guiver MD (2011) Enhancement of proton transport by nanochannels in comb-shaped copoly(arylene ether sulfone)s. Angew Chem Int Ed 50:9158–9161
8. Asano N, Aoki M, Suzuki S, Miyatake K, Uchida H, Watanabe M (2006) Aliphatic/aromatic polyimide ionomers as a proton conductive membrane for fuel cell applications. J Am Chem Soc 128:1762–1769
9. Fujimoto CH, Hickner MA, Cornelius CJ, Loy DA (2005) Ionomeric poly(phenylene) prepared by diels-alder polymerization: synthesis and physical properties of a novel polyelectrolyte. Macromolecules 38:5010–5016
10. Goto K, Rozhanskii I, Yamakawa Y, Otsuki T, Naito Y (2009) Development of aromatic polymer electrolyte membrane with high conductivity and durability for fuel cells. Polym J 41:95–104
11. Umezawa K, Oshima T, Yoshizawa-Fujita M, Takeoka Y, Rikukawa M (2012) Synthesis of hydrophilic-hydrophobic block copolymer ionomers based on polyphenylenes. ACS Macro Lett 1:969–972
12. Miyake J, Taki R, Mochizuki T, Shimizu R, Akiyama R, Uchida M, Miyatake K (2017) Design of flexible polyphenylene proton-conducting membrane for next-generation fuel cells. Sci Adv 3:eaao0476
13. Jouanneau J, Mercier R, Gonon L, Gebel G (2007) Synthesis of sulfonated polybenzimidazoles from functionalized monomers: preparation of ionic conducting membranes. Macromolecules 40:983–990
14. Chang Y, Brunello GF, Fuller J, Hawley M, Kim YS, Disabb-Miller M, Hickner MA, Jang SS, Bae C (2011) Aromatic ionomers with highly acidic sulfonate groups: acidity, hydration, and proton conductivity. Macromolecules 44:8458–8469
15. Weiber EA, Takamuku S, Jannasch P (2013) Highly proton conducting electrolyte membranes based on poly(arylene sulfone)s with tetrasulfonated segments. Macromolecules 46:3476–3485
16. Elabd YA, Hickner MA (2011) Block copolymers for fuel cells. Macromolecules 44:1–11
17. Miyahara T, Hayano T, Matsuno S, Watanabe M, Miyatake K (2012) Sulfonated polybenzophenone/poly(arylene ether) block copolymer membranes for fuel cell applications. ACS Appl Mater Interfaces 4:2881–2884
18. Okanishi T, Yoda T, Sakiyama Y, Miyahara T, Miyatake K, Uchida M, Watanabe M (2012) Durability of an aromatic block copolymer membrane in practical PEFC operation. Electrochem Commun 24:47–49
19. Miyake J, Mochizuki T, Miyatake K (2015) Effect of the hydrophilic component in aromatic ionomers: simple structure provides improved properties as fuel cell membranes. ACS Macro Lett 4:750–754
20. Mochizuki T, Uchida M, Uchida H, Watanabe M, Miyatake K (2014) Double-layer ionomer membrane for improving fuel cell performance. ACS Appl Mater Interfaces 6:13894–13899
21. Mochizuki T, Uchida M, Miyatake K (2016) Simple, effective molecular strategy for the design of fuel cell membranes: combination of perfluoroalkyl and sulfonated phenylene groups. ACS Energy Lett 1:348–352

Chapter 14
Development of Polymer Electrolyte Membranes for Solid Alkaline Fuel Cells

Shoji Miyanishi and Takeo Yamaguchi

Glossary

AEM	Anion-exchange membrane
BTMA	Benzyl trimethyl ammonium
DFT	Density functional theory
DSC	Differential scanning calorimetry
IEC	Ion-exchange capacity
HOMO	Highest occupied molecular orbital
LUMO	Lowest unoccupied molecular orbital
PEMFC	Proton exchange membrane fuel cell
PEM	Proton exchange membrane
PES	Polyethersulfone
PIM	Polymer intrinsic microporosity
PSF	Polysulfone
PPO	Polyphenylene oxide
QA	Quaternary ammonium
SAXS	Small-angle X-ray scattering
SAFC	Solid-state alkaline fuel cell

14.1 Introduction

Fuel cells are a promising, green alternative for alleviating the world's increasingly complex energy issues. Proton exchange membrane fuel cells (PEMFCs) possess intrinsic advantages of lower operating temperatures, portability, high-efficiency,

S. Miyanishi · T. Yamaguchi (✉)
Laboratory for Chemistry and Life Science, Institute of Innovative Research, Tokyo Institute of Technology, Midori-ku, Yokohama 226-8503, Japan
e-mail: yamag@res.titech.ac.jp

S. Miyanishi
e-mail: miyanishi.s.aa@m.titech.ac.jp

N. Nakashima (ed.), *Nanocarbons for Energy Conversion: Supramolecular Approaches*, Nanostructure Science and Technology,
https://doi.org/10.1007/978-3-319-92917-0_14

and zero CO_2 emissions. Currently, PEMFC technology offers good performance and high power densities obtainable with low catalyst loading. Nevertheless, the need to use expensive Pt-based catalysts and hydrogen cylinders or stations has limited widespread, real-world application of technology. Recently, solid-state alkaline fuel cells (SAFCs) using anion-exchange membranes (AEMs) has drawn increasing attention as alternatives to PEMFCs (Fig. 14.1) [1–5].

The main principle of SAFCs are similar to those found in PEMFCs, with the only difference being the use proton (H^+) or hydroxide (OH^-) as the ion transporting media. In SAFCs, water is produced at the anode and OH^- is transferred from cathode to anode. In contrast, water is produced at the cathode in PEMFCs, and a proton is transferred from anode to cathode. As such, the water management technology in SAFCs is vastly different from that in PEMFCs [6–8].

SAFCs offer many advantages, the most obvious being the use of cheaper, non-noble metals, and carbon alloys as catalysts. Since most metals are susceptible to corrosion in highly acidic environments, effective electrode catalysts in PEMFC are limited to only Pt and Pt-alloys. With the exception of Al or Zn, however, most metals are stable under the prevailing alkaline conditions often found in SAFCs, thus allowing for the use of various types of metal catalysts [9]. Other promising candidates in SAFC technology are carbon-alloy catalysts. In acidic media, the oxygen reduction reaction (ORR) of carbon alloys is generally lower than that of Pt-based catalysts, but the performance is comparable or even superior in alkaline media [10–12]. Enhanced kinetics of ORR activity is also observed for other metal catalysts under alkaline conditions. Another advantage of SAFC is the use of various fuels for electrochemical reactions. To date, only hydrogen fuel has been

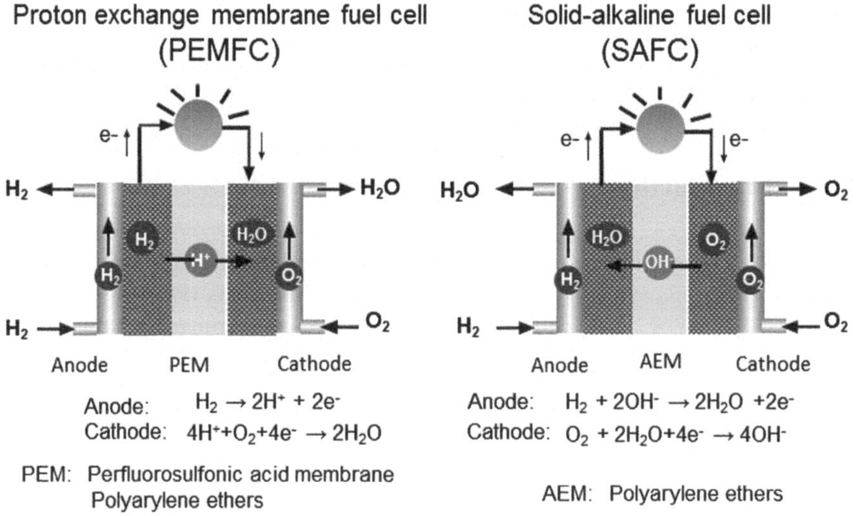

Fig. 14.1 Schematics of a proton exchange membrane fuel cell (PEMFC) and a solid-state alkaline fuel cell (SAFC)

commercially utilized in PEMFC because Pt-based catalysts are notoriously inactive in other fuels (e.g., methanol) [13, 14]. However, this is not the case on SAFCs, particularly, as it relates to the use of high-density liquid fuels such as methanol [14, 15], ethanol [14, 16], ethylene glycol [14], hydrazine [17], ammonia [18], and formate [19].

Despite these merits, real-world application of SAFC is still rather limited owing to the absence of high-performance membranes. Commercialization of PEMFC has been accomplished partly because of the development of high-performance perfluorosulfonic acid membranes, such as Nafion, which exhibit high proton conductivity and exceptional chemical durability when compared to other PEMs. This was attempted in real-world situations by performing simple modifications of the ion-exchange group from a sulfonic acid group to a quaternary ammonium (QA) cation of this perfluoro membrane [20]. However, this technique was abandoned after the resulting membrane showed low alkaline durability. Development of high-performance membrane, thus, remains a very important research subject for the real-world application of SAFC.

14.2 AEMs for Fuel Cell Applications

The typical structure of aromatic AEMs is shown in Fig. 14.2 Most AEMs are composed of an engineered plastic-based polymer backbone and an anion-exchange group. Although QA is the most popular anion-exchange site, imidazolium, phosphonium, guanidinium, and pyridinium can be used for the same purposes. Studies have shown that even though AEMs with an aliphatic polymer backbone (e.g., polyolefins, polyvinyl alcohols, and polystyrenes) [21–27] can be utilized for

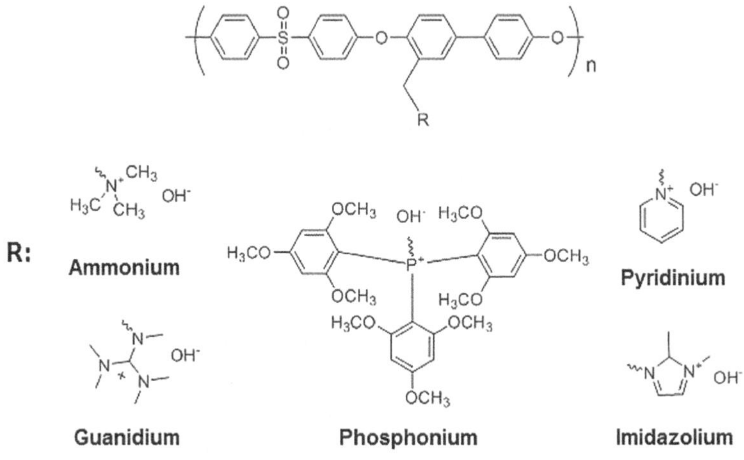

Fig. 14.2 Chemical structure of typical polyethersulfone AEMs

applications in salt purification or ion-exchange resin, these polymers are inadequate for fuel cell application because of their low chemical stability in oxidants and low thermal stability under the temperature conditions of low glass transition. As such, only the development and use of aromatic AEM will be discussed herein.

In order for a membrane to be considered suitable for real-world application in fuel cells, several properties must be present. First, since low ion conductivity of the membrane results in high ohmic resistance and decreased the performance of the cell, membranes with high ion conductivity are of utmost importance. Additionally, the membrane should have high chemical and thermal stability, given the fact that polymers with low glass transition temperature tend to be problematic in devices that operate at high temperatures. Thirdly, the membrane must also be very durable against oxidants, since it is not uncommon for highly reactive oxidants, such as hydrogen peroxide or OH radicals, to be formed during fuel cell operation. A fourth point to note is that a membrane must be mechanically strong and very stable in order to effectively suppress fuel crossover during operation of the fuel cell. This is especially important in liquid fuel devices, since cracking or the appearance of pinholes in the membrane can result in fuel crossover and drastically lower the cell's performance. Furthermore, a membrane's mechanical strength is also influential when it comes to the preparation and use of thinner membranes, which are especially beneficial for water and ion transport through the membrane. However, in order to maintain mechanical stability of the membrane, it is also important to suppress any swelling which may occur when the membrane is subjected to hydration.

Both PEMs and AEMs share many of the aforementioned traits, but it is worth noting that some important differences exist between the two. Firstly, the ion conductivity in AEMs is generally lower than in PEMs. Most of the recently developed PEMs show high conductivity with range from 0.1 to 1.0 S cm^{-1}, whereas anion conductivity in AEMs are generally lower by one order of magnitude (typically from 0.01 to 0.2 S cm^{-1}) [28]. This low conductivity may be caused by the lowered intrinsic self-diffusion of OH^- ions and reduced tendency of the QA hydroxide group to dissociate. Table 14.1 shows the ionic mobility of various cations and anions in water.

Ion mobility of proton and hydroxide is 36.23×10^{-8} m^2 s^{-1} v^{-1} and 20.64×10^{-8} m^2 s^{-1} v^{-1}, respectively. The higher mobility of protons and hydroxides when

Table 14.1 Ionic mobility of various cations and anions in water at 25 °C

Cation	Mobility (10^{-8} m^2 s^{-1} v^{-1})	Anion	Mobility (10^{-8} m^2 s^{-1} v^{-1})
H^+	36.23	OH^-	20.64
Na^+	5.19	F^-	5.74
K^+	7.62	Cl^-	7.92
NH_4^+	7.62	Br^-	8.09
$[N^+(CH_3)^4]$	4.65	I^-	7.96
$[N^+(C_2H_5)^4]$	3.38	CO_3^{2-}	7.18

compared to that of the other ions is thought to be the result of differences present in the mechanism of conduction, since these ions can be transported by both diffusion mechanisms (namely the Vehicle Mechanism) and via mediation of the hydrogen bonds that exist between the ions and water molecules (known as the Grotthuss Mechanism). However, it is important to note that the ionic mobility of OH^- is still two times lower than that of the proton, indicating that the exact processes through which protons and hydroxides are transported differ somewhat. In addition to this, the dissociation constant of QA salts are typically lower when compared to that of sulfonic acid. Dissociation of sulfonic acid in PEM can occur even with very small amounts of water due to high pKa ($pK_a \sim -1$); however, since the pK_b value of the QA is 4, the QA cation cannot dissociate if the content of water is too low. These reasons account for why most AEMs with high anion conductivity are extremely water absorbent and suffer from extensive swelling of the membrane. The anion conductivity of AEMs is generally measured in water, but the influence of humidity on the membrane tends to be significantly large because dissociation of the anion-exchange group becomes difficult at lower humidity.

Another point of difference between PEMs and AEMs is their chemical stability. AEMs are considerably unstable in hot alkaline solution and are prone to losing their conductivity, although a typical persulfonic acid PEM, Nafion, displays extremely high stability against chemical species because of its strong C–F bonds. PEMFCs with Nafion membranes are resilient during the long-term operation of the devices. In contrast, AEMs with a comparable level of chemical stability to Nafion have yet to be developed. Thus, the long-term durability of SAFCs has yet to be ascertained. The third point of difference relates to the issue of degradation of ion conductivity over time. Since the environmental atmosphere under which AEMs operate is slightly acidic due to existence of carbon dioxide, once the hydroxide anion is obtained by ion exchange of $NaOH_{(aq)}$ or $KOH_{(aq)}$, it is rapidly neutralized upon exposure to air in order to form the carbonate anion [29]. Carbonate anions can be only transferred by the Vehicle Mechanism; as such, the ion conductivity of AEMs tends to drop continuously over the lifetime of the device.

14.3 Synthesis of AEM

Generally, AEMs made of poly(arylene ether) can be synthesized in three steps as shown in Scheme 14.1 [30]. Firstly, polymerization is conducted via aromatic nucleophilic polycondensation of bisphenol and the bis-halogenated aryl monomers. Since this reaction is usually performed at high temperatures in an inorganic base, monomers with anion-exchange groups are not directly used due to its instability. Water must be carefully removed during reaction using azeotropic distillation in order to obtain high molecular weight polymers. Afterward, the resulting polymer can be functionalized through chloromethylation and successive quaternization to afford benzyl alkyl ammonium-modified AEMs. Controlling of the chloromethylation reaction is crucial in order to avoid the formation of

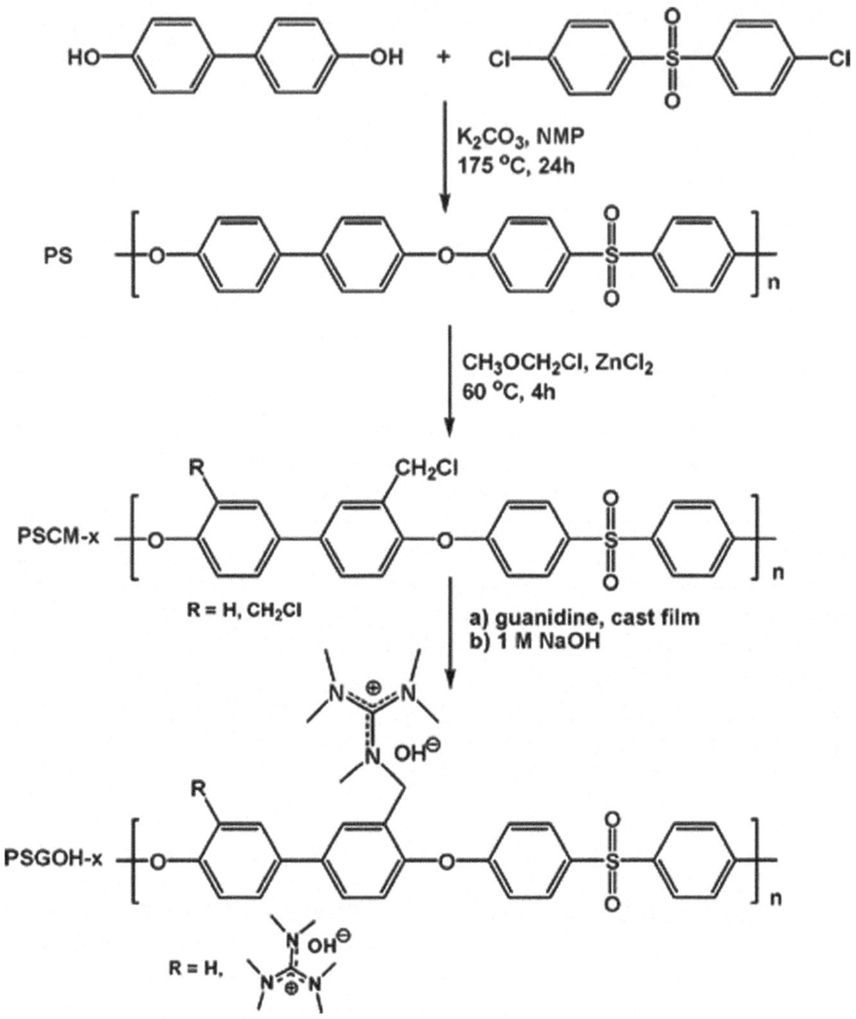

Scheme 14.1 Synthetic route for PS, PSCM-x, and PSGOOH-x [30]

cross-linked material. Finally, the chloride anion of the AEM is exchanged with the hydroxide anion by soaking the membrane in NaOH or KOH solution either before or after preparing the cast membrane.

AEMs made from poly(phenylene oxide) (PPO) can be synthesized via a slightly different procedure. Firstly, the precursor polymer is prepared using a Cu(I)-catalyzed oxidative polymerization of 2,6-dimethylphenol (this precursor polymer is also commercially available). The anion-exchange groups are attached through Wohl–Ziegler bromination of the methyl group in the polymer and followed by successive quaternization to afford QA-modified PPO AEMs [31].

Although most of the typical AEMs are synthesized through one of these two procedures, various types of poly(arylene ether) AEMs have been made recently using unique methods, including post-functionalization of the polymer through Suzuki–Miyaura Coupling [32–34], Click Chemistry [35–37], and lithiation [38] reactions to name a few. A. D. Mohanty et al. reported the successful preparation of an AEM by means of an iridium-catalyzed C–H borylation followed by palladium-catalyzed Suzuki Coupling as a type of polymer post-functionalization (Scheme 14.2) [33]. These highly efficient metal-catalyzed reactions allows for the synthesis of aromatic AEM materials under mild conditions and with short reaction times, thus minimizing the occurrence of side reactions along the polymer's chains (e.g., degradation and gelation). Side reactions are common during post-functionalization reactions that involve chloromethylation or benzylic bromination. Since both C–H borylation and Suzuki Coupling reactions give reliable conversions, controlling of the degree of functionalization of the desired QA moiety is relatively easy. This synthetic route could provide a way to incorporate bulky, more stable QA structures into AEMs through the use of the appropriate amine substrates during the Suzuki Coupling stage.

Zhang et al. reported the preparation of AEMs with pendent O-butyl trimethyl ammonium groups via direct polyacylation of pre-quaternized monomers (Scheme 14.3) [39]. Pre-quaternized monomers cannot be used in the general preparation of poly(arylene ether) AEMs due to the instability of the anion-exchange group at high temperatures under alkaline conditions. However, acid-catalyzed polymerization has the advantage of enabling direct preparation of AEMs in one step from the appropriate monomer, where the length of spacer and the species of the cationic group can be varied.

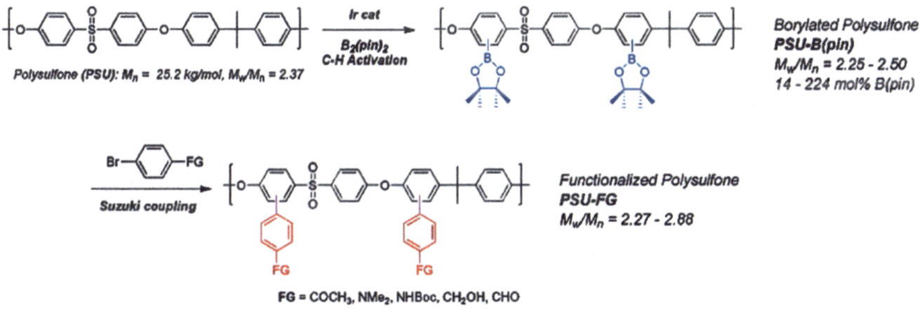

Scheme 14.2 Synthesis of trimethylbenzylammonium-functionalized polysulfone via C–H borylation and Suzuki Coupling [33]. © 2009 American Chemical Society

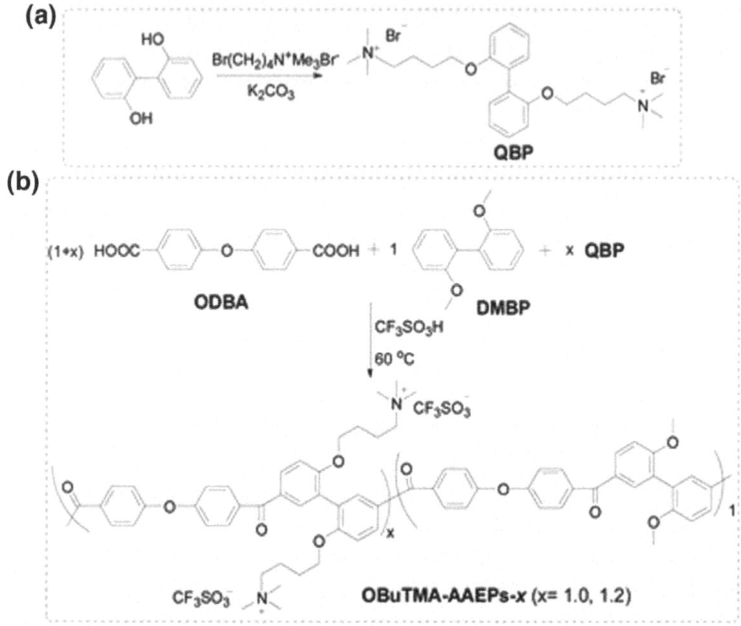

Scheme 14.3 Schematic synthesis of **a** the QBP monomer and **b** the OBuTMA–AAEPs-x polymer (x = 1.0, 1.2) [39]

14.4 Controlling the Solid-State Nanostructure in AEMs Through Targeted Molecular Design

Water uptake and the conductivity of hydroxide ions are particularly important for fuel cell applications. An ideally designed AEM would be focused on improving conductivity and controlling water uptake. So, far, most of the AEMs in existence that exhibit high hydroxide conductivity suffer from significant swelling, and are therefore unable to maintain their mechanical strength under hydration. To simultaneously achieve fast ion transport and membrane stability, it is necessary to construct hydrophilic ion channels that can efficiently move anions while still maintaining a strong hydrophobic domain. This is done to keep the membrane's mechanical strength high and to suppress any swelling associated with hydration. For both PEMs and AEMs, there are several approaches aimed at controlling the nanostructure of the membrane through more targeted molecular design.

As shown in Scheme 14.4 [40], Li et al. synthesized comb-shaped, hydroxide-conducting co-polymers containing hydrophobic grafting chains and functionalized polymer backbones with QA groups (PPO–DMHDA). Poly(phenylene oxide) (PPO) is an excellent choice as the polymer backbone because it is commercial available and has rather impressive thermal, mechanical, and chemical stability. Li was able to control the degree of grafting chain (x) density and

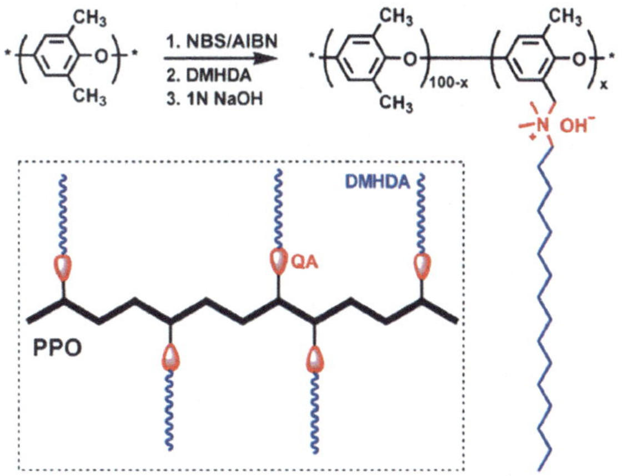

Scheme 14.4 Synthesis of comb-shaped polymers PPO–DMHDA-x containing quaternary ammonium (QA) groups [40]. © RSC publishing group

compared his results with those obtained using conventional benzyl trimethyl ammonium-modified PPO membranes (PPO–TMA) that had been synthesized without grafting chains. The nanostructure of PPO–DMHDAs is characterized using small-angle X-ray scattering (SAXS), in which clear peaks can be seen for all PPO–DMHDA samples but not for PPO–TMA samples. The periodicity *d* tends to rise when the content of the grafting chain is increased. This result is consistent with lamellar microstructures suggested by Li's comb-shaped architecture, although the large width of the peaks indicated that the arrangement of the phase-separated domains is only locally correlated. In Li's experiments, there was no proof that long-range ordered structures were ever formed. The self-assembled structure visible in the SAXS experiments shows the strong correlation between water uptake and ion transport. PPO–DMHDA-*x* membranes have much lower water uptake but comparable conductivity values can be seen for PPO–TMA (Table 14.2). These results indicate that construction of phase-separated ion channels is vital for fast ion conduction in AEMs operating under drier conditions.

In addition to controlling morphology via the use of self-assembled, comb-shaped polymers [41], the application of microphase separation of the block co-polymer is a promising approach for regulating the ion channels of AEMs.

Table 14.2 Properties of comb-shaped PPO–DMHDA-x and PPO–TMA-20 membranes [40]

Samples	IEC (meq g^{-1})	WU (wt%)[a]	λ	Δl (%)	σ (mS cm^{-1})[a]
PPO–DMHDA-20	1.08	8.2	5.6	2.6	7
PPO–DMHDA-30	1.48	13.5	5.2	4.2	14
PPO–DMHDA-40	1.67	16.2	5.4	5.0	23
PPO–DMHDA-55	1.92	20.4	5.9	5.6	35
PPO–TMA-20	1.39	25.9	10.4	7.4	5

[a]Measured in water at room temperature (20 °C)

Investigations on aromatic PEMs have shown that multi-block co-polymers, which contain highly localized, dense sulfonic acid groups in their hydrophilic blocks have unique phase-separated morphologies with interconnected proton transport pathways. These sulfonated multi-block co-polymer membranes are extremely conductive at high temperatures and show fuel cell performances comparable to those of Nafion under severe conditions. A similar strategy of utilizing sequential hydrophilic/hydrophobic structures with dense ion groups in the hydrophilic blocks also proves to be useful in designing the ideal AEMs [17, 42, 43].

As shown in Scheme 14.5, Tanaka et al. developed multi-block, co-poly(arylene ether) AEMs with QA groups (QPE) [17]. Tanaka was able to regulate the ion-exchange capacity (IEC) and control the unit length of the hydrophobic domains (X) and hydrophilic domains (Y) by controlling the rate of post-chloromethylation reaction, thereby creating a whole new series of membranes. The physical properties of the synthesized multi-block co-polymer (QPE), including anion conductivity, water uptake, microstructure, and alkaline stability, were investigated in detail and compared those attributed to the corresponding random co-polymer (rQPE).

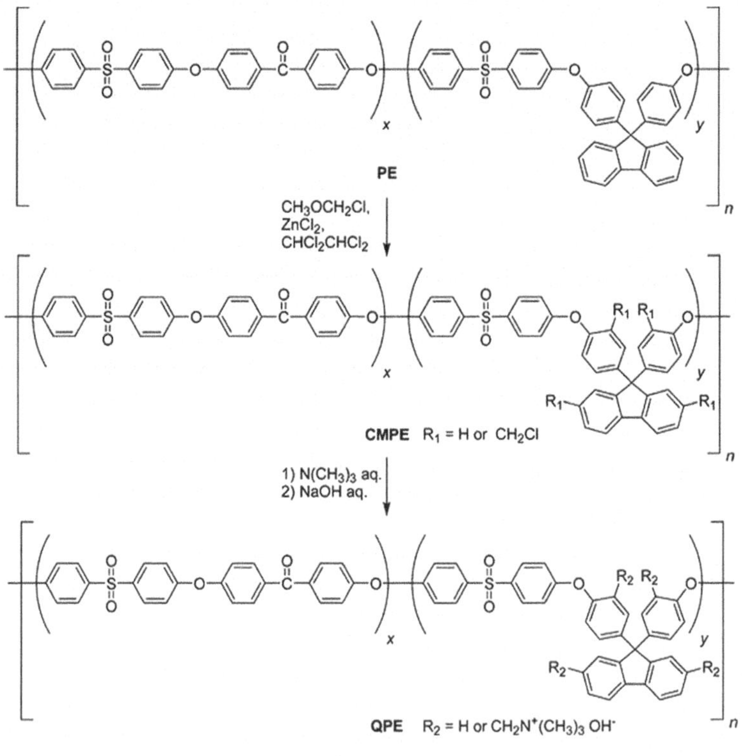

Scheme 14.5 Synthesis of multi-block co-polymer QPE [17]

A comparison of multi-block and random QPE membranes reveals that the multi-block QPE membranes, with the exception of the QPE-*X*8*Y*8 membranes, absorb less water (Fig. 14.3). In the case of the multi-block QPE membranes, lower water uptake and λ values associated with the number of absorbed water molecules per anion-exchange group are considered to be related to the development of hydrophilic/hydrophobic, phase-separated morphology. The QPE-*X*16*Y*11 membrane (IEC = 1.93 meq g^{-1}) had the highest conductivity (144 mS cm^{-1}) at 80 °C, which is approximately 3.2 times higher than that of the rQPE membrane (IEC = 1.88 meq g^{-1}) under the same conditions (Fig. 14.3). Multi-block QPE membranes generally have lower water uptake, but considerably higher conductivity than the corresponding rQPE membranes, thereby indicating that multi-block QPE membranes can utilize water molecules more efficiently for hydroxide ion transport than their counterparts. As already demonstrated for PEMs, these results suggest that the concept of applying multi-block structures with highly ionized hydrophilic blocks is effective for improving the ion conductivity of AEMs.

Several trials have been undertaken to construct microphase-separated nanostructures using multi-block segments or comb-shaped structures. These have proven beneficial in improving the ion conduction behavior of AEMs. However, it is important to note that anion-exchange functionality, such as the presence of QA groups or pyridinium groups, imparts weaker hydrophilic property to the membrane, especially when compared to the results seen in proton exchange membrane (PEM) with sulfonic acid groups. These polymers have been shown to have structural distributions with polydispersity. As such, it is difficult to construct

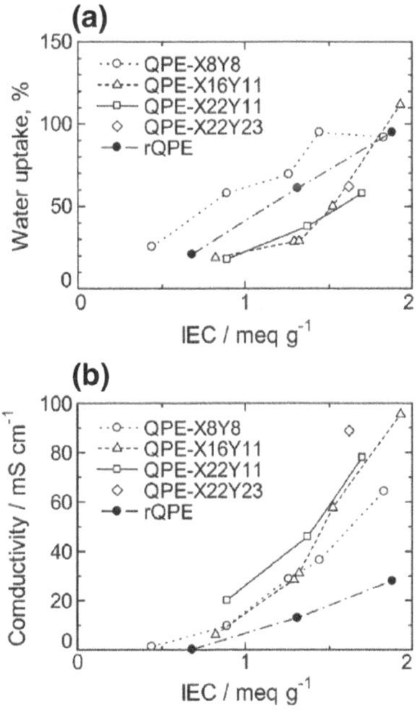

Fig. 14.3 **a** Water uptake at 30 °C. **b** Hydroxide ion conductivity at 40 °C of QPE membranes as a function of ion-exchange capacity (IEC) [17]. © 2012 American Chemical Society

long-range, well-defined phase-separated structures of both segments. Furthermore, if such phase-separated structures can be induced simply by the immiscible nature of the hydrophobic and hydrophilic segments, then the complexity of the micro-morphology can be changed through absorption of water, which changes the volume fraction of the hydrophilic segment. Therefore, even though the solid-state structure of a dried polymer is designed via such molecular designs, it is difficult to anticipate what its hydrated structure will be at a certain humidity and to analyze the relation between its structure and the membrane's properties in the hydrated state. To study the structures and properties of hydrated polymers and to obtain a suitable design strategy for high-performance AEMs, a model polymer electrolyte system with properly controlled, highly stabilized ion channels that are unaffected by the level of hydration in the membrane has been conducted via the use of semicrystalline AEMs.

Miyanishi et al. developed a new type of aromatic AEM for this purpose [44]. Self-assembling poly(3-alkylthiophene) was applied to spontaneously constructed bilayer lamellae structures that had ion channels composed of densely accumulated anion-exchange groups (as shown in Fig. 14.4). Long-ranged, well-ordered bilayer lamellae structures of the developed membrane (A2) were verified by the presence of sharp peaks in both SAXS and wide-angle X-ray diffraction. In contrast to microphase separation of block co-polymer polyelectrolytes in which miscibility of the hydrophobic and hydrophilic segments is the driving force of structural formation, these lamellae structures are induced through crystallization of the segment. Interestingly, these lamellae structures do not change during hydration with vapor and/or water, even though lamellae spacing becomes larger upon absorption of water (Fig. 14.5). Such structural behaviors are completely different from those typical for hydrophobic/hydrophilic block co-polymer electrolytes, whose

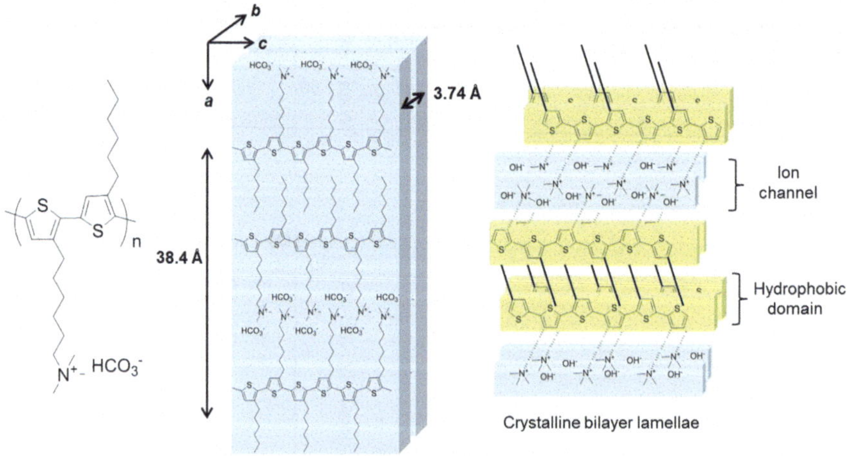

Fig. 14.4 Schematic of solid-state nanostructure of alternative poly(3-alkyalphiphene) AEM (A2) [44]

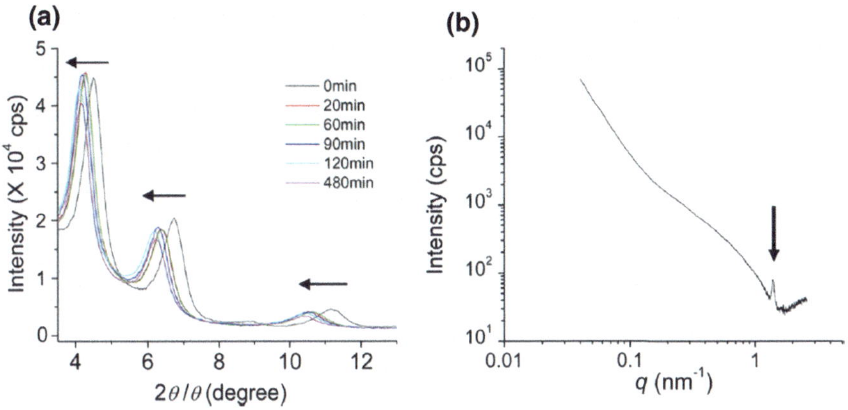

Fig. 14.5 **a** Time course of in situ, out-of-plane WAXRD of the A2 (HCO_3^-) membrane after introducing saturated vapor at room temperature. The arrows show the direction of peak shift. **b** SAXS of A2 (HCO_3^-) bulk sample in water at 25 °C. The arrows show (100) diffraction peaks derived from bilayer lamellae structures [44]. © 2015 American Chemical Society

structures usually change after absorbing water (lamellae, cylinder, sphere, etc.). This behavior can be understood by considering the semicrystalline nature of poly-3-alkylthiophenes. In these compounds, the polymers tend to crystalize due to packing of the alkyl chain and the π–π interactions that exist in the backbone. Their local structures can change relatively easily due to the flexible nature of alkyl chain. In addition to developing well-defined ion channels and the semicrystalline nature of these polymers, water uptake of these membranes is much lower than in typical AEMs that with the same IEC. In the aforementioned experiments, the anion conductivity (Br^-) of this membrane was a much higher value (56.2 mS cm^{-1}, 70 °C RH 100%) than that of the corresponding control polymer with a relatively disorderly structure (28.6 mS cm^{-1}, 70 °C RH 100%). The activation energy of this membrane was lower than that seen in the control polymer. The differences in ion conductivity and activation energies between A2 and its corresponding membrane with a disorderly structure becomes more pronounced when the humidity decreases, namely, there is a lower water content in the membrane. This is indicative of the benefits of A2's self-assembled, bilayer lamellae structures for fast anion transportation, especially in much drier conditions.

Although the conclusion obtained by this membrane is similar to those inferred for the aforementioned multi-block co-polymers or comb-shaped polymers, the semicrystalline nature of the membrane offers advantages for further improving the morphological stability of ion channels under hydration.

To achieve efficient ion conduction in AEMs, a few studies have recently used porous solid-state nanostructures. Although most membranes are packed to minimize free spaces in solid-state, resulting in no porosity formation, some of the polymers still tend to form molecular sieves in solid-state because the rotation and entanglement of the polymer chain is relatively restricted. These membranes are called as polymer intrinsic microporosity (PIM) [45]. Among various microporous polymers, PIM-1 and its analogs, first developed by McKeown et al., have attracted attention in gas separation applications because of their unique polymeric properties [46]. These materials are ladder-type polymers that feature a rigid and contorted backbone as well as high processability into a free-standing microporous membrane. Some of the studies reported expected that such microporous polymers would serve as a new motif for high-performance AEMs, when given the right anion-exchange groups.

As shown in Scheme 14.6, Ishiwari et al. reported the synthetic route and the anion-conducting properties of a trimethylbenzyl ammonium-appended PIM with a spirobiindane backbone [47]. In those investigations, the resulting PIM (1_{OH}) afforded a free-standing membrane with good OH^- conductivity (65 mS cm^{-1}) under 100% relative humidity at 80 °C as well as long-term durability in water. Similar approaches using this PIM design were tried by other researchers, including McKeown, Guiver, and Xu, and ultimately led to the development of AEMs with a Tröger's-base polymer backbone [48]. These novel AEMs exhibited very high hydroxide conductivity (164 mS cm^{-1} at 80 °C in water), despite having relatively low IEC (0.82 meq cm^{-1}). Although the exact relationship between these solid-state microporous structures and their anion conductivity still remains unknown, the molecular sieves in these polymers are thought to enhance the mobility of dissociated anions and also might help to improve water retention within the pore itself (Fig. 14.6).

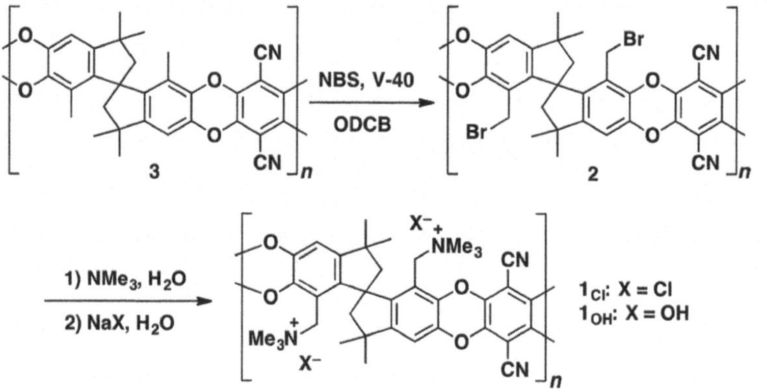

Scheme 14.6 Synthesis of spirobiindane-based PIMs ($\mathbf{1}_{Cl}$ and $\mathbf{1}_{OH}$) using V-40 and 1,1′-azobis (1-cyclo-hexanecarbonitrile) [47]

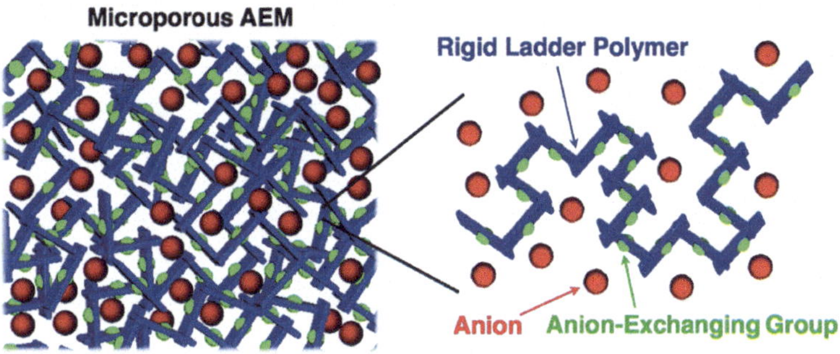

Fig. 14.6 Illustrations of a microporous AEM composed of a rigid ladder polymer with anion-exchanging groups [47]. © RSC publishing group

14.5 Alkaline Stability of AEM

The chemical stability of AEMs is also very important for practical device applications. It is well known that the alkaline stability of AEMs is relatively low, especially if the AEM is subjected to hot alkaline solutions for long periods of time. Under these conditions, the electrical resistance of the membrane increases whereas the IEC decreases [49]. This phenomenon can be reasonably explained by considering the degradation of the anion-exchange group at that particular stage. In fact, the QA is often converted to a nonionic compound after treatment in hot alkaline or organic base solution under specific conditions. The most famous of these reactions is the Hoffman Elimination, in which the β hydrogen of the QA is eliminated by OH^- in order to form olefin and tertiary amine [50]. In addition to this, it has also been reported that the QA can be decomposed by various mechanisms, including ylide-mediated reactions, a Stevens or a Sommlet–Hauer rearrangement [51, 52], and nucleophilic substitution reactions to name a few (Scheme 14.7).

14.6 Alkaline Stability of the Anion-Exchange Groups

Alkaline stability of anion-exchange group is very important for the development of durable AEMs. Mohanty et al. researched the alkaline stability of small molecular model anion-exchange groups [53] by preparing QA hydroxides from the corresponding QA halides via ion-exchange reactions with silver oxide (Fig. 14.7). The stability of the resulting QA hydroxides was noted in D_2O at 60 and 120 °C with 18-crown-6 as the internal standard. The degradation behavior was monitored via ^1H NMR and the corresponding degradation products were analyzed by GC-MS and 2D NMR.

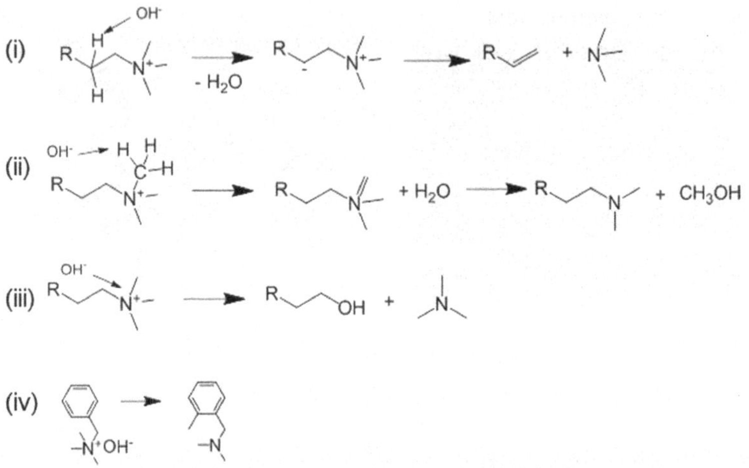

Scheme 14.7 Degradation pathways for QA cations during (i) Hoffman elimination (ii) Ammonium ylide-mediated reaction (iii) Nucleophilic substitution (iv) Rearrangement

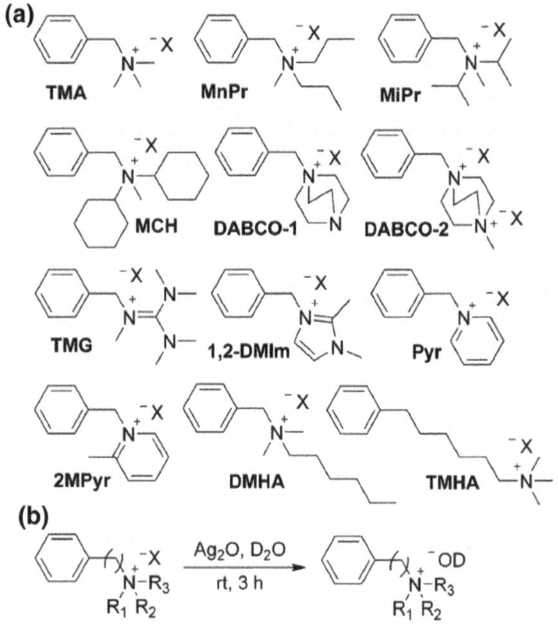

Fig. 14.7 **a** QA-model compounds prepared for alkaline stability studies, where X = Br or I. **b** Ag_2O-mediated ion-exchange reaction [53]. © RSC publishing group

The degradation rate of the aforementioned QA hydroxides is summarized in Fig. 14.8. From the results, it is clear that the DABCO-2 in which two cation groups co-exist in one molecule is very unstable, whereas DABCO-1 is relatively stable (Fig. 14.8a). Furthermore, benzylic-substituted resonance-stabilized cation structures (e.g., TMG, 1,2-DMIm, Pyr, and 2MPyr as shown in Fig. 14.7) are relatively unstable in alkaline conditions in comparison to TMA (Fig. 14.8a). The other trialkyl ammoniums, such as MnPr and MCH, are only slightly more stable than TMA (Fig. 14.8b). The most stable compound that has been examined thus far is TMHA. This is somewhat unexpected, given that TMHA is a famous substrate

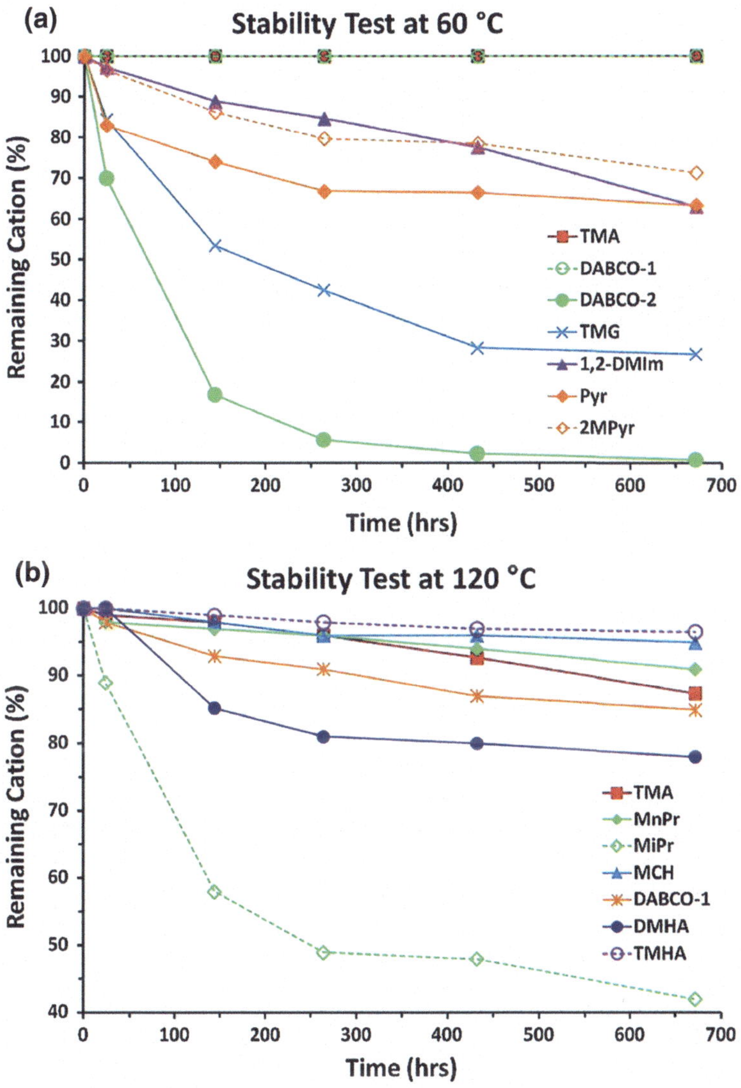

Fig. 14.8 Percentage of the remaining cations at **a** 60 °C for TMA, DABCO-1, DABCO-2, TMG, 1,2-DMIm, Pyr, and 2MPyr. **b** 120 °C for TMA, MnPr, MiPr, MCH, DABCO-1, DMHA, and TMHA [53]. Note that at 60 °C, the TMA and DABCO-1 datasets overlap at 100% of the remaining cations. © RSC publishing group

for Hoffman Elimination reactions. Through these studies, Mohanty was able to verify that even though the olefin compound derived from Hoffman elimination reaction was only observed at 140 °C, this compound was found to be stable at 120 °C in D_2O.

Among the investigated compounds, it was clear that BTMA was not as unstable as generally believed. Similarly, no degradation of the BTMA group was observed even after the compound was stirred in 1 M $NaOH_{(aq)}$ at 80 °C for one month.

14.7 Quantum Calculation for the Analysis of Alkaline Stability of Anion-Exchange Group

Researchers have been trying to establish the relationship between alkaline stability and the electronic structure of the anion-exchange group [54–57]. Using density functional theory (DFT) calculations, Pivovar et al. analyzed the alkaline stability of substituted trimethyl ammoniumcations and concluded that the alkyl chain length of the ammonium cation have a significant influence on alkaline stability [56]. Matsuyama et al. looked at the electronic state of the anion-exchange groups that had been reported by Mohanty (Figs. 14.7 and 14.9) with the aim of using quantum calculations to better understand alkaline degradation behavior [58].

It was found that a strong relationship existed between the lowest unoccupied molecular orbital (LUMO) energy and the alkaline stability of the anion-exchange groups (Fig. 14.10) [59]. Given that the hydroxide ion is an electron-rich nucleophile, it is able to interact with the vacant orbitals of other molecules. When compared to other vacant orbitals, the LUMO can more easily accept an electron because it has the lowest energy of all of the vacant orbitals. Moreover, electrons in the HOMO have the highest energy of all the occupied orbitals. Thus, electrons can be readily transferred from the HOMOs of all the hydroxide ions to the LUMOs of all the anion-exchange groups.

Figure 14.10 shows relation between the LUMO energy and the degradation rate of the anion-exchange groups. When a hydroxide ion and the model molecules interact, the lower LUMO energy value for the model molecule corresponds to more stabilized electron energy. As a result, model molecules with lower LUMO energies are decomposed easily. On the other hand, some molecules (e.g., 2 and 5) are decomposed relatively quickly despite having high LUMO energies. These molecules have π-type LUMOs localized on the anion-exchange groups, and since the HOMO isosurface of hydroxide ion is also π-type, some overlap occurs. Since the interactions between π–π orbitals are more extensive than those between π–σ orbitals, the faster rate of degradation can be attributed to greater overlap between the LUMO of the anion-exchange group and the HOMO of hydroxide ion. This explains why anion-exchange groups with benzylic-substituted resonance-stabilized cation structures (e.g., guanidines, imidazolium, and pyridinium) decompose faster than general QA cations. The stability of an anion-exchange group is affected by

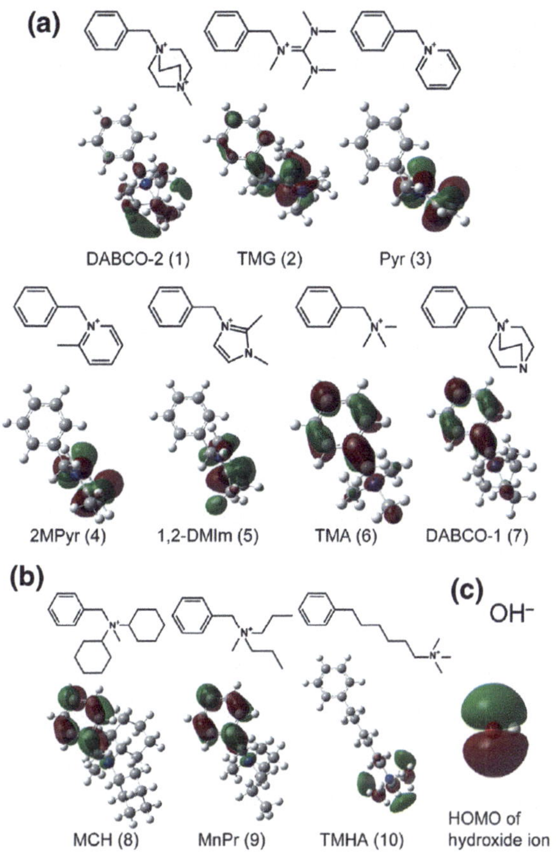

Fig. 14.9 Molecular structures and LUMO isosurfaces of the model molecules. Model molecules were subjected to decomposition tests at **a** 60 °C, **b** 120 °C, and **c** HOMO isosurface of hydroxide (isovalue of 0.04) as reported by Matsuyama [58]. © RSC publishing group

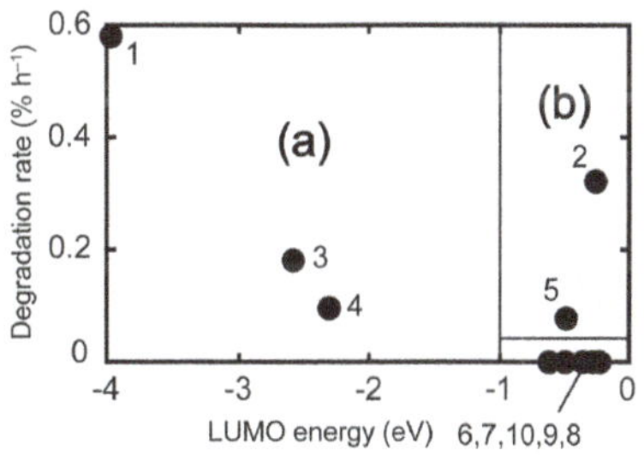

Fig. 14.10 The relationship between the LUMO energy and the degradation rate [58]. © RSC publishing group

both the molecule's electronic structure and various other kinetic factors (such as the ability of bulky structures to kinetically suppress the attacking of OH^- ion). A simple evaluation of LUMO is insufficient for discussing the durability of QA cations because of this complexity, but this technique is effective for rapid screening to establish durable QA cations.

14.8 Alkaline Stability of Anion-Exchange Group Attached in the Polymer

Although small molecular model ammonium is important in directly evaluating the stability of the anion-exchange groups, it is necessary to not to jump to the conclusion that the stability of anion-exchange group can be achieved through a simple attachment in the polymer system. Many studies found that the speed of degradation of QA cations was drastically accelerated in the polymer system when compared to small molecular QA cations, even though both the physical mobility and the reactivity of polymer were generally lower than those for small molecular compounds. This result may be explained by solvation and concentration effects.

Pivovar et al. experimented with the activation energy needed for degradation reactions by changing the dielectric constant around QA cations [54]. As shown in Fig. 14.11, it was found that the activation energy barrier for degradation of the tetramethyl ammonium increased linearly with dielectric constant. Since dielectric constant of water is very high (~80), the dielectric constant around a fully hydrated ammonium cation is naturally much higher than that of a less hydrated ammonium cation. As such, the degradation of ammonium cation can be suppressed through hydration.

This result is also consistent with the key principles of organic chemistry. Nucleophilic substitution (SN2-type) or elimination reactions (E2-type) are known to be drastically accelerated in polar aprotic solvents (such as diemthyl formamide, dimethylsulfoxide, and acetonitrile) but are less efficient in protonic solvents. In fact, the reaction kinetics of some types of nucleophilic reactions can change by orders of a million simply by changing solvent of the reaction. Protonic solvents such as water generally weaken the reactivity of anion and QA cations. QA cations in small molecular system are water soluble, and therefore fully hydrated, whereas the water solvation of the cation attached in the polymer is relatively limited. As

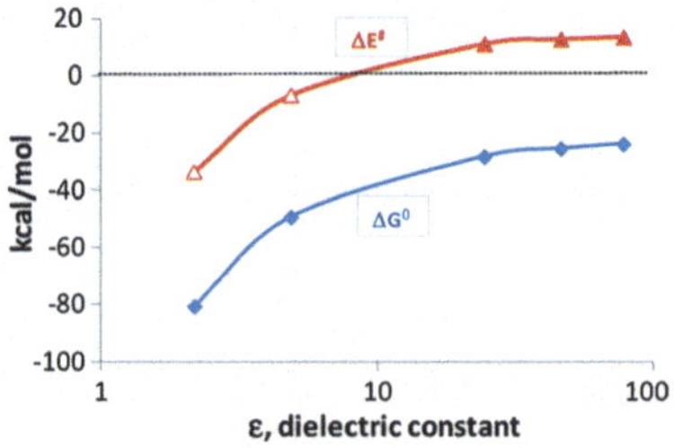

Fig. 14.11 Changes in the standard free energy (ΔG^0) and activation energy barriers ($\Delta E^{\#}$) for the degradation of tetramethylammonium as a function of the dielectric constant used in the PCM model as reported by Chempath [54]. Note that a $\Delta E^{\#}$ value below zero (represented by the open red triangles) indicates a reaction without activation barriers and is thus only limited by diffusion. © 2008 American Chemical Society

such, the stability of anion-exchange groups attached in the membrane tends to be much lower than that of small molecular ammonium compounds [60].

These solvation effects are also observed when looking at the differences in the stability between water and methanol. When solubility of the compound in water is a problem, testing the alkaline stability of AEMs or their model small molecular QA is sometimes conducted in methanol. In these cases, the speed of degradation is much faster in methanol than it is in water. Miyanishi's research focused on the degradation behavior of several low molecular model compounds of typical poly (arylene ether) AEMs (Fig. 14.12) [61]. They compared the speed of degradation for these compounds in water and alcohol solvents. As clearly seen in Fig. 14.13, alkaline degradation of the model compound is exponentially accelerated in methanol and ethanol in comparison to water. This indicated that stability of AEMs is of greater importance in methanol or ethanol liquid fuel cells than in typical oxygen/hydrogen fuel cells.

The degradation kinetics between small molecular QA cations and the ones attached to the polymer membrane are also influenced by the concentration of the hydroxide in the presence of a swollen membrane. Price et al. calculated effective concentration of hydroxide and water in polymer membranes [62] and found that there was a large decrease in the amount of water available per cation attached to a polymer film in comparison to the amount associated with small molecular ammonium groups. This fact correlates with the observed increase in hydroxide concentration on the swollen polymer film. Namely, even if AEMs are treated in 1 M $NaOH_{(aq)}$, the concentration of hydroxide anion around the QA cations in the polymer remain much higher than 1 M, which explains why materials that show minimal degradation when dissolved in solution often degrade when immobilized on water-soaked polymer films that are swollen from hydration.

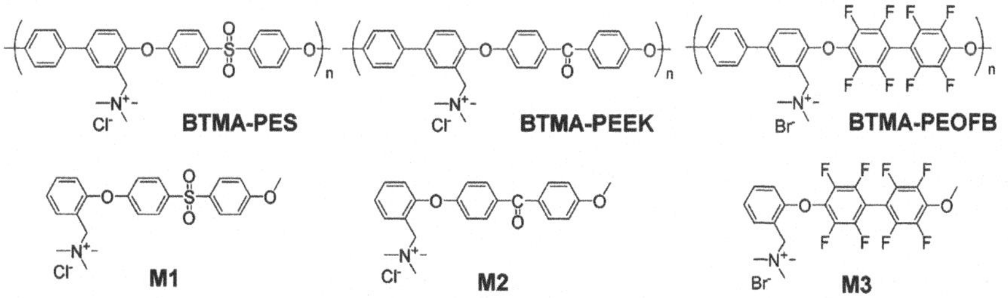

Fig. 14.12 Chemical structures of BTMA-modified PES, PEEK, and PEOFB, along with their corresponding model compounds (M1–M3) [61].© RSC publishing group

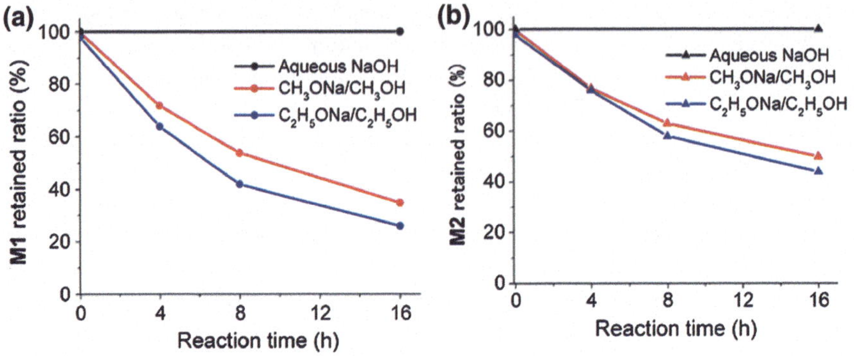

Fig. 14.13 **a** Degradation rates of M1 in 1 M aqueous NaOH (black), 1 M CH_3ONa/CH_3OH (red), and 1 M C_2H_5ONa/C_2H_5OH (blue) at 80 °C **b** degradation rates of M2 in 1 M aqueous NaOH (black), 1 M CH_3ONa/CH_3OH (red), and 1 M C_2H_5ONa/C_2H_5OH (blue) at 80 °C [61]. © RSC publishing group

14.9 Design of Alkaline Stable Anion-Exchange Group

Research has also been focused on the alkaline stability of the anion-exchange groups. Some of the anion-exchange groups are reported to be more stable than alkyl ammonium cations. However, there are few examples of anion-exchange groups which exhibit exceptional alkaline stability. Coates et al. reported that tetrakis(dialkylamino)phosphonium cation $[P(N(Me)Cy)]^+$ was very high stability in a 1 M $NaOD/CD_3OD$ solution (Fig. 14.14) [63] for up 20 days, which was a more strenuous operating condition when compared that of aqueous alkali. Individual crystals of methoxide anion formed $[P(N(Me)Cy)]^+ OMe^-$, which could be isolated because of its high stability. In light of these results, this anion-exchange group was applied to polyethylene AEMs, and it was found that the ion conductivity of the resulting membrane remained stable for 140 days in 1 M $NaOH_{(aq)}$ at 80 °C.

Fig. 14.14 Stability of $[BnNMe_3]^+$ and $[P(N(Me)Cy)_4]^+$ in 1 M $NaOD/CD_3OD$ at 80 °C as reported by Noonan [63]. © 2012 American Chemical Society

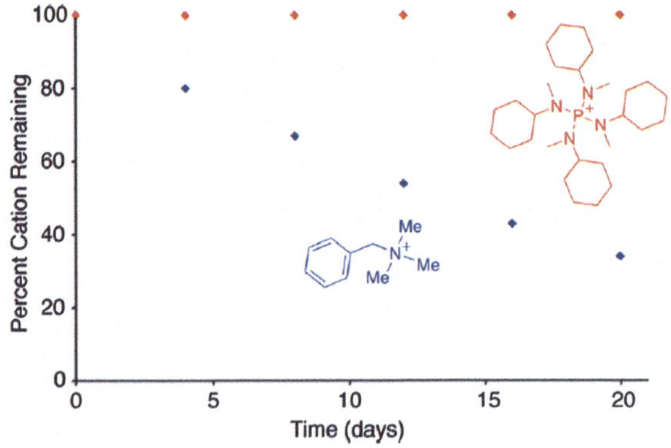

14.10 Backbone Degradation of Poly(Arylene Ether) AEMs

During the initial stages, the degradation of AEMs is believed to be caused by degradation of the anion-exchange groups. However, recent reports cast doubt on this. While investigating alkaline degradation behavior of QA-modified poly (fluorenyl ether sulfone ketone) AEMs (QA-PEEKS) and their corresponding imidazolium-modified AEMs (IM-PEEKS) [64], Chen et al. found that QA-PEEKS became mechanically brittle after treatment in alkaline solution (Fig. 14.15). The intrinsic viscosity of these membranes also decreased drastically, thereby hinting at the occurrence of backbone degradation in the membrane. A similar phenomenon was observed in other poly(arylene ether) AEMs, including benzyl trimethylammonium (BTMA) modified PES and PPO AEM.

Several studies were tried to explain this degradation behavior under alkaline conditions through spectroscopic analysis. Infrared (FTIR or Raman) and 1D ^{1}H NMR analysis is usually chosen for elucidating the chemical structure of AEMs after alkaline durability experiments. However, it is difficult to identify the individual chemical moieties using these methods due to broadness of the spectrum or variations in the chemical environment stemming from changes in the distribution of the polymer's molecular weight. Precise identification of the degraded chemical structure/moieties is essential in understanding the degradation mechanism of AEMs in detail. Arges et al. looked at the degradation products of polysulfone (PSF) AEMs after alkaline stability tests by using two-dimensional NMR spectroscopy [65] through the application of correlation spectroscopy (COSY) and heteronuclear multiple quantum correlation spectroscopy (HMQC) in an attempt to identify the degraded species derived from exposure of PSF-based AEMs to alkaline solutions. Using these methods, it can be confirmed that (i) PSF and chloromethylated PSF by themselves are stable in alkaline media and (ii) functionalization of PSF with relatively stable cation groups tethered to the benzyl

Fig. 14.15 Optical image of QA-PEEKS (left) and IM-PEEKS (right) after degradation tests [64]. © 2012 American Chemical Society

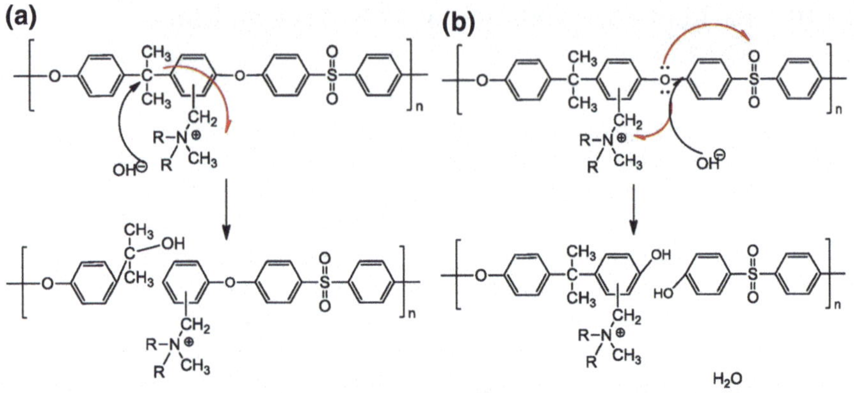

Fig. 14.16 Hydrolysis of **a** quaternary carbon **b** ether bonds in PSF AEMs containing relatively stable fixed cations as reported by Arges [65]. The red lines represent electron delocalization. © 2013 National Academy of Sciences

position of the diphenylpropane group in PSF resulted in backbone degradation upon exposure to alkaline solutions for both quaternary carbons and ether hydrolysis (Fig. 14.16).

14.11 The Relation Between Backbone and Cation Degradation in Poly(Arylene Ether) AEMs

Details of the degradation mechanism of PES-AEM were investigated using small molecular compound (M1) in order to account for the relationship between backbone degradation and anion-exchange group degradation of poly(arylene ether) AEMs [66]. Miyanishi et al. found that degradation products derived from simple degradation of anion-exchange group (N1, N2, or N3 in Scheme 14.8) were not detected after alkaline stability tests were conducted on M1. From ^{1}H NMR analysis of the degradation products, it is confirmed that the ether linkages in the backbone were much weaker than those in the QA groups. Scheme 14.8 details the alkaline degradation mechanism for M1. After cleavage of M1's ether linkage, the resulting 2-hydroxybenzyl trimethylammonium (D2) was very unstable in hot alkaline solution and rapidly converted to benzyl alcohol and tertiary amine through the formation of the highly reactive *o*-quinone methide. Several studies conducted on this highly reactive *o*-quinone methide and its precursor clearly support this explanation [67–70]. As such, cleavage of the ether bond not only results in degradation of polymer backbone, but also triggers the degradation of the anion-exchange groups.

In studies conducted with small molecular QA cations, BTMA itself is relatively stable. Both the backbone and the anion-exchange groups are decomposed in most

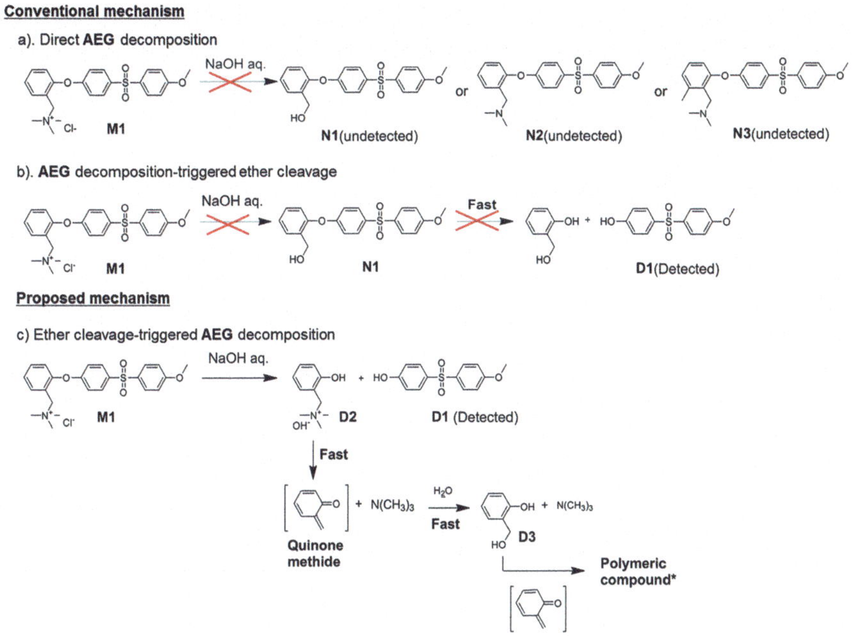

Scheme 14.8 Conventional and proposed degradation mechanisms for M1 in $NaOH_{(aq)}$ as reported by Miyanishi [66]. *Formed by the reaction between quinone methide and D3

benzyl alkylammonium modified poly(arylene ether) AEMs. Degradation of the anion-exchange group of typical benzyl alkyl ammonium-modified PES AEM may not be derived from direct degradation by OH^-, but rather from the cleavage of ether, resulting in the formation of a highly unstable quinone methide precursor. Given this, it can be said that degradation of the backbone and the anion-exchange groups occur simultaneously, and the mechanism outlined supports these experimental results.

Therefore, it can be inferred that stabilizing the ether linkages are key to enhancing poly(arylene ether) AEMs. Investigations were conducted on the stability of arylene ether linkage to verify that the ether linkages sandwiched by BTMA and a phenylene become relatively stable if the electron densities around linkages are increased (Figs. 14.17 and 14.18) [61]. PPO membranes were found to

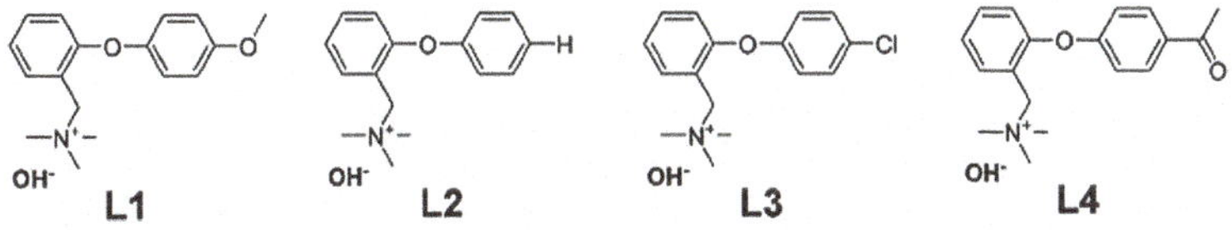

Fig. 14.17 Chemical structures of L1–L4 [61]. © RSC publishing group

be relatively more stable than PES or PEEK membranes in which the ether linkages were attached to electron withdrawing sulfone or ketone-substituted phenylene and benzyl ammonium [71].

14.12 Molecular Designs to Enhance the Alkaline Stability of Poly(Arylene Ether) AEMs

Several groups have been able to improve the alkaline stability of poly(arylene ether) AEMs by introducing alkyl spacers or spacing units between the backbone and the anion-exchange group [72, 73]. Dang et al. reported PPO AEMs functionalized with ammonium cations through a flexible alkyl spacer (PPO-7Q) (Scheme 14.9) [73]. The resulting membranes had much higher alkaline stability when compared to BTMA-modified PPO AEMs (PPO-1Q) (Fig. 14.19). The

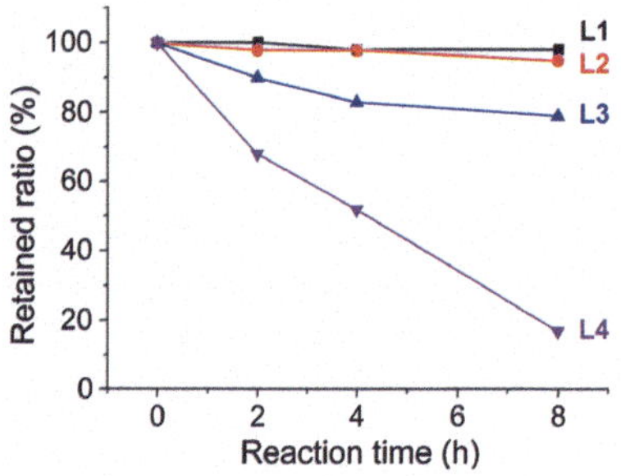

Fig. 14.18 Degradation rates of L1 (black), L2 (red), L3 (blue), and L4 (purple) in 2 M CD_3ONa/NaOD (CD_3OD:D_2O = 4:1) solution at 80 °C [61]. © RSC publishing group

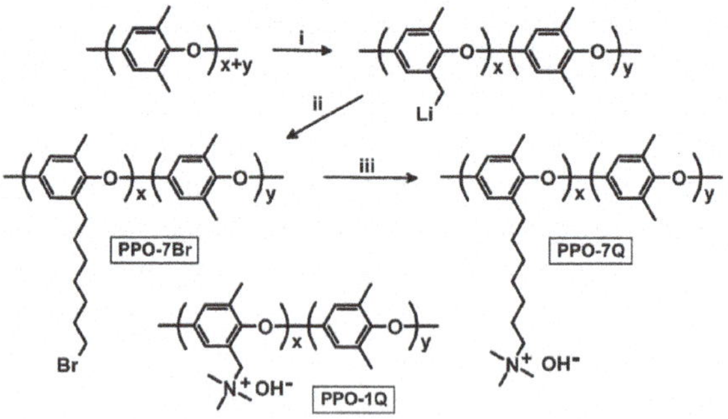

Scheme 14.9 Synthetic pathway of PPO modified with heptyltrimethyl-ammonium side chains (PPO-7Q) via (i) lithiation with n-BuLi, THF, −78 °C, (ii) bromoalkylation with 1,6-dibromohexane, THF, −78 °C, and (iii) quaternization using trimethylamine, NMP, 25 °C, NaOH as reported by Dang et al. [73]. PPO that has been modified with benzyltrimethylammonium groups (PPO-1Q) is used as a benchmark material

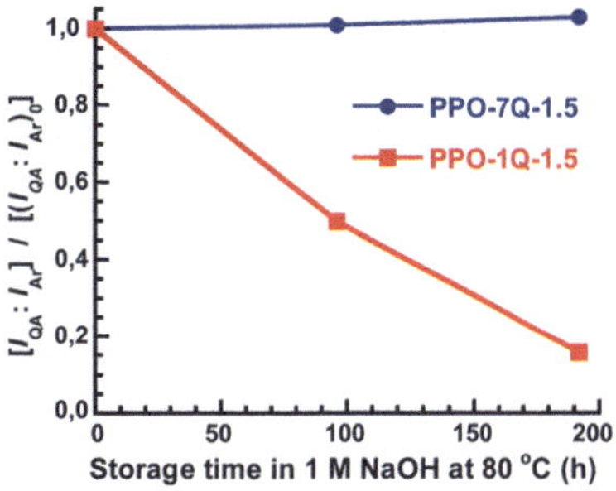

Fig. 14.19 Evolution of the ^{1}H NMR signal ratio of the QA protons and the aromatic protons [I_{QA}: I_{Ar}] during immersion in 1 M $NaOH_{(aq)}$ at 80 °C in comparison to the original ratio [(I_{QA} : I_{Ar})$_0$] as reported by Hai-Son Dan et al. [73]. © RSC publishing group

enhanced alkaline stability of PPO-7Q can be attributed to a stabilization of the anion-exchange group itself as opposed to the benzyl ammonium groups. The polymer backbone of this membrane can be further stabilized via the attachment of the anion-exchange groups through alkyl spacer. No detectable changes were observed in the membrane itself even after soaking in 1 M $NaOH_{(aq)}$ at 80 °C for 192 h.

These experimental results also supported the idea that the stability of both the backbone and the anion-exchange groups can be controlled simply by changing the electronic structure around the poly(arylene ether) linkages.

14.13 Design of Heteroatom-Free Aromatic AEMs

Since typical polymers for AEM applications have been prepared via nucleophilic aromatic substitution of aromatic diol and dihalide monomers, they unavoidably contain C–O linkages along the backbone. Although the aryl ether linkages allow for more freedom of rotation along the polymer backbone and improve both the solubility and the mechanical properties of the resulting polymers, the presence of such ether linkages makes AEMs potentially labile under alkaline conditions. Optimization of the electronic structure around the C–O linkages can enhance backbone stability. This can be achieved by changing the anion-exchange groups from benzyl ammonium cations to the groups attached through flexible spacer. However, in order to develop durable AEMs for long-term use at high temperatures, it is crucial to find alternative polymer backbones that are free of aryl ether bonds. In this section, studies aimed at improving said backbone stability are discussed.

As shown Fig. 14.20 [74, 75], Hibbs et al. first reported utilizing a poly(phenylene) backbone prepared via Diels–Alder polymerization for the synthesis of more durable AEMs. They evaluated and compared alkaline stability of the newly synthesized membrane (ADAPP) to that of both BTMA-modified PSF AEMs (PS-AEMs) and Tokuyama's membrane (AHA). The chemical stabilities of ADAPP, PS–AEM, and Tokuyama's AHA were then tested by immersing samples into aqueous solutions of NaOH (4 M) at 60 °C for 28 days and subjected them to mechanical agitation. The test results are shown in Fig. 14.21 After only a few days, the PS–AEM samples became extremely brittle and broke apart due to the

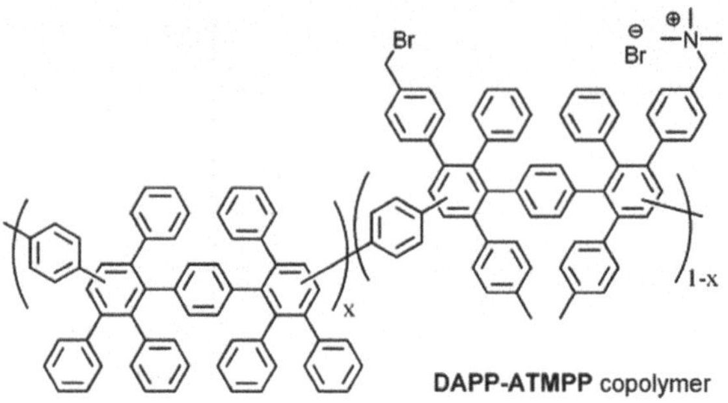

Fig. 14.20 Chemical structure of BTMA-modified branched polyphenylene AEM (DAPP-ATMPP) as reported by Hibbs et al. [74]. © 2009 American Chemical Society

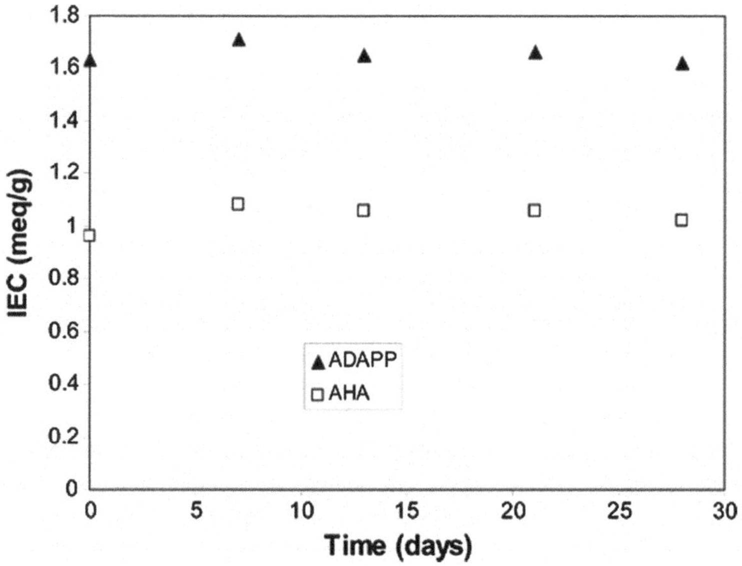

Fig. 14.21 Alkaline stability of DAPP-ATMPP and Tokuyama Membrane (AHA) in 4 M NaOH at 60 °C as reported by Hibbs [74]. © American Chemical Society

mechanical agitation. This embrittlement was attributed to degradation of the PSF backbone. However, the ADAPP and AHA membranes exhibited no changes in appearance or flexibility during the testing, nor were there any measurable changes in their IEC values. While this single set of test conditions is hardly enough to exactly duplicate the real-world operating conditions of an alkaline fuel cell, surprising stability of the BTAM groups on polyphenylene backbone indicated that ADAPP membranes may have a future in alkaline fuel cells.

Li et al. reported the development of new types of poly(phenylene alkylene) AEMs that were free of aryl ether linkages ($PBPA1^+$ and $PBPA2^+$). As shown in

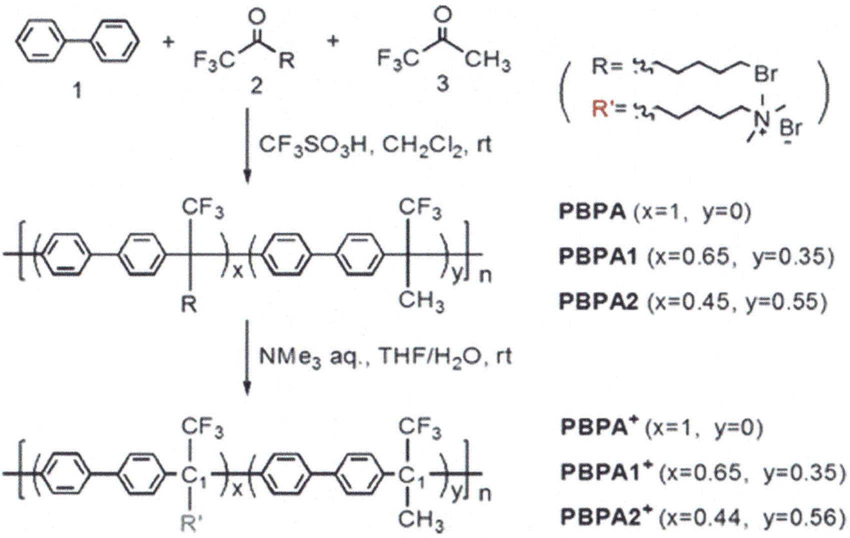

Scheme 14.10 Synthetic route for QA polymers $PBPA^+$ [76]

Scheme 14.10 [76, 77], these membranes could be synthesized via one-pot, acid-catalyzed Friedel–Crafts polycondensation between biphenyl and trifluoromethyl alkyl ketones. Although several types of Friedel–Craft reactions using acid catalysts have been attempted with the aim of constructing aryl-aryl linkages, most of these reactions were accompanied by unwanted side reactions. The synthetic procedure for this poly(phenylene alkylene) first developed by Guzmán-Gutiérrez et al. is a rare example of how to achieve selectivity and efficiency during polymerization [78, 79]. The synthesized $PBPA1^+$ and $PBPA2^+$ membranes maintained their IECs even after an alkaline test in 1 M NaOH solution at 80 °C for 30 days (Fig. 14.22). Furthermore, the hydroxide ion conductivity was also only negligibly different after the alkaline test, thereby indicating that these membranes also possessed good long-term thermochemical stability. These results confirmed that the AEM design concept of utilizing rigid polymer backbone structures without aryl ether bonds and adding QAs onto flexible alkyl tethered chains offered remarkable chemical stability even at high temperatures and under alkaline conditions.

Miyatake et al. developed new types of multi-block co-polymer AEMs (QPAF) in which the hydrophobic components were composed of perfluoroalkylene chains and the hydrophilic components were made of quaternized oligophenylene groups (Scheme 14.11) [80, 81]. No heteroatom linkages (such as ether, sulfone, and ketone groups) were included in the polymer main chains. The introduction of the highly hydrophobic perfluoroalkylene chains were found to improve the ion conductivity, mechanical properties, and solvent solubility of AEMs.

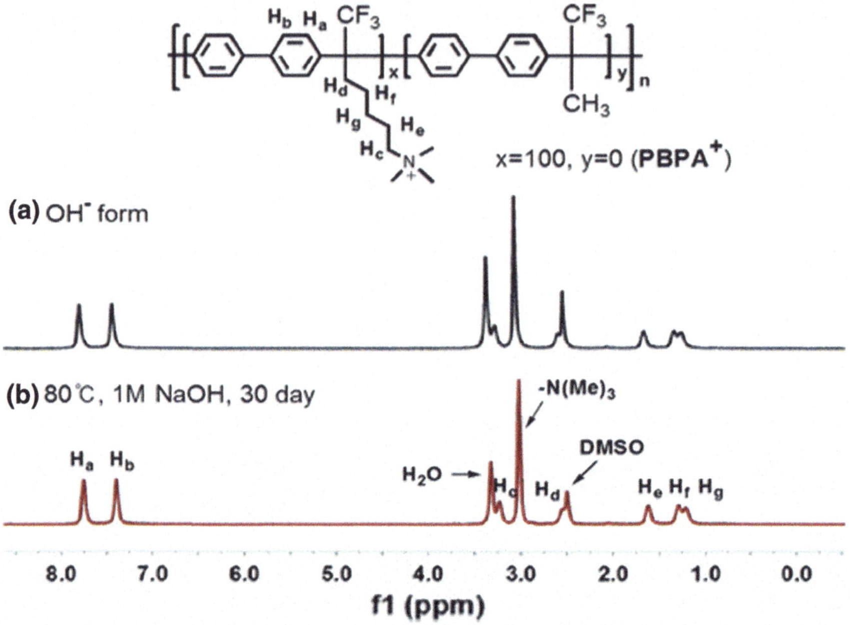

Fig. 14.22 ^{1}H NMR spectra of PBPA+ **a** before and **b** after stability tests [76]. © 2015 American Chemical Society

The QPAF multi-block co-polymers were compared to their multi-block co-polymer QPE-BI counterparts which were composed of poly(arylene ether) hydrophobic units and quaternized oligophenylene hydrophilic units [80]. It was revealed that QPAFs with BTMA cations (IEC = 1.26 meq g^{-1}) exhibited very high hydroxide conductivity (95.5 mS cm^{-1}, 80 °C) and conductivity that was double that of QPE-BIs (IEC = 1.30 meq g^{-1}, 44.4 mS cm^{-1}, 80 °C). When compared with their QPE-BI counterparts, elongation of the QPAF membrane was largely enhanced via the introduction of the perfluoroalkylene group. However, it should be said that this group negatively affected the thermomechanical properties of the membrane due to lowered glass transition temperatures. The QPAF membrane still kept its shape and bendability even after being soaked in alkaline solution at 80 °C, whereas typical poly(arylene ether) broke into pieces. This meant that having a heteroatom-free QPAF membrane was needed to stabilize the polymer backbone.

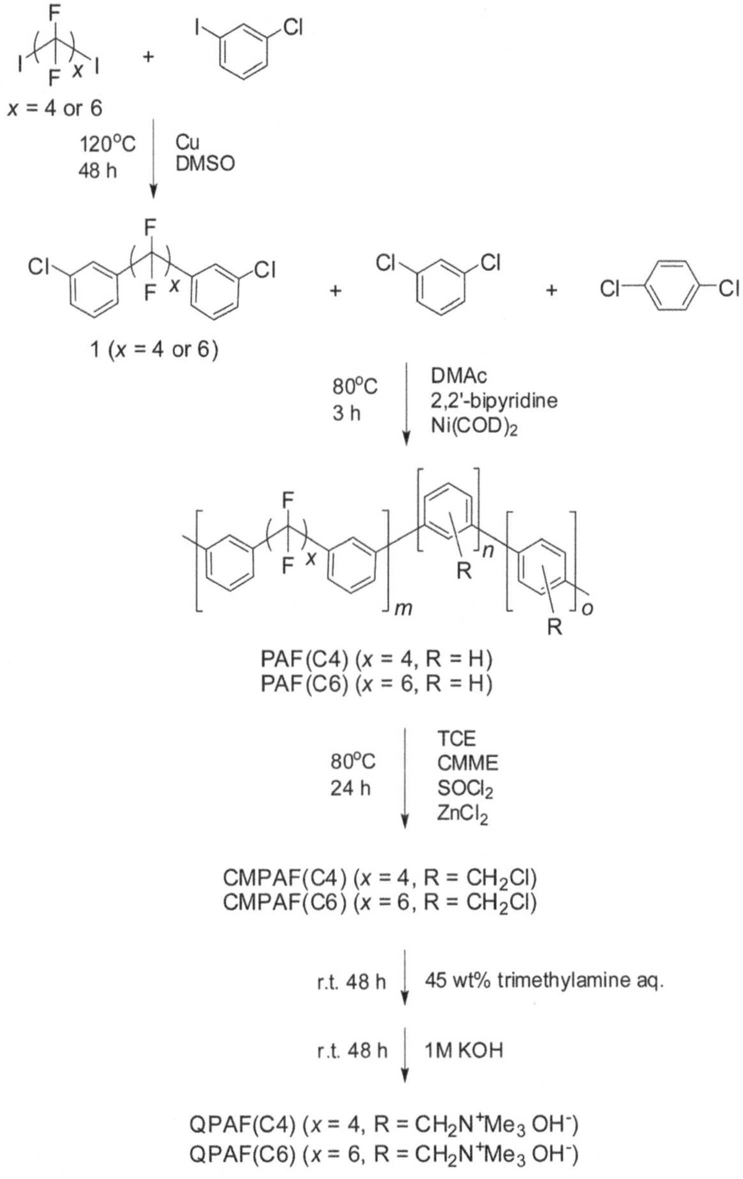

Scheme 14.11 Synthesis of QPAF membranes with various ammonium structures [80]

14.14 Pore-Filling Technology for AEMs

This section focuses on the numerous approaches to the development of new membranes as means of controlling the physical properties of AEMs, such as high hydroxide conductivity, chemical, and thermal durability, mechanical strength in both dry and hydrated state, and lower fuel crossover. Although the development of AEMs with all these necessities has proven to be extremely difficult, one idea has been to use pore-filling technology as an effective method for preparing high-performance membranes [82].

A pore-filled membrane is composed of two materials: a porous substrate with pore sizes on the sub-micrometer scale and a polymer that is used to fill the pores of the substrate as shown in Fig. 14.23. For AEM applications, the porous substrate must be completely stable in alkaline conditions and against any form of oxidant. In addition, the substrate must be mechanically strong in order to homogeneously suppress the inevitable swelling that occurs when the pores are filled with polyelectrolyte material. To date, several promising porous substrates that exhibit the aforementioned properties have been developed. For example, high density polyethylene porous substrate is commercially applied for several applications, including as a separator for electrochemical devices, since polyethylene is highly stable in alkaline solutions. Even though polyolefins are slightly unstable in the presence of highly active oxidants (such as hydroxy radicals), they are also used in this type of technology. This is because High-density polyethylene substrate can resist decomposition by Fenton's reagent due to the crystalline nature of the material. In addition, this substrate possesses a high mechanical strength due to large molecular weight of the polymer.

Pore-filling technology offers advantages when it comes to the preparing membrane made from polyelectrolytes that have low mechanical strength or become extremely swollen when immersed in water. Furthermore, the physical properties of membrane such as fuel permeability, conductivity, and chemical

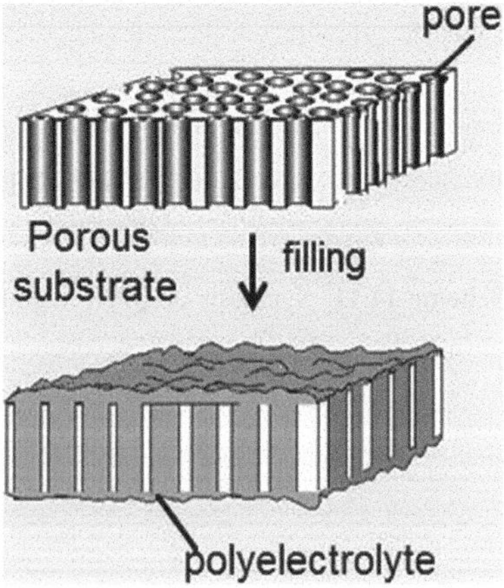

Fig. 14.23 Illustration of pore filling of an AEM

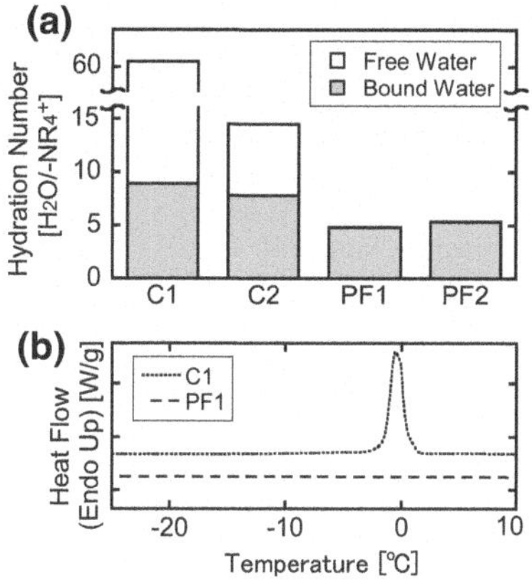

Fig. 14.24 **a** The water state and hydration number of each membrane, **b** thermogram of water fusion enthalpy of the membranes [84]. © Science Direct

durability can be effectively tuned by controlling the state of absorbed water surrounding the filled polyelectrolyte [83]. This technology enables researchers to develop high-performance membranes that exhibit all the requisite characteristics.

Jung et al. applied pore-filling technology to PES-AEMs by filling partially chloromethylated PES-AEMs (C1, C2) with cross-linked polyethylene porous substrates that were 17 μm thick [84]. After filling, the chloromethyl group of the AEM was cross-linked by *N,N,N′,N′*-tetramethyl-1,6-hexanediamine. The researchers then compared the water uptake, anion conductivity, and fuel crossover of the cast membranes (C1, C2), and their corresponding pore-filled membranes (PF1, PF2). IEC of both cast and pore-filled membranes were approximately 3.0 meq g^{-1} after being filled with *N,N,N′,N′*-tetramethyl-1,6-hexanediamine. Figure 14.24a shows hydration numbers (λ) of the membranes, whereas the number of free water estimated using differential scanning calorimetry (DSC) is shown in Fig. 14.24b.

Despite relatively humid conditions and the high IEC of the membrane, no free water is observed in the pore-filled membrane. This means that only bound water molecules exist in the membrane itself. Furthermore, total water absorption of pore-filled membrane is much lower than that of the corresponding cast membrane. Therefore, both the water state and the amount of swelling can be effectively controlled by filling the membrane with substrate.

The hydroxide ion conductivity (σ) for the aforementioned membranes as a function of temperature under humid conditions is shown in Fig. 14.25a. Interestingly, the hydroxide ion conductivity for membranes C1 and C2 are similar, although the free water content of C1 is 7.6 times higher than that of C2. The excessive free water does not contribute positively to hydroxide ion migration, probably because the concentration of the hydroxide ion is lowered by the presence of excess water. PF-AEMs, which only contain bound water molecules, show σ of 6.8 mS cm^{-1} for PF1 and 4.8 mS cm^{-1} for PF2 at 60 °C. That is to say, hydroxide ion conduction through bound water molecules without the presence of any free water molecules is possible in pore-filled membranes. In general, the hydroxide

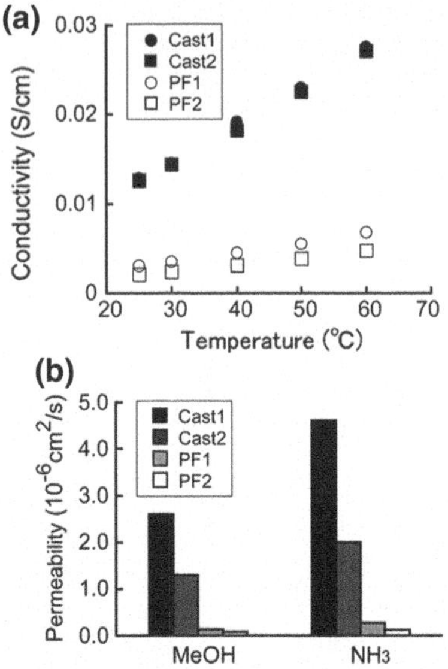

Fig. 14.25 **a** Hydroxide ion conductivity at various temperatures at R.H. 100%, **b** permeability of 10 wt% $methanol_{(aq)}$ and 10 wt% $ammonia_{(aq)}$ at 25 °C as reported by Jung [84]. © Science Direct

conductivities of PF-AEMs are about four to six times smaller than those of cast membranes. When these principles are transferred for the practical usage in fuel cells, the effect of the membrane's thickness must be taken into account. The mathematical product of area resistance, membrane thickness, and the reciprocal of conductivity must be calculated in order to get a more accurate picture of the processes occurring inside the fuel cell. Since a 17-µm-thick PF-AEM is much thinner than a 100-µm-thick cast membrane, the area resistance of PF1 and PF2 (0.25 Ω cm^2 and 0.36 Ω cm^2 at 60 °C, respectively) will be smaller than the value calculated for the cast membranes (which is 0.36 Ω cm^2 for C1 and 0.37 Ω cm^2 for C2). As such, preparation of thinner cast membranes is extremely difficult due to mechanical instability and excessive swelling.

As shown in Fig. 14.25b, permeability of methanol (MeOH) and ammonia (NH_3), both of which are potential candidates for liquid fuels in SAFCs, depend on the amount of free water molecules available in cast membranes. Larger amounts of water tend to cause greater permeability of MeOH or NH_3. In the aforementioned experiments, C1's permeability is almost twice that of C2's; this is because C1's λ is greater than C2's due to a smaller amount of bridging agent. When compared to cast membranes made from the same polymer, the fuel permeation of PF-AEMs is 17–38 times more restricted, whereas the σ is only one-third to one-twelfth times smaller. Therefore, PF-AEMs exhibit excellent performance even under the restrictions of MeOH and NH_3 permeation due to the ineffectiveness of transmitting MeOH and NH_3 via the bound water molecules. Thus, by filling the AEM with an extremely porous substrate, the water saturation levels of the membrane can be controlled, resulting in suppression of fuel crossover and excessive membrane swelling.

Controlling of water state by using pore-filled membranes is also beneficial for enhancing the alkaline stability of AEMs. Jung et al. further tried to reduce the hydroxide ion's ability to attack by suppressing its mobility via a pore-filled membrane [85]. As previously shown in Fig. 14.24a, cast membranes contain many free water molecules whereas pore-filling membranes only contain bound water molecules. Since the hydroxide ion in a pore-filled membrane is surrounded by structured water with low mobility, the reaction kinetics of the hydroxide can be suppressed.

Table 14.3 shows the self-diffusion coefficient in a humid environment with a cast membrane (C_H) and a pore-filled membrane (PF_H) at 25 °C. This coefficient is determined using Pulsed Field Gradient Spin Echo spectroscopy (PFGSE^{-1}NMR). The diffusion coefficient of water in the pore-filled membrane is 0.65×10^{-10} m^2 s^{-1}, which is one-fourteenth of the diffusion coefficient of the cast membrane. This indicates that water and hydroxide ion in the pore-filled membrane moves very slowly compared to that in the cast membrane. When the alkaline stability of the membrane (OH^- form) was tested in water at 100 °C, the cast membrane became larger and more fragile, shattering completely after only four hours of immersion.

Changes in the IEC that were observed during the tests are shown in Fig. 14.26. The IECs of the cast membranes decreased drastically over the course of one hour and then continued to decline moderately after two hours of immersion. On the other hand, the IECs of the pore-filled membranes exhibited no change during the course of two hours of immersion and then only began to decrease moderately after three hours of immersion. These results support the theory that the alkaline stability of pore-filled AEMs can be enhanced by controlling of water state of the membrane since doing so lowers the mobility and attacking ability of the hydroxide ions.

For pore-filling experiments, Jung et al. used typical PES-AEMs which had very low alkaline stability that did not meet the requirements for practical fuel cell application. On the other hand, Sailaja et al. focused on pore-filled technology using a new alkaline durable AEM [86]. As previously mentioned, BTMA cations by themselves are relatively stable in alkaline solutions, however, they exhibit instability if PSF or poly(arylene ether) groups are attached. As such, it is expected that a simple structure based on the benzyl ammonium group and benzene without a heteroatom segment would be promising as a very durable alkaline membrane. Based on this idea, researchers designed a new aromatic anionic electrolyte precursor which has three tertiary amino groups, 1,3,5-tri[*p*-(*N*,*N'*-dimethylaminomethyl)phenyl]benzene (TDAMPB). This monomer's tertiary amino groups can be

Table 14.3 Self-diffusion coefficient of water in fully saturated membranes at 25 °C as measured via PFGSE-^{1}H NMR [85]

Membrane	Pore-filling membrane (PF_H)	Cast membrane (C_H)
Self-diffusion coefficient/$\times 10^{-10}$ m^2 s^{-1}	0.65	9.16

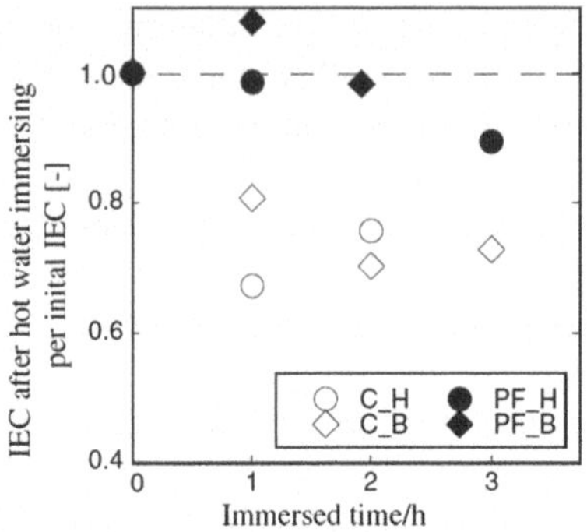

Fig. 14.26 Ion-exchange capacity ratio after immersing membranes in H_2O at 100 °C plotted against the immersion time of the membranes: cast membranes bridged with TMHDA (C_H, open circle) and *N,N,N′,N′*-tetranethyl-1,4-butanediamine (TMBDA) (C_B, open diamond) and pore-filled membranes bridged with *N,N,N′,N′*-tetramethyl-1,6-hexanediamine (TMHDA) (PF_H, solid circle), and TMBDA (PF_B, solid diamond) [85]

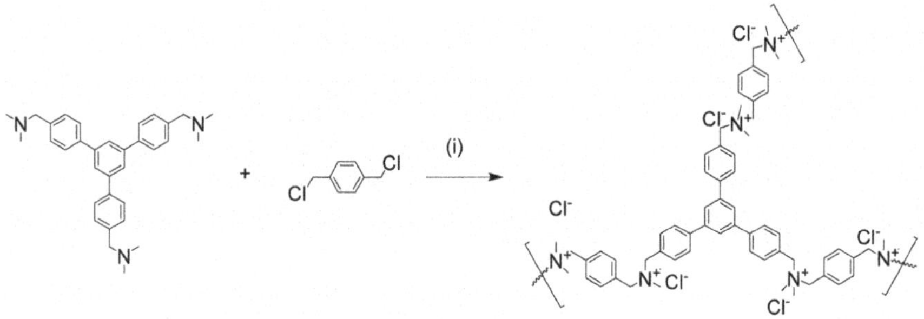

Scheme 14.12 Polymerization of TDAMPB and α,α′-dichloro-*p*-xylene [86]. Reagents and reaction conditions: (i) 60 °C in THF

converted to three benzyl trimethylammonium anion-exchange groups via in situ cross-linking with α,α′-dichloro-*p*-xylene. The result is an extensively cross-linked polyelectrolyte (PTDAMPB) with a high IEC value (IEC = 4.52), as shown in Scheme 14.12 For these experiments, poly(vinylbenzyltrimethyl) ammonium chloride (PVBTAC, IEC = 3.96), a simple but highly conductive anionic electrolyte, is used as the reference pore-filled membrane for comparison with pore-filled PTDAMPB membranes.

Although most conventional AEMs with high IEC and conductivity values suffer from problems due to excessive membrane swelling, the PTDAMPB membrane exhibited lower water uptake (19.9 wt% at 90% R.H.) in contrast to the PVBTAC membrane. In turn, the PVBTAC pore-filling membrane itself even has lower water uptake when compared to a typical AEM with a similar IEC. Table 14.4 shows the physical properties of the PTDAMPB and PVBTAC membranes. The area and thickness changes that occur during membrane swelling demonstrate the

Table 14.4 Physical properties of PTDAMPB and PVBTAC membranes [86]

Membrane	IEC[a] (meq g^{-1})	Water[b] uptake (wt%)	Ion conductivity[c] (mS Cm^{-1})		Methanol permeability (mol mm^{-2} s^{-1})
			Cl^-	OH^-	
PTDAMPB membrane	1.98	20	64	86	1.0×10^{-7}
PVBTAC membrane	2.26	30	69	91	1.2×10^{-6}

[a]Measured by titration
[b]At 80 °C under 90% R.H.
[c]70 °C under 100% R.H.

insignificance of changes in the dimensions of these types of membranes. The low water uptake of the resulting PTDAMPB membrane is thought to be derived from both the effect of filling and the extremely rigid, cross-linked molecular structure of the PTDAMPB polyelectrolyte itself. Both PTDAMPB (1370 kgf cm^{-2}) and PVBTAC (1048 kgf cm^{-2})-filled membranes possess excellent tensile strength (1370 kgf cm^{-2} for the PTDAMPB membrane and 1048 kgf cm^{-2} for the PVBTQAC membrane) when compared to that of typical AEMs. It is noted that preparation of the cast membrane of PTDAMPB is almost impossible due to the brittle nature of the membrane which arises from its extensively cross-linked structure. Therefore, the application of pore-filling technology makes it possible to prepare AEMs that would otherwise be impractical for real-world applications.

Despite the low water content, high ion conductivity values of 64 mS cm^{-1} and 86 mS cm^{-1} were obtained at 70 °C for the Cl^- and OH^- forms of the PTDAMPB membrane, respectively. Furthermore, there is very little change in the conductivity of the PTDAMPB membrane after soaking in 1 M $KOH_{(aq)}$ at 60 °C for 30 days. This is indicative of the alkaline stability of the membrane (Fig. 14.27a). Oxidative stability tests using Fenton solution (3 wt% H_2O_2, 3 ppm $FeSO_4$, 60 °C, 8 h) reveal that PTDAMPB AEMs experience insignificant changes in weight, whereas PVBTAC membranes are prone to gradual weight loss. In addition to the basic rigid aromatic configuration, the extensively cross-linked structure of PTDAMPB, which is analogous with typical thermosetting polymers, also significantly contributes to better oxidative stability of the PTDAMPB AEM when compared to PVBTACs with an aliphatic polymer backbone. The diffusion coefficient and permeability of methanol in PTDAMPB membranes are 2800 s, 4.6×10^{-10} cm^2 s^{-1} and 1.0×10^{-7} mol mm^{-2} s^{-1}, respectively. On the other hand, PVBTAC membranes exhibit a diffusion coefficient of 2.1×10^{-9} cm^2 s^{-1} and a permeability of 1.2×10^{-6} mol mm^{-2} s^{-1} at 25 °C (Fig. 14.27b). The permeability of the PTDAMPB membrane is $^1/_{10}$ that of the PVBTAC membrane, and as low as $^1/_{53}$ that of the Nafion 117 membrane.

With these results in mind, the ability to customize characteristics such as high ion conductivity, chemical durability, low fuel permeability, and simultaneous mechanical stability in both dry and hydrated state, mean that this new type of

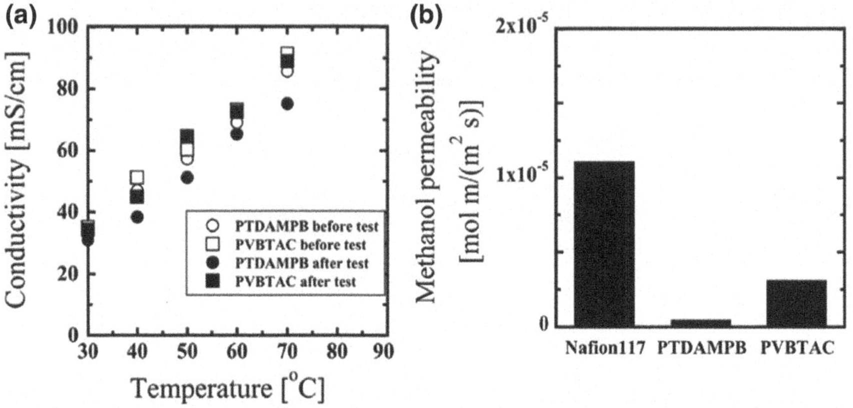

Fig. 14.27 **a** Hydroxide conductivity of the membrane before and after stability tests in 1 M $NaOH_{(aq)}$ at 60 °C for 30 days **b** methanol permeability of the membrane [86]. © RSC publishing group

membrane will have far-reaching influence in the future of AEM development. These novel design concepts for PTDAMPB membranes and their further application in pore-filling technology opens the door for countless real-world fuel cell applications.

References

1. Varcoe JR, Slade RCT (2005) Prospects for alkaline anion-exchange membranes in low temperature fuel cells. Fuel cells 5:187–200
2. Merle G, Wessling M, Nijmeijer K (2011) Anion exchange membranes for alkaline fuel cells: a review. J Membr Sci 377:1–35
3. Hickner MA, Herring AM, Coughlin EB (2013) Anion exchange membranes: current status and moving forward. J Polym Sci Part B: Polym Phys 51:1727–1735
4. Cheng J, He G, Zhang F (2015) A mini-review on anion exchange membranes for fuel cell applications: stability issue and addressing strategies. Inter J Hydrogen Energy 40:7348–7360
5. Maurya S, Shin SH, Kim Y, Moon SH (2015) A review on recent developments of anion exchange membranes for fuel cells and redox flow batteries. RSC Adv 5:37206–37230
6. Zhang H, Ohashi H, Tamaki T, Yamaguchi T (2012) Direction and management of water movement in solid-state alkaline fuel cells. J Phys Chem C 116:7650–7657
7. Zhang H, Ohashi H, Tamaki T, Yamaguchi T (2013) Water movement in a solid-state alkaline fuel cell affected by the anion-exchange pore-filling membrane properties. J Phys Chem C 117:16791–16801
8. Oshiba Y, Hiura J, Suzuki Y, Yamaguchi T (2017) Improvement in the solid-state alkaline fuel cell performance through efficient water management strategies. J Power Sources 345:221–226
9. Lu S, Pan J, Huang A, Zhuang L, Lu J (2008) Alkaline polymer electrolyte fuel cells completely free from noble metal catalysts. Proc Natl Acad Sci 105:20611–20614
10. Gong K, Du F, Xia Z, Durstock M, Dai L (2009) Nitrogen-doped carbon nanotube arrays with high electrocatalytic activity for oxygen reduction. Science 323:760–764

11. Qu L, Liu Y, Baek JB, Dai L (2010) Nitrogen-doped graphene as efficient metal-free electrocatalyst for oxygen reduction in fuel cells. ACS Nano 4:1321–1326
12. Li X, Popov BN, Kawahara T, Yanagi H (2011) Non-precious metal catalysts synthesized from precursors of carbon, nitrogen, and transition metal for oxygen reduction in alkaline fuel cells. J Power Sources 196:1717–1722
13. Matsuoka K, Iriyama Y, Abe T, Matusoka M, Ogumi Z (2005) Alkaline direct alcohol fuel cells using an anion exchange membrane. J Power Sources 150:27–31
14. Bambagioni V, Bianchini C, Marchionni A, Filippi J, Vizza F, Teddy J, Serp P, Zhiani M (2009) Pd and Pt–Ru anode electrocatalysts supported on multi-walled carbon nanotubes and their use in passive and active direct alcohol fuel cells with an anion-exchange membrane (alcohol = methanol, ethanol, glycerol). J Power Sources 190:241–251
15. Li L, Wang YX (2005) Quaternized polyethersulfone Cardo anion exchange membranes for direct methanol alkaline fuel cells. J Membr Sci 262:1–4
16. Hou HY, Sun GQ, He RH, Wu ZM, Sun BY (2008) Alkali doped polybenzimidazole membrane for high performance alkaline direct ethanol fuel cell. J Power Sources 182:95–99
17. Tanaka M, Fukasawa K, Nishino E, Yamaguchi S, Yamada K, Tanaka H, Bae B, Miyatake K, Watanabe M (2011) Anion conductive block poly(arylene ether)s: synthesis, properties, and application in alkaline fuel cells. J Am Chem Soc 133:10646–10654
18. Lan R, Tao SW (2010) Direct ammonia alkaline anion-exchange membrane fuel cells. Electrochem Solid State Lett 13:B83–B86
19. Bartrom AM, Haan JL (2012) The direct formate fuel cell with an alkaline anion exchange membrane. J Power Sources 214:68–74
20. Matsui K, Tobita E, Sugimoto K, Kondo K, Seita T, Akimoto A (1986) Novel anion exchange membranes having fluorocarbon backbone: preparation and stability. Appl Polym Sci 32:4137–4143
21. Vinodh R, Ilakkiya A, Elamathi S, Sangeetha D (2010) A novel anion exchange membrane from polystyrene (ethylene butylene) polystyrene: synthesis and characterization. Mater Sci Eng B 167:43–50
22. Zeng QH, Liu QL, Broadwell I, Zhu AM, Xing YX, Tu P (2010) Anion exchange membranes based on quaternized polystyrene-block-poly(ethylene-ran-butylene)-block-polystyrene for direct methanol alkaline fuel cells. J Membr Sci 349:237–243
23. Kumar M, Singh S, Shahi VK (2010) Cross-linked poly(vinyl alcohol)−poly(acrylonitrile-*co*-2-dimethylamino ethylmethacrylate) based anion-exchange membranes in aqueous media. J Phys Chem B 114:198–206
24. Xiong Y, Fang J, Zeng QH, Liu QL (2008) Preparation and characterization of cross-linked quaternized poly(vinyl alcohol) membranes for anion exchange membrane fuel cells. J Membr Sci 311:319–325
25. Xiong Y, Liu QL, Zhang QG, Zhu AM (2008) Synthesis and characterization of cross-linked quaternized poly(vinyl alcohol)/chitosan composite anion exchange membranes for fuel cells. J Power Sources 183:447–453
26. Zhamg M, Kim HK, Chalkova E, Mark F, Lvov SN, Chung TCM (2011) New polyethylene based anion exchange membranes (PE–AEMs) with high ionic conductivity. Macromolecules 44:5937–5946
27. Varcoe JR, Slade RCT, Yee ELH, Poynton SD, Driscoll DJ, Apperley DC (2007) Poly (ethylene-*co*-tetrafluoroethylene)-derived radiation-grafted anion-exchange membrane with properties specifically tailored for application in metal-cation-free alkaline polymer electrolyte fuel cells. Chem Mater 19:2686–2693
28. Zhang H, Shen PK (2012) Recent development of polymer electrolyte membranes for fuel cells. Chem Rev 112:2780–2832
29. John J, Hugar KM, Rivera-Meléndez J, Kostalik HA IV, Rus ED, Wang H, Coates GW, Abruña HD (2014) An electrochemical quartz crystal microbalance study of a prospective alkaline anion exchange membrane material for fuel cells: anion exchange dynamics and membrane swelling. J Am Chem Soc 136:5309–5322

30. Wang J, Li S, Zhang S (2010) Novel hydroxide-conducting polyelectrolyte composed of an poly(arylene ether sulfone) containing pendant quaternary guanidinium groups for alkaline fuel cell applications. Macromolecules 43:3890–3896
31. Li N, Leng Y, Hickner MA, Wang CY (2013) Highly stable, anion conductive, comb-shaped copolymers for alkaline fuel cells. J Am Chem Soc 135:10124–10133
32. Mohanty AD, Lee Y, Zhu BL, Hickner MA, Bae C (2014) Anion exchange fuel cell membranes prepared from C-H Borylation and Suzuki coupling reactions. Macromolecules 47:1973–1980
33. Jo TS, Kim SH, Shin J, Bae C (2009) Highly efficient incorporation of functional groups into aromatic main-chain polymer using iridium-catalyzed C−H activation and Suzuki−Miyaura reaction. J Am Chem Soc 131:1656–1657
34. Zhu L, Pan J, Christensen CM, Lin B, Hickner MA (2016) Functionalization of poly (2,6-dimethyl-1,4-phenylene oxide)s with hindered fluorene side chains for anion exchange membranes. Macromolecules 49:3300–3309
35. Ge Q, Ran J, Miao J, Yang Z, Xu T (2015) Click chemistry finds its way in constructing an ionic highway in anion-exchange membrane. ACS Appl Mater Interfaces 7:28545–28553
36. Li N, Guiver MD, Binder WH (2013) Towards high conductivity in anion-exchange membranes for alkaline fuel cells. Chemsuschem 6:1376–1383
37. Zhu L, Zimudzi TJ, Li N, Pan J, Lin B, Hickner MA (2016) Crosslinking of comb-shaped polymer anion exchange membranes *via* thiolene click chemistry. Polym Chem 7:2464–2475
38. Li N, Zhang Q, Wang C, Lee YM, Guiver MD (2012) Phenyltrimethylammonium functionalized polysulfone anion exchange membranes. Macromolecules 45:2411–2419
39. Zhang Z, Wu L, Varcoe J, Li C, Ong AL, Poynton S, Xu T (2013) Aromatic polyelectrolytes *via* polyacylation of pre-quaternized monomers for alkaline fuel cells. J Mater Chem A 1:2595–2601
40. Li N, Yan T, Li Z, Thurn-Albrecht T, Binder WH (2012) Comb-shaped polymers to enhance hydroxide transport in anion exchange membranes. Energy Environ Sci 5:7888–7892
41. Ran J, Wu L, Xu T (2013) Enhancement of hydroxide conduction by self-assembly in anion conductive comb-shaped copolymers. Polym Chem 4:4612–4620
42. Zhao Z, Wang J, Li S, Zhang S (2011) Synthesis of multi-block poly(arylene ether sulfone) copolymer membrane with pendant quaternary ammonium groups for alkaline fuel cell. J Power Sources 196:4445–4450
43. Dong X, Hou S, Mao H, Zheng J, Zhang S (2016) Novel hydrophilic-hydrophobic block copolymer based on cardo poly(arylene ether sulfone)s with *bis*-quaternary ammonium moieties for anion exchange membranes. J Membr Sci 518:31–39
44. Miyanishi S, Fukushima T, Yamaguchi T (2015) Synthesis and property of semicrystalline anion exchange membrane with well-defined ion channel structure. Macromolecules 48:2576–2584
45. Wu D, Xu F, Sun B, Fu R, He H, Matyjaszewski K (2012) Design and preparation of porous polymers. Chem Rev 12:3959–4015
46. Carta M, Malpass-Evans R, Croad M, Rogan Y, Jansen JC, Bernardo P, Bazzarelli F, McKeown NB (2013) An efficient polymer molecular sieve for membrane gas separations. Science 339:303–307
47. Ishiwari F, Sato T, Yamazaki H, Kondo JN, Miyanishi S, Yamaguchi T, Fukushima T (2016) An anion-conductive microporous membrane composed of a rigid ladder polymer with a spirobiindane backbone. J Mater Chem A 4:17655–17659
48. Yang Z, Guo R, Malpass-Evans R, Carta M, Mckeown NB, Guiver MD, Wu L, Xu T (2016) Highly conductive anion-exchange membranes from microporous Tröger's base polymers. Angew Chem Int Ed 55:11499–11502
49. Sata T, Tsujimoto M, Yamaguchi T, Matsusaki K (1996) Change of anion exchange membranes in an aqueous sodium hydroxide solution at high temperature. J Membr Sci 112:161–170
50. Cope AC, Trumbull ER (1960) Olefins from amines: the Hofmann elimination reaction and amine oxide pyrolysis. Org React 11:317–493

51. Stevens TS (1930) CCLXX—degradation of quaternary ammonium salts. Part II. J Chem Soc 2107–2119
52. Kantor SW, Hauser CR (1951) Rearrangements of benzyltrimethylammonium ion and related quaternary ammonium ions by sodium amide involving migration into the ring. J Am Chem Soc 73:4122–4131
53. Mohanty AD, Bae C (2014) Mechanistic analysis of ammonium cation stability for alkaline exchange membrane fuel cells. J Mater Chem A 2:17314–17320
54. Chempath S, Einsla BR, Pratt LR, Macomber CS, Boncella JM, Rau JA, Pivovar BS (2008) Mechanism of tetraalkylammonium headgroup degradation in alkaline fuel cell membranes. J Phys Chem C 112:3179–3182
55. Chempath S, Boncella JM, Pratt LR, Henson N, Pivovar BS (2010) Density functional theory study of degradation of tetraalkylammonium hydroxides. J Phys Chem C 114:11977–11983
56. Long H, Kim K, Pivovar BS (2012) Hydroxide degradation pathways for substituted trimethylammonium cations: a DFT study. J Phys Chem C 116:9419–9426
57. Wang W, Wang S, Xie X, Iv Y, Ramani V (2014) Density functional theory study of hydroxide-ion induced degradation of imidazolium cations. Int J Hydrogen Energy 39:14355–14361
58. Matsuyama K, Ohashi H, Miyanishi S, Ushiyama H, Yamaguchi T (2016) Quantum chemical approach for highly durable anion exchange groups in solid-state alkaline fuel cells. RSC Adv 6:36269–36272
59. Lin B, Dong H, Li Y, Si Z, Gu F, Yan F (2013) Alkaline stable C2-substituted imidazolium-based anion-exchange membranes. Chem Mater 25:1858–1867
60. Dekel DR, Amar M, Willdorf S, Kosa M, Dhara S, Diesendruck CE (2017) Effect of water on the stability of quaternary ammonium groups for anion exchange membrane fuel cell applications. Chem Mater 29:4425–4431
61. Miyanishi S, Yamaguchi T (2017) Analysis of the degradation mechanism of the polyarylene ether anion-exchange membrane for alkaline fuel cell and water-splitting cell applications. New J Chem 41:8036
62. Price SC, Ren X, Savage AM, Beyer FL (2017) Synthesis and characterization of anion-exchange membranes based on hydrogenated poly(norbornene). Polym Chem 8:5708–5717
63. Noonan KJT, Hugar KM, Kostalik HA IV, Lobkovsky EB, Abruña HD, Coates GW (2012) Phosphonium-functionalized polyethylene: a new class of base-stable alkaline anion exchange membranes. J Am Chem Soc 134:18161–18164
64. Chen D, Hickner MA (2012) Degradation of imidazolium- and quaternary ammonium-functionalized poly(fluorenyl ether ketone sulfone) anion exchange membranes. ACS Appl Mater Interfaces 4:5775–5781
65. Arges CG, Ramani V (2013) Two-dimensional NMR spectroscopy reveals cation-triggered backbone degradation in polysulfone-based anion exchange membranes. Proc Natl Acad Sci 110:2490–2495
66. Miyanishi S, Yamaguchi T (2016) Ether cleavage-triggered degradation of benzyl alkylammonium cations for polyethersulfone anion exchange membranes. Phys Chem Chem Phys 18:12009–12023
67. Modica E, Zanaletti R, Freccero M, Mella M (2001) Alkylation of amino acids and glutathione in water by *o*-quinone methide. Reactivity and selectivity. J Org Chem 66:41–52
68. Antonio MD, Doria F, Richter SN, Bertipaglia C, Mella M, Sissi C, Palumbo M, Freccero M (2009) Quinone methides tethered to naphthalene diimides as selective G-quadruplex alkylating agents. J Am Chem Soc 131:13132–13141
69. Wang P, Liu R, Wu X, Ma H, Cao X, Zhou P, Zhang J, Weng X, Xiao-Lian Zhang, Qi J, Zhou X, Weng L (2003) A potent, water-soluble and photoinducible DNA cross-linking agent. J Am Chem Soc 125:1116–1117
70. Chiang Y, Kresge AJ, Zhu Y (2002) Flash photolytic generation and study of *p*-quinone methide in aqueous solution. An estimate of rate and equilibrium constants for heterolysis of the carbon−bromine bond in *p*-hydroxybenzyl bromide. J Am Chem Soc 124:6349–6356

71. Amel A, Zhu L, Hickner M, Ein-Eli Y (2014) Influence of sulfone linkage on the stability of aromatic quaternary ammonium polymers for alkaline fuel cells. J Electrochem Soc 161: F615–F621
72. Yang Z, Zhou J, Wang S, Hou J, Wu L, Xu T (2015) A strategy to construct alkali-stable anion exchange membranes bearing ammonium groups *via* flexible spacers. J Mater Chem A 3:15015–15019
73. Dang HS, Weiber EA, Jannasch P (2015) Poly(phenylene oxide) functionalized with quaternary ammonium groups *via* flexible alkyl spacers for high-performance anion exchange membranes. J Mater Chem A 3:5280–5284
74. Hibbs MR, Fujimoto CH, Comelius CJ (2009) Synthesis and characterization of poly (phenylene)-based anion exchange membranes for alkaline fuel cells. Macromolecules 42:8316–8321
75. Fujimoto C, Kim DS, Hibbs M, Wrobleski D, Kim YS (2012) Backbone stability of quaternized polyaromatics for alkaline membrane fuel cells. J Membr Sci 423–424:438–449
76. Lee WH, Kim YS, Bae C (2015) robust hydroxide ion conducting poly(biphenyl alkylene)s for alkaline fuel cell membranes. ACS Macro Lett 4:814–818
77. Lee WH, Park EJ, Han J, Shin DW, Kim YS, Bae C (2017) Poly(terphenylene) anion exchange membranes. the effect of backbone structure on morphology and membrane property. ACS Macro Lett 6:566–570
78. Olvera LI, Guzmán-Gutiérrez MT, Zolotukhin MG, Fomine S, Cárdenas J, Ruiz-Trevino FA, Villers D, Ezquerra TA, Prokhorov E (2013) Novel high molecular weight aromatic fluorinated polymers from one-pot, metal-free step polymerizations. Macromolecules 46:7245–7256
79. Cruz AR, Hemandez MCG, Guzmán-Gutiérrez MT, Zolotukhin MG, Fomine S, Morales SL, Kricheldorf H, Wilks ES, Cárdenas J, Salmón M (2012) Precision synthesis of narrow polydispersity, ultrahigh molecular weight linear aromatic polymers by A_2 + B_2 nonstoichiometric step-selective polymerization. Macromolecules 45:6774–6780
80. Ono H, Miyake J, Shimada S, Uchida M, Miyatake K (2015) Anion exchange membranes composed of perfluoroalkylene chains and ammonium-functionalized oligophenylenes. J Mater Chem A 3:21779–21788
81. Mahmoud AMA, Elsaghier AMM, Otsuji K, Miyatake K (2017) High hydroxide ion conductivity with enhanced alkaline stability of partially fluorinated and quaternized aromatic copolymers as anion exchange membranes. Macromolecules 50:4256–4266
82. Yamaguchi T, Hua Z, Nakazawa T, Hara N (2007) An extremely low methanol crossover and highly durable aromatic pore-filling electrolyte membrane for direct methanol fuel cells. Adv Mater 19:592–596
83. Hara N, Ohashi H, Ito T, Yamaguchi T (2009) Rapid proton conduction through unfreezable and bound water in a wholly aromatic pore-filling electrolyte membrane. J Phys Chem B 113:4656–4663
84. Jung H, Fujii K, Tamaki T, Ohashi H, Ito T, Yamaguchi T (2011) Low fuel crossover anion exchange pore-filling membrane for solid-state alkaline fuel cells. J Membr Sci 373:107–111
85. Jung H, Ohashi H, Tamaki T, Yamaguchi T (2013) Improvement in thermal stability of anion-exchange membranes for fuel cell applications by controlling water state. Chem Lett 42:14–16
86. Sailaja GS, Miyanishi S, Yamaguchi T (2015) A durable anion conducting membrane with packed anion-exchange sites and an aromatic backbone for solid-state alkaline fuel cells. Polym Chem 6:7964–7973

Chapter 15
Carbon Nanotube-Based Enzymatic Biofuel Cells

Matsuhiko Nishizawa

15.1 Introduction

Enzymatic biofuel cells (BFCs) are a type of fuel cell in which enzymes are used as electrocatalysts to directly convert biochemical energy into electricity (Fig. 15.1) [1–7]. The oxidative reaction occurring at the enzyme-modified bioanode comprises of electro-oxidation of a desired fuel, e.g., fructose, glucose, lactate, etc., by highly selective enzymes to generate electrons. The cathode reaction commonly involves four-electron reduction of O_2 to H_2O. The extremely high reaction selectivity of enzymes, as compared with metallic catalysts such as Pt, eliminates the need for fuel purification and allows a separator-free design that consists of just a pair of enzyme-functionalized anode and cathode electrodes exposed to solutions containing both fuel and oxygen. Such a simple BFC system provides high flexibility in structural design and miniaturization, and as such, miniature enzymatic fuel cells could be an attractive power source for future wearable and implantable microdevices.

Recent enzyme-modified electrodes with high activity ($\sim$mA cm^{-2} level) have been achieved by tailoring nanocomposite of enzymes and carbon nanomaterials such as particulate carbon black and carbon nanotubes (CNTs). Since the carbon material is excellent in conductivity and chemical stability and also relatively inexpensive, it is attractive as an electrode material for BFC, and in particular, carbon nanomaterials provide a three-dimensional conductive network and high specific surface area, so larger amounts of enzyme could be immobilized [8–14]. In addition, since it is easy to modify molecules on the surface of carbon materials, it is expected that various electron transfer systems can be designed at the nanolevel and built into the enzyme electrode. In this chapter, we will introduce the method of making enzyme electrodes with CNTs and application of these electrodes to BFC devices.

M. Nishizawa (✉)
Tohoku University, 6-6-01 Aoba, Sendai 980-8579, Japan
e-mail: nishizawa@biomems.mech.tohoku.ac.jp

N. Nakashima (ed.), *Nanocarbons for Energy Conversion: Supramolecular Approaches*, Nanostructure Science and Technology,
https://doi.org/10.1007/978-3-319-92917-0_15

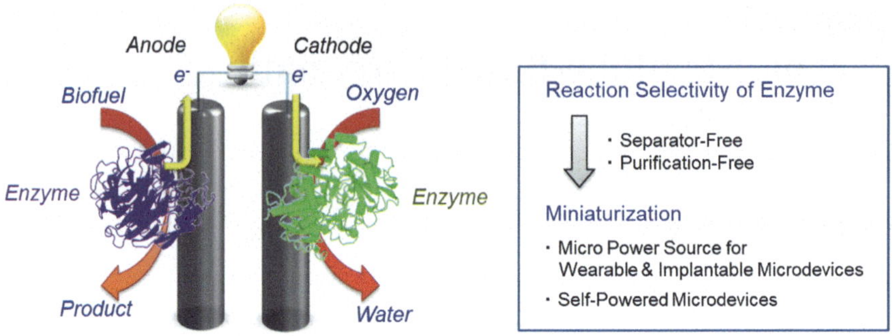

Fig. 15.1 Enzymatic biofuel cell

15.2 Carbon Nanotube-Based Enzyme Electrodes

In the recent BFC technology, CNTs have played an important role for interfacing enzymes with electronic circuitry. This section starts from the basic and attractive features of CNT as the material for enzyme electrode, followed by explanation of particular examples of the CNT-based enzyme electrodes: the film is composed of the ideally aligned CNT and enzyme [15–17], the flexible [18, 19] and stretchable [20] enzyme electrodes, and the printable enzyme electrode [21]. All these unique CNT-based enzyme electrodes show high performance in mA level.

15.2.1 Carbon Nanotube for Enzyme Electrodes

The CNT has many attractive features as an electrode material (Fig. 15.2): uniform nanowire morphology, biocompatibility, excellent conductivity [22], and high specific surface constituting of more than 1000 $m^2\ g^{-1}$, thus an attractive building block for the construction of highly porous three-dimensional nanostructured CNT electrodes [14]. Furthermore, the possibility to add appropriate functionalities via molecular modification enabled optimal tuning of such nanostructured electrodes. Many methods for making the CNT-based BFCs electrodes (in mm to cm scale) have been reported.

The most common and convenient CNT electrode is prepared by dropping and drying a CNT dispersion on a current collector. In order to obtain a homogeneous CNT dispersion, various solvents and additives such as surfactants have been combined with surface modification of CNTs [23, 24]. Using the spray-coating method, the film formation on a curved surface is possible [25]. It is also possible to deposit CNTs by electrophoretic deposition in dispersed aqueous solution of CNT with fixed charge introduced by surface treatment [26].

Reports on free-standing CNT-aggregated electrodes in the form of films and fibers are increasing [27–30]. An electrode called buckypaper is prepared by

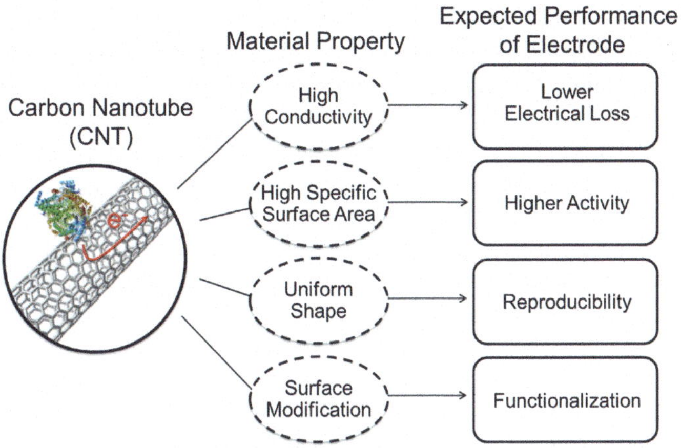

Fig. 15.2 CNT: attractive electrode material

vacuum filtration of CNT dispersions, followed by peeling off from the filter, and the high specific surface area of about 680 $m^2\ g^{-1}$ is obtained [28]. In the case of a thin CNT-deposited film, it becomes a transparent electrode by dissolution and removal of the filter [31]. A fibrous composite made of CNT and resin is produced by electrospinning method [32, 33]. By decomposing the resin component by appropriate heat treatment, the CNT fiber shows both a nanostructure (300 $m^2\ g^{-1}$) suitable for enzyme immobilization and microholes necessary for transporting fuel molecules [34]. Cosnier et al. reported that the compressing a powder mixture of CNT and enzyme [35] provides higher performance BFC electrodes.

The surface-treatments is important for both the preparation of CNT dispersion and the functionalization of CNTs. Oxygenated functional groups (such as quinones and hydroxyl groups), including carboxyl groups, are exposed when defects present at the side and terminal of CNT are treated with strong acid, and various molecular modification becomes possible. For example, modification of mediators and enzymes are performed. Attempts have also been made to attach apoenzyme GOD to the coenzyme FAD previously introduced to the CNT surface [36]. Molecular modification of adsorption type that does not require structure defects of CNTs is also performed, especially by utilizing the π–π interaction of aromatic molecules. As a representative example, there is a method of adsorbing a pyrene derivative having a NHS ester for the versatile modification of functional molecules including enzymes [37].

15.2.2 Enzyme/Carbon Nanotube Ensemble Films

A variety of nanoengineered carbon electrodes for biofuel cells have recently been developed in rapid succession due to their large specific surface area. However, all attempts to incorporate nanoengineered carbon electrodes have focused on prestructuring electrodes before enzyme modification. If the nanostructure of the electrode can be regulated in response to the enzyme to be immobilized, the resultant enzymatic ensemble would avoid the difficulty in post modification of enzymes. We have developed a method to achieve ideal enzyme electrodes having suitable intra-nanospace automatically regulated to the size of enzymes [15]. A carbon nanotube forest (CNTF) consisting of extremely long ($\sim$1 mm) single-walled CNTs was utilized as a 100% binder-free carbon film [38, 39]. When liquids are introduced into the CNTF (CNTs with a pitch of 16 nm) and dried, the CNTF shrinks to a near-hexagonal close-packed structure because of the surface tension of the liquids. Using an enzyme solution as the liquid, the CNTF is expected to dynamically entrap the enzymes during the shrinkage, as illustrated in Fig. 15.3. We used two model enzymes: D-fructose dehydrogenase (FDH) and Laccase (LAC), which can directly catalyze the oxidation of D-fructose and the reduction of dioxygen, respectively, without the presence of electron mediators [40]. The CNTF film was first treated by Triton X-100 and then immersed in a stirred buffer solution of FDH or LAC for 1 h to introduce enzymes into its void space, followed by drying on a carbon paper. The quantitative analysis of the amount of enzyme in a film indicated the structure of the ensemble as shown in Fig. 15.3a; enzyme molecules are aligned within the intraspace of the shrinked CNTF. The FDH-CNTF ensembles exhibited superior electrocatalytic activity for fructose oxidation by the current density of 16 mA cm^{-2} in stirred conditions. On the other hand, the LAC-entrapped CNTF cathode showed O_2 reduction with a maximum current density of $\sim$4 mA cm^{-2}. By connecting these anode and cathode through an external resistance, the fuel cell performance was evaluated in an O_2-saturated 200 mM fructose solution. The power density reached 1.8 mW cm^{-2} (at 0.45 V) in stirred condition, 84% of which could be maintained after continuous operation for 24 h.

Most of the redox enzymes including glucose oxidase (GOD) require electron mediator for their electrode reaction [41]. A conventional engineering for accelerating the electron transfer to the redox enzymes is the inclusive immobilization with mediator polymer matrices, such as Os-complex-pendant polymers, in which the successive electron exchange between the neighboring mediator groups

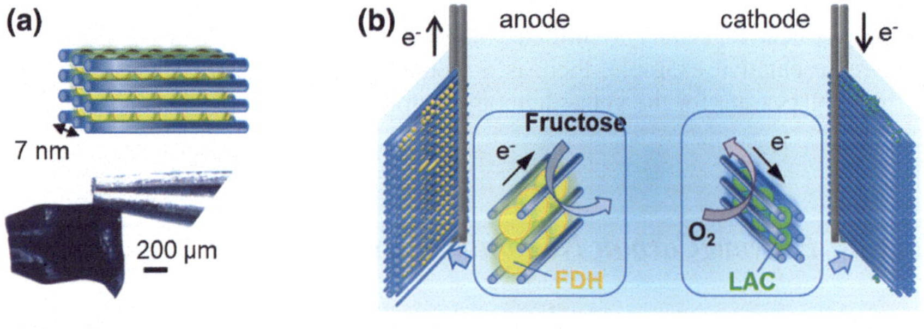

Fig. 15.3 CNTF-based enzyme electrodes for fructose/O_2 BFC [15]

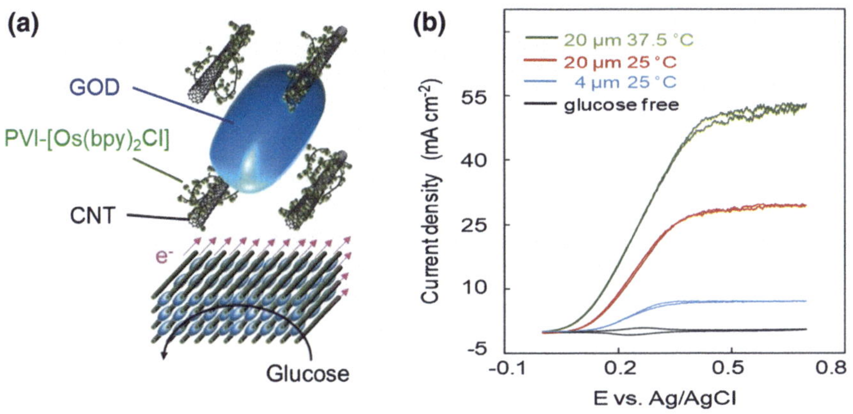

Fig. 15.4 CNTF-based enzyme electrodes for glucose oxidation. Reproduced from [16] with permission from John Wiley and Sons

connects the enzyme redox center and the electrode surface. We have developed [16] the stepwise process to construct molecular architecture with CNTF, polyvinylimidazole-[Os(bipyridine)$_2$Cl] (PVI-[Os(bpy)$_2$Cl]), and GOD, as illustrated in Fig. 15.4a. A part of the free imidazole groups of the mediator polymer would adsorb on CNT surfaces via π–π interaction, followed by electrostatic loading of enzyme GOD. Figure 15.4b shows the cyclic voltammograms of GOD/PVI-[Os(bpy)$_2$Cl]/CNTF ensemble films at 10 mV s^{-1} in a stirred 200 mM D-glucose PBS solution. The catalytic current for glucose oxidation increased in response to the thickness of CNTF films, indicating that also GOD can entirely penetrate inside the PVI-[Os(bpy)$_2$Cl]-modified CNTF films. The current density under stirred condition was enhanced to as high as 26.7 mA cm^{-2} by turning up the buffer temperature to 37.5 °C. Importantly, more than 90% of the electrode activity could be maintained even after 6 days storage in a buffer solution. The electron transfer turnover rate for the 20 μm-thick film was calculated from the current value at 25 °C (0.29 mA) to be ca. 650 s^{-1}, being comparable with that of GOD in bulk solution containing the natural electron acceptor O_2 (700 s^{-1}) at 25 °C [42]. These results indicate that most of ca. 3×10^{12} GOD units within a film could efficiently work to the fullest extent, presumably owing to the molecularly ordered structure of enzyme/mediator/electrode ensemble [16].

15.2.3 Carbon Fabric-Based Flexible Enzyme Electrodes

A flexible sheet-shaped biofuel cells that can be combined with advanced wearable electronics can be prepared using carbon fabric (CF) as the flexible, conductive base for the enzyme electrodes [18, 19]. The pre-modifications with CNTs were found to increase the specific surface area of the CF electrodes, resulting in effective enzyme immobilization for stable, high performance of the flexible enzyme electrodes.

For the preparation of FDH-anode, a strip of CF was first modified with multi-walled CNTs. The CNTs were pretreated by heating at 400 °C for 11 h and by immersing in mixed acid (H_2SO_4 + HNO_3 in a 1:3 ratio) for 5 h. The treated CNT were dispersed (10 mg ml^{-1} CNT) in water containing 0.5% Triton X-100 surfactant, and dropped on the CF strip and dried in air. After degassing the CNT-modified strip by immersion in a stirred McIlvaine buffer solution for more than 1 h under vacuum (0.09 MPa), the resulting hydrophilic CNT-modified strip was immersed in a stirred solution of FDH. Figure 15.5a shows cyclic voltammograms of the FDH/CNT/CF electrodes (thick line) at 10 mV s^{-1} in a stirred buffer solution containing 500 mM fructose. In comparison with the FDH/CF electrode prepared without CNTs (thin line), the increased specific surface area produced by CNT-modification obviously increased the current density two times. In fact, the measured double-layer capacitance of the CNT-modified electrodes has a two orders larger value (ca. 6.7 mF cm^{-2}) than that of the original CF (0.07 mF cm^{-2}). The maximum current of the optimized bioanode produced 28.0 mA cm^{-2} at 0.6 V using a 500 mM buffer.

The cathode strip for reduction of O_2 was prepared using the bilirubin oxidase (BOD), which is one of the multi-copper oxidases that can directly catalyze the four-electron reduction of O_2 to H_2O even without electron transfer mediators [43]. A 10 mg ml^{-1}, CNT solution was put on a CF strip and dried in air, followed by thoroughly washing out the surfactant by soaking in an ethanol solution for more than 1 h with stirring. The surface of the CNT-modified CF electrode was further modified with a 0.1 ml solution of 5 mg ml^{-1} BOD in a vacuum oven. The strip was additionally coated with an untreated CNT solution to make the surface

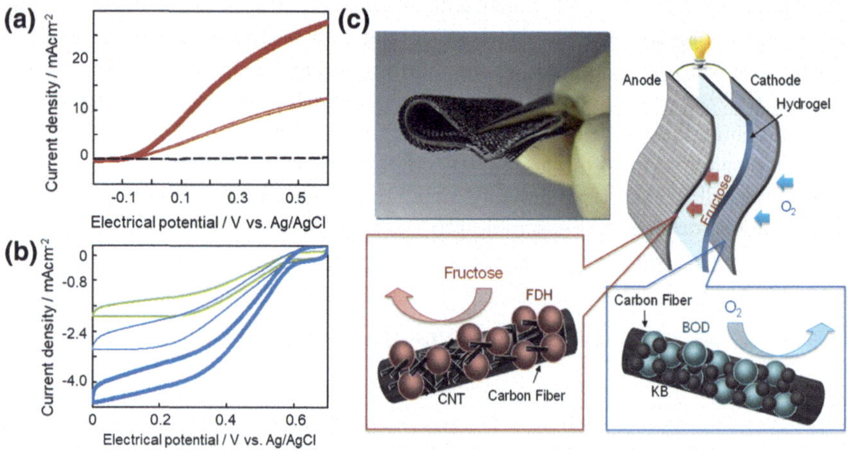

Fig. 15.5 The performances of flexible anode (**a**) and cathode (**b**) and the BFC assembly. Reproduced from [19] with permission from Elsevier

hydrophobic. Figure 15.5b shows cyclic voltammograms of a BOD-modified CF cathode at 10 mV s^{-1}. The electrode strip was put on an oxygenic pH 5.0 buffer solution so as to contact the solution by the BOD-modified face (green line). The reduction current density reaches ~1.9 mA cm^{-2} (at 0 V) by utilizing an oxygen supply from the ambient air through the CF. Moreover, an additional CNT coating onto the BOD-modified face of the CF strip enhanced the performance further to 2.9 mA cm^{-2} (thin blue line). Presumably, the hydrophobic nature of that coating controls excess penetration of solution into the CF electrode [18–20]. In addition to optimizing the performance of the bioanode fabric, the cathodic performance can be optimized by changing the buffer concentration; the maximum current was 4.6 mA cm^{-2} at 0 V using a 250 mM buffer condition (thick blue line).

As shown in Fig. 15.5c, the FDH-modified CF anode, and the BOD-modified gas-diffusion CF cathode were laminated to the opposite faces of a double-network (DN) hydrogel film (0.5 mm thick) [44, 45] made with 250 mM McIlvaine buffer solution (pH 5.0) containing 500 mM fructose. The enzyme-modified hydrophilic anode appeared to become moistened by blotting of the solution from the hydrogel layer. On the other hand, the O_2 reduction at the hydrophobic cathode proceeded at the three-phase boundary of the hydrogel–electrode interface. The cell performance was limited by the BOD-cathode because of its comparatively inferior activity to the FDH-anode fabric.

Figure 15.6a shows typical examples of the cell performance. The open-circuit voltage of the single cell was 0.74 V, which is similar to the difference between the potentials at which fructose oxidation and oxygen reduction start to occur in cyclic voltammograms (−0.14 V in Fig. 15.5a and 0.60 V in Fig. 15.5b, respectively). The maximum power density reached 0.95 mW cm^{-2} at 0.36 V, which was determined by the BOD-cathode fabric because of its comparatively inferior activity compared to the FDH-anode fabric. Importantly, the performance of the cell bent into a circle (red plots) is almost identical to that of a cell that was not bent (green plots). Such high flexibility originates in the superior mechanical strength of both the fabric bioelectrodes and the DN hydrogel. The booster cell was fabricated simply by lamination of anode/hydrogel/cathode sheets (blue plots). The open-circuit voltage of the laminated cell was over 2 V, which is roughly threefold that of a single cell. The maximum current was quite similar to that of the single cell. These results indicate that layered cells can be connected in series without suffering from short-circuit, even without packaging. Ionic isolation between the cells could be avoided by the hydrophobicity of gas-diffusion cathodes and the solid-like property of hydrogels. The laminated cell produced a maximum power of 0.64 mW at 1.21 V (2.55 mW cm^{-2}, 6.28 mW cm^{-3}); using this level of power, we were able to light the LEDs, as demonstrated in Fig. 15.6b.

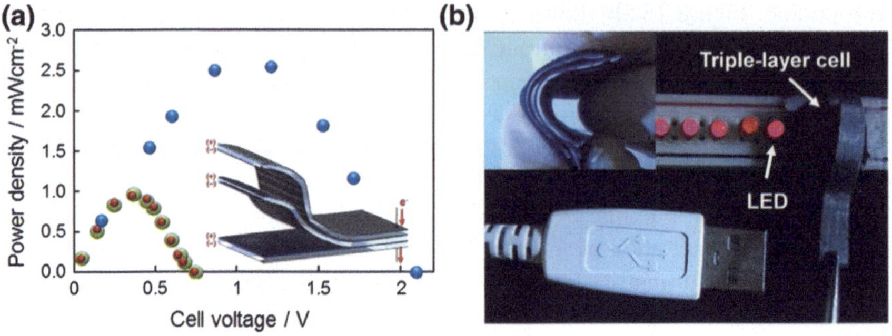

Fig. 15.6 The structure and performance of the layered BFC sheets. Reproduced from [19] with permission from Elsevier

15.2.4 Pantyhose-Based Stretchable Enzyme Electrodes

In addition, to the "flexibility" described above, the "stretchability" is desired for the BFC that will be used as a component of the wearable and patched devices. To make a stretchable BFC, we employ a nylon/polyurethane co-fiber textile, commercial stretchable pantyhose textile (ST, 0.18 mm thickness), as an underlying base material (Fig. 15.7a) [20]. This stretchable but insulating textile was modified with the two kinds of CNTs. One is a super-growth single-walled carbon nanotube (SCNT), which is extremely long (200 μm in average) and shows higher conductivity than conventional CNTs [39]. SCNTs were modified on the virgin ST to achieve high conductivity of ~200 mS/cm. Another CNT used is a comparatively short (5 μm in average) multi-walled carbon nanotube (MCNT) to obtain large specific surface area, which is necessary for enzyme immobilization.

Figure 15.7b, c shows a photograph of the textile electrode under 50% stretching, which is the largest strain possible at the junction part of the body, and electrical conductance during successive stretching at the elongation rate of 4 mm/s.

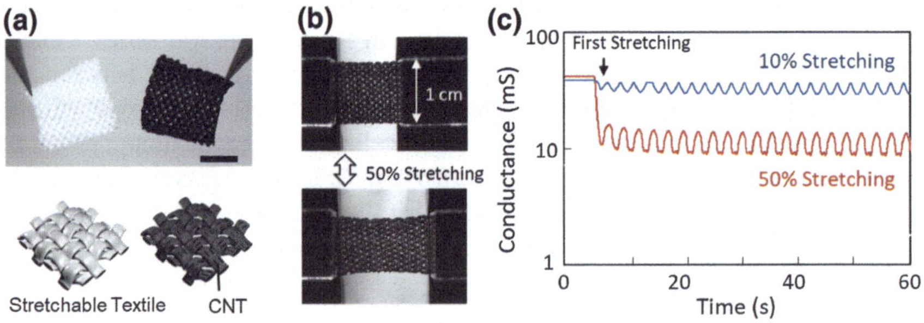

Fig. 15.7 Photographs and illustrations of virgin and CNT-modified ST strips (the scale bar indicates 5 mm), and the electrical conductance during successive stretching. Reproduced from [20] with permission from Elsevier

The first 50% stretching decreased the conductance to the one-third of the original value. Such large decrease in conductance did not occur in the case of 10% stretching. SEM observation indicated the partial breaking of the CNT bundles at the junction between the ST fibers, probably due to slipping of the ST fibers during stretching. Importantly, this partial breaking of the conductive network occurs only on the first stretching, and the textile electrode showed stable conductive behavior in the subsequent cycles with reversible 30% fluctuation that corresponds to resistance changes of 30 Ω.

The FDH-modified anode textile and the BOD-modified cathode textile were laminated to the opposite faces of the DN hydrogel sheet (0.5 mm thick) containing 200 mM fructose (Fig. 15.8a). The MCNT/SCNT/ST strip was modified by FDH and BOD through almost the same procedures used for the CF-based flexible electrodes. Only difference was that the strip of BOD electrode was additionally coated with the MCNT dispersed in ethanol with 1.0 wt% polytetrafluoroethylene

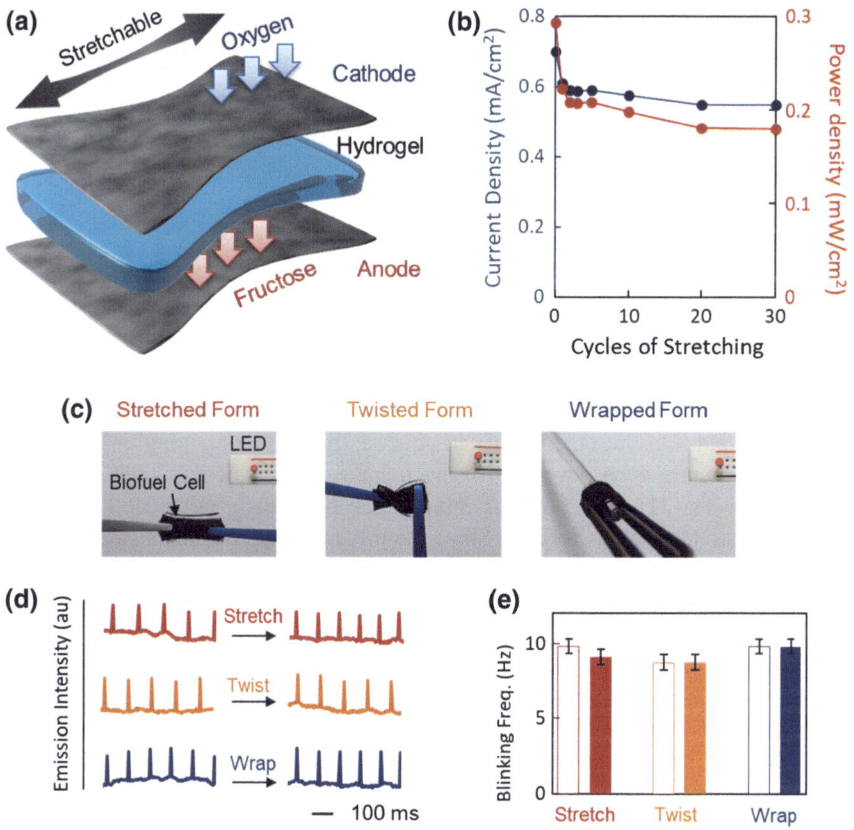

Fig. 15.8 The schematic of a stretchable BFC sheet and the deviation in the cell performances on successive 50% stretching, twisting, and wrapping. Reproduced from [20] with permission from Elsevier

(PTFE) as a binder. Although these electrodes were just stacked together with the hydrogel sheet without any adhesive materials, the assembled cell was mechanically stable enough for our measurements. The maximum current density and the power density of the cell were 1.0 mA/cm^2 and 0.25 mW/cm^2, respectively, which were determined by the BOD-cathode because of its comparatively inferior performance compared to the FDH-anode. Figure 15.8b shows that the current and power densities (with an external load of 1.2 kΩ) decreased to 0.6 mA/cm^2 and 0.2 mW/cm^2, respectively, and was stable during the 50% stretching cycles. The stretchability of the textile-based BFC originates from the flexible meshwork structure of the textile; the base fibers (nylon/polyurethane co-fiber) themselves are not significantly expanded during the stretching of the textile. Therefore, the CNTs and enzymes are stable on the fibers except for the part of junctions, and can maintain their performance even under continuous stretching.

As shown in the photos in Fig. 15.8c, the deviations of cell power during deformations were evaluated using a LED device, of which blinking frequency (Fig. 15.8d) is positively correlating with the cell power [16, 46]. The blinking frequency (namely, cell power) derived from Fig. 15.8d was almost constant during the 50% stretching, the 180° twisting, and wrapping on a glass tube (0.7 cm diameter). The blinking frequency in each form was almost the same as the original state, which indicates that the developed cell was stretchable and flexible enough to operate in various forms.

15.2.5 Printable Enzyme Electrode

The nanocarbon-based enzyme electrodes generally requires two-step preparation processes, because modification with enzymes is incompatible with harsh conditions for the fabrication of nanostructured electrodes such as heating and use of organic solvent. On the other hand, an aqueous mixture containing enzymes and CNTs has recently been developed using trehalose as a biocompatible dispersing agent for CNTs [21]. This aqueous mixture can be directly printed on insulating flexible substrates such as textiles. Since the mixture also contains a small amount of binder molecules, the printed electrode maintains structural integrity even in aqueous solution, and it shows catalytic activity in mA cm^{-2} level. The versatility of this single-step printing process for arbitrary patterning of enzyme electrodes has been proved for FDH and GOD.

As shown in Fig. 15.9a, the newly developed aqueous mixture is composed of CNT, trehalose, enzyme, and two kinds of binders. A single-walled super-growth CNT (SGCNT, 200 μm average length) [38, 39] was used. Trehalose was used as an additive for freeze-drying enzyme [47], which we also found serves as a dispersing agent for CNTs to give an enzyme/CNT aqueous mixture without surfactant. The printed mixture was immediately freeze-dried for preservation, and rehydrated on its use (Fig. 15.9b). As drawn in the schematic illustration of Fig. 15.9c, the structure and activity of the electrode were expected to be

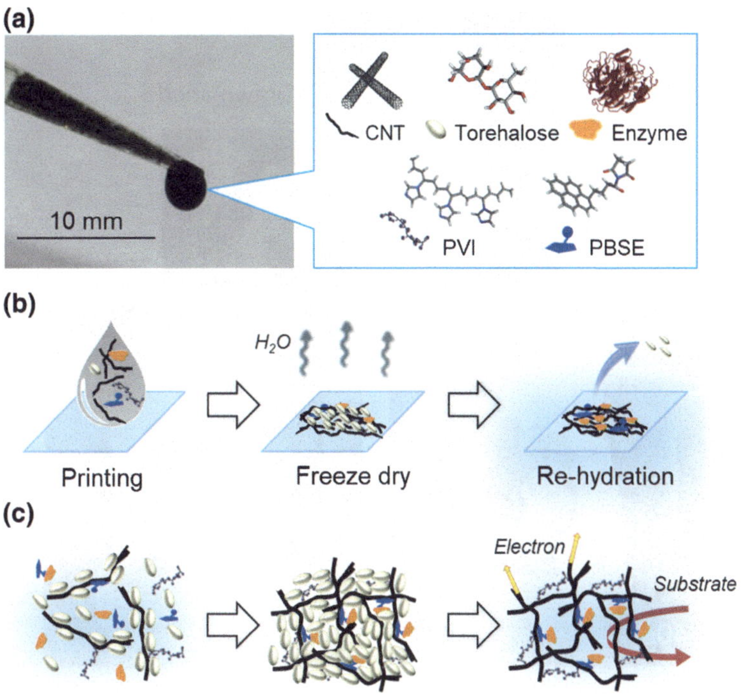

Fig. 15.9 The process for preparing enzyme electrodes: printing the mixture, freeze-drying, and rehydration for the use [21]

maintained even after rehydration by the effects of two molecular binders in the mixture: polyvinyl imidazole (PVI) and 1-pyrenebutanoic acid succinimide ester (PBSE).

The rehydrated film electrode that is composed of FDH performed biocatalytic activity in a stirred 0.2 M fructose solution with the maximum current density of 1.3 mA cm^{-2} at 0.6 V. The presence of PVI in the mixture was necessary to maintain the film structure. In fact, the film prepared without PVI immediately collapses on dipping into the solution. The free imidazole groups of PVI could attach to the CNT surface via π–π interaction and bind CNTs together [48]. The stability of the electrode performance was ensured by PBSE that attaches to the CNT surface and covalently binds to the amino group of FDH [49]. When the electrode film was prepared without addition of PBSE, the electrode activity was lower and soon decreased surely due to leakage of FDH from the electrode film to the solution.

Printing enzyme electrode is essential for the development of the paper-based and textile-based biodevices. Figure 15.10 shows a complex drawing of the FDH electrode (a shadow picture of a lady) made by the present FDH/CNT mixture with a stencil mask. Importantly, the conventional two-step fabrication of an enzyme

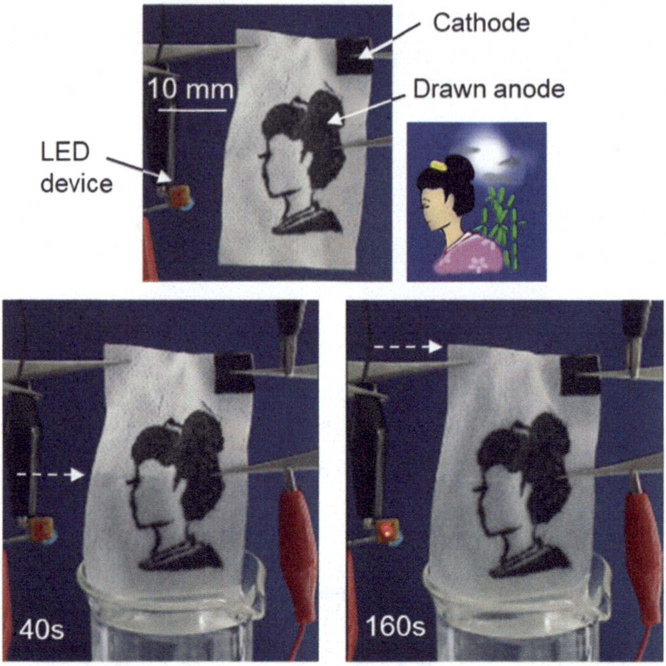

Fig. 15.10 Demonstration of a biofuel cell constructed from a drawn FDH-anode and the BOD-cathode [21]

electrode is not suitable for preparing the complicated patterns of enzyme electrode. In order to prove that the drawn FDH electrode serves as an anode of a biofuel cell, a piece of BOD-modified carbon fabric for O_2 reduction as a cathode was put on the upper right corner of the textile, and these electrodes were electrically connected with metallic tweezers to drive a LED device [16, 46]. At 40 s, after dipping the lower edge of the textile to an electrolyte solution containing fructose, the moving front of the permeating electrolyte reached the middle part of the textile without significant disruption of the drawn electrode. At 160 s, the electrolyte reached the cathode, and the LED started to blink, demonstrating the drawn electrode served as the anode of a fructose/O_2 biofuel cell.

15.3 Enzymatic Biofuel Cell Devices

There are a variety of applications for BFCs, including the implantation for power generation from biological fluids. The first BFC operating in a living organism was in a grape [50], followed by the implantation into a rat [51], a snail [52], and a clam [53]. Cosnier et al. made further developments in rat implantation for longer term (for 3 months) [35]. On the other hand, attempts to mount BFCs on wearable devices such as contact lenses [54, 55] and patches [56, 57] have been progressing. In this section, we will introduce our recent developments of BFC-based miniature devices.

15.3.1 Insertion Devices for Direct Power Generation from Biofluids

The high reaction selectivity of enzyme catalysts provides unique advantages of BFCs, including the possibility of direct power generation from carbohydrates in biofluids, such as juices or blood. Power generation from real living organisms, however, entails the consideration that natural organisms are generally covered by a skin. One possible solution is to use a needle-shaped insertion anode to approach fuels through the skins. Another important consideration is that oxygen in many organisms is limited to a lower concentration than that of sugars. Taking this consideration into account, an insertion BFC was assembled with a needle anode and a gas-diffusion cathode designed to be exposed to atmospheric air, as illustrated in Fig. 15.11a [58]. The anode and cathode are assembled using a PDMS chamber and an ion-conducting agarose hydrogel (pH 5.0) as the inner matrix. The insertion BFC was combined with an LED device [16, 46]. In practice, the LED blinks at a higher frequency with an increase in the fructose concentration, and as a result, the concentration of fructose within the grape was estimated to be roughly 20–40 mM (Fig. 15.11b, c). In fact, a separate measurement revealed that the true concentration of fructose in the grape was 35 mM.

An insertion BFC device for power generation from blood was also developed with a needle glucose anode prepared by co-immobilization of glucose dehydrogenase (GDH), diaphorase (Dp), and vitamin K_3-pendant poly-L-lysine [58]. The fuel cell performance was evaluated by inserting the device into a vein of a rabbit ear and reached 130 μW cm^{-2} at 0.56 V. An anti-biofouling modification of the needle tip with 2-methacryloyloxyethyl phosphoryl-choline (MPC)-polymer was required, however, to stabilize the cell performance.

Fig. 15.11 Insertion BFC for sensing of sugar in a grape. Reproduced from [58] with permission from the Royal Society of Chemistry

Fig. 15.12 Insertion EBFC using an enzyme/CNTF anode. Reproduced from [16] with permission from John Wiley and Sons

The enzyme-CNT ensemble film explained in Sect. 15.2.2 can be used for constructing insertion BFC devices [16]. As shown in Fig. 15.12, a piece of GOD/ PVI-[Os(bby)$_2$Cl]/CNTF film anode (Fig. 15.12a) was wound on one electric lead of the LED device, whose blinking interval is inversely proportional to the flowing current. The other lead was connected to the BOD-modified CF air-breathing cathode. The blinking interval of the LED upon inserting the device to a grape was coincident with that for the extracted juice, proving that this device could serve as a sugar indicator by simply being inserted into a grape. In this case, the leaked juice was ionically connecting the anode and cathode, instead of the intra matrix (hydrogel) of the previous needle-based insertion BFC device in Fig. 15.12. The woundable enzyme-CNT film is free-standing and thus can be readily replaced when the original is degraded.

15.3.2 Patch Devices with Built-in EFC

The electric current-induced iontophoresis has been used to enhance the efficiency of transdermal dosing [59]. However, the electrically controlled skin patches need to be wired to an external power source that is usually heavy, rigid, and hazardous. The integration of a lightweight and safe power source would produce a more patient-compliant iontophoresis patch. The enzyme-modified CFs explained in Sect. 15.2.3 contained the first completely organic iontophoresis patch with a built-in EFC, as shown in Fig. 15.13 [56]. A FDH-modified CF anode and a BOD-modified O_2-diffusion CF cathode will be connected with the stretchable PEDOT/PU internal resistor, combined with hydrogel films containing fructose and the chemical to be delivered, and covered with an O_2 permeable medical tape. The BFC patch shows the maximum current density of $\sim$300 μA cm^{-2} even at a small external load because of the presence of the skin, of which resistance was $\sim$700 Ω. As a reference, the resistance of human skin on the arms and the back of hands of three men was 270–360 Ω. Therefore, to regulate the transdermal current,

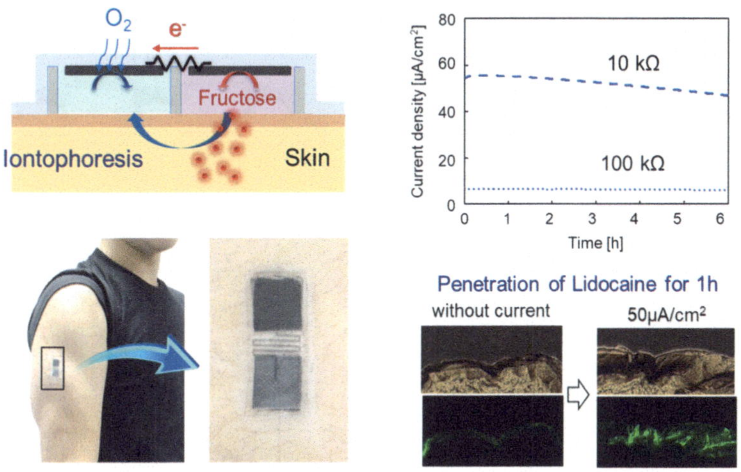

Fig. 15.13 Totally organic iontophoresis patch with built-in BFC. Reproduced from [56] with permission from John Wiley and Sons

regardless of the skin condition, the resistance of the built-in resistor is to be sufficiently larger than that of skin (>1 kΩ). BFC patches, placed on pigskin and under a variety of resistance conditions, maintained their activity for 6 h, indicating that the patches could possibly be used overnight. The cell voltages of the patches with the 100 kΩ resistor (red line) and the 10 kΩ resistor (blue line) were approximately 0.7 V and 0.55 V, respectively, indicating that the patches were generating transdermal currents of ~10 μA cm^{-2} and ~50 μA cm^{-2}. The cell voltages showed only slight, reversible decreases when a hand was flexed, as a consequence of minor change in performance of BFC under the flexed condition. The transdermal ionic current is known to be accompanied by a water flow from the anode to the cathode, which is generated by the preferential movement of mobile cation instead of a fixed anion, e.g., keratin, in the stratum corneum of the skin. This osmotic water flow can assist the electrophoretic movement of small molecules into the skin. We examined the iontophoretic penetration of lidocaine (painkiller) and ascorbyl glucoside (skin-care compound) using pigskins.

Wound healing and closure are elaborately organized biological processes in which cells migrate and proliferate at and around the wound site [60]. It has long been known that an endogenous electric current arises at the wound site (Fig. 15.14a) [61]. This "wound current", together with other biological signals such as secreted cytokines, is suggested to enhance cell migration and proliferation in wound healing. And therefore, the use of external electrical stimulation to cure chronic wounds and ulcers has been approved by the US Government in 2002 [62].

The totally stretchable BFC patch can serve as a "bioelectric plaster" that fits tightly to the skin to promote wound healing by generating ionic current above the wound on the skin (Fig. 15.14b) [57]. The bioelectric plaster consists of enzymatic electrodes, an elastic conductive resistor, a hydrogel, and medical adhesive tape.

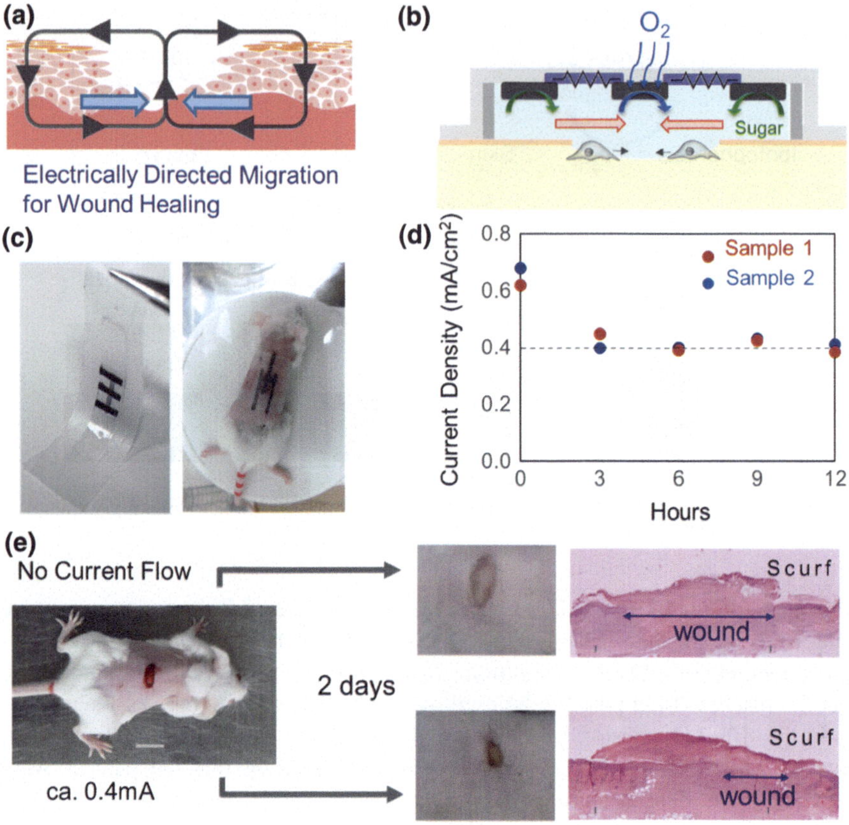

Fig. 15.14 Development of bioelectric bandage for wound healing [57]

The enzymatic electrode is made of a piece of carbon fiber fabric coated with carbon nanotubes, on which a redox enzyme (FDH for an anode, BOD for a cathode) was immobilized by spontaneous physical adsorption in buffer solution. The large specific surface area and high conductivity of the carbon nanotube-coated carbon fiber fabric enable the fabrication of enzymatic electrodes with high current density. The elastic conductive resistor was made of the composite of PEDOT and PU. The DN hydrogel sheet was adopted as a reservoir of a buffer and fuel, an ion transport path, and a stretchable contact material with skin. The thickness of hydrogel was fixed at 0.5 mm, which is the minimum possible thickness for sufficient mechanical strength, good fitting to skin, and a wet environment for the electrodes (Fig. 15.14c). All these components are assembled and fixed on the skin by wrapping them in stretchable O_2-permeable medical adhesive tape.

The stability of output current of the bioelectric plaster was studied under the conditions with varied buffer concentrations and external resistances, and the condition was optimized to the use of a hydrogel with 200 mM citrate buffer solution of pH 5 with 400 mM fructose and a stretchable resistor of 10 kΩ as it maintains the electric current of around 0.4 mA cm^{-2} over 12 h (Fig. 15.14d). An oval hole of ~8 mm width (length perpendicular to a mouse body axis) and ~4 mm height (length parallel to a body axis) was punched on a back skin of a female ICR mouse to cause artificial wounding. In this wound, both the epidermis and dermis were removed, and this can be considered as a model of delayed wound healing in a large wound. The bioelectric plaster was replaced every 12 h. Figure 15.14e shows a typical results indicating that the patch with the ionic current showed more pronounced decrease in the size of wound than the case without the current (only a hydrogel sheet), which suggested that wound closure by the proliferation and migration of the cells was effectively promoted by the ionic current of the bioelectric plaster. It is worthwhile to note that this demonstrates the first application of ionic current of BFCs to direct manipulation of biological functions of a living body.

15.4 Future Outlook

In this chapter, we described the importance of CNTs in the engineering of BFCs. The nanostructure of the CNT electrodes determines the quality of enzyme electrode and brought the recent advancement of BFC performance in mA cm^{-2}. The CNT-based flexible, stretchable, and printable enzyme electrodes will be applied to wearable self-powered devices. In order to prepare a porous CNT electrode with further reproducibility, there is still room for strict control of the dispersion/aggregation of CNTs by surface modification. It can be expected that the progress in the technology to control the nanostructure of CNT electrodes and to fruition in the remarkable performance improvement of BFC devices.

References

1. Luckarift HR, Atanassov PB. Johnson GR (eds) (2014) Enzymatic fuel cells: from fundamentals to applications. Wiley
2. Rasmussen M, Abdellaoui S, Minteer SD (2016) Enzymatic biofuel cells: 30 years of critical advancements. Biosens Bioelectron 76:91–102
3. Barton SC, Gallaway J, Atanassov P (2004) Enzymatic biofuel cells for implantable and microscale devices. Chem Rev 104:4867–4886
4. Heller A (2004) Miniature biofuel cells. Phys Chem Chem Phys 6:209–216
5. Cooney MJ, Svoboda V, Lau C, Martin G, Minteer SD (2008) Enzyme catalyzed biofuel cells. Energy Environ Sci 1:320–337
6. Meredith MT, Minteer SD (2012) Biofuel cells: enhanced enzymatic bioelectrocatalysis. Annu Rev Anal Chem 5:157–179
7. Kwon CH, Lee SH, Choi YB, Lee JA, Kim SH, Kim HH, Spinks GM, Wallace GG, Lima MD, Kozlov ME, Baughman RH, Kim SJ (2014) High-power biofuel cell textiles from woven biscrolled carbon nanotube yarns. Nat Commun 5:3928

8. Arechederra R, Minteer SD (2009) Nanomaterials in biofuel cells. American Scientific Publishers, Stevenson Ranch, CA, pp 287–299
9. Che AF, Germain V, Cretin M, Cornu D, Innocent C, Tingry S (2011) Fabrication of free-standing electrospun carbon nanofibers as efficient electrode materials for bioelectrocatalysis. New J Chem 35:2848–2853
10. Minteer SD (2012) Nanobioelectrocatalysis and its applications in biosensors, biofuel cells and bioprocessing. Top Catal 55:1157–1161
11. Walcarius A, Minteer SD, Wang J, Lin Y, Merkoçi A (2013) Nanomaterials for bio-functionalized electrodes: recent trends. J Mater Chem B 1:4878–4908
12. Wen D, Deng L, Zhou M, Guo S, Shang L, Xu G, Dong S (2010) A biofuel cell with a single-walled carbon nanohorn-based bioanode operating at physiological condition. Biosens Bioelectron 25:1544–1547
13. Yan Y, Zheng W, Su L, Mao L (2006) Carbon nanotube-based glucose/O_2 biofuel cells. Adv Mater 18:2639–2643
14. Cosnier S, Holzinger M, Goff AL (2014) Recent advances in carbon nanotube-based enzymatic fuel cells. Front Bioeng Biotechnol 2:45
15. Miyake T, Yoshino S, Yamada T, Hata K, Nishizawa M (2011) Self-regulating enzyme-nanotube ensemble films and their application as flexible electrodes for biofuel cells. J Am Chem Soc 133:5129–5134
16. Yoshino S, Miyake T, Yamada T, Hata K, Nishizawa M (2013) Molecularly ordered bioelectrocatalytic composite inside a film of aligned carbon nanotubes. Adv Energy Mater 3:60–64
17. Ogawa Y, Yoshino S, Miyakea T, Nishizawa M (2014) Surfactant-assisted direct electron transfer between multi-copper oxidases and carbon nanotube-based porous electrodes. Phys Chem Chem Phys 16:13059–13062
18. Haneda K, Yoshino S, Ofuji T, Miyake T, Nishizawa M (2012) Sheet-shaped biofuel cell constructed from enzyme-modified nanoengineered carbon fabric. Electrochim Acta 82:175–178
19. Miyake T, Haneda K, Yoshino S, Nishizawa M (2013) Flexible, layered biofuel cells. Biosens Bioelectron 40:45–49
20. Ogawa Y, Takai Y, Kato Y, Kai H, Miyake T, Nishizawa M (2015) Stretchable biofuel cell with enzyme-modified conductive textiles. Biosens Bioelectron 74:947–952
21. Kai H, Ogawa Y, Chen HY, Hata K, Nishizawa M (in press) Austin J Biosens Bioelectron
22. Smart SK, Cassady AI, Lu GQ, Martin DJ (2006) The biocompatibility of carbon nanotubes. Carbon 44:1034–1047
23. Inam F, Yan H, Reece MJ, Peijs T (2008) Dimethylformamide: an effective dispersant for making ceramic-carbon nanotube composites. Nanotechnology 19:195710
24. Ishibashi A, Nakashima N (2006) Individual dissolution of single-walled carbon nanotubes in aqueous solutions of steroid or sugar compounds and their raman and near-IR spectral properties. Chem Eur J 12:7595–7602
25. Kaempgen M, Duesberg GS, Roth S (2005) Transparent carbon nanotube coatings. Appl Surf Sci 252:425–429
26. Abe Y, Tomuro R, Sano M (2005) Highly efficient direct current electrodeposition of single-walled carbon nanotubes in anhydrous solvents. Adv Mater 17:2192–2194
27. Whitby RL, Fukuda T, Maekawa T, James SL, Mikhalovsky SV (2008) Geometric control and tuneable pore size distribution of buckypaper and buckydiscs. Carbon 46:949–956
28. Vohrer U, Kolaric I, Haque MH, Roth S, Detlaff-Weglikowska U (2004) Carbon nanotube sheets for the use as artificial muscles. Carbon 42:1159–1164
29. Hennrich F, Lebedkin S, Malik S, Tracy J, Barczewski M, Rösner H, Kappes M (2002) Preparation, characterization and applications of free-standing single walled carbon nanotube thin films. Phys Chem Chem Phys 4:2273–2277
30. Hussein L, Urban G, Krüger M (2011) Fabrication and characterization of buckypaper-based nanostructured electrodes as a novel material for biofuel cell applications. Phys Chem Chem Phys 13:5831–5839

31. Wu Z, Chen Z, Du X, Logan JM, Sippel J, Nikolou M, Kamaras K, Reynolds JR, Tanner DB, Hebard AF, Rinzler AG (2004) Transparent, conductive carbon nanotube films. Science 305:1273–1276
32. Ko F, Gogotsi Y, Ali A, Naguib N, Ye H, Yang GL, Li C, Willis P (2003) Electrospinning of continuous carbon nanotube-filled nanofiber yarns. Adv Mater 15:1161–1165
33. Vigolo B, Penicaud A, Coulon C, Sauder C, Pailler R, Journet C, Bernier P, Poulin P (2000) Macroscopic fibers and ribbons of oriented carbon nanotubes. Science 290:1331–1334
34. Gao F, Viry L, Maugey M, Poulin P, Mano N (2010) Engineering hybrid nanotube wires for high power biofuel cells. Nat Commun 1:2
35. Cosnier S Le, Goff A, Holzinger M (2014) Towards glucose biofuel cells implanted in human body for powering artificial organs. Electrochem Commun 38:19–23
36. Patolsky F, Weizmann Y, Willner I (2004) Long-range electrical contacting of redox enzymes by SWCNT connectors. Angew Chem Int Edit 43:2113–2117
37. Chen RJ, Zhang Y, Wang D, Dai H (2001) Noncovalent sidewall functionalization of single-walled carbon nanotubes for protein immobilization. J Am Chem Soc 123:3838–3839
38. Hiraoka T, Izadi-Najafabadi A, Yamada T, Futaba DN, Yasuda S, Tanaike O, Hatori H, Yumura M, Iijima S, Hata K (2010) Compact and light supercapacitor electrodes from a surface-only solid by opened carbon nanotubes with 2200 m^2 g^{-1} surface area. Adv Funct Mater 20:422–428
39. Ata S, Kobashi K, Yumura M, Hata K (2012) Mechanically durable and highly conductive elastomeric composites from long single-walled carbon nanotubes mimicking the chain structure of polymers. Nano Lett 12:2710–2716
40. Kamitaka Y, Tsujimura S, Setoyama N, Kajino T, Kano K (2007) Fructose/Dioxygen biofuel cell based on direct electron transfer-type bioelectrocatalysis. Phys Chem Chem Phys 9:1793–1801
41. Heller A, Feldman B (2010) Electrochemistry in diabetes management. Acc Chem Res 43:963–973
42. Bourdillon C, Demaille C, Gueris J, Moiroux J, Saveant JM (1993) A fully active monolayer enzyme electrode derivatized by antigen-antibody attachment. J Am Chem Soc 115:12264–12269
43. Tsujimura S, Kamitaka Y, Kano K (2007) Diffusion-controlled oxygen reduction on multi-copper oxidase-adsorbed carbon aerogel electrodes without mediator. Fuel Cells 7:463–469
44. Gong JP (2010) Why are double network hydrogels so tough? Soft Matter 6:2583–2590
45. Wu ZL, Gong JP (2011) Hydrogels with self-assembling ordered structures and their functions. NPG Asia Mater 3:57–64
46. Hanashi T, Yamazaki T, Tsugawa W, Ferri S, Nakayama D, Tomiyama M, Ikebukuro K, Sode K (2009) BioCapacitor-a novel category of biosensor. Biosens Bioelectron 24:1837–1842
47. Dráber P, Sulimenko V, Sulimenko T, Dráberová E (2014) Stabilization of protein by freeze-drying in the presence of trehalose: a case study of tubulin. protein downstream processing: design, development and application of high and low-resolution methods. Methods Mol Biol 1129:443–458
48. Roman T, Dino WA, Nakanishi H, Kasai H (2006) Amino acid adsorption on single-walled carbon nanotubes. Eur Phys J 38:117–120
49. Hickey DP, Reid RC, Milton RD, Minteer SD (2016) A self-powered amperometric lactate biosensor based on lactate oxidase immobilized in dimethylferrocene-modified LPEI. Biosens Bioelectron 77:26–31
50. Mano N, Heller A (2003) A miniature membraneless biofuel cell operating at 0.36 V under physiological conditions. J Electrochem Soc 150:A1136–A1138
51. Cinquin P, Gondran C, Giroud F, Mazabrard S, Pellissier A, Boucher F, Alcaraz JP, Gorgy K, Lenouvel F, Mathe S, Porcu P, Cosnier S (2010) A glucose biofuel cell implanted in rats. PLoS ONE 5:e10476
52. Halámková L, Halámek J, Bocharova V, Szczupak A, Alfonta L, Katz E (2012) Implanted biofuel cell operating in a living snail. J Am Chem Soc 134:5040–5043

53. Szczupak A, Halámek J, Halámková L, Bocharova V, Alfonta L, Katz E (2012) Living battery-biofuel cells operating in vivo in clams. Energy Environ Sci 5:8891–8895
54. Falk M, Andoralov V, Blum Z, Sotres J, Suyatin DB, Ruzgas T, Arnebrant T, Shleev S (2012) Biofuel cell as a power source for electronic contact lenses. Biosens Bioelectron 37:38–45
55. Reid RC, Minteer SD, Gale BK (2015) Contact lens biofuel cell tested in a synthetic tear solution. Biosens Bioelectron 68:142–148
56. Ogawa Y, Kato K, Miyake T, Nagamine K, Ofuji T, Yoshino S, Nishizawa M (2015) Organic transdermal iontophoresis patch with built-in biofuel cell. Adv Healthcare Mater 4:506–510
57. Kai H, Yamauchi T, Ogawa Y, Tsubota A, Magome T, Miyake T, Yamasaki K, Nishizawa M (2017) Accelerated wound healing on skin by electrical stimulation with a bioelectric plaster. Adv Healthcare Mater 6:1700465–1700470
58. Miyake T, Haneda K, Nagai N, Yatagawa Y, Ohnami H, Yoshino S, Abe T, Nishizawa M (2011) Enzymatic biofuel cells designed for direct power generation from biofluids in living organisms. Energy Environ Sci 4:5008–5012
59. Saluja S, Kasha PC, Paturi J, Anderson C, Morris R, Banga AK (2013) A novel electronic skin patch for delivery and pharmacokinetic evaluation of donepezil following transdermal iontophoresis. Int J Pharm 453:395–399
60. Gurtner GC, Werner S, Barrandon Y, Longaker MT (2008) Wound repair and regeneration. Nature 453:314–321
61. Zhao M (2009) Electrical fields in wound healing—an overriding signal that directs cell migration. Semin Cell Dev Biol 20:674–682
62. Kloth LC (2005) Electrical stimulation for wound healing: a review of evidence from in vitro studies, animal experiments, and clinical trials. Int J Lower Extrem Wounds 4:23–44

Chapter 16
Improved Synthesis of Graphene-Like Materials and Their Application

Yuta Nishina

16.1 Preface: Graphene Production

Graphene research was awarded the Nobel Prize in Physics in 2010, therefore, research on graphene has been developed based on physics. Graphene is not only the lightest and strongest material, but also has excellent thermal conductivity and electrical conductivity than anything. For this reason, graphene is considered to be one of the most versatile materials among substances existing on the earth, and research on the application of graphene has been extensively conducted.

To understand the potential use of graphene, it is necessary to understand basic characteristics of materials. In 2004, graphene was manufactured by a very simple method by peering graphite with scotch tape. This graphene was sufficient for the elucidation of fundamental physical properties and verification of the operating principle of devices using graphene. On the other hand, in order to apply graphene to an actual device, graphene of large area and high quality of centimeter-to-meter size is indispensable. Unfortunately, however, it is impossible to prepare large-size graphene by this method. In addition, since graphene is a perfect two-dimensional material, its surface area is 2,630 m^2 per gram (1,315 m^2 on some substrate), therefore, it might be possible to use graphene in a large area, but using large weight is extremely difficult; you may recognize that it will be 131,500 m^2 for only 1 kg of graphene. In other words, how to make, store, and use graphene in three dimensions is important in practical application. To disperse in solution would be one solution of the problem of two-dimensional materials.

Chemical vapor deposition (CVD) is being studied as a large-area manufacturing technology of graphene having relatively high purity and crystallinity. The CVD method is a bottom-up method in which graphene crystals are grown by thermal

Y. Nishina (✉)
Research Core for Interdisciplinary Sciences, Okayama University, Okayama, Japan
e-mail: nisina-y@cc.okayama-u.ac.jp

N. Nakashima (ed.), *Nanocarbons for Energy Conversion: Supramolecular Approaches*, Nanostructure Science and Technology,
https://doi.org/10.1007/978-3-319-92917-0_16

decomposition of a source gas, such as methane or alcohol on a substrate (i.e. copper foil) at a high temperature of about 1000 °C. In the CVD method, although it is not difficult to increase the area of graphene, improvement of homogeneity (defectless structure) of the crystal, cost of the metal substrate, and transfer technique, etc., are the subjects to be solved. At present, it is possible to synthesize single-crystal graphene of several cm sizes by the CVD method.

On the other hand, this chapter focuses on the mass production method of graphene-like materials from graphite. Graphite can be obtained both naturally and artificially, and is inexpensive and stable. However, graphite cannot be easily peeled off, except scotch tape method. For this reason, peeling is performed by weakening the interaction between the graphene layers by chemical treatments, such as electrolysis, chemical functionalization, and oxidation. In this process, defects are introduced into the graphene. That is, it is difficult to obtain highly crystalline graphene. Nevertheless, there are excellent advantages in cost and mass productivity when graphite is used as a raw material and applying chemistry-based treatments.

16.2 Electrochemical Production of Graphene

It has long been known that a metal salt penetrates (intercalates) between the layers of graphite when a metal salt is contained in an electrolytic solution and electric potential is applied for a graphite electrode. Many reports mainly used an organic solvent having a large potential window [1, 2], but inorganic electrolyte, such as hydrochloric acid, can also be used [3]. In recent years, attention has been paid to peel graphite using such an electrochemical method. This is because it is not necessary to use a strong oxidizing agent, and there is a possibility of producing less defective graphene.

16.2.1 In Water

In water, oxygen functional groups are introduced by reactive oxygen species generated by electrolysis of water. One of the experimental procedures is as follows: a graphite electrode and a platinum electrode are inserted into a 0.1 mol/L sulfuric acid aqueous solution, and a peeled product having a yield of ca. 50% can be obtained by applying a voltage (ca. 10 V, and water decomposes to generate oxy radicals (OH radicals or O radicals)). It is thought that by adding these radicals to the edge of graphite, the edge exfoliates and intercalation of sulfate ions are promoted. After that, sulfate ions are decomposed between graphite layers to generate gas such as SO_2, and peeling occurs by the gas pressure (Fig. 16.1).

Li et al. confirmed that graphite was damaged by sulfuric acid, and attempted to prevent it by adjusting pH to 1.2 by adding KOH aqueous solution [4]. After

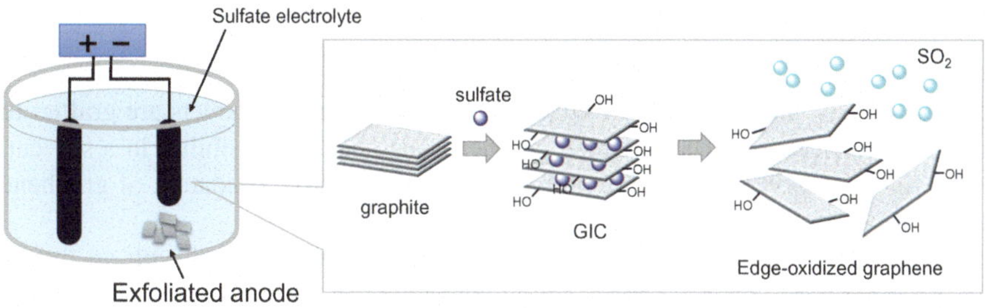

Fig. 16.1 Electrochemical peeling of graphite in water

applying a voltage of +2.5 V for 1 min, then applying +10 V and −10 V, uniform peeling of few layers of graphene with an average thickness of 2 nm was obtained. Although this graphene is slightly oxidized, oxygen functional groups can be removed and the electrical conductivity can be improved after heating at 450 °C. Müllen et al. reported the yield of graphene is reduced to 7% when sulfuric acid with a high concentration of ca. 5.0 mol/L is used [5]. This is because when the amount of sulfuric acid is high, the amount of oxygen radicals generated decreases and partial oxidation of graphite does not occur. Conversely, when the sulfuric acid concentration is as low as 0.01 mol/L, sulfate-ion intercalation does not occur sufficiently, and the yield is as low as 2% even over 2 h for electrolysis. Also, when only concentrated sulfuric acid is used, peeling by oxidation does not occur and intercalation into graphite takes place and only expansion occurs. These results suggest that the composition of the electrolyte is very important. The Raman spectra of the peeled product showed the intensity ratio of D band and G band (I_D/I_G) is about 0.4, indicating that there are many defects. Furthermore, Müllen et al. reported a method of electrochemically peeling graphite with as little oxidation as possible [6]. If sulfate (ammonium sulfate, potassium sulfate, sodium sulfate, etc.) is used in place of sulfuric acid and a voltage is applied under neutral conditions, graphite of 85% or more is peeled off to 1–3 layers with a size of 5 μm or larger. The oxygen content of graphene obtained by this method is 5.5% or less, which is closer to graphene. The type of salt has a great influence on peeling, with particular importance being on the anion side; sulfate is better than chlorates and nitrates. This is because 1.42 and 0.92 V are required to reduce chlorate and nitrate to Cl_2 and NO_2, while sulfate is reduced to SO_2 at 0.20 V.

The graphene can be dispersed in a polar organic solvent such as DMF and NMP. For the application of electrode, the graphene is coated on metal foil. A graphene film coated with a thickness of 6 nm shows a light transmission of 91% and a sheet resistance of 24.2 kΩ/sq. When the thickness is increased to 16 nm, the transmittance decreases to 80%, but the resistivity lowers to 7.56 kΩ/sq. In addition, when the graphene film is treated at 300 °C, the resistance of each film lowers to 7.61 and 1.81 kΩ/sq.

Production of graphene by such electrolysis can also be carried out using pencils for both the anode and the cathode [7].

16.2.2 In Ionic Liquid

By electrolysis, the ionic liquid itself and its decomposition products are grafted to the graphite electrode. The obtained graphene has good dispersibility in a solvent. Also, since no oxygen atoms are present in the ionic liquid, oxidation of graphene does not occur unless water is present.

Luo et al. inserted two graphite rods into an electrolyte mixed with an imidazolium ionic liquid having various alkyl chains and an equal amount of water and applied a voltage of 15 V at room temperature for 6 h [8]. It is believed that ionic liquids generate radicals by electrolysis, and the radicals are added to the edges of graphite. The obtained product was dispersed in DMF, and then coated to form a thin film. AFM analysis revealed that the thickness of the graphene was 1.1 nm. This is considerably larger than the ideal graphene thickness (0.34 nm) due to the attachment of long-chain alkyl groups derived from ionic liquids. However, there are many questions about the mechanism of radical addition to the graphite electrode. In a very similar system, Loh et al. attempted to elucidate the mechanism of graphite peeling [9]. If radicals are generated at the cathode, the radical will react with the graphite of the cathode and the cathode should also be peeled off, however, only anode was peeled off. Also, because the electrolysis is done at 15 V in the presence of water, decomposition of water should occur before radical formation from the ionic liquid. Therefore, Loh et al. proposed that hydroxy radicals generated at the anode, which oxidize the anode graphite.

Even in electrolysis in an ionic liquid, the product varies greatly depending on the type of ion (BF_4^- or Cl^-), water content (10–90%), and kind of graphite (HOPG or graphite rod). Since BF_4^- has a higher oxidation potential than water, hydroxyl radicals, and oxygen radicals are generated and oxidation occurs. On the other hand, in ionic liquids having Cl^- whose oxidation potential is lower than that of water, Cl_3^- generates first, and then intercalation occurs, thereafter, hydroxyl radicals and oxygen radicals attack the detached positions, as a result, larger graphene is obtained. To obtain graphene without oxidization, it is desirable to perform the electrolysis of graphite using an ionic liquid in a nonaqueous system.

16.3 Chemical Functionalization

In order to stably disperse graphene in solution, it is necessary to chemically modify its surface and edge. Although the electrolysis of graphite described in the above section also induced chemical modification of graphene, this section describes none-electrochemical methods. Non-covalent bonds and covalent bonds are used as chemical modification of graphene, and the former is less likely to lose the potential properties of graphene. Particularly, in chemical modification by in-plane covalent functionalization, the sp^2 hybridized carbon of graphene becomes sp^3 hybridized

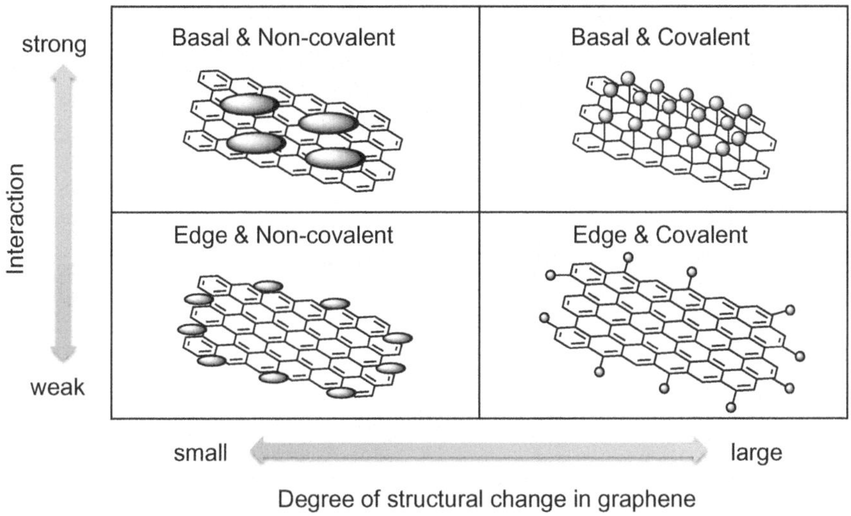

Fig. 16.2 Classification of chemical modification of graphene

carbon, and the characteristic properties of graphene such as high electric conductivity are lost (Fig. 16.2).

16.3.1 Functionalization by Non-covalent Interaction

Since benzene is a unit structure of graphene, it interacts with an aromatic compound by π–π interaction. For this reason, an aromatic compound having a functional group, which is highly compatible with a solvent and graphite (and graphene) is used as dispersing reagent [10]. Derivatives of fluorinated benzene, naphthalene, anthracene, triphenylene, coronene, phthalocyanine, porphyrin, and perylene promote the stabilization of graphene in solvent (Fig. 16.3a). Among such aromatic compounds, a pyrene skeleton bearing different numbers, positions, and types of functional groups are used as graphene dispersant (Fig. 16.3b). The graphene conversion rate is evaluated by mixing with graphite under sonication. In recent years, it has been reported that simple compounds such as long-chain alkyl amines, carboxylic acids, alcohols, and even alkanes promote the graphene formation from graphite. These are not aromatic compounds, therefore, van der Waals interaction can also enhance the stabilization of graphene in solvent, and hence, the choice of dispersants has now widened [11].

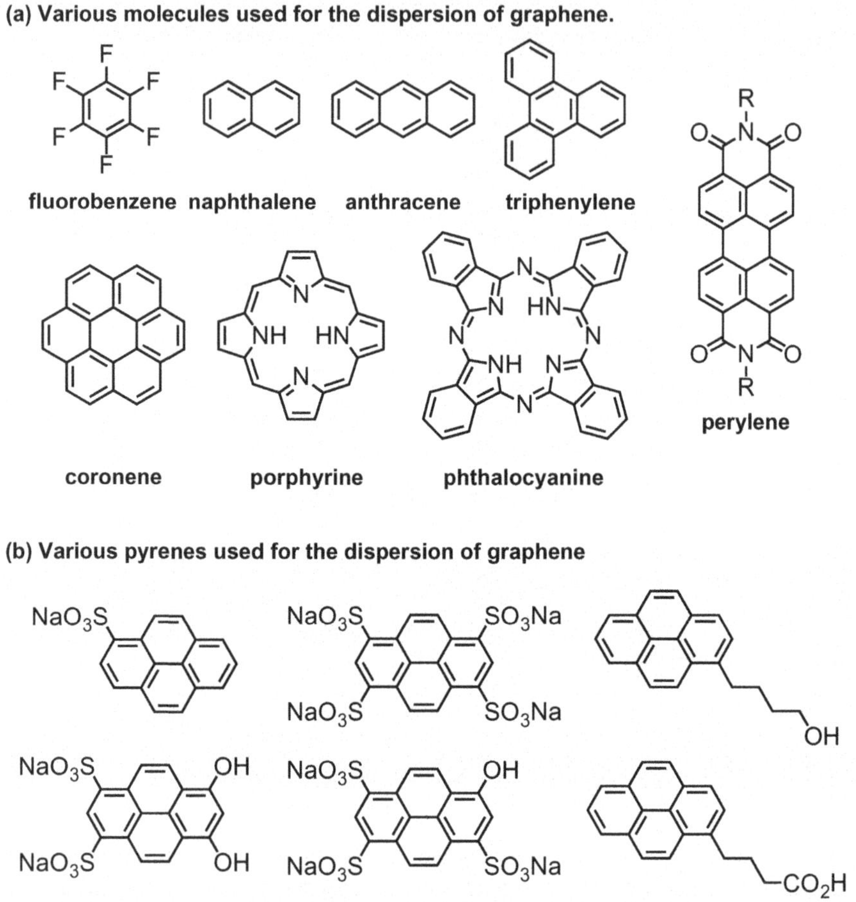

Fig. 16.3 Aromatic compounds for exfoliation and dispersion of graphite

16.3.2 Functionalization by Covalent Interaction

The edges of graphene terminated with oxygen functional groups such as carbonyl groups, epoxy groups, carboxyl groups, etc., can be reactive sites for further functionalization [12]. The carboxyl group on the graphene is functionalized with alcohol or amine to produce an ester or amide (Fig. 16.4i) [13–15]. Covalent bond formation of an amine to an epoxy group is also effective. Metachlorobenzoic acid (mCPBA) is used for edge selective epoxidation of graphite. For such epoxidized graphite, a ring-opening nucleophilic addition reaction takes place and functional groups can be introduced (Fig. 16.4ii) [16]. In addition, adjacent carbonyl groups react with diamines to form nitrogen-containing aromatic rings (Fig. 16.4iii) [17].

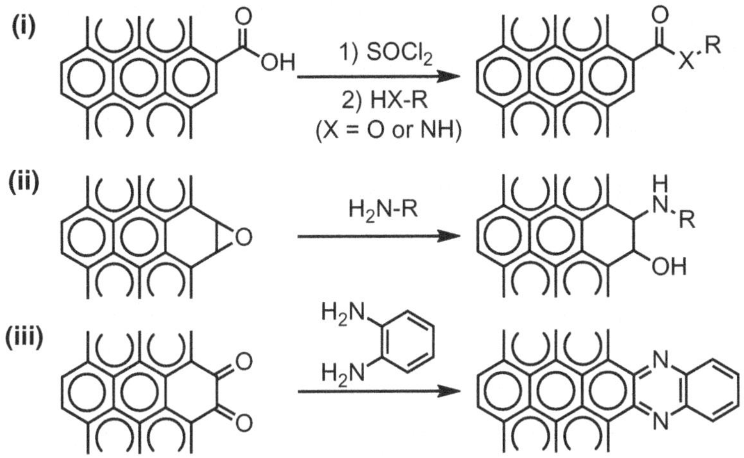

Fig. 16.4 Covalent bond formation of oxygen functional groups at the edge of graphene

Cycloaddition to graphene plane is another type of modification method by covalent bond formation. There are mainly four methods and are as follows: (1) [2 + 1] cycloaddition forming a three-membered ring, (2) [2 + 2] cycloaddition forming a four-membered ring, (3) [3 + 2] cycloaddition to form a five-membered ring, and (4) a [4 + 2] cycloaddition to form a six-membered ring.

Carbene is a reactive species of [2 + 1] cycloaddition; the carbene's vacant orbital (LUMO) interacts with the graphene's π-orbital (HOMO), and the carbene's electron pair (HOMO) interacts with graphene's π*-orbital (LUMO). As a result, three-membered ring is formed (Fig. 16.5i) [18]. Also, nitrene reacts with graphite to form an aziridine framework. Highly reactive singlet nitrene is formed by thermal decomposition or photolysis of the azide group, and cycloaddition reaction occurs on the graphene sp^2 carbon to give a three-membered ring (Fig. 16.5ii) [19]. Alkenes and dienes form four- or six-membered ring via the Diels–Alder reaction (Fig. 16.5iii) [20]. In recent years, various benzynes can be generated by simple procedures. Benzyne is added to the alkene moiety of graphene to give a four-membered ring (Fig. 16.5iv) [21]. Azomethine ylide, which is generally prepared from an amino acid and aldehyde through condensation and decarboxylation, is used as a reactive species for [3 + 2] cycloaddition to form a five-membered nitrogen-containing ring (Fig. 16.5v) [22].

In addition, Friedel–Crafts-type reaction [23, 24], reaction with organometallic reagent [25], silane coupling reaction [26], etc., have been reported, however, structural analysis is sometimes insufficient, and structure of the products are uncertain. In addition, the correlation between functional groups to be introduced and physical properties of graphene has not been comprehensively studied. This field is still under development.

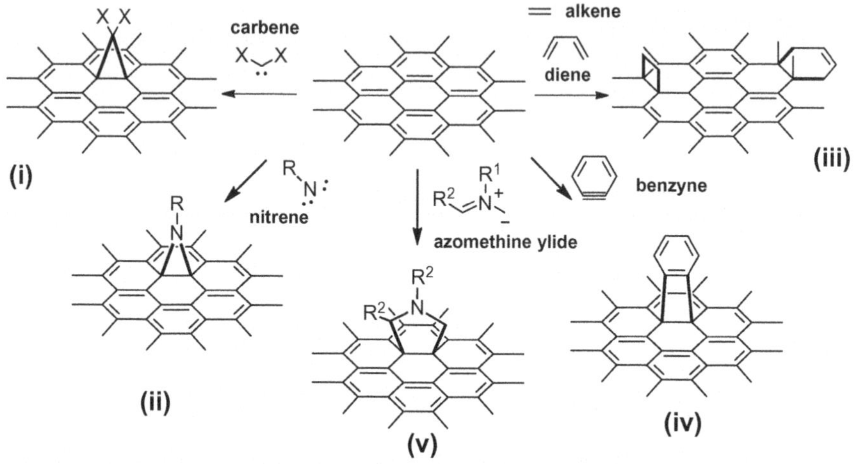

Fig. 16.5 Covalent bond formation reactions on graphene plane

16.4 Oxidation

Graphene oxide, which is often regarded as an analog of graphene, has oxygen functional groups on most of the carbon atoms, therefore, its physical properties are largely different from those of pure graphene. For example, electrical conductivity and thermal conductivity are very low. Also, holes are opened during oxidation process and/or the sheet breaks and becomes smaller. In order to improve the physical properties required for graphene, such as electric conductivity, such damages should be suppressed. Nonetheless, graphene oxide is attracting great attention because mass production is much easier than other processes. Using inexpensive graphite and optimizing the oxidation process, industrial production of graphene oxide is underway.

There are roughly three methods of graphene oxide production, which are as follows: Brodie's method, Staudenmaier's method, and Hummers' method. Depending on the fabrication method, the amount and type of oxygen functional group are changed, and the physical properties of the graphene oxide also change. For example, graphene oxide prepared by the Hummers' method shows excellent performance in the adsorption of metal ions, in contrast, the Brodie's method is superior in conductivity after reduction. Each preparation method is given in detail below.

16.4.1 Brodie's Method

Brodie's method is one of the oldest methods for the oxidation of graphite [27]. Potassium chlorate is used as an oxidizing agent in concentrated nitric acid to oxidize graphite. Compared with other oxidation methods, Brodie's method is thought that damage to graphene is smaller and easier to obtain a graphene-like structure by reduction. In the method reported by B. C. Brodie in 1859, oxygen content can be increased by repeating oxidation treatment. Recently, the process was optimized and become easier than the original method. For example, after adding graphite, potassium chlorate, and fuming nitric acid, and leaving it at room temperature for 24 h, nitric acid is added and the reaction mixture is heated at 60 °C, which reduces the oxidation time required for 3–4 days in the original method. By increasing the amount of potassium chlorate and prolonging the reaction time, it is also possible to complete the oxidation in one step. By slightly changing the reaction conditions, the degree and efficiency of oxidation changes to give graphene oxide with different characters.

16.4.2 Staudenmaier's Method

In the general Brodie's method, it is necessary to perform repetitive oxidation in order to increase the oxygen content. By adding concentrated sulfuric acid to the conditions of Brodie's method, it becomes possible to oxidize in one step and the process can be simplified [28]. The formation of nitroium ion (NO_2^+) from sulfuric acid and nitric acid would enhance the oxidation of graphite, but the detailed mechanism is unknown. However, as mentioned in Sect. 16.4.1, the Brodie's method has also improved to sufficiently oxidize in a short time in a single step. Further study is necessary to properly verify the effect of sulfuric acid.

16.4.3 Hummers' Method

Unlike the Brodie's method and Staudenmaier's method, Hummers' method developed in 1958 uses potassium permanganate as oxidant [29]. Since potassium permanganate quickly reacts with graphite exothermically, it is added into the reaction mixture by splitting under cooling. Since scale-up is easier than other methods, Hummers' method is the most frequently used in recent years. The synthesis method first reported by W. S. Hummers is shown below.

1. Place 100 g of graphite and 50 g of sodium nitrate into the flask, add 2.3 L of sulfuric acid and stir with ice cooling.
2. Add 300 g of potassium permanganate gradually. At this time, the temperature of the reaction solution should not exceed 20 °C.

3. Heat the mixture in a water bath at 35 °C for 30 min.
4. After 30 min, add 4.6 L of water and leave in a water bath at 98 °C for 15 min.
5. Dilute the reaction mixture to 14 L with water.
6. Add 3% hydrogen peroxide to the resulting solution until foaming no longer occurs.
7. Remove impurities by filtration.

Normally, graphene oxide causes clogging of filter, so it is very difficult to perform filtration. In addition, the reaction time is as short as 30 min. Therefore, there is a possibility that only the surface of graphite is oxidized by this original method.

16.4.4 Improved Hummers' Method

It has been shown that oxidation efficiency is improved by pre-oxidizing graphite, and is used for the synthesis of single-layer graphene oxide in a short time [30]. In this method, since pre-oxidation of graphite is carried out, sodium nitrate is unnecessary in the main oxidation procedure, there is an advantage that the formation of harmful gas (NOx) during the main oxidation can be suppressed. In addition, the amount of unexfoliated graphite portion decreased, and the yield of graphene oxide increased. Although it is not described in detail what kind of effects the graphite pre-oxidation has, it is anticipated that affinity of graphite, sulfuric acid and/or potassium permanganate, formation rate of graphite intercalation compound, etc., might be improved by the pre-oxidation.

In 2010, another improved method using phosphoric acid instead of sodium nitrate was reported [31]. With this method, generation of harmful gas such as NOx can also be suppressed. However, the amount of sulfuric acid and potassium permanganate used in this method is much larger than the original method, therefore, a simple comparison cannot be done.

Besides these, some papers say that by changing the graphite pretreatment method (mixing it with manganese oxide, irradiating with microwaves, and expanding), or changing additives (water, phosphoric acid, nitric acid, etc.), oxidation of graphite is enhanced. However, these are empirical results and many are not chemically explained. Therefore, comprehensive investigation for the preparation method of graphene oxide reported so far was performed to eliminate unnecessary operations and reagents, and clarify the minimum necessary condition. As a result, it was found that only the potassium permanganate and sulfuric acid are required for preparing graphene oxide from graphite. The reaction time was found to be sufficient for 2 h. This is not only empirical but also supported by chemical analysis of the oxidation reaction in situ. That is, the process of disordering the laminated structure of graphite and the process of reducing manganese (oxidizing graphite), were traced by in situ X-ray diffraction analysis and X-ray absorption spectroscopy analysis. Through these analyses, the optimum conditions under which graphene oxide was formed could be determined. Additionally, the effect of reaction temperature was investigated. If the reaction temperature is too high, oxidation proceeds excessively to generate carbon dioxide, as a result, graphene oxide having many defects is formed. In contrast, the oxidation reaction hardly progresses at 10 °C or lower. As a result, it was found that graphene oxide with few carbon defects can be obtained by carrying out the reaction at around 35 °C.

By applying the optimized method, graphene oxide can be safely produced at a scale of 500 g or more even in the laboratory. In addition, since the ratio of reagents, concentration, temperature, and reaction time are determined, continuous flow process can be achieved [32].

16.5 Mass Production of Graphene Oxide

Mass production of pure graphene by scotch tape exfoliation and CVD on some substrate are quite difficult, therefore, scientists have focused on the production of graphene-like materials in liquid phase or powder state.

Ball milling in the presence of solvent [33] and under CO_2 atmosphere [34], microwave irradiation [35], and mixer treatment [36] have been investigated. Here, the production method of graphene oxide, which has a great potential for real application [37], is introduced. Previously, mass production of graphene oxide is reported, however, the experiment was performed in only 15 g scale [38].

16.5.1 Oxidation Process

For the oxidation of graphite, a strong oxidizing agent is required. Since the reaction uses sulfuric acid as a solvent, there is little worry of fire, but there is a possibility of explosion. For example, Mn_2O_7 generated by the reaction of $KMnO_4$ and sulfuric acid may explode at higher than 55 °C. Furthermore, Mn_2O_7 is volatile and gradually vaporizes as a purple gas even at room temperature. Of course, these gases are toxic and must be carried out under suitable exhaust facilities. The oxidation reaction of graphite is an exothermic reaction, and it is dangerous if the overreaction occurs. Therefore, the reaction must be carried out at low temperature. Graphite does not undergo any reaction without an oxidizing agent, on the other hand, graphene oxide is reactive due to the presence of oxygen functional groups; even sulfuric acid, a solvent for the reaction, can react with graphene oxide. Further, there is a possibility that the remaining Mn_2O_7 reacts with the generated graphene oxide during the reaction, and an excessive oxidation reaction may proceed. For example, diols may be dehydrated by concentrated sulfuric acid to form epoxides, which may rearrange to ketones. Also, diols may undergo direct oxidation to diketones. Furthermore, carbon–carbon bonds of hydroxyl ketones and diketones can be oxidatively cleaved (Fig. 16.6). Therefore, it is difficult to obtain exactly the

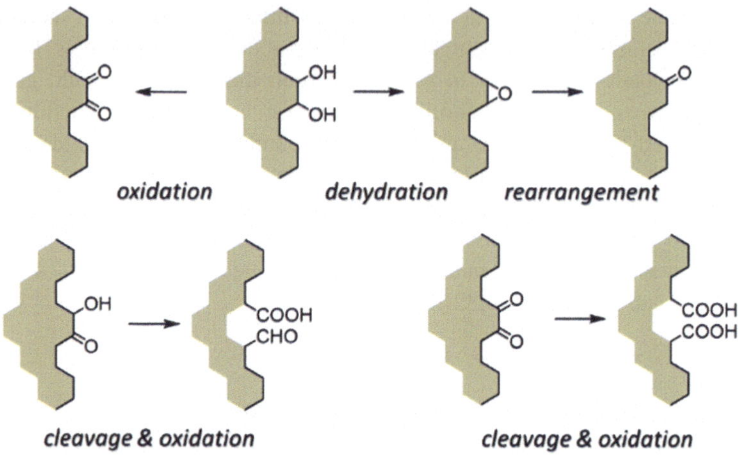

Fig. 16.6 Possible side reactions on graphene oxide

same graphene oxide with high reproducibility unless the amount of oxidizing agent, reaction temperature, reaction time, amount of water present, etc., are properly controlled.

Also, the kind of graphite greatly affects the physical properties of graphene oxide. Small graphite produces small graphene oxide, but large graphite does not always produce large graphene oxide. This is because the crystallinity of graphite is also an important factor in controlling the properties of graphene oxide. Even if the particle size is large, graphite having low crystallinity and a large amount of amorphous component tends to generate small graphene oxide.

16.5.2 *Quenching*

The oxidation reaction stops when water is added. When water is added, the reduced manganese species elutes from the interlayer of graphene oxide into the liquid, and causes a homogenization reaction with remaining high valence manganese species. Therefore, oxidation must be finished before adding water. A method of water addition is also important. Since the reaction mixture contains concentrated sulfuric acid, the temperature of the mixture is increased when water is added. When water is added to the mixture in a short time, it boils up; if a large amount of unreacted Mn_2O_7 remains, it is extremely dangerous. Therefore, it is necessary to add water while confirming the temperature of the reaction mixture. Since sulfuric acid has a specific gravity larger than water (1.8 g/cm^3), water will float to the upper layer and heat generation may occur suddenly when mixed, therefore, great care should be taken for this step. In contrast to adding water into the reaction solution, there is also a method of pouring the reaction mixture into water. This method is easier to control the temperature, and it does not cause the local temperature to increase. However, there is a problem that two reactors are required, and there is a danger of pouring concentrated sulfuric acid reaction mixture into another reactor.

16.5.3 Purification

It is necessary to remove sulfuric acid and manganese salts from the crude reaction solution. Graphene oxide can be precipitated by centrifugation, therefore, it is often purified by repeated centrifugation. Purification efficiency depends on the rejection rate of the solution. For example, if centrifugation is performed under the condition that the rejection rate is 80%, 20% of impurities (sulfuric acid and manganese) remain. In order to reduce the amount of impurities to the 0.01% or less of the original, it is necessary to repeat centrifugation about five times ($0.2 \times 0.2 \times 0.2$ 0.2×0.2). Because of this, a lot of waste liquid comes out. Actually, graphene oxide swells as impurities are removed, so the rejection rate gradually decreases, and the number of centrifugation and the amount of wastewater increase is estimated.

Therefore, attention is focused on filtering for purification, but graphene oxide swells and gelation occurs as purification progresses as described above, and clogging of the filter occurs. To solve this problem, a method of washing with hydrochloric acid and acetone has been proposed. Since hydrochloric acid is effective for removing a metal salt and is a volatile acid, it can be removed by drying even if a small amount of hydrochloric acid remains. Also, acetone has the effect of agglomerating graphene oxide and prevents gelation. Since acetone is also volatile, it can be removed by drying even if a small amount remains. It is also possible to add another solvent to graphene oxide after washing with acetone, and redisperse it in the solvent. By purifying in this manner, the amount of metal in graphene oxide can be reduced to 0.1 wt% or less. To further reduce the amount of impurities, it is necessary to redisperse and perform centrifugation. When the degree of oxidation is very high, or when graphene oxide having a very small size is synthesized, precipitation does not occur as the centrifugal separation is repeated. For such highly dispersible graphene oxide, sulfuric acid and manganese salt may be removed by dialysis.

16.5.4 Exfoliation and Storage

Single-layer graphene oxide is definitely not obtained only by the above operation, and it is necessary to exfoliate by sonication after purification. The concentration of graphene oxide purified by centrifugation is around 3 wt%. To obtain a concentration higher than 3 wt%, evaporation of water is necessary. However, if 3 wt% of graphene oxide is subjected to ultrasonic wave, gelation occurs as exfoliation proceeds. Single-layered graphene oxide cannot be obtained at a concentration higher than 1.5 wt%.

In addition, when graphene oxide is dried as it is by heating or under reduced pressure, the sheets are strongly bonded to each other and become like paper, making it difficult to redisperse. Although irradiation with ultrasonic waves can redisperse to a certain extent, it takes a long time, and the structure of graphene oxide may change during the process. Therefore, it is desirable to dry it while maintaining a high surface area state such as by freeze-drying, spray drying, and supercritical drying. As a result of such drying, sponge-like graphene oxide powder can be obtained, which is readily redispersed.

Graphene oxide is reduced not only by physical stimuli such as light and heat but also by organic molecules and metals. Therefore, it is desirable to store in cool and dark place. If you place the dispersion in a transparent bottle in the room, agglomeration by reduction will occur in several days. Also, since the dried graphene oxide powder has hygroscopic property, it is desirable to keep it tightly closed.

16.6 Conclusion and Future Remarks

Developing a method for making graphene on a practical scale is very attractive in academia and industry. For this purpose, it is desirable to prepare graphene from graphite via a chemical method, because graphite is low cost and easy to obtain. With regard to graphene oxide, it is technically possible to prepare on a large scale, but the current problem is improvement of processes such as purifying and securing of safety, therefore, the research stage is now shifting from academia to industry.

Unfortunately, it is impossible to obtain graphene with high crystallinity, which has been studied in the field of physics by the chemical methods. Therefore, the level of physical properties for graphene obtained by the chemical methods should be somewhat lower than ideal graphene. More unfortunately, miniaturized graphite is sometimes published in scientific papers and commercialized as graphene. For the future study of graphene and its practical application, a balance between compromise and deterrence seems to be necessary.

References

1. Billaud D, El Haouari A, Gerardin R (1989) Synthesis and characterization of ternary $FeCl_4^-CH_3NO_2$-graphite intercalation compounds. Synth Met 29:241–246
2. Billaud D, El Haouari A (1989) Electrointercalation of tetrahedral $AlCl_4^-$ and $GaCl_4^-$ anions into graphite. Synth Met 34:79–83
3. Shioyama H, Crespin M, Seron A et al (1993) Electrochemical oxidation of graphite in an aqueous medium: intercalation of $FeCl_4^-$. Carbon 31(1):223–226

4. Su CY, Lu AY, Xu YP et al (2011) High-quality thin graphene films from fast electrochemical exfoliation. ACS Nano 5(3):2332–2339. https://doi.org/10.1021/nn200025p
5. Parvez K, Li RJ, Puniredd SR et al (2013) Electrochemically exfoliated graphene as solution-processable, highly conductive electrodes for organic electronics. ACS Nano 7 (4):3598–3606. https://doi.org/10.1021/nn400576v
6. Parvez K, Wu Z, Li R et al (2014) Exfoliation of graphite into graphene in aqueous solutions of inorganic salts. J Am Chem Soc 136(16):6083–6091. https://doi.org/10.1021/ja5017156
7. Liu JL, Yang HP, Zhen SG et al (2013) A green approach to the synthesis of high-quality graphene oxide flakes via electrochemical exfoliation of pencil core. RSC Adv 3:11745–11750. https://doi.org/10.1039/C3RA41366G
8. Liu N, Luo F, Wu HX et al (2008) One-step ionic-liquid-assisted electrochemical synthesis of ionic-liquid-functionalized graphene sheets directly from graphite. Adv Funct Mater 18 (10):1518–1525. https://doi.org/10.1002/adfm.200700797
9. Lu J, Yang JX, Wang JZ et al (2009) One-pot synthesis of fluorescent carbon nanoribbons, nanoparticles, and graphene by the exfoliation of graphite in ionic liquids. ACS Nano 3 (8):2367–2375. https://doi.org/10.1021/nn900546b
10. Ciesielski A, Samori P (2014) Graphene *via* sonication assisted liquid-phase exfoliation. Chem Soc Rev 43:381–398. https://doi.org/10.1039/C3CS60217F
11. Haar S, Bruna M, Lian JX et al (2016) Liquid-phase exfoliation of graphite into single- and few-layer graphene with α-functionalized alkanes. J Phys Chem Lett 7(14):2714–2721. https://doi.org/10.1021/acs.jpclett.6b01260
12. Eigler S, Hirsch A (2014) Chemistry with graphene and graphene oxide—challenges for synthetic chemists. Angew Chem Int Ed 53(30):7720–7738. https://doi.org/10.1002/anie.201402780
13. Feng L, Yang X, Shi X et al (2013) Polyethylene glycol and polyethylenimine dual-functionalized nano-graphene oxide for photothermally enhanced gene delivery. Small 9(11):1989–1997. https://doi.org/10.1002/smll.201202538
14. Robinson JT, Tabakman SM, Liang Y et al (2011) Ultrasmall reduced graphene oxide with high near-infrared absorbance for photothermal therapy. J Am Chem Soc 133(17):6825–6831. https://doi.org/10.1021/ja2010175
15. Tian B, Wang C, Zhang S et al (2011) Photothermally enhanced photodynamic therapy delivered by nano-graphene oxide. ACS Nano 5(9):7000–7009. https://doi.org/10.1021/nn201560b
16. Mei KC, Rubio N, Costa PM et al (2015) Synthesis of double-clickable functionalised graphene oxide for biological applications. Chem Commun 51:14981–14984. https://doi.org/10.1039/C5CC05412E
17. Qi B-P, Hu H, Bao L et al (2015) An efficient edge-functionalization method to tune the photoluminescence of graphene quantum dots. Nanoscale 7:5969–5973. https://doi.org/10.1039/C5NR00842E
18. Strom AT, Dillon EP, Hamilton CE et al (2010) Nitrene addition to exfoliated graphene: a one-step route to highly functionalized graphene. Chem Commun 46:4097–4099. https://doi.org/10.1039/C001488E
19. Quintana M, Spyrou K, Grzelczak M et al (2010) Functionalization of graphene *via* 1,3-dipolar cycloaddition. ACS Nano 4(6):3527–3533. https://doi.org/10.1021/nn100883p
20. Li J, Li M, Zhou L-L et al (2016) Click and patterned functionalization of graphene by Diels-Alder reaction. J Am Chem Soc 138:7448–7451. https://doi.org/10.1021/jacs.6b02209
21. Zhong X, Jin J, Li S et al (2010) Aryne cycloaddition: highly efficient chemical modification of graphene. Chem Commun 46:7340–7342. https://doi.org/10.1039/C0CC02389B
22. Georgakilas V, Bourlinos AB, Zboril R et al (2010) Organic functionalisation of graphenes. Chem Commun 46:1766–1768. https://doi.org/10.1039/B922081J
23. Choi EK, Jeon IY, Bae SY et al (2010) High-yield exfoliation of three-dimensional graphite into two-dimensional graphene-like sheets. Chem Commun 46:6320–6322. https://doi.org/10.1039/C0CC00753F

24. Jeon IY, Yu D, Bae SY et al (2011) Formation of large-area nitrogen-doped graphene film prepared from simple solution casting of edge-selectively functionalized graphite and its electrocatalytic activity. Chem Mater 23:3987–3992. https://doi.org/10.1021/cm201542m
25. Yuan C, Chen W, Yan L (2012) Amino-grafted graphene as a stable and metal-free solid basic catalyst. J Mater Chem 22:7456–7460. https://doi.org/10.1039/C2JM30442B
26. Li Z, Wang R, Young RJ et al (2013) Control of the functionality of graphene oxide for its application in epoxy nanocomposites. Polymer 54:6437–6446
27. Brodie BC (1859) XIII. On the atomic weight of graphite. Philos Trans R Soc Lond 149:248–259. https://doi.org/10.1098/rstl.1859.0013
28. Staudenmaier L (1898) Verfahren zur Darstellung der Graphitsäure. Ber Dtsch Chem Ges 31:1481–1487. https://doi.org/10.1002/cber.18980310237
29. Hummers WS Jr, Offeman RE (1958) Preparation of graphitic oxide. J Am Chem Soc 80 (6):1339–1339. https://doi.org/10.1021/ja01539a017
30. Kovtyukhova NI, Ollivier PJ, Martin BR et al (1999) Layer-by-layer assembly of ultrathin composite films from micron-sized graphite oxide sheets and polycations. Chem Mater 11 (3):771–778. https://doi.org/10.1021/cm981085u
31. Marcano DC, Kosynkin DV, Berlin JM et al (2010) Improved synthesis of graphene oxide. ACS Nano 4(8):4806–4814. https://doi.org/10.1021/nn1006368
32. Morimoto N, Suzuki H, Takeuchi Y et al (2017) Real-time, in situ monitoring of the oxidation of graphite: lessons learned. Chem Mater 29(5):2150–2156. https://doi.org/10.1021/acs.chemmater.6b04807
33. Lin T, Chen J, Bi H et al (2013) Facile and economical exfoliation of graphite for mass production of high-quality graphene sheets. J Mater Chem A 1:500–504
34. Jeon IY, Choi HJ, Jung SM et al (2013) Large-scale production of edge-selectively functionalized graphene nanoplatelets via ball milling and their use as metal-free electrocatalysts for oxygen reduction reaction. J Am Chem Soc 135(4):1386–1393. https://doi.org/10.1021/ja3091643
35. Shin Y, Lee J, Yang J et al (2014) Mass production of graphene quantum dots by one-pot synthesis directly from graphite in high yield. Small 10(5):866–870. https://doi.org/10.1002/smll.201302286
36. Hernandez Y, Nicolosi V, Lotya M et al (2008) High-yield production of graphene by liquid-phase exfoliation of graphite. Nat Nanotechnol 3:563–568. https://doi.org/10.1038/nnano.2008.215
37. Tung VC, Allen MJ, Yang Y et al (2009) High-throughput solution processing of large-scale graphene. Nat Nanotechnol 4:25–29. https://doi.org/10.1038/nnano.2008.329
38. Sun L, Fugetsu B (2013) Mass production of graphene oxide from expanded graphite. Mater Lett 109:207–210

Chapter 17
Alcoholic Compounds as an Efficient Energy Carrier

Takashi Fukushima, Sho Kitano, Masaaki Sadakiyo and Miho Yamauchi

17.1 Electrochemical Reduction of Oxalic Acid to Glycolic Acid

17.1.1 Introduction

The high impact of increase of CO_2 concentration in the atmosphere on climate change has been seriously recognized. The intergovernmental panel on climate change attributed climate change to a large increase in atmospheric CO_2 concentration over the past 40 years, of which 78% was caused by fossil fuel combustion and industrial processes [1]. Therefore, the efficient utilization of renewable energies is indispensable for the sustainable development of our human society [2]. Recently, the efficiency of electric power generation using renewable energies such as solar photovoltaic, wind turbine and wave has been greatly improved. However, such electricity, i.e. renewable electricity, is not widely used because of its supply instability. Energy carriers [3], which are high-energy chemicals produced via hydrogenation or electroreduction using low-carbon technologies, have recently attracted substantial attention as storage and transportation media for renewable electricity. Electrochemically produced hydrogen is believed to be the cleanest

T. Fukushima · S. Kitano · M. Sadakiyo · M. Yamauchi (✉)
International Institute for Carbon-Neutral Energy Research, Kyushu University, Motooka 744, Nishi-Ku, Fukuoka 819-0395, Japan
e-mail: yamauchi@i2cner.kyushu-u.ac.jp

T. Fukushima
e-mail: fukushima@i2cner.kyushu-u.ac.jp

S. Kitano
e-mail: skitano@i2cner.kyushu-u.ac.jp

M. Sadakiyo
e-mail: sadakiyo@i2cner.kyushu-u.ac.jp

N. Nakashima (ed.), *Nanocarbons for Energy Conversion: Supramolecular Approaches*, Nanostructure Science and Technology,
https://doi.org/10.1007/978-3-319-92917-0_17

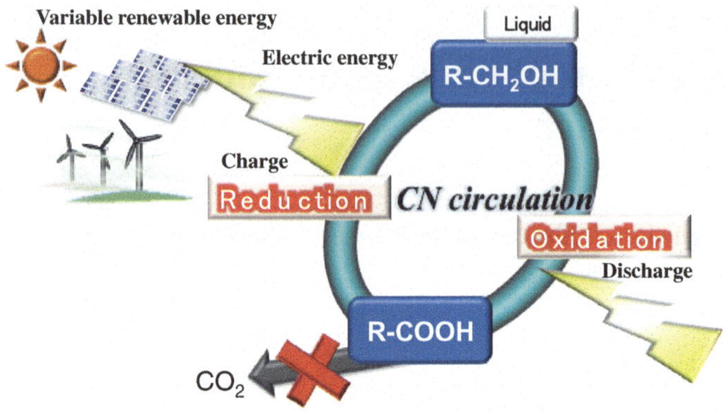

Fig. 17.1 Schematic image of carbon-neutral power circulations based on alcohol/acid redox reactions

energy carrier due to zero emission of greenhouse gases both in its production and combustion for the energy generation. However, fatal drawbacks of utilization of H_2 as an energy carrier still remain. Namely, gaseous and chemically active H_2 fold an infeasibility of widespread distribution due to its low volumetric energy density (13 MJ m^{-3}) compared to conventional liquid fossil fuel such as gasoline (33,400 MJ m^{-3}), which makes difficulties in long-time storage and long-distance transport. H_2 transport using a cryogenic liquid hydrogen tanker or a massive hydrogen pipeline system is expected to be a possible solution but requests a high cost and extra consumption of energy. On the other hand, various H_2 carriers such as organic hydrides [4], NH_3 [5], amides [6] and formate converted from CO_2 [7] have been proposed as an H_2 carrier, and some have achieved the efficiencies demanded for practical use. In this regard, liquid energy carriers offer great merits by considering feasible manageabilities. Alcohols have recently emerged as a candidate of an energy carrier because of their low explodability and low corrosiveness, in addition to the high energy density in a liquid or solution. Thus, we advocate utilization of alcoholic compounds as an energy carrier for renewable electricity and power generation from an alcohol using a fuel cell. Recently, we have proposed an idea for power generation from an alcoholic compound without CO_2 emission via highly selective oxidation of the alcohol to carboxylic acid whereas complete oxidation of an alcohol generates CO_2. Considering that CO_2 capture from the air involves considerable cost, carboxylic acids, which can be collected as a solution, is a feasible waist to reuse. If a carboxylic acid is electrochemically reduced into an alcohol, energy carriers are recyclable. Then, we focused on the carbon-neutral (CN) energy circulation via selective redox reactions between an alcohol and carboxylic acid as illustrated in Fig. 17.1 [8].

17.1.2 Glycolic Acid as an Energy Carrier

We take notice of glycolic acid (**GC**, $HOCH_2–CO_2H$) as an energy carrier. **GC** is a monovalent alcohol having 8,600 MJ m^{-3} of volumetric energy capacity in the case of selective oxidation to oxalic acid (**OX**, $(CO_2H)_2$), a divalent carboxylic acid, which is formed via four-electron oxidation of **GC** in the following equation:

$$HCOO–CH_2OH + H_2O \rightarrow HCOO–COOH + 4H^+ + 4e^- \qquad (17.1)$$

Through the reverse reaction of Eq. (17.1), we can reproduce **GC**. It should be noted that **OX** can be produced from biomass-derived resources, e.g. cellulose [9]. **GC** is thus considered to be a promising environmentally friendly carrier, which is produced from bio-derived materials and water by the help of renewable electricity. In this process, water is used as a hydrogen source in the electroreduction. Considering that industrially utilized hydrogen is produced via fuel reforming, which emits CO_2, electroreduction of carboxylic acid is significantly clean process for the production of alcoholic compounds. Creation of the highly selective electro-catalysts is a key to realizing the CN system. However, limited knowledge has been accumulated about both the selective the electroreduction of carboxylic acids into alcohols and selective electro-oxidation of alcohols to date.

17.1.3 TiO_2 Catalyst to Electrochemically Reduce Oxalic Acid to Glycolic Acid

Carboxylic acids are stable chemicals due to electron conjugation and low chemical activity assignable to low electrophilicity of carboxyl carbon [10]. Usually, hydrogenation of carboxylic acid using molecular hydrogen to produce an alcohol is achieved under relatively severe conditions (240–380 °C, 21 bar of hydrogen atmosphere) [11, 12]. Limited numbers of results concerning electrocatalysts for the reduction of carboxylic acids have been reported so far. Then, we tested various kinds of materials, such as Al, Ti, V, Ni, Cu, Zr, Nb, Mo, Sn, Pb and Pt, for **OX** electroreduction and found that only Ti foil calcined in the air exhibits the electrochemical **OX** reduction ability. However, the main product was glyoxylic acid (GO) that is a two-electron reduction aldehyde compound and a small amount of **GC** was formed [8, 13]. We noticed that the catalytic activity was enhanced by the calcination of Ti foil, which indicated that TiO_2 acts as an active site for **OX** reduction. To increase the catalytic activity, we tried to apply TiO_2 particles with a high specific surface area, namely porous TiO_2 spheres (PTSs), for **OX** reduction. PTSs were prepared from layered protonated titanate (LPT) according to previous report [14]. The preparation procedure and transmission electron microscopy images for LPT and PTS are provided in Fig. 17.2.

PTSs prepared at calcination temperatures greater than 500 °C were mainly composed of a highly crystalline anatase phase (>99 wt%) and characterized with a

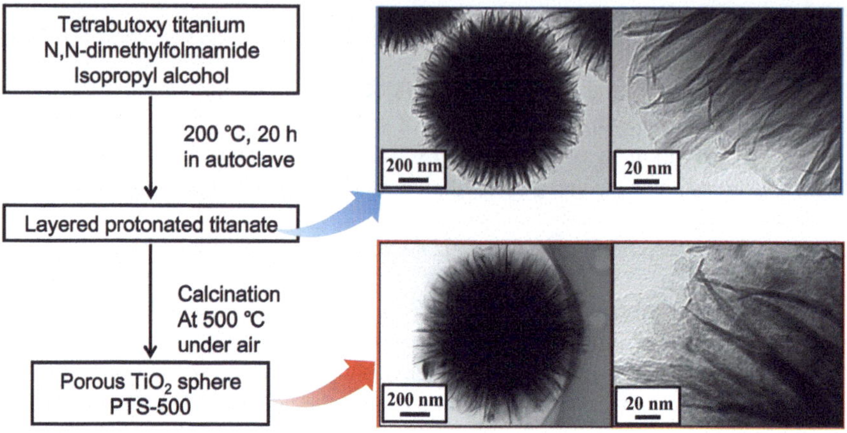

Fig. 17.2 TEM images for layered protonated titanate and porous TiO_2 sphere prepared by calcination at 500 °C

large surface area (43.2–82.1 m^2/g). The catalytic activity of PTSs for **OX** electroreduction was examined by chronoamperometry (CA). Electrodes were prepared by drying a methanol suspension of a PTS dropped onto calcined Ti foil, followed by calcination at various temperatures. The product yield and Faradaic efficiency of the PTS/Ti foil electrodes increased with increase of the calcination temperature up to 500 °C. However, the product yield and Faradaic efficiency on the electrode treated above 500 °C sharply deteriorated, e.g. PTS prepared via calcination at 600 °C (PTS-600) showed a minimal product yield (0.5%), although the PTS structures were proved to be nearly unchanged based on powder X-ray diffraction results. Active and inactive PTSs, i.e. PTS-500 and PTS-600, were characterized using X-ray photoemission spectroscopy, N_2 gas adsorption and classical **OX** adsorption methods. However, a significant difference between these catalysts was not confirmed. UV-vis spectra gave us a hint to answer the question. Absorption edge for PTS-600 was shifted to the longer wavelength side, indicating that PTS-600 contains TiO_2 phases with a narrower band gap, such as a rutile phase, compared to the gap of anatase TiO_2. Anatase- and rutile-type TiO_2 were reported to exhibit characteristic electron energy-loss spectral (EELS) charts. We produced phase distribution maps of the anatase and rutile types on PTS-500 and -600 via the spectral structure recognition of each phase, as shown in Fig. 17.3. The EELS mapping represents that PTS-500 is made of anatase-type TiO_2 (Fig. 17.3a, b), whereas PTS-600 includes a rutile-type TiO_2, especially around the surface (Fig. 17.3c, d). To confirm our assumption, we examined the catalytic activity of commercial rutile nanoparticles and observed that both product yield and Faradaic efficiency on rutile nanoparticles were considerably low compared to those on PTS-500. Thus, we concluded that the active catalytic site is anatase TiO_2 surface.

Why does anatase TiO_2 serve as a highly efficient and selective electrocatalyst to convert **OX** into **GC**? The conduction-band bottom level of anatase-type TiO_2 is

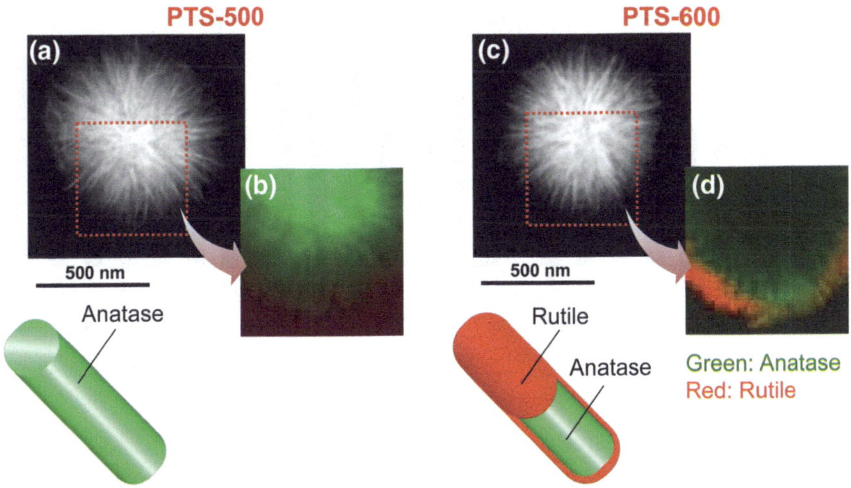

Fig. 17.3 Scanning transmission electron microscopy (STEM) images for PTS-500 (**a**) and PTS (**c**), and electron energy-loss spectral (EELS) maps for PTS-500 (**b**) and PTS-600 (**d**)

known to lie at a position 0.2 V more negative than that of rutile-TiO_2. Electrons provided by the electrode are possibly accumulated in the conduction-band bottom of TiO_2. Considering that electrons located at a higher energy level have a lower work function, electrons introduced in conduction-band bottom of anatase TiO_2 can have greater reducibility compared to those in rutile TiO_2. Therefore, the high catalytic ability of anatase TiO_2 is attributed to its energy state.

Catalytic conditions such as temperature and **OX** concentration over PTS-500 were optimized. Catalytic performances of PTS-500 at various temperatures are shown in Fig. 17.4. PTS-500 exhibited relatively high Faradaic efficiency (70–95%) under mild conditions, e.g. at 50 °C and in the potential range from −0.5 to −0.7 V versus RHE. Noted that the aqueous solution used was highly acidic (pH = 2.1). Generally speaking, in aqueous media, H_2 evolution is a major reaction. Surprisingly, hydrogen evolution is suppressed on PTS-500, which is possibly attributable to the conversion of Ti^{4+} sites to Ti^{3+} sites under application of electric field. The Ti^{3+} sites are stable under acidic condition and are favourable for the adsorption of a hydrophilic carboxyl group analogously to the behaviour of the photogenerated hydrophilicity of TiO_2 [15], leading to the high Faradaic efficiency for **OX** reduction, i.e. suppression of hydrogen evolution.

17.1.4 Photo-assisted Hydrogenation of Oxalic Acid

Photoelectrochemical water splitting using a photoelectrode, such as oxide [16–21], (oxy)nitride [22–25] and sulphide [26], is expected as a green process to produce hydrogen from water using renewable solar energy and provide an effective

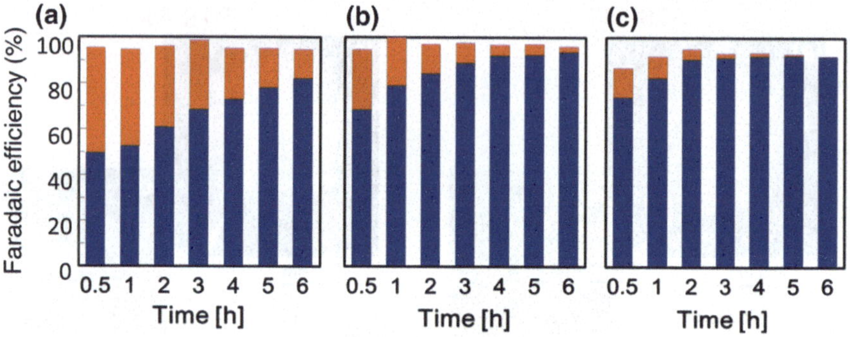

Fig. 17.4 Faradaic efficiency for the production of GO (orange) and **GC** (blue) at −0.7 V versus RHE on PTS-500 in **OX** electroreduction at **a** 24, **b** 40 and **c** 50 °C

approach for energy conversion and storage. The utilization of the hydrogen photocatalytically generated from water is expected to accelerate electrochemical reactions for **OX** reduction due to photovoltaic effects. Therefore, the electrochemical reduction of **OX** with the assistance of light energy achieves the highly efficient **GC** production acid without fossil-derived hydrogen, i.e. a light-energy-driven alcohol synthesis, as shown in Fig. 17.5. This process is an excellent environmentally friendly alcohol production route since it progresses under milder conditions (0.1 MPa, <100 °C) than those of hydrogenation reactions. The equations proceeding in each electrode are shown below:

$$\text{Cathode: } (COOH)_2 + 4\,H^+ + 4\,e^- \rightarrow HOOC - CH_2OH + H_2O \quad +0.13\text{ V versus RHE} \tag{17.2}$$

$$\text{Anode: } 2H_2O \rightarrow 4\,H^+ + 4\,e^- + O_2 \quad +1.23\text{ V versus RHE} \tag{17.3}$$

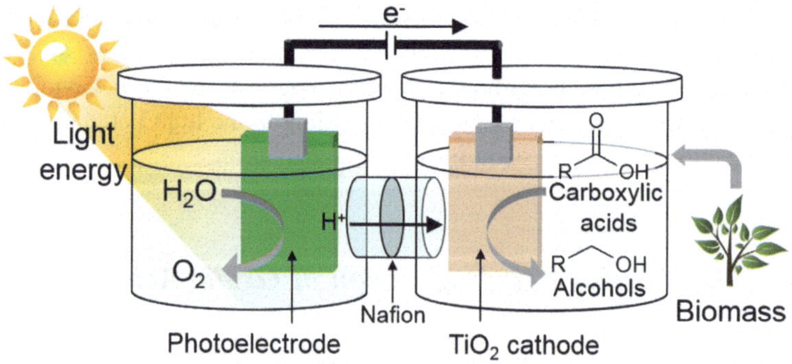

Fig. 17.5 Fabricated light-assisted alcohol electrosynthesis system from carboxylic acid and water

$$\text{Overall: } (COOH)_2 + H_2O \rightarrow HOOC{-}CH_2OH + O_2 \quad 1.1\ V \qquad (17.4)$$

A light-assisted electrochemical alcohol production system was constructed by applying oxide semiconductor photoelectrodes as an anode for water oxidation to provide protons and electrons for the hydrogenation of a carboxylic acid on a cathode, which enables the direct light energy conversion into low-carbon alcoholic chemicals. We applied gallium oxide (Ga_2O_3) [27, 28], bismuth vanadate ($BiVO_4$) [29] and tungsten trioxide (WO_3) [18, 30] electrodes, which are known as highly active photocatalysts for water oxidation, to oxidize water and to supply electrons and protons for the electrochemical reduction of **OX** on an anatase-type TiO_2 cathode [31].

17.1.5 Performance of Oxide Photoelectrodes for Applied Bias Decrease

Commercial powders of Ga_2O_3 and WO_3 were applied in the electrochemical experiments. The $BiVO_4$ powders were synthesized by using the homogeneous precipitation method [32]. Two-electrode system was employed using two-compartment glass cells, where the cathode and anode were separately fixed in each cell. Figure 17.6 shows the current–voltage curves for the electroreduction of **OX** using Ga_2O_3, $BiVO_4$ and WO_3 photoelectrodes in the dark and under irradiation of UV–visible light ($\lambda > 200$ nm). Current for the reaction using all three anodes was observed by applying external bias above 2.1 V in the dark. The results indicated that extra 1.0 V bias in addition to 1.1 V of theoretical voltage which is required to begin the **GC** production through **OX** reduction and water oxidation was needed to overcome the overpotential without light irradiation. Under light irradiation, minimum biases applied to flow reaction current in the system were significantly decreased compared with those observed in the dark, indicating that UV–visible light energy absorbed over photoelectrodes enhanced to apply external bias.

The onset potential for the reaction current flow under irradiation of UV–visible light depended on the photoelectrodes; the order of onset bias was as follows: $WO_3 < BiVO_4 < Ga_2O_3$. This order corresponded to the photovoltaic performance of photoelectrodes, i.e. $WO_3 > BiVO_4 > Ga_2O_3$. An observed decrease of the onset biases for **OX** reduction is associated with the photovoltaic performance, which depends on the amount of photoexcited electrons and holes and recombination probabilities between such excited species, and coincides with the activity of photoelectrodes for oxidation of water. Photoexcitation probabilities for the formation of active species strongly depend on the photoabsorption spectrum range for the photoelectrodes [33]. A wider absorption range of the photoelectrodes is expected to enable more efficient photoexcitation, resulting in formation of a larger amount of excited electrons and holes. All excited active species are not consumed by reactions since the recombination between some parts of excited electrons and

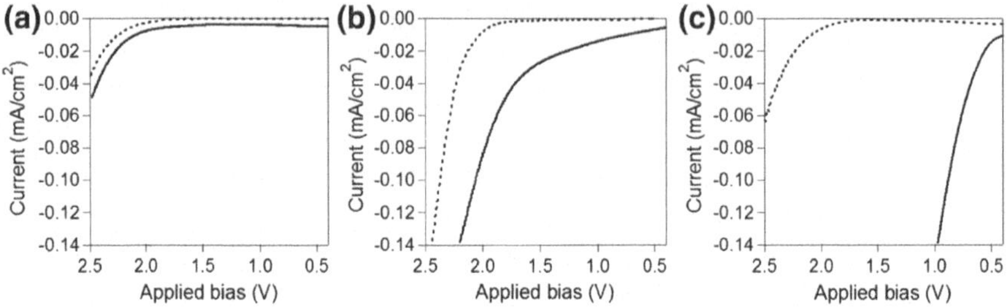

Fig. 17.6 Current–voltage curves for electroreduction of **OX** using a two-electrode system employing a TiO_2 cathode and **a** Ga_2O_3, **b** $BiVO_4$ and **c** WO_3 photoelectrodes under UV–visible light ($\lambda > 200$ nm) irradiation (solid line) or in the dark (broken line)

holes occurs before the water oxidation proceeds [34]. Lower recombination probability achieves higher reaction probability of holes and water molecules, leading to higher activity of water oxidation and a larger amount of available excited electrons, i.e. highly active photovoltaic performance [35]. Therefore, the photoelectrode, which shows a wide photoabsorption property and low recombination probability, exhibits large bias decrease ability due to the high activity for water oxidation and photovoltaic performance.

The Ga_2O_3 electrode showed the lowest bias decrease performance under light irradiation (0.15 V, as shown in Fig. 17.6) since the narrowest range of photoabsorption spectrum of the Ga_2O_3 photoanode up to 270 nm corresponds to the smallest amount of excited electrons and holes as illustrated in Fig. 17.7, thus, resulting in the low photovoltaic performance. Although the $BiVO_4$ electrode exhibited the widest absorption spectrum up to 550 nm among all the photoelectrodes, **OX** reduction required a relatively high bias application, i.e. 1.7 V, compared with 0.7 V, which was shown for the WO_3 electrode. It is known that the $BiVO_4$ photoelectrode shows a high recombination probability due to its poor mobility for excited electrons (Fig. 17.7) [36–39]. This result clearly indicates that the recombination probability is more dominant for water oxidation over the $BiVO_4$ anode. By contrast, the WO_3 photoanode, absorbing photons at wavelengths less than 480 nm, showed the highest bias decrease ability among all the anodes, as shown in Fig. 17.7. Therefore, the largest decrease of applied bias for electrochemical reduction of **OX** observed on the WO_3 photoelectrode was attributed to the relatively wide absorption spectrum and low recombination probability for efficient water oxidation and high photovoltaic performance [18].

17.1.6 Effect of Irradiated Light Wavelength on the WO_3 Photoanode Performance

Effects of the wavelength of irradiation light on the WO_3 electrode were investigated by employing a two-electrode system. Figure 17.8 shows the current–voltage

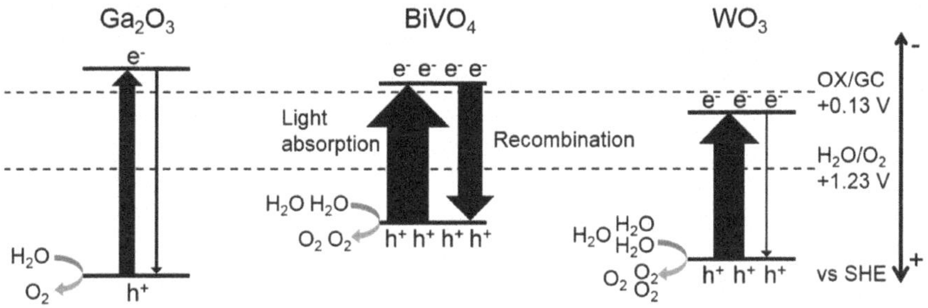

Fig. 17.7 Schematic illustrations for energy diagrams and photovoltaic performances on the photoelectrodes under irradiation of UV–visible light

curves both in the dark and under light irradiation with various wavelengths (UV–visible light with $\lambda > 200$ and 300 nm, visible light with $\lambda > 400$ nm). Onsets of reaction current were observed at 0.5 V under light irradiation with all wavelengths, which are 0.6 V smaller than the theoretical bias required for **GC** production through **OX** reduction and water oxidation, 1.1 V, whereas the onset potentials were observed at 2.1 V in the dark, suggesting that 1.6 V of bias decrease was achieved by light irradiation. Light irradiation with $\lambda > 200$ and 300 nm covers a wide energy range of photoabsorption property of WO_3. Therefore, the rate determining step for electrochemical reduction of **OX** under irradiation of UV–visible light is not the water oxidation on the WO_3 anode, where holes with sufficient positive potentials are formed [40], but the electrochemical reduction of **OX** on the cathode, suggesting that the irradiation of UV–visible light with $\lambda > 300$ nm efficiently works in the system. Furthermore, almost the same onset potential of reaction current was observed under irradiation of visible light, indicating that the WO_3 photoanode efficiently work for the bias decrease under visible light irradiation. The system has applicability under both the irradiation of solar light and only visible light.

17.1.7 OX Reduction Under Various Conditions

Chronoamperometric **OX** reduction was conducted while applying an external bias of 1.0 or 1.5 V using a two-electrode system with the TiO_2 cathode and the WO_3 photoanode for 2 h under irradiation of UV–visible light with $\lambda > 300$ nm, visible light with $\lambda > 400$ nm or in the dark. Figure 17.9 shows the Faradaic efficiencies determined based on the amount of GO and **GC** formed under various conditions. No products were produced in the dark even for an external bias below 1.5 V, which is coincident with the result for no reductive current below 1.5 V in the current–voltage curve as shown in Fig. 17.8, indicating that the system does not work with applying bias below 1.5 V without irradiation. Meanwhile, GO and **GC** production was observed under irradiations of both UV–visible and visible light,

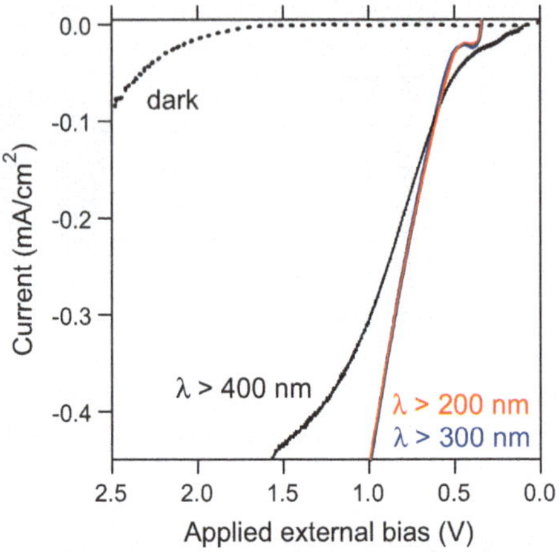

Fig. 17.8 Current–voltage curves of electrochemical reduction of **OX** using a two-electrode system comprising a TiO_2 cathode and a WO_3 photoelectrode under light irradiation, with wavelength >200 nm (red), >300 nm (blue) or >400 nm (black), or in the dark (broken line)

indicating that **OX** reduction was promoted with the support of light energy irradiated on the WO_3 photoelectrode through an electrochemical reaction system. Faradaic efficiencies for GO and **GC** production observed under the irradiation of UV–visible light were higher than those under the irradiation of visible light. The GO and **GC** formed with an 80% Faradaic efficiency in total with applying a bias of 1.5 V under UV–visible light irradiation. The anode potential when applying a bias of 1.5 V under UV–visible light irradiation was measured to be 1.05 V versus RHE, which is 0.18 V more negative than the theoretical potential of water oxidation, 1.23 V versus RHE, indicating that the photogenerated hole oxidized water molecules [41, 42].

Figure 17.10 shows the Faradaic efficiencies calculated based on the amount of GO, **GC** and gaseous products, i.e. H_2 that generated in the cathode cell in **OX** reduction and O_2 that generated in water oxidation in the anode cell when applying a bias of −1.5 V under UV–visible light irradiation. Faradaic efficiencies for products in each cell achieved 100%, indicating that all electrons formed from water oxidation were used for the electrochemical reduction of **OX** or water to produce GO, **GC** and H_2 and other reactions did not occur such as dissolution of electrodes.

In conclusion, **GC** production through electroreduction of **OX** and oxidation of water using a light-assisted electrochemical reaction system was successfully conducted. Irradiation of UV–visible light to the WO_3 photoanode enables a drastic decrease of minimal bias, i.e. 0.5 V, which is 0.6 V smaller than the theoretical bias potential, i.e. 1.1 V, required for **GC** production via **OX** reduction and water oxidation. The green synthetic process for the production of an alcoholic compound from an organic acid procurable from biomass via electro-oxidation of water with the assistance of light energy is achievable.

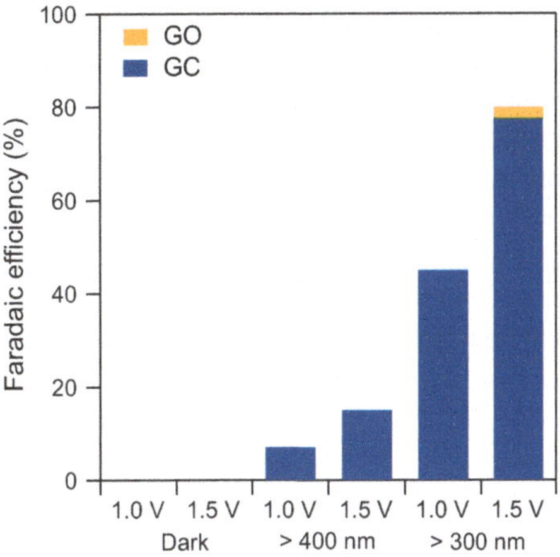

Fig. 17.9 Faradaic efficiencies calculated based on the amount of GO and **GC** formed in a two-electrode system comprising a TiO_2 cathode and a WO_3 photoelectrode for 2 h by applying an external bias of 1.0 or 1.5 V under irradiation of UV–visible light with $\lambda > 300$ nm, visible light with $\lambda > 400$ nm or in the dark

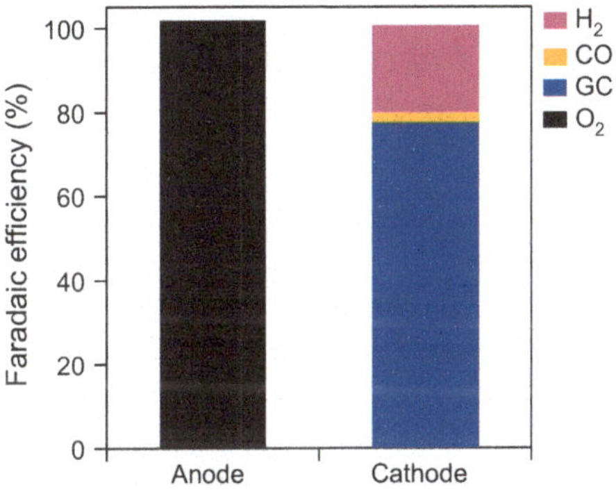

Fig. 17.10 Faradaic efficiencies calculated based on the amount of GO, **GC** and H_2 generated in the cathode cell and O_2 formed in the anode cell when applying a bias of 1.5 V under the irradiation of UV–visible light ($\lambda > 300$ nm)

17.2 Electrosynthesis Cell for Continuous Power Storage in Glycolic Acid

17.2.1 Introduction

Electrolyzers can store electric power into various chemicals, such as hydrogen and alcohols, through the conversion of these chemicals on the electrodes. An important role of the electrolyzers is a temporal storage of intermissive electric power generated from renewable energies, such as wind power and solar cell. Electrolyzers can be roughly classified into two types: The first is a batch-wise two-compartment cell (Fig. 17.11a) [43] and the second is a flow-type polymer electrolyte cell (Fig. 17.11b) [44]. Batch-wise two-compartment cell must include additional electrolytes in the reaction solution (e.g. KOH, Na_2SO_4, etc.), while the polymer electrolyte cell does not need it. In the case of hydrogen production through electrolysis of water, both types of electrolyzers are acceptable for continuous operation of the electrolysis because the gas product (hydrogen) is automatically separated from the reaction solution and the reactant (water) is easily refilled. In contrast, liquid or soluble solid chemicals, such as alcohols, can be continuously produced only by the flow-type polymer electrolyte cell because the products always remain in the reaction solution that includes electrolyte as impurity, and thus it should be separated by further experimental manipulation.

As described above, efficient electrosynthesis of alcohols from ubiquitous oxidized chemicals, i.e. carboxylic acids, will be one of the important issues for efficient power storage. **GC**, made from **OX**, can be one of the candidates as next-generation energy carriers because of its potentially high volumetric capacity (<417 Ah l^{-1} (at 60 °C)). In this chapter, we focus on continuous production of **GC** through an electrochemical reduction of **OX** solution using a novel polymer electrolyte electrolyzer, named as polymer electrolyte alcohols electrosynthesis cell (PEAEC) [45].

17.2.2 Fabrication of Membrane Electrode Assembly for PEAEC

A novel membrane electrode assembly (MEA) for the PEAEC was fabricated by employing the TiO_2 catalysts on cathode, as described above, due to its high over potential for hydrogen production which is the dominant side reaction of the electrochemical reduction of **OX** aqueous solution. The MEA should be composed of the cathode for **OX** reduction, proton-exchange membrane and anode for water oxidation. The cathode in the PEAEC is also required to have electron conductivity and substrate diffusivity. Therefore, Ti mesh or Ti felt, to which the reaction solution is permeable, is preferable for electrode material for the cathode. Since rigid linking between electrode and catalysts is also indispensable, direct growth of the porous TiO_2 on the electrode using Ti ions is one of the suitable methods to

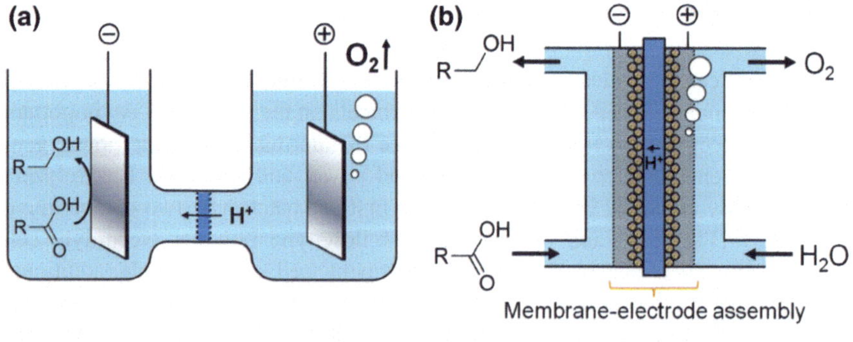

Fig. 17.11 **a** Batch-wise two-compartment cell and **b** flow-type polymer electrolyte cell for alcohol electrosynthesis

fabricate the cathode for the PEAEC (Fig. 17.12). Thus, the novel cathode was fabricated by a two-step hydrothermal reaction: the first step is heating of the Ti at 220 °C for various time (e.g. for 12 h) in aqueous NaOH solution to form $H_2Ti_2O_5 \cdot H_2O$ and the second step is to heat the $H_2Ti_2O_5 \cdot H_2O$-coated Ti at 200 °C for 24 h in water to form anatase TiO_2 grown on the Ti mesh (TiO_2/Ti-M) (Fig. 17.12) [46].

After the two-step hydrothermal reaction, fibrous TiO_2 was formed on the surface of Ti lines. From the results of XRD measurements with the sample prepared through different reaction times (Fig. 17.13), it was clear that the amount of the anatase TiO_2 catalysts could be controlled just by changing the reaction time of the first-step hydrothermal treatment.

The effect of the hydrothermal reaction time, i.e. amount of the anatase TiO_2, on catalytic performance for the electrochemical reduction of **OX** was clarified by CA using a batch-wise two-compartment electrolyzer with a three electrode system. Figure 17.14 shows **OX** conversion and Faradaic efficiency after the CA for 2 h at an applied potential of 0.76 V (vs. RHE) using each TiO_2/Ti-M electrode using 0.03 M **OX** solution with 0.2 M Na_2SO_4 (pH = 2.1). The conversion of **OX** increased with increasing the hydrothermal reaction time below 12 h, indicating that the amount of TiO_2 catalyst on the Ti line is important to accelerate the catalytic reduction of **OX** in this region. Faradaic efficiency for the target product (**GC**) also drastically increased with increasing the hydrothermal reaction time, suggesting that the metallic Ti surface, uncovered by TiO_2, preferably generates hydrogen gas as a by-product. Above 12 h, the conversion of **OX** and Faradaic efficiency are almost independent of the hydrothermal reaction time in the first step, indicating that the extra amount of TiO_2 on the electrode does not contribute to increasing catalytic performance. This might be because TiO_2 crystals located far from the metal Ti electrode do not work as electrocatalysts due to high electrical resistivity. Note that the excess amount of TiO_2 (e.g. 24 h of the first step) is not preferable for fabrication of MEA because it causes a peel-off of the cathode electrode from Nafion after a hot press, meaning that 12 h is an optimal

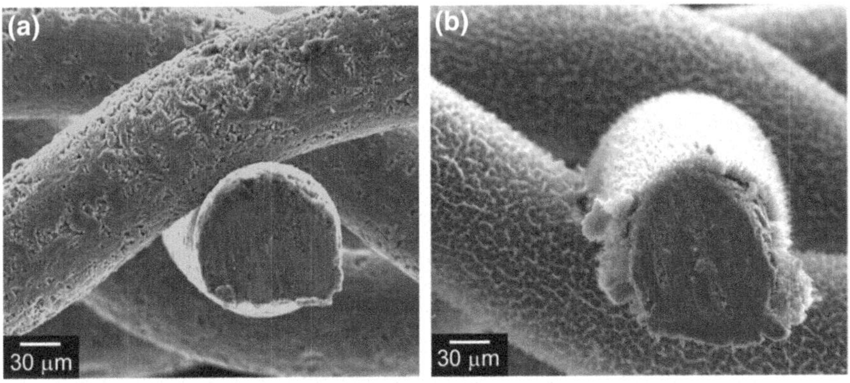

Fig. 17.12 **a** SEM image of the cross section of the Ti mesh electrode and **b** TiO_2/Ti-M electrode after the two-step hydrothermal reaction

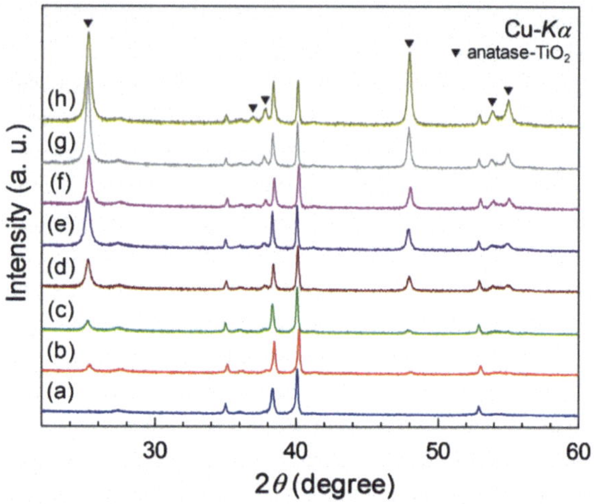

Fig. 17.13 XRD patterns of **a** Ti mesh and TiO_2/Ti-M with the first-step hydrothermal reaction time for **b** 3, **c** 6, **d** 12, **e** 18, **f** 24, **g** 48 and **h** 72 h (second step is fixed to 24 h)

hydrothermal reaction time in the first step for fabrication of the TiO_2/Ti-M electrode for the PEAEC.

Optimum reaction temperature for the catalysis with the prepared TiO_2/Ti-M electrode (12 h for the first-step hydrothermal reaction) was revealed by a linear sweep voltammetry (LSV) at 25, 50, 60, and 70 °C (Fig. 17.15) using aqueous solutions of Na_2SO_4 with and without **OX** substrate. The TiO_2/Ti-M cathode showed different LSV curves in the presence or absence of **OX** at 50–60 °C, indicating that **OX** reduction preferably occurs on the anatase TiO_2 catalyst rather than the hydrogen evolution reaction in this temperature region. Note that the overpotential for the hydrogen evolution is very low above 70 °C, which is not preferable for the selective **OX** reduction.

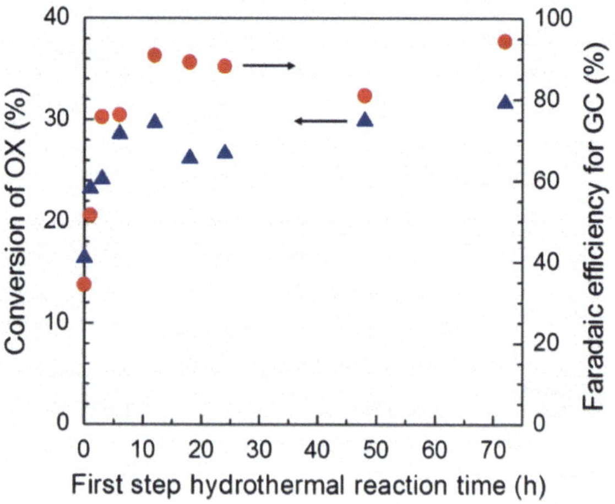

Fig. 17.14 Dependence of (triangle) **OX** conversion and (circle) Faradaic efficiency on the first-step hydrothermal reaction time, after the CA at 0.76 V (vs. RHE) for 2 h using the TiO_2/Ti-M electrode

This synthetic method, producing TiO_2 microcrystals on Ti lines, is applicable not only for the Ti mesh but for a Ti felt having higher surface area. Figure 17.16 shows SEM images before and after the optimized two-step hydrothermal treatment (first step: 12 h, second step: 24 h). TiO_2 microcrystals were also formed on the surface of Ti lines of Ti felt (TiO_2/Ti-F), while it has almost ten times higher surface area (BET surface area of Ti mesh and Ti felt electrode are 2.4 and 19.6 m^2 g^{-1}, respectively).

IrO_2 can be used as an anode catalyst for water oxidation reaction. The anode electrode was prepared the by hand painting method with the IrO_2 powder and a gas-diffusion carbon electrode [47]. The catalyst ink was prepared by sonication of a mixture containing IrO_2 powder, slight amount of Nafion solution and mixed solvent of 2-propanol and water (1:1) for several tens of minutes. The catalyst ink was then painted on the gas-diffusion carbon electrode and dried under air at ambient temperature.

The MEA for the PEAEC was prepared by a hot press at 120 °C with the TiO_2/Ti-M or TiO_2/Ti-F electrode as a porous cathode, Nafion 117 as a solid polymer electrolyte, and gas-diffusion carbon electrode loading IrO_2 (IrO_2/C) as an anode for water oxidation.

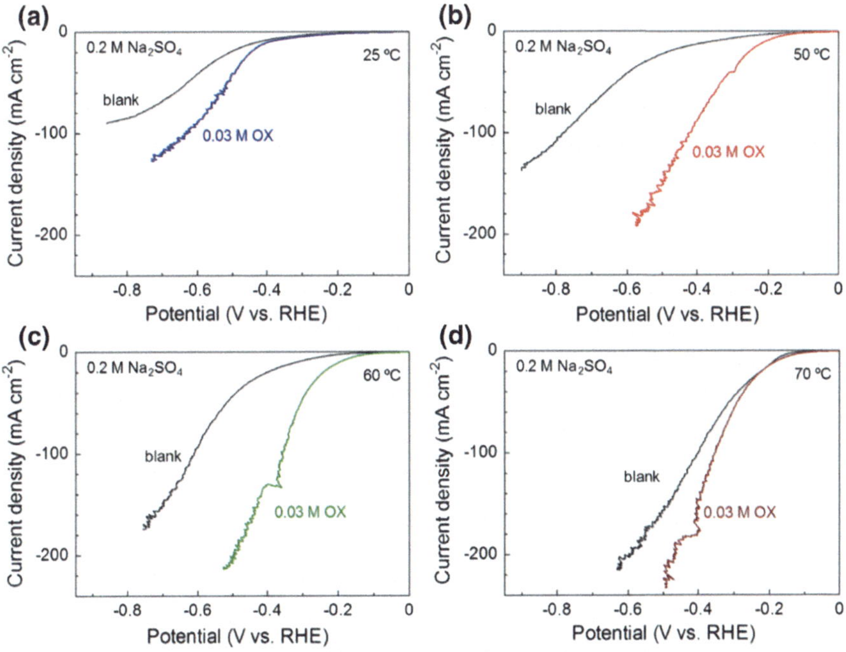

Fig. 17.15 LSV curves of the TiO_2/Ti-M (the first-step reaction: 12 h) cathode at **a** 25, **b** 50, **c** 60 and **d** 70 °C

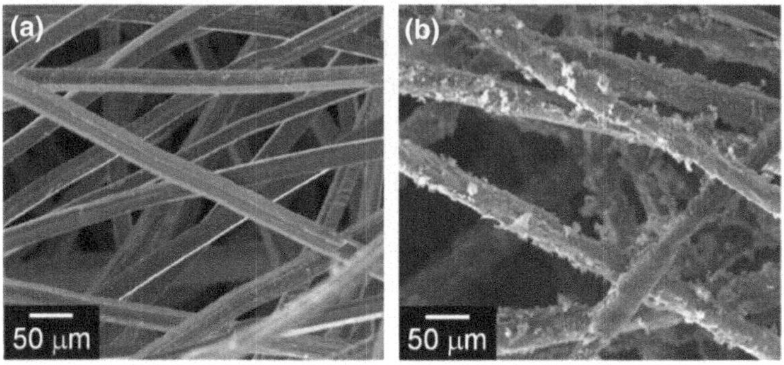

Fig. 17.16 SEM images of **a** Ti felt and **b** TiO_2/Ti-F

17.2.3 Continuous Operation of Electrochemical Production of Glycolic Acid from Oxalic Acid Using PEAEC

The PEAEC was constructed by attaching the MEA between carbon electrodes having flow channels with silicone rubbers as gaskets (Fig. 17.17). The performance of the prepared PEAEC for the continuous electrochemical production of **GC** from **OX** was evaluated by controlled-potential electrolysis of **OX** (0.03 M aqueous solution with no additional electrolyte, reaction area: 4 cm^2, flow rate: 0.5 ml min^{-1}). The product could be separated and quantified by high-performance liquid chromatography (HPLC) after applying various voltages (from 1.8 to 3.0 V). Figure 17.18 shows the temperature dependence of the **OX** conversion and Faradaic efficiency at each applied voltage (TiO_2/Ti-M was used). Both **OX** conversion and Faradaic efficiency for the carbon products completely depend on the reaction temperature. **OX** conversion at a lower applied voltage (below 2.4 V) significantly increased with increasing temperature, suggesting that the overpotential for the **OX** reduction decreases with increasing temperature. On the other hand, Faradaic efficiency decreases, especially above 70 °C, suggesting the increase of hydrogen production, which is ascribed to lowering of overpotential for the hydrogen production above 70 °C. These observations are almost consistent with the results of the LSV measurements and showed that the optimal temperature for PEAEC operation is around 60 °C. In this condition, the maximum Faradaic efficiency for **GC** achieved 69.4% at 2.0 V. The conversion of **OX** did not reach 100% but was 39.4% at 2.4 V even at the optimal temperature of 60 °C. It is also clear that production of an intermediate product, GO, becomes almost zero above 60 °C, while a considerable amount of GO (≈7% FE) was observed at 25 °C.

Optimization of the performance of the PEAEC was performed by changing various factors, such as reaction area (1–25 cm^2), flow rate (0.1–2 ml min^{-1}), and **OX** concentration (0.01–1 M). Figure 17.19 shows the dependence of **OX** conversions on these factors. The **OX** conversion monotonically decreased with increasing flow rate and **OX** concentration, but increased with the electrode area, which clearly suggests that the reaction rate on the cathode is not high enough to achieve 100% conversion of **OX**. It should be noted that the Faradaic efficiency for **GC** increases with increasing **OX** concentration of the reaction solution, indicating that the selective reduction of **OX**, rather than hydrogen evolution, preferably occurs in **OX** solution with higher concentrations, e.g. 1 M solution. From these results, it is clear that higher **OX** conversion with higher Faradaic efficiency for **GC** production using the PEAEC would be achieved by applying a large reaction area, low flow rate and high concentration of **OX** solution. The amount of IrO_2 on the anode was also varied for optimization. We found that 10 mg cm^{-2} of IrO_2 seems to show a maximum performance of the PEAEC. It should also be noted that 10 mg cm^{-2} seemed to be the maximum amount of catalyst in our hand painting method because of a peel-off of anode paper from the MEA after a hot press.

The difference in performance between the TiO_2/Ti-M and TiO_2/Ti-F cathodes was also tested using the PEAEC. Figure 17.20 shows a comparison of the performance between TiO_2/Ti-M and TiO_2/Ti-F cathode in the PEAEC with the same condition (**OX** concentration: 1 M, reaction area: 4 cm^2, flow rate: 0.5 ml min^{-1}, IrO_2 catalyst: 10 mg cm^{-2}, temperature: 60 °C). The TiO_2/Ti-F showed higher **OX** conversion (26.6% at 3.0 V) than TiO_2/Ti-M (17.1% at 3.0 V), indicating that the reaction rate on the TiO_2/Ti-F cathode is faster than that on TiO_2/Ti-M. This might be ascribed to the higher surface area of TiO_2/Ti-F (19.6 m^2 g^{-1}) than that of TiO_2/Ti-M (2.4 m^2 g^{-1}). The Faradaic efficiency of TiO_2/Ti-F for **GC** is also higher than that of TiO_2/Ti-M at all applied voltages and achieves 90.2% at 2.0 V at maximum. Theoretical electrolysis voltage for the PEAEC producing **GC** from **OX** (E_{GC}) could be described as 1.1 V due to the standard redox potentials of **GC** production from **OX** (Eq. 17.1) and water oxidation (Eq. 17.2). Efficiency of energy storage in **GC** (η_{GC}) can be calculated using the following equation:

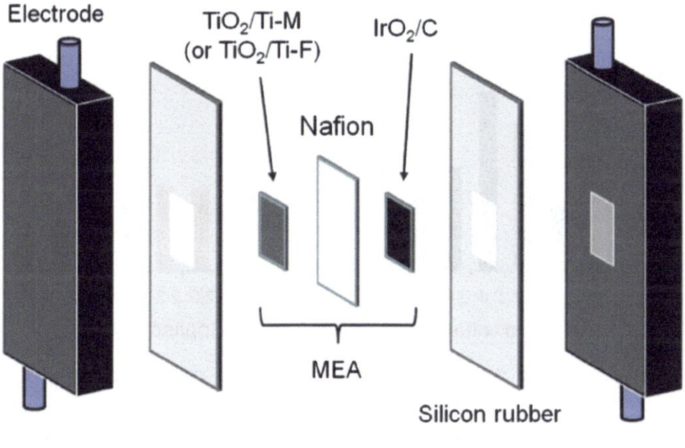

Fig. 17.17 Assembly of building blocks for the PEAEC

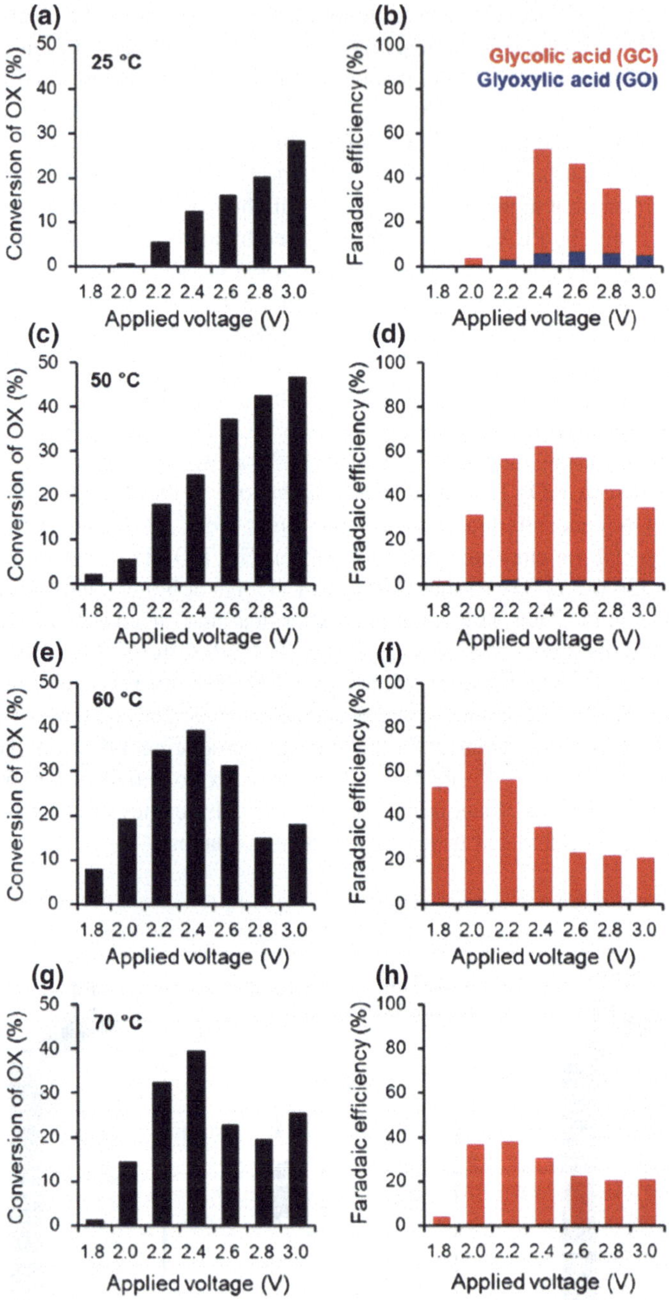
(a)
25 °C
Conversion of OX (%)
Applied voltage (V)
(b)
Glycolic acid (GC)
Glyoxylic acid (GO)
Faradaic efficiency (%)
Applied voltage (V)
(c)
50 °C
Conversion of OX (%)
Applied voltage (V)
(d)
Faradaic efficiency (%)
Applied voltage (V)
(e)
60 °C
Conversion of OX (%)
Applied voltage (V)
(f)
Faradaic efficiency (%)
Applied voltage (V)
(g)
70 °C
Conversion of OX (%)
Applied voltage (V)
(h)
Faradaic efficiency (%)
Applied voltage (V)

◂**Fig. 17.18** Performances of **OX** reduction through the PEAEC with TiO_2/Ti-M cathode operating with a reaction area of 4 cm^2, flow rate of 0.5 ml min^{-1} and **OX** concentration of 0.03 M. Conversions of **OX** at **a** 25, **c** 50, **e** 60 and **g** 70 °C and Faradaic efficiencies at **b** 25, **d** 50, **f** 60 and **h** 70 °C at each applied voltage. Red and blue colours correspond to **GC** and **GO**, respectively

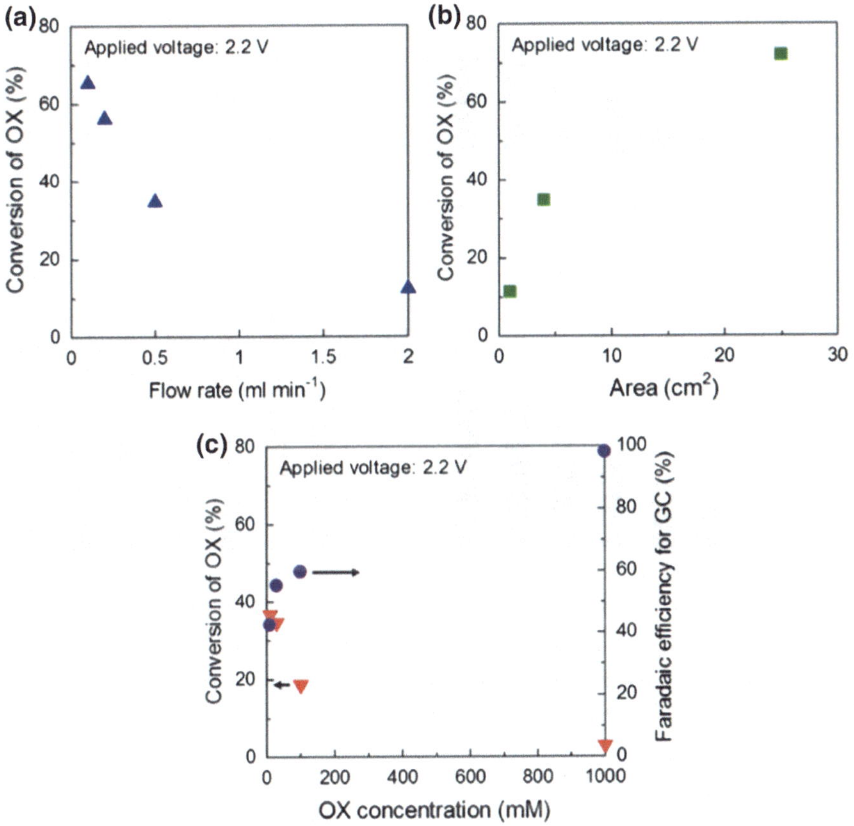

Fig. 17.19 Dependences of **OX** conversion or Faradaic efficiency on **a** flow rate of reaction solution (reaction area: 4 cm^2, **OX** concentration: 0.03 M), **b** reaction area (flow rate: 0.5 ml min^{-1}, **OX** concentration: 0.03 M) and **c OX** concentration (reaction area: 4 cm^2, flow rate: 0.5 ml min^{-1}) under the operation of the PEAEC at 60 °C

$$\eta_{GC} = \frac{E_{GC} \times F_{GC}}{E_{appl}} \quad (17.5)$$

where E_{appl} is the applied voltage, and F_{GC} is the Faradaic efficiency for **GC** production. In this condition, the η_{GC} of the PEAEC with TiO_2/Ti-F electrode leads to 49.6% as a maximum.

According to the above experiments, an optimized PEAEC was constructed with TiO_2/Ti-F by employing the following conditions: **OX** concentration = 1 M,

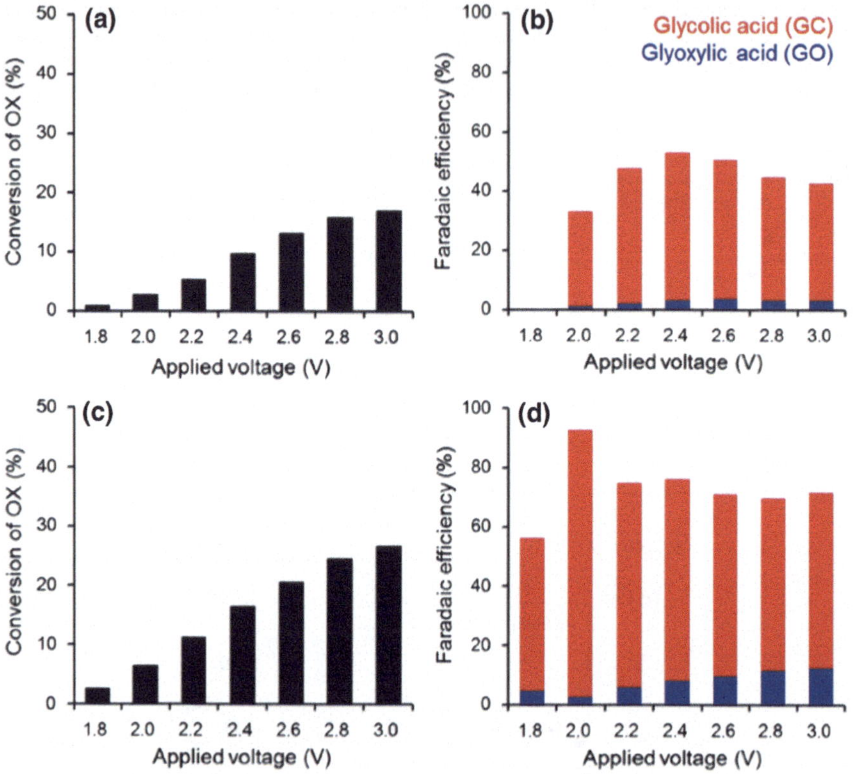

Fig. 17.20 Performances of the PEAEC at 60 °C with TiO_2/Ti-M or TiO_2/Ti-F cathode (reaction area: 4 cm^2, flow rate: 0.5 ml min^{-1}, **OX** concentration: 1 M). **a** Conversion of **OX** and **b** Faradaic efficiency in the case of TiO_2/Ti-M. **c** Conversion of **OX** and **d** Faradaic efficiency in the case of TiO_2/Ti-F

reaction area = 25 cm^2, flow rate = 0.1 ml min^{-1}, IrO_2 = 10 mg cm^{-2}, and temperature = 60 °C. Figure 17.21 shows **OX** conversion, Faradaic efficiency and current density of the PEAEC. **OX** conversion finally reached almost 100% (99.8% at 3.0 V applied voltage) with moderate Faradaic efficiency (31.9% for **GC** production at 3.0 V), while 1 M **OX** solution was used, which is an almost saturated solution at room temperature (solubility of **OX** at 20 °C is around 1.01 M) [48]. Note that the Faradaic efficiency for GO production was vanishingly small (0.4% at 3.0 V). η_{GC} is calculated to be 26.5% at maximum (E_{appl} = 1.8 V, F_{GC} = 43.4%) in this case. Lowering the applied voltage and increasing the Faraday efficiency would be important for further increase in the efficiency for energy storage. Figure 17.21 also shows the current density of the PEAEC. The maximum average current density was 53.8 mA cm^{-2} at 3.0 V. The power density for energy storage in **GC** (P_{GC}) can be expressed with the equation,

$$P_{GC} = E_{GC} \times I \times F_{GC} \quad (17.6)$$

in which I is the current density of the PEAEC. Considering that the I and F_{GC} at 3.0 V are 53.8 mA cm^{-2} and 31.9%, respectively, the power density for the energy storage in **GC** is calculated to be 18.9 mW cm^{-2} at maximum. Water electrolyzers [44, 49], which continuously produce hydrogen as an energy carrier, and redox flow batteries [50, 51], which continuously convert electrical energy into chemical energy of catholyte (and anolyte) stored outside the cell, are some of the competitive devices for the PEAEC. The current density of our PEAEC (<60 mA cm^{-2}) is approximately one order lower than that of commercial water electrolyzers (250–1000 mA cm^{-2}) [49]. Considering that **GC** solution is storable in a common container under ambient conditions without liquefaction at low temperatures or under high pressures, **GC** has a significant advantage as an energy carrier compared with chemically active and gaseous hydrogen. Furthermore, if we could recognize the **OX** solution as a flowable electron pool for the energy storage, the theoretical

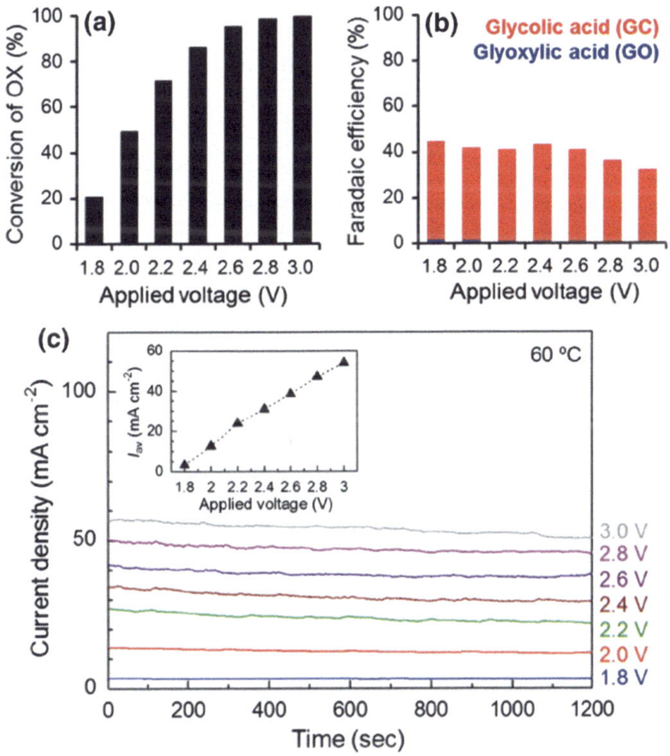

Fig. 17.21 **a** **OX** conversion, **b** Faradaic efficiency and **c** current density during the operation of the PEAEC equipped with TiO_2/Ti-F cathode (**OX** concentration: 1 M, reaction area: 25 cm^2, flow rate: 0.1 ml min^{-1}, IrO_2: 10 mg cm^{-2}, temperature: 60 °C). Inset shows average current density (I_{av}) at each applied voltage

volumetric capacity of the flowed 1 M **OX** solution can be calculated as 107 Ah l^{-1} (four-electron reduction of **OX** to form **GC**). This value is approximately 50 times higher than that of hydrogen gas (2.2 Ah l^{-1} (SATP)) and almost double that of vanadium-based catholyte (60 Ah l^{-1}) [50], implying that conversion of the carboxylic acid group into a hydroxyl group (alcohol) with the PEAEC has great potential as a flow-type energy storage device. It should be noted that the **GC** can generate electricity on a fuel cell accompanied by the production of **OX**, as we previously reported [43]. In addition, as noted above, the theoretical capacity of **GC** solution produced through a four-electron reduction of saturated aqueous solution of **OX** at the operation temperature ($\approx$3.89 M at 60 °C) is around 417 Ah l^{-1}, which is one order higher than that of a typical redox flow battery and 190 times higher than that of hydrogen. Thus, we believe that the PEAEC producing alcohols would be a significant device for energy storage, while some of the issues such as the low energy efficiency (our PEAEC < 27%, redox flow battery $\approx$ 80% [51]), i.e. low Faradaic efficiency for **GC**, and high overpotential should be solved in the future.

17.3 Power Generation from Alcoholic Compounds

17.3.1 Introduction

Alcohols have attracted considerable recent interest as energy carrier, because they offer multiple advantages over hydrogen, including ease of transportation, storage and handling as well as higher energy density. To generate electric power from alcohols efficiently, tremendous efforts have been made to improve direct alcohol fuel cells (DAFCs) using methanol [52–54], ethanol [55, 56], ethylene glycol [57, 58] and glycerol [59] as fuels. In the development of anode catalysts for DAFCs, most efforts have been directed towered complete electro-oxidation of alcohols to generate CO_2, because it provides higher oxidation current than partially electro-oxidation of alcohols. However, renewing alcoholic compounds by direct electroreduction of inherently stable CO_2 is still challenging at present, and this makes it difficult to realize a CN cycle. On the other hand, we have achieved selective production of **GC** by electroreduction of **OX**, which is partially oxidized form of **GC**, under mild conditions as described in Sect. 17.1. Therefore, **GC** has a potential to be an electrochemically renewable alcoholic fuel, and catalyst for selective electro-oxidation of **GC** to **OX** is required to complete a CN energy circulation using **GC/OX** redox couple.

17.3.2 Search for Catalyst Materials for GC Electro-oxidation

To the best of our knowledge, there are no reports on electrocatalysts for the oxidation of **GC**. Thus, we examined the catalytic performances of various transition metals, including Ti, V, Fe, Co, Ni, Cu, Zn, Zr, Nb, Mo, Rh, Pd, Ag, Ta, W, Re, Ir, Pt and Au, for the electrochemical oxidation of **GC** [60]. The catalytic activities of these metals were evaluated by performing cyclic voltammetry (CV) measurements using plates or wires of these metals as a working electrode in the presence and absence of **GC**. The CVs recorded with the metal working electrodes in 0.5 M Na_2SO_4 aqueous solution containing and not containing 0.1 M **GC** are shown in Figs. 17.22, 17.23 and 17.24.

Most of the non-noble metals such as V, Fe, Co, Cu, Zn, Mo, Re and W exhibited large anodic current (peak current: 5–700 mA) in their CVs (Fig. 17.22). After the CV measurement, the metal plates became smaller. In addition, the electrolyte solution after the CV measurements of Fe, Co and Cu exhibited characteristic colours originated from their ions. These observations indicate that oxidation of these metals causes elution of the metal ions into the electrolyte solution. Intriguingly, the anodic current in CVs of Fe, Co, Cu, Zn, Mo and W was increased in the presence of **GC**. Thus, these metals may have certain ability to catalyze electro-oxidation of **GC** although elution of the metal occurs, which is a disadvantage on these catalysts.

The remaining non-noble metals, i.e. Ti, Zr, Nb, Ta and Ni, displayed lower anodic current (peak current: 20 μA–60 mA) in the first cycle of the CVs compared to the oxidation current on the catalysts introduced above, and the anodic current reduced extremely in the second and the third CV cycle (Fig. 17.23), suggesting the oxide film formation on the metal surface.

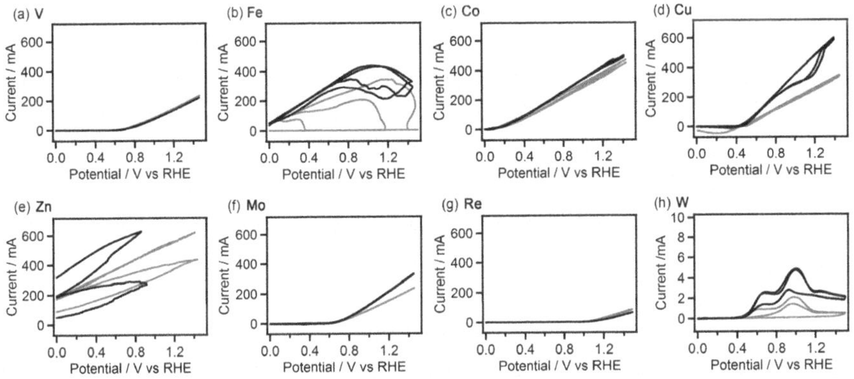

Fig. 17.22 CVs (two-cycle scan) of V (**a**), Fe (**b**), Co (**c**), Cu (**d**), Zn (**e**), Mo (**f**), Re (**g**) and W (**h**) recorded in 0.1 M **GC** + 0.5 M Na_2SO_4 (black line) and 0.5 M Na_2SO_4 (grey line). The current range of graphs (**a–g**) is fixed to –50 to 700 mA

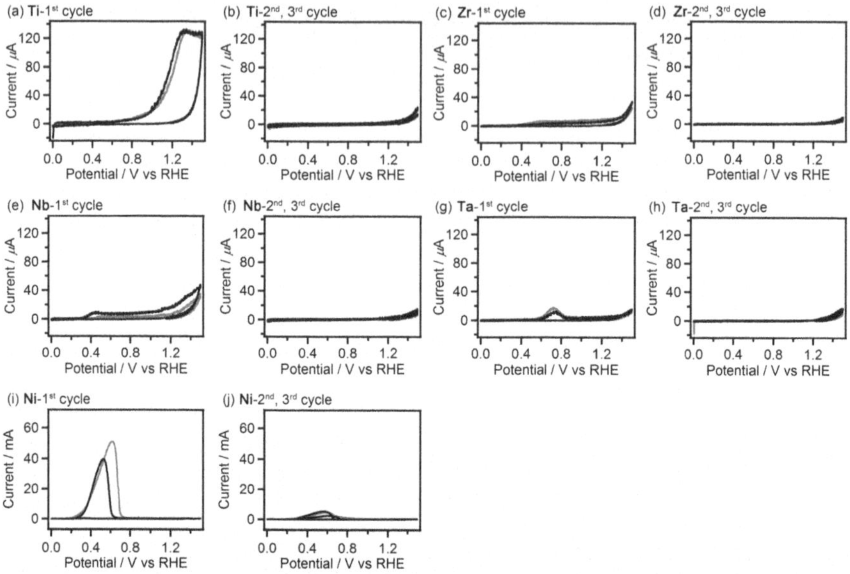

Fig. 17.23 CVs (three-cycle scan) of Ti (**a**, **b**), Zr (**c**, **d**), Nb (**e**, **f**), Ta (**g**, **h**) and Ni (**i**, **j**) recorded in 0.1 M **GC** + 0.5 M Na_2SO_4 (black line) and 0.5 M Na_2SO_4 (grey line). The first (**a**, **c**, **e**, **g**, **i**) and the second, third (**b**, **d**, **f**, **h**, **j**) scan cycles of CVs are depicted separately. The current range of graphs (**a**–**h**) is fixed to –20 to 140 μA

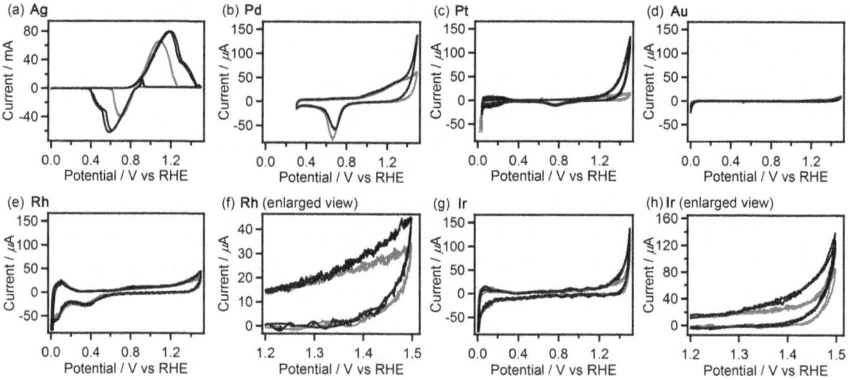

Fig. 17.24 CVs (two-cycle scan) of Ag (**a**), Pd (**b**), Pt (**c**), Au (**d**), Rh (**e**, **f**) and Ir (**g**, **h**) recorded in 0.1 M **GC** + 0.5 M Na_2SO_4 (black line) and 0.5 M Na_2SO_4 (grey line). The current range of graphs (**b**–**e**, **g**) is fixed to –80 to 700 μA

On the other hand, the anodic current in the absence of **GC** on the noble metals except for Ag, i.e. Pd, Pt, Au, Rh and Ir, is very small (peak current: <80 μA, Fig. 17.24b–h), and this is consistent with the high resistance to self-oxidation on these noble metals. Ag showed the large anodic waves (peak current: 60–80 mA) in the forward (positive) scans of the CVs accompanied by the corresponding cathodic waves in the reverse (negative) scans (Fig. 17.24a). These redox waves can be explained by the Ag/Ag_2O redox reaction [61–63]. In the CVs recorded with the electrode of Rh, Pd, Ir and Pt, the anodic current at potential more positive than 1.2 V versus RHE was certainly increased in presence of **GC**, indicating that electrochemical **GC** oxidation is catalyzed on these metals. The results of CV measurements presented in this section suggest that Rh, Pd, Ir and Pt have preferable features, including high resistance to self-oxidation and catalytic activity, for the electrochemical oxidation of **GC**, and Pt shows the highest catalytic performance among all examined metals.

17.3.3 Performance of Pt/C in Electro-oxidation of GC

As described above, Pt shows the highest catalytic activity for electrochemical oxidation of **GC** in the various transition metals we tested. We, hence, investigate catalytic properties of a carbon-supported Pt catalyst (Pt/C) and appropriate conditions for the electrochemical **GC** oxidation [60]. Firstly, we examined catalytic performance of Pt/C for electro-oxidation of **GC** in both acidic and alkaline media. Figure 17.25 compares CVs of Pt/C loaded carbon felt electrode recorded in acidic (0.5 M **GC** + 0.5 M Na_2SO_4, pH 2.3, (a)) and alkaline (0.5 M **GC** + 20 wt% KOH, pH 14, (b)) media at 50 °C. The catalytic current for the **GC** oxidation in acidic and alkaline media arose at 1.1 and 0.3 V versus RHE, respectively, and increased constantly with potential to reach 150 mA and 500 mA at 1.5 V, respectively. A number of studies have demonstrated that the high concentration of OH^- in the electrolyte and adsorbed OH on the Pt surface significantly promote the deprotonation of alcohols, and thus greatly reduce energy barrier of alcohol oxidation reaction [64–66]. This explains the substantial reduction of the onset potential and the increase of catalytic current for the **GC** oxidation in alkaline media. In acidic condition, the catalytic current observed on the reverse (negative) scan was definitely lower than that obtained on the forward (positive) scan. (Fig. 17.25a). This indicates poisoning of the catalyst caused by the adsorption of CO generated by oxidative degradation of **GC** on the catalyst surface (vide infra), and the anodic peak around 0.6 V on the reverse scan is attributable to the desorption of CO.

To test the durability of the Pt/C catalyst and determine the oxidation products of **GC**, CA experiments were conducted in both acidic (0.5 M **GC** + 0.5 M Na_2SO_4, pH 2.3) and alkaline (0.5 M **GC** + 20 wt% KOH, pH 14) media at 50 °C [60]. It is interesting to note that Pt/C catalyst maintained the high activity (490–770 mA) during the CA experiment in alkaline media (Fig. 17.26b), while the catalyst was significantly deactivated in early stage of the CA experiment, up to 80 s, and the reaction proceeded with low catalytic performance (45–65 mA) in acidic media (Fig. 17.26a). As a result, 385 C and 4300 C of charge passed across the electrode during the CA experiment in acidic and alkaline media, respectively. Figure 17.27a,

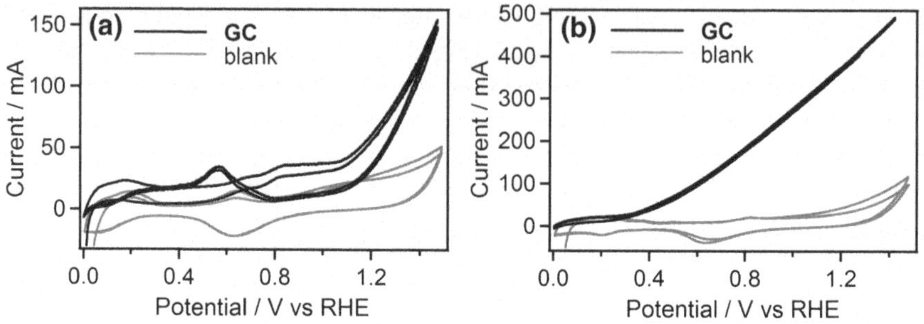

Fig. 17.25 CVs of Pt/C in 0.5 M **GC** + 0.5 M Na_2SO_4 (**a**) and 0.5 M **GC** + 20 wt% KOH (**b**) at 50 °C

b shows the products of the CA experiments and Faradaic efficiency for each product, and Fig. 17.28 illustrates the pathway for electro-oxidation of **GC** as inferred from the reports dealing with ethylene glycol electro-oxidation [64–70]. The major products of **GC** oxidation in acidic solution were CO_2 and formic acid (90.9 and 7.2% Faradaic efficiency, respectively). The dominant formation of the C_1 compounds indicates the accelerated occurrence of the C–C bond cleavage of the reacting molecules followed by further oxidation of the C_1 intermediates generating adsorbed CO [64, 69]. This result supports our consideration for the catalyst deactivation observed in the CV and the CA experiments in acidic media. On the other hand, electrochemical **GC** oxidation in alkaline media gave **OX** and CO_2 (47.8 and 52.1% Faradaic efficiency, respectively). These results suggest that the use of alkaline media is favourable for the selective four-electron oxidation of **GC**, i.e. no CO_2 production, including higher catalytic activity and stability.

To achieve selective electro-oxidation of **GC** to **OX**, effects of the applied potential (1.2 or 1.5 V vs. RHE), temperature (40 or 50 °C) and alkali cations (K^+ or Li^+) on the product selectivity were examined in alkaline media [60]. Figure 17.27b–d reveal that the reaction temperature gives greater influence on the product selectivity rather than applied potential. The alkali cations made a

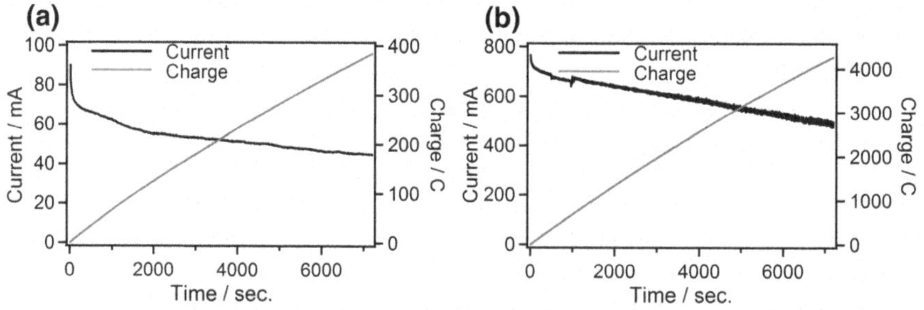

Fig. 17.26 CA curves of Pt/C in 0.5 M **GC** + 0.5 M Na_2SO_4 (**a**) and 0.5 M **GC** + 20 wt% KOH (**b**) at 50 °C

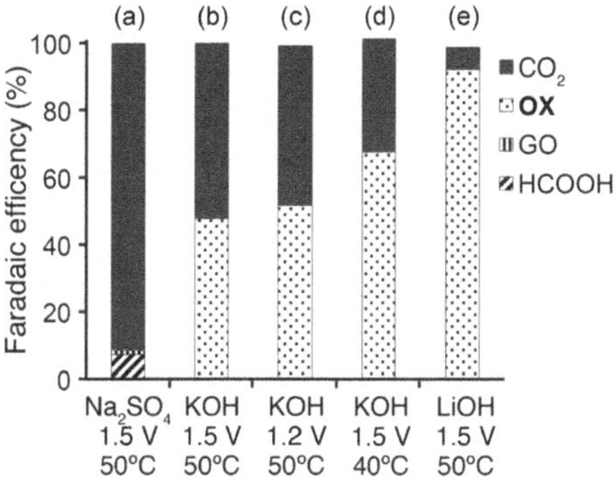

Fig. 17.27 Faradaic efficiency for CO_2, **OX**, GO and HCOOH in **GC** electro-oxidation **a** at 1.5 V in 0.5 M **GC** and 0.5 M Na_2SO_4 at 50 °C, **b** at 1.5 V in 0.5 M **GC** and 20 wt%, i.e. 3.56 M KOH at 50 °C, **c** at 1.2 V in 0.5 M **GC** and 20 wt% KOH at 50 °C, **d** at 1.5 V in 0.5 M **GC** and 20 wt% KOH at 40 °C and **e** at 1.5 V in 0.5 M **GC** and 3.56 M LiOH at 50 °C

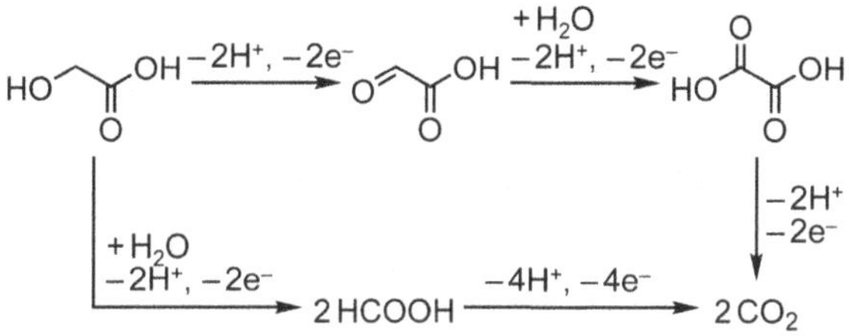

Fig. 17.28 Possible pathway for **GC** oxidation

significant impact on the product selectivity. The CA experiment in LiOH electrolyte resulted in high Faradaic efficiency of 92.2% for **OX** production (Fig. 17.27e), which is approximately twice as high as the yield with K^+ ions (Fig. 17.27b). Garcia-Arez et al. proposed that alkali cations can form an ion pair with anions adsorbed on Pt surface [71]. It is well known that alkali cations form a cluster-like structure with water molecules ($M^+(H_2O)_X$) in aqueous solutions, and a larger effective charge of Li^+ facilitates the formation of a solvated cluster. The non-covalent interactions between OH species adsorbed on the surface, i.e. OH_{ad}, and solvated cations inactivate OH_{ad} and block the catalyst surface, decreasing thus the amount of surface sites needed for the C–C bond breaking [67]. Thus, the smaller amount of the active surface site in LiOH solution realizes the high selectivity for **OX** production [72]. We found that the selective oxidation of **GC** to **OX** on Pt/C more favourably proceeds in alkaline media. In particular, electro-oxidation of **GC** in LiOH solution resulted in dominant formation of **OX**, i.e. 92.2% of Faradaic efficiency.

We conducted a series of experiments as a proof of concept of a CN power circulation system using solutions of an alcohol and an acid. We hope that the CN systems contribute to realization of a sustainable society driven by renewable energy resources.

References

1. Stocker TF (2014) Climate change 2013: the physical science basis: Working Group I contribution to the Fifth assessment report of the Intergovernmental Panel on Climate Change. Cambridge University Press
2. Turner JA (1999) A realizable renewable energy future. Science 285(5428):687–689
3. Züttel A, Borgschulte A, Schlapbach L (2011) Hydrogen as a future energy carrier. Wiley
4. Alhumaidan F, Cresswell D, Garforth A (2011) Hydrogen storage in liquid organic hydride: producing hydrogen catalytically from methylcyclohexane. Energy Fuels 25(10):4217–4234
5. Klerke A, Christensen CH, Nørskov JK, Vegge T (2008) Ammonia for hydrogen storage: challenges and opportunities. J Mater Chem 18(20):2304–2310
6. S-i Orimo, Nakamori Y, Eliseo JR, Züttel A, Jensen CM (2007) Complex hydrides for hydrogen storage. Chem Rev 107(10):4111–4132
7. Hull JF, Himeda Y, Wang W-H, Hashiguchi B, Periana R, Szalda DJ, Muckerman JT, Fujita E (2012) Reversible hydrogen storage using CO_2 and a proton-switchable iridium catalyst in aqueous media under mild temperatures and pressures. Nat Chem 4(5):383–388
8. Yamauchi M, Ozawa N, Kubo M (2016) Experimental and quantum chemical approaches to develop highly selective nanocatalysts for CO_2-free power circulation. Chem Rec 16(5):10
9. Yosuke Muranaka AI, Hasegawa Isao, Mae Kazuhiro (2013) Selective production of valuable chemicals from biomass by two-step conversion combining pre-oxidation and hydrothermal degradation. Chem Eng J 234:189–194
10. Dub PA, Ikariya T (2012) Catalytic reductive transformations of carboxylic and carbonic acid derivatives using molecular hydrogen. ACS Catal 2(8):1718–1741
11. Onyestyak G, Harnos S, Klebert S, Stolcova M, Kaszonyi A, Kallo D (2013) Selective reduction of acetic acid to ethanol over novel Cu_2In/Al_2O_3 catalyst. Appl Catal A-Gen 464:313–321. https://doi.org/10.1016/j.apcata.2013.05.042
12. Rachmady W, Vannice MA (2000) Acetic acid hydrogenation over supported platinum catalysts. J Catal 192(2):322–334. https://doi.org/10.1006/jcat.2000.2863
13. Watanabe R, Yamauchi M, Sadakiyo M, Abe R, Takeguchi T (2015) CO_2-free electric power circulation via direct charge and discharge using the glycolic acid/oxalic acid redox couple. Energy Environ Sci 8(5):1456–1462
14. Wu HB, Lou XWD, Hng HH (2012) Synthesis of uniform layered protonated titanate hierarchical spheres and their transformation to anatase TiO_2 for lithium-ion batteries. Chem Eur J 18(7):2094–2099
15. Wang R, Hashimoto K, Fujishima A, Chikuni M, Kojima E, Kitamura A, Shimohigoshi M, Watanabe T (1997) Light-induced amphiphilic surfaces. Nature 388(6641):431
16. Fujishima A, Honda K (1972) Electrochemical photolysis of water at a semiconductor electrode. Nature 238(5358):37–38. https://doi.org/10.1038/238037a0
17. Le Formal F, Tetreault N, Cornuz M, Moehl T, Gratzel M, Sivula K (2011) Passivating surface states on water splitting hematite photoanodes with alumina overlayers. Chem Sci 2 (4):737–743. https://doi.org/10.1039/c0sc00578a
18. Amano F, Li D, Ohtani B (2010) Fabrication and photoelectrochemical property of tungsten (VI) oxide films with a flake-wall structure. Chem Commun 46(16):2769–2771. https://doi.org/10.1039/b926931b
19. Iwashina K, Kudo A (2011) Rh-doped $SrTiO_3$ photocatalyst electrode showing cathodic photocurrent for water splitting under visible-light irradiation. J Am Chem Soc 133 (34):13272–13275. https://doi.org/10.1021/ja2050315
20. Gu J, Yan Y, Krizan JW, Gibson QD, Detweiler ZM, Cava RJ, Bocarsly AB (2014) p-type CuRhO2 as a self-healing photoelectrode for water reduction under visible light. J Am Chem Soc 136(3):830–833. https://doi.org/10.1021/ja408876k
21. Qin D-D, Li Y-L, Wang T, Li Y, Lu X-Q, Gu J, Zhao Y-X, Song Y-M, Tao C-L (2015) Sn-doped hematite films as photoanodes for efficient photoelectrochemical water oxidation. J Mater Chem A 3(13):6751–6755. https://doi.org/10.1039/c4ta06872f

22. Minegishi T, Nishimura N, Kubota J, Domen K (2013) Photoelectrochemical properties of $LaTiO_2N$ electrodes prepared by particle transfer for sunlight-driven water splitting. Chem Sci 4(3):1120–1124. https://doi.org/10.1039/c2sc21845c
23. Maeda K, Higashi M, Siritanaratkul B, Abe R, Domen K (2011) SrNbO2 N as a water-splitting photoanode with a wide visible-light absorption band. J Am Chem Soc 133 (32):12334–12337. https://doi.org/10.1021/ja203391w
24. Higashi M, Domen K, Abe R (2012) Highly stable water splitting on oxynitride TaON photoanode system under visible light irradiation. J Am Chem Soc 134(16):6968–6971. https://doi.org/10.1021/ja302059g
25. Kibria MG, Zhao S, Chowdhury FA, Wang Q, Nguyen HPT, Trudeau ML, Guo H, Mi Z (2014) Tuning the surface Fermi level on p-type gallium nitride nanowires for efficient overall water splitting. Nat Commun 5. ARTN 3825. https://doi.org/10.1038/ncomms4825
26. Kato T, Hakari Y, Ikeda S, Jia QX, Iwase A, Kudo A (2015) Utilization of metal sulfide material of (CuGa)(1-x)Zn2xS2 solid solution with visible light response in photocatalytic and photoelectrochemical solar water splitting systems. J Phys Chem Lett 6(6):1042–1047. https://doi.org/10.1021/acs.jpclett.5b00137
27. Hisatomi T, Miyazaki K, Takanabe K, Maeda K, Kubota J, Sakata Y, Domen K (2010) Isotopic and kinetic assessment of photocatalytic water splitting on Zn-added Ga2O3 photocatalyst loaded with Rh2-yCryO3 cocatalyst. Chem Phys Lett 486(4–6):144–146. https://doi.org/10.1016/j.cplett.2010.01.006
28. Sakata Y, Matsuda Y, Yanagida T, Hirata K, Imamura H, Teramura K (2008) Effect of metal ion addition in a Ni supported Ga(2)O(3) photocatalyst on the photocatalytic overall splitting of H(2)O. Catal Lett 125(1–2):22–26. https://doi.org/10.1007/s10562-008-9557-7
29. Kudo A, Omori K, Kato H (1999) A novel aqueous process for preparation of crystal form-controlled and highly crystalline $BiVO_4$ powder from layered vanadates at room temperature and its photocatalytic and photophysical properties. J Am Chem Soc 121 (49):11459–11467. https://doi.org/10.1021/ja992541y
30. Sayama K, Mukasa K, Abe R, Abe Y, Arakawa H (2001) Stoichiometric water splitting into H-2 and O-2 using a mixture of two different photocatalysts and an IO3-/I-shuttle redox mediator under visible light irradiation. Chem Commun 23:2416–2417
31. Kitano S, Yamauchi M, Hata S, Watanabe R, Sadakiyo M (2016) Hydrogenation of oxalic acid using light-assisted water electrolysis for the production of an alcoholic compound. Green Chem 18(13):3700–3706
32. Yu JQ, Kudo A (2006) Effects of structural variation on the photocatalytic performance of hydrothermally synthesized $BiVO_4$. Adv Funct Mater 16(16):2163–2169. https://doi.org/10.1002/adfm.200500799
33. Kudo A, Miseki Y (2009) Heterogeneous photocatalyst materials for water splitting. Chem Soc Rev 38(1):253–278. https://doi.org/10.1039/b800489g
34. Murakami SY, Kominami H, Kera Y, Ikeda S, Noguchi H, Uosaki K, Ohtani B (2007) Evaluation of electron-hole recombination properties of titanium(IV) oxide particles with high photocatalytic activity. Res Chem Intermediat 33(3–5):285–296. https://doi.org/10.1163/156856707779238612
35. Chen XB, Shen SH, Guo LJ, Mao SS (2010) Semiconductor-based photocatalytic hydrogen generation. Chem Rev 110(11):6503–6570. https://doi.org/10.1021/cr1001645
36. Liang YQ, Tsubota T, Mooij LPA, van de Krol R (2011) Highly improved quantum efficiencies for thin film $BiVO_4$ photoanodes. J Phys Chem C 115(35):17594–17598. https://doi.org/10.1021/jp203004v
37. Abdi FF, Savenije TJ, May MM, Dam B, van de Krol R (2013) The origin of slow carrier transport in $BiVO_4$ thin film photoanodes: a time-resolved microwave conductivity study. J Phys Chem Lett 4(16):2752–2757. https://doi.org/10.1021/jz4013257
38. Zhong M, Hisatomi T, Kuang YB, Zhao J, Liu M, Iwase A, Jia QX, Nishiyama H, Minegishi T, Nakabayashi M, Shibata N, Niishiro R, Katayama C, Shibano H, Katayama M, Kudo A, Yamada T, Domen K (2015) Surface modification of CoOx loaded $BiVO_4$

photoanodes with ultrathin p-type NiO layers for improved solar water oxidation. J Am Chem Soc 137(15):5053–5060. https://doi.org/10.1021/jacs.5b00256
39. Kim TW, Choi KS (2014) Nanoporous $BiVO_4$ photoanodes with dual-layer oxygen evolution catalysts for solar water splitting. Science 343(6174):990–994. https://doi.org/10.1126/science.1246913
40. Scaife DE (1980) Oxide semiconductors in photoelectrochemical conversion of solar-energy. Sol Energy 25(1):41–54. https://doi.org/10.1016/0038-092x(80)90405-3
41. Hashimoto K, Irie H, Fujishima A (2005) TiO_2 photocatalysis: a historical overview and future prospects. Jpn J Appl Phys 1 44(12):8269–8285. https://doi.org/10.1143/jjap.44.8269
42. Nosaka Y (2011) Solar cells and photocatalysts. Comprehensive nanoscience and technology, vol 1: Nanomaterials, pp 571–605
43. Watanabe R, Yamauchi M, Sadakiyo M, Abe R, Takeguchi T (2015) CO_2-free electric power circulation via direct charge and discharge using the glycolic acid/oxalic acid redox couple. Energy Environ Sci 8(5):1456–1462. https://doi.org/10.1039/c5ee00192g
44. Millet P, Andolfatto F, Durand R (1996) Design and performance of a solid polymer electrolyte water electrolyzer. Int J Hydrogen Energy 21(2):87–93. https://doi.org/10.1016/0360-3199(95)00005-4
45. Sadakiyo M, Hata S, Cui XD, Yamauchi M (2017) Electrochemical production of glycolic acid from oxalic acid using a polymer electrolyte alcohol electrosynthesis cell containing a porous TiO_2 catalyst. Sci Rep 7. https://doi.org/10.1038/s41598-017-17036-3
46. Wu WQ, Rao HS, Xu YF, Wang YF, Su CY, Kuang DB (2013) Hierarchical oriented anatase TiO_2 nanostructure arrays on flexible substrate for efficient dye-sensitized solar cells. Sci Rep 3. https://doi.org/10.1038/srep01892
47. Ma S, Luo R, Moniri S, Lan YC, Kenis PJA (2014) Efficient electrochemical flow system with improved anode for the conversion of CO_2 to CO. J Electrochem Soc 161(10):F1124–F1131. https://doi.org/10.1149/2.1201410jes
48. Apelblat A, Manzurola E (1987) Solubility of oxalic, malonic, succinic, adipic, maleic, malic, citric, and tartaric-acids in water from 278.15-K to 338.15-K. J Chem Thermodyn 19(3):317–320. https://doi.org/10.1016/0021-9614(87)90139-x
49. Zeng K, Zhang DK (2010) Recent progress in alkaline water electrolysis for hydrogen production and applications. Prog Energy Combust Sci 36(3):307–326. https://doi.org/10.1016/j.pecs.2009.11.002
50. Chen HN, Zou QL, Liang ZJ, Liu H, Li Q, Lu YC (2015) Sulphur-impregnated flow cathode to enable high-energy-density lithium flow batteries. Nat Commun 6. https://doi.org/10.1038/ncomms6877
51. Skyllas-Kazacos M, Chakrabarti MH, Hajimolana SA, Mjalli FS, Saleem M (2011) Progress in flow battery research and development. J Electrochem Soc 158(8):R55–R79. https://doi.org/10.1149/1.3599565
52. Tiwari JN, Tiwari RN, Singh G, Kim KS (2013) Recent progress in the development of anode and cathode catalysts for direct methanol fuel cells. Nano Energy 2(5):553–578. https://doi.org/10.1016/j.nanoen.2013.06.009
53. Gomez JCC, Moliner R, Lazaro MJ (2016) Palladium-based catalysts as electrodes for direct methanol fuel cells: a last ten years review. Catalysts 6(9). https://doi.org/10.3390/catal6090130
54. Antolini E (2017) Pt-Ni and Pt-M-Ni (M = Ru, Sn) anode catalysts for low-temperature acidic direct alcohol fuel cells: a review. Energies 10(1). https://doi.org/10.3390/en10010042
55. An L, Zhao TS, Li YS (2015) Carbon-neutral sustainable energy technology: direct ethanol fuel cells. Renew Sustain Energy Rev 50:1462–1468. https://doi.org/10.1016/j.rser.2015.05.074
56. Wang Y, Zou SZ, Cai WB (2015) Recent advances on electro-oxidation of ethanol on Pt- and Pd-based catalysts: from reaction mechanisms to catalytic materials. Catalysts 5(3):1507–1534. https://doi.org/10.3390/catal5031507

57. An L, Chen R (2016) Recent progress in alkaline direct ethylene glycol fuel cells for sustainable energy production. J Power Sources 329:484–501. https://doi.org/10.1016/j.jpowsour.2016.08.105
58. Serov A, Kwak C (2010) Recent achievements in direct ethylene glycol fuel cells (DEGFC). Appl Cataly B-Environ 97(1–2):1–12. https://doi.org/10.1016/j.apcatb.2010.04.011
59. Simoes M, Baranton S, Coutanceau C (2012) Electrochemical valorisation of glycerol. Chemsuschem 5(11):2106–2124. https://doi.org/10.1002/cssc.201200335
60. Fukushima T, Kitano S, Hata S, Yamauchi M (2018) Carbon neutral energy cycles using alcohols. Sci Technol Adv Mater 19(1):142–152. https://doi.org/10.1080/14686996.2018.1426340
61. Dirkse TP, Devries DB (1959) The effect of continuously changing potential on the silver electrode in alkaline solutions. J Phys Chem 63(1):107–110. https://doi.org/10.1021/j150571a028
62. Teijelo ML, Vilche JR, Arvia AJ (1982) Complex potentiodynamic response of silver in alkaline electrolytes in the potential range of the Ag/Ag_2O couple. J Electroanal Chem 131 (JAN):331–339
63. Wan Y, Wang XL, Liu SY, Li YB, Sun H, Wang Q (2013) Effect of electrochemical factors on formation and reduction of silver oxides. Int J Electrochem Sci 8(12):12837–12850
64. BeltowskaBrzezinska M, Luczak T, Holze R (1997) Electrocatalytic oxidation of mono- and polyhydric alcohols on gold and platinum. J Appl Electrochem 27(9):999–1011. https://doi.org/10.1023/a:1018422206817
65. Simoes M, Baranton S, Coutanceau C (2010) Electro-oxidation of glycerol at Pd based nano-catalysts for an application in alkaline fuel cells for chemicals and energy cogeneration. Appl Catal B-Environ 93(3–4):354–362. https://doi.org/10.1016/j.apcatb.2009.10.008
66. Xin L, Zhang Z, Qi J, Chadderdon D, Li W (2012) Electrocatalytic oxidation of ethylene glycol (EG) on supported Pt and Au catalysts in alkaline media: reaction pathway investigation in three-electrode cell and fuel cell reactors. Appl Catal B-Environ 125:85–94. https://doi.org/10.1016/j.apcatb.2012.05.024
67. Christensen PA, Hamnett A (1989) The oxidation of ethylene-glycol at a platinum-electrode in acid and base—an insitu FTIR study. J Electroanal Chem 260(2):347–359. https://doi.org/10.1016/0022-0728(89)87149-9
68. Chang SC, Ho YH, Weaver MJ (1991) Applications of real-time FTIR spectroscopy to the elucidation of complex electroorganic pathways—electrooxidation of ethylene-glycol on gold, platinum, and nickel in alkaline-solution. J Am Chem Soc 113(25):9506–9513. https://doi.org/10.1021/ja00025a014
69. Wieland B, Lancaster JP, Hoaglund CS, Holota P, Tornquist WJ (1996) Electrochemical and infrared spectroscopic quantitative determination of the platinum-catalyzed ethylene glycol oxidation mechanism at CO adsorption potentials. Langmuir 12(10):2594–2601. https://doi.org/10.1021/la9506943
70. Bambagioni V, Bevilacqua M, Bianchini C, Filippi J, Marchionni A, Vizza F, Wang LQ, Shen PK (2010) Ethylene glycol electrooxidation on smooth and nanostructured Pd electrodes in alkaline media. Fuel Cells 10(4):582–590. https://doi.org/10.1002/fuce.200900120
71. Garcia-Araez N, Climent V, Rodriguez P, Feliu JM (2010) Thermodynamic evidence for K+-SO42-ion pair formation on Pt(111). New insight into cation specific adsorption. Phys Chem Chem Phys 12(38):12146–12152. https://doi.org/10.1039/c0cp00247j
72. Sitta E, Batista BC, Varela H (2011) The impact of the alkali cation on the mechanism of the electro-oxidation of ethylene glycol on Pt. Chem Commun 47(13):3775–3777. https://doi.org/10.1039/c0cc05353h

Chapter 18
Nanocarbons in Li-Ion Batteries

Seok-Kyu Cho, JongTae Yoo and Sang-Young Lee

18.1 Introduction

The rapidly emerging ubiquitous energy era, including smart portable electronics, the internet of things (IOTs), and electric vehicles, has driven a relentless pursuit of high-energy-density/high-safety rechargeable power sources with reliable electrochemical performance [1–4]. Furthermore, the inconsistent energy supply from renewable resources (i.e., solar and wind power) strongly demands for reliable energy storage systems, which balances between energy demand and supply therefrom. Of various rechargeable energy storage systems reported to date, lithium-ion batteries (LIBs) have undoubtedly occupied a predominant position, since the commercialization of LIBs by Sony in 1991. The LIB system shows the superiority in gravimetric and volumetric energy density (Fig. 18.1 and Table 18.1) compared to other traditional rechargeable batteries (e.g., lead-acid, Ni-Cd, and Ni-MH batteries) [5]. No memory effect (compared to Ni-Cd batteries) and environmental benignity (over lead-acid batteries) have accelerated the application expansion of LIBs.

Meanwhile, the ever-increasing demand for advanced power sources with higher energy density and/or versatile form factors strongly pushes us to search for new battery materials and structures beyond current state-of-the-art LIBs [6–8]. Recent progress in nanoscience and nanotechnology suggests opportunities to develop

S.-K. Cho · S.-Y. Lee (✉)
Department of Energy Engineering, School of Energy and Chemical Engineering, Ulsan National Institute of Science and Technology (UNIST), Ulsan 44919, Republic of Korea
e-mail: syleek@unist.ac.kr

J. Yoo
Center of General R&D Coordination, Korea Institute of S&T Evaluation and Planning (KISTEP), Seoul 06775, Republic of Korea
e-mail: jtyoo@kistep.re.kr

N. Nakashima (ed.), *Nanocarbons for Energy Conversion: Supramolecular Approaches*, Nanostructure Science and Technology,
https://doi.org/10.1007/978-3-319-92917-0_18

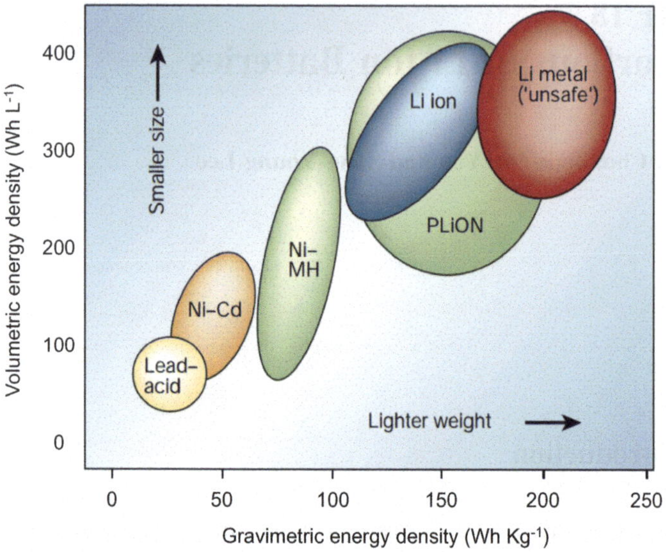

Fig. 18.1 Comparison in gravimetric/volumetric energy density of different battery technologies. Reproduced with permission from Ref. [1]. Copyright 2001, Nature Publishing Group

novel electrode materials and architectures for next-generation LIBs. Notably, nanocarbon materials have garnered considerable attention as a promising battery element to enrich electrodes' chemistry [9–11]. Among the nanocarbon families (e.g., 0D fullerene, 1D carbon nanotube (CNT), 2D graphene, 3D graphite and nanodiamond, etc.) with diverse dimensions, the CNT and graphene exhibit outstanding electrical and mechanical properties, large surface area, and unique structural characteristics, which are favorable for superior electrochemical performance and flexibility/design diversity of LIBs [10–13].

In this chapter, we aim to provide a brief overview of recent advances of nanocarbons in LIBs, which covers their use in conventional systems and potential in new application fields.

18.2 Chemistry and Electrochemical Principles of Li-Ion Batteries

The Li-ion cells are typically assembled with a positive electrode (called as "cathode") and a negative electrode (called as "anode"), a separator membrane (i.e., porous polymer membranes), and liquid electrolytes composed of carbonate solvents and lithium salts (Fig. 18.2). Under external electrical connection between the electrodes, electrochemical reaction of the electrode materials produces electrons and ions, resulting in the charge/discharge reaction of LIBs. Specifically, during

Table 18.1 Comparison between different battery technologies. Reproduced with permission from Ref. [5]. Copyright 2015, Elsevier

Specifications	Lead-acid	NiCd	NiMH	Li-ion		
				Cobalt	Manganese	Phosphate
Specific energy density (W h/kg)	30–50	45–80	60–120	150–190	100–135	90–120
Internal resistance (mΩ)	<100	100–300	200–300	150–300	25–75	25–50
	12 V pack	6 V pack	6 V pack	7.2 V	Per cell	Per cell
Cycle life (80% discharge)	200–300	1000	300–500	500–1000	500–1000	1000–2000
Fast charge time	8–16 h	1 h typical	2–4 h	2–4 h	1 h or less	1 h or less
Overcharge tolerance	High	Moderate	Low	Low. Cannot tolerate tickle charge		
Self-discharge/month (room temp) (%)	5	20	30	<10		
Cell voltage (nominal) (V)	2	1.2	1.2	3.6	3.8	3.3
Charge cutoff voltage (V/cell)	2.4	Full charge detection by voltage signature		4.2		3.6
	Float 2.25					
Discharge cutoff voltage (V/cell, 1C)	1.75	1		2.50–3.00		2.8
Peak load current	5C	20C	5C	>3C	>30C	>30C
Best result	0.2C	1C	0.5C	>1C	<10C	<10C
Charge temperature	−20 to 50 °C (−4 to 122 °F)	0 to 45 °C (32–113 °F)		0 to 45 °C (32–113 °F)		
Discharge temperature	−20 to 50 °C (−4 to 122 °F)	−20 to 65 °C (−4 to 49 °F)		−20 to 60 °C (−4 to 140 °F)		
Maintenance requirement	3–6 months (topping charge)	30–60 days (discharge)	60–90 days (discharge)	Not required		
Safety requirements	Thermally stable	Thermally stable, fuse protection common		Protection circuit mandatory		
In use since	Late 1800s	1950	1990	1991	1996	1999
Toxicity	Very high	Very high	Low	Low		

charge process, the electrons and ions are released from the cathode materials, then the liberated electrons and ions move toward anode materials, respectively, through the external circuit and liquid electrolyte. This electrochemical conversion generates the difference in chemical potential between the cathode and anode, indicating that the external electrical energy is stored in the form of chemical energy in the electrode materials. In contrast, the discharge process proceeds via spontaneous chemical reactions of the electrode materials without the external electrical energy supply, until the redox reactions reach the electrochemical equilibrium. Through these electrochemical reactions at the electrodes, the stored chemical energy is released and utilized as power for electronic devices.

The stored electricity in the LIBs can be quantitatively estimated by various parameters, including voltage, capacity, and gravimetric/volumetric energy density. The cell voltage between the electrodes (i.e., cathodes and anodes) is theoretically calculated via the total change of Gibbs free energy ($\Delta E = -\Delta G/nF$), at the electrode active materials. The battery capacity is defined as amount of charge generated in the electrochemical reaction. The amount of electrical energy (i.e., energy

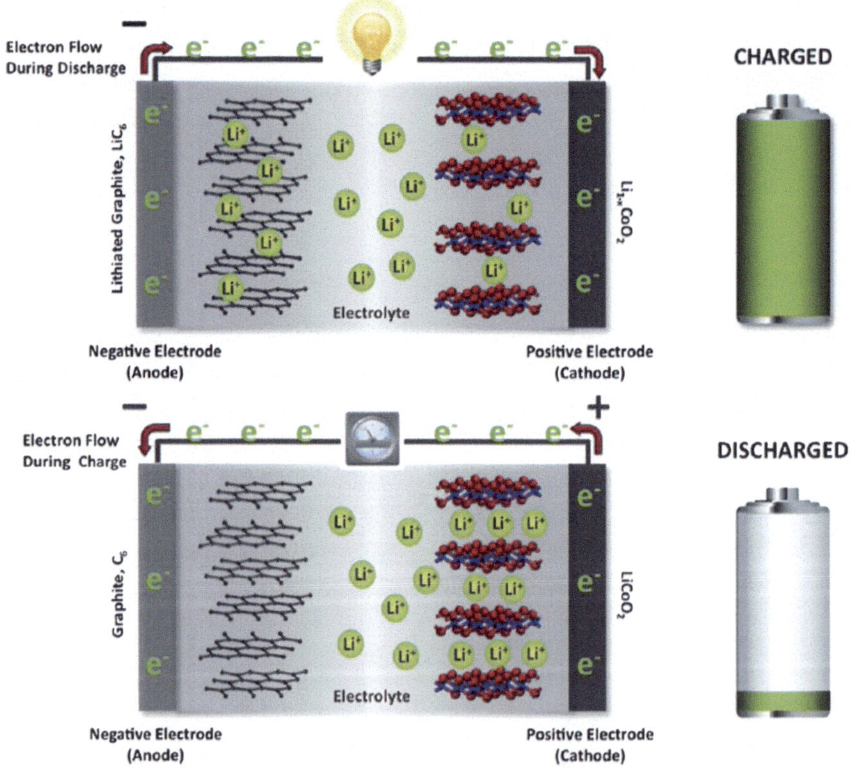

Fig. 18.2 Schematic illustration of operating principle (i.e., charge and discharge) of conventional LIBs. Reproduced with permission from Ref. [4]. Copyright 2012, Royal Society of Chemistry

density) is calculated by integrating the cell potential and capacity, and the values are described as gravimetric energy density (W h kg^{-1}) and volumetric energy density (W h L^{-1}).

In addition to the energy storage, cyclability (i.e., charge/discharge repeatability), rate performance (i.e., charge/discharge current density), and abuse tolerance (i.e., safety) should be also considered. The cyclability represents the reversibility of electrochemical reaction during the repeated charge/discharge process, where the higher capacity retention indicates the better cyclability. The rate (also called "C-rate") performance describes the electrochemical kinetics of the cells. The C-rate is calculated as a function ($C = 1/h$, $C_p = i \times h$) of the charge/discharge current and the time required to fully charge and discharge them, wherein C is the C-rate, h is the time required to fully charge and discharge, i is the current, and C_p is the capacity. The higher C-rate indicates the higher applied current, and thus shorter time required for full charge/discharge. The high-rate performance LIBs is greatly desired especially for use in electric vehicles (EVs). The reliable electrochemical performance under the high C-rate condition can be achieved by facilitating electronic/ionic conduction in the cells. Note that fast charging is a big challenge to be urgently addressed in EV batteries.

18.3 Li-Ion Storage in Nanocarbons

18.3.1 Graphitic Carbon

Since the commercialization of LIBs, which are based on lithium cobalt oxide (LCO)-graphite paired rocking-chair system, the graphite has been utilized as the most widely used anode material. The graphite exhibits superior specific capacities (=372 mA h g^{-1}) and lower redox potentials (<0.2 V vs. Li/Li^+) than other metal oxide-based anode materials (Table 18.2) [14–16]. Moreover, the structural stability of the lithiated graphite provides the decent cycle performance than alloy-based materials (e.g., silicon, aluminum, and tin). Driven by the electrochemical reduction of the carbon host, the lithium ions are desolvated from the liquid electrolyte and migrate between graphene layers, leading to the storage of the intercalated lithium ions in a form of Li_xC_n compounds. The maximum lithium content stored in the graphite is known as LiC_6 (i.e., one lithium per a benzene ring unit). During the lithium intercalation process, two adjacent graphene layers existing in AB-stacked conformation allow the carbon-layers shift to AA-stacked conformation, thus the Li-C_6-Li-C_6-chain is arranged along the c-axis with slight increase (10.3%) in carbon interlayer distance (Fig. 18.3a).

The in-plane distribution of the lithium on the LiC_6 is elucidated from a staging formation. At the low concentration of guest species (i.e., lithium), the occupation of the neighboring unit is avoided, thus a periodic array of unoccupied layer gaps is shown in the intercalation process [15–19]. Figure 18.3b shows a schematic

Table 18.2 Comparison in basic physical/electrochemical properties of various anode materials. Reproduced with permission from Ref. [14]. Copyright 2012, Elsevier

Materials	Li	C	$Li_4Ti_5O_{12}$	Si	Sn	Sb	Al
Density	0.53	2.25	3.5	2.3	7.3	6.7	2.7
Lithiated phase	Li	LiC_6	$Li_7Ti_5O_{12}$	$Li_{4.4}Si$	$Li_{4.4}Sn$	Li_3Sb	LiAl
Theoretical specific capacity (mA h g^{-1})	3862	372	175	4200	994	660	993
Volume change (%)	100	12	1	420	260	200	96
Potential versus Li/Li^+	0	0.05	1.6	0.4	0.6	0.9	0.3

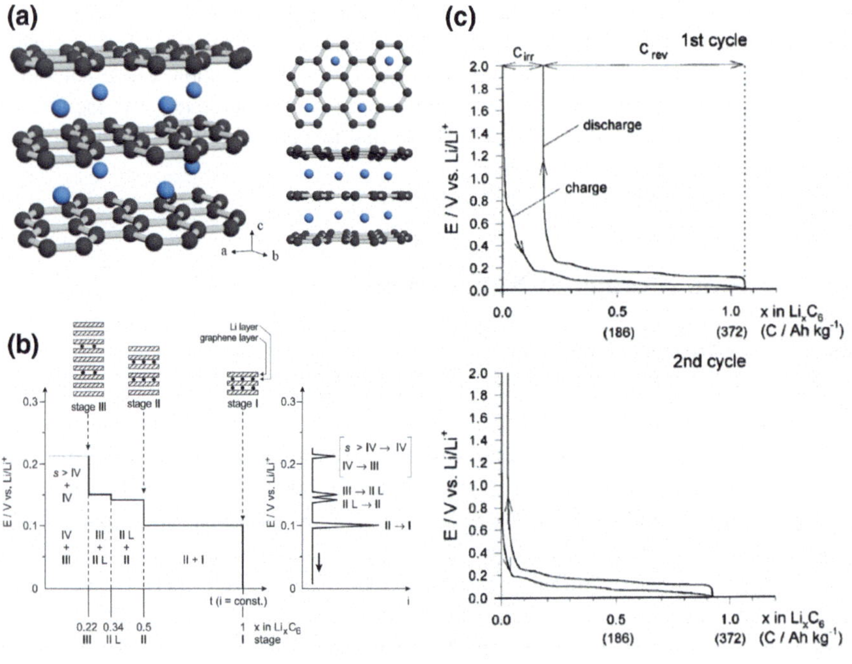

Fig. 18.3 **a** Schematic illustration of crystal structure of LiC_6 with AA stacking sequence of graphite. **b** Stage formation during lithium intercalation into graphite. **c** Galvanostatic charge/discharge profiles of graphite in first and second cycle. C_{irr} and C_{rev} is the irreversible and reversible capacity, respectively. Reproduced with permission from Ref. [15]. Copyright 1998, Wiley

illustration for the LiC_6 staging formation as a function of the potential during the galvanostatic lithiation of graphite, wherein the plateaus indicate two-phase coexistent regions. Due to the different lithium packing density, five different stages (s = IV, III, II L, II, and I) are sequentially developed during the lithiation. As expected, the final stage (s = I) or fully lithiated state of graphite is LiC_6.

Figure 18.3c shown the experimental galvanostatic charge/discharge curves for Li^+ intercalation into graphite. Note that the first lithiation capacity of the graphite significantly exceeds its theoretical specific capacity of LiC_6. The subsequent deintercalation recovers only about 80–90% of the lithiation charge. In the second and subsequent cycles, however, charge consumption is lower and charge recovery rate is close to 100% in accordance with the repeated lithiation/delithiation, of which the capacity is called "reversible capacity". The excess charge consumed in the first cycle is generally ascribed to solid–electrolyte interphase (SEI) formation on the graphite surfaces, which is a charge/electrolyte-consuming side reaction in the first few cycles [20] and the corresponding charge loss is frequently called "irreversible capacity". The Li-conducting and electron-insulating SEI layer prevents continuous decomposition of lithium source and electrolyte, therefore the SEI design is crucial to the cell performance.

18.3.2 Partially-Graphitic Carbon

Although the graphite provides the good cycle performance, its insufficient capacity (=372 mA h g^{-1}) compared to other lithium-storage materials (e.g., 3862 mA h g^{-1} for lithium and 4200 mA h g^{-1} for silicon) has remained a formidable challenge. To overcome this limitation, partially-graphitic carbons have been investigated as a promising alternative that can exceed the maximum lithium contents of graphite (i.e., crystalline LiC_6 conformation) [21–28]. Partially-graphitic carbonaceous materials are composed of crystalline (with planar hexagonal network) and amorphous (without far-reaching crystallographic order in the c-direction) phase (Fig. 18.4a). Grain size and composition between the crystalline and amorphous region depend on precursor materials and manufacturing conditions (e.g., temperature, heating time, gas flow, and so on) [21–27]. Depending on the structural characteristics, they are classified into following two types: graphitizable (called as "soft carbon") and non-graphitizable (called as "hard carbon") carbons. The formation of the soft/hard carbons is determined from carbonization temperature (i.e., graphitization) of precursors. Soft carbons are generally synthesized over 1000 °C and possess the disordered turbostratic structure. The additional heat treatment allows further graphitization of the turbostratic disorders, where the fraction of ordered graphitic layers increases [23]. In contrast, the morphology of the hard carbons, which are synthesized at the lower temperatures (500–1000 °C) compared to the soft carbons, is hardly changed after the additional heating [23].

The as-synthesized, graphitized structures and additional post-heating influence their lithiation behavior and resultant capacities. Soft carbons (e.g., turbostratic carbons, cokes, and carbon blacks) exhibit the lower specific capacity than that of the graphite [23], of which lower stoichiometric factor ($x < 1$, Li_xC_6) is due to the wrinkled/buckled segments that are unfavorable for lithium intercalation site and hinders shift to AA stacking by amorphous phases. Figure 18.4b shows the first lithiation/delithiation cycle of soft carbons. The lithium intercalation to the soft

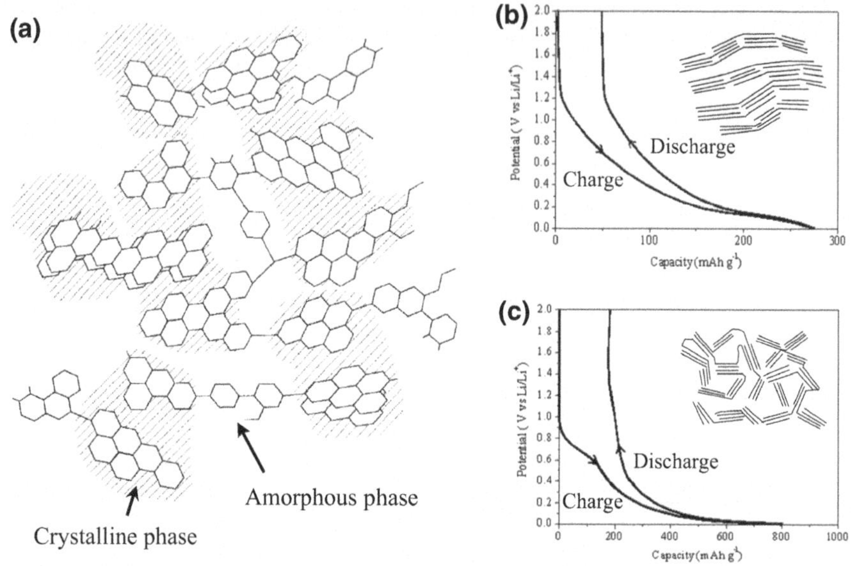

Fig. 18.4 **a** Schematic illustration of non-graphitic carbon. Reproduced with permission from Ref. [15]. Copyright 1998, Wiley. Galvanostatic charge/discharge profiles of **b** soft carbon and **c** hard carbon. Their conceptual illustration is shown in insets. Reproduced with permission from Ref. [16]. Copyright 2009, Wiley

carbon begins at the higher voltage (at ~1.2 V vs. Li/Li^+) compared to the graphite (at 0.1 V vs. Li/Li^+) with drawing fluent curves, moreover, the plateaus (at 0.1 V vs. Li/Li^+) hardly appear. Such ambiguous plateaus and lower capacities indicate that the turbostratic disorders largely provide the lower stoichiometric sites for lithium storage, although partial graphitic region contributes to the formation of crystalline LiC_6 similar to graphite. On the other hand, the reversible capacity (=400–2000 mA h g^{-1}) of the hard carbons is much larger than the theoretical capacity of the graphite, representing that the stoichiometric factor for lithium storage is significantly enhanced in the hard carbons (x = 1.2–5, Li_xC_6) as shown in Fig. 18.4c.

Dahn et al. explained the intercalation behavior in soft and hard carbons as follows [23]. The lithiation capacity and the degree of hysteresis tend to be improved in accordance with the increase in the hydrogen content of the carbons, indicating that the lithium adsorbs to hydrogen-adjacent sites on both sides of single-layered sheets that are arranged like a "house of cards" [15, 16, 23, 25]. In additions, to explain the high specific capacity of partially-graphitic carbons, several models have been proposed. The extra volume (sometimes, nanosized cavities [32]) emerged from the disordered structures could additionally store lithium [27]. This hypothesis has been supported from the higher gravimetric capacity of partially-graphitic carbons than graphite, while the difference in the volumetric one is negligible [28]. Sato et al. suggested that lithium occupation to the nearest neighboring sites in intercalated layers contributes to the capacity rise [29]. Meanwhile, a hypothesis based on the formation of lithium multilayers has been proposed by Yazami et al. [30]. On the other hand, Peled et al. elucidated that the extra capacity was attributed to mildly-oxidized sites in the partially-graphitic carbons, in which the edge planes between two adjacent crystallites and/or in the vicinity of defects and impurities allow the accommodation for lithium [31].

18.3.3 Fullerene

Several efforts have been undertaken to utilize fullerenes as lithium-storage materials in LIBs [33, 34]. However, in liquid electrolytes, the electrochemically reduced C_{60} (up to five electrons per a C_{60}) is dissolved into the electrolytes, which thus remains a formidable challenge to its application [33]. To avoid the dissolution issue, Armand et al. employed a PEO-based gel-polymer electrolyte and investigated the lithiation behavior in the fullerene [34]. The lithiated fullerene finally forms a unique structure of $Li_8(4Li^+)C_{60}{}^{4-}$ at maximum capacity, indicating that the extra four lithium per a C_{60} can be bound without generating cluster anions by electron transfer. Although this unprecedented behavior remains unresolved yet, further studies will provide a clue to develop high-capacity lithium-storage materials. Meanwhile, the hydrogenation ($x \leq 60$ for $C_{60}H_x$ and $x \leq 70$ for $C_{70}H_x$) of fullerenes shows an enhancement in the lithium storage and electrochemical stability [35]. The lithium adsorbs to hydrogen-adjacent sites similar to the partially-graphitic carbons as described above.

18.3.4 Carbon Nanotube

Carbon nanotubes (CNTs) have received significant attention as one of the most promising nanomaterials since Iijima's report in 1991 [36]. The CNTs mainly comprising carbon atoms display one-dimensional (1D) cylindrical shapes as if graphene sheets are rolled-up, which are categorized depending on a number of layers (Fig. 18.5a); single-walled (SWCNTs), double-walled (DWCNT), and multiwalled carbon nanotubes (MWCNTs). Moreover, the SWCNTs are further classified into two types (i.e., metallic and semiconducting SWCNTs) in terms of electronic properties, which depends on the chirality representing the rolling direction of the sheet [37]. The CNTs have been widely investigated in energy application due to their outstanding electronic [38]/thermal [39] properties, mechanical strength [40, 41], and high aspect ratio [36]. Moreover, the bundle formation of CNTs, owing to the strong van der Waals interaction among them [42], generates specific nano-interfaces (e.g., interstitial channels and external grooves), where the intense interaction with various molecules (e.g., gases [43–45], small alkanes [46, 47], linear aromatic molecules [48], and ions [49–52]) enables their wide uses in molecular storage applications. The lithium-storage behavior/mechanism at the nano-interfaces has been extensively investigated for energy storage applications. Theoretical calculations predicted that remarkable lithium storage (>1116 mA h g^{-1}) beyond the LiC_2 stoichiometry can be attainable via the nano-interfaces, which is significantly larger than that (=372 mA h g^{-1}) of the graphite [50–52] (Fig. 18.5c). However, experimental studies failed to fulfill the ideal capacity. Figure 18.5c shows the first lithiation step of the CNTs. The voltage plateau at ~0.8 V is ascribed to the formation of SEI layers on the CNT surfaces, due to the large specific surface area than other spherical carbon particles, the large consumption of current (i.e., irreversible capacity close to 1000 mA h g^{-1}) is observed. Nevertheless, the higher reversible capacities than the graphite are shown in the CNTs [50–65], which is possibly attributed to the presence of graphitic disorders (e.g., micro-cavities, edge sites, and defects) generated during the synthetic process of CNTs. In addition, several approaches have been undertaken to further improve the reversible capacities through the partial etching of tube-caps/side-walls [51]. However, the contribution of the nano-interfaces remains unclear because of their critically poor capacities compared to the theoretical value (>1116 mA h g^{-1}). Therefore, significant attentions should be devoted to elucidating the lithiation mechanism, which include the effect of nano-interfaces (especially, interstitial sites of close-packed CNT bundles [50–52]) and diffusion behavior [56, 57, 59, 61].

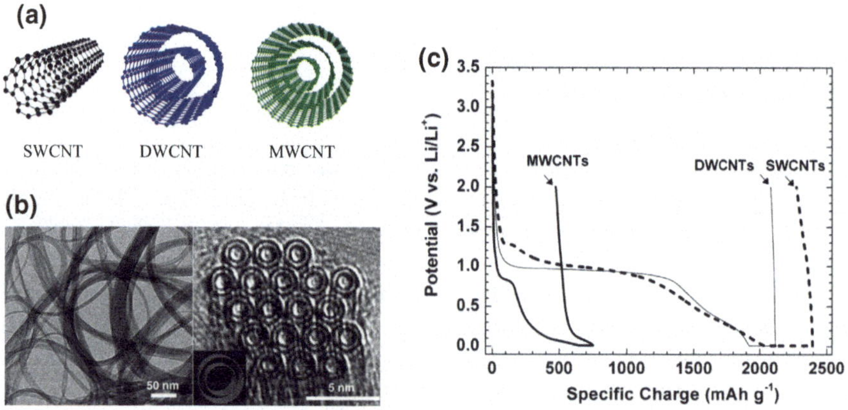

Fig. 18.5 **a** Schematic illustration of single-walled (SWCNT), double-walled (DWCNT), and multi-walled carbon nanotube (MWCNT). **b** TEM images of bundled DWCNTs and their cross-sectional view. Reproduced with permission from Ref. [60]. Copyright 2006, Wiley. **c** Galvanostatic charge/discharge profiles of SWCNT, DWCNT, and MWCNT. Reproduced with permission from Ref. [63]. Copyright 2009, Elsevier

18.3.5 Graphene

A single layer virtually cut out of the graphite structure is graphene, which is truly a two-dimensional (2D) single aromatic layer. The graphene has also attracted considerable attention due to its high specific surface areas (theoretical value (both sides) = $\sim$2600 $m^2\ g^{-1}$), outstanding electrical/thermal properties, and mechanical strength comparable to CNT [66–68]. Notably, the superior chemical stability and the wide electrochemical window make it a promising material for battery applications especially in lithium storage [69–78]. The lithium can be bound to both sides of graphene sheets. Moreover, open planar space sometimes facilitates multilayer adsorption of lithium. Therefore, the lithiation capacity in the graphene is expected to significantly increase compared to the graphite. However, synthetic graphenes typically exist in the shape of multilayer (10–20 layers)-stacked structures, due to the strong van der Waals force between the sheets, thus further works for improving the reversible capacities are needed for practical applications. Meanwhile, it was revealed that the extra capacity (with superior long-term cycling stability) could be obtained through introducing the nanopores/cavities, defects, oxygen-containing functional groups, and other heteroatom sites [72–78]. Regarding the nanopores/cavities, Honma et al. reported nanospace-defined graphene hybrids and their effects on electrochemical performance. The d-spacing of graphene nanosheet (GNS) was controlled by incorporating other nanocarbon materials such as CNT (GNS/CNT) and C_{60} (GNS/C_{60}) during exfoliation and reassembly of the GNS (Fig. 18.6a) [73]. As a consequence, the d-spacing of the hybrids shifted as follows: 0.365 (for the GNS) < 0.4 (for the GNS/CNT) < 0.42 nm (for GNS/C_{60}), which was well consistent with their capacities (540 (for the GNS) < 730 (for the GNS/CNT) < 784 mA h g^{-1} (for GNS/C_{60})) as shown in Fig. 18.6b, c.

18.4 Use of Nanocarbons in Li-Ion Batteries

18.4.1 Conductive Agents

One of the most essential prerequisite to improve electrochemical performance of LIBs is the development of well-networked electron-conducting pathway inbetween electrode materials. The use of nanocarbons (especially CNT and graphene) has obvious advantages over conventional carbon additives such as carbon blacks and acetylene blacks, due to their exceptional electrical conductivity and structural characteristics. The CNT and graphene featuring the high aspect ratio enable the well-developed electrical networks than spherical conductive carbons due to their long-range interconnectivity. Enormous efforts have been made to employ the nanocarbons as conductive agents [79–86] in LIB electrodes, with a focus on mixing methods such as hand-milling with a mortar [80, 81], ball-milling [79, 85],

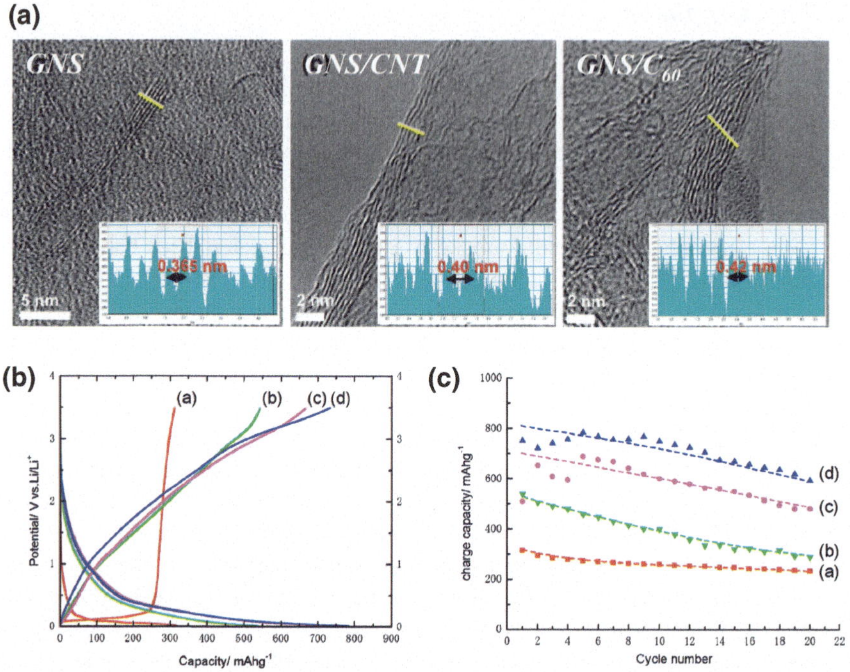

Fig. 18.6 **a** Cross-sectional TEM images of GNS families (GNS, GNS/CNT, and GNS/C_{60} from left) with almost the same numbers (5–6) of graphene stacking layers. **b** Galvanostatic charge/discharge profiles and **c** cycle performance of **a** graphite, **b** GNS, **c** GNS/CNT, and **d** GNS/C_{60}. Reproduced with permission from Ref. [73]. Copyright 2008, American Chemical Society

and sonication-mediated solution mixing [81]. The nanocarbons-containing LIB electrodes show significant improvement in reversible capacities than those with conventional conductive carbons, due to the well-interconnected electrical networks. The advantageous effects of the nanocarbons become more pronounced at higher charge/discharge current densities, where the nanocarbon percolation boosts the electrical conduction without hampering ionic transport, leading to remarkable improvement in electrochemical kinetics. The energy storage at high charge and discharge rates is a key technology for EV applications [86]. The use of nanocarbons will contribute to fast charging/discharging of LIBs for the EVs.

However, complete replacement of conductive agents (from spherical conductive carbons to CNTs) cannot always guarantee improvement of electrochemical performance. Dresselhaus et al. reported that optimal combination of spherical carbons and CNTs leads to the lowest electrical resistivity of the electrode (1 wt% CNTs with 2 wt% acetylene blacks), where the 0D/1D-combined structure acts as electrical/thermal/structural linking segments and provides suitable porosity, thereby shortening the electrolyte absorption time [85]. In addition, the acetylene blacks on outer surface of active materials can act as effective binders and suppress the unwanted electrolysis of the electrolyte solution. Therefore, rational combination of functional nanocarbons is crucial for their use as a conductive agent of LIBs.

18.4.2 Current Collectors

The current collector acts as a medium to allow electron transport between electrode materials and external circuit. Also, it contributes to mechanical/dimensional stability of the electrodes. Considering the abovementioned roles, high electronic conductivity and electrochemical stability are needed for the current collector. Among various materials, metal-based current collectors (e.g., aluminum and copper) have been commercially used in conventional LIBs. However, insufficient adhesion and contact area between the metal-based current collector and electrode materials give rise to the undesired growth of interfacial resistance and thereby degradation of cell performance. Moreover, the metal-based current collector occupies $\sim$30% of total electrode weight [87] in LIB cells, exerting a negative influence on specific energy density. Therefore, much attention should be paid to develop reliable alternatives. Among various materials, carbon-based current collectors (e.g., graphite sheets, carbon fiber mats, and carbon papers) have received significant attention due to the superior chemical stability and lightweight. In particular, the high electrical conductivity and structural benefits (i.e., large active surface area and percolation network) of the nanocarbons (especially, CNT and graphene) are beneficial for both the current collector and electrodes [88–92].

The rapidly approaching era of ubiquitous energy devices with design diversity has driven continuous demands for flexible and stretchable power sources. However, in the conventional LIBs, slurry-coated electrode materials are easily detached from the metal-based current collectors upon the repeated mechanical deformation due to the weak interfacial adhesion, eventually causing the rapid capacity fading and sometimes internal short-circuit-triggered safety failure. Nanocarbon-based current collectors have gained significant attention as an effective way to address the aforementioned issue because their mechanical strength and tolerance could contribute to the battery deformability. Lee et al. reported the highly flexible free-standing LIBs incorporating the CNT/cellulose nanofiber (CNF)-integrated architecture (heterolayered 1D nanomats) [93]. These LIBs were fabricated through vacuum-assisted filtration of the CNF and CNT/active-materials dispersions (Fig. 18.7a), where the CNT-containing electrodes ($LiFePO_4$ (LFP) for cathode or $Li_4Ti_5O_{12}$ (LTO) for anode) were assembled with the CNF-based separator (separator/electrode assembly (SEA) architecture). The resultant LIBs showed the superior flexibility (see Fig. 18.7b) without incorporating polymeric binders. As a consequence, the highly interconnected conduction path (for electrons) and the well-developed pores (for ions) enabled the significant improvement in the electrochemical performance and energy density (Fig. 18.7c). Moreover, such reliable performance was maintained after the repeated mechanical deformation (Fig. 18.7d).

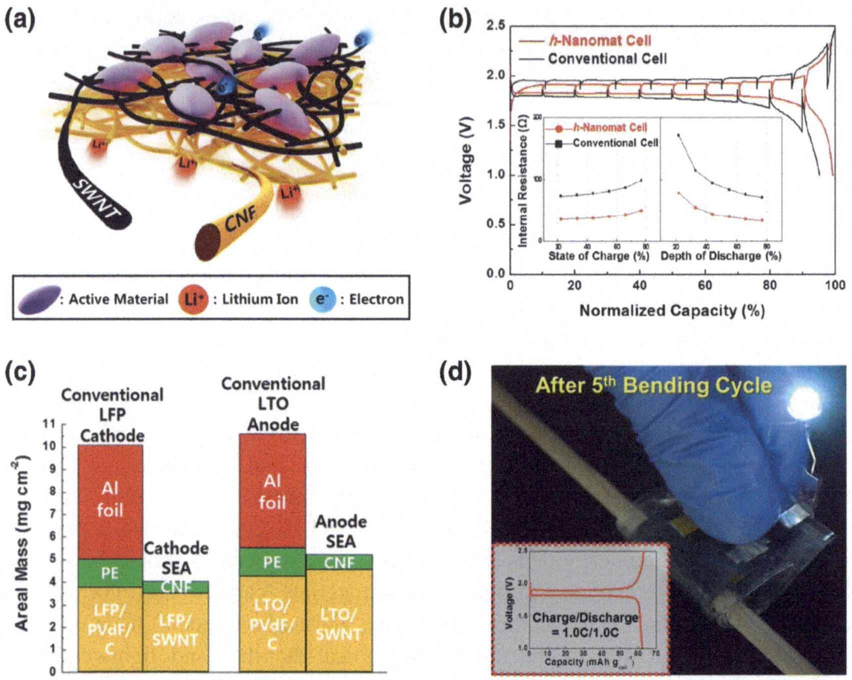

Fig. 18.7 **a** Schematic illustration of 1D nanobuilding block(CNF/SWCNT)-enabled structural uniqueness of SEA. **b** GITT profiles of cells, where the inset shows the variation of internal cell resistance as a function of state of charge and depth of discharge. **c** Comparison of areal mass between conventional electrodes and SEAs. **d** A photograph showing the electrochemical activity of h-nanomat cell after the 5 bending/unbending cycles (rod diameter = 5 mm), where the inset shows the corresponding charge/discharge profile at charge/discharge current density = 1.0/1.0 C. Reproduced with permission from Ref. [93]. Copyright 2014, American Chemical Society

18.4.3 Composites with Electrode Materials

Driven by exceptional structural/electrical properties, the nanocarbons could be used as a promising material for LIB electrodes. The agglomeration of electrode active materials and the mechanical rupture of electrodes are effectively mitigated due to the binding effect of the nanocarbons, providing the superior long-term stability of the LIBs. Moreover, they can reduce the amount of polymeric binders and metallic current collectors, thus beneficially contributing to specific capacities and energy densities. Here, we introduce several examples of nanocarbon composites (hybridized with metal oxides, sulfur, and silicon [88–90]) and their structural design for high-performance electrodes (e.g., porous sponge [94–97], core-shell [98, 99], and yolk shell [100]).

18.4.3.1 Sponge-Like Electrodes

The good mechanical strength and interconnected structure of the CNT and graphene enable the formation of free-standing electrodes in the form of aerogels and sponges, where the well-developed porous structure effectively facilitates ionic conduction as well as nanocarbon-induced electronic conduction. Moreover, the uniformly anchored active materials on the nanocarbon surface with the high specific surface area increase the electrochemically active sites, leading to significant improvement in the electrochemical kinetics and capacities [94–97]. The free-standing porous electrodes comprising nanocarbons and active materials are typically fabricated through direct/template growth and sol-gel assembly.

Mullen et al. reported 3D graphene/Fe_3O_4 sponges and their effects on the electrochemical performance as LIB anodes [94]. During the hydrothermal reaction of as-prepared graphene oxide (GO)/Fe_3O_4 suspensions, the Fe_3O_4 particles were anchored onto oxygen-containing crosslinking sites of the GO, then simultaneous reduction reaction of the GO to the graphene eventually formed the 3D porous graphene/Fe_3O_4 sponges (Fig. 18.8a). The hierarchical graphene aerogels showed the good cycling performance and delivered a favorable capacity of 363 mA h g^{-1} at a current density of 15 C (=4800 mA g^{-1} (charging time = $\sim$4 min)), while conventional graphene/Fe_3O_4 composites exhibited much smaller capacity of 150 mA h g^{-1} (Fig. 18.8b). It is expected that the well-connected 3D networks and porous structures, respectively, contribute to accelerating the electronic and ionic conductions, which mitigates the charge transfer resistance as shown in the electrochemical impedance spectroscopy (EIS) result (Fig. 18.8c).

Rubloff et al. also reported MWCNT/V_2O_5 sponge electrodes [95], which were fabricated through atomic layer deposition (ALD) of V_2O_5 onto MWCNT sponges. The MWCNT frame played multifunctional roles as the current collector, conductive agent, and mechanical binder (Fig. 18.9a). The V_2O_5 layer (4–6 nm, see Fig. 18.9b) homogeneously coated on the MWCNT sponge exhibited the superior specific capacity (818 μA h cm^{-2} at 1.0 C) than that (1.8 μA h cm^{-2}) of the planar V_2O_5 film deposited on a metal-based current collector (i.e., stainless steel disk), moreover, such beneficial effect was pronounced at a high rate of 50 C (155 μA h cm^{-2} for areal capacity, 21.7 mW cm^{-2} for power density).

Advantageous roles of the nanocarbon frames are similarly observed in post LIB systems (especially, lithium–sulfur (Li-S) batteries). Although the Li-S system has received significant attention owing to its high theoretical capacity (1672 mA h g^{-1}) and environmental benignity, serious drawbacks of the Li-S system hindered its practical applications. The intrinsically low electronic conductivity of sulfur and continuous dissolution/loss of sulfur species (i.e., lithium polysulfides) cause the initial capacity drop and subsequent fading thereof, moreover, the breakage of electrode structure due to the large volume change during the repeated charge/discharge has limited the long-term use of the Li-S batteries. Interestingly, nanocarbon-based porous structures (i.e., sponges and aerogels) provide ideal architectures for Li-S electrodes as follows: (1) high electrical conductivity, (2) ionic conduction through porous structures, and (3) chemical affinity with

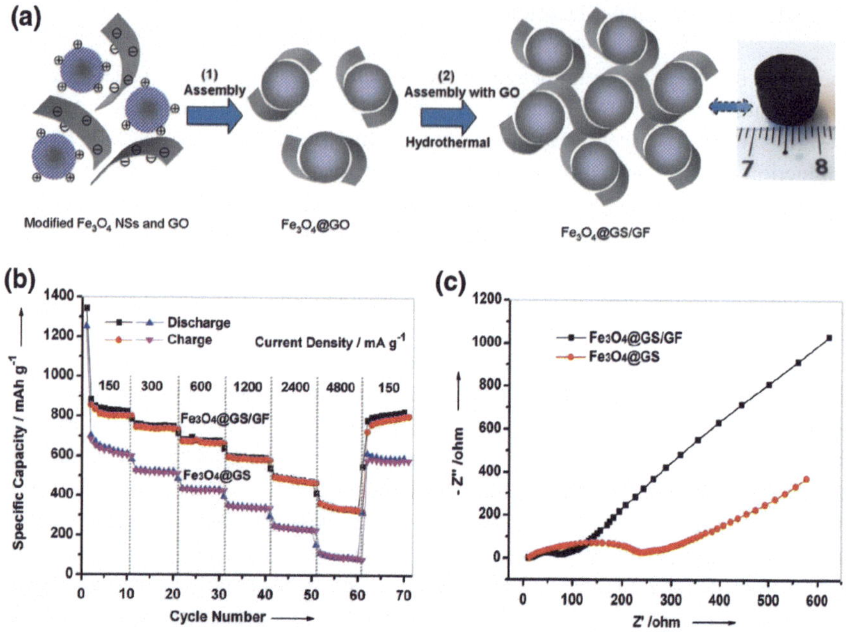

Fig. 18.8 **a** Fabrication process and photograph of Fe_3O_4@GS/CF sponge electrode. **b** Rate capability and **c** Nyquist plots of Fe_3O_4@GS/CF and Fe_3O_4@GS electrodes. Reproduced with permission from Ref. [94]. Copyright 2013, Wiley

lithium polysulfides to prevent their dissolution. Manthiram et al. developed free-standing graphene/sulfur sponge cathodes doped with nitrogen (N) and sulfur (S) atoms [96]. The GO/thiourea was hydrothermally reacted and freeze-dried. The sponge electrode exhibited high capacity (=1157 mA h g^{-1} at 0.2 C) and improvement in the long-term cycling stability (822 mA h g^{-1} after 100 cycles) without binders and metal current collectors. Considering that the N/S-doped graphene sponge electrodes provide favorable adsorption sites for the sulfur species, the synergistic effect between the graphene sponge and heteroatom doping enables notable improvement in the capacity and cyclability (Fig. 18.10).

Li et al. developed CNT/graphene-based 3D architectures via a solvothermal reduction of sulfur-impregnated CNT/GO hydrogels and followed by freeze-drying process [97], in which well-developed porous structures enabled high sulfur contents (~80%). The CNT-bridged graphene network dramatically improved the electrical conductivity (324.7% higher than the graphene-based electrode). In addition, bimodal mesoporous structures (pore size = 3.5 and 32.1 nm) accelerated the ionic conduction and thereby electrochemical kinetics, leading to the improvement in rate capability (1217 and 653.4 mA h g^{-1} at 0.1 and 4.0 C, respectively) and coulombic efficiency than that of the graphene sponge cathode.

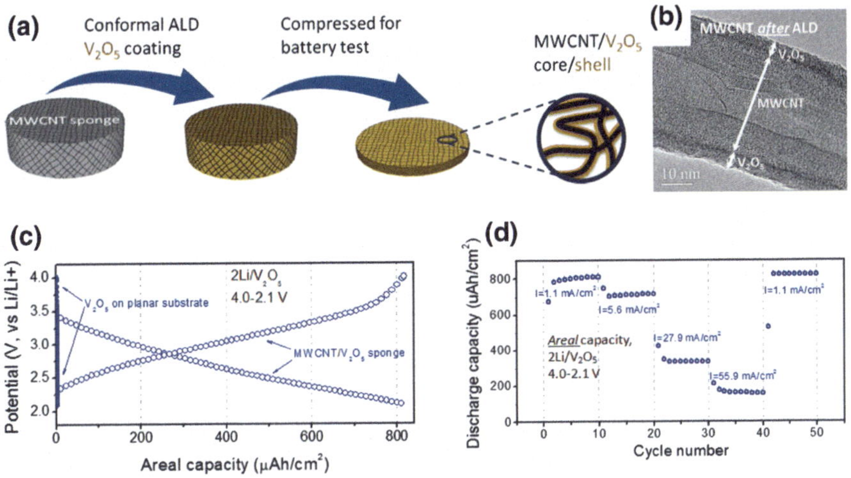

Fig. 18.9 **a** Schematic representation of fabrication for V_2O_5-coated MWCNT sponges. **b** TEM image of coaxial structures of V_2O_5/MWCNT. **c** Charge/discharge profiles (second cycle) of sponge electrode and planar electrode in the voltage range of 2.1–4.0 V versus Li/Li^+ with 1.0 C current. **d** Rate capability of V_2O_5/MWCNT sponge electrode in the voltage range of 2.1–4.0 V versus Li/Li^+. Reproduced with permission from Ref. [95]. Copyright 2012, American Chemical Society

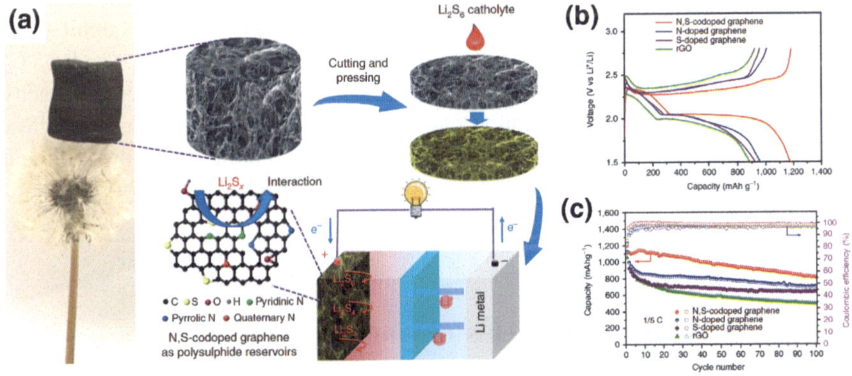

Fig. 18.10 **a** Schematic illustration of N, S-codoped graphene/sulfur sponge electrode. **b** Galvanostatic charge/discharge profiles (second cycle) and **c** cycle performance of the rGO, S-doped graphene, N-doped graphene and N, S-codoped graphene electrodes at 0.2 C rate. Reproduced with permission from Ref. [96]. Copyright 2015, Nature Publishing Group

18.4.3.2 Core-Shell Structures

The nanocarbons have been extensively investigated as shell substances for electrode active materials (especially, silicon) that suffer from low electrical conductivity and high volume expansion during charge/discharge process. Although the silicon has attracted wide attention owing to the ultra-high theoretical capacity (4200 mA h g^{-1}) and natural abundance, its large volume change (~300%) and low electronic conductivity hindered its practical applications. Here, we introduce nanocarbon-based core-shell structures, with a particular focus on application to silicon anodes.

Sohn et al. synthesized nanosize Si particles coated with ultra-thin (5–10 nm) amorphous carbon shells (n-Si@C) through calcination of poly(vinyl alcohol) (PVA)-coated Si particles at 750 °C [98]. The n-Si@C exhibited good capacity retention (1800 mA h g^{-1} after 50 cycles) than that (500 mA h g^{-1}) of pristine nano Si particles (n-Si), possibly due to the carbon-derived electrical conductivity and buffering effect against to volume change of the Si (Fig. 18.11). In addition, Choi et al. reported the evolved Si@C architecture in a form of hierarchical fibers via the electrospinning and subsequent carbonization [99]. In details, the precursor fiber comprising the Si/poly(methyl methacrylate) (PMMA)-core and polyacrylonitrile (PAN)-shell was fabricated through the dual-electrospinning, then the core and shell polymers were respectively evaporated and carbonized after the calcination. The obtained hollow carbon fiber filled with n-Si (Si loading: ~0.6 mg cm^{-2}) showed the improved reversible capacity (1305 mA h g^{-1} at 0.1 C), rate capability, and cycling stability (Fig. 18.12).

Cui et al. prepared pomegranate-inspired Si@C yolk shell structure [100]. SiO_2-coated Si nanoparticles (Si@SiO_2) were self-assembled in a water-in-oil emulsion, then the secondary particles (1–10 μm) were sequentially coated by resorcinol-formaldehyde resin and carbonized. Finally, the Si@C yolk shell was formed after chemical etching of the SiO_2 layer. The secondary Si architecture showed the significant improvement in specific capacity (2350 mA h g^{-1} at 0.05 C) and capacity retention (97% after 1000 cycles), moreover, the areal capacity reached up to 3.1 mA h cm^{-2} (areal mass loading = 3.12 mg cm^{-2}). The hollow carbon shell is expected to be favorable for mitigating the pulverization of Si and continuous SEI formation, resulting in the good electrochemical performance as described above (Fig. 18.13).

18.5 New Application Fields of Nanocarbons

18.5.1 Flexible Batteries

Conventional LIBs are usually rigid and heavy, which is not suitable for flexible/wearable electronic devices with various form factors. The conventional electrodes are fabricated by casting electrode slurry on top of metal current collectors followed by drying and pressing. Thus, the electrode active particles are easily delaminated from the current collector upon the repeated deformation stress, resulting in loss of electrochemical performance. Electrode fabrication based on the nanocarbons can be suggested as a new solution for flexible and deformable electrodes.

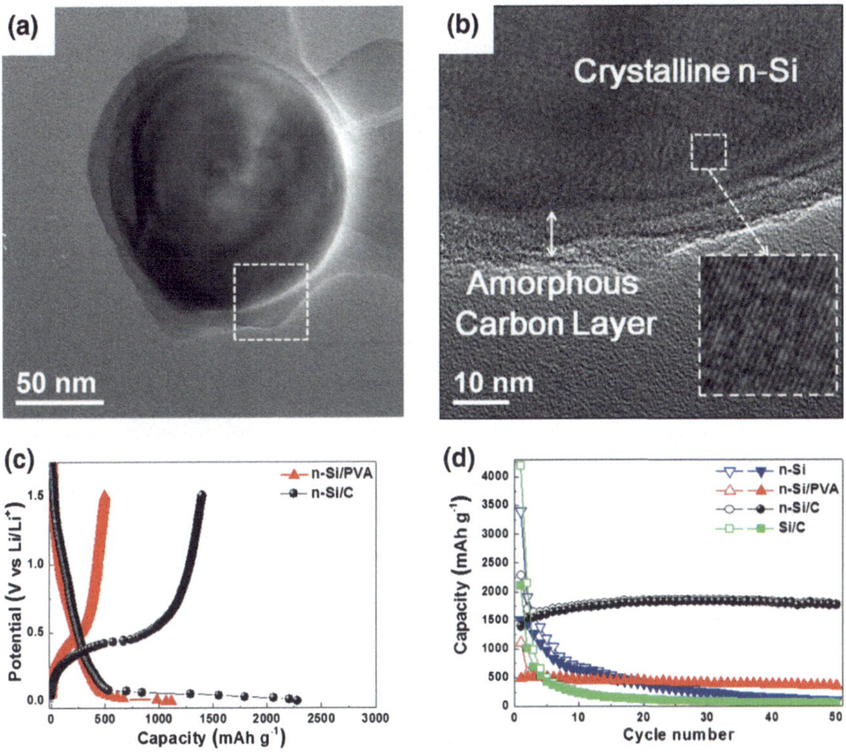

Fig. 18.11 **a** Bright field TEM image and **b** HRTEM image of n-Si/C. **c** Galvanostatic charge/discharge profiles of n-Si/PVA and n-Si/C. **d** Comparison of cyclability of n-Si, n-Si/PVA, n-Si/C, and Si/C electrodes at constant current of 100 mA g^{-1}. Reproduced with permission from Ref. [98]. Copyright 2012, Elsevier

Weaving or knotting is a promising technology for fabrication of thin and flexible electrodes. Carbon fiber textile or cloth exhibits excellent mechanical/electrical properties, which can replace conventional metal current collectors. Recently, many conductive textile-based electrodes have been reported. They are prepared by coating of cotton/polymer textile with various nanocarbons or carbonization of the polymer-based cloth. Textile LIBs with high areal mass loading were developed using the 3D porous textile current collector (Fig. 18.14a, b) [101]. The textile current collector-containing electrode was fabricated by simple dip-coating of polyester textile onto CNT dispersion, followed by coating of electrode slurry. The flexible textile current collector showed highly improved electronic conductivity of ~1300 S cm^{-1}, leading to better electrochemical performance than that of flat metal current collectors. As shown in Fig. 18.14c, the overpotential for LFP-containing 3D textile electrode was 50 mV, but that of Al foil was 100 mV, which was attributed to excellent ion transport from the porous electrode structure and also from the facile electron transport through the embedded CNT (Fig. 18.14d).

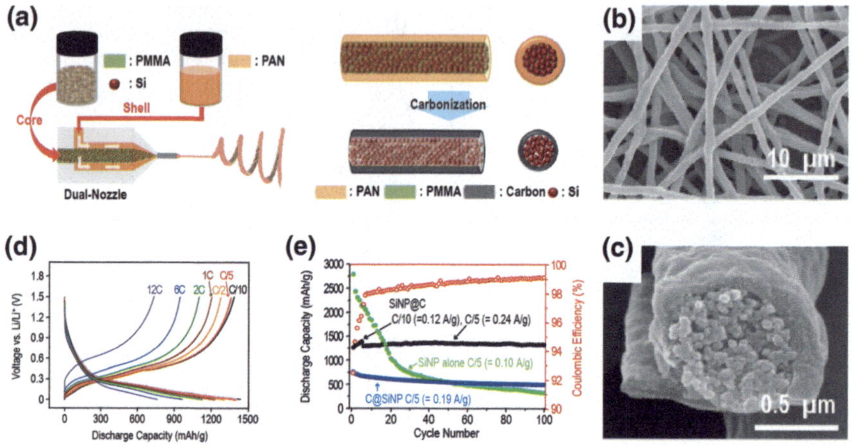

Fig. 18.12 **a** Schematic illustration of electrospinning and subsequent carbonization steps for 1D fiber Si/C electrode. **b** SEM images of Si@C fiber electrode and **c** cross-sectional view. **d** Rate and **e** cycle performance of Si@C at 0.2 C rate. Reproduced with permission from Ref. [99]. Copyright 2012, American Chemical Society

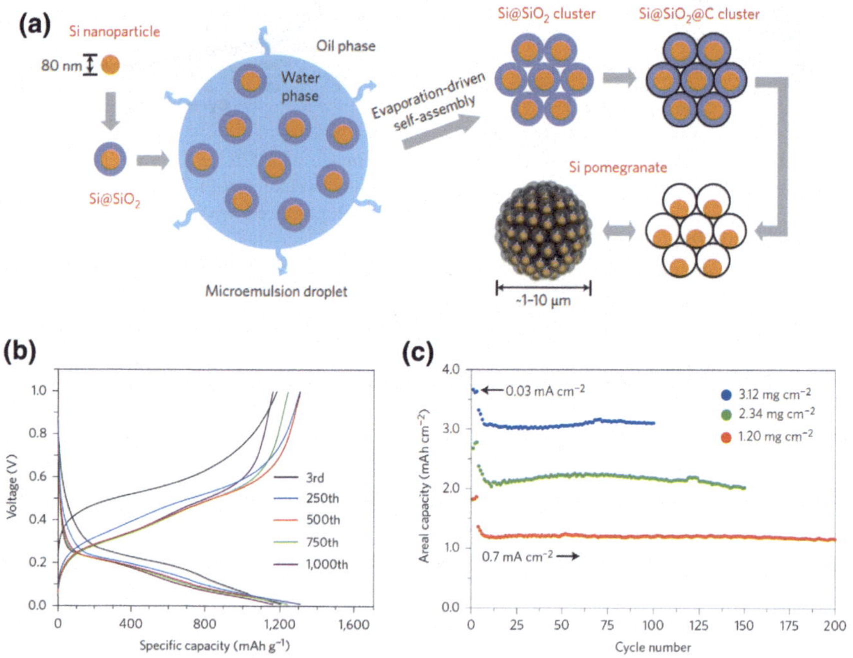

Fig. 18.13 **a** Schematic illustration of fabrication for silicon pomegranates. **b** Voltage profiles of the yolk shell Si@C for third, 250th, 500th, 750th, and 1000th cycle. **c** High areal mass loading test of the yolk shell Si@C. Reproduced with permission from Ref. [100]. Copyright 2014, Nature Publishing Group

The form factor of batteries plays a viable role for their application in flexible electronic devices. The above-described sponge/textile-based configurations as well as wire-shaped electrodes were explored as promising candidates. Peng et al. demonstrated twisted, aligned MWCNT/Si composite wire electrode for potential use as LIB anode [102]. The wire-shaped electrode was fabricated through spinning of Si deposited MWCNT arrays, which exploit the high specific capacity of Si, good conductivity of MWCNT, and interstitial buffer space for volume change of the Si. The mechanically deformable wire electrode exhibited a high specific capacity of 1970 mA h g^{-1} after 50 cycles, which was higher than pure Si with 1648 mA h g^{-1} at 30 cycles (Fig. 18.15). Furthermore, Peng and coworkers also reported stretchable LIB with MWCNT/LTO or LMO yarns-type electrodes [103].

The aligned arrays of nanostructured CNTs [104] and several LIB-based materials [105–108] were investigated in LIBs, due to their high specific surface area and mechanical flexibility. Yu et al. reported a flexible and self-supported carbon coated LTO (LTO-C) nanotube array electrode [108]. ZnO nanorod arrays were used as a template of TiO_2 nanotube arrays and then LTO nanotube arrays

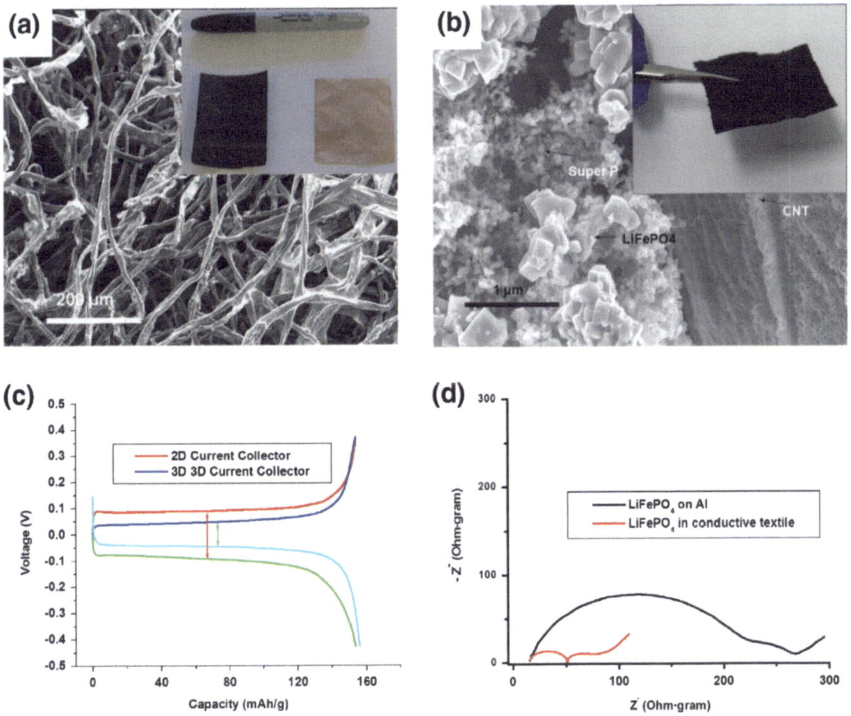

Fig. 18.14 SEM images of **a** CNT-coated conductive textile and **b** interface between LFP and conductive materials after electrode slurry coating. **c** Voltage profiles and **d** EIS analysis of 3D textile electrode and pristine electrode. Reproduced with permission from Ref. [101]. Copyright 2011, Wiley

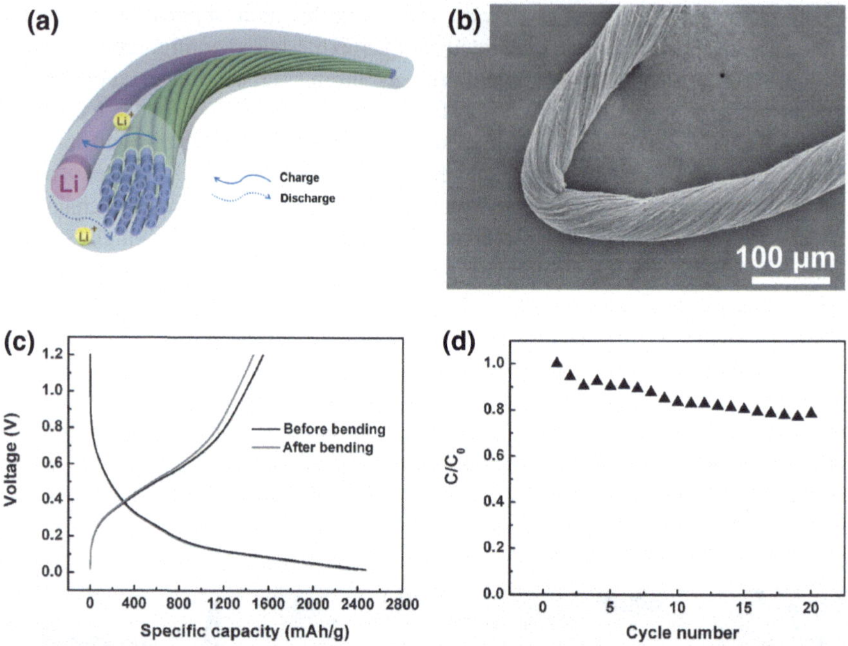

Fig. 18.15 **a** Schematic illustration of wire-type half-cell based on aligned MWCNT/Si fiber electrode. **b** SEM image of bent MWCNT/Si fiber electrode. **c** Voltage profiles of MWCNT/Si fiber electrode half-cell before/after 100 bending deformation. **d** Cycle performance of MWCNT/Si fiber electrode after 100 bending deformation at current density of 2 A g^{-1}. Reproduced with permission from Ref. [102]. Copyright 2014, Wiley

were obtained by lithiation and finally, LTO-C nanotube array electrodes were fabricated by carbonization of adsorbed glucose. The LTO-C nanotube array electrode exhibited good rate capability (80 mA h g^{-1} at 100 C) and long cycle life (7% capacity loss after 500 cycles at 10 C), which were attributed to the increased specific surface area and electronic conductivity via the inner and outer surface coating of carbon layers (Fig. 18.16).

18.5.2 Paper Batteries

Lignocellulose, due to its natural abundance, environmental friendliness, recyclability, and dimensional stability, has been widely used as a reliable/sustainable basic constituent for a vast variety of materials/devices in our daily lives [109–111]. Recently, driven by its exceptional physical/chemical properties, lignocellulose has vigorously extended its potential application toward rapidly growing industrial fields including mobile/intelligent electronics, displays, and sensors, in addition to traditional use in paper and packaging substances [112–117]. More notably, as a

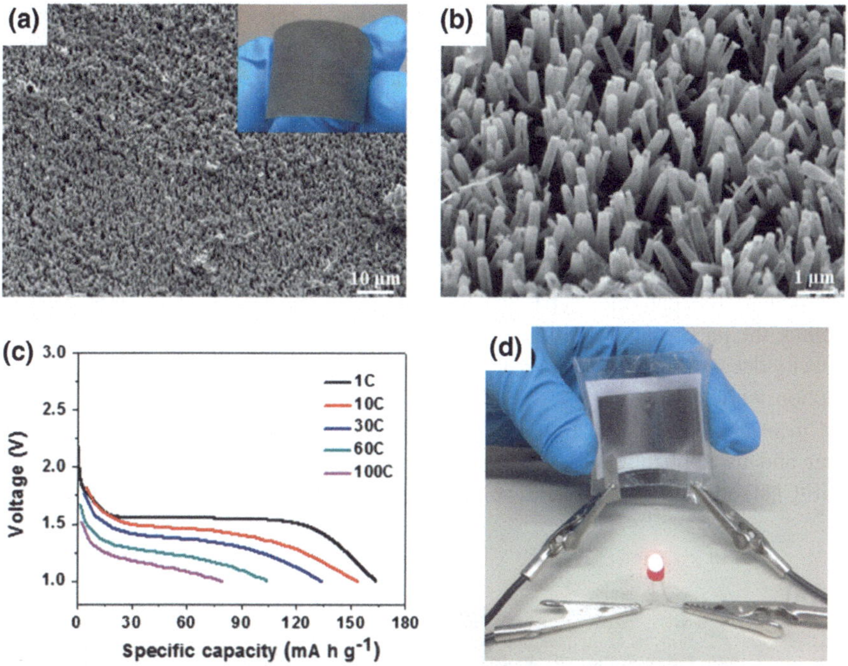

Fig. 18.16 SEM images of LTO-C nanotube arrays electrode in **a** low and **b** high magnification. **c** Galvanostatic charge/discharge profiles at different rate from 1 to 100 C. **d** Digital image of red LED powered by flexible LTO-C nanotube battery under bent state. Reproduced with permission from Ref. [108]. Copyright 2014, American Chemical Society

promising building element for next-generation flexible paper energy storage systems, lignocellulose has been extensively investigated. CNF, which are a main constituent of cellulose and characterized with a high aspect ratio, have brought significant benefits to design/fabrication of energy storage systems which lie far beyond those achievable with traditional battery technologies. To ensure electronic network as a prerequisite for cellulose-based battery electrodes, the cellulose is combined with various conducting materials such as CNTs, graphenes, conductive metal oxides, metal nanowires, and conductive polymers through molecular mixing or physical integration.

Cui et al. reported a fully-integrated flexible lithium-ion paper battery [118]. Conductive SWCNT films were used as current collectors for both anodes and cathodes. The electrode active materials were coated on the surface of SWNT films and the physically integrated electrodes were delaminated from a stainless steel substrate. The highly porous paper acts as both a mechanical support and also a separator membrane. Due to the flexibilities of CNT current collectors, the flexible paper LIB successfully lighted up an LED device even after being folded (Fig. 18.17b). The flexible paper battery also showed excellent cycle performance.

The superiority of this paper LIB over the previously reported commercial flexible cells was depicted in Fig. 18.17d as a function of total thickness and energy density.

Leijonmarck et al. developed the flexible single-unit paper battery, where cathodes, separator membranes, and anodes were seamlessly integrated [119]. Notably, CNFs was utilized both as a binder and a separator membrane. The paper battery was made through a traditional papermaking process by sequential filtration of aqueous anode/separator/cathode dispersions. After the sequential paper-making process, thin (250 μm) and flexible paper battery were fabricated. The cross-sectional SEM image showed three tightly integrated layers, which were assigned to the cathode, separator, and anode, respectively. After hundreds of cycling, the paper battery exhibited the meaningful level of capacity retention. The capacity decay is believed to originate from loss of cyclable lithium or parasitic side reaction with water.

Cui et al. reported the lightweight conductive CNF paper coated with Si thin layer and its application as LIB anodes [120]. Mechanically flexible, conductive CNF/CNT aerogel paper could be cut into different sizes and shapes with sheet

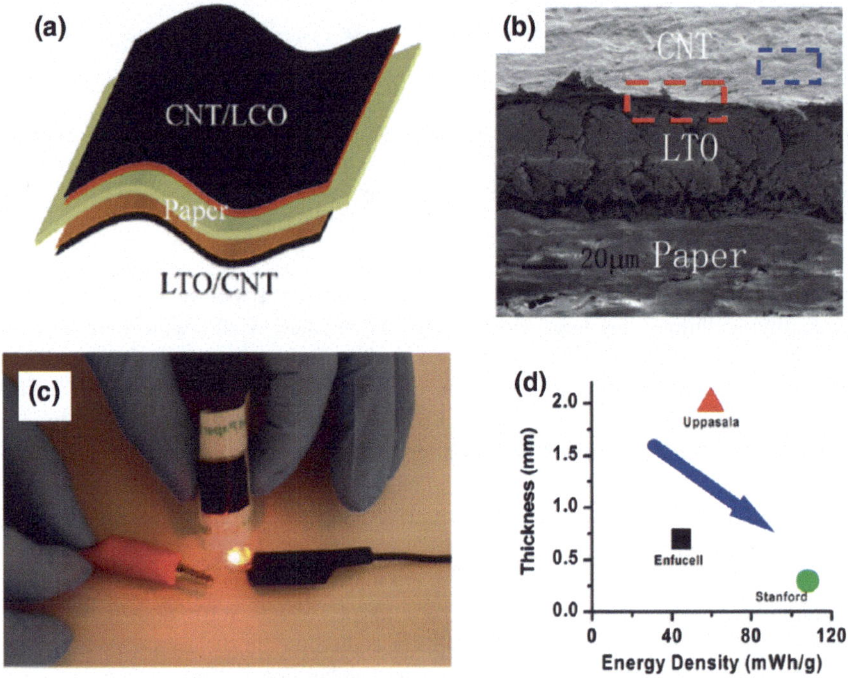

Fig. 18.17 **a** Schematic illustration of paper battery with laminated LCO/CNT and LTO/CNT on both sides of the paper substrate. **b** Cross-sectional SEM image of laminated paper battery. **c** Digital image of red LED powered by flexible paper battery. **d** Comparison in thickness and energy density between the paper batteries. Reproduced with permission from Ref. [118]. Copyright 2010, American Chemical Society

resistance of ~80 Ω sq^{-1}. After the plasma-enhanced CVD deposition, the Si thin layer-coated paper was still flexible (Fig. 18.18b). The cycling performance of the paper anode was shown in Fig. 18.18c, d. The discharge capacity retention was still 83% (compared to second cycle) and 77% (compared to first cycle) after 100 cycles. The 3D porous structure can accommodate volume change of Si during repeated charge/discharge. In comparison to conventional metal current collectors, the CNF/CNT conductive paper shows good mechanical tolerance and thus withstands volume expansion of Si.

In addition to the aforementioned SEA architecture, Lee et al. developed the hetero-nanonet (HN) paper batteries based on 1D building blocks of CNFs/MWCNTs [121]. The HN paper electrodes consisted of CNF/MWCNT hetero-nanonets integrating active materials and microporous CNF separator membranes. In contrast to conventional slurry casting-based electrodes, the paper electrodes were fabricated using vacuum-assisted filtration analogous to traditional papermaking process (Fig. 18.19a). Because of its 1D building blocks-driven flexible characteristics, the self-standing, ultrathick (thickness ≈ 1400 μm) LFP

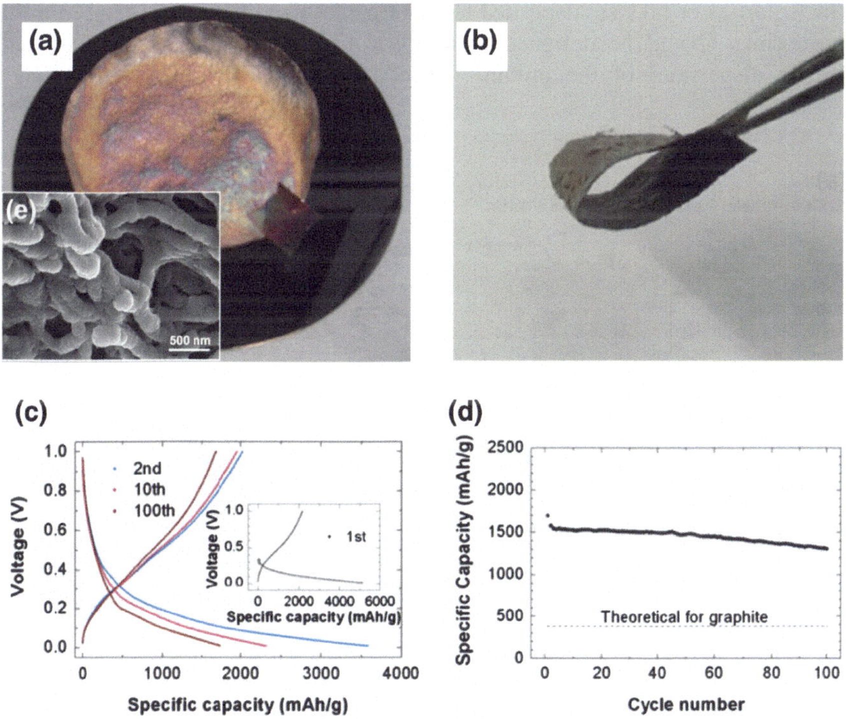

Fig. 18.18 **a** Digital image (inset = SEM image) of Si-coated conductive nanopaper. **b** Digital image of bent Si-conductive nanopaper. **c** Voltage profiles and **d** cycle performance of Si-coated paper electrode at 0.2 C rate. Reproduced with permission from Ref. [120]. Copyright 2013, Elsevier

cathode was successfully fabricated without disrupting its dimensional stability. By comparison, for the conventional LFP cathode prepared by typical slurry coating process, the serious problem in structural integrity (specifically, detachment of electrode components from an aluminum current collector) was observed even at the cathode thickness of ≈500 μm. Such a novel electrode structure brought significant improvement in the mechanical flexibility, reaction kinetics, and electrode mass loading. More notably, the metallic current collectors-free CNF/MWNT heteronets allowed the multiple stacking of paper electrodes in series, leading to the user-tailored, ultrathick (i.e., high-mass loading) electrodes with reliable electrochemical performance. In addition, the heteronet-enabled mechanical compliance of the paper electrodes, in combination with readily deformable CNF separators, allowed the fabrication of paper crane batteries via origami folding technique.

18.5.3 *Printable Batteries*

The printing technology is a facile and reproducible manufacturing process in which slurries or inks are deposited to make predesigned patterns [122, 123]. The slurries/inks should be designed to fulfill requirements (such as rheology and particle dispersion) of the printing process. Development of printed batteries

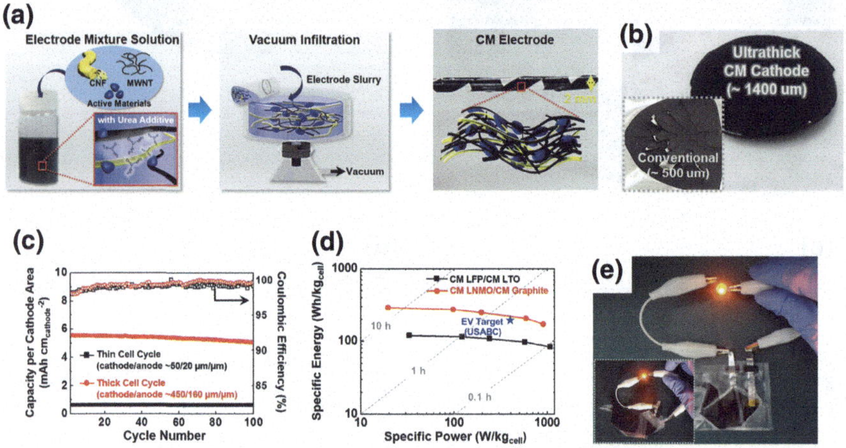

Fig. 18.19 **a** Schematic representations and photographs depicting the overall fabrication of paper electrodes. **b** A photograph depicting the self-standing, dimensionally tolerable ultrathick (thickness ≈ 1400 μm) paper electrode. Inset image shows the poor structural stability of the conventional cathode (thickness ≈ 500 μm). **c** Cycle performance of single-unit paper cell with different active material loading. **d** Ragone plot of single-unit paper cell, where the gravimetric energy/power densities of cells are determined on the basis of cell mass. **e** Digital image showing the electrochemical activity of paper crane cell. Reproduced with permission from Ref. [121]. Copyright 2015, Wiley

involves the design and fabrication of battery component slurries/inks. Most studies of the printed batteries have been devoted to the development of printed electrodes. However, in order to reach an ultimate goal of so-called “all-printed-batteries”, printed separator membranes and printed electrolytes should be also developed along with the printed electrodes.

A primary step for developing printed electrodes is the preparation of electrode slurry/inks, which should be tailored to ensure process compatibility with printing techniques [123]. Key requirements for the printed electrode slurry/inks include: (1) dispersion state of components, (2) rheological properties (i.e., viscosity and viscoelasticity) tuned for specific printing processes, (3) structural/dimensional stability after printing (e.g., cohesion between particles, adhesion with substrates and mechanical tolerance upon external stress). Among various electrode components, electrically conductive additives are essentially employed to facilitate electron transport in electrodes. Of readily available conductive additives, carbonaceous substances such as CNTs and graphenes have been extensively investigated due to their high electrical conductivity and also affordable electrochemical capacitance which is often beneficial for electrical double layer capacitors (EDLCs) [124]. A major challenge facing the carbonaceous substances is their dispersion in water and other polar solvents. In particular, for CNT cases, strong inter-tube affinity and intrinsic hydrophobicity pose a stringent problem for securing good dispersion state [124, 125]. In most studies, the dispersion of CNTs has been achieved with an assistance of CNT-dispersing agents such as SDBS (sodium dodecylbenzene sulfonate), and surface functionalization of CNTs. An important note is that these additional procedures for CNT dispersion should not impair intrinsic electrical conductivity of CNTs.

To develop printed electrodes with design versatility, the current collectors should be also printable in addition to the other printed battery components. Currently widespread current collectors in commercial electrodes are metallic foils, however, their predetermined shape and mechanical stiffness pose impediments to diversifying design and form factors of the printed batteries. Unfortunately, only a few works have been reported on the printed current collectors. Cui et al. presented the CNT-coated conductive paper current collector, which showed a sheet resistance of ~ 1 ohm sq^{-1} [126]. Another approach was the utilization of a spray coating method to produce a conductive CNT current collector. The spray coating solution consisted of CNT/carbon black (=8/2 (w/w)) in NMP solvent [127]. The sheet resistance of the resulting spray-coated CNT current collector was found to be as low as ~ 10 ohm sq^{-1}. In addition to the CNT current collector, a spray-coated metal current collector was prepared using Cu (copper) inks. This result underscores the potential applicability of metallic inks as a printed current collector for use in printed batteries.

Ajayan et al. reported the painted battery using a spray printing technique [127]. The battery components, including the electrodes, separator membranes, and current collectors, were prepared in the form of paint solutions and then spray-coated sequentially to construct the painted battery. The paints of the aforementioned battery components were sprayed through shadow masks with the predetermined geometries, while temperature of substrates was controlled in the range of 90–120 °C to remove solvents in the painted components. The painted battery provided an areal capacity of >1 mA h cm^{-2} with a working voltage of ~2.3 V. The battery was fabricated with a large area of 25 cm^2, demonstrating the scalability of the adopted spraying process. Also, the painted battery was expected to be integrated with a photovoltaic panel to develop energy conversion/storage hybrid devices for various outdoor applications (Fig. 18.20).

Lee et al. demonstrated solid-state flexible supercapacitors which were fabricated directly on conventional A4 paper using a household desktop inkjet printer [128]. The new supercapacitors looked like typical inkjet-printed letters or patterns commonly found in office documents. To attain high-resolution print images on A4 paper, a CNF nanomat was inkjet-printed on A4 paper in advance as a primer layer. The Ag nanowires were introduced onto the SWCNT/AC electrodes in order to improve electrical conductivity of the resulting inkjet-printed electrodes, which was followed by the SWCNT-assisted photonic interwelding of the Ag nanowire networks. The ([BMIM][BF_4]/ETPTA) solid-state electrolyte was applied to the inkjet-printed electrodes via the same inkjet printing, followed by the UV-crosslinking process for solidification. The inkjet-printed supercapacitors exhibited the reliable electrochemical performance over 2000 cycles as well as the good mechanical flexibility. The inkjet-printed supercapacitors were easily connected in series or in parallel without the extra aid of metallic interconnects, thereby enabling the user-customized control of the cell voltage and capacitance. More notably, the inkjet-printed supercapacitors with the computer-designed artistic patterns and letters were fabricated and

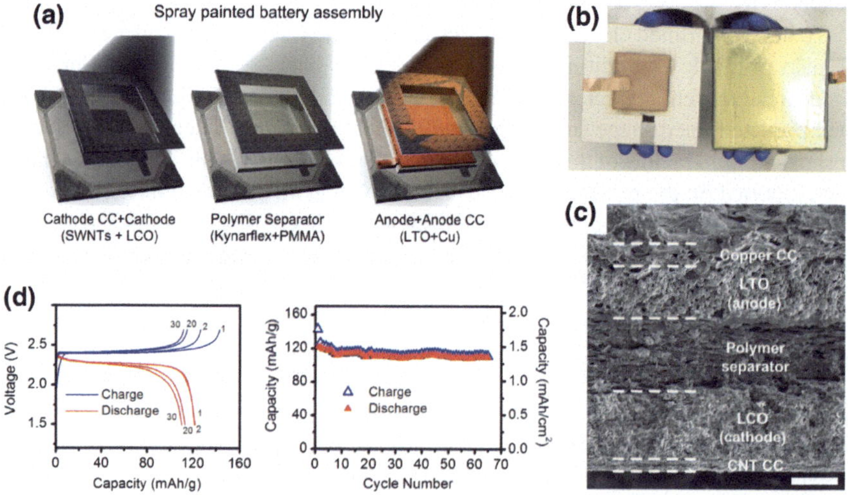

Fig. 18.20 a Schematic representation of sequential spraying-based battery fabrication. b Glazed ceramic tile with spray pained Li-ion cell (left) before and (right) after packaging. c Cross-sectional SEM image of the spray painted full cell showing its multilayered structure. d Voltage profiles and cycle performance of the painted battery at C/8 rate. Reproduced with permission from Ref. [127]. Copyright 2012, Nature Publishing Group

also aesthetically unitized with other inkjet-printed art images and smart glass cups, underscoring their exceptionally versatile aesthetics and potential applicability as a new class of object-tailored power sources for use in forthcoming IOTs.

18.6 Perspective and Outlook

In summary, we have described the current status and challenges of nanocarbons, with a particular focus on their applications to lithium-storage anode materials, conductive agents, current collectors, and structure-directing substance for electrodes. Enormous efforts have been undertaken to utilize the nanocarbons (i.e., partially-graphitic carbons, CNTs, and graphenes) as a potential alternative to graphite anode materials. However, there is a significant gap between the theoretical calculation and experimental results, which thus needs in-depth understanding of lithiation mechanism as a prerequisite, including interfacial phenomena and diffusion behavior on CNT and graphene surfaces. Due to the outstanding electrical/thermal conductivity, mechanical strength, and structural benefits (i.e., high aspect ratio and specific surface area), the nanocarbons (especially, CNTs and graphenes) are highly desirable for use in LIB electrodes and current collectors. The unwanted agglomeration of electrode active materials and the mechanical rupture of electrodes are effectively mitigated due to binding effect of the nanocarbons, providing the superior long-term stability of the LIBs. Moreover, they can reduce the content of polymeric binders and metallic current collectors in electrodes, eventually contributing to the increase in specific capacities and energy densities. The structural/electrochemical uniqueness allow the realization of flexible and free-standing electrodes with reliable electrochemical performance. Future research direction of the nanocarbons in energy storage systems will be dedicated to application extension to new battery components as a promising building block and understanding of their physicochemical phenomena at bulk and interface. We envision that the nanocarbons will hold great promise as a versatile and effective element in various energy storage systems, in addition to LIB materials presented herein.

Acknowledgements This work was supported by the Basic Science Research Program (2018R1A2A1A05019733) and Wearable Platform Materials Technology Center (2016R1A5A1009926) through the National Research Foundation of Korea (NRF) funded by the Ministry of Science, ICT and future Planning. J. Yoo and S.-Y. Lee are co-correspondence.

References

1. Tarascon J-M, Armand M (2001) Issues and challenges facing rechargeable lithium batteries. Nature 414:359–367
2. Tollefson J (2008) Car industry: charging up the future. Nat News 456:436–440
3. Scrosati B, Garche J (2010) Lithium batteries: status, prospects and future. J Power Sources 195:2419–2430

4. Thackeray MM, Wolverton C, Isaacs ED (2012) Electrical energy storage for transportation—approaching the limits of, and going beyond, lithium-ion batteries. Energy Environ Sci 5:7854–7863
5. Diouf B, Pode R (2015) Potential of lithium-ion batteries in renewable energy. Renew Energy 76:375–380
6. Bauer S (2013) Flexible electronics: sophisticated skin. Nat Mater 12:871–872
7. Lee S-Y, Choi K-H, Choi W-S, Kwon YH, Jung H-R, Shin H-C, Kim JY (2013) Progress in flexible energy storage and conversion systems, with a focus on cable-type lithium-ion batteries. Energy Environ Sci 6:2414–2423
8. Gogotsi Y (2014) Materials science: energy storage wrapped up. Nature 509:568–570
9. Chabot V, Higgins D, Yu A, Xiao X, Chen Z, Zhang J (2014) A review of graphene and graphene oxide sponge: material synthesis and applications to energy and the environment. Energy Environ Sci 7:1564–1596
10. Liu W, Song MS, Kong B, Cui Y (2017) Flexible and stretchable energy storage: recent advances and future perspectives. Adv Mater 29:1603436
11. Sun H, Xu Z, Gao C (2013) Multifunctional, ultra-flyweight, synergistically assembled carbon aerogels. Adv Mater 25:2554–2560
12. Baughman RH, Zakhidov AA, De Heer WA (2002) Carbon nanotubes—the route toward applications. Science 297:787–792
13. Geim AK (2009) Graphene: status and prospects. Science 324:1530–1534
14. Wu H, Cui Y (2012) Designing nanostructured Si anodes for high energy lithium ion batteries. Nano Today 7:414–429
15. Winter M, Besenhard JO, Spahr ME, Novák P (1998) Insertion electrode materials for rechargeable lithium batteries. Adv Mater 10:725–763
16. Kaskhedikar NA, Maier J (2009) Lithium storage in carbon nanostructures. Adv Mater 21:2664–2680
17. Besenhard JO, Fritz HP (1983) The electrochemistry of black carbons. Angew Chem Int Ed 22:950–975
18. Magerl A, Zabel H, Anderson I (1985) In-plane jump diffusion of Li in LiC_6. Phys Rev Lett 55:222
19. Langer J, Epp V, Heitjans P, Mautner F, Wilkening M (2013) Lithium motion in the anode material LiC_6 as seen via time-domain ^{7}Li NMR. Phys Rev B 88:094304
20. Verma P, Maire P, Novák P (2010) A review of the features and analyses of the solid electrolyte interphase in Li-ion batteries. Electrochim Acta 55:6332–6341
21. Yata S et al (1994) Structure and properties of deeply Li-doped polyacenic semiconductor materials beyond C_6Li stage. Synth Met 62:153–158
22. Takami N, Satoh A, Hara M, Ohsaki T (1995) Rechargeable lithium-ion cells using graphitized mesophase-pitch-based carbon fiber anodes. J Electrochem Soc 142:2564–2571
23. Dahn JR, Zheng T, Liu Y, Xue J (1995) Mechanisms for lithium insertion in carbonaceous materials. Science 270:590–590
24. Tatsumi K, Akai T, Imamura T, Zaghib K, Iwashita N, Higuchi S, Sawada Y (1996) ^{7}Li-nuclear magnetic resonance observation of lithium insertion into mesocarbon microbeads. J Electrochem Soc 143:1923–1930
25. Liu Y, Xue J, Zheng T, Dahn J (1996) Mechanism of lithium insertion in hard carbons prepared by pyrolysis of epoxy resins. Carbon 34:193–200
26. Lee S-J, Nishizawa M, Uchida I (1999) Fabrication of mesophase pitch carbon thin film electrodes and the effect of heat treatment on electrochemical lithium insertion and extraction. Electrochim Acta 44:2379–2383
27. Mochida I, Ku C-H, Korai Y (2001) Anodic performance and insertion mechanism of hard carbons prepared from synthetic isotropic pitches. Carbon 39:399–410
28. Ohzuku T, Iwakoshi Y, Sawai K (1993) Formation of lithium-graphite intercalation compounds in nonaqueous electrolytes and their application as a negative electrode for a lithium ion (shuttlecock) cell. J Electrochem Soc 140:2490–2498

29. Sato K, Noguchi M, Demachi A, Oki N, Endo M (1994) A mechanism of lithium storage in disordered carbons. Science 264:556–559
30. Yazami R, Deschamps M (1995) High reversible capacity carbon-lithium negative electrode in polymer electrolyte. J Power Sources 54:411–415
31. Peled E, Menachem C, Bar-Tow D, Melman A (1996) Improved graphite anode for lithium-ion batteries chemically bonded solid electrolyte interface and nanochannel formation. J Electrochem Soc 143:L4–L7
32. Tokumitsu K, Fujimoto H, Mabuchi A, Kasuh T (2000) Structural characterization and electrochemical properties of non-graphitizable carbons for a lithium ion battery. J Power Sources 90:206–213
33. Seger L, Wen LQ, Schlenoff JB (1991) Prospects for using C_{60} and C_{70} in lithium batteries. J Electrochem Soc 138:L81–L82
34. Chabre Y et al (1992) Electrochemical intercalation of lithium into solid fullerene C_{60}. J Am Chem Soc 114:764–766
35. Loutfy RO, Katagiri S (2002) Fullerene materials for lithium-ion battery applications. Perspectives of fullerene nanotechnology. Springer, Berlin, pp 357–367
36. Iijima S (1991) Helical microtubules of graphitic carbon. Nature 354:56–58
37. Bachilo SM, Strano MS, Kittrell C, Hauge RH, Smalley RE, Weisman RB (2002) Structure-assigned optical spectra of single-walled carbon nanotubes. Science 298:2361–2366
38. Ebbesen T, Lezec H, Hiura H, Bennett J, Ghaemi H, Thio T (1996) Electrical conductivity of individual carbon nanotubes. Nature 382:54–56
39. Kim P, Shi L, Majumdar A, McEuen P (2001) Thermal transport measurements of individual multiwalled nanotubes. Phys Rev Lett 87:215502
40. Wong EW, Sheehan PE, Lieber CM (1997) Nanobeam mechanics: elasticity, strength, and toughness of nanorods and nanotubes. Science 277:1971–1975
41. Yu M-F, Lourie O, Dyer MJ, Moloni K, Kelly TF, Ruoff RS (2000) Strength and breaking mechanism of multiwalled carbon nanotubes under tensile load. Science 287:637–640
42. Fujigaya T, Nakashima N (2008) Methodology for homogeneous dispersion of single-walled carbon nanotubes by physical modification. Polym J 40:577–589
43. Jakubek ZJ, Simard B (2004) Two confined phases of argon adsorbed inside open single walled carbon nanotubes. Langmuir 20:5940–5945
44. Jiang J, Sandler SI (2004) Nitrogen and oxygen mixture adsorption on carbon nanotube bundles from molecular simulation. Langmuir 20:10910–10918
45. Heroux L, Krungleviciute V, Calbi MM, Migone AD (2006) CF_4 on carbon nanotubes: physisorption on grooves and external surfaces. J Phys Chem B 110:12597–12602
46. Kondratyuk P, Wang Y, Johnson JK, Yates JT (2005) Observation of a one-dimensional adsorption site on carbon nanotubes: adsorption of alkanes of different molecular lengths. J Phys Chem B 109:20999–21005
47. Cannon JJ, Vlugt TJ, Dubbeldam D, Maruyama S, Shiomi J (2012) Simulation study on the adsorption properties of linear alkanes on closed nanotube bundles. J Phys Chem B 116:9812–9819
48. Yoo J, Fujigaya T, Nakashima N (2013) Molecular recognition at the nanoscale interface within carbon nanotube bundles. Nanoscale 5:7419–7424
49. Wang J, Chu H, Li Y (2008) Why single-walled carbon nanotubes can be dispersed in imidazolium-based ionic liquids. ACS Nano 2:2540–2546
50. Zhao J, Buldum A, Han J, Lu JP (2000) First-principles study of Li-intercalated carbon nanotube ropes. Phys Rev Lett 85:1706–1709
51. Meunier V, Kephart J, Roland C, Bernholc J (2002) Ab initio investigations of lithium diffusion in carbon nanotube systems. Phys Rev Lett 88:075506
52. Song B, Yang J, Zhao J, Fang H (2011) Intercalation and diffusion of lithium ions in a carbon nanotube bundle by ab initio molecular dynamics simulations. Energy Environ Sci 4:1379–1384

53. Gao B, Kleinhammes A, Tang X, Bower C, Fleming L, Wu Y, Zhou O (1999) Electrochemical intercalation of single-walled carbon nanotubes with lithium. Chem Phys Lett 307:153–157
54. Maurin G, Bousquet C, Henn F, Bernier P, Almairac R, Simon B (1999) Electrochemical intercalation of lithium into multiwall carbon nanotubes. Chem Phys Lett 312:14–18
55. Frackowiak E, Gautier S, Gaucher H, Bonnamy S, Beguin F (1999) Electrochemical storage of lithium in multiwalled carbon nanotubes. Carbon 37:61–69
56. Gao B et al (2000) Enhanced saturation lithium composition in ball-milled single-walled carbon nanotubes. Chem Phys Lett 327:69–75
57. Claye AS, Fischer JE, Huffman CB, Rinzler AG, Smalley RE (2000) Solid-state electrochemistry of the Li single wall carbon nanotube system. J Electrochem Soc 147:2845–2852
58. Yang Z, Wu H (2001) Electrochemical intercalation of lithium into carbon nanotubes. Solid State Ionics 143:173–180
59. Eom J, Kwon H, Liu J, Zhou O (2004) Lithium insertion into purified and etched multi-walled carbon nanotubes synthesized on supported catalysts by thermal CVD. Carbon 42:2589–2596
60. Kim YA et al (2006) In situ Raman study on single- and double-walled carbon nanotubes as a function of lithium insertion. Small 2:667–676
61. Wang XX, Wang JN, Chang H, Zhang YF (2007) Preparation of short carbon nanotubes and application as an electrode material in Li-ion batteries. Adv Func Mater 17:3613–3618
62. Landi BJ, Ganter MJ, Schauerman CM, Cress CD, Raffaelle RP (2008) Lithium ion capacity of single wall carbon nanotube paper electrodes. J Phys Chem C 112:7509–7515
63. Chew SY et al (2009) Flexible free-standing carbon nanotube films for model lithium-ion batteries. Carbon 47:2976–2983
64. Yoon S, Lee S, Kim S, Park K-W, Cho D, Jeong Y (2015) Carbon nanotube film anodes for flexible lithium ion batteries. J Power Sources 279:495–501
65. Ramos-Sanchez G, Chen G, Harutyunyan A, Balbuena P (2016) Theoretical and experimental investigations of the Li storage capacity in single-walled carbon nanotube bundles. RSC Adv 6:27260–27266
66. Novoselov KS et al (2004) Electric field effect in atomically thin carbon films. Science 306:666–669
67. Dikin DA et al (2007) Preparation and characterization of graphene oxide paper. Nature 448:457–460
68. Balandin AA, Ghosh S, Bao W, Calizo I, Teweldebrhan D, Miao F, Lau CN (2008) Superior thermal conductivity of single-layer graphene. Nano Lett 8:902–907
69. Sun Y, Wu Q, Shi G (2011) Graphene based new energy materials. Energy Environ Sci 4:1113–1132
70. Novoselov KS, Fal V, Colombo L, Gellert P, Schwab M, Kim K (2012) A roadmap for graphene. Nature 490:192–200
71. Huang X, Qi X, Boey F, Zhang H (2012) Graphene-based composites. Chem Soc Rev 41:666–686
72. Gerouki A, Goldner M, Goldner R, Haas T, Liu T, Slaven S (1996) Density of states calculations of small diameter single graphene sheets. J Electrochem Soc 143:L262–L263
73. Yoo E, Kim J, Hosono E, H-s Zhou, Kudo T, Honma I (2008) Large reversible Li storage of graphene nanosheet families for use in rechargeable lithium ion batteries. Nano Lett 8:2277–2282
74. Pan D, Wang S, Zhao B, Wu M, Zhang H, Wang Y, Jiao Z (2009) Li storage properties of disordered graphene nanosheets. Chem Mater 21:3136–3142
75. Wang G, Shen X, Yao J, Park J (2009) Graphene nanosheets for enhanced lithium storage in lithium ion batteries. Carbon 47:2049–2053
76. Pollak E, Geng B, Jeon K-J, Lucas IT, Richardson TJ, Wang F, Kostecki R (2010) The interaction of Li^+ with single-layer and few-layer graphene. Nano Lett 10:3386–3388

77. Uthaisar C, Barone V (2010) Edge effects on the characteristics of Li diffusion in graphene. Nano Lett 10:2838–2842
78. Wu Z-S, Ren W, Xu L, Li F, Cheng H-M (2011) Doped graphene sheets as anode materials with superhigh rate and large capacity for lithium ion batteries. ACS Nano 5:5463–5471
79. Varzi A, Täubert C, Wohlfahrt-Mehrens M, Kreis M, Schütz W (2011) Study of multi-walled carbon nanotubes for lithium-ion battery electrodes. J Power Sources 196:3303–3309
80. Guoping W, Qingtang Z, Zuolong Y, MeiZheng Q (2008) The effect of different kinds of nano-carbon conductive additives in lithium ion batteries on the resistance and electrochemical behavior of the $LiCoO_2$ composite cathodes. Solid State Ionics 179:263–268
81. Mun J et al (2014) New dry carbon nanotube coating of over-lithiated layered oxide cathode for lithium ion batteries. J Mater Chem A 2:19670–19677
82. Wu Z et al (2014) Depolarized and fully active cathode based on $Li(Ni_{0.5}Co_{0.2}Mn_{0.3})O_2$ embedded in carbon nanotube network for advanced batteries. Nano Lett 14:4700–4706
83. Li X, Kang F, Bai X, Shen W (2007) A novel network composite cathode of $LiFePO_4$/multiwalled carbon nanotubes with high rate capability for lithium ion batteries. Electrochem Commun 9:663–666
84. Li X, Kang F, Shen W (2006) A comparative investigation on multiwalled carbon nanotubes and carbon black as conducting additive in $LiNi_{0.7}Co_{0.3}O_2$. Electrochem Solid-State Lett 9: A126–A129
85. Sotowa C et al (2008) The reinforcing effect of combined carbon nanotubes and acetylene blacks on the positive electrode of lithium-ion batteries. Chemsuschem 1:911–915
86. Kang B, Ceder G (2009) Battery materials for ultrafast charging and discharging. Nature 458:190–193
87. Johnson BA, White RE (1998) Characterization of commercially available lithium-ion batteries. J Power Sources 70:48–54
88. Guo Z, Zhao Z, Liu H, Dou S (2005) Electrochemical lithiation and de-lithiation of MWNT–Sn/SnNi nanocomposites. Carbon 43:1392–1399
89. Fu Y, Ma R, Shu Y, Cao Z, Ma X (2009) Preparation and characterization of SnO_2/carbon nanotube composite for lithium ion battery applications. Mater Lett 63:1946–1948
90. Zhu X, Zhu Y, Murali S, Stoller MD, Ruoff RS (2011) Nanostructured reduced graphene oxide/Fe_2O_3 composite as a high-performance anode material for lithium ion batteries. ACS Nano 5:3333–3338
91. Lee SW, Yabuuchi N, Gallant BM, Chen S, Kim B-S, Hammond PT, Shao-Horn Y (2010) High-power lithium batteries from functionalized carbon-nanotube electrodes. Nat Nanotechnol 5:531–537
92. Patil N, Aqil A, Ouhib F, Admassie S, Inganäs O, Jérôme C, Detrembleur C (2017) Bioinspired redox-active catechol-bearing polymers as ultrarobust organic cathodes for lithium storage. Adv Mater 29:1703373
93. Choi K-H et al (2014) Heterolayered, one-dimensional nanobuilding block mat batteries. Nano Lett 14:5677–5686
94. Wei W, Yang S, Zhou H, Lieberwirth I, Feng X, Müllen K (2013) 3D graphene foams cross-linked with pre-encapsulated Fe_3O_4 nanospheres for enhanced lithium storage. Adv Mater 25:2909–2914
95. Chen X, Zhu H, Chen Y-C, Shang Y, Cao A, Hu L, Rubloff GW (2012) MWCNT/V_2O_5 core/shell sponge for high areal capacity and power density Li-ion cathodes. ACS Nano 6:7948–7955
96. Zhou G, Paek E, Hwang GS, Manthiram A (2015) Long-life Li/polysulphide batteries with high sulphur loading enabled by lightweight three-dimensional nitrogen/sulphur-codoped graphene sponge. Nat Commun 6:7760
97. He J, Chen Y, Li P, Fu F, Wang Z, Zhang W (2015) Three-dimensional CNT/graphene–sulfur hybrid sponges with high sulfur loading as superior-capacity cathodes for lithium–sulfur batteries. J Mater Chemy A 3:18605–18610

98. Hwa Y, Kim W-S, Hong S-H, Sohn H-J (2012) High capacity and rate capability of core–shell structured nano-Si/C anode for Li-ion batteries. Electrochim Acta 71:201–205
99. Hwang TH, Lee YM, Kong B-S, Seo J-S, Choi JW (2012) Electrospun core–shell fibers for robust silicon nanoparticle-based lithium ion battery anodes. Nano Lett 12:802–807
100. Liu N, Lu Z, Zhao J, McDowell MT, Lee H-W, Zhao W, Cui Y (2014) A pomegranate-inspired nanoscale design for large-volume-change lithium battery anodes. Nat Nanotechnol 9:187–192
101. Hu L, La Mantia F, Wu H, Xie X, McDonough J, Pasta M, Cui Y (2011) Lithium-ion textile batteries with large areal mass loading. Adv Energy Mater 1:1012–1017
102. Lin H et al (2014) Twisted aligned carbon nanotube/silicon composite fiber anode for flexible wire-shaped lithium-ion battery. Adv Mater 26:1217–1222
103. Ren J et al (2014) Elastic and wearable wire-shaped lithium-ion battery with high electrochemical performance. Angew Chem 126:7998–8003
104. Goyal A, Reddy AL, Ajayan PM (2011) Flexible carbon nanotube–Cu_2O hybrid electrodes for li-ion batteries. Small 7:1709–1713
105. Li Y, Tan B, Wu Y (2008) Mesoporous Co_3O_4 nanowire arrays for lithium ion batteries with high capacity and rate capability. Nano Lett 8:265–270
106. Zeng W, Zheng F, Li R, Zhan Y, Li Y, Liu J (2012) Template synthesis of $SnO_2/\alpha\text{-}Fe_2O_3$ nanotube array for 3D lithium ion battery anode with large areal capacity. Nanoscale 4:2760–2765
107. Liao J-Y, Higgins D, Lui G, Chabot V, Xiao X, Chen Z (2013) Multifunctional TiO_2-C/MnO_2 core-double-shell nanowire arrays as high-performance 3D electrodes for lithium ion batteries. Nano Lett 13:5467–5473
108. Liu J, Song K, van Aken PA, Maier J, Yu Y (2014) Self-supported $Li_4Ti_5O_{12}$–C nanotube arrays as high-rate and long-life anode materials for flexible Li-ion batteries. Nano Lett 14:2597–2603
109. Klemm D, Heublein B, Fink HP, Bohn A (2005) Cellulose: fascinating biopolymer and sustainable raw material. Angew Chem Int Ed 44:3358–3393
110. Einchhorn S et al (2010) Review: current international research into cellulose nanofibres and composites. J Mater Sci 45:1–33
111. Siró I, Plackett D (2010) Microfibrillated cellulose and new nanocomposite materials: a review. Cellulose 17:459–494
112. Habibi Y, Lucia LA, Rojas OJ (2010) Cellulose nanocrystals: chemistry, self-assembly, and applications. Chem Rev 110:3479–3500
113. Moon RJ, Martini A, Nairn J, Simonsen J, Youngblood J (2011) Cellulose nanomaterials review: structure, properties and nanocomposites. Chem Soc Rev 40:3941–3994
114. Giese M, Blusch LK, Khan MK, MacLachlan MJ (2015) Functional materials from cellulose-derived liquid-crystal templates. Angew Chem Int Ed 54:2888–2910
115. Li Y et al (2015) Nanocellulose as green dispersant for two-dimensional energy materials. Nano Energy 13:346–354
116. Shafiee H et al (2015) Paper and flexible substrates as materials for biosensing platforms to detect multiple biotargets. Sci Rep 5:8719
117. Klemm D, Kramer F, Moritz S, Lindström T, Ankerfors M, Gray D, Dorris A (2011) Nanocelluloses: a new family of nature-based materials. Angew Chem Int Ed 50:5438–5466
118. Hu L, Wu H, La Mantia F, Yang Y, Cui Y (2010) Thin, flexible secondary Li-ion paper batteries. ACS Nano 4:5843–5848
119. Leijonmarck S, Cornell A, Lindbergh G, Wågberg L (2013) Single-paper flexible Li-ion battery cells through a paper-making process based on nano-fibrillated cellulose. J Mater Chem A 1:4671–4677
120. Hu L, Liu N, Eskilsson M, Zheng G, McDonough J, Wågberg L, Cui Y (2013) Silicon-conductive nanopaper for Li-ion batteries. Nano Energy 2:138–145
121. Cho SJ et al (2015) Hetero-nanonet rechargeable paper batteries: toward ultrahigh energy density and origami foldability. Adv Func Mater 25:6029–6040

122. Tian D, Song Y, Jiang L (2013) Patterning of controllable surface wettability for printing techniques. Chem Soc Rev 42:5184–5209
123. Sousa RE, Costa CM, Lanceros-Méndez S (2015) Advances and future challenges in printed batteries. Chemsuschem 8:3539–3555
124. Lawes S, Riese A, Sun Q, Cheng N, Sun X (2015) Printing nanostructured carbon for energy storage and conversion applications. Carbon 92:150–176
125. Aleeva Y, Pignataro B (2014) Recent advances in upscalable wet methods and ink formulations for printed electronics. J Mater Chem C 2:6436–6453
126. Hu L, Choi JW, Yang Y, Jeong S, La Mantia F, Cui L-F, Cui Y (2009) Highly conductive paper for energy-storage devices. Proc Natl Acad Sci 106:21490–21494
127. Singh N et al (2012) Paintable battery. Sci Rep 2:481
128. Choi K-H, Yoo J, Lee CK, Lee S-Y (2016) All-inkjet-printed, solid-state flexible supercapacitors on paper. Energy Environ Sci 9:2812–2821

Chapter 19
Nanocarbons and Their Composite Materials as Electrocatalyst for Metal–Air Battery and Water Splitting

Suyeon Hyun, Arumugam Sivanantham and Sangaraju Shanmugam

19.1 Introduction

Carbon is one of the well-known and most earth-abundant elements since antiquity. In the earth crust, it exists as a diamond, graphite, coal, etc. [1, 2]. The carbon materials with nanoscale dimensions are called as "nanocarbon materials" in the materials world. The intensive research has been done in the past few decades and found the various nanocarbon allotropes, including fullerenes, graphene, single-walled carbon nanotubes (SWCNT) [3]. The C_{60}, $C_{70,}$ and C_{84} are the well-known fullerene structures, and the representative species is C_{60}. Kroto et al. reported that the 60 carbon atoms in C_{60} have 12 five-membered and 20 six-membered rings. The fullerenes are an excellent electron acceptor [4]. In the year 1991, Iijima observed the existence of nanometer-sized carbon nanotubes (CNT) with the cylindrical structure like rolled graphene sheet [5]. The same sp^2 carbons only occur in the CNT as like fullerenes. In general, the CNTs exhibit the two different natures such as semiconducting and metallic. Specifically, when the normal fabrication methods are used, the two-thirds of the CNTs have semiconducting nature and the remaining amount of one-third is metallic CNTs due to the random carbon sheet rolling [6].

Similarly, graphene is another nanocarbon material, which consists of an only six-membered ring with sp^2 carbons. It has a sheet structure with two dimensions, which can be extracted from the graphite using a simple synthesis procedure. At the first time, Novoselov et al. successfully separated the graphite into graphene sheets in 2004 and in the year 2010 they received the Nobel prize for their remarkable achievement [7]. Interestingly, the graphene has specific characteristics of

S. Hyun · A. Sivanantham · S. Shanmugam (✉)
Department of Energy Science & Engineering, Daegu Gyeongbuk Institute of Science & Technology (DGIST), 50-1, Sang-Ri, Hyeonpung-Myeon, Dalseong-Gun, Daegu 42988, Republic of Korea
e-mail: sangarajus@dgist.ac.kr

N. Nakashima (ed.), *Nanocarbons for Energy Conversion: Supramolecular Approaches*, Nanostructure Science and Technology,
https://doi.org/10.1007/978-3-319-92917-0_19

electronic, chemical, and mechanical properties. Moreover, it has 100 times higher electron mobility than the silicon. So, the electrical property is one of the most remarkable characteristic features of graphene [8]. Another hard allotrope of carbon is diamond, which has excellent thermal conductivity and insulation properties. Widely, it is used as a cutter and polisher tools in the industrial applications [9].

After the findings of the CNTs, fullerenes and their derivatives, the field of nanocarbon materials has achieved remarkable scientific and technological improvements. Importantly, the expanded nanocarbon frameworks show the unique and advanced properties of excellent charge-carrier property, narrow bandgap, more mechanical strength, and flexibility [10, 11]. Nanocarbon materials are widely used in practical use particularly for the energy applications due to its higher surface area (>1000 $m^2\ g^{-1}$), stability, low toxicity, and density (weight). So, the nanocarbon materials have their distinctive electronic and physicochemical properties [12]. However, these features of the carbon are changeable by the functionality of the different functional groups on the surface of the nanocarbons. As discussed earlier, the nanocarbon has two different properties of semiconductor and metallic. Also, it exhibits both p-type and n-type semiconducting property, which mainly depends on the charge carriers present in it. However, most of the as-synthesized carbons contain more oxygen functional groups and it has the p-type charge carriers as a major part [13].

The molecular or ionic doping/de-doping with the injection of charge carriers, the nanocarbon properties are easily tunable, and their Seebeck coefficient also can be altered [14, 15]. Meanwhile, the large surface area of the sp^2 planes of the nanocarbon promotes the efficient adsorption of different species such as ions, organic and inorganic materials. The Vander Waals interaction with the electron system in the nanocarbon is the key reason for the efficient adsorption of various substances on its surface. Mostly, their Π-electron systems support to make the Π–Π interaction and charge-transfer interactions between the species and sp^2 planes. For instance, the species which can accept the electron from the nanocarbon can afford the p-type nanocarbon. The n-type nanocarbon exit, when the electrons are injected from the surface adsorbed substances. The partial charge-transfer interaction between two materials has been extensively studied using the Raman and Electron spin resonance (ESR) analyses [16, 17].

In general, the acid or base treatment has been carried out to introduce the charge-carrier species on the nanocarbon surface using various chemicals. Also, the sodium or potassium metals has been attached to the nanocarbon sp^2 planes by chemical doping and try to form the n-type nanocarbon. However, at the ambient conditions for the practical applications, those doped nanocarbon are not good enough due to its poor stability and conductivity [18]. So, in recent days the researchers introduce the nitrogen (N) species in the nanocarbon during the preparations using various N-containing organic molecules. Mostly, the following methods are widely used to prepare the nanocarbon materials, such as

- Electrospinning-assisted pyrolysis
- In situ hydrothermal method

- Dry-autoclaving method (Swagelok)
- Electrodeposition
- Precipitation, etc.

The various synthetic techniques of nanocarbon materials are discussed in detail in the following section of nanocarbon catalysts fabrication. Hence, the unique structures of nanocarbon with their novel properties have drawn much attention not only on their basic science; also have been studied extensively as a future suitable candidate for the various potential fields such as energy storage, production, and conversion devices (battery, water electrolyzer, fuel cell, solar cell, etc.), electronics (field-effect transistors, thermoelectronics, etc.) and biomedical engineering, etc. Mostly, the advanced energy storage and conversion devices of battery and water electrolyzer, including fuel cell, solar cell, field-effect transistors, and thermoelectronics [19, 20].

19.2 Fabrication of Nanocarbon Catalysts

19.2.1 Hydrothermal Synthesis

Hydrothermal method is a kind of wet-chemical approach that includes heterogeneous reaction in aqueous solution over 100 °C and the pressure of 1 bar as shown in Fig. 19.1 [21]. To endure the pressure included in the hydrothermal process, the autoclave is usually used. The advantages of the hydrothermal method over other types of crystal growth include the possibilities to create crystalline phases which are not stable at the melting point [22]. Also, materials which have a high vapor pressure near melting points can be synthesized by the hydrothermal method. The method is particularly suitable for the growth of large good-quality crystals while maintaining their composition. In the nanostructured carbon materials point of view, it is a simple and green one-step procedure towards functionalized carbon nanostructures with the controlled surface area, pore size, and shape [23]. Moreover, this technique enables the large-scale production of many energy storage materials in a controlled manner.

The nanostructures and morphologies of the as-synthesized carbon materials can be tuned by changing the concentrations of the precursor solution. Sun et al. reported the synthesis of low-dimensional nanocarbon, especially one-dimensional (1D) rods and fibers via hydrothermal method assisted by surfactant-templating followed by carbonization process [24]. To get desire architectures of nanocarbon, they used resorcinol and hexamethylene tetramine (HMT) as the monomers, and Pluronic F127 surfactant was applied as the structure-directing agent. Thus, the synthesized nanocarbon is used as an anode which represented a good rate capability compared to that of the commercial artificial graphite for the rechargeable lithium-ion batteries. Prabu et al. applied a simple single-step solvothermal technique for the development of transition metal oxide–carbon composite material, which is composed of chrysanthemum flower-like $NiCo_2O_4$ incorporated

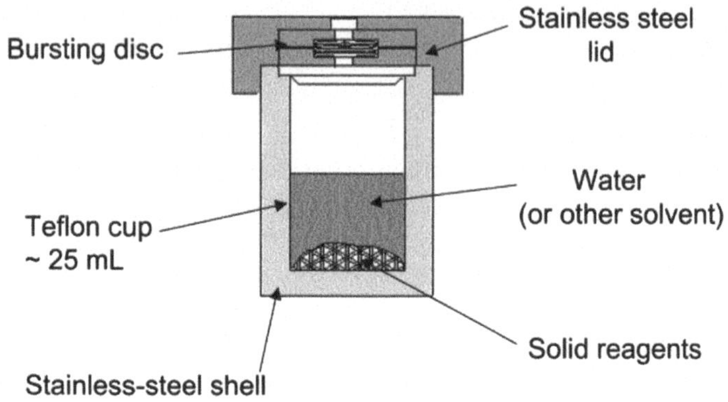

Fig. 19.1 Illustration of Teflon-lined stainless steel autoclave for hydrothermal synthesis

nitrogen-doped graphene oxide (NCO/N-rGO), and it was applied as a bifunctional air cathode for primary and rechargeable lithium–oxygen/zinc–air batteries. In a typical synthetic procedure, the stoichiometric amounts of nickel acetate, cobalt acetate, graphene oxide powder, and ammonium hydroxide were dissolved in the ethanol. The as-prepared solution is refluxed at 80 °C for 10 h and then transferred into a Teflon autoclave, which was heated in an electric oven at 150 °C for 6 h. The obtained product was separated from the impurities by centrifugation, and the product was further calcinated at 250 °C in open air for 12 h [25].

19.2.2 Dry-Autoclaving Method (Swagelok)

It is a simple and an easy method for the preparation of hollow nitrogen-doped carbon spheres as well as other compounds compared to different methodologies (Fig. 19.2). Also, it is quite straightforward, environmental harmfulness, high reproducibility, solvent-free, and an inexpensive method [26]. Recently, heteroatom-doped carbon materials have been widely studied by researchers as non-precious metal-based catalysts to replace the Pt/C or RuO_2, IrO_2, etc., which are the state-of-art oxygen electrocatalysts so far. By substituting of a heteroatom with nitrogen (N), sulfur (S), phosphorus (P), or boron (B) into the carbon framework, the electrical property and the chemical activity can be changed, thereby enhancing the oxygen redox reaction. In particular, since the higher electronegativity of N (3.04) than that of the carbon atom (2.55), nitrogen doping can form a positive charge density on adjacent carbon atoms, it results in enhancement of electron donor tendency, electrical conductivity, basicity, adsorption of acidic gases and the wettability of carbon in aqueous solution [27–29]. Such N-functionalized carbon materials have drawn much attention in various energy storage and conversion application.

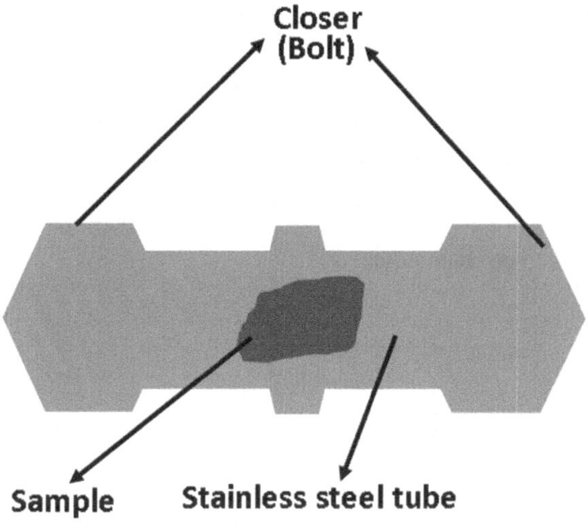

Fig. 19.2 Scheme of the basic setup of the dry-autoclaving method using Swagelok

Sanetuntikul et al. developed metal coordinated with nitrogen-doped carbon (M-N-C) as oxygen reduction reaction catalyst, for example, hollow nitrogen-doped carbon spheres (HNCSs) by adopting commercial Swagelok union cells followed by acid treatment to remove the oxide core. A single precursor of iron(III) diethylenetriaminepentaacetate, which acts as a source of metal, nitrogen, and carbon is annealed at 700 °C under autogenic pressure, and the Fe_0/Fe_xO_y filled carbon shell is obtained. The obtained product was stirred in concentrated HCl for 24 h at room temperature then finally HNCSs catalyst is obtained by centrifugation [29]. For another example, Ganesan et al. developed cobalt-sulfide nanoparticles grown in situ on nitrogen and sulfur co-doped graphene oxide as an efficient bifunctional electrocatalyst for oxygen reduction and evolution reactions by solid-state thermolysis approach which includes a single-step direct pyrolysis of the graphene oxide and cobalt thiourea complex in Swagelok cells [30]. By modifying the thermolysis temperature, the coupling effect between CoS_2 nanoparticles and the graphene oxide surface as well as the amount of doping and functional groups of nitrogen and sulfur in the carbon and its surface area can be controlled.

19.2.3 Electrospinning Technique

The electrospinning is a mass-productive, simple, versatile, and low-cost fabrication approach (Fig. 19.3). Moreover, synthesis of carbon with plenty of N content and the porosity can be achieved easily without the help of hard or soft templates compared to general synthetic strategies which require further time-consuming carbonization process [31]. Commonly, N-containing polymers such as polyacrylonitrile (PAN), acetonitrile, melamine, polypyrrole (PPy), polyaniline (PANi), and polyvinyl pyrrolidone (PVP) have been utilized as a precursor materials, and the subsequent carbonization process is conducted under inert atmosphere [32]. To

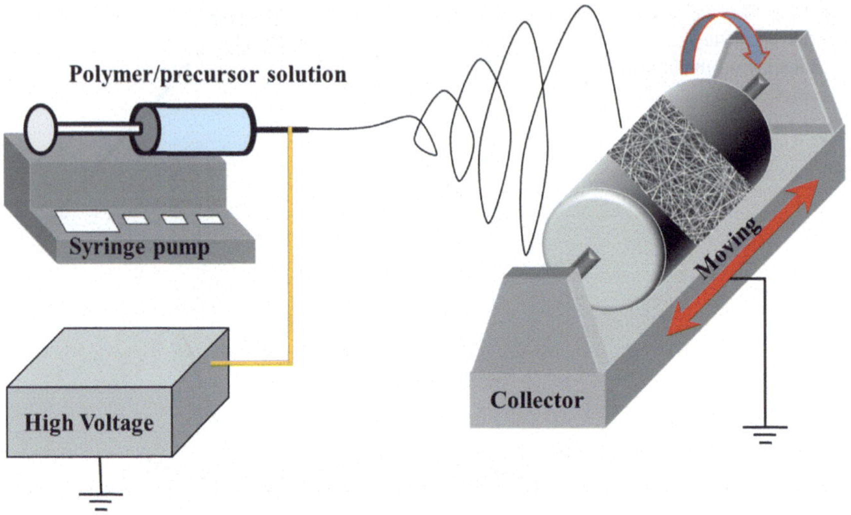

Fig. 19.3 Schematic illustration of the basic configuration of electrospinning method

obtain highly porous carbon structures with large surface area, a post-activation process using KOH treatment can also be applied [33].

Sivanantham et al. introduced self-redox CeO_2 supported Co in nitrogen-doped carbon nanorods (Co-CeO_2/N-CNR) by an electrospinning method and utilized as an efficient bifunctional catalyst for both oxygen reduction and evolution reaction [34]. Polyacrylonitrile, cobalt (II) and cerium (III) acetylacetonates were prepared with a ratio of 8:1:1 then added in dimethylformamide as the precursors for electrospinning. The electrospun fiber mat is gathered on the aluminum foil at the applied voltage of 12 ± 2 kV, with the distance of 10 cm between the solution filled the syringe and the aluminum foil. The mat was removed from the aluminum foil and then annealed in the tubular furnace at 300 °C for 3 h in the ambient air atmosphere and subsequent pyrolysis step at various temperatures of 850 and 900 °C with time duration from 1 to 2 h in the inert atmosphere to obtain the targeted catalysts. Based on the high-resolution XPS N1s spectra of Co-CeO_2/N-CNR catalysts, they confirmed that the carbon nanorods possess less-energy bandgap with enhanced conductivity toward electron transfer. Also, the presence of structural defects and the amorphous nature of carbon might be the main reason for superior oxygen redox catalytic activity.

19.3 Nanocarbons for Metal–Air Batteries

Metal–air batteries which follow a new concept of working principles compared to the lithium-ion batteries (LIBs) with conventional intercalation reaction mechanism have been considered as a promising energy storage technology [35]. In particular, taking the most representative examples, Zn–air batteries and Li–air batteries have drawn much attention because of their extremely high energy densities ($\sim$3458

W h kg^{-1} (Li_2O_2) for Li–air battery and 2288 W h kg^{-1} for zinc-air battery), low-cost, and eco-friendly operation compatible with EV [36]. Compared to conventional LIBs system, the metal–air battery is composed of a catalytically active oxygen reduction reaction (ORR) and an oxygen evolution reaction (OER) bifunctional electrode instead of an intercalation material as a cathode [37]. Metal–air batteries generate electricity through a redox reaction between the metal and the oxygen gas in the air. In general, during a discharge process, the metal anode is oxidized, and the electrons are transferred through the external circuit. Meanwhile, at the cathode side, oxygen accepts the electrons from the anode and is reduced to oxygen-containing species, so-called ORR happens. Finally, the oxygen-reduced species and dissociated cations combine to different metal oxides depending on the environment system. In contrast, the charging process follows the reverse procedure of discharging process. Especially at the cathode, an oxygen evolution reaction (OER) occurred and metal ions are electroplated at the anode side.

Basically, metal–air batteries can be categorized into two types depending on the electrolytes of the system. One is composed of aqueous electrolyte, which is not sensitive to moisture while the other is a water-sensitive system containing aprotic-based electrolyte [35, 38]. Metals, such as Zn, Fe, Na, K, Ca, Al, and Cd is most suitable for aqueous environment system. Among them, Zn–air batteries are the low-cost system with high efficiency (\$160 kW h^{-1}) compared to all other metal–air batteries. Hence they already utilized as a practical primary battery with high capacities and applied to hearing aids. Moreover, Zn–air batteries possess high theoretical gravimetric energy density as well (2,288 W h kg^{-1}) to satisfy the DOE target of 2,100 W h kg^{-1} for electric vehicle application. Also, zinc metal has numerous advantages, for example, low material cost, abundance, low equilibrium potential, environmental friendliness, and long shelf life, so on [38, 39].

The zinc–air battery consists of three main parts: zinc metal as anode, air electrode as a cathode which is the combination of catalytic active layer and gas diffusion layer, and a separator (Fig. 19.4a). The overall cell reaction in alkaline electrolyte-based zinc–air cell are followed [37, 40]:

$$\textbf{Anode}: Zn \leftrightarrow Zn^{2+} + 2e^-$$

$$Zn^{2+} + 4OH^- \leftrightarrow Zn(OH)_4^{2-} \quad E^0 = 1.25\,\text{V}$$

$$Zn(OH)_4^{2-} \leftrightarrow ZnO + H_2O + 2OH^-$$

$$\textbf{Cathode}: 1/2O_2 + H_2O + 2e^- \leftrightarrow 2OH^- \quad E^0 = 0.4\,\text{V}$$

$$\textbf{Overall reaction}: \text{Zn} + 1/2O_2 \leftrightarrow \text{ZnO} \quad E^0 = 1.65\,\text{V}$$

In brief, during discharge, oxygen gas diffuses into the porous cathode by the difference in oxygen pressure between the inside and the outside of battery cell. Consequently, the catalyst at the air cathode catalyzes the oxygen reduction reaction (ORR) and forms hydroxyl ions in alkaline solution. This process occurs at a three-phase boundary, which involves catalyst (solid), electrolyte (liquid), and oxygen (gas). Hence, it is so called as a three-phase reaction. The generated OH^-

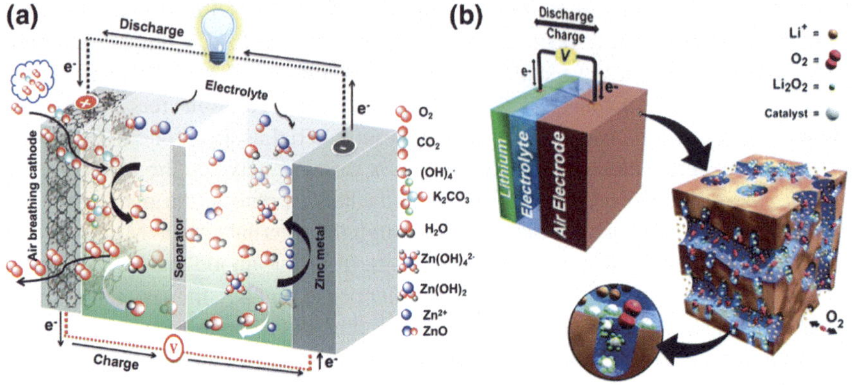

Fig. 19.4 Schematic illustration of cell figurations and the working principles of **a** Zinc–air battery in an alkaline aqueous electrolyte under ambient air and **b** Li–O_2 battery in alkaline aqueous and a nonaqueous electrolyte. Reproduced with permission from [41]. Copyright (2012) American Chemical Society

reacts with zinc ions and completes the cell reaction with 1.65 V of an equilibrium standard potential. Nonetheless, the practical working potential of the Zn–air battery is lower than 1.65 V due to an occurrence of cell overpotential, combining activation, ohmic, and concentration loss [42]. The activity loss of zinc–air cell mainly stems from the instability of air cathode when charging step happened and low efficiency of oxygen reduction and evolution reactions [43].

Meanwhile, the dendrite formation on the surface of zinc metal anode which leads a short lifetime of battery operation is also an important issue to be handled. Therefore, (1) the optimization of air electrode architecture, which is chemically stable and highly active for both OER and ORR, and (2) applying protective layer on the Zn metal surface or making Zn metal alloy or adding some additives to the electrolytes have been intensely studied to solve these problems. The point to note when manufacturing the air electrode is again, it should have a high porosity with volume area as well as excellent OER/ORR catalytic properties. However, highly porous architecture can easily be mechanically damaged by forming oxygen molecules during the charging process of the battery cell [43].

There are two types of Li–O_2 batteries being investigated, the nonaqueous and aqueous system [41] and its cell figuration is presented in Fig. 19.4b. Despite of some merits related with an aqueous system such as inexpensive, high ion conductivity, and easiness of expansion, the facile operation is being prohibited due to the sensitivity to absorb CO_2 from ambient air and evaporation of electrolyte through the open structure of cathode [44]. The fundamental electrochemical reactions are shown in the following:

Anode: $Li \leftrightarrow Li^{+} + e^{-}$

Cathode: $2Li^{+} + 2e^{-} + O_2 \leftrightarrow Li_2O_2 \quad E^0 = 2.96\,\mathrm{V}\ vs.\ Li/Li^{+}$ (Non-aqueous)

$O_2 + 2H_2O + 4e^{-} \leftrightarrow 4OH^{-} \quad E^0 = 3.43\,\mathrm{V}\ vs.\ Li/Li^{+} \quad (Aqueous)$

Overall reaction: $2Li^{+} + O_2 \leftrightarrow Li_2O_2 \quad E^0 = 2.96\,\mathrm{V}\ vs.\ Li/Li^{+}\,(Non-aqueous)$

Also, considering the electrochemical reactions in an aqueous electrolyte, which involve water as active reagent, the preliminary results reveal that the theoretical energy density of Li–O_2 battery operated under the nonaqueous system is higher than that of an aqueous Li–O_2 battery system. As a result, nonaqueous Li–air batteries have widely studied around the world in recent years. Similar to Zn–air batteries, however an oxygen electrocatalyst is one of the most challenging issues among other challengeable problems in Li–O_2 battery, such as incompatibility with electrolyte and air coming from lithium metal anode, instability of electrolytes in oxygen-rich electrochemical conditions, etc.

Modern air electrode is composed of activated carbon with a porous architecture and the hydrophobic film, such as polytetrafluoroethylene (PTFE) as a wet-proofing agent, which is permeable to only air but not moisture [45]. Even though precious metal-based electrocatalysts, such as Ru, Pt, and their alloys possess superior activity for ORR and OER, they include high manufacturing fee and poor stability during cycling. To achieve the target value of energy densities required for practical metal–air batteries, inexpensive, abundance, lightweight, and porous material should be chosen for the air electrode. In this regard, carbon itself has been considered a highly attractive candidate as the practical cathode material. In particular, carbon could catalyze the oxygen reduction reaction very efficiently during discharge with high electrical conductivity, leading high energy efficiency of metal–air battery [46–49].

19.3.1 Metal-Free Nanocarbon-Based Electrodes

Recently, the nanocarbon-based structures, such as carbon nanotubes, carbon nanofibers, and graphene have been investigated as air electrodes for metal–air batteries [50–53]. Many attempts have been made to utilize carbon-based electrode for the practical application by tuning the various surface properties including surface area, surface functional groups, degree of disorder, pore distribution, and the porosity. Especially, heteroatom-doped carbon materials with electroneutrality breaks and charge relocation have been intensely studied due to their tunable surface chemistry and fast electron transfer ability. It was already reported in many literatures that nanocarbon materials doped with nitrogen could be an efficient alternatives to Pt ORR catalyst [46, 48]. The incorporation of nitrogen into the graphitic network of nanocarbon in the forms of graphitic, pyridinic, or pyrrolic-nitrogen multiplies the extent of defectiveness and edge-plan sites, thereby enhancing the ORR catalytic activity.

Also, by introducing second heteroatom, e.g., B, S, P, O, or Si, etc., into the nitrogen-doped carbon nanomaterials, the ORR activity can be further increased by modulating the electronic properties and the surface polarity of nanocarbon [49].

Inspired by this point, in 2015, Zhang et al. developed three-dimensional nitrogen and phosphorus co-doped mesoporous nanocarbon (NPMCs) foams, which possess a large surface area of 1,663 m^2 g^{-1} and excellent bifunctional electrocatalytic properties including both OER & ORR, thus enhanced battery performance and long-term durability can be achieved (Fig. 19.5) [54]. For the synthesis of NPMCs, they followed solution polymerization followed by template-free pyrolysis of polyaniline (PANi) aerogels in the presence of phytic acid at the temperature of 1000 °C. The NPMCs catalyst consists of graphitic carbon domains that are co-doped with N and P atoms from the PANi and phytic acid, respectively, with a large number of edge-like graphitic nanostructures. The ORR and OER activity of NPMCs are examined using three-electrode system. It shows a high ORR onset potential of 0.94 V and a half-wave potential of 0.85 V versus RHE comparable to those of Pt/C. Furthermore, it exhibits a lower onset potential than that of OER electrode based on RuO_2, however, still lower current densities at higher potentials are obtained ($\sim$1.8 V at 10 mA cm^{-2}). When applied

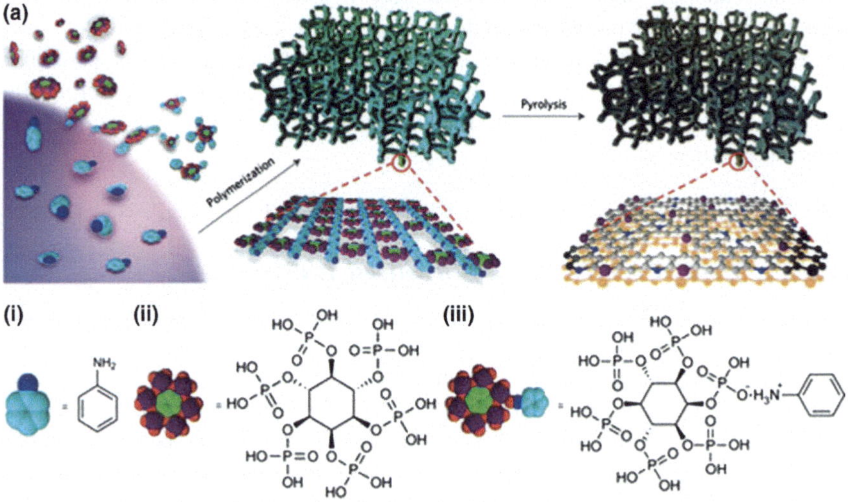

Fig. 19.5 **a** Preparation process of the N and P co-doped porous carbon (NPMCs) electrocatalysts. An aniline (**i**) phytic acid (**ii**) complex (**iii**) is formed (for clarity, only one of the complexed anilines is shown for an individual phytic acid), followed by oxidative polymerization into a three-dimensional PANi hydrogel crosslinked with phytic acids. The PANi hydrogel is freeze-dried into an aerogel and pyrolyzed in Ar to produce the NPMCs. Adapted from Zhang et al. with permission from Nature publishing group [Nature Nanotechnology], copyright (2015) [54]

NPMCs as a bifunctional air electrode in primary and rechargeable Zn–air batteries, it exhibits an open-circuit voltage (OCV) of ~1.48 V and an energy density of ~835 W h kg_{Zn}^{-1}. Also, the peak power density of ~55 mW cm^{-2} as well as excellent durability for over 240 h of operation after two mechanical recharges are achieved.

Further research efforts have been directed toward the development of new metal-free nanocarbon-based ORR/OER catalysts. For instance, Tang et al. reported bifunctional graphene catalysts with numerous topological defects through the carbonization of natural gelatinized sticky rice, and it is one of the best metal-free bifunctional carbon catalysts ever reported [55]. Based on the DFT calculations, they have first reported that edge-induced topological defects with nitrogen incorporated can deliver superior activity for both ORR and OER. The overpotential gap is achieved with a value of only 0.90 V, an excellent ORR performance exceeding to Pt/C regarding both activity and stability, and a remarkable OER activity (1.67 V at 10 mA cm^{-2}) comparable with Ir/C in 0.1 M KOH. Furthermore, a prototype Zn–air battery cell based on the NGM catalysts is constructed, and it achieves a high OCV of 1.42 V, a current density of ≈6.0 mA cm^{-2} and a peak power density of ≈3.0 mW cm^{-2}, respectively, which are comparable to those of a Pt/C electrode.

Nonetheless, enlarging the active catalytic sites of heteroatom-doped carbon still remains challenging. One of the effective ways is the addition of metal species into the heteroatom-doped carbon frameworks as active additives, which can be utilized in metal–air batteries [56]. From this point of view, Chen et al. recently developed a new way of involving S for the first time to achieve atomic dispersion of Fe-N_x species on N and S co-decorated hierarchical carbon layers, resulting in single-atom bifunctional OER/ORR catalysts with higher efficiency [57]. Taking advantage from the plenty of atomically dispersed highly active Fe-N_x sites and unique hierarchical structure to increase active site exposure, the obtained electrocatalyst denoted as S, N-Fe/N/C-CNT shows excellent OER/ORR electrocatalytic activity under alkaline conditions. Especially, further applying S, N-Fe/N/C-CNT into a rechargeable Zn–air battery presents an open-circuit voltage of 1.35 V, a lower charge overpotential compared to Pt/C, and higher stability (no visible voltage drop over 100 cycles).

Apart from the physical and electrochemical properties which are demanded in an air cathode, management of total electrode weight and the space in electronics are also essential to enhance the battery performance. Typically, an electrode fabricated with so-called "brick-and-mortar" route contains some ancillary materials including polymeric binders and catalyst supports. Due to the presence of inactive ancillaries, it provides an excess weight to the final electrode for almost 40% and increases the interfacial resistance caused by the reduction of effective active sites and the insulating binders [58]. Also, the decomposition or the oxidation of additives can cause the detachment of some active materials from the electrode surfaces over extended cycling [59, 60]. Keeping this in mind, the electrode should be fabricated with a minimized utilization of additives. In this regard, one of the effective ways to enhance the performance of metal–air batteries is to design an integration air cathode, which is free from the usage of the binder.

For example, carbon nanofibers directly grown on a porous alumina substrate can also be used as binder-free electrodes in Li–air battery provides gravimetric energies of about 2500 Wh $kg^{-1}_{electrode}$ at the power of about 100 W kg^{-1} even though an alumina substrate itself is inactive [51]. Instead of using an inert substrate, the oxygen electrocatalysts directly grown on the conductive current collector is the best way to get an entirely binder-free air electrode which can efficiently facile the electron transfer between active materials and the current collector. The current collectors consist of graphitic carbons such as carbon cloth, and carbon paper can mainly reduce the total weight of electrode, as the employment of metallic mesh current collectors still occupy a significant portion of the weight of traditional air cathode. The modified graphitic carbon acts as a diffusion layer and bifunctional oxygen electrode with a high electrochemical catalytic activity and electrical conductivity. Furthermore, it can simplify the three components into one part only, thus, the modified graphite carbon can be a potential candidate for the air electrode [61].

Liu et al. fabricated free-standing and flexible nitrogen-doped nanoporous carbon nanofiber film (NCNF) via scalable method through pyrolysis of electrospun polyimide [62]. The obtained electrodes possess a large specific surface area of 1249 m^2 g^{-1}, the high electrical conductivity of 147 S m^{-1}, and moderate tensile strength (1.89 MPa) as well as tensile modulus (0.31 GPa). In addition, it shows superior electrocatalytic activities for both OER (onset potential = 1.43 V vs. RHE, potential = 1.84 V @ 10 mA cm^{-2}) and ORR (onset potential = 0.97 V vs. RHE; limiting current density = 4.7 mA cm^{-2}), respectively. Based on these excellent properties, the air electrode was utilized in various types of zinc-air battery, such as, primary liquid Zn–air battery, rechargeable liquid Zn–air battery, and particularly, flexible all-solid-state rechargeable Zn–air battery. The OCV and the maximum power density of the Zn–air battery applying NCNF catalyst were found to be as high as 1.48 V and 185 mW cm^{-2}, respectively, exceeding those of the Pt/C-based counterpart. A slightly lower charge–discharge voltage gap was observed for the Zn–air battery with the NCNF air cathode compared to that of Pt/C air cathode, indicating a good rechargeability. In particular, after 500 cycling operation (about 83 h operation), the NCNF air cathode shows a slight performance loss with a small increase in the voltage gap by 0.13 V, whereas Pt/C demonstrates a significantly higher increase in the voltage gap (0.38 V) under the same condition. The observed performance decay for the NCNF may due to the carbon oxidization and the loss of active sites. Apart from liquid Zn–air battery, all-solid-state Zn–air micro batteries without binder and current collector are fabricated and exposed to ambient air. Series-connected all-solid-state Zn–air microbatteries can be used to power a blue LED (3.0 V) without apparent decay in performance even when the cells are bent by human hand movements, indicating possibilities for the all-solid-state NCNF cathode based Zn–air microbatteries to be utilized in flexible and wearable optoelectronic devices.

In the nonaqueous based Li–O_2 cell, graphene nanosheets result in high discharge capacity exceeding 8000 mA h g^{-1}. Furthermore, comparing discharge characteristics of Li–air cells with different carbon material based cathodes, it is revealed that the mesopore volume of carbon materials gives a large contribution to the capacity. For example, Guo et al. fabricated a 3D electrode that consists of mesoporous/macroporous carbon sphere arrays [63]. The ordered mesoporous channels and hierarchical structures provide enough space for discharge products, which improving the electrochemical performance of $Li–O_2$ batteries.

In point of fact, the most reported non-precious metal-based OER catalysts are consisted of transition metal oxides with the support of carbon materials, and the use of only carbon nanomaterials as bifunctional catalysts has rarely been discussed to date.

19.3.2 Metal Oxide/Nanocarbon-Based Electrodes

Carbon can also be hybridized with other materials to improve electrode performance. For instance, in our group, the 1D $NiCo_2O_4$ nanostructures with carbon was successfully synthesized [64]. To further enhance the battery performance, the morphology was tuned using a simple electrospinning technique. Usually, one-dimensional (1D) nanostructure affords efficient transport pathways for electrons and ions. The 1D $NiCo_2O_4$ cathode can deliver an oxygen electrode potential ($E_{oxygen} = E_{OER} - E_{ORR}$) of 0.84 V and a discharge capacity of $\sim$580 mA h g^{-1} for deep discharge process. Additionally, an overall overpotential for the battery charge and discharge after 50 cycles of 1D $NiCo_2O_4$ was $\sim$0.84 V compared with the initial value of $\sim$0.7 V with a small overpotential increase of $\sim$0.14 V. Our results show that the manipulation of the architecture of $NiCo_2O_4$ plays a role in enhancing the bifunctional activity without employing additional supports.

In the other hand, Liu et al. developed $Li–O_2$ battery based on binder-free air cathode catalyst made of hierarchical $NiCo_2O_4$ nanosheets grown on carbon nanofiber films ($NiCo_2O_4$@CNFs) via a hydrothermal method followed by heat treatment (Fig. 19.6a, b) [65]. Typically, porous carbon nanofiber films were obtained by electrospinning followed by carbonization. Thanks to the integration of an efficient bifunctional catalyst with rationally designed low-dimensional ultrathin nanosheet structures, lightweight carbon networks with high conductivity, and binder-free design, $NiCo_2O_4$@CNFs achieves extraordinary high energy density of 2110 W h kg^{-1} and excellent cycling durability for over 350 cycles (tested at a current density of 200 mA g^{-1} with a fixed specific capacity of 1000 mA h g^{-1}) as shown in Fig. 19.6c.

Similarly, we developed a new hybrid material, 1D $LaTi_{0.65}Fe_{0.35}O_{3-\delta}$ perovskite nanoparticle anchored on nitrogen-doped carbon nanorods (LTFO-C) by an electrospinning method, and applied as an advanced hierarchical air cathode for rechargeable metal–air batteries shown in Fig. 19.7. It is reported that perovskite nanostructures possess noticeable physicochemical properties, with excellent electrochemical stability, and eco-friendliness. Also, the enhanced electrocatalytic activities toward both OER and ORR can be explained by the electron-occupancy of, e.g., orbitals and high covalence of metal–oxygen bonds [66, 67]. Figure 19.7a,

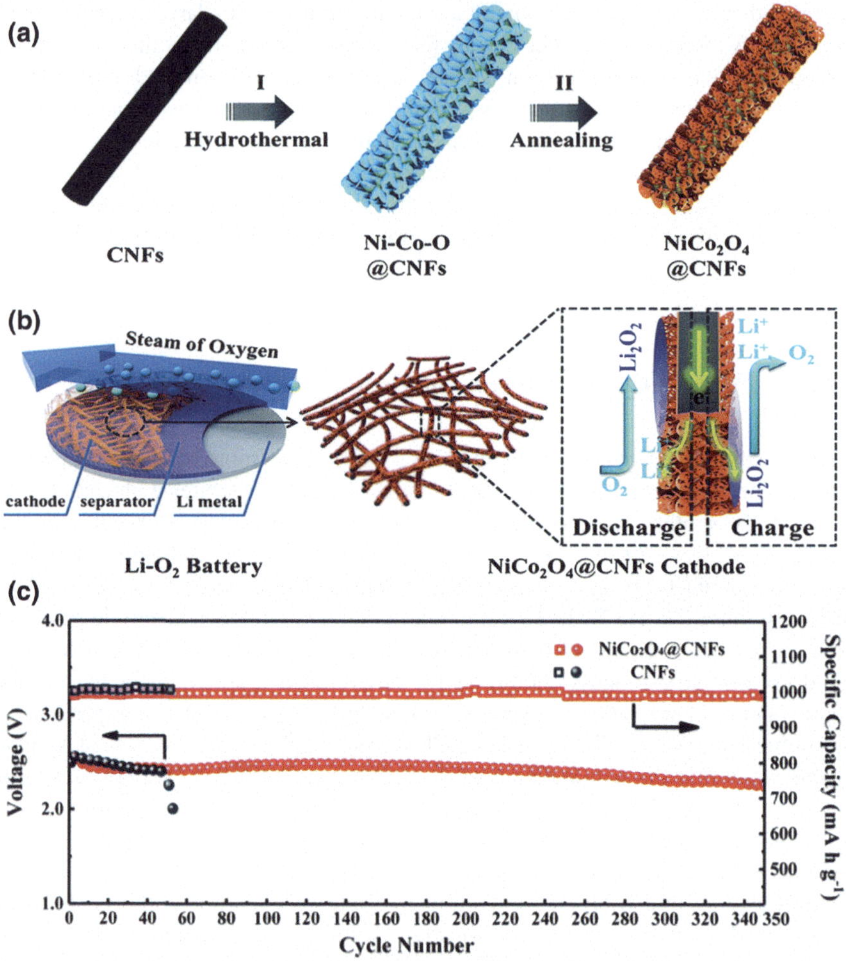

Fig. 19.6 **a** Schematic representation of the synthesis of the $NiCo_2O_4$@CNFs air cathode; **b** structure of the rechargeable Li–O_2 battery with $NiCo_2O_4$@CNFs air cathode **c** the terminal discharge voltage versus cycle number for Li–O_2 batteries fabricated with $NiCo_2O_4$@CNFs and CNFs. Reproduced from Ref. [65] with permission of The Royal Society of Chemistry

b show the LTFO-C is composed of ultrafine LTFO nanoparticles with an average size of 10 nm, uniformly distributed at the surface and also embedded in the NCNR ($\sim$200 $\pm$ 20 nm). The optimized LTFO-C electrode shows an excellent high stability and electrocatalytic performance toward both primary and rechargeable zinc–air batteries even operated in an ambient air environment compared to the commercial catalysts of Pt/C and IrO_2. In detail, The LTFO-C catalyst shows a plateau discharge voltage of 1.16 V with a specific capacity of 440 mA h g^{-1} for over 12 h operation, whereas the NCNR presents a plateau discharge voltage of 1.13 V with a specific discharge capacity of 230 mA h g^{-1} for 8 h operation (Fig. 19.7c). It

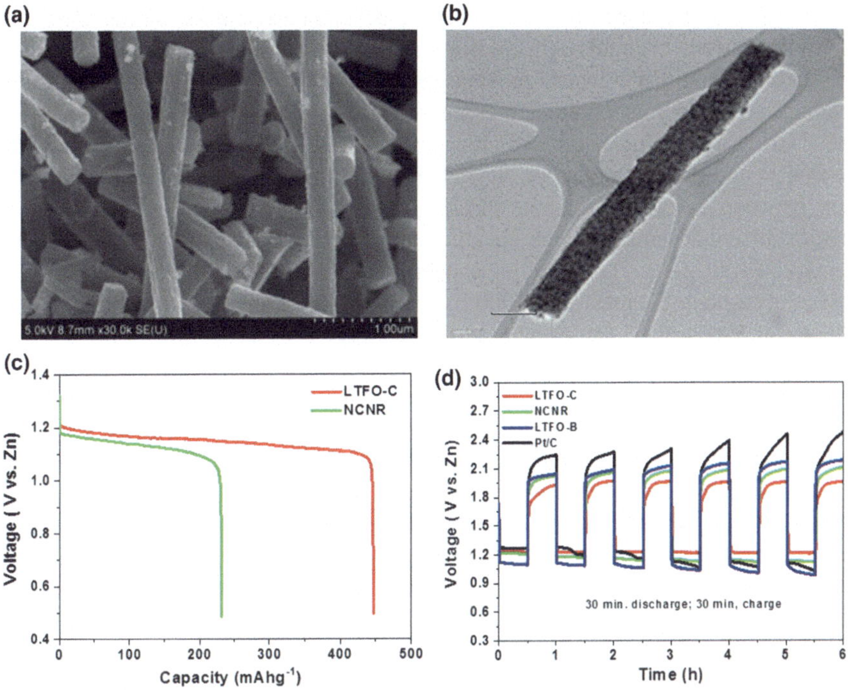

Fig. 19.7 **a** FE-SEM images and **b** FE-TEM images of one-dimensional LTFO-C. **c** Discharge profiles of primary zinc–air battery with the one-dimensional of NCNR and the perovskite LTFO catalyst-embedded in carbon nanorod electrodes as air cathode and polished zinc plate as an anode that consumes ambient air. (Current density: 5 A g^{-1}, cutoff voltage: 0.5 V vs. Zn.) **d** galvanostatic charge–discharge curves of various catalysts at a current density of 5 A g^{-1} in long (1 h) interval cycles. Reproduced from Ref. [67], Copyright (2015), with permission from Elsevier

should be noted that all the specific capacity of the primary zinc–air battery is normalized to the used mass of zinc metal. When assembled a rechargeable Zn–air cell, the potential gap does not change with increasing cycle number up to 6 cycles for LTFO-C-based cathode, however, the commercial IrO_2-based cathode degraded quickly with a significant increase in the charge potential for IrO_2 (Fig. 19.7d). Moreover, the charge–discharge voltage profile of LTFO-C shows significantly reduced overpotential over 60 cycles, and the voltage gap slightly increases from 0.57 to 0.83 V, while the precious catalyst-based electrode quickly degraded with a considerable increment in overall overpotential (not shown).

Further, we also prepared Co-CoO nanoparticles embedded in mesoporous N-doped carbon nanorod (Co-CoO/N-CNR) as air cathode via an electrospinning for $Li–O_2$ battery application [68]. While designing the cathode, diffusion of oxygen, optimization of pore size, and distribution of cathode structures are the critical issues as the ability of porous materials to accommodate the formation of discharge products is closely related with a gravimetric energy density

(Wh $kg^{-1}_{discharged}$). Based on this aspect, the developed one-dimensional Co-CoO/N-CNR electrode occupies uniform network nanostructure and huge voids between the carbon nanofibers, which prevents the pore blockage during discharge and facilitates oxygen transfer. Also, making a hybrid structure with cobalt metal/cobalt metal oxide nanoparticles on CNR reduces the overall overpotential for both OER and ORR and delays the carbon degradation, minimizing the unwanted side reaction between Li_2O_2 and carbon surface during charge. Therefore, it leads to a significant enhancement of the specific capacity (10,555 mA h g^{-1} at a current density of 100 mA g^{-1}) and cycling capability (over 86 cycles without any capacity loss at a vast cutoff capacity of 1000 mA h g^{-1}). Moreover, even at a high current density of 500 mA g^{-1}, an excellent metal-battery performance of 7514 mA h g^{-1} is achieved.

For the other example, we have tried developing a hybrid material composed of $CoMn_2O_4$ nanoparticles covalently coupled with nitrogen-doped reduced graphene oxide (denoted as CMO/N-rGO) as a highly active and durable air electrode for zinc–air battery via hydrothermal approach [69]. The coupling interaction of semiconducting metal oxides and carbon-based materials can create more active sites and superior chemical stability, preventing the detachment and agglomeration of metal oxide nanoparticles on the graphene surface. The CMO/N-rGO catalyst presented the ORR onset potential of 0.90 V and half-wave potential at −3 mA cm^{-2} of 0.80 V, respectively. Meanwhile, in case of OER, it could afford a current density of 10 mA cm^{-2} at a potential of 1.66 V. So the overall potential for oxygen electrode to catalyze both OER, and ORR was only 0.86 V, which outperforms over the commercial catalyst of Pt/C. Based on the excellent oxygen catalytic activities, zinc–air battery cell is assembled with CMO/N-rGO cathode, and it showed an OCV around 1.5 V and the charge–discharge potential gap is maintained to 0.71 V until 100 cycle's battery operation. This work suggests that the combination of N-doped reduced graphene oxide and mixed valence spinel material is favorable to be an oxygen electrode which can be applied to Zn–air batteries.

19.3.3 Metal Chalcogenide/Nanocarbon-Based Electrodes

Also, the CoS_2 nanoparticles supported on a nitrogen and sulfur co-doped graphene oxide (CoS_2(400)/N, S-GO) was fabricated by solid-state thermolysis at 400 °C using Swagelok from cobalt thiourea complex and the graphene oxide as precursors [70]. The hybrid CoS_2(400)/N, S-GO exhibits low overpotential gap about 0.78 V versus Zn with remarkable and robust charge and discharge profile. Also, the CoS_2(400)/N, S-GO shows discharge specific capacity of 767 mA h g^{-1}, which almost achieved the theoretical capacity of 819.8 mA h g^{-1}, and stability up to 7.5 h battery operation. This work demonstrated that the synergistic effect of N, S-GO coupling reaction which improves the discharge capacity dramatically was stem from a strong interaction between CoS_2 (400) particles to the N, S-GO surface.

Meng and his co-worker synthesized in situ coupling of ultrafine $Co_{0.85}Se$ and nitrogen-doped carbon ($Co_{0.85}Se$@NC) by one-step selenization of metal-organic frameworks (zeolitic imidazolate framework-67) as a trifunctional catalyst for overall water-splitting and Zn–air batteries (Fig. 19.8a) [71]. Recently, metal–organic framework (MOF)-derived materials have drawn large interest, as both the precursor and template can provide abundant channels to transport the electron/ion efficiently, and heteroatom-doped carbon incorporates with metal species via in situ carbonization process to further increase active sites, conductivity, and adsorptivity [72]. Thanks to the synergistic effect between $Co_{0.85}Se$ and NC, with conductive carbon network and porous nature, $Co_{0.85}Se$@NC shows high oxygen evolution reaction (OER) catalytic activity with a very low overall overpotential of 0.80 V at 10 mA cm^{-2} current density, remarkable durability, and a high faradaic efficiency. Furthermore, Co0.85Se@NC can also efficiently catalyze hydrogen evolution reaction (HER) and oxygen reduction reaction (ORR). Specifically, when it is applied as air cathode for Zn–air batteries, $Co_{0.85}Se$@NC presents a current density of 186 mA cm^{-2} at a voltage of 1.0 V, and the peak power density of $Co_{0.85}Se$@NC is as high as 268 mW cm^{-2} at 0.64 V, which is much higher than those of $Co_{0.85}Se$ (139 mW cm^{-2}) and Pt/C (173 mW cm^{-2}) shown in Fig. 19.8b, c. Specifically, $Co_{0.85}Se$@NC presents a discharging potential superior to $Co_{0.85}Se$, and comparable to the Pt/C during the ORR, while for the charging process, $Co_{0.85}Se$@NC outperforms the commercial Pt/C electrode. Also, the voltage gap of charge–discharge curves for a $Co_{0.85}Se$@NC electrode at 20 mA cm^{-2} is only 0.88 V, which is much smaller than those of $Co_{0.85}Se$ (1.13 V) and commercial Pt/C catalyst (0.96 V) (Fig. 19.8d). At the same time, it also shows high cycling performance with no obvious voltage drop for over 180 cycles within 30 h, whereas an apparent increase in both the charge and discharge voltages is detected after 120 cycles in 20 h for both $Co_{0.85}Se$ and Pt/C (Fig. 19.8e). Therefore, the bifunctional $Co_{0.85}Se$@NC air cathode is feasible in practical Zn–air batteries with enhanced electrocatalytic performance and cyclic stability.

19.4 Nanocarbon-Based Electrode Catalyst for Water Splitting

The water electrolysis (WE) is an emerging technology in the production of sustainable hydrogen (H_2) energy. The distinct advantage of WE is that it can use the earth abundant water (H_2O) molecule as a resource to produce the H_2 energy without much contribution to the greenhouse effect. It is an electrochemical process, and the electrical energy acts as a driving force. In a water-splitting process, the H_2O is split into the oxygen and hydrogen molecules at the anode and cathode electrodes, respectively. Based on the kind of electrolyte usage and the charge carrier (OH^-, H^+, O^{2-}), WE is classified into three main categories such as alkaline [73], polymer electrolyte membrane (PEM) [74], and solid oxide electrolyzes

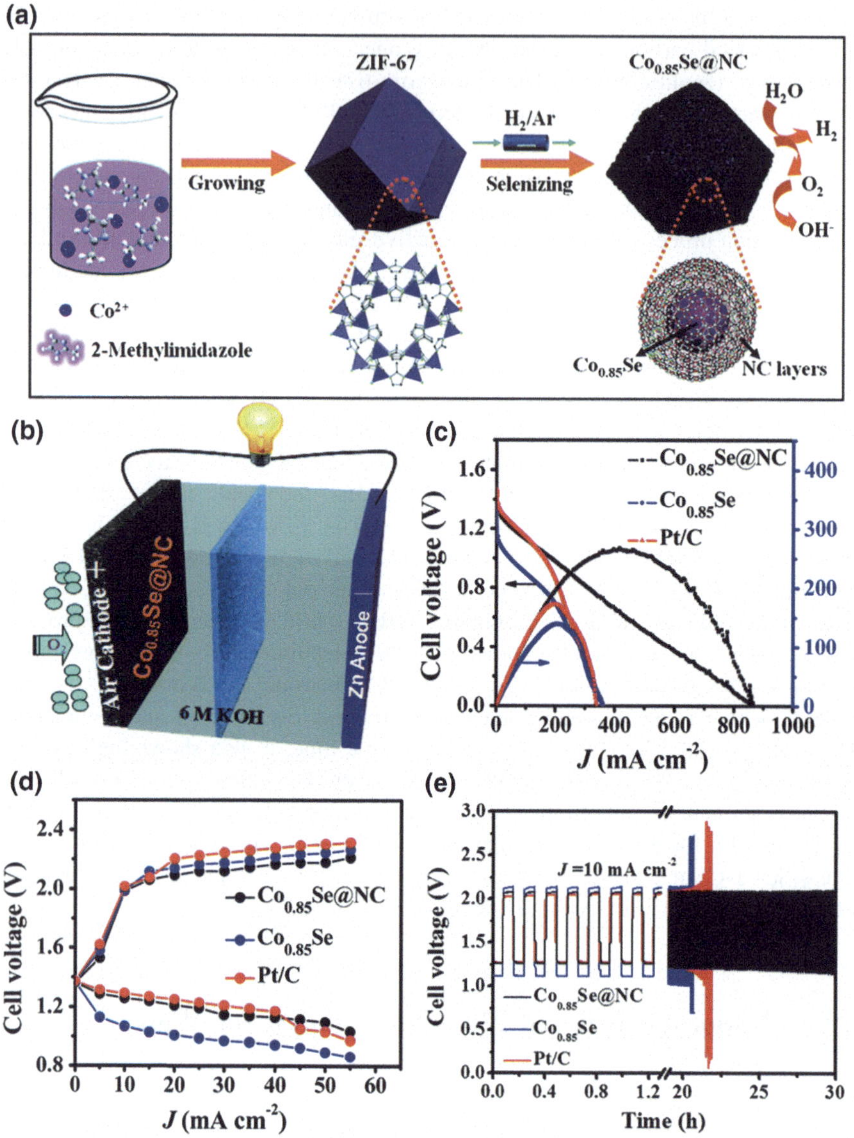

Fig. 19.8 **a** Illustration of synthesis procedure for $Co_{0.85}Se$@NC electrocatalyst. **b** Schematic of rechargeable Zn–air battery based on $Co_{0.85}Se$@NC air cathode. **c** Polarization curves and the corresponding power density plots of primary Zn–air batteries using $Co_{0.85}Se$@NC, $Co_{0.85}Se$, and commercial Pt/C catalyst as the air-breathing cathode. **d** Charge–discharge curves of the rechargeable Zn–air batteries **e** cycling performance of rechargeable Zn–air batteries (10 min cycle period). Reproduced from Ref. [71] with permission of The Royal Society of Chemistry

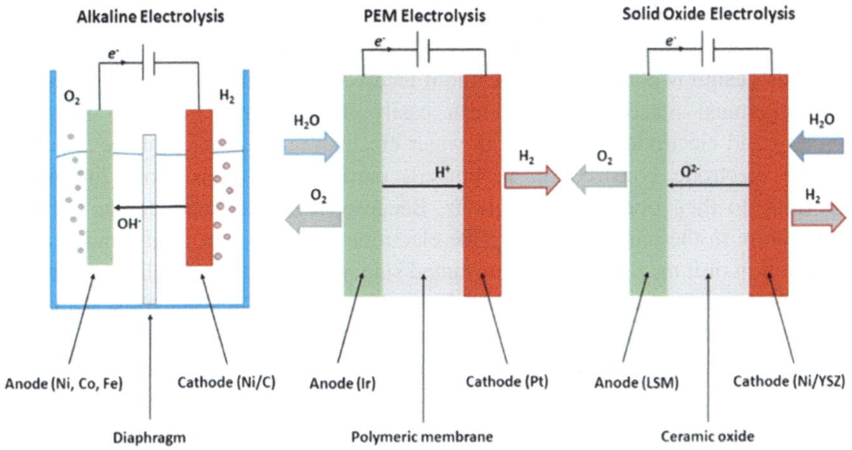

Fig. 19.9 Operating principles of different water electrolysis system (Alkaline, PEM and solid oxide). OER occurs at the anode side, HER occurs at the cathode side. An O^{2-} is the conducting electrolyte in solid oxide electrolysis with a nickel/yttria-stabilized zirconia (cathode) and a lanthanum strontium manganite (anode). Reproduced from Ref. [76], Copyright (2017), with permission from Elsevier

(SOE) [75]. All the water electrolysis systems operating principles are presented in Fig. 19.9.

The anode and cathode reactions are given as follows:

Alkaline electrolysis:

$$\text{Anode: } 4OH^{-} \rightarrow O_2 + 2H_2O + 4e^{-}$$
$$Cathode\text{: } 4H_2\text{O} + 4e^{-} \rightarrow 4OH^{-} + 2H_2$$

PEM electrolysis:

$$Anode\text{: } 2H_2\text{O} \rightarrow 4H^{+} + 4e^{-} + O_2$$
$$\text{Cathode: } 4H^{+} + 4e^{-} \rightarrow 2H_2$$

Solid oxide electrolysis:

$$\text{Anode: } O^{2-} \rightarrow \frac{1}{2}O_2 + 2e^{-}$$
$$Cathode\text{: } H_2\text{O} + 2e^{-} \rightarrow H_2 + O^{2-}$$

The reactions mentioned above are the key reactions in the water splitting and which can mainly decide the cell efficiency [76]. Interestingly, the above-mentioned electrochemical reactions happen on the electrode material surface. So, the most important component to construct the electrolysis cell is the adequate electrode materials and which should not support the unwanted side reactions for avoiding the cell efficiency loss and impurities in H_2 gas, and it must only try to enhance the necessary electrochemical reactions of OER and HER [77]. So, the electrode materials design is not an easy task, and it received much attention worldwide. The electrochemical reactions in WE can easily deviate from electrode materials property and electrode design. So, in water electrolysis the choice of the materials for the electrode construction is limited, particularly it has more limitations according to their operating conditions. Because, if the cell operates at a higher temperature in the harsh electrolyte the electrode materials can easily undergo the deactivation or it may support the unwanted side reactions. In particular, most of the electrode materials have been easily subjected to the corrosion at anodic potentials due to the oxidation or reduction of electrode materials [78].

Each electrolyzer has their own pros and cons by their operating temperature and types of electrode materials. Initially, the acid aqueous electrolyte-based electrolysis has been widely used, but nowadays the alkaline WE has more attraction due to the advantages and freedom in the selection of electrode materials [77]. Since in the acid environments, the limitations on electrode design are very high, and widely the noble metals such as Ru, Ir, and Pt and their derivatives have been considered as the suitable active sites due to the high tolerance property. So, it does not require the expensive acid-resistant electrode materials to avoid the self-corrosion in the acid environments [79]. Most importantly, the cheaper and various kinds of non-precious electrode materials are easily utilized in the alkaline WE. The reduction of electrolysis unit cost is the ultimate goal of the industrial WE.

Normally, the metal electrodes are suitable for the gas evolving process. According to the reaction kinetics, the nickel-related electrode materials have shown reliable performance in the caustic electrolytes. So, both anode and cathode in the industrial alkaline WE are typically nickel-coated iron, stainless steel, or nickel-based electrode materials and the cell operates in 6–9 mol/L concentration of potassium hydroxide solution with the operating temperature of 60–80 °C. Unfortunately, during the electrolysis, the electrode materials undergo the extensive deactivation due to the high initial electrochemical activity. This electrode deactivation within the short period of electrolysis has strongly affected the water electrolysis current/cell efficiency [80]. Initially, the nickel deactivation has been stopped by dissolving the V_2O_5 in the electrolyte solution. They found that the dissolved V_2O_5 reactivates the deactivated nickel efficiently in the cathode. Even though, the electrocatalytic activity was not beyond the fresh nickel electrode material [81]. Then, the cobalt- and tungsten-based ionic activators have been utilized in the electrolyte to obtain the enhanced hydrogen production using the nickel foam electrode materials. Interestingly, they reduced 15% of the energy requirements per unit hydrogen production as compared with the non-activated system [82]. Also, the addition of iron in the manganese-molybdenum oxides has enhanced the electrode stability [83]. Furthermore, the active layer coating on the oxides materials, alloy kind of materials formation, including the doping of precious metal and their oxides with non-precious electrode materials have been studied to improve the different parameters of the electrode for the enhanced the WE activity and stability as well [84, 85, 86].

Recently, to increase the electrical conductivity, roughness, surface area, and porosity of the electrocatalysts, the heteroatoms-doped carbon-based electrode materials have been used widely as anode and cathode or both in the overall WE. However, the stability of the carbon is not sufficient for the higher cell voltage WE. Also, to improve the catalytic active sites and stability of the electrode materials, the transition metal, metal oxides, metal chalcogenides, metal pnictides are combined with carbon and introduced as a composite in the WE. Furthermore, the Li and La metal have been doped with core electrode materials [87, 88]. Furthermore, the nanostructured materials with the enlarged surface area and unique electrical property also attained much attention in WE. Because, the increased surface area has the more active sites and reduces the electrolyzer operating current density, including the 25% overpotential reduction with 20% less-energy consumption [89]. Also, the nanostructured materials have increased the charge transfer of the applied potential to the electrocatalysts surface during WE [90]. Apart from that the electrocatalysts design and electrode preparation also play the key role in the electrode activity and stability. Because, during the electrocatalytic performance the electrode should facilitate the electrolyte diffusion and escape the formed gas bubbles. So, several electrocatalysts have been designed to reduce the water electrolysis overpotential and also to increase the stability of industrial WE.

To sum up, most importantly the electrode materials give the significant influence in the cell overpotential and current density. The physical and chemical modifications of electrode materials improve the water electrolysis activity and stability. By doping or coating of different elements, the electronic and surface property of electrodes can be determined and the electrocatalytic performance is decided. Widely, the nanostructured materials have multiple advantages toward the enhanced and stable water electrolysis in the industrial-scale H_2 production. Also, the transition metals incorporation in the nanostructures could improve the electrode materials activity.

19.5 Nanocarbon Electrodes in Water Electrolysis

The emerging electrocatalytic materials of nanocarbon and their composites have great attraction and significant interest due to their unique properties and structural integrity toward the multiple energy applications. Carbon is earth-abundant, low-cost material and it has a tunable molecular structure, extremely high surface area, more roughness, excellent electrical conductivity, and excellent surface tolerance in the harsh conditions [91, 92]. Remarkably, the various traditional nanocarbon materials of nanorods, nanotubes, graphene, carbon black are widely introduced as efficient electrode materials for the high-performance electrocatalytic activities in water electrolysis [93].

19.5.1 Heteroatoms-Doped Carbon Nanostructures

The pristine nanocarbon materials are not electrochemically active, or mostly it has poor activity due to the insufficient active sites. So, it is widely used as a current collector such as carbon cloth, glassy carbon, and carbon paper [94]. However, the nanocarbon electrode materials activity and stability have been increased by changing their morphology, chemical composition, and surface functional groups using several advanced chemicals and physical modification approach. Mostly, the chemical modification may strongly influence in the electronic state density of the nanocarbon, and it enhances the electrocatalytic activity [95, 96]. For instance, the oxygen functionalization on the nanocarbon also tries to modify the electronic property of nanocarbon surface. The chemical modification has been done by the doping of different heteroatoms in the nanocarbon chemical structure. However, the doping of heteroatom in the structure of nanocarbon gives much more influence in the electrocatalytic activity than that of functionalization. Because, the surface functionalized groups may only have weak adsorption on the surface of the nanocarbon, unlike the doping of heteroatom in the crystal structure which has affected the whole nanocarbon chemical structure. So, the single heteroatom (N, B, P, S, I, etc.) has been doped in the nanocarbon and strongly modified their electronic structure, and then the heteroatom-doped nanocarbon is introduced as an efficient electrocatalytic material [97]. Moreover, the secondary heteroatom doping has further disturbed the single-heteroatom-doped nanocarbon structure for the enhanced electrocatalytic activity [98]. Overall, the two essential factors for the performance of the carbon-based electrocatalysts are: (i) the chemical structure modification by the heteroatom(s) doping in the nanocarbon; (ii) The nanocarbon surface area and porosity, particularly micro and mesoporosity.

For example, the N-doping in the nanocarbon has been studied as a most efficient electrocatalyst for overall WE. The introduced N-heteroatom in the nanocarbon determines the intrinsic property of the active sites for the electrocatalytic activity. Most of the N-doped carbon nanomaterials are widely prepared from the N-rich polymer precursors assisted by pyrolysis method and used as alternative electrode materials for the OER and HER electrocatalysts in alkaline electrolyte [99]. Generally, the three major kinds of N species exist in the carbon matrix such as, pyridinic-N, pyrrolic-N and graphitic-N. Among all, the pyridinic-N plays a crucial role in OER. At the first step, the adjacent carbon atoms to pyridinic-N atoms are positively charged due to the electron-withdrawing nature of pyridinic-N atom in the carbon matrix [100]. So, the positive charge attained carbon atoms in the carbon matrix acts as an active site and which facilitates the adsorption of OER intermediates of OH^- and OOH^-. This intermediates adsorption by the positively charged carbon atoms is the first rate-determining step in the OER. From this, all the positively charged carbon atoms in the carbon matrix provide the support for the recombination of adjacent two O_{ads} additives [101–103]. So, the carbon atoms also act as active sites for the OER because of the N-doping in the nanocarbon. Also, the pyridinic-N may be another possible active site for the OER, because the density states of the nitrogen in the carbon matrix (sp^2) locates very close to the Fermi level, so the active nitrogen sites also possible during the OER [104, 105]. However, according to the density functional theory (DFT) and first-principle calculation, the real active sites are not yet confirmed for the OER. Similarly, the single-heteroatom of N-doped graphene, N-doped carbon nanotube, N-doped coaxial carbon nanocables are also exhibited the promising OER activity in the water splitting [106–108]. Also, the N-doped nanocarbon materials are the most responsible for the HER electrocatalytic activity. As next to the pyridinic-N the carbon atoms act as active sites for the water molecules adsorption and which is the initial rate-determining step for the HER in basic media [109]. It means, the heteroatoms doping in the carbon matrix also show their strong influence in the HER rate-determining steps.

For instance, the simply synthesized N-doped multiwall carbon nanotubes (NMWNT) have studied as overall WE electrocatalyst by Davodi et al. As-synthesized 20-NMWCNT catalyst shows the superior OER current density of 10 mA cm^{-2} with an overpotential of 360 and 320 mV in 0.1 and 1 M KOH electrolyte, respectively. Moreover, the Tafel slope linearity and small values reveal that the possibility of higher conductivity with fast electron transfer. Also, the NMWCNT catalyst requires an overpotential of 340 mV to reach the HER current density of −10 mA cm^{-2} in 0.1 M KOH. It reveals that the N-functionalities in the MWCNT plays a key role in the OER and HER activities enhancement, particularly the pyridinic-N acts as the real active site as compared with other nitrogen species. Also, the NMWCNT exhibits the excellent electrochemical stability during the long-term WE [110]. Further, Balogan et al. developed the monolithic 3D N-doped porous carbon cloth (3D NiD-PCC) as shown in Fig. 19.10a with the high surface area (135 m^2 g^{-1}) and explored as another metal-free and self-standing water-splitting electrodes (Fig. 19.10b, c). It requires the OER overpotential of 360 mV to afford the 10 mA cm^{-2}. The overall WE current density of 10 mA cm^{-2} has achieved with the cell voltage of 1.85 V, which is quite higher as compared to the reported values (Fig. 19.10d, e) [111]. So, Zhang et al. reported that the polydopamine analog derived defect-rich porous carbon acting as an efficient bifunctional metal-free nanocarbon electrode for overall water splitting. This defect-rich porous nanocarbon has a large surface area with micro- and meso-porosity. Besides, the N-rich precursor provides the more pyridinic-N in the carbon matrix, which is most responsible for the bifunctional activity in the water splitting. It requires the low cell voltage of 1.74 V to deliver the current density of 10 mA cm^{-2}. It has significantly lower cell voltage as compared with different metal-based electrode materials. Furthermore, a self-powered WE system with Zn–air battery also demonstrated successfully [112].

This single-heteroatom doping and their influence on the OER and HER activities and stability provide a new way to design dual-heteroatoms-doped nanocarbon electrode materials. So, co-doping of N with other heteroatoms can improve the electrocatalytic activity significantly due to the synergistic effect between the two different heteroatoms in the carbon matrix. The sulfur (S), boron (B) and phosphorous (P) atoms are considered as secondary heteroatoms, which are

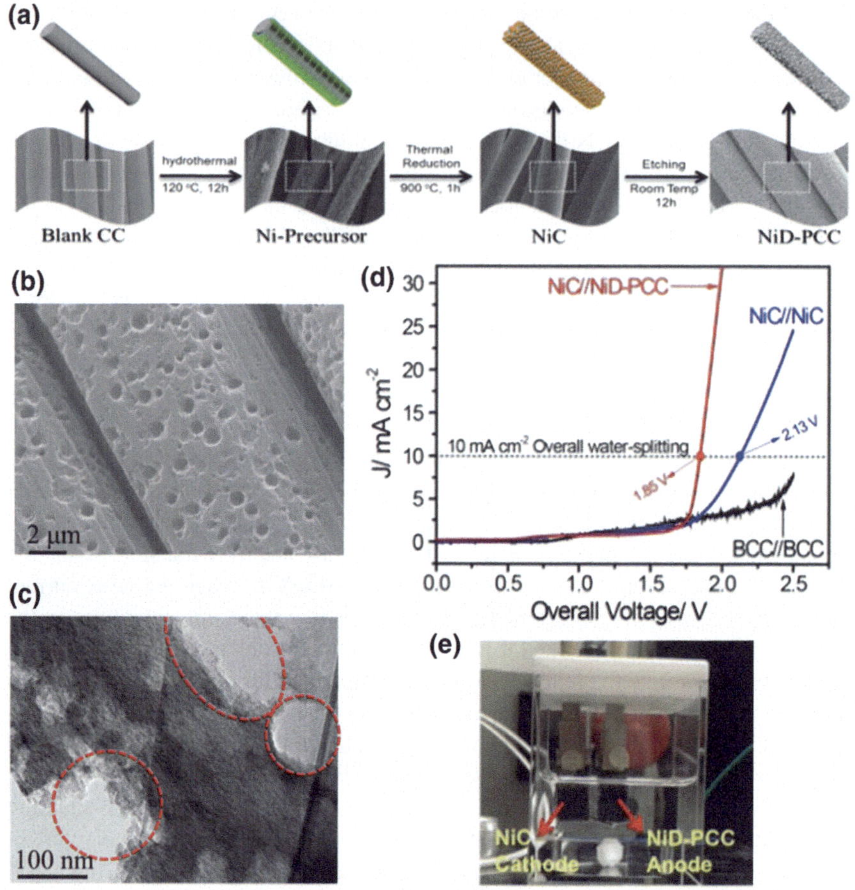

Fig. 19.10 **a** Schematic representation of monolith 3D NiD-PCC synthesis. **b**, **c** SEM and TEM images of the NiD-PCC. Overall water splitting activities of the various alkaline electrolyzers. **d** Polarization plots in 1 M KOH (scan rate of 10 mV s^{-1}). **e** Optical photographic image shows the hydrogen and oxygen bubbles generation from the electrodes. Reproduced from Ref. [111] with permission of The Royal Society of Chemistry

doped with N-doped nanocarbon materials. The dual-heteroatoms-doped carbon nanostructures also prepared from the single/multiple polymer precursors with high-temperature pyrolysis or hydrothermal method. For instance, the N and P co-doped graphene/carbon nanosheets tested as enhanced OER and HER nanocarbon electrode materials. Comparatively, the electrocatalytic activity of dual-heteroatoms-doped nanocarbon materials is much enhanced than single-heteroatom-doped nanocarbon materials. The possible reported reason for this enhanced electrocatalytic activity is that the additional doping of P in N-doped nanocarbon may enhance the carbon atoms charge delocalization as well as asymmetric spin density together. So, these altered carbon atoms are trying to promote the N-doping mostly at the edge of the carbon matrix, which can increase the more active sites and also provide the triggering synergistic effect [113–115]. In addition to this, when the extra support can be provided to the dual-heteroatom-doped nanocarbon, for example, the carbon fiber paper supported N and P elements-doped graphitic-carbon nitrides attaining the higher OER activity as well as stability. Because, the support increases the charge transfer and also exposes the highly active surface area, including the synergistic effect of dual heteroatoms [116].

For that, the polydopamine inspired dual heteroatom-doped carbon nanotubes using graft and pyrolyze method. As-synthesized N, S co-doped carbon nanotube shows the improved bifunctional activity in 1 M KOH electrolyte. In overall water splitting the N, S co-doped nanocarbon electrode material needs 2.03 V to the solar-to-fuel current density of 10 mA cm^{-2} in 1 M KOH solution. Almost all the carbon atoms serve as active sites to adsorb the H* and OOH* due to the dual doping, particularly S atoms [117]. Also, the N, S-co-doped graphitic sheets with stereoscopic holes (SHG) was prepared by Hu et al., and reported as metal-free electrocatalyst s for both OER and HER in 0.1 M KOH. This newly prepared porous carbon has the larger surface area of 576 m^2 g^{-1}. Also, this 900 °C annealed electrocatalyst show the more graphitization (I_G/I_D = 1.29) with defects which is confirmed by Raman analysis, and it can improve the electrical conductivity. During OER in 0.1 M KOH, the SHG electrocatalyst shows the OER potential of 1.56 V at the anodic current density of 10 mA cm^{-2} with lower Tafel slope value of 71 mV dec^{-1}. In the cathodic activity of HER, the same electrocatalyst needs −0.31 V potential to provide the −10 mA cm^{-2} current density with smaller Tafel slope value (112 mV cm^{-2}). In overall WE, it requires 1.70 V cell voltage to deliver the WE current density of 10 mA cm^{-2} in 1 M KOH with 30 h stable activity. Hence, the stereoscopic holes in the electrocatalysts ensure the larger surface area, abundant active sites with good electron/electrolyte transport during the electrochemical reactions [118]. Interestingly, the tri-heteroatoms (N, P, and O) co-doping in porous graphite carbon-oxidized carbon cloth has reported as full water-splitting electrocatalysts. In 1 M KOH electrolyte, the WE shows a very low cell voltage of 1.66 V at the current density of 10 mA cm^{-2}, which is quite low as compared with above-mentioned WE system [119].

Beyond the heteroatom doping effect on OER in water splitting, the two different nanocarbon structures interaction also play the key role in the electrocatalytic activity and stability. As, all the N-doped nanocarbon may have good electrocatalytic activity, but it suffers from the poor electron transfer during the OER activity, so it is also a kind of barrier for the electrochemical applications. For that case, the active carbon materials need the proper support by good interaction. For example, the graphitic carbon nitride nanosheet is an N-doped carbon structure and good for electrocatalytic applications, but the charge transfer is poor during the reaction. So, the carbon nanotube or other conductive carbon materials may provide a support and form the three-dimensional porous composite. It can show an excellent OER and stronger durable electrocatalyst in the base environment (pH = 13) due to the Π-Π stacking and strong electrostatic interaction between nanosheets and nanotubes [120]. As we know, the OER takes place in the positive potential, and it has the slow kinetics. So, the material stability is also a significant problem in the water splitting. The major current density loss or overpotential comes from the anode side due to the easy electrode material deactivation or corrosion. Interestingly, the heteroatom-doped nanocarbon shows the excellent stability with activity in the harsh electrolyte environments. Recently, the N-doped nanocarbon electrode materials have been prepared and reported as metal-free electrocatalyst with remarkable long-term stability [121]. Also, the coupled graphitic-carbon nitride with nitrogen-doped graphene shows a good HER activity with the low overpotential and Tafel slope as compared with the various metallic electrocatalysts. From the experimental results and DFT calculations, it is clear that the activity originates from the coupling of intrinsic chemical and electronic properties. This strong coupling leads the synergistic effect and promotes the proton efficiently for the enhanced HER activity [122]. Also, the N and S co-doped nanoporous graphene is one of the best metal-free nanocarbon electrodes for the HER. Because the N and S co-doped graphene demonstrates that the catalytically active sites are associated with the doped heteroatoms with around defects in the graphene. Furthermore, the S-heteroatom is not influenced in the HER activity directly, which creates the coupling effect with co-doped N atoms in the graphene [123]. The P co-doped with N in the graphene also provides the synergistic effect towards the HER activity. Here, both the heteroatoms together activate the neighbor carbon atoms in the graphene sheet by tailoring the valance orbital energy levels of the carbon and induce the more adsorption of water molecules. Also, the C-N and C-P in the graphene matrix enhance the stability of carbon without any corrosion and easy oxidation [124].

19.5.2 Transition Metal-Doped Carbon Nanostructures

Furthermore, the modified nanocarbon by various heteroatoms can easily couples with different inorganic metal active sites, and the nanocarbon composite exhibits more active sites toward the enhanced OER and HER activities, including their synergistic effect during the electrochemical analysis. It is mainly due to the existence of a chemical bond between the nanocarbon and metal, which enhances efficient charge-transfer between them and also increasing the electrochemical stability of the catalysts in the OER and HER [125]. For example, the cobalt-embedded porous N-doped nanofibers prepared by the electrospun method has introduced as an efficient catalyst for the overall WE in 1 M KOH. The morphology analysis confirms the co-existence of metallic-Co and N elements in the nanofibers. The Raman and Brunauer–Emmett–Teller (BET) analyses reveal that the metal introduced nanostructure has more defects ($I_D/I_G = 1.17$) and much higher surface area of 356 $m^2\ g^{-1}$ as compared with the pristine nanostructures. It shows the OER overpotential of 285 mV at 10 mA cm^{-2} with lower Tafel slope value of 73 mV dec^{-1}. In the case of HER, the −10 mA cm^{-2} current density has achieved with an overpotential of 249 mV (92 mV dec^{-1}). Finally, the constructed WE produce the water-splitting current density of 10 mA cm^{-2} with the cell voltage of 1.66 V. Zhao et al. concluded that the obtained excellent electrochemical performance (both OER and HER) is due to the unique encapsulated metallic-Co with porous nanostructures [126]. Also, the nitrogen and cobalt co-doped CNFs provide more active sites with synergistic effect on enhanced catalytic activity and stability as well. Similarly, Hou et al. has pyrolyzed the graphene oxide (GO) with zeolitic imidazolate framework (ZIF-67) and obtained a nitrogen-doped graphene/cobalt-embedded porous carbon polyhedron (N/Co-doped PCP//NRGO) hybrid electrocatalyst, followed by the etching (partially) of metallic-Co. It has unique advantages of dodecahedral shape and more Co–N moieties. From the Raman spectrum, the higher I_G/I_D value (1.12) of this hybrid catalyst confirms the existence of more defects with a BET surface area of 375 m^2 g^{-1}. The hybrid exhibits the superior water-splitting activity with an OER potential of 1.66 V (at 10 mA cm^{-2}), due to the synergistic effect of N/Co-doped porous carbon polyhedron and N-doped reduced graphene oxide nanostructures [127]. Furthermore, the bi-metal incorporation in the carbon nanostructure also increases the number of multiple active sites and tuning the electronic states of an existed metals/carbon. Accordingly, the Ni and Co metals are coated on a porous nitrogen-doped carbon thin film by using two-step pulsed laser deposition method and reported as a bifunctional OER and HER electrocatalysts in 1 M KOH (Fig. 19.11a). The peaks shift (1 eV) in the high-resolution XPS Co2p spectrum confirms the interaction between two different metals of Co and Ni in the carbon nanostructures (Fig. 19.11b, c). As-prepared $Ni_{0.5}Co_{0.5}$/NC hybrid electrocatalyst exhibits the HER overpotential of 176 mV at −10 mA cm^{-2}. Besides, it shows various double layer capacitance values by varying the Ni and Co ratio, as a result, the change in OER and HER overpotential have been noticed (Fig. 19.11d). The same electrocatalyst attains the OER current density of 10 mA cm^{-2} with overpotential of 300 mV. The remarkable bifunctional activity (cell voltage of 1.75 V @ 10 mA cm^{-2}) is due to the synergetic coupling interaction of nickel-cobalt active sites with the porous N-carbon framework [128]. Similarly, the metal-organic frameworks (MOFs)-derived FeNi alloy nanoparticles encapsulated in n-doped carbon nanotubes also exhibits the water splitting OER overpotential of 300 mV at the anodic current density of 10 mA cm^{-2} in 1 M KOH [129].

19.5.3 Transition Metal Oxide–Carbon Composite Nanostructures

The metal oxide supported on nanocarbon is considered as promising electrocatalysts for efficient and stable WE. The metal-based electrocatalysts show the excellent HER activity than OER. To realize the overall WE, it is necessary to improve the OER overpotential and their kinetics by introducing the oxide-based active sites. Accordingly, the nickel, iron, and cobalt-based oxides are considering

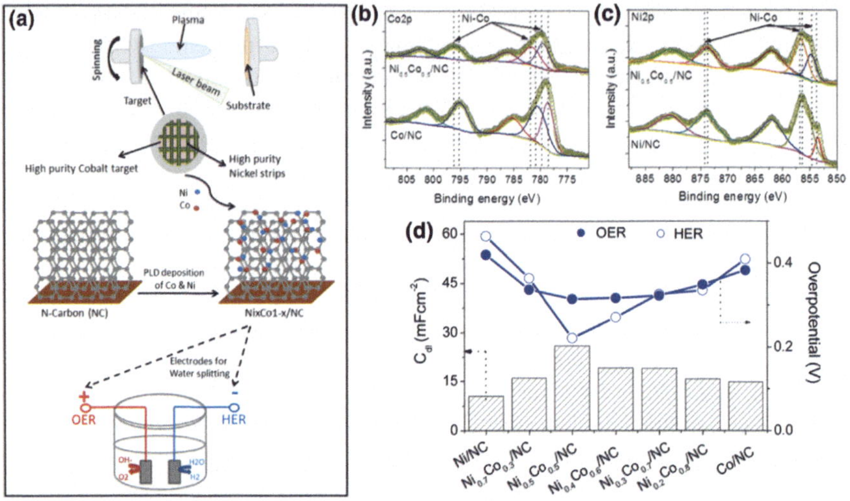

Fig. 19.11 **a** synthesis process of Ni_xCo_{1-x}/NC catalytic electrodes and applied in overall water splitting. **b** High-resolution Ni2p spectra of Ni/NC and $Ni_{0.5}Co_{0.5}$/NC. **c** High-resolution Co2p spectra of Co/NC and $Ni_{0.5}Co_{0.5}$/NC. **d** Roughness factor and the HER/OER overpotential relationships (at 10 mA cm^{-2}). Reproduced from Ref. [128] with permission of The Royal Society of Chemistry

as the most promising OER catalysts, particularly with carbon support. Jin et al. synthesized the cobalt–cobalt oxide/N-doped carbon hybrids (CoO_x@CN) and reported their bifunctionality. Here, the obtained hybrids consist of Co, CoO, and Co_3O_4 which show the HER overpotential of 231 mV at a current density of 10 mA cm^{-2} and low charge-transfer resistance of 41 Ω. The excellent activity is attributed form the high electrical conductivity of carbon and the synergistic effect between the metallic-Co and Co oxides. The carbon encapsulation also provides the strong stability to the active sites. Furthermore, the same catalysts require the OER overpotential of 260 mV to reach the current density of 10 mA cm^{-2} in 1 M KOH and which is due to the presence of more Co^{2+} proportion in the hybrids. The constructed alkaline WE cell needs only 1.55 V to deliver the overall water splitting current density of 20 mA cm^{-2} [130]. Moreover, the higher intrinsic OER and HER activities are expected from the bimetallic oxides than the single metal-based oxide electrocatalysts. Also, the electrochemical activities of those electrocatalysts have been modified by controlled their electronic and valence states. For example, Li et al. reported a highly ordered carbon-coated Co–Mn oxide nanoparticle superlattice (CoMnO@CN) as an efficient bifunctional electrocatalyst for the water splitting. The as-prepared and uniform CoMnO nanoparticles have the smaller bandgap, stronger and weaker adsorption of OH and O_2 molecules, including an excellent OER reagent during electrochemical activities than that of single metal oxides of CoO and MnO. The authors claimed that the obtained activities are mainly attributed to the synergistic effect of bimetals and N-doped carbon species.

Also, the surrounding N-doped carbon framework acts as catalytic active sites for HER and it provides an efficient pathway for electron/ion transport to the embedded CoMnO nanoparticles. Besides, the ordered superlattice has increased the catalytic sites and minimizes the adjacent CoMoO nanoparticles aggregation/dissolution. Therefore, the CoMnO@CN superlattice structure serves as an efficient OER and HER electrocatalysts in alkaline electrolyte. Finally, the bifunctional electrode was coupled with the photovoltaic cell for solar water splitting, and it shows continuous activity for almost 5 days with a solar-to-hydrogen conversion efficiency of 8% [131].

Moreover, the rational design of bi-junction nanostructured catalyst of metallic nickel-decorated transition metal oxides nanosheets vertically grown on ceria film (ceria/Ni-TMO) is synthesized by Long et al. (Fig. 19.12a). They claimed that the as-synthesized catalyst has an ideal electronic structure, abundant active sites with a strong synergistic effect between carbon cloth, ceria, and Ni-TMO. This electrocatalyst-introduced WE system requires the small cell voltage of 1.58 V to produce the water splitting current density of 10 mA cm^{-2} (Fig. 19.12b), including the

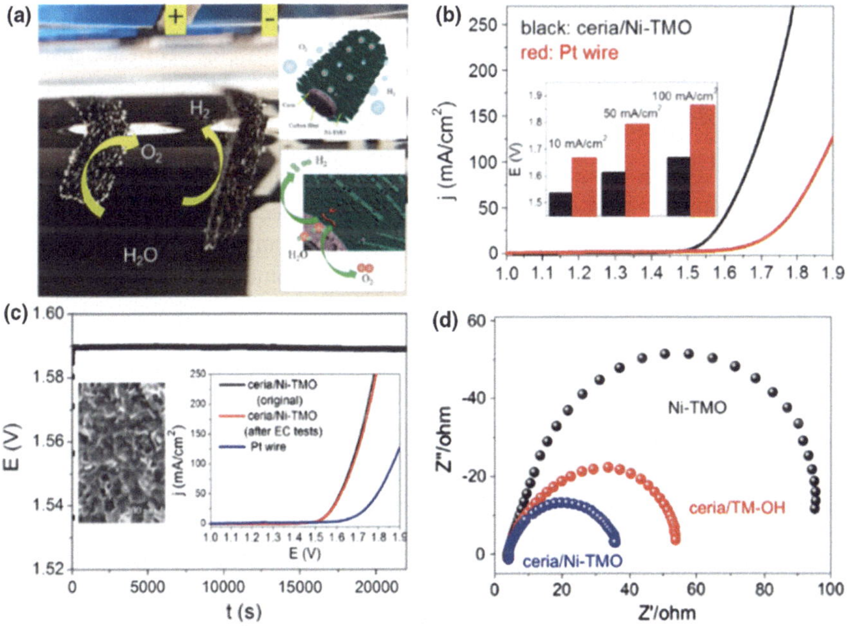

Fig. 19.12 Overall water-splitting system using the ceria/Ni-TMO/CC as both anode and cathode. **a** Digital image (left) displays H_2 and O_2 production on ceria/Ni-TMO electrodes at 1.60 V and schematic description of the catalysts (inset). **b** LSV curves of water splitting by using ceria/Ni-TMO and Pt nanowire in a two-electrode setup in 1 M KOH. **c** Durability (chronopotentiometry) test of ceria/Ni-TMO in full water splitting at 10 mA/cm^{-2}). SEM image (C inset) and polarization curve (C inset) of ceria/Ni-TMO after the stability test. **d** EIS Nyquist plots of various catalyst at the potential of 1.45 V (vs. RHE). Reproduced with permission from [132]. Copyright (2018) American Chemical Society

excellent stability (Fig. 19.12c). Through this work, the authors stated that (i) the idealized electronic structure of metal ions obtained from the introduced Fe, Mn, and Ni and which enhances the intrinsic activity, (ii) the hierarchical nanostructure increases the active surface area and an intimated heterostructure promotes the electron transfer between Ni-TMO and ceria film (Fig. 19.12d), (iii) the unique property of oxygen storage-release in ceria can favor the intermediates adsorption and desorption during the WE [132]. Another efficient method to improve the catalytic activity of oxide catalysts is an electrochemical transformation of nanoparticles ($\sim$20 nm) to ultra-small nanoparticles with a diameter of $\sim$2–5 nm by lithium metal-induced conversion reactions and this approach was reported by Wang et al. First they are grown the CoO nanoparticles with a diameter of $\sim$20 nm on the carbon fibers (CoO/CNFs). Then, to reduce the size of CoO, the CoO/CNFs catalysts was assembled in the lithium-ion pouch cell and they are carried out the galvanostatic lithiation–delithiation process for several cycles in 1 M $LiPF_6$ in 1:1 w/w ethylene carbonate/diethyl carbonate electrolyte (Fig. 19.13a–f). Likewise, the various oxides of NiO/CNFs, Fe_3O_4/CNFs and the mixed oxides of $NiFeO_x$/CNFs have been prepared by using the similar approach to improve their catalytic activity and stability in both 0.1 and 1 M KOH electrolyte. All the catalysts show the promising electrocatalytic activities, particularly $NiFeO_x$/CNFs exhibits the remarkable and lower OER potential of 1.51 V at 200 mA cm^{-2} current density in 1 M KOH with small Tafel slope value of 31.5 mV dec^{-1}. Also, the same catalyst needs only 88 mV overpotential to deliver the HER current density of −10 mA cm^{-2}. By using this catalyst, the overall WE current density of 10 mA cm^{-2} was obtained with the lowest cell the voltage of 1.51 V with 200 h long-term stability. It reveals that the excellent electrical interaction among nanoparticles and the larger surface area with many catalytic active sites (Fig. 19.13g, h) are the most responsible for the promising bifunctionality and stability of the catalyst [133].

19.5.4 Transition Metal Chalcogenides/Pnictides-Carbon Composite Nanostructures

Recently, the metal-sulfides, selenides or phosphides-carbon composites/hybrids have received a great attention in the alkaline water splitting applications, including excellent performance and durability as compared with above-discussed electrocatalysts. Because, all of them have the optimal and favorable bandgap energy, which can try to support both the OER and HER activities. Also, comparatively these catalysts exhibit the high anti-corrosion nature in the harsh electrolyte environments, which can improve their oxidative/dissolution (in electrolyte) ability. Accordingly, tremendous progress has been made to develop the most promising bifunctional catalyst, which can efficiently replace the precious metal-based

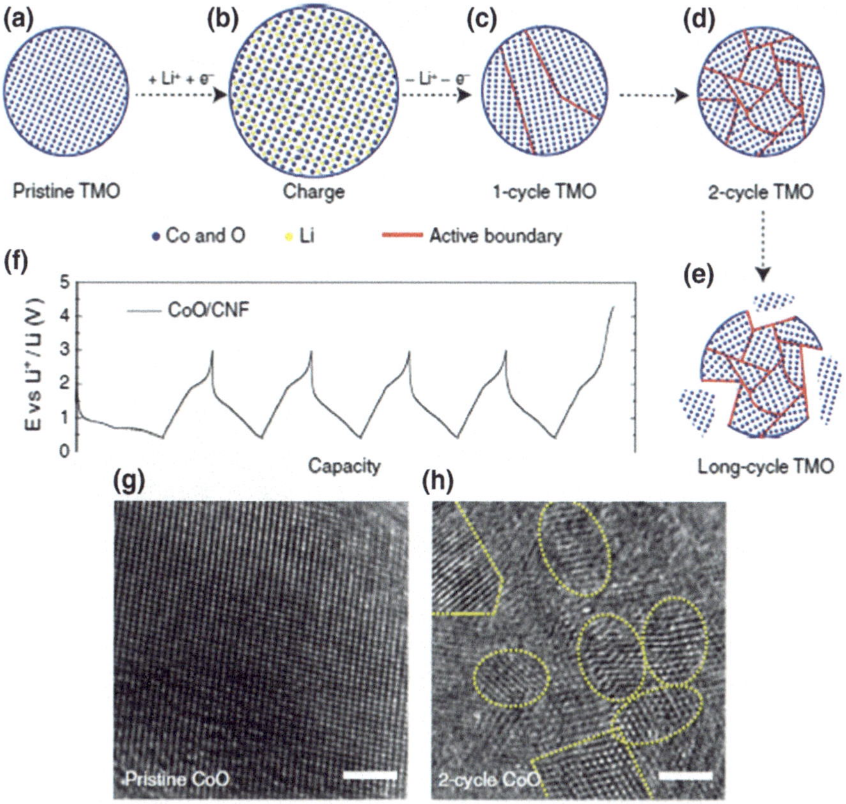

Fig. 19.13 Schematic illustration of evolution of TMO morphology under galvanostatic cycles. **a–e** Gradual change in the TMO particles from single crystalline to ultra-small interconnected crystalline NPs. **f** The galvanostatic cycling profile of CoO/CNF. **g** High-resolution TEM image of pristine CoO. **h** High-resolution TEM image of CoO after second cycle. Adapted from Wang et al. with permission from Nature publishing group [Nature Communication], copyright (2015) [133]

electrodes in the real scale WE. Few examples are going to discuss in the following section.

An integrated three-dimensional (3D) array electrocatalyst (carbon paper/carbon tubes/cobalt-sulfide) has been introduced as a cost-effective electrode for both OER and HER in alkaline solution by Wang et al. This electrode has been prepared by using multiple synthesis steps including ZnO growth and etching on the CP substrate. As-developed electrodes show the lower OER potential of 1.536 V (Fig. 19.14a) for solar fuel synthesis (@ 10 mA cm^{-2}) with more favorable kinetics (72 mV dec^{-1}) (Fig. 19.14b). Also, in the HER activity, it shows only −0.190 V at the current density of −10 mA cm^{-2} in 1 M KOH (Fig. 19.14a), including good stability (Fig. 19.14c). The two-electrode conventional WE with the same electrode catalyst displays the overall cell voltage of 1.743 V, which is 64 mV higher than the precious electrode-based WE. Also, this system shows more stability than that

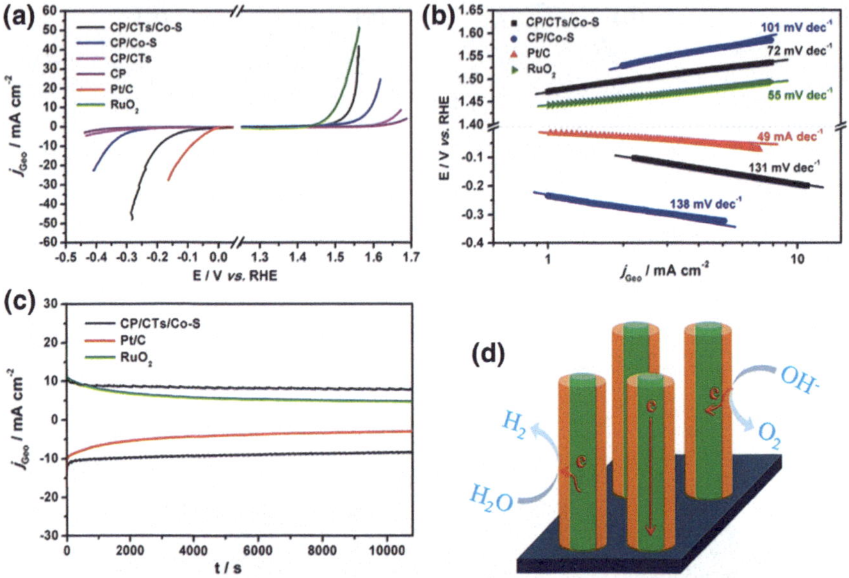

Fig. 19.14 **a** Polarization curves for HER and OER in 1 M KOH (scan rate of 2 mV s^{-1}). **b** Corresponding Tafel plots. **c** Chronoamperometric durability tests of CP/CTs/Co-S, Pt/C, and RuO_2 electrodes at a fixed HER and/or OER potentials. **d** Schematic representation of the operating principle of CP/CTs/Co-S. Reproduced with permission from [134]. Copyright (2016) American Chemical Society

of precious electrode-based system. Obtained stable and outstanding bifunctionality originates from the unique 3D arrays architecture with more exposure and utilization of vast active sites, perfect electron transport in the vertical carbon nanotubes on the carbon paper and fast gas movement in the porous structure (Fig. 19.14d) [134]. Similarly, Sun et al. studied the overall water splitting activity for the MOFs derived-$CoSe_2$ nanoparticles anchored on carbon fibers ($CoSe_2$/CF) in the same electrolyte environment. Two-step synthesized electrode has the catalyst loading of 2.9 mg cm^{-2}. Interestingly, the electrode exhibits the OER current density of 10 mA cm^{-2} at overpotential of 297 mV. Also, the smaller Tafel slope (41 mV dec^{-1}) than the RuO_2/CF catalyst, it explains that the non-precious electrode has more favorable OER kinetics/rate-determining step than the precious electrode. In addition, the same electrode with the same loading delivers the HER current density of −10 mA cm^{-2} with an overpotential of 95 mV, which is very less as compared with many other reported catalysts. Finally, the self-made two-electrode WE system with this $CoSe_2$/CF electrode gives the overall electrolysis current density of 10 mA cm^{-2} at the cell voltage of 1.63 V, including excellent durability in alkaline electrolyte conditions. Here, the formed catalyst has more rough surface nature, which can utilize more electrolytes and increase the active sites exposer to adsorb the reaction intermediates from the electrolyte [135].

Furthermore, to improve the intrinsic activity and anti-oxidation of mono-metal sulfide, the Fe has doped in NiS_2 and developed on the carbon fiber (Fe-NiS_2/CF) by using sulfurization method. Because the active oxygen species formation during the in situ electrochemical tuning at the OER potential may inhabit the decoupling of their intrinsic activity. So, Yu et al. mainly focus the OER activity and stability on this catalyst. The developed Fe-NiS_2/CF catalyst shows the improved OER activity with a lower overpotential of 243 mV at 10 mA cm^{-2}, and it is quite lower than the mono-metal sulfide and selenide catalysts as discussed above. Also, it shows the good OER stability even at 100 mA cm^{-2} current density for over 30 h with only 24 mV increment. For the overall, WE system was constructed with the same catalyst and tested their overall water splitting cell potential, which delivers the 10 mA cm^{-2} current density with reasonable cell voltage with good stability [136]. Comparatively, the bi-metal-based chalcogenides/pnictides are improving the electrocatalytic activities than that of single metal-based catalysts. Accordingly, Liu et al. explained the bifunctional activity of $NiCo_2S_4$ nanowire arrays on the carbon cloth substrate ($NiCo_2S_4$/CC) and compared with $NiCo_2O_4$/CC catalyst. In 1 M KOH, this obtained catalyst produces a promising OER and HER current density of 100 mA cm^{-2} at an overpotential of 340 and 305 mV, respectively, which is quite less than $NiCo_2O_4$/CC. Importantly, in the overall water splitting the $NiCo_2O_4$/CC catalyst need 300 mV (1.71 V) higher overpotential than $NiCo_2S_4$/CC catalyst (1.68 V). Because, the in situ grown nanowire arrays have more electrochemical surface with bimetal active sites and direct electron pathways for the fast electron transfer within the catalyst [137].

Also, some researchers introduced the cobalt-molybdenum sulfides/phosphides as a remarkable catalyst with enhanced OER and HER activities to the overall water splitting. For example, an amorphous phosphorus incorporated cobalt-molybdenum sulfide with the support of carbon cloth (P-CoMoS/CC) has reported by Ray et al. In the bifunctional activity test, the catalyst exhibits the lower overpotential with higher mass activity in 1 M KOH. Such electrode-based overall water splitting system split the water and produce the current density of 10 mA cm^{-2} at a cell voltage of 1.54 V (Fig. 19.15a, b). As-prepared catalyst shows the very low OER and HER charge-transfer resistances of 8.41 and 6.99 Ω, respectively (Fig. 19.15c, d), and very high electrochemical surface area (Fig. 19.15e). Also, the significant activity and very good electron conductivity are due to the more unsaturated active sites presents in the P-CoMoS as compared with P-CoS catalyst [138].

Recently, Wang et al. fabricated the self-supported $NiMoP_2$ nanowires on carbon cloth ($NiMoP_2$/CC) by using in situ P/O exchange process (Fig. 19.16) and explored as an efficient and durable catalyst for full water splitting. Here, the $NiMoP_2$ nanowire consists of small nanocrystallites, which provides the large amounts of active sites for the enhanced catalytic activities, including more defects/dislocations. Also, each nanocrystallite is strongly interconnected and promote the electron movements during the water-splitting reactions. The $NiMoP_2$ nanowire catalyst delivers the HER and OER current density of 100 mA cm^{-2} at 199 and 330 mV overpotential. In two-electrode system, it shows the cell voltage of 1.67 V at 10 mA cm^{-2}. The reason behind this excellent activity and stability are: (a) in

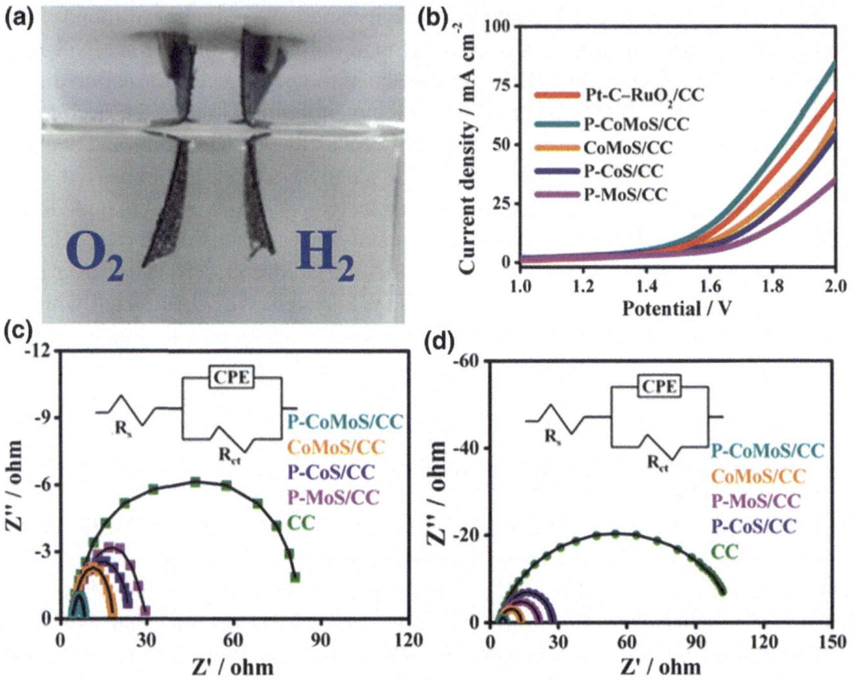

Fig. 19.15 **a** Overall water splitting (Digital image). **b** Overall water splitting polarization curves of different two electrode systems in 1 M KOH. Electrochemical impedance spectroscopy (Nyquist plots) of the various catalysts at a potential of **c** 1.50 (OER) and **d** −0.2 V (HER) versus RHE. Reproduced with permission from [138]. Copyright (2017) American Chemical Society

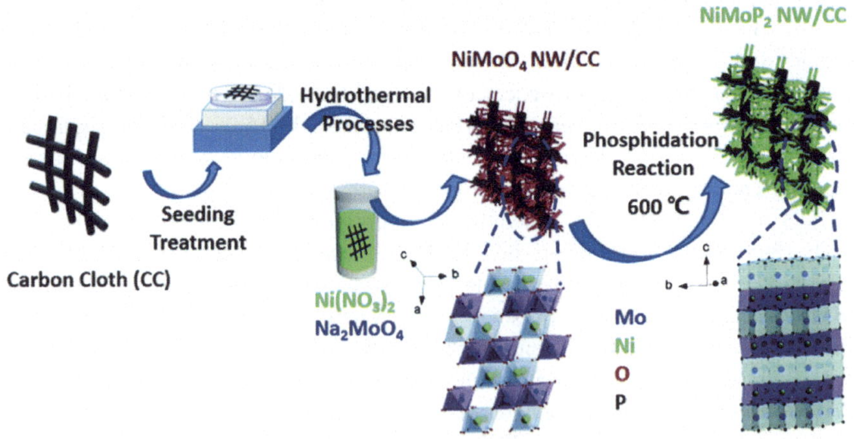

Fig. 19.16 Schematic illustration of the synthesis of $NiMoP_2$ nanowires on carbon cloth ($NiMoP_2$ NW/CC). Reproduced from Ref. [139] with permission of The Royal Society of Chemistry

ternary metal phosphide, the Ni and Mo elements merging together has integrated the oxygen-containing substance; (b) the direct growth on 3D carbon cloth using in situ method increases the active surface and it is binder-free hierarchical architecture. So, the fast electron and mass transport with less disintegration tendency during the reactions; (c) more electroactive sites availability in grain boundary nanowire structure ($NiMoP_2$). Among all the above discussed catalysts, it shows the remarkable OER and HER overpotential in alkaline electrolyte [139].

Apart from this, carbon-coated non-precious metals have been reported as electrocatalyst in acid conditions. For example, carbon-shell-coated FeP nanoparticles has reported as the robust HER electrocatalyst in acid electrolyte environment. The transition metal-based electrocatalysts are not stable in the acid conditions. But, the carbon-shell protects the core active sites of FeP from the oxidation and hold the stable catalytic activity for a long-time [140].

19.6 Summary

Hence, the nanocarbon composites/hybrids are considered as the key non-precious electrode catalysts in metal–air batteries and water splitting for alternative energy conversion and production due to their bifunctional electrochemical activity and excellent stability. So, the widely available electrode materials are helping to attain the cost-effective energy conversion and production.

References

1. Zhu H, Wei J, Wang K, Wu D (2009) Applications of carbon materials in photovoltaic solar cells. Sol Energy Mater Sol Cells 93:1461–1470
2. Bernardi M, Lohrman J, Kumar PV, Kirkeminde A, Ferralis N, Grossman JC, Ren S (2012) Nanocarbon-based photovoltaics. ACS Nano 6:8896–8903
3. Nonoguchi Y, Nakano M, Murayama T, Hagino H, Hama S, Miyazaki K, Matsubara R, Nakamura M, Kawai T (2016) Simple salt-coordinated n-type nanocarbon materials stable in air. Adv Funct Mater 26:3021–3028
4. Kroto HW, Heath JR, O'Brien SC, Curl RF, Smalley RE (1985) C60: Buckminsterfullerene. Nature 318:162–163
5. Iijima S (1991) Helical microtubules of graphitic carbon. Nature 354:56–58
6. Kim J, Page AJ, Irle S, Morokuma K (2012) Dynamics of local chirality during SWCNT growth: armchair versus zigzag nanotubes. J Am Chem Soc 134:9311–9319
7. Novoselov KS, Geim AK, Morozov SV, Jiang D, Zhang Y, Dubonos SV, Grigorieva IV, Firsov AA (2004) Electric field effect in atomically thin carbon films. Science 306:666–669
8. Bolotin KI, Sikes KJ, Jiang Z, Klima M, Fudenberg G, Hone J, Kim P, Stormer HL (2008) Ultrahigh electron mobility in suspended graphene. Solid State Commun 146:351–355
9. Branson BT, Beauchamp PS, Beam JC, Lukehart CM, Davidson JL (2013) Nanodiamond nanofluids for enhanced thermal conductivity. ACS Nano 7:3183–3189
10. Kishi N, Kuwahara S, Kobayashi K, Ramesh P (2014) Recent advances in nanocarbon materials. J Nanomater 2014:1–2

11. Sekiguchi A, Tanaka F, Saito T, Kuwahara Y, Sakurai S, Futaba DN, Yamada T, Hata K (2015) Robust and soft elastomeric electronics tolerant to our daily lives. Nano Lett 15:5716–5723
12. Chen W, Duan L, Zhu D (2007) Adsorption of polar and nonpolar organic chemicals to carbon nanotubes. Environ Sci Technol 41:98295
13. Avouris P (2002) Molecular electronics with carbon nanotubes. Acc Chem Res 35:1026–1034
14. Takenobu T, Takano T, Shiraishi M, Murakami Y, Ata M, Kataura H, Achiba Y, Iwasa Y (2003) Stable and controlled amphoteric doping by encapsulation of organic molecules inside carbon nanotubes. Nat Mater 2:683–688
15. Yanagi K, Kanda S, Oshima Y, Kitamura Y, Kawai H, Yamamoto T, Takenobu T, Nakai Y, Maniwa Y (2014) Tuning of the thermoelectric properties of one-dimensional material networks by electric double layer techniques using ionic liquids. Nano Lett 14:6437–6442
16. Chen Y, Chen J, Hu H, Hamon MA, Itkis ME, Haddon RC (1999) Solution-phase EPR studies of single-walled carbon nanotubes. J Chem Phys Lett 299:532–535
17. Saito R, Hofmann M, Dresselhaus G, Jorio A, Dresselhaus MS (2011) Raman spectroscopy of graphene and carbon nanotubes. Adv Phys 60:413–550
18. Lee RS, Kim HJ, Fischer JE, Thess A, Smalley RE (1997) Conductivity enhancement in single-walled carbon nanotube bundles doped with K and Br. Nature 388:255–257
19. Li X, Guard LM, Jiang J, Sakimoto K, Huang JS, Wu J, Li J, Yu L, Pokhrel R, Brudvig GW, Beigi SI, Hazari N, Taylor AD (2014) Controlled doping of carbon nanotubes with metallocenes for application in hybrid carbon nanotube/Si solar cells. Nano Lett 14:3388–3394
20. Javey A, Guo J, Wang Q, Lundstrom M, Dai H (2003) Ballistic carbon nanotube field-effect transistors. Nature 424:654–657
21. Albrecht R (1985) The role of hydrothermal synthesis in preparative chemistry. Angew Chem Int Ed Engl 24:1026–1040
22. Feng W, Xiaogang L (2009) Recent advances in the chemistry of lanthanide-doped upconversion nanocrystals. Chem Soc Rev 38:976–989
23. Baccile N, Antonietti M, Titirici MM (2010) One-step hydrothermal synthesis of nitrogen-doped nanocarbons: albumine directing the carbonization of glucose. Chemsuschem 3:246–253
24. Sun Q, Zhang XQ, Han F, Li WC, Lu AH (2012) Controlled hydrothermal synthesis of 1D nanocarbons by surfactant-templated assembly for use as anodes for rechargeable lithium-ion batteries. J Mater Chem 22:17049–17054
25. Moni P, Hyun S, Vignesh A, Shanmugam S (2017) Chrysanthemum flower-like $NiCo_2O_4$–nitrogen doped graphene oxide composite: an efficient electrocatalyst for lithium–oxygen and zinc–air batteries. Chem Commun 53:7836–7839
26. Ramakrishnan P, Park SG, Shanmugam S (2015) Three-dimensional hierarchical nitrogen-doped arch and hollow nanocarbons: morphological influences on supercapacitor applications. J Mater Chem A 3:16242–16250
27. Pol VG, Pol SV, Gedanken A (2011) Dry autoclaving for the nanofabrication of sulfides, selenides, borides, phosphides, nitrides, carbides, and oxides. Adv Mater 23:1179–1190
28. Zhao Z, Li M, Zhang L, Dai L, Xia Z (2015) Design principles for heteroatom-doped carbon nanomaterials as highly efficient catalysts for fuel cells and metal–air batteries. Adv Mater 27:6834–6840
29. Sanetuntikul J, Chuaicham C, Choi YW, Shanmugam S (2015) Investigation of hollow nitrogen-doped carbon spheres as non-precious $Fe–N_4$ based oxygen reduction catalysts. J Mater Chem A 3:15473–15481
30. Ganesan P, Prabu M, Sanetuntikul J, Shanmugam S (2015) Cobalt sulfide nanoparticles grown on nitrogen and sulfur codoped graphene oxide: an efficient electrocatalyst for oxygen reduction and evolution reactions. ACS Catal 5:3625–3637

31. Huang ZM, Zhang YZ, Kotaki M, Ramakrishna S (2003) A review on polymer nanofibers by electrospinning and their applications in nanocomposites. Compos Sci Technol 63:2223–2253
32. Ramakrishnan P, Shanmugam S (2014) Electrochemical performance of carbon nanorods with embedded cobalt metal nanoparticles as an electrode material for electrochemical capacitors. Electrochim Acta 125:232–240
33. Wang J, Kaskel S (2012) KOH activation of carbon-based materials for energy storage. J Mater Chem 22:23710–23725
34. Sivanantham A, Ganesan P, Shanmugam S (2017) A synergistic effect of Co and CeO_2 in nitrogen-doped carbon nanostructure for the enhanced oxygen electrode activity and stability. Appl Catal B. https://doi.org/10.1016/j.apcatb.2017.08.063
35. Wang ZL, Xu D, Xu JJ, Zhang XB (2014) Oxygen electrocatalysts in metal–air batteries: from aqueous to nonaqueous electrolytes. Chem Soc Rev 43:7746
36. Lee JS, Kim ST, Cao R, Choi NS, Liu M, Lee KT, Cho J (2011) Metal–air batteries with high energy density: Li–air versus Zn–air. Adv Energy Mater 1:34–50
37. Rahman MA, Wang X, Wen C (2013) High energy density metal-air batteries: a review. J Electrochem Soc 160:1759–1771
38. Abraham KM, Jiang Z (1996) A polymer electrolyte-based rechargeable lithium/oxygen battery. J Electrochem Soc 143:1–5
39. Li Y, Gong M, Liang Y, Feng J, Kim JE, Wang H, Hong G, Zhang B, Dai H (2013) Advanced zinc–air batteries based on high-performance hybrid electrocatalysts. Nat Commun 4:1805
40. McLarnon FR, Cairns EJ (1991) The secondary alkaline zinc electrode. J Electrochem Soc 138:645–656
41. Shao Y, Park S, Xiao J, Zhang JG, Wang Y, Liu J (2012) Electrocatalysts for nonaqueous lithium–air batteries: status, challenges, and perspective. ACS Catal 2:844–857
42. Sapkota P, Kim H (2009) Zinc–air fuel cell, a potential candidate for alternative energy. J Ind Eng Chem 15:445–450
43. Toussaint G, Stevens P, Akrour L, Rouget R, Fourgeot F (2010) Development of a rechargeable zinc–air battery. ECS Trans 28:25–34
44. Cheng F, Chen J (2012) Metal-air batteries: from oxygen reduction electrochemistry to cathode catalysts. Chem Soc Rev 41:2172–2192
45. Wang X, Sebastian PJ, Smit MA, Yang H, Gamboa SA (2003) Studies on the oxygen reduction catalyst for zinc–air battery electrode. J Power Sources 124:278
46. Shanmugam S, Osaka T (2011) Efficient electrocatalytic oxygen reduction over metal-free nitrogen doped carbon nanocapsules. Chem Commun 47:4463–4465
47. Jo G, Shanmugam S (2012) Single step synthetic approach for boron-doped carbons as non-precious catalyst for oxygen reduction in alkaline medium. Electrochem Commun 25:101–104
48. Shanmugam S, Momma T, Osaka T (2014) Enhanced oxygen reduction activities of Pt supported on nitrogen-doped carbon nanocapsules. Electrochim Acta 137:41–48
49. Sanetuntikul J, Hang T, Shanmugam S (2014) Hollow nitrogen-doped carbon spheres as efficient and durable electrocatalysts for oxygen reduction. Chem Commun 50:9473–9476
50. Wei Q, Fu Y, Zhang G, Sun S (2017) Rational design of carbon-based oxygen electrocatalysts for zinc–air batteries. Curr Opin Electrochem 4:45–59
51. Mitchell R, Gallant B, Thompson C, Horn Y (2011) All-carbon-nanofiber electrodes for high-energy rechargeable Li–O_2 batteries. Energy Environ Sci 4:2952–2958
52. Xiao J, Mei D, Li X, Xu W, Wang D, Graff GL, Bennett WD, Nie Z, Saraf LV, Aksay IA, Liu J, Zhang JG (2011) Hierarchically porous graphene as a lithium–air battery electrode. Nano Lett 11:5071–5078
53. Xu Y, Zhang Y, Guo Z, Ren J, Wang Y, Peng H (2015) Flexible, stretchable, and rechargeable fiber-shaped zinc–air battery based on cross-stacked carbon nanotube sheets. Angew Chem 54:15390–15394

54. Zhang J, Zhao Z, Xia Z, Dai L (2015) A metal-free bifunctional electrocatalyst for oxygen reduction and oxygen evolution reactions. Nat Nanotechnol 10:444–452
55. Tang H, Zeng Y, Liu D, Qu D, Luo J, Binnemans K, Vos DD, Fransaer J, Qu D, Sun SG (2016) Dual-doped mesoporous carbon synthesized by a novel nanocasting method with superior catalytic activity for oxygen reduction. Nano Energy 26:131–138
56. Tang C, Wang HF, Chen X, Li BQ, Hou TZ, Zhang B, Zhang Q, Titirici MM, Wei F (2016) Topological defects in metal-free nanocarbon for oxygen electrocatalysis. Adv Mater 28:6845–6851
57. Chen P, Zhou T, Xing L, Xu K, Tong Y, Xie H, Zhang L, Yan W, Chu W, Wu C, Xie Y (2016) Atomically dispersed iron–nitrogen species as electrocatalysts for bifunctional oxygen evolution and reduction reactions. Angew Chem 56:610–614
58. Zhu AL, Duch D, Roberts GA, Li SX, Wang H, Duch K, Bae E, Jung KS, Wilkinson D, Kulinich SA (2015) Increasing the electrolyte capacity of alkaline Zn–air fuel cells by scavenging zincate with Ca(OH)2. ChemElectroChem 2:134–142
59. Li JC, Hou PX, Zhao SY, Liu C, Tang DM, Cheng M, Zhang F, Cheng HM (2016) A 3D bi-functional porous N-doped carbon microtube sponge electrocatalyst for oxygen reduction and oxygen evolution reactions. Energy Environ Sci 9:3079–3084
60. Fan Y, Ida S, Staykov A, Akbay T, Hagiwara H, Matsuda J, Kaneko K, Ishihara T (2017) Ni-Fe nitride nanoplates on nitrogen-doped graphene as a synergistic catalyst for reversible oxygen evolution reaction and rechargeable Zn–Air battery. Small 13:1700099
61. Pan J, Xu YY, Yang H, Dong Z, Liu H, Xia BY (2018) Advanced architectures and relatives of air electrodes in Zn–Air batteries. Adv Sci. https://doi.org/10.1002/advs.201700691
62. Liu Q, Wang Y, Dai L, Yao J (2016) Scalable fabrication of nanoporous carbon fiber films as bifunctional catalytic electrodes for flexible Zn–Air batteries. Adv Mater 28:3000–3006
63. Guo Z, Zhou D, Dong X, Qiu Z, Wang Y, Xia Y (2013) Ordered hierarchical mesoporous/macroporous carbon: a high-performance catalyst for rechargeable Li–O_2 batteries. Adv Mater 25:5668–5672
64. Prabu M, Kriangsak K, Shanmugam S (2014) Hierarchical nanostructured $NiCo_2O_4$ as an efficient bifunctional non-precious catalyst for rechargeable zinc–air battery. Nanoscale 6:3173–3181
65. Liu G, Zhang L, Wang S, Ding LX, Wang H (2017) Hierarchical $NiCo_2O_4$ nanosheets on carbon nanofiber films for high energy density and long-life Li–O_2 batteries. J Mater Chem A 5:14530–14536
66. Prabu M, Ramakrishnan P, Nara H, Momma T, Osaka T, Shanmugam S (2014) zinc–air battery: understanding the structure and morphology changes of graphene supported $CoMn_2O_4$ bifunctional catalysts under practical rechargeable conditions. ACS Appl Mater Interfaces 6:16545–16555
67. Prabu M, Ramakrishnan P, Ganesan P, Manthiram A, Shanmugam S (2015) LaTi0.65Fe0.35O3-δ nanoparticle-decorated nitrogen-doped carbon nanorods as an advanced hierarchical air electrode for rechargeable metal-air batteries. Nano Energy 15:92–103
68. Hyun S, Shanmugam S (2017) Mesoporous Co-CoO/N-CNR nanostructure as high-performance air cathode for lithium-oxygen batteries. J Power Sources 354:48–56
69. Prabu M, Ramakrishnan P, Shanmugam S (2014) $CoMn_2O_4$ nanoparticles anchored on nitrogen-doped graphene nanosheets as bifunctional electrocatalyst for rechargeable zinc–air battery. Electrochem Commun 41:59–63
70. Ganesan P, Prabu M, Shanmugam S (2015) Nitrogen and sulfur co-doped graphene supported cobalt sulfide nanoparticles as an efficient air cathode for zinc–air battery. Electrochim Acta 183:63–69
71. Meng T, Qin J, Wang S, Zhao D, Mao B, Cao M (2017) *In situ* coupling of $Co_{0.85}Se$ and N-doped carbon via one-step selenization of metal–organic frameworks as a trifunctional catalyst for overall water splitting and Zn–air batteries. J Mater Chem A 5:7001–7014

72. Aijaz A, Fujiwara N, Xu Q (2014) From Metal–organic framework to nitrogen-decorated nanoporous carbons: high CO_2 uptake and efficient catalytic oxygen reduction. J Am Chem Soc 136:6790–6793
73. Zeng K, Zhang D (2010) Recent progress in alkaline water electrolysis for hydrogen production and applications. Prog Energy Combust Sci 36:307–326
74. Carmo M, Fritz DL, Mergel J, Stolten D (2013) A comprehensive review on PEM water electrolysis. Int J Hydrogen Energy 38:4901–4934
75. Hauch A, Ebbesen SD, Jensen SH, Mogensen M (2008) Highly efficient high temperature electrolysis. J Mater Chem 18:2331–2340
76. Sapountzi FM, Gracia JM, Weststrate CJ (KJ), Fredriksson HOA, Niemantsverdriet JW (Hans) (2017) Electrocatalysts for the generation of hydrogen, oxygen and synthesis gas. Prog Energy Combust Sci 58:1–35
77. Zoulias E, Varkaraki E, Lymberopoulos N, Christodoulou CN, Karagiorgis GN (2004) A review on water electrolysis. TCJST 4:41–71
78. Leroy RL (1983) Industrial water electrolysis: present and future. Int J Hydrogen Energy 8:401–417
79. Sasaki K, Li HW, Hayashi A, Yamabe J, Ogura T, Lyth SM (2016) Hydrogen energy engineering: a Japanese perspective. Springer, Japan
80. Mauera AE, Kirk DW, Thorpe SJ (2007) The role of iron in the prevention of nickel electrode deactivation in alkaline electrolysis. Electrochim Acta 52:3505–3509
81. Abouatallah RM, Kirk DW, Thorpe SJ, Graydon JW (2001) Reactivation of nickel cathodes by dissolved vanadium species during hydrogen evolution in alkaline media. Electrochim Acta 47:613–621
82. Nikolic VM, Tasic GS, Maksic AD, Saponjic DP, Miulovic SM, Kaninski MPM (2010) Raising efficiency of hydrogen generation from alkaline water electrolysis energy saving. Int J Hydrogen Energy 35:12369–12373
83. Ghany NAA, Kumagai N, Meguro S, Asami K, Hashimoto K (2002) Oxygen evolution anodes composed of anodically deposited Mn/Mo/Fe oxides for seawater electrolysis. Electrochim Acta 48:21–28
84. Huot JY, Trudeau ML, Schulz R (1991) Low hydrogen overpotential nanocrystalline Ni-Mo cathode for alkaline water electrolysis. J Electrochem Soc 138:1316–1321
85. Conway BE, Bai L (1986) H_2 evolution kinetics at high activity Ni-Mo-Cd electrocoated cathodes and its relation to potential dependence of sorption of H*. Int J Hydrogen Energy 11:533–540
86. Stojic DL, Grozdic TD, Kaninski MPM, Maksic AD, Simic ND (2006) Intermetallics as advanced cathode materials in hydrogen production via electrolysis. Int J Hydrogen Energy 31:841–846
87. Hamdani M, Pereira MIS, Douch J, Addi AA, Berghoute Y, Mendonca MH (2004) Physicochemical and electrocatalytic properties of Li-Co_3O_4 anodes prepared by chemical spray pyrolysis for application in alkaline water electrolysis. Electrochim Acta 49:1555–1563
88. Singh RN, Mishra D, Anindita Sinha ASK, Singh A (2007) Novel electrocatalysts for generating oxygen from alkaline water electrolysis. Electrochem Commun 9:1369–1373
89. Kim S, Koratkar N (2006) Water electrolysis activated by Ru nanorod array electrodes. Appl Phys Lett 88:263106–263108
90. Kamat PV (2007) Meeting the clean energy demand: Nanostructure architectures for solar energy conversion. J Phys Chem C 111:834–2860
91. Zhang J, Zhao Z, Xia Z, Dai L (2015) A metal-free bifunctional electrocatalyst for oxygen reduction and oxygen evolution reactions. Nat Nanotechnol 10:444–452
92. Qu K, Zheng Y, Zhang X, Davey K, Dai S, Qiao SZ (2017) Promotion of electrocatalytic hydrogen evolution reaction on nitrogen-doped carbon nanosheets with secondary heteroatoms. ACS Nano 11:7293–7300
93. Wang DW, Su D (2014) Heterogeneous nanocarbon materials for oxygen reduction reaction. Energy Environ Sci 7:576–591

94. Zou X, Huang X, Goswami A, Silva R, Sathe BR, Mikmekov E, Asefa T (2014) Cobalt-embedded nitrogen-rich carbon nanotubes efficiently catalyze hydrogen evolution reaction at all pH values. Angew Chem Int Ed 53:4372–4376
95. Zeng M, Li Y (2015) Recent advances in heterogeneous electrocatalysts for the hydrogen evolution reaction. J Mater Chem A 3:14942–14962
96. Zhou W, Zhou Y, Yang L, Huang J, Ke Y, Zhou K, Li L, Chen S (2015) N-doped carbon-coated cobalt nanorod arrays supported on a titanium mesh as highly active electrocatalysts for the hydrogen evolution reaction. J Mater Chem A 3:1915–1919
97. Zhou W, Jia J, Lu J, Yang L, Hou D, Li G, Chen S (2016) Recent developments of carbon-based electrocatalysts for hydrogen evolution reaction. Nano Energy 28:29–43
98. Choi CH, Chung MW, Parka SH, Woo SI (2013) Additional doping of phosphorus and/or sulfur into nitrogen-doped carbon for efficient oxygen reduction reaction in acidic media. Phys Chem Chem Phys 15:1802–1805
99. Zhao Y, Nakamura R, Kamiya K, Nakanishi S, Hashimoto K (2013) Nitrogen-doped carbon nanomaterials as non-metal electrocatalysts for water oxidation. Nat Commun 4:2390
100. Kim H, Lee K, Woo SI, Jung Y (2011) On the mechanism of enhanced oxygen reduction reaction in nitrogen-doped graphene nanoribbons. Phys Chem Chem Phys 13:17505–17510
101. Yeo BS, Bell AT (2011) Enhanced activity of gold-supported cobalt oxide for the electrochemical evolution of oxygen. J Am Chem Soc 133:5587–5593
102. Zhao Y, Watanabe K, Hashimoto K (2012) Poly(bis-2,6-diaminopyridinesulfoxide) as an active and stable electrocatalyst for oxygen reduction reaction. J Mater Chem 22:12263–12267
103. Rao CV, Cabrera CR, Ishikawa Y (2010) In Search of the active site in nitrogen-doped carbon nanotube electrodes for the oxygen reduction reaction. J Phys Chem Lett 1:2622–2627
104. Czerw R, Terrones M, Charlier JC, Blase X, Foley B, Kamalakaran R, Grobert N, Terrones H, Tekleab D, Ajayan PM, Blau W, Ruhle M, Carroll DL (2001) Identification of electron donor states in N-doped carbon nanotubes. Nano Lett 1:457–460
105. Hou Z, Wang X, Ikeda T, Terakura K, Oshima M, Kakimoto M (2013) Electronic structure of N-doped graphene with native point defects. Phys Rev B 87:165401
106. Lin Z, Waller GH, Liu Y, Liu M, Wong CP (2013) Simple preparation of nanoporous few-layer nitrogen-doped graphene for use as an efficient electrocatalyst for oxygen reduction and oxygen evolution reactions. Carbon 53:130–136
107. Tian GL, Zhao MQ, Yu D, Kong XY, Huang JQ, Zhang Q, Wei F (2014) Nitrogen-doped graphene/carbon nanotube hybrids: in situ formation on bifunctional catalysts and their superior electrocatalytic activity for oxygen evolution/reduction reaction. Small 10:251–2259
108. Tian GL, Zhang Q, Zhang B, Jin YG, Huang JQ, Su DS, Wei F (2014) Toward full exposure of "active sites": nanocarbon electrocatalyst with surface enriched nitrogen for superior oxygen reduction and evolution reactivity. Adv Funct Mater 24:5956–5961
109. Li J, Yin S, Dong F, Cen W, Chu Y (2017) Tailoring active sites via synergy between graphitic and pyridinic N for enhanced catalytic efficiency of a carbocatalyst. ACS Appl Mater Interfaces 9:19861–19869
110. Davodi F, Tavakkoli M, Lahtinen J, Kallio T (2017) Straightforward synthesis of nitrogen-doped carbon nanotubes as highly active bifunctional electrocatalysts for full water splitting. J Catal 353:19–27
111. Balogun MS, Qiu W, Yang H, Fan W, Huang Y, Fang P, Li G, Ji H, Tong Y (2016) A monolithic metal-free electrocatalyst for oxygen evolution reaction and overall water splitting. Energy Environ Sci 9:411–3416
112. Zhang Z, Yi Z, Wang J, Tian X, Xu P, Shib G, Wang S (2017) Nitrogen-enriched polydopamine analoguederived defect-rich porous carbon as a bifunctional metal-free electrocatalyst for highly efficient overall water splitting. J Mater Chem A 5:17064–17072

113. Li R, Wei Z, Gou X (2015) Nitrogen and phosphorus dual-doped graphene/carbon nanosheets as bifunctional electrocatalysts for oxygen reduction and evolution. ACS Catal 5:4133–4142
114. Choi CH, Park SH, Woo SI (2012) Binary and ternary doping of nitrogen, boron, and phosphorus into carbon for enhancing electrochemical oxygen reduction activity. ACS Nano 6:7084–7091
115. Wang X, Li X, Zhang L, Yoon Y, Weber PK, Wang H, Guo J, Dai H (2009) N-doping of graphene through electrothermal reactions with ammonia. Science 324:768–771
116. Ma TY, Ran J, Dai S, Jaroniec M, Qiao SZ (2015) Phosphorus-doped graphitic carbon nitrides grown in situ on carbon-fiber paper: flexible and reversible oxygen electrodes. Angew Chem Int Ed 54:4646–4650
117. Qu K, Zheng Y, Jiao Y, Zhang X, Dai S, Qiao SZ (2017) Polydopamine-inspired, dual heteroatom-doped carbon nanotubes for highly efficient overall water splitting. Adv Energy Mater 7:1602068
118. Hu C, Dai L (2017) Multifunctional carbon-based metal-free electrocatalysts for simultaneous oxygen reduction, oxygen evolution, and hydrogen evolution. Adv Mater 29:1604942
119. Lai J, Li S, Wu F, Saqib Luque MR, Xu G (2016) Unprecedented metal-free 3D porous carbonaceous electrodes for full water splitting. Energy Environ Sci 9:1210–1214
120. Ma TY, Dai S, Jaroniec M, Qiao SZ (2014) Graphitic carbon nitride nanosheet–carbon nanotube three-dimensional porous composites as high-performance oxygen evolution electrocatalysts. Angew Chem Int Ed 53:7281–7285
121. Jia N, Weng Q, Shi Y, Shi X, Chen X, Chen P, An Z, Chen Y (2017) N-doped carbon nanocages: Bifunctional electrocatalysts for the oxygen reduction and evolution reactions. Nano Res. https://doi.org/10.1007/s12274-017-1808-8
122. Zheng Y, Jiao Y, Zhu Y, Li LH, Han Y, Chen Y, Du A, Jaroniec M, Qiao SZ (2014) Hydrogen evolution by a metal-free electrocatalyst. Nat Commun 5:3783
123. Ito Y, Cong W, Fujita T, Tang Z, Chen M (2015) High catalytic activity of nitrogen and sulfur co-doped nanoporous graphene in the hydrogen evolution reaction. Angew Chem Int Ed 54:2131–2136
124. Zheng Y, Jiao Y, Li LH, Xing T, Chen Y, Jaroniec M, Qiao SZ (2014) Toward design of synergistically active carbon-based catalysts for electrocatalytic hydrogen evolution. ACS Nano 8:5290–5296
125. Li Q, Cao R, Cho J, Wu G (2014) Nanocarbon electrocatalysts for oxygen reduction in alkaline media for advanced energy conversion and storage. Adv Energy Mater 4:1301415
126. Zhao Y, Zhang J, Li K, Ao Z, Wang C, Liu H, Sun K, Wang G (2016) Electrospun cobalt embedded porous nitrogen doped carbon nanofibers as an efficient catalyst for water splitting. J Mater Chem A 4:12818–12824
127. Hou Y, Wen Z, Cui S, Ci S, Mao S, Chen J (2015) An advanced nitrogen-doped ggaphene/cobalt-embedded porous carbon polyhedron hybrid for efficient catalysis of oxygen reduction and water splitting. Adv Funct Mater 25:872–882
128. Bayatsarmadi B, Zheng Y, Russo V, Ge L, Casari CS, Qiao SZ (2016) Highly active nickel–cobalt/nanocarbon thin films as efficient water splitting electrodes. Nanoscale 8:18507–18515
129. Tao Z, Wang T, Wang X, Zheng J, Li X (2016) MOF-derived noble metal free catalysts for electrochemical water splitting. ACS Appl Mater Interfaces 8:35390–35397
130. Jin H, Wang J, Su D, Wei Z, Pang Z, Wang Y (2015) In situ cobalt–cobalt oxide/N-doped carbon hybrids as superior bifunctional electrocatalysts for hydrogen and oxygen evolution. J Am Chem Soc 137:2688–2694
131. Li J, Wang Y, Zhou T, Zhang H, Sun X, Tang J, Zhang L, Enizi AMA, Yang Z, Zheng G (2015) Nanoparticle superlattices as efficient bifunctional electrocatalysts for water splitting. J Am Chem Soc 137:14305–14312
132. Long X, Lin H, Zhou D, An Y, Yang S (2018) Enhancing full water-splitting performance of transition metal bifunctional electrocatalysts in alkaline solutions by tailoring CeO_2-transition metal oxides-Ni nanointerfaces. ACS Energy Lett 3:290–296

133. Wang H, Lee HW, Deng Y, Lu Z, Hsu PC, Liu Y, Lin D, Cui Y (2015) Bifunctional non-noble metal oxide nanoparticle electrocatalysts through lithium-induced conversion for overall water splitting. Nat Commun 6:7261
134. Wang J, Zhong H, Wang Z, Meng F, Zhang X (2016) Integrated three-dimensional carbon paper/carbon tubes/cobalt-sulfide sheets as an efficient electrode for overall water splitting. ACS Nano 10:2342–2348
135. Sun C, Dong Q, Yang J, Dai Z, Lin J, Chen P, Huang W, Dong X (2016) Metal–organic framework derived $CoSe_2$ nanoparticles anchored on carbon fibers as bifunctional electrocatalysts for efficient overall water splitting. Nano Res 9:2234–2243
136. Yu C, Huang H, Zhou S, Han X, Zhao C, Yang J, Li S, Guo W, An B, Zhao J, Qiu J (2018) An electrocatalyst with anti-oxidized capability for overall water splitting. Nano Res. https://doi.org/10.1007/s12274-017-1964-x
137. Liu D, Lu Q, Luo Y, Sun X, Asiri AM (2015) $NiCo_2S_4$ nanowires array as an efficient bifunctional electrocatalyst for full water splitting with superior activity. Nanoscale 7:15122–15126
138. Ray C, Lee SC, Sankar KV, Jin B, Lee J, Park JH, Jun SC (2017) Amorphous phosphorus-incorporated cobalt molybdenum sulfide on carbon cloth: an efficient and stable electrocatalyst for enhanced overall water splitting over entire pH values. ACS Appl Mater Interfaces 9:37739–37749
139. Wang X, Chen H, Xu Y, Liao J, Chen B, Rao H, Kuang D, Su C (2017) Self-supported $NiMoP_2$ nanowires on carbon cloth as an efficient and durable electrocatalyst for overall water splitting. J Mater Chem A 5:7191–7199
140. Chung DY, Jun SW, Yoon G, Kim H, Yoo JM, Lee KS, Kim T, Shin H, Sinha AK, Kwon SG, Kang K, Hyeon T, Sung YE (2017) Large-scale synthesis of carbon-shell-coated FeP nanoparticles for robust hydrogen evolution reaction electrocatalyst. J Am Chem Soc 139:6669–6674

Chapter 20
Applications of Carbon Nanotubes in Solar Cells

Feijiu Wang and Kazunari Matsuda

20.1 Introduction

Recently, carbon nanotubes (CNTs) have been used in many types of solar cells with high photovoltaic performance [1–7]. Initially, the CNTs were incorporated into solar cells as electron acceptors in organic photovoltaic (OPV) devices. Single-walled carbon nanotubes (SWCNTs) form a heterojunction with the employed polymer, poly(3-hexylthiophene) used for the effective dissociation of excitons (bound electron–hole pairs) and the creation of long-lived free charges for the carrier extraction. Suzuki et al. reported the integration of CNTs into dye-sensitized solar cells (DSSCs) as a counter electrode material, where the SWCNTs were deposited onto Teflon membrane filter cells, resulting in a device with a promising photovoltaic conversion efficiency (PCE) of 4.5% [8]. Moreover, CNTs have been incorporated into photoelectrochemical cells as additives in donor–acceptor pairs in conjunction with quantum dots, fullerenes, and many other photoactive polymers [9, 10].

Progressive studies of CNTs combined with Si, which is a well-understood semiconductor used in electronic devices, have recently been conducted to elucidate the mechanisms of photocarrier generation, exciton dissociation, and electron and hole transport. The typical architecture of a CNT and Si heterojunction (CNT/Si) device is similar to that of a single-junction crystalline Si solar device except that *p*-type or *n*-type Si is replaced with a thin CNT film to form *p*-type CNT/*n*-type Si (*p*-CNT/*n*-Si) or *n*-type CNT/*p*-type Si (*n*-CNT/*p*-Si) solar cells. The Si in the cells mainly absorbs photons and generates electron–hole pairs, which diffuse to the

F. Wang · K. Matsuda (✉)
Institute of Advanced Energy, Kyoto University, Uji, Kyoto 611-0011, Japan
e-mail: matsuda@scl.kyoto-u.ac.jp

F. Wang
JST, ERATO, Itami Molecular Nanocarbon Project, Nagoya University, Chikusa, Nagoya 464-8602, Japan

N. Nakashima (ed.), *Nanocarbons for Energy Conversion: Supramolecular Approaches*, Nanostructure Science and Technology,
https://doi.org/10.1007/978-3-319-92917-0_20

space-charge region. The electron–hole pairs are separated into free charge carriers (electrons and holes) under the driving force of the built-in voltage [11]. Through improvements in CNT films stemming from improvements in fabrication and chemical modification processes, the photovoltaic conversion efficiency (PCE) of CNT/Si solar cells has been improved from 1.3 to 17% in *p*-CNT/*n*-Si and from 0.01 to 4.62% in *n*-CNT/*p*-Si photovoltaic devices within less than a decade [12–15]. Notably, the *p*-CNT/*n*-Si cells show better performance than the *n*-CNT/*p*-Si devices because of their greater open circuit voltage (V_{oc}) and fill factor (FF), suggesting that the *p*-CNTs are more promising for use in solar cells.

Organic–inorganic hybrid perovskite solar cells (PSCs), which are a new class of photovoltaic (PV) device, have shown excellent photovoltaic performance (a certified efficiency of 22.1%) thus far [16]. However, one of the critical problems with this type of solar cell is poor stability under ambient and light illumination conditions. The devices typically include a hole transport layer (HTL) between a perovskite-structured light absorber and a hole-collecting electrode for efficient charge extraction. However, the performance of HTLs composed of either poly (triarylamine) (PTAA) or 2,2′,7,7′-tetrakis-(*N*,*N*-di-*p*-methoxyphenylamine)-9,9′-spirobifluorene (spiro-OMeTAD) commonly used in organic–inorganic hybrid PSCs decays within a very short time under ambient conditions. CNTs are potential alternative HTL materials for PSCs. Cells fabricated with SWCNTs wrapped with P3HT as a hole transporter exhibit a power conversion efficiency (PCE) as high as 15.3% with better stability than cells with PTAA or spiro-OMeTAD [17]. In previous work, we designed a hole transport/electron blocking layer based on an SWCNT network film combined with graphene oxide (GO) and poly(methyl methacrylate) (PMMA) to reduce shunt recombination and improve hole-selective transport. A PCE of 13.3% was attained in PSCs with a SWCNT/GO/PMMA layer, which is substantially greater than the PCE of the corresponding hole-transport-free device (3.3%) [18]. This SWCNT/GO/PMMA layer also improved the device stability by inhibiting the interaction between ambient moisture and the perovskite layer. In addition to their application as a hole transport material, CNTs have also been used as flexible transparent electrodes to replace the fluorine-doped tin oxide (FTO) substrates used for the film devices in PSCs [19].

CNTs have also been used as light absorber and photocarrier generation materials in solar cells. Yang et al. used semiconducting SWCNTs asymmetrically contacted with Pd and Sc (or Y) electrodes. Ohmic contact between the Pd electrode and the valence band (VB) of the SWCNTs as well as between Sc (or Y) and the conduction band (CB) contributed to barrier-free carrier extraction from the SWCNTs under illumination and resulted in a PCE of 0.11% in the fabricated solar cells [21]. The performance of polychiral semiconducting SWCNT/fullerene-based solar cells also suggests that the SWCNTs participate in both light absorption and carrier generation [23]. Moreover, Zhou et al. reported that a plasmonic nanostructure effectively promotes the photoelectric response in CNT-based devices [20]. A large number of advanced properties have stimulated interesting studies on solar cells with CNTs. We here provide an overview of the CNT-based solar cells, with particular emphasis on CNT/Si- and CNT-based PSCs.

In this review, we survey the roles of CNTs in solar cells and in CNT/Si- and CNT-based PSCs, in particular. In Sect. 20.2, we discuss the fundamentals of solar cells and the general evolution of CNT-based solar cells. In Sect. 20.3, we focus on CNT/Si solar cells, where the roles of CNTs are discussed in detail. In Sect. 20.4, PSCs in which CNTs are incorporated as electrodes, HTLs, or effective additives are demonstrated. In Sect. 20.5, we review the important research related to organic solar cells and sensitized solar cells with incorporated CNTs. Finally, we summarize the current state of CNT-based photovoltaic devices and future challenges facing the advancement of CNT-based optoelectronic devices.

20.2 CNTs in Solar Cells

20.2.1 Principle of Solar Cells

A photovoltaic device is designed to convert incident solar energy into electric power. The photons of the solar light with energy beyond the absorption edge are absorbed by the semiconducting materials. The electrons in the CB are excited to the VB, and holes are generated in the VB (Fig. 20.1). The electrons in the CB quickly decay to the lowest available energy state of the CB, whereas the holes in the VB decay to the highest energy state of the VB on a timescale on the order of 100 fs (10^{-13} s). The cooled electrons and holes are separated by the built-in electric field, and the electrons and holes as carriers are fed to the external circuit as photogenerated current. Note that the band gap is very important for maintaining the excited electrons at a sufficiently high energy level long enough to be ejected by the electrodes. The performance of a photovoltaic device is strongly dependent on the absorbing materials and on how effectively the carriers are extracted by the external circuit.

In photovoltaic devices, the junction between two electronically different materials introduces an electrostatic dipole field at the interface. The electrostatic dipole field induces a built-in voltage (V_d), which drives the photoexcited electrons

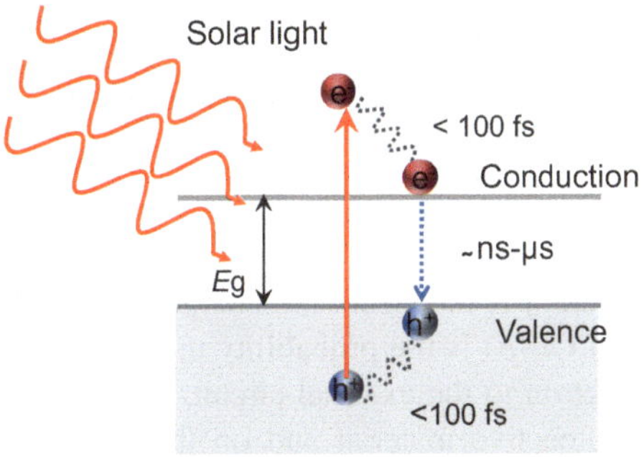

Fig. 20.1 Schematic of light absorption and carrier generation in a semiconductor with a bandgap of E_g. The electron in the conduction band is excited to the valence band, creating a hole in the valence band. The electrons (holes) thermalize to the band edges and are then extracted by the electrode under the built-in electric field

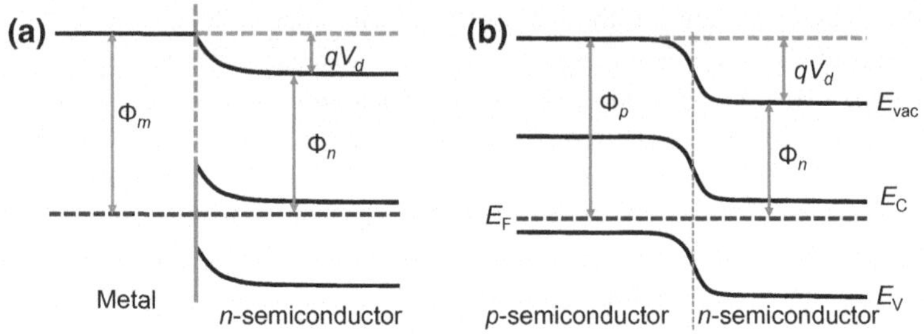

Fig. 20.2 **a** Band diagram of an *n*-type semiconductor/metal junction in equilibrium under dark conditions. **b** Band diagram of the *p*–*n* junction in equilibrium

and holes in opposite directions. A built-in voltage can be generated by either a Schottky (semiconductor/metal) or a *p*–*n* (*p*-type semiconductor/*n*-type semiconductor) junction. A band diagram of *n*-type semiconductor/metal Schottky junction is presented in Fig. 20.2a, where the work function (Φ_n) of the semiconductor is smaller than that of the metal (Φ_m). The Fermi levels are realigned because of carriers exchanged across the junction when the interfaces (metal and semiconductor) are connected to each other, resulting in a barrier difference of $\Phi_m - \Phi_n$ between the semiconductor and the metal. If no surface states exist, the electrostatic field generated by the built-in voltage V_d directly corresponds to the energy difference of $\Phi_m - \Phi_n$. The potential difference V_d is shared between the two materials according to their dielectric permittivities. A *p*–*n* junction, as another important heterojunction, is formed between *p*-type and *n*-type semiconductors. Figure 20.2b shows a schematic of a *p*–*n* junction. A dipole electric field is established at the heterojunction interface because of the difference in work functions, which drives photoexcited electrons toward the *n*-type semiconductor side and holes toward the *p*-type semiconductor side. In photovoltaic applications, a *p*–*n* junction has advantages over a Schottky barrier junction with low defect densities and a large built-in voltage.

The excited electron–hole pairs in the photovoltaic process are transported, collected, and extracted to an external circuit. A mechanism for charge carrier separation is required, and the intrinsic potential asymmetry for electrons and holes drives the electrons (holes) away from their point of origin in the solar cell. The separated electrons and holes as the carriers are extracted and transported to an external circuit to generate the photocurrent. The photocurrent density of the device is described as

$$J_{\mathrm{sc}} = q \int \mathrm{QE}(E) b_{\mathrm{s}}(E) dE, \tag{20.1}$$

where q is the electronic charge, $b_\mathrm{s}(E)$ is the incident spectral photon flux density, and QE(E) is the probability that an incident photon with energy E will deliver an electron to the external circuit, which depends on the absorption coefficient of the photoactive material and on the efficiency of charge separation and charge collection. A dark current $I_{\mathrm{dark}}(V)$, which acts in the opposite direction to the photocurrent, is generated by the potential difference between the terminals of the cell. Most of the cells exhibit diode behaver under dark conditions. The dark current $J_{\mathrm{dark}}(V)$ flowing across the diode under an applied voltage V is given by

$$J_{\mathrm{dark}}(V) = J_0(\mathrm{e}^{qV/nk_{\mathrm{B}}T} - 1), \tag{20.2}$$

where J_0 is a constant, n is an ideality factor, k_{B} is Boltzmann's constant, and T is the absolute temperature.

The overall current–voltage characteristic can be approximated as the sum of a short-circuit current and the dark current (Fig. 20.3). Therefore, the net photocurrent density in the cell is

$$J = J_0\left(\mathrm{e}^{qV/nk_{\mathrm{B}}T} - 1\right) - J_{\mathrm{sc}} \tag{20.3}$$

At open circuit, $J_{\mathrm{dark}}(V) = J_{\mathrm{sc}}$. We obtain an equation for the V_{oc}:

$$V_{\mathrm{oc}} = \frac{kT}{q}\ln\left(\frac{J_{\mathrm{sc}}}{J_0} + 1\right). \tag{20.4}$$

In an actual solar cell, the resistance of materials, the resistance between contacts, and the leakage current result in power losses. These resistances are electrically equivalent to two parasitic resistances in series (R_{s}) and in parallel (R_{sh}) (inset of Fig. 20.3). The current density is described as follows if we include the resistances:

$$J = J_0\left(\mathrm{e}^{q(V - JAR_s)/nk_BT} - 1\right) + \frac{V - JAR_s}{AR_{sh}} - J_{\mathrm{sc}} \tag{20.5}$$

where A is the effective area of the solar cells. Figure 20.3 shows a typical J–V curve based on Eq. (20.5). The power density of the solar cell is defined by $P = JV$ in the bias range from 0 to V_{oc}. The maximum value of P is the cell's

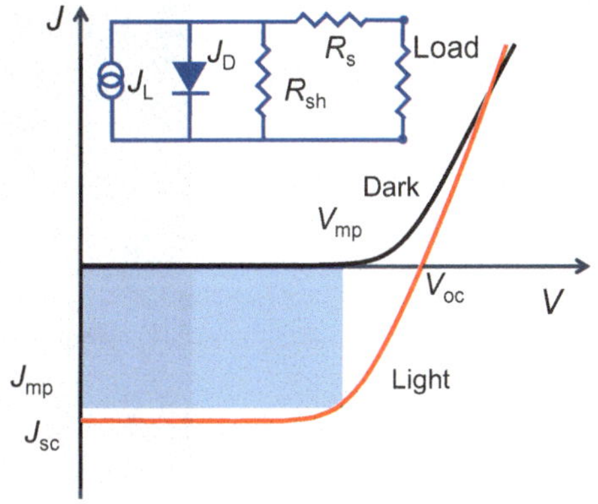

Fig. 20.3 Current–voltage curves of a diode under light and dark conditions. The inset shows the equivalent circuit of a solar cell with both series and shunt resistances

operating point, at which the voltage and current density are labeled as V_m and J_m, respectively. The FF is defined as

$$\mathrm{FF} = \frac{J_m V_m}{J_{sc} V_{oc}}. \tag{20.6}$$

The efficiency (η) of a solar cell is the power density delivered at the operating point as a fraction of the incident light power density (P_{in}):

$$\eta = \frac{J_m V_m}{P_{in}} = \frac{J_{sc} V_{oc} \mathrm{FF}}{P_{in}}, \tag{20.7}$$

where J_{sc}, V_{oc}, FF, and η are the important parameters used to characterize the performance of solar cells.

20.2.2 *Roles of CNTs in Solar Cells*

20.2.2.1 Carrier Generation

The photocarrier generation in semiconducting SWCNTs (s-SWCNT) has been extensively studied over the past decade. Figure 20.4 shows a schematic of photocarrier generation in an s-SWCNT. Freitag et al. investigated the infrared-laser-excited photoconductivity of individual s-SWCNTs [21]. The mechanism of photocurrent generation involves resonant excitation of the E_{22} exciton state in an s-SWCNT. In particular, the phenomenon of multiple electron–hole pair generation has also been reported in SWCNTs, where efficient multiple carrier generation from a single hot E_{22} exciton occurs [22]. This interesting property, in principle, could contribute to a higher PCE that overcomes the Shockley–Queisser limit, which is defined on the basis of the assumption that an absorbed single photon generates a

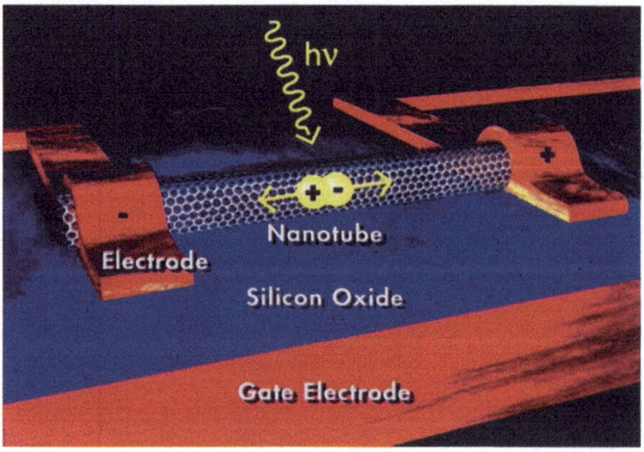

Fig. 20.4 Schematic of photocarrier generation in an s-SWCNT. Reprinted with permission from Ref. [21]. Copyright 2003 American Chemical Society

single electron–hole pair (exciton) in a semiconductor. Shockley and Queisser used a detailed balance calculation to derive a fundamental limit of photovoltaic conversion efficiency (PCE) of a single-junction solar cell [23].

The photocurrent quantum efficiency (QE), which is the most important factor in assessing the incident light conversion to electric power, was calculated as initial values of 1–5% [24]. Aspitarte et al. reported that a *p–n* junction device with a suspended carbon nanotube with a diameter of 2–3 nm could yield a photocurrent QE of 20–40% [25]. Aspitarte et al. speculated that the CNTs were excited by high photon energies at the fourth sub-band (S_{44}) and fifth sub-band (S_{55}) states and that large-diameter CNTs contribute to the high QE, which is consistent with the increase of QE with increasing nanotubes diameter and decreasing exciton binding energy [26]. Lee et al. demonstrated the photovoltaic effect in individual SWCNTs using a *p–n* junction diode, which demonstrated a PCE of 0.2% [27]. Calculations predicts that a high PCE of CNT solar cells will be achieved by further improvements of the quality of SWCNTs and of device architecture in the future [28].

Recently, researchers have developed CNT/Si heterojunction solar cells by taking advantage of the high mobility, conductivity, and transparency of CNTs. Ong et al. reported the photocurrent response of an SWCNT/Si heterojunction solar cell fabricated using SWCNTs with a narrow chirality distribution [29]. They observed a small photocurrent response peak at the E_{11} exciton band, which demonstrates that the SWCNTs contribute to the photoconversion process not only as a charge separator/transporter/collector but also as a light absorber in the SWCNT/Si heterojunction solar cell.

20.2.2.2 Carrier Transport Material

Because of their high carrier mobility, CNTs show great promise for use as carrier transport materials in optoelectronic devices. The initial success in the application of CNTs in solar cells was achieved with organic solar cells in which the CNTs were used as electron acceptors wrapped with a conjugated polymer [30]. However, the CNTs yielded relatively low PCEs when used as electron acceptors in photovoltaic devices because the Fermi level between the polymer and the nanotubes realigned to energetically favor hole transfer from the polymer to the nanotubes [31, 32]. Dabera et al. applied CNTs as the hole extraction layer in organic solar cells, resulting in improved device performance [33].

In principle, the CNTs in organic solar cells mainly play the role of a "bridge" to facilitate carrier transfer among the organic semiconducting molecules [34]. Three important factors for carrier transport have been identified:

(1) Metallic CNTs (m-CNTs) are not appropriate for improving the performance of OPV devices because they induce short circuits, undergo aggregation, introduce impurities, and exhibit low charge mobility when introduced into organic solar cells [7, 35]. The carrier mobility is dominated by Schottky barriers from the junctions

between the m-CNTs and the semiconductors (M–S junctions), leading to poor solar cell performance. This result is similar to that reported by Topinka et al. for m- and s-CNT Schottky barriers [36]. However, the chirality separated m- and s-SWCNTs exhibit higher performance than the mixed m- and s-SWCNTs [37]. The contribution of s-CNTs to the electron charge transfer has been elucidated through ab initio calculations of polymer/CNT heterojunctions [38]. Moreover, the CNTs serve as an effective HTL in CNT/Si and PSCs [18].

(2) The morphology and structure of CNTs greatly affect their carrier transport properties. Highly oriented and isolated SWCNTs coated with a P3HT layer exhibit high carrier transfer ability [39], which strongly suggests that controlling the dispersion of SWCNTs in the polymer is a critical factor affecting photovoltaic performance.

(3) The chemical doping or surface modification of CNTs can considerably improve their carrier transport ability because the carrier density can be improved, reducing the resistance in each individual CNT. Nogueira et al. attempted to mitigate the aggregation of SWCNTs through covalent functionalization with thiophene moieties and achieved a PCE of 1.48% using functionalized SWCNTs and P3OT polymer under 1.55 mW/cm^2 illumination [40].

20.2.2.3 Transparent Conducting Electrodes

CNTs are considered promising transparent conducting electrode (TCE) materials because they are fabricated from an abundant raw material; exhibit excellent conductivity, high specular transmittance over a wide spectral range, superior flexibility, and high chemical stability; and are cost-effective to fabricate [41–44]. Ago et al. first reported the replacement of indium tin oxide (ITO) electrodes with CNTs in polymer solar cells. They observed an enhanced external quantum yield (EQE), which they attributed to the interpenetrating network between the polymer and the MWCNTs and to a higher built-in electric field resulting from the higher work function of the MWCNTs (5.1 eV) compared with that of ITO [45]. Wu et al. demonstrated that conductive thin films made from high-purity SWCNTs represent a promising alternative to ITO [46]. An improved PCE was reported when a SWCNT layer was substituted for the common ITO electrode [47]. However, the photovoltaic performance was very low because of the high sheet resistance of the SWCNT film.

An ideal TCE should exhibit very high light transmission in a broad range from the UV to the NIR region while simultaneously exhibiting low sheet resistance. However, a trade-off exists between these two parameters. Ellmer et al. presented a useful method to judge the properties of a transparent electrode on the basis of the ratio between its dc electrical conductivity (σ_{dc}) and its optical conductivity (σ_{OP}), which is expressed as [48]

$$\frac{\sigma_{dc}}{\sigma_{OP}} = \frac{1}{2R_{sheet}} \frac{\left(\frac{\mu_0}{\varepsilon_0}\right)^{1/2}}{T'^{-1} - 1}, \tag{20.8}$$

where μ_0 and ε_0 are the free-space permeability and free-space permittivity, respectively, T' is the transmittance, and R_{sheet} is the sheet resistance. We can calculate the σ_{dc}/σ_{OP} ratio by measuring T' and R_{sheet} and obtain a value for comparison with the properties of a CNT film. Different σ_{dc}/σ_{OP} ratios are required depending on the application. Solar cells usually require high transmittance and low sheet resistance of the CNT films to ensure efficient energy harvesting and transport. A transfer-printing method was used to produce low sheet resistance CNT films (200 Ω/sq) with a high optical transmission of 85% in the visible range [49]. Muramoto et al. fabricated an SWCNT/Si solar cell with a PCE of 10.4% by utilizing a low sheet resistance (268 Ω/sq) and high-transmission (90%) CNT film deposited via vacuum filtration. An SWCNT/Si solar cell with the same window area and similar transmittance (88%) was fabricated using spray-coated CNTs; the resultant film exhibited a very high sheet resistance (7600 Ω/sq) and resulted in a low PCE of 0.83% [50]. These results strongly suggest that the properties of CNTs are critical factor affecting the performance of CNT-based solar cells.

Yu et al. [1] fabricated CNT TCEs by various approaches and compared their performance. To realize the practical applications of CNT electrodes in optoelectronic devices, especially solar cells, further modification of CNTs toward low resistance but high transparency is needed. The chemical doping of CNTs has been extensively studied and could drastically improve their electrode performance. Numerous *p*-type dopants (e.g., NO_2, Br_2, HNO_3, $SOCl_2$, HCl, $AuCl_3$, $FeCl_3$, metallic particles, and metal oxides) and *n*-type dopants (e.g., poly(ethyleneimine) (PEI), benzyl viologen, and potassium) have been reported. In practice, the *p*-type dopants are more promising because the CNTs are slightly *p*-doped by oxygen under ambient conditions [51]. In addition to chemical doping, the hybridization of CNTs with other materials as electrodes has also been investigated. The hybrid electrodes tend to exhibit improved stability and conductivity.

20.3 Carbon Nanotube/Silicon Solar Cells

Solar cells based on CNT and Si (CNT/Si) heterostructures have been extensively studied as model devices [52–54]. PCEs greater than 10% have been realized in CNT/Si solar cells by many groups, and a record efficiency of 17% has been reported (see Table 20.1). These large improvements in the performance of the CNT/Si solar cell devices have mainly resulted from various postprocessing techniques and modified device architectures; examples include improvements in the quality of CNT films [55], carrier doping of CNT films by infiltration of acid [50], controlling the electronic junctions [56, 57], and coating the CNT/Si surface with an

Table 20.1 Selected important results of CNT/Si heterojunction solar cells. Note that Si represents *n*-type doped Si unless otherwise noted and that rf-SWCNTs refers to redox-functionalized SWCNTs [59]

Solar cell structure	J_{sc} (mA/cm^2)	V_{oc} (V)	FF (%)	PCE (%)	R_s (Ω cm^2)	Active area (cm^2)	Note
DWCNT/Si [15]	13.8	0.5	19	1.31	14.4–98	0.49	H_2O_2 and HCl rinsed
DWCNT/Si [60]	26	0.54	53	7.4	6.4	0.49	H_2O_2 and HCl rinsed
Aligned DWCNT/Si [61]	33.4	0.54	58	10.5	1.72	0.12	Well-aligned nanotubes
FCCVD-SWCNT/Si [55]	30.2	0.59	70	12.5	1.41	0.008	Without doping
SWCNT/textured Si [62]	33.1	0.55	57.3	10.4	–	0.0314	
Pristine-SWCNT/Si [63]	29.7	0.54	68	10.8	2.48	0.09	
MWCNT/Si [64]	30.2	0.55	49	8.1	0.44	0.04	
$SOCl_2$-doped SWCNT/Si [53]	26.5	0.49	35	4.5	4.00	0.25	
HNO_3/$AuCl_3$-doped SWCNT/Si [52]	28.6	0.53	74.1	11.2	–	0.09	Super-acid slide coating method
Acid-doped SWCNT/Si [65]	36.3	0.53	72	13.8	2.25	0.09	Wet state
Cp_2Co-doped SWCNT/*p*-Si [12]	24.2	0.43	45	4.62	–	0.09	
HF/AgNWs/SWCNT/*p*-Si [66]	24.8	0.50	61	7.53	–	0.09	High PCE of the PV cell with *p*-Si
AgNWs/SWCNT/Si [67]	31.0	0.51	69	10.8	–	0.49	
GO/CNT/spiro-OMeTAD/Si [68]	26.15	0.55	66	9.49	5.48	0.087	Without doping and antireflective layer
Electronic-induced SWCNT/Si [57]	29.8	0.55	73	12.0	–	0.08	Ionic liquid gated in devices
TiO_2/acid-doped SWCNT/Si [69]	25.32	0.63	52	10.1	–	2.15	Large area
PMMA/acid-doped aligned DWCNTs/Si [70]	35.6	0.60	62	13.1	2.00	0.13	
rf-SWCNT/Si [59]	36.1	0.54	72.3	14.1	0.51	0.09	Highly stable
TiO_2/acid-doped SWCNT/Si [58]	32	0.61	77	15.1	–	0.15	
MoO_x-coated SWCNT/Si [14]	36.6	0.59	78	17.0	1.92	0.008	Record PCE

antireflection layer [58]. Some important research and the obtained photovoltaic performances of various CNT/Si solar cells are summarized in Table 20.1.

20.3.1 Effects of CNT Films

The physical properties of CNT films strongly affect the photovoltaic performance of CNT/Si heterostructure solar cells. The CNT films have been demonstrated to serve as a semitransparent carrier transport layer in CNTs/Si solar cell devices [71]. Hence, one of the major factors limiting the photovoltaic efficiency of previously reported CNT solar cells is likely related to the carrier transport properties of the CNT films.

Wei et al. [15] first reported the application of CNTs in planar solar cells based on a double-walled carbon nanotube (DWCNT)/Si heterojunction in 2007. An *n*-Si wafer with a window of insulating layer was used to pick up a floating DWCNT film for fabricating DWCNT/Si heterojunction solar cells. The nanotubes function as both a photocarrier generator and a collection layer in the fabricated devices. Figure 20.5a shows the prepared nanotube network film, and (b) displays the photovoltaic performance of a fabricated DWCNT/Si heterojunction solar cell. A PCE of only 1.3% was achieved in the DWCNT/Si heterojunction solar cell. This low PCE was caused by a high sheet resistance, which resulted in a low short-circuit current and a low FF. The same group later achieved a high PCE of 7.4% by reducing the thickness of native silicon oxide on the *n*-Si window [60]. Di et al. reported aligned DWCNT films (Fig. 20.5c), which were drawn from a spinnable nanotube array, for solar cell fabrication [61]. They compared both aligned and random CNTs and concluded that aligned CNTs prefer to densely cover the Si substrate to form a high CNT–Si junction density, whereas the random CNTs result in discrete junctions between the CNTs and Si (Fig. 20.5d). As a result, the aligned-CNT-film-based solar cells showed a substantially higher PCE than the randomly dispersed CNT-based devices (Fig. 20.5e). Because of the superior performance of a densely covered CNT network deposited onto a Si substrate to achieve a high DWCNT–Si junction density, a PCE of 10.5% was achieved without chemical treatments or other postprocess modifications.

SWCNTs show a competitive conversion efficiency even in the absence of chemical modifications. Kozawa et al. fabricated solar cells by directly coating a CoMoCAT (6,5)-rich SWCNT solution onto *n*-type Si and demonstrated that the PCE was strongly dependent on the thickness of the SWCNT film [54]. The highest efficiency of 2.4% achieved at a transmittance of approximately 70% was mainly attributed by the balance between the effective *p–n* junction area and the light intensity based on an equivalent circuit model. The authors noted that the photocurrent of the devices was mainly generated in the *n*-Si layer. With the further optimization, a PCE of 3.4% was achieved by the same group [50]. This relatively low PCE value was attributed to the high sheet resistance and low transmittance of the SWCNTs. Our group improved the PCE of SWCNT/Si solar cells by using a floating-catalyst chemical vapor deposited SWCNT (FC-CVD SWCNT) network

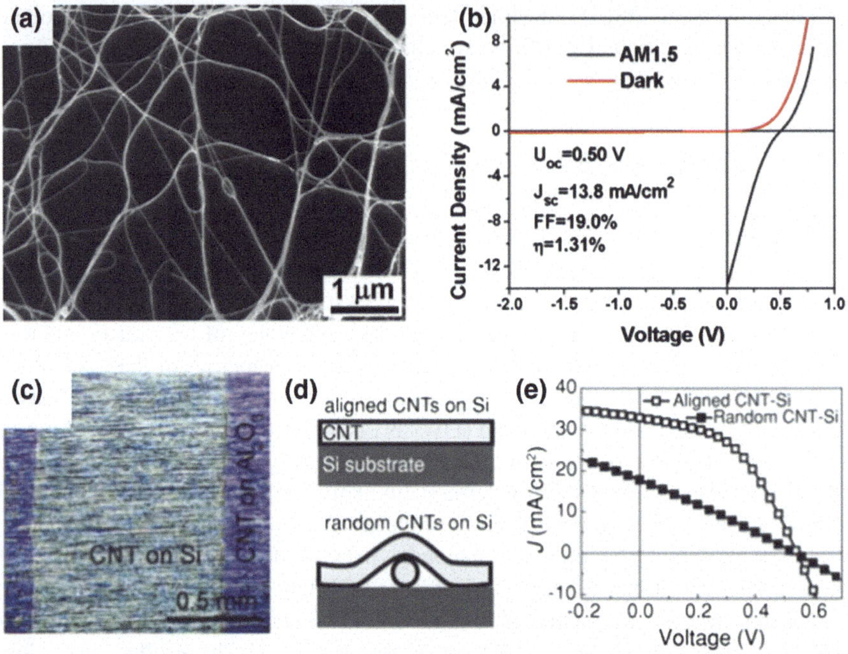

Fig. 20.5 **a** Scanning electron microscopy image of a DWCNT film. **b** *J–V* curves of DWCNT/Si solar cells. (a and b) adapted with permission from Ref. [15]. Copyright 2007 American Chemical Society. **c** Optical image of aligned CNTs on an Si substrate. **d** Schematic of the connection between the aligned and random CNTs. **e** *J–V* curves of the aligned-CNT/*n*-Si solar cells and random-CNT/Si solar cells. (**c–e**) Reprinted with permission from Ref. [61]. Copyright 2013 John Wiley & Son, Inc

film, which exhibits a higher σ_{dc}/σ_{OP} than spray-cast SWCNT films (Fig. 20.6b) [55, 72]. The fabricated devices showed increased PCEs with a decrease of the active window area (Fig. 20.6d), which indicates that the sheet resistance of the SWCNT films plays a critical role in determining the solar cells' performance. The SWCNT-thickness-dependent PCE for the different active window areas also indicates that thicker SWCNTs are preferred in solar cells with a larger window size to retard the decay of the PCE. These results suggest that we should carefully select suitable CNT films for a designated window size in solar cells to achieve the best performance. With the optimization of both the SWCNT film thickness and the window size of solar cells, the highest PCEs of 12.5, 11.6, 10.6, and 7.2% were achieved with devices with 1-, 1.5-, 2-, and 3-mm windows, respectively (Fig. 20.6d). Moreover, the high reproducibility and stability of the photovoltaic performance of these devices in air were demonstrated (Inset of Fig. 20.6c). These insights reveal promising strategies for further improving the PCE of cells and fabricating large-scale CNT solar cells with high PCEs.

MWCNTs have also been used for CNTs/Si solar cells because they are less expensive than SWCNTs. Castrucci et al. reported a MWCNT/Si device in which

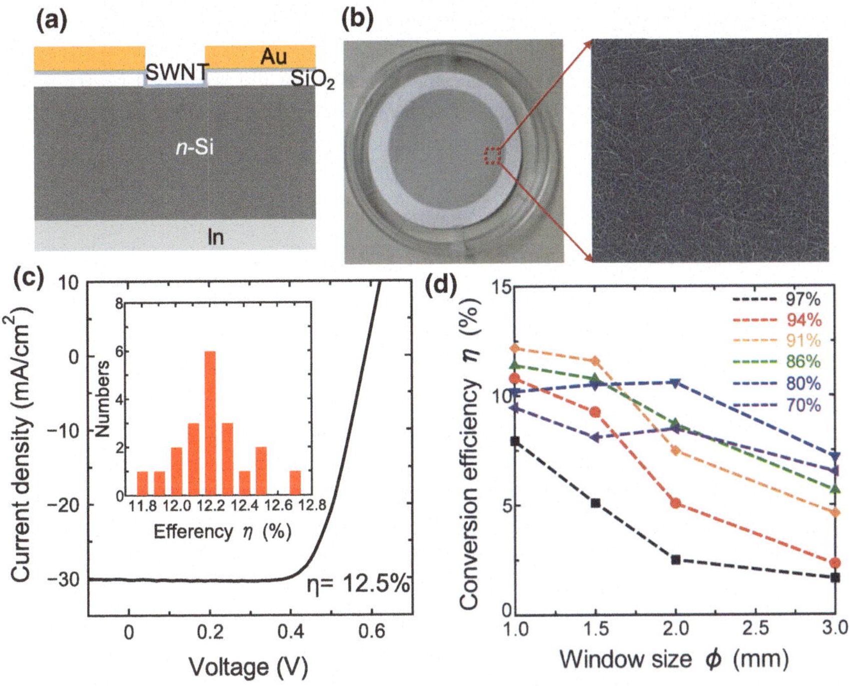

Fig. 20.6 **a** Schematic of the structure of a CNT/Si solar cell. **b** Optical micrograph and scanning electron microscopy image of an SWCNT film ($T' \approx 91\%$). **c** Best solar cell performance of SWCNT/Si solar cells using FC-CVD SWCNT films. The inset shows the reproducibility of the SWCNT/Si solar cells. **d** PCEs as a function of window size for devices fabricated using SWCNT films with various thicknesses. Reprinted with permission from Ref. [55]. Copyright 2014 American Chemical Society

MWCNTs served as both a photocurrent generator and a semitransparent electrode [73]. However, the reported PCE was very low. Nicola et al. reported MWCNT/Si heterojunction solar cells with a high reproducible PCE of 8 ± 1% by tailoring the optical and electrical properties of CNTs [64]. The heterojunction between the MWCNTs and Si extended the spectral absorption from the near-infrared to the near-UV region, which facilitated charge dissociation and the transport of electrons to the Si layer.

A comparison of the properties of MWCNTs and SWCNTs and their selection for further applications is important because many types of CNTs have been used in solar cells. Grace et al. used suspensions of single-walled, double-walled, and multiwalled CNTs for a reliable comparison of the differences in cell properties [74]. The σ_{dc}/σ_{OP} ratios for all of the films were approximately the same, and a detailed comparison of the cells was carried out after they were treated with HF and $SOCl_2$. Their comparison shows that cells with SWCNTs and DWCNTs exhibited similar PCEs that differed by less than 0.5%, whereas the MWCNT-based devices exhibited lower PCEs. Grace et al. also noted that large-diameter SWCNT films tended to exhibit the best performance in the CNT/Si solar cells. Harris et al. further demonstrated that, when chirality purified SWCNTs were used in devices, different-type and -chirality SWCNTs strongly affected the solar cell performance [37]. These results strongly suggest that SWCNT properties such as purity, chirality, type, and diameter will strongly affect device performance.

20.3.2 Chemically Functionalized CNTs

The PCE of CNT/Si solar cells can be improved to greater than 10% by the selection of appropriate CNTs. However, the application of pristine CNTs in solar cells is still restricted by the mixtures of both s- and m-CNTs and poorly conductive CNT bundles. Chemical modification effectively improves their properties for solar cell applications. The conductivity of CNT films is determined by the conductivity of the individual CNTs and by the contact resistance between CNTs. Chemical doping improves both aspects by shifting the Fermi level to increase the charge carrier density and reducing the Schottky barrier height between s- and m-SWCNTs [7]. Numerous efforts to modify CNTs through chemical doping to improve the performance of solar cells have been reported, including modification with acids, metal oxides, gold, and ionic liquid electrolytes.

Li et al. used $SOCl_2$ to modify SWCNT films for an SWCNT/Si solar cell that showed a 45% improvement in PCE because of modification of the Fermi level and increases in the carrier concentration and mobility [75]. Hall-effect measurements demonstrated that the carrier density increased from 3.1×10^{15} to 4.6×10^{17} cm^{-2} after the $SOCl_2$ treatment. Another effective doping method was reported by Jia et al.; their process is shown in Fig. 20.7a. The HNO_3 infiltrated into the CNT network and into the interface of the CNT/Si heterojunction, resulting in an increase of the PCE of doped CNT/Si cells to 13.8 from 6.2% in the pristine cell (Fig. 20.7c). They also investigated the acid doping effect on the performance of CNT/Si solar cells and concluded that both a decrease in sheet resistance and an increase in heterojunction density after HNO_3 doping contributed to an increase of the PCE of the solar cells [65]. Nicola et al. also reported the fabrication of photovoltaic devices with CNTs treated using HNO_3 vapors for 10 s; the devices exhibited a PCE as high as 12% and a record internal quantum efficiency of 100% [76].

In addition to acids, gold(III) chloride ($AuCl_3$) has also been reported to enhance the PCE of CNT/Si solar cells. Li et al. prepared SWCNT films by a super-acid sliding method and transferred them onto patterned Si wafers to fabricate SWCNT/Si solar cells [77]. A high PCE of 11.5% was achieved using SWCNTs treated with the nitric acid and $AuCl_3$ because of a large increase in the J_{sc}. The Au has been proposed to function as a precursor for Cl adsorption to effectively improve electron extraction from the CNTs [78].

Although the wet state doping with acids can increase the performance of solar cells, the stability of such devices is problematic because of the volatility of the

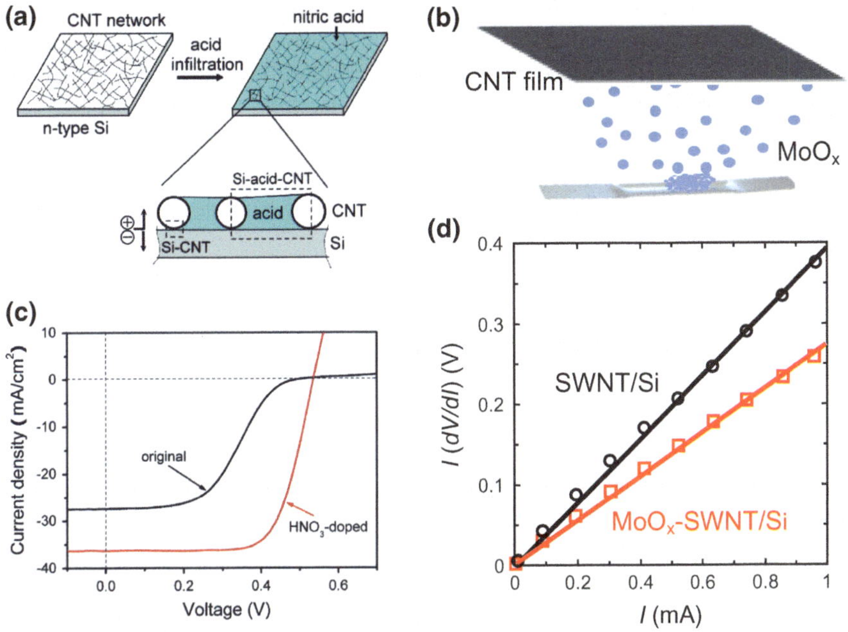

Fig. 20.7 **a** Illustration of the CNTs/Si solar cell and infiltration of nitric acid into the CNT network to form Si-acid-CNT photoelectrochemical units at the interface. **b** Process of vacuum evaporation of MoO_x onto a CNT film. **c** *J*–*V* curves of the CNTs/Si solar cell. (**a** and **c**) Reprinted with permission from Ref. [65]. Copyright 2011, American Chemical Society. **d** Series resistance before and after MoO_x coating. Reprinted with permission from Ref. [14]. Copyright 2015, Nature Publishing Group

acids. The dry-state doping was introduced as an alternative method for improving both performance and stability. We deposited a layer of molybdenum oxide (MoO_x) onto SWCNTs, which can effectively dope carriers into the CNTs (Fig. 20.7b) [14]. Figure 20.7d shows the $I(dV/dI)$ plots for the pristine-SWCNT/Si and MoO_x-SWCNT/Si (SWCNTs with MoO_x coating) solar cells as a function of the current (I). The series resistance (R_s) of the solar cells without and with the MoO_x coating was evaluated from the slope of this plot, and the resistance of the MoO_x-SWCNT/Si solar cell (244 Ω) was substantially lower than that of the pristine-SWCNT/Si (360 Ω). This decrease in series resistance was also confirmed independently by the two-probe method, indicating that the MoO_x could effectively dope carriers into the CNTs. This result is also consistent with the results of a previous report [79].

Chemical doping can also change the electronic properties of CNTs from *p*-type to *n*-type. CNT films functionalized with PEI to form *n*-type semiconductors have been reported [13]. Li et al. introduced cobaltocene (Cp_2Co) to improve the *n*-type (electron) doping of CNT films, resulting in a PCE of 4.62% with assistance from the current-stimulated oxygen removal process [12]. Li et al. emphasized that the oxygen removal process is critical for *n*-type CNTs and reported a record PCE of

7.53% for SWCNT/*p*-Si cells when an HF treatment was used to remove the oxygen [66]. We reported that zinc oxide (ZnO)-assisted β-nicotinamide adenine dinucleotide (NADH) enabled effective *n*-type doping into SWCNTs, which we subsequently used to fabricate *n*-SWCNT/*p*-Si solar cells. The PCE of these cells increased from 1.3% in the absence of ZnO to 4%. This increased performance arises from a reduction in the loss of solar light reflectance, the efficient extraction of electrons to the contact, and a decrease in the series resistance of the SWCNTs [14].

20.3.3 Interfacial Engineering

The performance of a solar cell is not only determined by the materials quality but also by the structure of the devices. In particular, the interface between different layers can result in a large recombination loss in the solar cells. In CNT/Si solar cells, interfacial optimization between the CNTs and the Si was achieved through comparing the CNT/Si solar cells with and without the native silicon oxide layer. A low PCE of 1.3% was obtained for CNT/Si solar cells with a native silicon oxide layer [15]. In the oxide-free CNT/Si solar cells, more efficient heterojunctions were formed between the CNTs and the Si, resulting in a higher PCE of 7.4% [60]. Jia et al. also reported the CNT/Si solar cells with a thin oxide layer (1 nm) grown by immersing the CNT/Si heterojunction into acid to create a CNT-oxide-Si structure. A PCE of 10.1% was achieved, which is more than 1% greater than that achieved with a structure without a thin oxide layer (PCE = 8.9%) [80]. The oxide layer in the solar cells could suppress the dark saturation current, inhibiting recombination and leading to an increase of both the V_{oc} and the minority carrier lifetime. An organic interlayer between the SWCNTs and the Si also could minimize the recombination loss at the interface [68]. Spiro-OMeTAD was used as an interlayer between a GO/SWCNT/Si heterojunction. The authors speculated that the organic interlayer improves the diode properties and increases the Schottky barrier height, resulting in a large V_{oc}. Moreover, the spiro-OMeTAD interlayer could improve the stability of devices by avoiding the formation of SiO_2 on the Si substrate under ambient conditions, which could drastically reduce the performance if the SiO_2 thickness exceeded 2 nm [68].

Wadhwa et al. reported a novel method for improving CNTs/Si solar cells through electronic junction control using an ionic liquid gate of 1-ethyl-3-methylimidazolium bis(trifluoromethylsulfonyl)imide (EMI-BTI) (Fig. 20.8a, b). The device exhibited a PCE of 8.5%, which was dynamically and reversibly adjustable between 4 and 11% via the ionic liquid gating [56]. The mechanism of the electronic junction control was explained by considering the gate-induced modulation of the CNTs' Fermi level and the gate-modulated control of the interface dipole at the junction (Fig. 20.8c). The interface dipole also could be controlled by inserting materials between the SWCNTs and the electrode of Au via the effective hole extraction layer. The deposition of MoO_x between the SWCNT

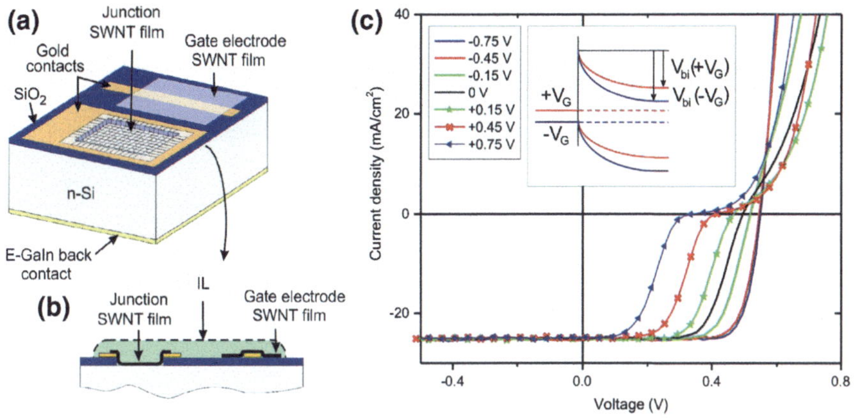

Fig. 20.8 **a** Structure of an electrically gated nanotube/Si solar cell. **b** Cross-section of the solar cell structure. **c** *J–V* curves of the device with variation of the gate voltage. The inset shows the change in built-in potential V_{bi} due to more or less Si band bending with the gate voltage. Reprinted with permission from Ref. [56]. Copyright 2010, American Chemical Society

and the Au electrode in SWCNT/Si solar cells was reported to modify the SWCNT surface potential and to optimize the SWCNT/Au band alignment. The change in the band alignment due to the insertion of the MoO_x layer promoted photocarrier extraction from the SWCNTs to the Au electrode. The MoO_x could enhance the photovoltaic performance by preventing current leakage and consequent counter-diode formation [81]. SWCNT/Si solar cells with and without insertion of the MoO_x layer between the SWCNT network and the Au have been reported. Figure 20.9a shows the performance of both pristine-SWCNT/Si and MoO_x-SWCNT/Si solar cells. The pristine-SWCNT/Si solar cell shows a J_{sc} of 29.4 mA/cm^2, an FF of 65%, and a V_{oc} of 0.58 V, which results in a PCE of 11.2%. For the MoO_x-SWCNT/Si solar cell, the J_{sc}, FF, and V_{oc} were 31 mA/cm^2, 78%, and 0.59 V, respectively, and an improved PCE of 14.2% was attained. The enhancement in the PCE mainly arises from a reduction of the energy barrier (Fig. 20.9b, c) for charge carrier (hole) extraction and efficient transport over the interface between the SWCNT film and the Au contact, which increases the J_{sc} and FF values. Because of the interfacial modification, both the recombination loss and the contact resistance of the MoO_x-SWCNT/Si solar cells were reduced.

20.3.4 Optical Absorption in Solar Cells

The large difference in the refractive index between air and Si results in a substantial power loss by a high reflectance of approximately 30–40% in the range from the UV to the visible region [14]. Thus, a strategy to improve the optical absorption is quite important to improving the performance. A mature technology is

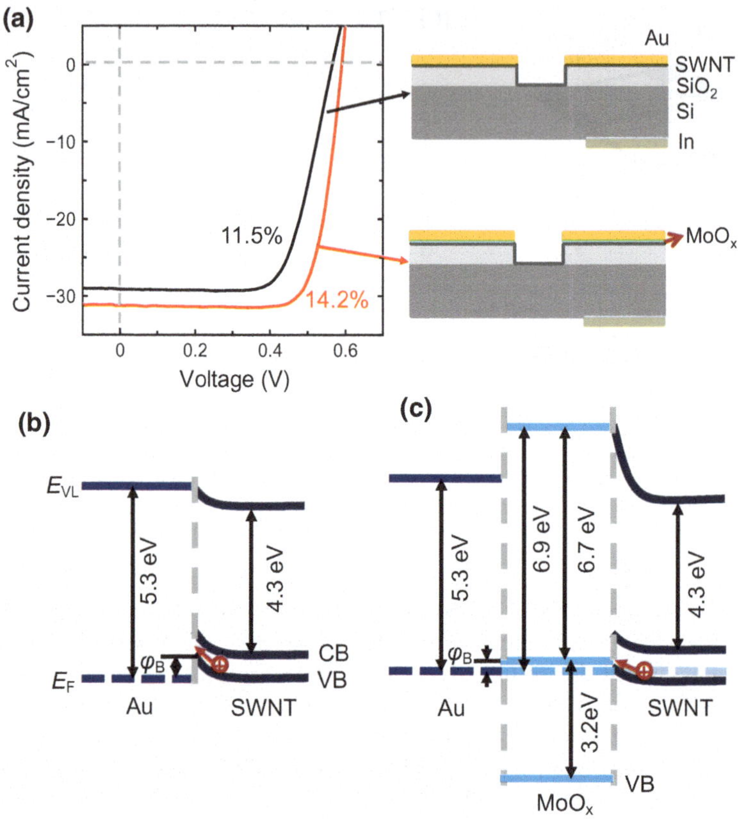

Fig. 20.9 **a** *J–V* curves of solar cells without and with an MoO_x layer. Schematics of the device structures are shown in the figure. **b** Bandgap alignments of the pristine-SWCNT/Au and **c** the MoO_x-SWCNT/Au interfaces. Reprinted with permission from Ref. [14]. Copyright 2015, Nature Publishing Group

to etch the Si surface with an alkaline solution to form a random pyramid to improve the light scattering and absorption. Muramoto et al. reported, for the first time, microsized pyramids formed on a CNT/Si active window using a dilute sodium hydroxide solution [62]. The pyramid effectively improved the incident light absorption, leading to a high PCE of 10.4%, which is a substantial improvement (~27%) compared with that without pyramid textures. This substantial improvement of PCEs mainly arises from an increase of the J_{sc}. Moreover, the polymer coating on the CNT films to improve the optical absorption has also been investigated. The PMMA was used in the CNT/Si solar cells as an antireflection layer, which reduces the reflectance up to less than 10% in the wavelength range from 600 to 1000 nm and results in an increase of the J_{sc} from 24.3 to 35.6 mA/cm^2 [70]. Other polymers such as PDMS and polystyrene have also been reported as antireflection coatings for CNT/Si solar cells [82].

Numerous researchers are interested in the polymers used in CNT/Si solar cells because they not only reduce the reflectance to increase the PCE but also protect the CNTs and the Si from moisture and oxygen to improve device stability [82]. In addition to polymers, metal oxides have also been used to enhance the optical absorption of CNT/Si solar cells. Shi et al. reported the effect of TiO_2 uniformly dispersed onto the SWCNT/Si solar cells to reduce the solar light reflection (Fig. 20.10a) [58]. The J_{sc} of the cell improved from 24.4 to 32.2 mA/cm^2 after the

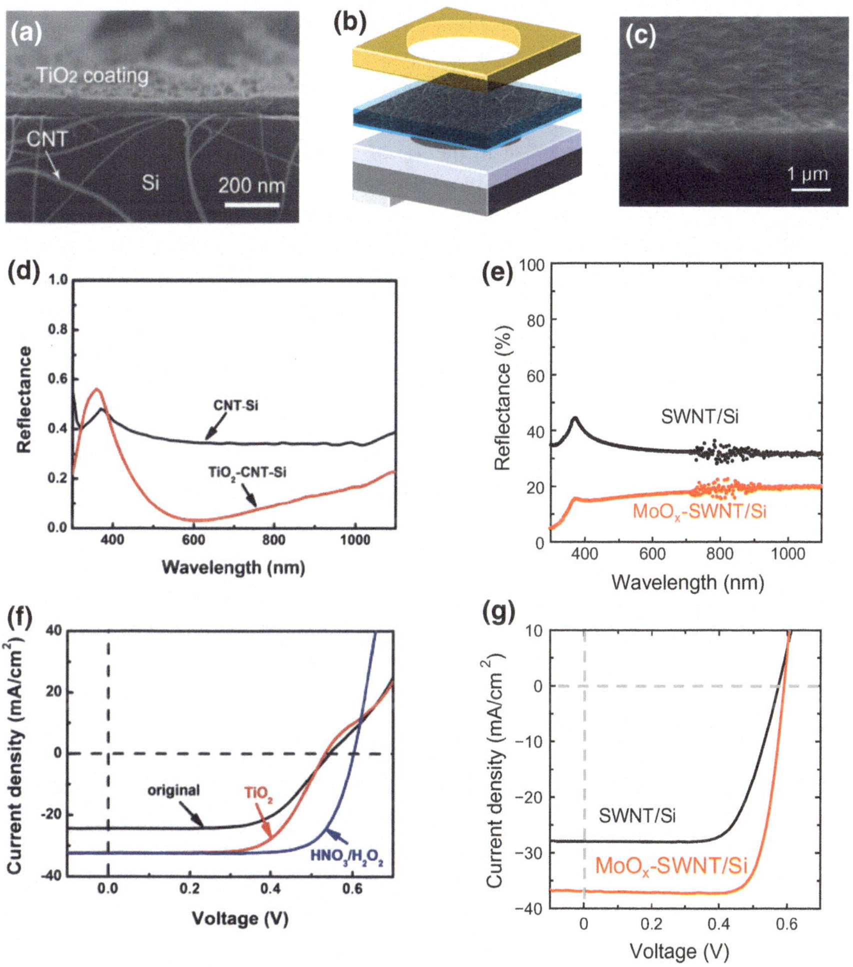

Fig. 20.10 **a**, **c** Scanning electron microscopy image of CNT/Si solar cells coated with TiO_2 and MoO_x layers, respectively. **b** Structure of a CNT/Si solar cell an MoO_x layer between the CNT film and the gold electrode. **d**, **e** Reflectance spectra of CNT/Si without and with TiO_2 and MoO_x layers, respectively. **f**, **g** The *J–V* curves of solar cells. (**a**, **d**, and **f**) Reprinted with permission from Ref. [58]. Copyright 2012, Nature Publishing Group. (**b**, **c**, **e**, and **g**) Reprinted with permission from Ref. [14]. Copyright 2015, Nature Publishing Group

TiO_2 coating was applied; the enhanced J_{sc} is very similar to the calculated values before (23.8 mA/cm^2) and after (31.7 mA/cm^2) the antireflection layer coating was applied (Fig. 20.10f). Note that the optimized layer thickness of TiO_2 for best optical absorption is 68 nm, in agreement with the reported value (70 nm) to minimize the light reflection (Fig. 20.10d).

MoO_x can also contribute to an increase of the optical absorption in solar cells (Fig. 20.10b, c) [14]. The reflectance spectra of the SWCNT/Si solar cells without and with an MoO_x layer are shown in Fig. 20.10e. The reflectivity (R) of the SWCNT/Si was approximately 30% and was reduced to less than 20% over the entire spectral region from 300 to 1100 nm by the MoO_x coating. The J_{sc} of the device, therefore, was improved from 29 to 35 mA/cm^2. The substantial increase in the short-circuit current density is due to the deposition of the MoO_x on the active area of the cell as an antireflection layer, contributing higher PCE than without MoO_x (Fig. 20.10g). The increase in the J_{sc} of 5.9 mA/cm^2 was calculated on the basis of the difference in reflectivity of the pristine-SWCNT/Si and the MoO_x-SWCNT/Si; this increase in the J_{sc} was approximately the same as the experimentally obtained value of 6 mA/cm^2. The role of the MoO_x layer was to reduce the refractive index mismatch between the Si and the air because MoO_x has an intermediate refractive index (n_0 = 2.2) [83]. The increase of the optical absorption in the MoO_x-SWCNT/Si solar cell reduces the loss of solar light and increases the photocarrier generation because the MoO_x layer operates as an antireflection layer.

20.4 Carbon Nanotubes in Perovskite Solar Cells

The organo-lead halide perovskites $CH_3NH_3PbX_3$ (X = Cl, Br, I) have attracted much interest since the first report of their use in solar cells [84]. Their direct bandgaps, large absorption coefficients, and high charge carrier mobilities render them very promising photoactive layers in photovoltaic devices. PSCs constructed from mesoporous scaffolds, perovskite light absorbers (such as $CH_3NH_3PbI_3$), and organic HTLs have exhibited outstanding performance [85]. The continuous improvement and engineering of interfaces [86, 87], solvents [88], and materials [89] have led to a substantial increase in the PCEs of PSCs from the first reported value of 3.8% to greater than 20% [16, 84]. However, many problems remain unsolved for the realistic application of organo-lead halide PSCs, such as the instability of both the perovskite and the organic transport layer when exposed to the atmosphere. By contrast, CNTs, being a chemically stable, flexible, and highly conductive material, have the prospect of wide application as the flexible electrode, HTL, and additives in the PSCs. A schematic of the typical device structure includes FTO, a dense TiO_2 layer, a mesoporous TiO_2 layer, the perovskite, a HTM, and a metal electrode, as shown in Fig. 20.11a. The CNTs could replace the HTM in the PSCs, as shown in Fig. 20.11b. Although different operating principles of PSCs have been reported, the most widely accepted one is similar to the operating principle of DSSCs, where the electron injection from the dye molecules is

Table 20.2 Performance of PSCs functioned by CNTs

Configuration of PSC	J_{sc} (mA cm^{-2})	V_{oc} (V)	FF (%)	PCE (%)	Note
FTO/cTiO$_2$/TiO$_2$/MAPbI$_3$/Al2O$_3$/ SWCNT + carbon [92]	21.26	1.01	69	14.7	CNTs doped in carbon electrode
FTO/c-TiO$_2$/TiO$_2$/MAPbI$_3$/Al$_2$O$_3$/ I-CSCNT/PMMA [93]	17.22	0.853	71	10.54	
FTO/c-TiO$_2$/TiO$_2$/MAPbI$_3$/CNT [94]	15.46	0.88	51	6.87	First publication on the use of CNTs in PSCs
FTO/c-TiO$_2$/TiO$_2$/(FAPbI$_3$)0.85 (MAPbI$_3$)0.15/SWCNT [95]	20.3	0.97	46	9.1	
FTO/c-TiO$_2$/TiO$_2$/(FAPbI$_3$)0.85 (MAPbI$_3$)0.15/spiro-OMeTAD/ SWCNT [95]	20.3	1.1	61	13.6	Without metal electrode
FTO/c-TiO$_2$/TiO$_2$/MAPbI$_3$/CNT [96]	18.54	0.703	60	7.83	
FTO/c-TiO$_2$/TiO$_2$/Cs$_5$(MA$_{0.17}$FA$_{0.83}$) Pb(I$_{0.83}$Br$_{0.17}$)$_3$/spiro-OMeTAD/ SWCNT [97]	21	1.12	71	16.6	Very high PCE
Ti/TNT/MAPbI$_3$/spiro-OMeTAD/ CNTs [98]	14.36	0.99	68	8.31	Ti-foil-based flexible devices
FTO/c-TiO$_2$/TiO$_2$/MAPbI$_3$/MWCNT [99]	15.6	0.88	75	10.3	Negligible hysteresis
ITO/c-TiO$_2$/MAPbI$_3$/CNT [100]	15.9	0.91	65.6	9.49	Fiber-type solar cells
FTO/c-TiO$_2$/TiO$_2$/MAPbI$_3$/ CNTs@P3HT [101]	22.7	0.91	65	13.43	
FTO/cTiO$_2$/TiO$_2$/MAPbI$_3$/SWCNT/ GO/PMMA/Au [18]	20.1	0.95	61	11.7	All carbon for hole transport
FTO/c-TiO$_2$/TiO$_2$/MAPbBr$_3$/CNTs/ PMMA/Au [102]	5.86	1.31	75	5.76	
FTO/c-TiO$_2$/Al$_2$O$_3$/perovskite/P3HT/ SWCNT-spiro-OMeTAD(stratified)/ Ag [103]	21.4	1.02	71	15.4	
FTO/c-TiO$_2$/MAPbI$_3$/spiro-OMeTAD/ MWCNTs/Au [104]	21.6	1.1	69.2	15.1	
FTO/c-TiO$_2$/MAPbI$_3$/s-SWCNTs/Au [105]	20.8	1.07	73	16.1	Semiconducting carbon nanotubes
FTO/c-TiO$_2$/TiO$_2$/MAPbI$_3$/ spiro-OMeTAD/ spiro-OMeTAD + MWCNTs/Au [106]	22.13	0.76	62	10.42	
FTO/c-TiO$_2$/TiO$_2$/MAPbI$_3$/ SWCNT-PhOMe/P3HT/Au [107]	22	0.854	62	11.6	
FTO/c-TiO$_2$/MAPbI$_3$/ spiro-OMeTAD + f-SWCNTs/Ag [108]	16.6	1	58	9.6	(7,6)-enriched SWCNTs
FTO/c-TiO$_2$/TiO$_2$/MAPbI$_3$/ P3HT-BCN(1 w%)/Au [109]	18.75	0.86	52	8.3	Doped in HTL

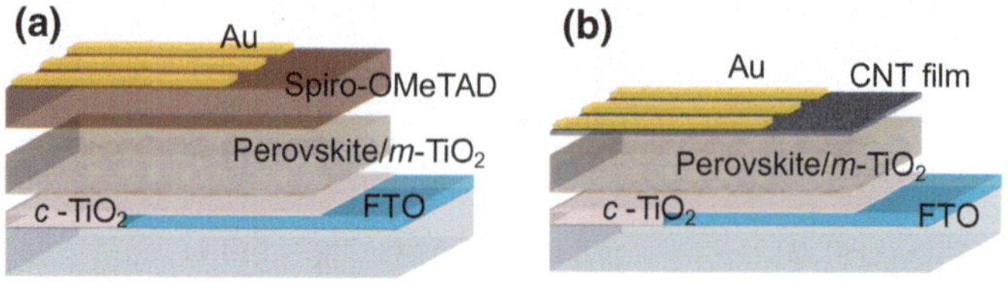

Fig. 20.11 **a** Structure of a typical PSC with a spiro-OMeTAD HTL. **b** Structure of a PSC with a CNT HTL

driven by the energy level gradient between the TiO_2 and dye interface [90]. The detailed mechanism has been reported in other papers [90, 91]. Here, we mainly focus on recent progress in elucidating the roles of CNTs in PSCs. Table 20.2 presented major results of PSCs by using CNTs.

20.4.1 CNT-Based Electrodes

CNTs have been extensively used as electrode materials in PSCs because they can realize a high PCE with greater stability than PSCs with a metal electrode even in the absence of an expensive organic HTM. In the first studies where CNTs were used as the counter electrode in mesostructured PSCs, a free-standing SWCNT film was laminated onto the perovskite layer as a cathode (Fig. 20.12a) [94]. In the resulting device, the CNT film plays the roles of both hole transporter and electrode, and the obtained PCE of the PSCs with a CNT cathode was 6.87% (Fig. 20.12b), which is higher than that of an otherwise identical PSC with an Au electrode (5.14%). This result indicates that CNTs can outperform precious metals as electrode materials in PSCs. Moreover, the efficiency of the PSC fabricated with the CNT cathode was improved to 9.9% when the CNTs were doped with spiro-OMeTAD. Wei et al. reported that MWCNTs could be used to replace the metal as the back electrode in PSCs [99]. They achieved a PCE of 12.67% with a very high FF of 0.80, which is attributed to the superior conductivity of the MWCNTs and low recombination loss at the $CH_3NH_3PbI_3$/MWCNT interface. The devices with MWCNTs display a negligible hysteresis of only 1.53%, which is much smaller than those of devices with spiro-OMeTAD and HTM-free PSCs. Wei et al. attributed the very low hysteresis in the devices with MWCNTs to the extraordinary charge transport along the MWCNTs.

In addition to improving the solar cell efficiency, upgrading the CNT properties is another effective approach to improving solar cell performance. The use of chirality-sorted CNTs in PSCs would bring a remarkable improvement in performance because the chirality-sorted CNTs could improve the carrier selectivity [110]. Another simple method is chemically doping CNTs to improve their conductivity, thereby enabling a high charge extraction rate. Thus, the chemical

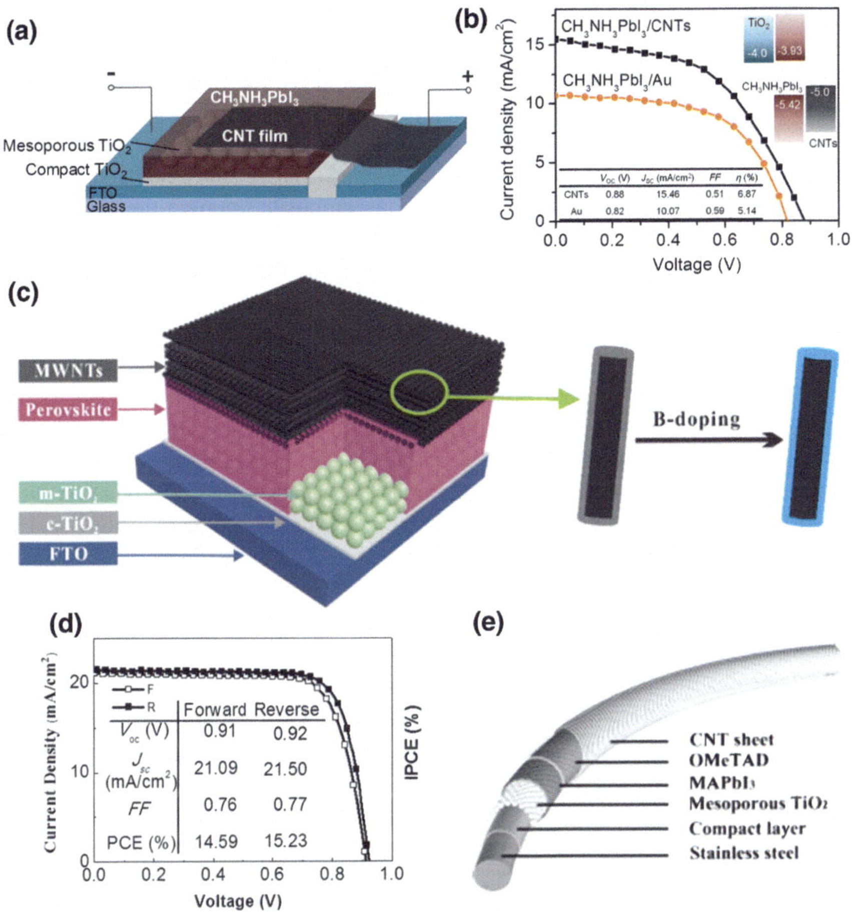

Fig. 20.12 **a** Schematic of a PSC with a CNT film. **b** *J–V* curves. (**a**, **b**) Adapted with permission from Ref. [94]. Copyright 2014, American Chemical Society. **c** Schematic of B-doped MWCNT solar cells. **d** *J–V* curves. (**c**, **d**) Adapted with permission from Ref. [111]. Copyright 2017, American Chemical Society. **e** Structure of a fiber solar cells with CNTs as electrode. Reprinted with permission from Ref. [112]. Copyright 2014, John Wiley & Son, Inc

functionalization of CNTs to further improve the performance of solar cells is a promising approach. Zheng et al. doped boron (B) into CNTs for use as the counter electrode in a PSC [111]. The devices fabricated using the B-MWCNTs as the back electrode for hole collection and transport showed a PCE (14.6%) substantially greater than that of devices without B-MWCNTs (10.7%). They further improved the PCE to 15.23% (Fig. 20.12d) by depositing a thin Al_2O_3 layer between the TiO_2 and the perovskite layer; the Al_2O_3 layer functions as an insulating barrier to avoid possible leakage between the m-TiO_2 and the B-MWCNTs and further reduces the charge carrier losses by restraining the charge recombination processes

(Fig. 20.12c). The PSCs with MWCNTs show negligible hysteresis behavior and also impressive durability and stability, as also reported by other groups [99]. The SWCNT films infiltrated with spiro-OMeTAD were used as electrodes in PSCs, which yielded a PCE greater than 16% with better stability than PSCs with metal electrodes [97]. Li et al. introduced SWCNTs into a hole-collecting carbon electrode composed of graphite and carbon black. Their study showed that the addition of 0.05 wt% SWCNTs to the carbon composite improve the PCE of the device from 9.9 to 14.7%, mainly by improving the V_{oc} and the FF [92].

CNTs are considered an ideal transparent electrode material for flexible devices because of their extraordinary network properties. Jeon et al. fabricated inverted PSCs using HNO_3-functionalized SWCNTs as a front transparent conductive electrode to replace the commercial ITO transparent electrode. They achieved a PCE of 5.38% in PSCs on polyethylene terephthalate (PET) substrates [19]. Wang et al. demonstrated a solid-state, flexible PSC based on an electrode composed of Ti foil and transparent CNTs, where the CNTs were laminated on top of a perovskite layer as the hole collector and the transparent electrode for light illumination [98]. The best PCE of 8.31% was attained using spiro-OMeTAD covered by CNTs. The device's tolerance to mechanical bending was tested, and the device was found to operate at approximately 85% of its initial PCE after 100 bending cycles. A coaxial-fiber-shaped PSC with excellent flexibility was fabricated using stainless steel wires with a TiO_2 blocking layer, $CH_3NH_3PbI_3$, spiro-OMeTAD, and CNTs as the anode, absorber, HTL, and cathode, respectively (Fig. 20.12e) [112]. The CNTs function as a flexible and conductive electrode in the flexible solar cell, which yields a PCE of 3.3%. A PCE of 6.8% was realized by improving the coverage of the perovskite layer, which prevents direct contact between the electron transport material and the CNT sheet electrode. Because of the high series resistance of the CNT electrode, Ag was introduced to increase its conductivity; a 7.1% of PCE was attained upon optimization of the Ag thickness [113]. With further improvements in the perovskite crystal size, a 9.49% PCE was attained because of a substantial improvement of the lifetime of charge carriers [100]. The fabricated devices maintained better than 90% of their initial PCE after 500 bending cycles at a bending curvature radius of 7.5 mm. More interestingly, the photovoltaic parameters J_{sc}, V_{oc}, and FF were independent of the incident light illumination angle because of the symmetric coaxial structure of the devices.

20.4.2 CNT-Based Carrier-Selective Layer

Spiro-OMeTAD, PTAA, and P3HT are effective HTMs in PSCs. However, the high cost and poor long-term stability of many organic hole conductors prompts the further development of hole collection and transport layers. CNTs are an ideal alternative material because of their high stability and high hole conductivity. The initial use of CNTs as a hole-collecting layer was reported by Habisreutinger et al., who directly deposited the polymer-wrapped SWCNTs onto the perovskite surface and then coated them with an inert polymer layer [17]. The polymer-wrapped SWCNTs act as a carrier-selective transport layer, which results in a PCE as high as 15.3%. The SWCNTs effectively extract the photogenerated holes from the perovskite to the cathode, while the inert polymer matrix could eliminate shunting pathways and reduce recombination losses between the perovskite film and the metal electrode. The best performance device yielding a PCE of 15.4% with a low series resistance of 1.8 Ω/cm^2 was achieved by replacing the inert polymer with an organic matrix (spiro-OMeTAD). The photovoltage decay of devices with the spiro-OMeTAD and the SWCNTs/spiro-OMeTAD demonstrated that the charge extraction rate was approximately one order of magnitude greater when the CNTs were present [103].

The CNT network filling with PMMA without HTM for improving the stability of devices has also been reported [33]. Figure 20.13a exhibits the architecture of $CH_3NH_3PbI_3$/SWCNT/GO/PMMA PSC devices, which could yield a J_{sc} of 20.1 mA/cm^2, a V_{oc} of 0.95 V, and an FF of 61%, resulting in a PCE (η) of 11.7% (Fig. 20.13e). The SWCNT layer functions as an efficient carrier dissociation and hole extraction layer. The photoluminescence (PL) and time-resolved PL of perovskite and perovskite/SWCNT thin films were compared to elucidate the charge extraction properties. The PL intensity in the perovskite/SWCNT is drastically quenched, and the time-resolved PL shows shorter carrier lifetime for the perovskite/SWCNT than for a perovskite layer without the CNTs, which suggests that the SWCNT layer plays an effective role in charge carrier dissociation and extraction across the interface between the perovskite and the SWCNTs. However, the SWCNTs, whose bandgap is narrower than that of $CH_3NH_3PbI_3$, cannot block the electron leakage, resulting in recombination losses at the interface. By contrast, the much larger bandgap of the GO compared with that of the perovskite and the SWCNTs may enable the GO to serve as an effective electron blocking layer. Figure 20.13c shows the energy levels of a perovskite/SWCNT/GO solar cell, where the bandgaps for $CH_3NH_3PbI_3$ and s-SWCNT films (diameter: 1.6 nm) were approximately 1.6 and 0.9 eV, respectively [14, 114]. The energy barrier between the perovskite and the GO layer effectively limits the electron diffusion, preventing carrier recombination loss at the anode and increasing the V_{oc} value. The stability of the perovskite/SWCNT/GO/PMMA solar cell is much better than that of the perovskite/spiro-OMeTAD solar cell because of the inert PMMA layer, which functions as an effective barrier to moisture and oxygen penetration and prevents degradation of the organo-lead perovskite.

Ihly et al. reported highly enriched s-SWCNT films as a hole extraction layer for PSCs, where the enriched s-SWCNT films provide a material-specific spectroscopic signature that enables tracking of the charge extraction and recombination and enables very long-lived charge separation [105]. The carrier transport dynamics in the $CH_3NH_3PbI_3$ layer with the SWCNTs were demonstrated; the process is shown in Fig. 20.13b. The processes of both photogenerated carriers in the pristine $CH_3NH_3PbI_3$ layer extracted by the SWCNTs and extracted carrier back transfer or recombination at the SWCNT/perovskite interface were investigated using time-resolved spectroscopy. Figure 20.13d shows that the generated hole was

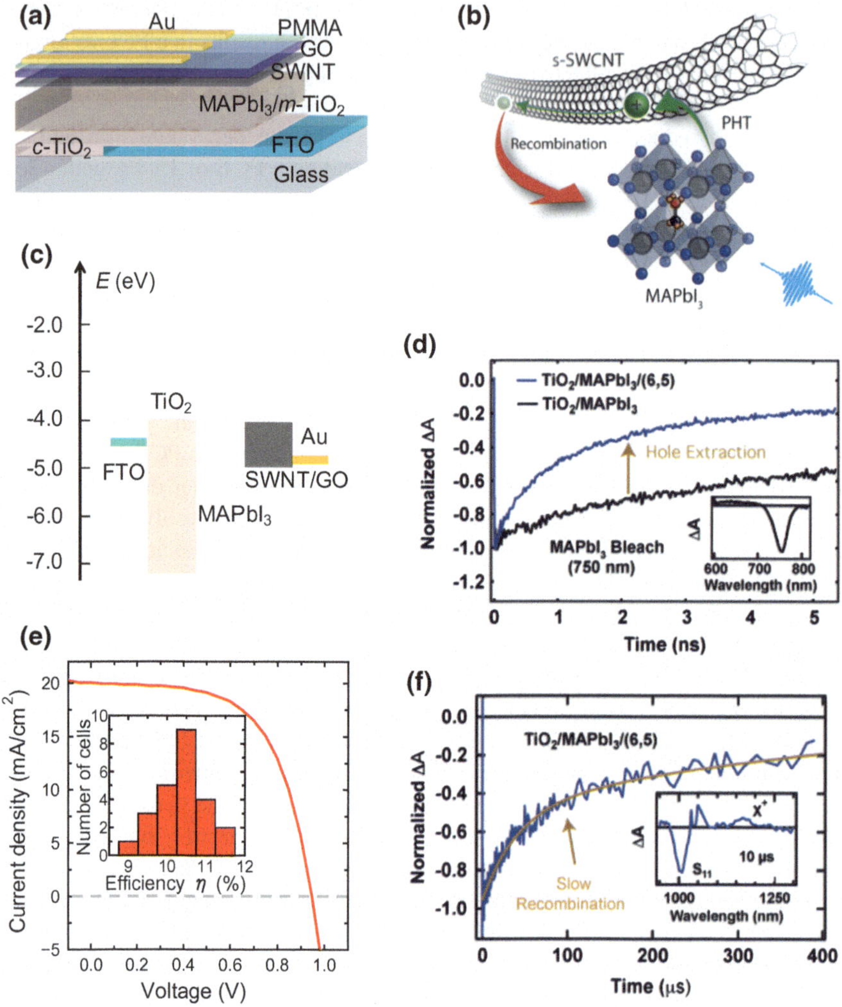

Fig. 20.13 **a** Schematic of a perovskite/SWCNT/GO/PMMA solar cell. **c** Band alignment and **e** typical *J–V* curve of a perovskite/SWCNT/GO solar cell. The inset shows a histogram of the device performance for 25 perovskite/SWCNT/GO/PMMA solar cells. (**a**, **c**, and **e**) Reprinted with permission from Ref. [18]. Copyright 2016, The Royal Society of Chemistry. **b** Fast photoinduced hole transfer is observed from $MAPbI_3$ to s-SWCNTs, whereas the recombination is slow. **d** Tracking hole extraction at the SWCNT–$MAPbI_3$ interface using the $MAPbI_3$ bleaching signals at 750 nm (inset). **f** Tracking hole back transfer and recombination at the SWCNT–$MAPbI_3$ interface using the (6,5) SWCNT bleaching signals at 1000 nm (inset). Reprinted with permission from Ref. [105]. Copyright 2016, The Royal Society of Chemistry

extracted from the perovskite to the s-SWCNTs extremely fast (sub-picosecond after carrier generation). Once transferred into the SWCNTs, the holes show a long lifetime, which is several orders of magnitude longer than that in the perovskite layer (Fig. 20.13f). These results indicate that s-SWCNTs establish the appropriate interfacial energetics to facilitate fast charge extraction and very slow recombination. A 5-nm-thick SWCNT film was used as the HTL between the perovskite and doped spiro-OMeTAD to enhance the charge extraction, resulting in an improvement of the PCE from 14.7 to 16.5%.

Wang et al. used MWCNT/spiro-OMeTAD as the HTL in PSCs to improve the devices' performance [106]. The electron lifetime was improved by the MWCNTs, resulting in a low recombination in the PSCs with the MWCNTs. They also demonstrated that MWCNTs in the HTL could reduce current loss during the charge transfer process and result in an enhanced J_{sc}. Acid-treated MWCNTs for improving device performance were reported by the same group [115]. A PCE of 8.7% was achieved because the acid treatment improved the dispersibility of MWCNTs because of the π–π interactions between the –COOH groups and the spiro-OMeTAD.

20.4.3 CNTs as Additives

The CNTs have been also used in organic solar cells and DSSCs as dopants to improve the carrier dissociation, extraction, and transport [116]. This strategy has also been applied in PSCs. Batmunkh et al. incorporated SWCNTs into the TiO_2 nanofiber (NF) photoelectrode for the PSCs, which significantly improved the PCE of the devices [3]. The SWCNTs provide fast electron transport pathways within the TiO_2 NF photoelectrode, resulting in an increase in the J_{sc}. Simultaneously, the energy level of the photoelectrodes was optimized by the introduction of SWCNTs, which improved the V_{oc} of the devices. Furthermore, SWCNTs introduced into the TiO_2 NFs reduced the hysteresis effect and improved the stability of the PSCs. A best PCE of 14.03% was achieved using an optimized concentration of SWCNTs both in the compact TiO_2 layer and in the photoelectrode. In subsequent work, the same group [117] reported a best PCE of 16.11% for PSCs with 0.10 wt% SWCNTs incorporated into their TiO_2 photoelectrode. The obtained PCE is significantly higher than the highest PCE of 13.53% for a control PSC fabricated without SWCNTs in its photoelectrode.

The SWCNTs functionalized with oligophenylenevinylenes drastically improved the performance and reproducibility of the PSCs. Cai et al. used CNTs to improve the charge transport properties in the P3HT hole transport material for PSCs [109]. The SWCNTs effectively enhance the crystallization of the polymer, whereas the MWCNTs enhance its conductivity. The CNTs used in the studies were a mixture of SWCNTs and MWCNTs for structural reasons; these mixtures were dubbed "bamboo-like carbon nanotubes." The performance of the P3HT solutions with low concentrations of CNTs was investigated in devices, and the highest PCE obtained was 8.3% for the device with a CNT loading of 1 wt%. The improvement resulting from the incorporation of the CNT additives was ascribed to the improved crystallinity and grain size of P3HT for effective hole extraction and transport, which is consistent with the findings of previous reports [118].

Gatti et al. reported the incorporation of both organic functionalized SWCNTs and reduced GO in the P3HT matrix for PSCs [107]. A PCE of ~11% was attained in a device with 2 wt% SWCNTs in the P3HT blend used as a hole extraction layer. The observed results indicate that the improvement of the performance could be linked with the enhancement of local contacts at the hybrid interfaces, such as the perovskite/HTM and the HTM/top electrode interfaces, leading to an effective extraction of the photogenerated carriers from the HTM in the presence of CNTs. They also noted that the stability of the solar cells was remarkably improved by the incorporation of CNT-doped P3HT.

The enriched semiconducting (7,6)-SWCNTs with chemical functionalization were used to dope the spiro-MeOTAD as the HTL in PSCs, which yielded a maximum PCE of 9.6% [108]. MWCNTs were also used as additives to improve the conductivity of spiro-OMeTAD; however, because of back-electron transfers, cell performance suffered when the CNTs were allowed to contact the perovskite layer directly [104]. The authors coated an additional layer of spiro-OMeTAD between the perovskite and the MWCNT/spiro-OMeTAD hybrid layer as a barrier to improve the performance of PSCs. When the concentration of CNTs was controlled and the structures were optimized, the loading of 2 wt% of MWCNTs in the spiro-OMeTAD solution improved the PCE from 12.8 to 15.1%.

The CNTs have also been doped into the perovskite layer for improving the grain size and reducing recombination losses. Bag et al. reported the incorporation of MWCNTs into the perovskite layer, which drastically reduced the recombination losses and the charge accumulation at the perovskite/PEDOT:PSS interface [119]. Zhang et al. demonstrated that the incorporation of sulfonate CNTs could enhance the grain size and reduce the grain boundary of the perovskite layer [120].

20.5 CNTs in Other Solar Cells

20.5.1 Organic Photovoltaics

The typical structure of an OPV device is shown in Fig. 20.14a; it is composed of a transparent conductive electrode (ITO or FTO glass), an HTL, an active layer, and a back electrode with a lower-work-function metal such as Al. In the OPV systems, CNT films have been incorporated into devices as the transparent anode, the cathode, the hole extraction layer, and the electron acceptor [4, 7]. Chaudhary et al. incorporated CNT films into OPVs and found that the use of SWCNTs as the anode or HTL improved the devices' performance [10]. These findings indicate that the SWCNTs can be made suitable for use in the HTL either through tuning the nanotube Fermi level [31] or through *p*-type doping effects. Barnes et al. reported using an SWCNT film as the transparent electrode to replace both the ITO electrode

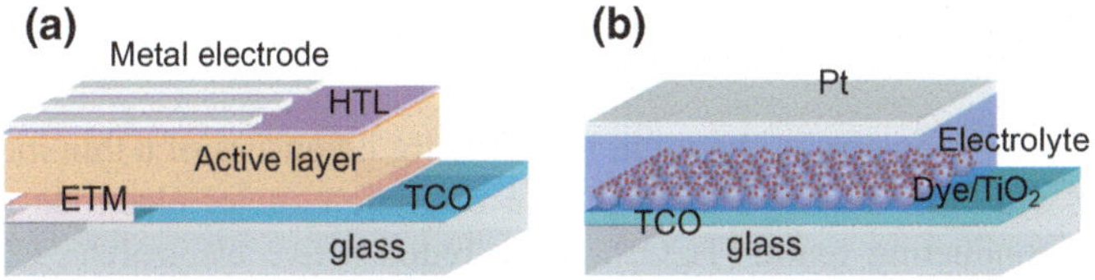

Fig. 20.14 **a** Structures of organic photovoltaics; ETM and HTL represent the electron transport material and HTL, respectively. **b** Structure of dye-sensitized solar cells

and the HTL. However, Lagemaat et al. found that an HTL was still required for efficient charge extraction [121]. Nevertheless, the SWCNTs have displayed several advantages such as the flexibility of nanotubes much better than conventional transparent electrodes of ITO and FTO [49]. However, the CNTs could be also used as the electron acceptor in polymer solar cells, which always exhibit poor performance [30, 122–124]. The observed low PCEs are caused by large connections between m-CNTs and polymers; these connects result from the stronger electrostatic interaction induced by the charge transfer, which results in large recombination and inefficient performance [38]. Moreover, the use of CNTs as an additive dispersed in P3HT to enhance the P3HT:PCBM solar cell performance has been reported [125]; this approach increases the hole mobility and reduces carrier recombination [126].

The utilization of SWCNTs as additives in PTB7:PCBM OPVs results in a high PCE of 6.3%, suggesting that the SWCNTs can be used as additives in modern OPV blends. The SWCNTs were utilized as light-harvesting materials in OPV devices for the exciton dissociation and the electron transfer [127, 128]. The authors initially used a polyfluorene sorting route to achieve a monochiral dispersion of SWCNTs and a PCE of ~1% [4]. An EQE of 43% was achieved but only at the sharp van Hove singularities of SWCNTs, which could harvest the energy only in a fraction of the solar spectrum [129]. Thus, a polychiral distribution of s-SWCNTs was used to broaden the solar light-harvesting, resulting in an average certified PCE of 2.5% and a highest PCE of 3.1% [130]. Moreover, Ihly et al. demonstrated that the driving force and reorganization energy could result in efficient photoinduced electron transfer between the SWCNTs and fullerene derivatives, indicating that further improvements to the SWCNT/acceptor systems could be realized through the selection of appropriate of SWCNT donor materials and fullerene derivatives [131]. Notably, the CNT active layer in the solar cells strongly absorbs light in the near-infrared region, which enables the harnessing of a portion of the solar spectrum that is currently inaccessible to many other thin-film OPV technologies.

20.5.2 Sensitized Solar Cells

A typical DSSC geometry is shown in Fig. 20.14b; it consists of a thin mesoporous TiO_2 layer with adsorbed dye molecules, a catalytic counter electrode [Pt-coated transparent conducting oxide (TCO)], and a redox-active electrolyte. CNTs have been used as the transparent conductive film in either the counter electrode or the photoanode. Aitola et al. deposited a SWCNT film onto a PET substrate as a counter electrode (ITO and Pt free) in the DSSC, resulting in a PCE of only 0.74%. The PCE of 1.53% was obtained using nitrogen (N)-doped CNTs as the counter electrode, which generates free electrons for easy reduction of the electrolyte and enhances the performance. Recently, Arbab et al. reported a textile fabric counter electrode containing activated charcoal (AC)-doped MWCNTs for N719-dye DSSCs [132]. They confirmed that the carbon composite exhibited a high electrocatalytic activity and showed a low charge transfer resistance, which resulted in a high PCE of 7.29%. A reference solar cell prepared using an FTO counter electrode exhibited a PCE of 7.16%, suggesting that the carbon fabric composite electrode is promising as a counter electrode for DSSCs. The CNTs function as an alternative electrode material to conventional TCOs in Pt-based counter electrodes for DSSCs, where the CNTs and the Pt salts can be dispersed well with the assistance of poly (oxyethylene)-backboned polyimide. The fabricated DSSC showed a PCE of 6.96%, which is similar to PCE of the cell fabricated with a Pt/FTO counter electrode [133].

CNTs are a promising candidate to replace the catalyst layer (Pt) in DSSCs [134–136]. Both SWCNTs and MWCNTs have been investigated as the catalyst layer in devices, which show comparable performance to the Pt-based cells [8, 137]. These results are ascribed to the excellent electrocatalytic activity, low charge transfer resistance, and high conductivity of CNTs. The presence of defects in CNTs is one of the critical factors for the electron transfer to enhance the catalytic rate at the counter electrode–electrolyte interface [137]. Thus, the "bamboo"-structured MWCNTs with abundant defects was used as the counter electrode material of a DSSC, which yielded a high PCE of 7.67%. Moreover, highly purified and well-aligned CNTs on an FTO substrate as the counter electrode for a DSSC have been reported; the resultant devices show a high PCE of 10.04%, which is even higher than the PCEs of devices fabricated with benchmark Pt-counter electrodes of 8.8% [138]. These results strongly suggest that CNTs are a promising material for use in counter electrodes to realize higher PCEs in DSSCs.

CNT films are also a potential alternative to TCO-coated glass used as the photoanode in DSSCs. Wei et al. reported a flexible solid-state DSSC with CNT-coated PET as the transparent electrode [139]. Because of the high sheet resistance (250 Ω/sq) and low transmittance ($T' = 65\%$), the PCE of the resultant devices was less than 0.5%. Du et al. improved the PCE of DSSCs to 2.5% by employing the electrodeposited CNT film ($R_s = 470$ Ω/sq at 86% transmittance) as the photoanode [140]. Compared with the PCE of 3.7% for ITO/PET-based devices ($15 \leq R_s \leq 70$ Ω/sq at $T' \approx 85\%$), the CNT-based photoanode is promising because the optoelectronic properties of CNTs could be much improved. Kyaw et al. proposed TiO_x coating onto sulfuric-acid-doped CNT films to reduce the charge recombination between the nanotubes and the electrolyte [141]. A PCE of 1.8% was achieved in a DSSC with a TiO_x/CNT electrode, which was substantially higher than that of devices fabricated using CNT films without TiO_x (PCE of 0.004%).

CNTs have been incorporated into the semiconducting layer of DSSCs to reduce recombination and improve electron transport. Sawatsuk et al. reported a high PCE of 10.29% for CNT/TiO_2 photoelectrode-based DSSCs, which is higher than the PCE of benchmark cells with a TiO_2-only electrode (6.31%) [142]. The concentration of CNTs doped into the semiconducting layer plays a critical role in determining the performance of the DSSCs. Grace et al. reported that typical concentrations with and without chemically modified CNTs range from 0.1 to 0.3 wt% and from 0.01 to 0.03 wt%, respectively [143]. Dang et al. achieved an impressive result by first doping the TiO_2 to semiconducting SWCNTs in the DSSCs and achieved a record PCE of 10.6% [144]. The s-SWCNTs with a band gap could transport the electrons from the conduction band to the electrode quickly without recombination. Moreover, the electron diffusion length was also improved after incorporation of the s-SWCNTs. These results indicate that the collection of electrons from dye molecules to the electrode can be more effective if s-SWCNTs are incorporated into the semiconducting layer.

20.6 Summary and Future Outlook

CNTs have many superior structural, electronic, and optical properties that render them suitable for solar cell applications. First, they have been extensively used as transparent conductive electrodes in several types of solar cells because of their excellent conductivity, superior flexibility with low brittleness, and high transmittance over a broad spectral range. Although solar cells with a CNT electrode usually show a slightly lower PCE than those fabricated using convectional electrodes, they are still very promising as transparent electrodes because the properties of the CNTs can be further improved. Second, the CNTs also play the role of an HTL in the photovoltaic devices, especially PSCs, where the CNTs enable the rapid extraction of holes with suppression of recombination and raise the quasi-Fermi level in the perovskite, thereby increasing the driving force for electron injection into the TiO_2 [105]. The CNTs associated with GO or spiro-OMeTAD for both the HTL and the electron blocking layer could further improve the performance. However, the CNT HTL improves the stability of solar cells, which is strongly required for the commercialization of devices. Third, CNTs can be doped into the photoelectrode to improve the electron injection to the TiO_2 and can be doped into the active layer of a solar cell to effectively dissociate the excitons and to collect and transport the carriers to the external circuit. The s-SWCNTs with a bandgap that enables the effective transport of electrons from the conduction band to the electrode quickly improve the diffusion length of the holes. Finally, the photocurrent generation in s-SWCNTs is quite interesting for solar cells. Theoretical estimations suggest that the PCE of CNT *p–n* junction devices will increase with decreasing tube diameter, irrespective of the exciton diffusion length [47]. Moreover, carrier multiplication in the CNTs is, in principle, highly promising for solar cells whose performance exceeds the thermodynamic limitation.

Although CNTs have been intensively studied and have been shown to exhibit numerous superior properties, many problems toward the industrial application of CNTs in solar cells remain to be solved. First, most of the CNTs used in investigations are mixtures of both s- and m-nanotubes, which causes Schottky barriers because of their different bandgaps and work functions, resulting in a high contact resistance. Moreover, in some specific applications, high-quality single-chirality nanotubes are strongly required. Second, most of the synthesis methods of CNTs are energy-intensive, and producing large area CNT films is still expensive. Moreover, the issue of long-term stability in CNT-based devices remains unresolved. In particular, chemically doped CNTs do not maintain adequate long-term performance for use in solar cell applications. Also, UV light could de-dope or cause traps in the CNTs, which would result in high sheet resistance. Thus, the controlled and precise synthesis of high-quality single-chirality CNTs via a low-cost fabrication process will be important not only for solar cells but also for many other devices. Despite the great challenges, the future of CNTs is very promising because of their unique properties for many applications.

References

1. Yu L, Shearer C, Shapter J (2016) Recent development of carbon nanotube transparent conductive films. Chem Rev 116:13413–13453. https://doi.org/10.1021/acs.chemrev.6b00179
2. Li X, Lv Z, Zhu H (2015) Carbon/silicon heterojunction solar cells: state of the art and prospects. Adv Mater 27:6549–6574. https://doi.org/10.1002/adma.201502999
3. Batmunkh M, Macdonald TJ, Shearer CJ, Bat-Erdene M, Wang Y, Biggs MJ, Parkin IP, Nann T, Shapter JG (2017) Carbon nanotubes in TiO_2 nanofiber photoelectrodes for high-performance perovskite solar cells. Adv Sci 4:1600504. https://doi.org/10.1002/advs.201600504
4. Barbero DR, Stranks SD (2016) Functional single-walled carbon nanotubes and nanoengineered networks for organic- and perovskite-solar-cell applications. Adv Mater 28:9668–9685. https://doi.org/10.1002/adma.201600659
5. Hecht DS, Hu L, Irvin G (2011) Emerging transparent electrodes based on thin films of carbon nanotubes, graphene, and metallic nanostructures. Adv Mater 23:1482–1513. https://doi.org/10.1002/adma.201003188
6. Liu Y, Wang S, Peng L-M (2016) Toward high-performance carbon nanotube photovoltaic devices. Adv Energy Mater 6:1600522. https://doi.org/10.1002/aenm.201600522
7. Ratier B, Nunzi J-M, Aldissi M, Kraft TM, Buncel E (2012) Organic solar cell materials and active layer designs—improvements with carbon nanotubes: a review. Polym Int 61:342–354. https://doi.org/10.1002/pi.3233

8. Suzuki K, Yamaguchi M, Kumagai M, Yanagida S (2003) Application of carbon nanotubes to counter electrodes of dye-sensitized solar cells. Chem Lett 32:28–29. https://doi.org/10.1246/cl.2003.28
9. Brown P, Takechi K, Kamat PV (2008) Single-walled carbon nanotube scaffolds for dye-sensitized solar cells. J Phys Chem C 112:4776–4782. https://doi.org/10.1021/jp7107472
10. Chaudhary S, Lu H, Müller AM, Bardeen CJ, Ozkan M (2007) Hierarchical placement and associated optoelectronic impact of carbon nanotubes in polymer-fullerene solar cells. Nano Lett 7:1973–1979. https://doi.org/10.1021/nl070717l
11. Tune DD, Hennrich F, Dehm S, Klein MFG, Glaser K, Colsmann A, Shapter JG, Lemmer U, Kappes MM, Krupke R, Flavel BS (2013) The role of nanotubes in carbon nanotube–silicon solar cells. Adv Energy Mater 3:1091–1097. https://doi.org/10.1002/aenm.201200949
12. Li X, Guard LM, Jiang J, Sakimoto K, Huang J-S, Wu J, Li J, Yu L, Pokhrel R, Brudvig GW, Ismail-Beigi S, Hazari N, Taylor AD (2014) Controlled doping of carbon nanotubes with metallocenes for application in hybrid carbon nanotube/Si solar cells. Nano Lett 14:3388–3394. https://doi.org/10.1021/nl500894h
13. Li Z, Saini V, Dervishi E, Kunets VP, Zhang J, Xu Y, Biris AR, Salamo GJ, Biris AS (2010) Polymer functionalized n-type single wall carbon nanotube photovoltaic devices. Appl Phys Lett 96:033110. https://doi.org/10.1063/1.3284657
14. Wang F, Kozawa D, Miyauchi Y, Hiraoka K, Mouri S, Ohno Y, Matsuda K (2015) Considerably improved photovoltaic performance of carbon nanotube-based solar cells using metal oxide layers. Nat Comun 6:6305. https://doi.org/10.1038/ncomms7305
15. Wei J, Jia Y, Shu Q, Gu Z, Wang K, Zhuang D, Zhang G, Wang Z, Luo J, Cao A, Wu D (2007) Double-walled carbon nanotube solar cells. Nano Lett 7:2317–2321. https://doi.org/10.1021/nl070961c
16. National Revewable Energy Laboratory (NREL) (2016). http://www.nrel.gov/ncpv/images/efficiency_chart.jpg. Accessed Sept 2016
17. Habisreutinger SN, Leijtens T, Eperon GE, Stranks SD, Nicholas RJ, Snaith HJ (2014) Carbon nanotube/polymer composites as a highly stable hole collection layer in perovskite solar cells. Nano Lett 14:5561–5568. https://doi.org/10.1021/nl501982b
18. Wang F, Endo M, Mouri S, Miyauchi Y, Ohno Y, Wakamiya A, Murata Y, Matsuda K (2016) Highly stable perovskite solar cells with an all-carbon hole transport layer. Nanoscale 8:11882–11888. https://doi.org/10.1039/c6nr01152g
19. Jeon I, Chiba T, Delacou C, Guo Y, Kaskela A, Reynaud O, Kauppinen EI, Maruyama S, Matsuo Y (2015) Single-walled carbon nanotube film as electrode in indium-free planar heterojunction perovskite solar cells: investigation of electron-blocking layers and dopants. Nano Lett 15:6665–6671. https://doi.org/10.1021/acs.nanolett.5b02490
20. Zhou C, Wang S, Sun J, Wei N, Yang L, Zhang Z, Liao J, Peng L-M (2013) Plasmonic enhancement of photocurrent in carbon nanotube by Au nanoparticles. Appl Phys Lett 102:103102. https://doi.org/10.1063/1.4794937
21. Freitag M, Martin Y, Misewich JA, Martel R, Avouris P (2003) Photoconductivity of single carbon nanotubes. Nano Lett 3:1067–1071. https://doi.org/10.1021/nl034313e
22. Gabor NM, Zhong Z, Bosnick K, Park J, McEuen PL (2009) Extremely efficient multiple electron-hole pair generation in carbon nanotube photodiodes. Science 325:1367–1371. https://doi.org/10.1126/science.1176112
23. Shockley W, Queisser HJ (1961) Detailed balance limit of efficiency of p-n junction solar cells. J Appl Phys 32:510–519. https://doi.org/10.1063/1.1736034
24. Malapanis A, Perebeinos V, Sinha DP, Comfort E, Lee JU (2013) Quantum efficiency and capture cross section of first and second excitonic transitions of single-walled carbon nanotubes measured through photoconductivity. Nano Lett 13:3531–3538. https://doi.org/10.1021/nl400939b
25. Aspitarte L, McCulley DR, Minot ED (2016) Photocurrent quantum yield in suspended carbon nanotube p–n junctions. Nano Lett 16:5589–5593. https://doi.org/10.1021/acs.nanolett.6b02148

26. Kazaoui S, Cook S, Izard N, Murakami Y, Maruyama S, Minami N (2014) Photocurrent quantum yield of semiconducting carbon nanotubes: dependence on excitation energy and exciton binding energy. J Phys Chem C 118:18059–18063. https://doi.org/10.1021/jp500105f
27. Lee JU (2005) Photovoltaic effect in ideal carbon nanotube diodes. Appl Phys Lett 87:073101. https://doi.org/10.1063/1.2010598
28. Stewart DA, Léonard F (2005) Energy conversion efficiency in nanotube optoelectronics. Nano Lett 5:219–222. https://doi.org/10.1021/nl048410z
29. Pang-Leen O, William BE, Igor AL (2010) Hybrid solar cells based on single-walled carbon nanotubes/Si heterojunctions. Nanotechnology 21:105203. https://doi.org/10.1088/0957
30. Kymakis E, Alexandrou I, Amaratunga GAJ (2003) High open-circuit voltage photovoltaic devices from carbon-nanotube-polymer composites. J Appl Phys 93:1764–1768. https://doi.org/10.1063/1.1535231
31. Dissanayake NM, Zhong Z (2011) Unexpected hole transfer leads to high efficiency single-walled carbon nanotube hybrid photovoltaic. Nano Lett 11:286–290. https://doi.org/10.1021/nl103879b
32. Lan F, Li G (2013) Direct observation of hole transfer from semiconducting polymer to carbon nanotubes. Nano Lett 13:2086–2091. https://doi.org/10.1021/nl400395c
33. Dabera GDMR, Jayawardena KDGI, Prabhath MRR, Yahya I, Tan YY, Nismy NA, Shiozawa H, Sauer M, Ruiz-Soria G, Ayala P, Stolojan V, Adikaari AADT, Jarowski PD, Pichler T, Silva SRP (2013) Hybrid carbon nanotube networks as efficient hole extraction layers for organic photovoltaics. ACS Nano 7:556–565. https://doi.org/10.1021/nn304705t
34. Lee JM, Park JS, Lee SH, Kim H, Yoo S, Kim SO (2011) Selective electron- or hole-transport enhancement in bulk-heterojunction organic solar cells with N- or B-doped carbon nanotubes. Adv Mater 23:629–633. https://doi.org/10.1002/adma.201003296
35. Park S, Vosguerichian M, Bao Z (2013) A review of fabrication and applications of carbon nanotube film-based flexible electronics. Nanoscale 5:1727–1752. https://doi.org/10.1039/c3nr33560g
36. Topinka MA, Rowell MW, Goldhaber-Gordon D, McGehee MD, Hecht DS, Gruner G (2009) Charge transport in interpenetrating networks of semiconducting and metallic carbon nanotubes. Nano Lett 9:1866–1871. https://doi.org/10.1021/nl803849e
37. Harris JM, Semler MR, May S, Fagan JA, Hobbie EK (2015) Nature of record efficiency fluid-processed nanotube–silicon heterojunctions. J Phys Chem C 119:10295–10303. https://doi.org/10.1021/acs.jpcc.5b02626
38. Kanai Y, Grossman JC (2008) Role of semiconducting and metallic tubes in P3HT/carbon-nanotube photovoltaic heterojunctions: density functional theory calculations. Nano Lett 8:908–912. https://doi.org/10.1021/nl0732777
39. Ham M-H, Paulus GLC, Lee CY, Song C, Kalantar-zadeh K, Choi W, Han J-H, Strano MS (2010) Evidence for high-efficiency exciton dissociation at polymer/single-walled carbon nanotube interfaces in planar nano-heterojunction photovoltaics. ACS Nano 4:6251–6259. https://doi.org/10.1021/nn1019384
40. Patyk RL, Lomba BS, Nogueira AF, Furtado CA, Santos AP, Mello RMQ, Micaroni L, Hümmelgen IA (2007) Carbon nanotube–polybithiophene photovoltaic devices with high open-circuit voltage. Phys Status Solidi (RRL) 1:R43–45. https://doi.org/10.1002/pssr.200600057
41. Zhou Y, Hu L, Grüner G (2006) A method of printing carbon nanotube thin films. Appl Phys Lett 88:123109. https://doi.org/10.1063/1.2187945
42. Kim S, Yim J, Wang X, Bradley DDC, Lee S, deMello JC (2010) Spin- and spray-deposited single-walled carbon-nanotube electrodes for organic solar cells. Adv Funct Mater 20:2310–2316. https://doi.org/10.1002/adfm.200902369
43. De S, Lyons PE, Sorel S, Doherty EM, King PJ, Blau WJ, Nirmalraj PN, Boland JJ, Scardaci V, Joimel J, Coleman JN (2009) Transparent, flexible, and highly conductive thin films based on polymer−nanotube composites. ACS Nano 3:714–720. https://doi.org/10.1021/nn800858w

44. Jo JW, Jung JW, Lee JU, Jo WH (2010) Fabrication of highly conductive and transparent thin films from single-walled carbon nanotubes using a new non-ionic surfactant via spin coating. ACS Nano 4:5382–5388. https://doi.org/10.1021/nn1009837
45. Ago H, Kugler T, Cacialli F, Salaneck WR, Shaffer MSP, Windle AH, Friend RH (1999) Work functions and surface functional groups of multiwall carbon nanotubes. J Phys Chem B 103:8116–8121. https://doi.org/10.1021/jp991659y
46. Wu Z, Chen Z, Du X, Logan JM, Sippel J, Nikolou M, Kamaras K, Reynolds JR, Tanner DB, Hebard AF, Rinzler AG (2004) Transparent, conductive carbon nanotube films. Science 305:1273–1276. https://doi.org/10.1126/science.1101243
47. Pasquier AD, Unalan HE, Kanwal A, Miller S, Chhowalla M (2005) Conducting and transparent single-wall carbon nanotube electrodes for polymer-fullerene solar cells. Appl Phys Lett 87:203511. https://doi.org/10.1063/1.2132065
48. Ellmer K (2012) Past achievements and future challenges in the development of optically transparent electrodes. Nat Photon 6:809–817. https://doi.org/nphoton.2012.282/metrics
49. Rowell MW, Topinka MA, McGehee MD, Prall H-J, Dennler G, Sariciftci NS, Hu L, Gruner G (2006) Organic solar cells with carbon nanotube network electrodes. Appl Phys Lett 88:233506. https://doi.org/10.1063/1.2209887
50. Wang F, Kozawa D, Miyauchi Y, Hiraoka K, Mouri S, Matsuda K (2013) Enhancement mechanism of the photovoltaic conversion efficiency of single-walled carbon nanotube/Si solar cells by HNO_3 doping. Appl Phys Express 6:102301. https://doi.org/10.7567/apex.6.102301
51. Ma X, Adamska L, Yamaguchi H, Yalcin SE, Tretiak S, Doorn SK, Htoon H (2014) Electronic structure and chemical nature of oxygen dopant states in carbon nanotubes. ACS Nano 8:10782–10789. https://doi.org/10.1021/nn504553y
52. Jung Y, Li X, Rajan NK, Taylor AD, Reed MA (2013) Record high efficiency single-walled carbon nanotube/silicon p–n junction solar cells. Nano Lett 13:95–99. https://doi.org/10.1021/nl3035652
53. Li Z, Kunets VP, Saini V, Xu Y, Dervishi E, Salamo GJ, Biris AR, Biris AS (2009) Light-harvesting using high density p-type single wall carbon nanotube/n-type silicon heterojunctions. ACS Nano 3:1407–1414. https://doi.org/10.1021/nn900197h
54. Kozawa D, Hiraoka K, Miyauchi Y, Mouri S, Matsuda K (2012) Analysis of the photovoltaic properties of single-walled carbon nanotube/silicon heterojunction solar cells. Appl Phys Express 5:042304. https://doi.org/10.1143/apex.5.042304
55. Wang F, Kozawa D, Miyauchi Y, Hiraoka K, Mouri S, Ohno Y, Matsuda K (2014) Fabrication of single-walled carbon nanotube/Si heterojunction solar cells with high photovoltaic performance. ACS Photonics 1:360–364. https://doi.org/10.1021/ph400133k
56. Wadhwa P, Liu B, McCarthy MA, Wu Z, Rinzler AG (2010) Electronic junction control in a nanotube-semiconductor schottky junction solar cell. Nano Lett 10:5001–5005. https://doi.org/10.1021/nl103128a
57. Wadhwa P, Seol G, Petterson MK, Guo J, Rinzler AG (2011) Electrolyte-induced inversion layer schottky junction solar cells. Nano Lett 11:2419–2423. https://doi.org/10.1021/nl200811z
58. Shi E, Zhang L, Li Z, Li P, Shang Y, Jia Y, Wei J, Wang K, Zhu H, Wu D, Zhang S, Cao A (2012) TiO_2-coated carbon nanotube-silicon solar sells with efficiency of 15%. Sci Rep 2:884. https://doi.org/10.1038/srep00884
59. Cui K, Qian Y, Jeon I, Anisimov A, Matsuo Y, Kauppinen EI, Maruyama S (2017) Scalable and solid-state redox functionalization of transparent single-walled carbon nanotube films for highly efficient and stable solar cells. Adv Energy Mater 7:1700449. https://doi.org/10.1002/aenm.201700449
60. Jia Y, Wei J, Wang K, Cao A, Shu Q, Gui X, Zhu Y, Zhuang D, Zhang G, Ma B, Wang L, Liu W, Wang Z, Luo J, Wu D (2008) Nanotube–silicon heterojunction solar cells. Adv Mater 20:4594–4598. https://doi.org/10.1002/adma.200801810
61. Di J, Yong Z, Zheng X, Sun B, Li Q (2013) Aligned carbon nanotubes for high-efficiency schottky solar cells. Small 9:1367–1372. https://doi.org/10.1002/smll.201202995

62. Muramoto E, Yamasaki Y, Wang F, Hasegawa K, Matsuda K, Noda S (2016) Carbon nanotube-silicon heterojunction solar cells with surface-textured Si and solution-processed carbon nanotube films. RSC Adv 6:93575–93581. https://doi.org/10.1039/c6ra16132d
63. Cui K, Anisimov AS, Chiba T, Fujii S, Kataura H, Nasibulin AG, Chiashi S, Kauppinen EI, Maruyama S (2014) Air-stable high-efficiency solar cells with dry-transferred single-walled carbon nanotube films. J Mater Chem A 2:11311–11318. https://doi.org/10.1039/c4ta01353k
64. De Nicola F, Salvato M, Cirillo C, Crivellari M, Boscardin M, Scarselli M, Nanni F, Cacciotti I, De Crescenzi M, Castrucci P (2016) Record efficiency of air-stable multi-walled carbon nanotube/silicon solar cells. Carbon 101:226–234. https://doi.org/10.1016/j.carbon.2016.01.099
65. Jia Y, Cao A, Bai X, Li Z, Zhang L, Guo N, Wei J, Wang K, Zhu H, Wu D, Ajayan PM (2011) Achieving high efficiency silicon-carbon nanotube heterojunction solar cells by acid doping. Nano Lett 11:1901–1905. https://doi.org/10.1021/nl2002632
66. Li X, Huang J-S, Nejati S, McMillon L, Huang S, Osuji CO, Hazari N, Taylor AD (2014) Role of HF in oxygen removal from carbon nanotubes: implications for high performance carbon electronics. Nano Lett 14:6179–6184. https://doi.org/10.1021/nl502401c
67. Li X, Jung Y, Huang J-S, Goh T, Taylor AD (2014) Device area scale-up and improvement of SWNT/Si solar cells using silver nanowires. Adv Energy Mater 4:1400186. https://doi.org/10.1002/aenm.201400186
68. Yu L, Batmunkh M, Grace T, Dadkhah M, Shearer C, Shapter J (2017) Application of a hole transporting organic interlayer in graphene oxide/single walled carbon nanotube-silicon heterojunction solar cells. J Mater Chem A 5:8624–8634. https://doi.org/10.1039/c7ta01782k
69. Xu W, Wu S, Li X, Zou M, Yang L, Zhang Z, Wei J, Hu S, Li Y, Cao A (2016) High-efficiency large-area carbon nanotube-silicon solar cells. Adv Energy Mater 6:1600095. https://doi.org/10.1002/aenm.201600095
70. Li R, Di J, Yong Z, Sun B, Li Q (2014) Polymethylmethacrylate coating on aligned carbon nanotube-silicon solar cells for performance improvement. J Mater Chem A 2:4140–4143. https://doi.org/10.1039/c3ta14625a
71. Jeong S, Garnett EC, Wang S, Yu Z, Fan S, Brongersma ML, McGehee MD, Cui Y (2012) Hybrid silicon nanocone-polymer solar cells. Nano Lett 12:2971–2976. https://doi.org/10.1021/nl300713x
72. D-m Sun, Timmermans MY, Tian Y, Nasibulin AG, Kauppinen EI, Kishimoto S, Mizutani T, Ohno Y (2011) Flexible high-performance carbon nanotube integrated circuits. Nat Nano 6:156–161. https://doi.org/10.1038/nnano.2011.1
73. Paola C, Claudia S, Del Silvano G, Manuela S, Luca C, Mirko S, Bernard D, Alessandra C, De Maurizio C (2011) Light harvesting with multiwall carbon nanotube/silicon heterojunctions. Nanotechnology 22:115701. https://doi.org/10.1088/0957
74. Grace T, Yu L, Gibson C, Tune D, Alturaif H, Al Othman Z, Shapter J (2016) Investigating the effect of carbon nanotube diameter and wall number in carbon nanotube/silicon heterojunction solar cells. Nanomater 6:52. https://doi.org/10.3390/nano6030052
75. Li Z, Kunets VP, Saini V, Xu Y, Dervishi E, Salamo GJ, Biris AR, Biris AS (2008) SOCl2 enhanced photovoltaic conversion of single wall carbon nanotube/n-silicon heterojunctions. Appl Phys Lett 93:243117. https://doi.org/10.1063/1.3050465
76. De Nicola F, Salvato M, Cirillo C, Crivellari M, Boscardin M, Passacantando M, Nardone M, De Matteis F, Motta N, De Crescenzi M, Castrucci P (2017) 100% internal quantum efficiency in polychiral single-walled carbon nanotube bulk heterojunction/silicon solar cells. Carbon 114:402–410. https://doi.org/10.1016/j.carbon.2016.12.050
77. Li X, Jung Y, Sakimoto K, Goh T-H, Reed MA, Taylor AD (2013) Improved efficiency of smooth and aligned single walled carbon nanotube/silicon hybrid solar cells. Energy Environ Sci 6:879–887. https://doi.org/10.1039/c2ee23716d

78. Kim SM, Kim KK, Jo YW, Park MH, Chae SJ, Duong DL, Yang CW, Kong J, Lee YH (2011) Role of anions in the $AuCl_3$-doping of carbon nanotubes. ACS Nano 5:1236–1242. https://doi.org/10.1021/nn1028532
79. Hellstrom SL, Vosgueritchian M, Stoltenberg RM, Irfan I, Hammock M, Wang YB, Jia C, Guo X, Gao Y, Bao Z (2012) Strong and stable doping of carbon nanotubes and graphene by MoO_x for transparent electrodes. Nano Lett 12:3574–3580. https://doi.org/10.1021/nl301207e
80. Jia Y, Li P, Gui X, Wei J, Wang K, Zhu H, Wu D, Zhang L, Cao A, Xu Y (2011) Encapsulated carbon nanotube-oxide-silicon solar cells with stable 10% efficiency. Appl Phys Lett 98:133115. https://doi.org/10.1063/1.3573829
81. Irwin MD, Buchholz DB, Hains AW, Chang RPH, Marks TJ (2008) p-Type semiconducting nickel oxide as an efficiency-enhancing anode interfacial layer in polymer bulk-heterojunction solar cells. Proc Natl Acad Sci USA 105:2783–2787. https://doi.org/10.1073/pnas.0711990105
82. Yu L, Tune DD, Shearer CJ, Shapter JG (2015) Implementation of antireflection layers for improved efficiency of carbon nanotube–silicon heterojunction solar cells. Sol Energy 118:592–599. https://doi.org/10.1016/j.solener.2015.06.014
83. Dimitrov V, Sakka S (1996) Electronic oxide polarizability and optical basicity of simple oxides. J Appl Phys 79:1736–1740. https://doi.org/10.1063/1.360962
84. Kojima A, Teshima K, Shirai Y, Miyasaka T (2009) Organometal halide perovskites as visible-light sensitizers for photovoltaic cells. J Am Chem Soc 131:6050–6051. https://doi.org/10.1021/ja809598r
85. Im J-H, Jang I-H, Pellet N, Grätzel M, Park N-G (2014) Growth of $CH_3NH_3PbI_3$ cuboids with controlled size for high-efficiency perovskite solar cells. Nat Nanotechnol 9(11):927–932. https://doi.org/10.1038/nnano.2014.181
86. Zhou H, Chen Q, Li G, Luo S, T-b Song, Duan H-S, Hong Z, You J, Liu Y, Yang Y (2014) Interface engineering of highly efficient perovskite solar cells. Science 345:542–546. https://doi.org/10.1126/science.1254050
87. Wang F, Shimazaki A, Yang F, Kanahashi K, Matsuki K, Miyauchi Y, Takenobu T, Wakamiya A, Murata Y, Matsuda K (2017) Highly efficient and stable perovskite solar cells by interfacial engineering using solution-processed polymer layer. J Phys Chem C 121:1562–1568. https://doi.org/10.1021/acs.jpcc.6b12137
88. Jeon NJ, Noh JH, Kim YC, Yang WS, Ryu S, Seok SI (2014) Solvent engineering for high-performance inorganic–organic hybrid perovskite solar cells. Nat Mater 13:897–903. https://doi.org/10.1038/nmat4014
89. Jeon NJ, Noh JH, Yang WS, Kim YC, Ryu S, Seo J, Seok SI (2015) Compositional engineering of perovskite materials for high-performance solar cells. Nature 517:476–480. https://doi.org/10.1038/nature14133
90. Laban WA, Etgar L (2013) Depleted hole conductor-free lead halide iodide heterojunction solar cells. Energy Environ Sci 6:3249–3253. https://doi.org/10.1039/c3ee42282h
91. Chen H, Yang S (2017) Carbon-based perovskite solar cells without hole transport materials: the front runner to the market? Adv Mater 29:1603994. https://doi.org/10.1002/adma.201603994
92. Li H, Cao K, Cui J, Liu S, Qiao X, Shen Y, Wang M (2016) 14.7% efficient mesoscopic perovskite solar cells using single walled carbon nanotubes/carbon composite counter electrodes. Nanoscale 8:6379–6385. https://doi.org/10.1039/c5nr07347b
93. Luo Q, Ma H, Zhang Y, Yin X, Yao Z, Wang N, Li J, Fan S, Jiang K, Lin H (2016) Cross-stacked superaligned carbon nanotube electrodes for efficient hole conductor-free perovskite solar cells. J Mater Chem A 4:5569–5577. https://doi.org/10.1039/c6ta01715k
94. Li Z, Kulkarni SA, Boix PP, Shi E, Cao A, Fu K, Batabyal SK, Zhang J, Xiong Q, Wong LH, Mathews N, Mhaisalkar SG (2014) Laminated carbon nanotube networks for metal electrode-free efficient perovskite solar cells. ACS Nano 8:6797–6804. https://doi.org/10.1021/nn501096h

95. Aitola K, Sveinbjornsson K, Correa-Baena J-P, Kaskela A, Abate A, Tian Y, Johansson EMJ, Gratzel M, Kauppinen EI, Hagfeldt A, Boschloo G (2016) Carbon nanotube-based hybrid hole-transporting material and selective contact for high efficiency perovskite solar cells. Energy Environ Sci 9:461–466. https://doi.org/10.1039/c5ee03394b
96. Gopi CVVM, Venkata-Haritha M, Prabakar K, Kim H-J (2017) Low-temperature easy-processed carbon nanotube contact for high-performance metal- and hole-transporting layer-free perovskite solar cells. J Photochem Photobiol A 332:265–272. https://doi.org/10.1016/j.jphotochem.2016.09.003
97. Aitola K, Domanski K, Correa-Baena J-P, Sveinbjörnsson K, Saliba M, Abate A, Grätzel M, Kauppinen E, Johansson EMJ, Tress W, Hagfeldt A, Boschloo G (2017) High temperature-stable perovskite solar cell based on low-cost carbon nanotube hole contact. Adv Mater 29:1606398. https://doi.org/10.1002/adma.201606398
98. Wang X, Li Z, Xu W, Kulkarni SA, Batabyal SK, Zhang S, Cao A, Wong LH (2015) TiO2 nanotube arrays based flexible perovskite solar cells with transparent carbon nanotube electrode. Nano Energy 11:728–735. https://doi.org/10.1016/j.nanoen.2014.11.042
99. Wei Z, Chen H, Yan K, Zheng X, Yang S (2015) Hysteresis-free multi-walled carbon nanotube-based perovskite solar cells with a high fill factor. J Mater Chem A 3:24226–24231. https://doi.org/10.1039/C5TA07714A
100. Qiu L, He S, Yang J, Jin F, Deng J, Sun H, Cheng X, Guan G, Sun X, Zhao H, Peng H (2016) An all-solid-state fiber-type solar cell achieving 9.49% efficiency. J Mater Chem A 4:10105–10109. https://doi.org/10.1039/c6ta03263j
101. Zheng X, Chen H, Wei Z, Yang Y, Lin H, Yang S (2016) High-performance, stable and low-cost mesoscopic perovskite ($CH_3NH_3PbI_3$) solar cells based on poly(3-hexylthiophene)-modified carbon nanotube cathodes. Front Optoelectron 9:71–80. https://doi.org/10.1021/acsami.5b07703
102. Li Z, Boix PP, Xing G, Fu K, Kulkarni SA, Batabyal SK, Xu W, Cao A, Sum TC, Mathews N, Wong LH (2016) Carbon nanotubes as an efficient hole collector for high voltage methylammonium lead bromide perovskite solar cells. Nanoscale 8:6352–6360. https://doi.org/10.1039/c5nr06177f
103. Habisreutinger SN, Leijtens T, Eperon GE, Stranks SD, Nicholas RJ, Snaith HJ (2014) Enhanced hole extraction in perovskite solar cells through carbon nanotubes. J Phys Chem Lett 5:4207–4212. https://doi.org/10.1021/jz5021795
104. Lee J, Menamparambath MM, Hwang J-Y, Baik S (2015) Hierarchically structured hole transport layers of spiro-OMeTAD and multiwalled carbon nanotubes for perovskite solar cells. Chemsuschem 8:2358–2362. https://doi.org/10.1002/cssc.201403462
105. Ihly R, Dowgiallo A-M, Yang M, Schulz P, Stanton NJ, Reid OG, Ferguson AJ, Zhu K, Berry JJ, Blackburn JL (2016) Efficient charge extraction and slow recombination in organic-inorganic perovskites capped with semiconducting single-walled carbon nanotubes. Energy Environ Sci 9:1439–1449. https://doi.org/10.1039/c5ee03806e
106. Wang J, Li J, Xu X, Bi Z, Xu G, Shen H (2016) Promising photovoltaic application of multi-walled carbon nanotubes in perovskites solar cells for retarding recombination. RSC Adv 6:42413–42420. https://doi.org/10.1039/c6ra04743b
107. Gatti T, Casaluci S, Prato M, Salerno M, Di Stasio F, Ansaldo A, Menna E, Di Carlo A, Bonaccorso F (2016) Boosting perovskite solar cells performance and stability through doping a poly-3(hexylthiophene) hole transporting material with organic functionalized carbon nanostructures. Adv Funct Mater 26:7443–7453. https://doi.org/10.1002/adfm.201602803
108. Miletić T, Pavoni E, Trifiletti V, Rizzo A, Listorti A, Colella S, Armaroli N, Bonifazi D (2016) Covalently functionalized SWCNTs as tailored p-type dopants for perovskite solar cells. ACS Appl Mater Interfaces 8:27966–27973. https://doi.org/10.1021/acsami.6b08398
109. Cai M, Tiong VT, Hreid T, Bell J, Wang H (2015) An efficient hole transport material composite based on poly(3-hexylthiophene) and bamboo-structured carbon nanotubes for high performance perovskite solar cells. J Mater Chem A 3:2784–2793. https://doi.org/10.1039/C4TA04997G

110. Arnold MS, Green AA, Hulvat JF, Stupp SI, Hersam MC (2006) Sorting carbon nanotubes by electronic structure using density differentiation. Nat Nanotechnol 1:60–65. https://doi.org/10.1038/nnano.2006.52
111. Zheng X, Chen H, Li Q, Yang Y, Wei Z, Bai Y, Qiu Y, Zhou D, Wong KS, Yang S (2017) Boron doping of multiwalled carbon nanotubes significantly enhances hole extraction in carbon-based perovskite solar cells. Nano Lett 17:2496–2505. https://doi.org/10.1021/acs.nanolett.7b00200
112. Qiu L, Deng J, Lu X, Yang Z, Peng H (2014) Integrating perovskite solar cells into a flexible fiber. Angew Chem Int Ed 53:10425–10428. https://doi.org/10.1002/anie.201404973
113. Qiu L, He S, Yang J, Deng J, Peng H (2016) Fiber-shaped perovskite solar cells with high power conversion efficiency. Small 12:2419–2424. https://doi.org/10.1002/smll.201600326
114. Yamada Y, Nakamura T, Endo M, Wakamiya A, Kanemitsu Y (2014) Photocarrier recombination dynamics in perovskite $CH_3NH_3PbI_3$ for solar cell applications. J Am Chem Soc 136:11610–11613. https://doi.org/10.1021/ja506624n
115. Wang J, Li J, Xu X, Xu G, Shen H (2016) Enhanced photovoltaic performance with carbon nanotubes incorporating into hole transport materials for perovskite solar cells. J Electron Mater 45:5127–5132. https://doi.org/10.1007/s11664-016-4724-x
116. Batmunkh M, Biggs MJ, Shapter JG (2015) Carbon nanotubes for dye-sensitized solar cells. Small 11:2963–2989. https://doi.org/10.1002/smll.201403155
117. Batmunkh M, Shearer CJ, Bat-Erdene M, Biggs MJ, Shapter JG (2017) Single-walled carbon nanotubes enhance the efficiency and stability of mesoscopic perovskite solar cells. ACS Appl Mater Interfaces 9:19945–19954. https://doi.org/10.1021/acsami.7b04894
118. Ren S, Bernardi M, Lunt RR, Bulovic V, Grossman JC, Gradečak S (2011) Toward efficient carbon nanotube/P3HT solar cells: active layer morphology, electrical, and optical properties. Nano Lett 11:5316–5321. https://doi.org/10.1021/nl202796u
119. Bag M, Renna LA, Jeong SP, Han X, Cutting CL, Maroudas D, Venkataraman D (2016) Evidence for reduced charge recombination in carbon nanotube/perovskite-based active layers. Chem Phys Lett 662:35–41. https://doi.org/10.1016/j.cplett.2016.09.004
120. Zhang Y, Tan L, Fu Q, Chen L, Ji T, Hu X, Chen Y (2016) Enhancing the grain size of organic halide perovskites by sulfonate-carbon nanotube incorporation in high performance perovskite solar cells. Chem Commun 52:5674–5677. https://doi.org/10.1039/c6cc00268d
121. Jvd Lagemaat, Barnes TM, Rumbles G, Shaheen SE, Coutts TJ, Weeks C, Levitsky I, Peltola J, Glatkowski P (2006) Organic solar cells with carbon nanotubes replacing In_2O_3: Sn as the transparent electrode. Appl Phys Lett 88:233503. https://doi.org/10.1063/1.2210081
122. Kazaoui S, Minami N, Nalini B, Kim Y, Hara K (2005) Near-infrared photoconductive and photovoltaic devices using single-wall carbon nanotubes in conductive polymer films. J Appl Phys 98:084314. https://doi.org/10.1063/1.2113419
123. Arranz-Andrés J, Blau WJ (2008) Enhanced device performance using different carbon nanotube types in polymer photovoltaic devices. Carbon 46:2067–2075. https://doi.org/10.1016/j.carbon.2008.08.027
124. Geng J, Zeng T (2006) Influence of single-walled carbon nanotubes induced crystallinity enhancement and morphology change on polymer photovoltaic devices. J Am Chem Soc 128:16827–16833. https://doi.org/10.1021/ja065035z
125. Pradhan B, Batabyal SK, Pal AJ (2006) Functionalized carbon nanotubes in donor/acceptor-type photovoltaic devices. Appl Phys Lett 88:093106. https://doi.org/10.1063/1.2179372
126. Mallajosyula AT, Iyer SSK, Mazhari B (2011) Increasing the efficiency of charge extraction limited poly-(3-hexylthiophene):[6] phenyl C61 butyric acid methyl ester solar cells using single walled carbon nanotubes with metallic characteristics. J Appl Phys 109:124908. https://doi.org/10.1063/1.3598081
127. Bindl DJ, Wu M-Y, Prehn FC, Arnold MS (2011) Efficiently harvesting excitons from electronic type-controlled semiconducting carbon nanotube films. Nano Lett 11:455–460. https://doi.org/10.1021/nl1031343

128. Bindl DJ, Safron NS, Arnold MS (2010) Dissociating excitons photogenerated in semiconducting carbon nanotubes at polymeric photovoltaic heterojunction interfaces. ACS Nano 4:5657–5664. https://doi.org/10.1021/nn1012397
129. Shea MJ, Arnold MS (2013) 1% solar cells derived from ultrathin carbon nanotube photoabsorbing films. Appl Phys Lett 102:243101. https://doi.org/10.1063/1.4811359
130. Gong M, Shastry TA, Xie Y, Bernardi M, Jasion D, Luck KA, Marks TJ, Grossman JC, Ren S, Hersam MC (2014) Polychiral semiconducting carbon nanotube–fullerene solar cells. Nano Lett 14:5308–5314. https://doi.org/10.1021/nl5027452
131. Ihly R, Mistry KS, Ferguson AJ, Clikeman TT, Larson BW, Reid O, Boltalina OV, Strauss SH, Rumbles G, Blackburn JL (2016) Tuning the driving force for exciton dissociation in single-walled carbon nanotube heterojunctions. Nat Chem 8:603–609. https://doi.org/10.1038/nchem.2496
132. Arbab AA, Sun KC, Sahito IA, Memon AA, Choi YS, Jeong SH (2016) Fabrication of textile fabric counter electrodes using activated charcoal doped multi walled carbon nanotube hybrids for dye sensitized solar cells. J Mater Chem A 4:1495–1505. https://doi.org/10.1039/c5ta08858e
133. Liu C-T, Wang Y-C, Dong R-X, Wang C-C, Huang K-C, Vittal R, Ho K-C, Lin J-J (2012) A dual-functional Pt/CNT TCO-free counter electrode for dye-sensitized solar cell. J Mater Chem 22:25311–25315. https://doi.org/10.1039/c2jm35102a
134. Murakami TN, Grätzel M (2008) Counter electrodes for DSC: application of functional materials as catalysts. Inorg Chim Acta 361:572–580. https://doi.org/10.1016/j.ica.2007.09.025
135. Costa RD, Lodermeyer F, Casillas R, Guldi DM (2014) Recent advances in multifunctional nanocarbons used in dye-sensitized solar cells. Energy Environ Sci 7:1281–1296. https://doi.org/10.1039/c3ee43458c
136. Yun S, Hagfeldt A, Ma T (2014) Pt-free counter electrode for dye-sensitized solar cells with high efficiency. Adv Mater 26:6210–6237. https://doi.org/10.1002/adma.201402056
137. Lee WJ, Ramasamy E, Lee DY, Song JS (2009) Efficient dye-sensitized solar cells with catalytic multiwall carbon nanotube counter electrodes. ACS Appl Mater Interfaces 1:1145–1149. https://doi.org/10.1021/am800249k
138. Nam JG, Park YJ, Kim BS, Lee JS (2010) Enhancement of the efficiency of dye-sensitized solar cell by utilizing carbon nanotube counter electrode. Scr Mater 62:148–150. https://doi.org/10.1016/j.scriptamat.2009.10.008
139. Di W, Husnu Emrah U, Dongxue H, Qixian Z, Li N, Gehan A, Tapani R (2008) A solid-state dye-sensitized solar cell based on a novel ionic liquid gel and ZnO nanoparticles on a flexible polymer substrate. Nanotechnology 19:424006. https://doi.org/10.1088/0957
140. Du J, Bittner F, Hecht DS, Ladous C, Ellinger J, Oekermann T, Wark M (2013) A carbon nanotube-based transparent conductive substrate for flexible ZnO dye-sensitized solar cells. Thin Solid Films 531:391–397. https://doi.org/10.1016/j.tsf.2012.12.051
141. Kyaw AKK, Tantang H, Wu T, Ke L, Peh C, Huang ZH, Zeng XT, Demir HV, Zhang Q, Sun XW (2011) Dye-sensitized solar cell with a titanium-oxide-modified carbon nanotube transparent electrode. Appl Phys Lett 99:021107. https://doi.org/10.1063/1.3610488
142. Sawatsuk T, Chindaduang A, Sae-kung C, Pratontep S, Tumcharern G (2009) Dye-sensitized solar cells based on TiO_2–MWCNTs composite electrodes: performance improvement and their mechanisms. Diamond Relat Mater 18:524–527. https://doi.org/10.1016/j.diamond.2008.10.052
143. Peng H, Li Q, Chen T (2017) Industrial applications of carbon nanotubes. In: Grace T, Shearer C, Tune D, Yu L, Batmunkh M, Biggs MJ, ALOthman ZA, Shapter JG (eds) Use of carbon nanotubes in third-generation solar cells, 1st edn. Elsevier
144. Dang X, Yi H, Ham M-H, Qi J, Yun DS, Ladewski R, Strano MS, Hammond PT, Belcher AM (2011) Virus-templated self-assembled single-walled carbon nanotubes for highly efficient electron collection in photovoltaic devices. Nat Nanotechnol 6:377–384. doi:10.1038/nnano.2011.50

Chapter 21
Photon Energy Up-conversion in Carbon Nanotubes

Yuhei Miyauchi

21.1 Introduction

Single-walled carbon nanotubes (SWNTs) [1] (Fig. 21.1) are nanostructures in which graphene is rolled up in a cylindrical shape with a diameter of the order of 1 to several nanometers and a length of several hundred nanometers or more. Their physical properties are determined by their structures. In particular, their band gaps can be in the range from zero (metal) to more than 1 eV (semiconductor), depending on their diameter and the roll-up direction [2]. The structure of a nanotube can be specified by two integers (n, m) called chiral indices; for example, the indices (6, 5) refer to a semiconducting SWNT with a diameter of about 0.8 nm. Semiconducting SWNTs show photoluminescence (PL) that originates from the recombination of excitons [3–5] (hydrogen-like bound states of electrons and holes via Coulomb interaction that are generated via photon absorption). The emission wavelength of semiconducting SWNTs with a diameter of the order of 0.6–1.3 nm is within the near-infrared wavelength range of about 800–1600 nm [5, 6].

Attempts to apply near-infrared PL of SWNTs for luminescent imaging of the deep inside of the living animal's body [7–12] and brain tissues [13] have recently attracted much attention. Using the PL of SWNTs in the near-infrared region of about 1000–1700 nm, which is called NIR-II, the observable depth in biological tissues can be maximized via a combination of relatively small light absorption of water constituting the living body and low light scattering of the biological tissues for the near-infrared light. Thus far, this concept has been demonstrated, for example, by the injection of aqueous dispersion of SWNTs into blood vessels of small animals such as mice [7, 8], which enables the luminescence imaging of the vessels and organs in the deep inside of the small animal's body. A long-term monitoring of nitrogen oxide generation in living mice using the PL of SWNTs has

Y. Miyauchi (✉)
Institute of Advanced Energy, Kyoto University, Kyoto, Japan
e-mail: miyauchi@iae.kyoto-u.ac.jp

N. Nakashima (ed.), *Nanocarbons for Energy Conversion: Supramolecular Approaches*, Nanostructure Science and Technology,
https://doi.org/10.1007/978-3-319-92917-0_21

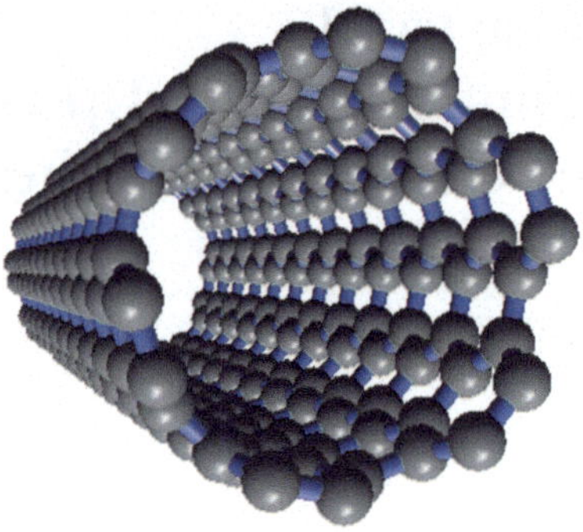

Fig. 21.1 Schematic of a single-walled carbon nanotube

also been proposed [12]. However, near-infrared light with the wavelengths of 1000 nm or more can hardly be detected using conventional silicon-based image sensors. Therefore, these applications require to newly introduce an expensive compound semiconductor-based near-infrared image detector, which has prevented the widespread application of SWNTs as luminescent probes for deep-tissue PL imaging.

21.2 Up-conversion Photoluminescence Phenomenon

PL phenomena found in typical materials follow Stokes' law of luminescence [14] (empirical rule that the wavelength of emission is longer than that of excitation light). In contrast, up-conversion PL (UCPL) [15] (also called anti-Stokes PL) is a luminescence phenomenon that does not follow Stokes' law. In the up-conversion light-emission process, light emission with a wavelength shorter than that of the light irradiated on the material can be obtained. The wavelength and energy of a photon are inversely proportional to each other. Therefore, in a UCPL process, the photons of higher energy are emitted from the material compared to the photon energy that was initially given to it.

Figure 21.2 schematizes the difference between (a) normal Stokes PL and (b) UCPL. In the usual Stokes light-emission process, electrons in the material absorb the energy of the irradiated high-energy (short wavelength) photons and are launched into a high-energy state; then, they release heat to lose energy and finally emit lower energy (long wavelength) photons than the irradiated photons. In contrast, in the case wherein up-conversion luminescence that is capable of obtaining high-energy (short wavelength) luminescence via light irradiation with low-energy (long wavelength) photons exists, electrons in the material absorb low-energy photons and are excited to an intermediate energy level; thereafter, they are excited to a higher energy state (energy up-conversion) by obtaining additional energy via some internal mechanism. Then, the electrons (or excitons) experience the process of releasing energy as light from a higher energy state.

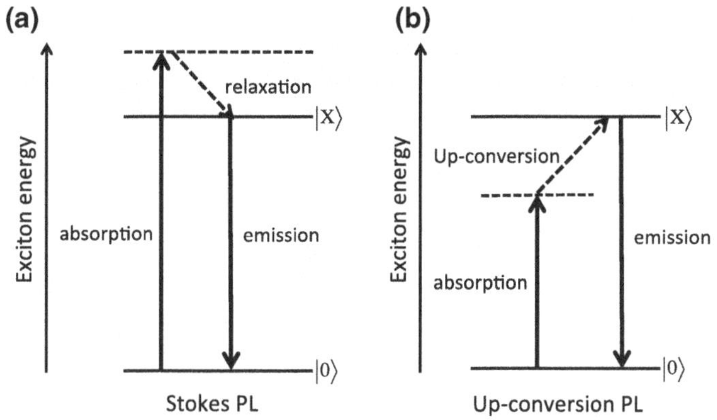

Fig. 21.2 Energy diagram of **a** Stokes PL and **b** up-conversion PL. The horizontal dashed lines represent the intermediate levels that are initially photoexcited in each process. $|X\rangle$ and $|0\rangle$ are the excited states that emit PL and ground states, respectively

Such a unique light-emission phenomenon may seem to be strange at first glance. However, special materials that exhibit outstanding UCPL are equipped with specific mechanisms that enable efficient UCPL. For example, in ceramics doped with rare earth elements, UCPL occurs when two or more photons are absorbed in multiple stages because of the discrete energy level structure of rare earth ions [15]. Some organic dye molecules can emit photons with higher energy than incident photons via a process called triplet–triplet annihilation in which electrons can be excited to a higher energy singlet state via the diffusion-collision of molecules in the excited triplet state [16, 17]. Furthermore, in some semiconductors, thermal up-conversion occurs owing to the excitation of photoexcited electrons to a high-energy state by receiving thermal (phonon) energy in solids [18, 19].

Several unique applications have been proposed for materials exhibiting efficient UCPL. For example, energy harvesting in the unused light wavelength region below the bandgap of silicon in photoelectric conversion [16, 17, 20], biological tissue imaging using the advantage of UCPL with strongly suppressed autofluorescence [21], application of up-conversion multicolor luminescence to color display [22], application to laser [23], and optical cooling of solid using thermally pumped UCPL of semiconductors [18, 24] have been proposed.

21.3 Up-conversion Photoluminescence of Carbon Nanotubes

As already mentioned, the biological tissues have high transparency for the light in the near-infrared wavelength region (wavelength of about 700–1700 nm); if a material that generates efficient UCPL in this region can be realized, it is optimum for biological imaging through the deep tissue of the living body with suppressed autofluorescence. Thus, developing new up-conversion luminescent probes in this wavelength range has been considered to be an important issue [21]. Recently, we found that when light with long wavelength (low energy) of about 1100–1200 nm is irradiated to an aqueous dispersion of (6, 5) SWNTs, UCPL with a short wavelength (high energy) of 950–1000 nm can be efficiently obtained [25]. Figure 21.3 shows the near-infrared UCPL of a nanotube dispersion photographed using a silicon-based electron multiplying (EM) CCD camera. SWNTs dispersed in D_2O using a surfactant were used as the sample. Light with a wavelength of about 1100 nm (photon energy of about 1.13 eV) was irradiated to the dispersion contained in the quartz cuvette, and light with a wavelength of 950–1000 nm (photon energy of about 1.24–1.3 eV) was selected using optical filters and photographed from the front. Although the wavelength of the irradiated light (1100 nm) is sufficiently longer than the light wavelength (950–1000 nm) observed, the light emission of 1000 nm or less could be clearly observed.

Figure 21.4 shows the spectra of Stokes PL (referred to as PL in the figure) [under irradiation with a wavelength of 568 nm (photon energy 2.18 eV)] and UCPL [under light irradiation with a wavelength of 1100 nm (photon energy 1.13 eV)] of the SWNTs dispersion. For comparison, the vertical axis is normalized. The peak of the UCPL feature was observed at almost the same wavelength

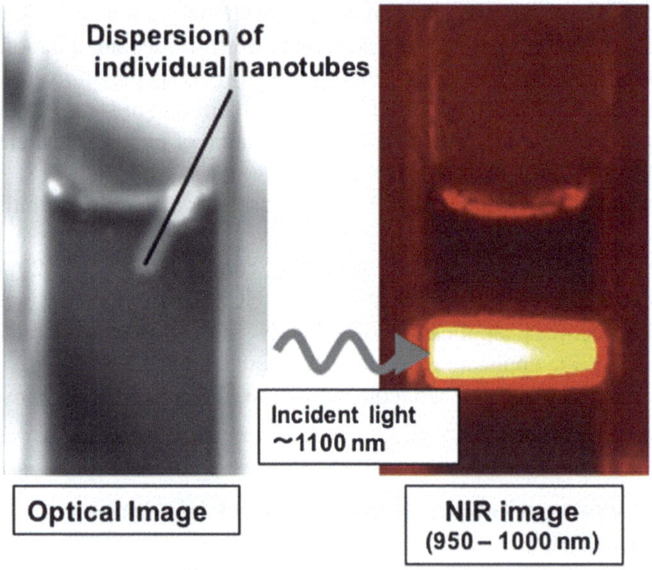

Fig. 21.3 Optical image of aqueous dispersion of SWNTs in a quartz cell (left) and UCPL image (Right). The excitation power density was 200 mW/cm^2. Reproduction from [25] (Licensed under CC BY 4.0) https://creativecommons.org/licenses/by/4.0/

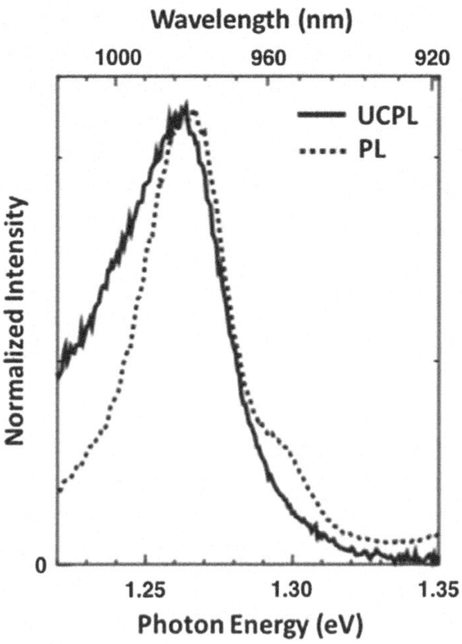

Fig. 21.4 Comparison of Stokes PL (dotted curve) and UCPL spectra (solid curve) of SWNTs dispersion. The photon energy of emission is shown on the lower horizontal axis, and the corresponding optical wavelength is shown on the upper horizontal axis. Reproduction from [25] (Licensed under CC BY 4.0) https://creativecommons.org/licenses/by/4.0/

(photon energy) as in the case of normal PL from (6, 5) SWNTs. Furthermore, recent microscopic spectroscopy observations have confirmed that each individual SWNT emits UCPL [26].

21.4 Mechanism of the Up-conversion

First, the possibilities of multiphoton absorption or anti-Stokes Raman scattering are conceivable as mechanisms for the up-conversion light emission. Therefore, we measured the dependences of excitation energy and excitation strength for verifying the possibility of the above mechanisms [25]. As a result, these possibilities were denied from the following experimental results (i) and (ii): (i) Excitation intensity dependence showing weak saturation behavior was observed. This eliminates the possibility of coherent multiphoton absorption because in the case of multiphoton absorption, excitation intensity dependence should exhibit nonlinear behavior of either second or higher order. (ii) Even when the excitation photon energy changed, the energy of the emission peak remained almost constant. From this result, the possibility of being anti-Stokes Raman scattering is excluded, because the energy difference between emission and excitation photons should be constant in the case of Raman scattering.

If neither multiphoton absorption nor Raman scattering takes place, the next mechanism to be suspected is the possibility of energy up-conversion caused by the absorption of the thermal (phonon) energy. If the energy up-conversion is caused by the thermal (phonon) energy absorption, the up-conversion emission intensity is expected to exhibit characteristic temperature dependence. Therefore, we examined the temperature dependence of the up-conversion emission intensity. Figure 21.5 shows the results of measurements performed at different temperatures. Figure 21.5a shows up-conversion emission spectra measured at the temperatures of 281 and 297 K; even with a slight temperature difference, if the temperature is high, the up-conversion emission intensity increases. Figure 21.5b is a plot of Stokes PL intensity (peak integrated intensity) (I_{PL}) and UCPL intensity (peak integrated intensity) (I_{UCPL}) as functions of temperature, respectively. As shown in Fig. 21.5b, I_{PL} and I_{UCPL} show completely different temperature dependence.

Figure 21.5c shows the semilogarithmic plot of the up-conversion emission intensity normalized by the Stokes emission intensity (I_{UCPL}/I_{PL}) under the same temperature condition as a function of reciprocal temperature (Arrhenius plot). The dotted line in Fig. 21.5c shows the dependency that is well reproduced using the Arrhenius equation [$\propto \exp(-E_a/k_B T)$] when the activation energy E_a is about 120 meV (where k_B and T are Boltzmann constant and temperature, respectively). The experiment in Fig. 21.5 is performed under the excitation energy of 1.13 eV, and the energy gain (difference in excitation and emission photon energy) in UCPL is about 130 meV. Therefore, the result in Fig. 21.5c conforms to the Arrhenius equation for the activation energy E_a, which is almost equal to the energy gain. This strongly suggests that UCPL of SWNTs is enabled by the absorption of thermal energy (phonons). Considering a large energy gain of 120 meV or more, there is a high possibility that optical phonons with high energies more than 100 meV are involved.

One of the remaining mysteries is the identity of the intermediate energy level (the level shown by the horizontal dotted line in Fig. 21.2b), which is the starting point of the energy up-conversion process. The near-infrared photoexcitation energy ($\sim$1.1 eV or less) used for the observation of UCPL is sufficiently lower than the lowest free exciton energy (E_{11} at $\sim$1.26 eV) in (6, 5) SWNTs; however,

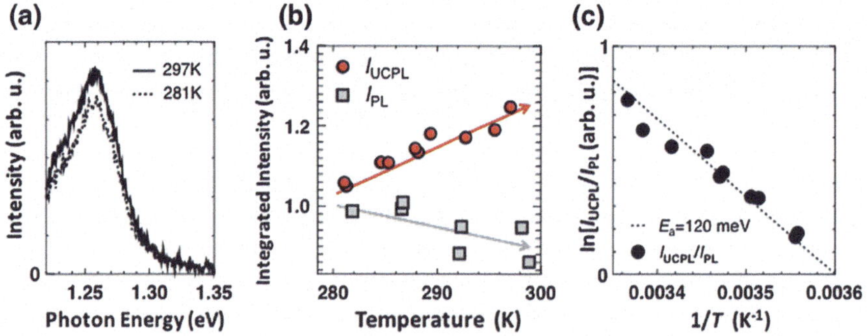

Fig. 21.5 **a** Comparison of UCPL spectra at temperatures of 297 and 281 K. **b** Temperature dependence of UCPL intensity (I_{UCPL}) and Stokes PL intensity (I_{PL}). Arrows are guides to the eye. **c** Arrhenius plot of I_{UCPL}/I_{PL}. The dotted line is a fit according to Arrhenius equation with the activation energy E_a = 120 meV. Reproduction from [25] (Licensed under CC BY 4.0) https://creativecommons.org/licenses/by/4.0/

in this energy region, there is no intrinsic exciton level that can be excited by light. In principle, the anti-Stokes process accompanied by nonresonant excitation to the virtual level and phonon absorption could occur. However, the excitation intensity dependence of the up-conversion emission showed a weak saturation behavior [25], and it is rather suspected that there is a real state that causes absorption saturation because of the state-filling effect.

As a candidate for such a low-energy intermediate level, optically active mid-gap localized states [27–36] that could be introduced by the formation of point defects on the nanotube wall could be considered. Such a localized state is known to be introduced by either oxygen-atom doping [27, 29, 31, 34] or sp^3 defect generation [30, 33, 35, 37]. Although, the SWNTs used in the experiments shown in Figs. 21.3, 21.4, and 21.5 were not subjected to special defect introduction treatment, due to accidental defect generation in the sample preparation process, the possibility of introducing a certain type of mid-gap localized state is conceivable. Therefore, the hypothesis that the localized state originating from the point defect functions as an intermediate state serving as the starting point of the UCPL is worthy of verification.

To verify the aforementioned hypothesis, we intentionally introduced point defects (oxygen-atom dope [27] and sp^3 defect [30]) by applying the chemical treatments to carbon nanotubes; owing to this, changes in UCPL spectra were observed. Figure 21.6a shows the Stokes PL spectrum (2.18 eV excitation) of (6, 5) SWNTs in which localized states were introduced by oxygen-atom doping on the nanotube wall. As shown in Fig. 21.6a, in SWNTs after the localized state introduction, new emission peaks appear around 1.05–1.15 eV because of the formation of optically active localized exciton states. Figure 21.6b shows the UCPL spectrum at the excitation photon energy of 1.08 eV before and after the introduction of the localized states. The UCPL intensity after the introduction of the localized state is clearly larger than that before the introduction. Experiments using various excitation light energies have confirmed that the enhancement rate of the UCPL intensity is the greatest around the excitation photon energy of 1.08 eV [25]. Since this photon energy substantially coincides with the energy of the luminescence peak newly appeared at ~ 1.1 eV after the introduction of the localized state shown in Fig. 21.6a, the result strongly suggests that the localized exciton states function as an intermediate state at the starting point of the energy up-conversion process. The enhancement of the UCPL by intentional defect introduction suggests a possibility of *engineering* the efficiency of UCPL by controlling the density and nature of the localized state that acts as an intermediate state [36].

From the detailed analysis, the quantum efficiency of the up-conversion from the intermediate localized level to the free exciton level has been estimated to be on the order of 10^{-1} at room temperature for an energy gain of 120 meV [25]. However, it is still unclear as to how such a large energy upward conversion of about five times or more than the energy of the room temperature (about 26 meV) is possible with such efficiency under room temperature conditions. With regard to this question, we have reported that a quantitatively consistent explanation is possible by considering a model incorporating the one-dimensional diffusion motion of excitons on an

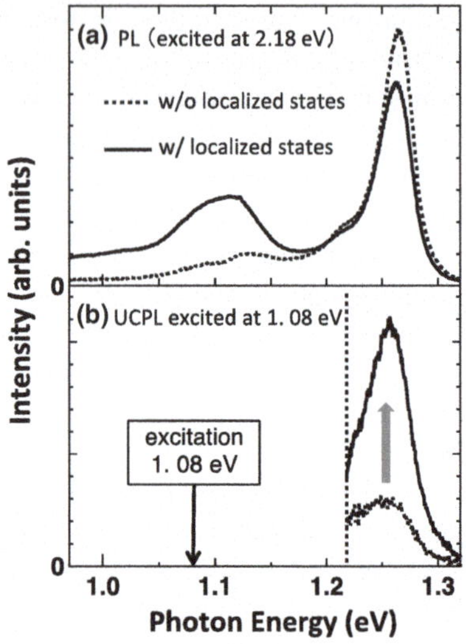

Fig. 21.6 **a** Comparison of Stokes PL spectra before and after introduction of localized state by oxygen-doping procedure. Excitation photon energy was 2.18 eV. **b** Comparison of UCPL spectra before and after the introduction of the localized states. The excitation photon energy was 1.08 eV. Reproduction from [25] (Licensed under CC BY 4.0) https://creativecommons.org/licenses/by/4.0/

SWNT [25]. Although details are omitted herein, after the excitons are photogenerated at the localized intermediate level, they are up-converted to the one-dimensional free exciton state by the absorption of thermal energy (phonons), and low-energy acoustic phonon scattering immediately induces rapid diffusive exciton migration along the nanotube axis direction, and most of the excitons that are up-converted can be immediately transported to positions spatially separated from the localized state (Fig. 21.7). Once the excitons are up-converted to the free exciton states, the possibility of these excitons returning to the original localized state can be reduced. This unique exciton dynamics may significantly suppress the process of returning the excitons to the initial localized state; as a consequence,

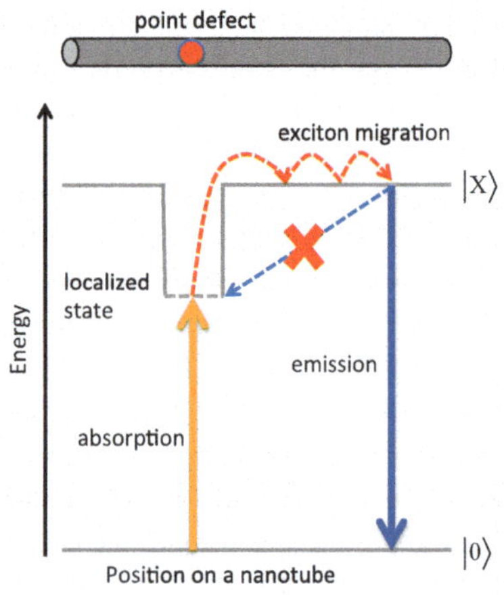

Fig. 21.7 Schematic of the photoexcitation at a localized state induced by a point defect, exciton up-conversion to the intrinsic free exciton state, its migration along the nanotube axis, and light emission from the free exciton state at the higher energy than the excitation photon energy

exciton energy up-conversion that jumps over the energy difference more than five times of the thermal energy of the room temperature can be realized with relatively high efficiency.

21.5 Outlook

At the end of this chapter, the applicability of the UCPL of SWNTs will be discussed by focusing on the results that we have examined so far. With regard to the application of luminescence of SWNTs, two major directions have been proposed. One is the application as a light source in near-infrared optoelectronics [6, 38–42], whereas the other application corresponds to biosensing or imaging [7–13, 43]. We expect that the discovery of the efficient UCPL of SWNTs has an important impact, particularly on the latter. By utilizing the PL of SWNTs in the NIR-II biological window under irradiation with excitation light at a wavelength of about 800 nm (photon energy of about 1.55 eV), successful PL imaging of blood vessels and organs in deep tissues have been reported [7, 8, 11]. However, for photoexcitation at around 800 nm, autofluorescence of living tissues cannot be avoided up to about 1050 nm. Thus, for high contrast observation, luminescence signals in the wavelength range shorter than $\sim$1100 nm must be removed using optical filters.

However, as described above, to perform PL imaging in the long wavelength region exceeding 1100 nm, image sensors based on a widely used silicon semiconductor (such as CCD and CMOS sensors) cannot be used; an expensive compound semiconductor-based detector should be introduced to perform PL imaging. Using UCPL of SWNTs, it is possible to swap the wavelength regions of excitation and emission light for avoiding the above problem. For example, using an inexpensive continuous wave solid-state laser with a wavelength of 1064 nm as a light source (since UCPL of SWNTs is a one-photon process, an expensive ultrashort pulse laser as commonly used in multiphoton microscopy is unnecessary), a silicon-based camera can be used for the image acquisition because UCPL in SWNTs (for example, (6, 5) nanotubes) has a wavelength of free exciton emission of less than 1000 nm. Using SWNTs with smaller diameters such as (5, 4) nanotubes [44], it is expected that UCPL can also be generated at the wavelength of around 800 nm with excitation light in the wavelength range of $\sim$850–1000 nm; even a standard photomultiplier tube can be used to detect UCPL for this case.

Herein, we demonstrate the results of UCPL imaging of SWNTs embedded in a mock tissue [25] to confirm the potential applicability of the UCPL for the deep-tissue imaging. Figure 21.8a shows an image of a sample in which the SWNTs dispersion is confined in a “7” shape (a region surrounded by dotted lines) in a transparent resin spread over the bottom of a glass dish. In addition, as shown in Fig. 21.8b, a white mock tissue (1% Intralipid® [8] with light scattering parameters similar to those of a biological tissue) is placed on the surface to embed the SWNTs inside the mock tissue.

In the visible image (Fig. 21.8b), the mock tissue is completely opaque, and the nanotube dispersion sample embedded behind it or the background honeycomb grid

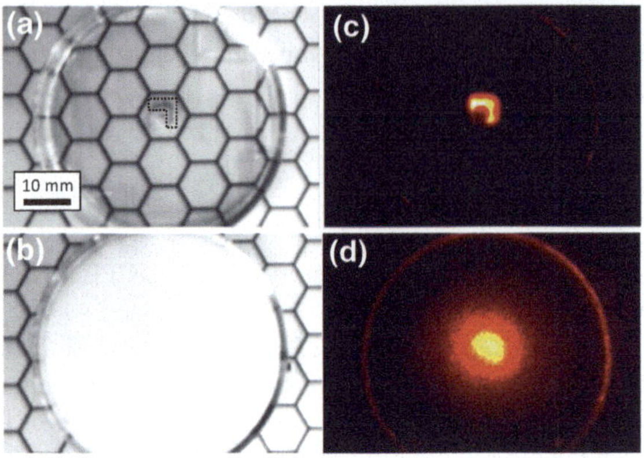

Fig. 21.8 **a** Optical image of a nanotube dispersion (a region surrounded by a black dotted line) encapsulated in a resin spread on the bottom of a transparent glass dish. **b** An image of the glass dish covered with light scatterers imitating living tissue. **c** A UCPL image of the sample (without light scatterer) in (**a**). **d** UCPL image of SWNTs dispersion placed behind light scatterer with thickness of 4 mm. The wavelength of the incident light in (**c**) and (**d**) was 1064 nm, photographed using a silicon-based EMCCD camera. The excitation light power density was about 6 mW/cm^2. Reproduction from [25] (Licensed under CC BY 4.0) https://creativecommons.org/licenses/by/4.0/

pattern (printed on a paper for comparison) could not be observed. Figure 21.8c shows a UCPL image of the sample before it was covered by the mock tissue. Here, incident light with a wavelength of 1064 nm (photon energy: 1.165 eV) was irradiated, and an emission image with a wavelength of 950–1000 nm was acquired using a silicon EMCCD camera. The UCPL from SWNTs that is placed in the center of the glass dish is clearly observed ("7" shape). Figure 21.8d shows an up-conversion emission image after coating with the 4-mm-thick mock tissue, and the figure shows the UCPL from the SWNTs embedded behind the mock tissue. These results suggest the usefulness of the UCPL of SWNTs in imaging deep inside light scatterers such as a living body.

21.6 Summary and Remaining Issues

In this chapter, the recent discovery of efficient UCPL phenomenon in SWNTs, its mechanism, and its applicability to deep-tissue bioimaging were outlined. The UCPL is a new optical function of SWNTs triggered by the existence of defect-induced localized states that lead to conversion of thermal energy to exciton energy, which consequently enable photon energy up-conversion in SWNTs. Although the basic mechanism of the UCPL phenomena in SWNTs has been proposed and confirmed as outlined in this chapter, the precise value of the up-conversion quantum yield and its dependence on the nanotube structure still remain to be clarified. An optimal engineering method to enhance the efficiency of the UCPL also remains to be developed. Based on the various findings introduced in this chapter, the next challenge will be to further improve the efficiency of the UCPL of SWNTs and to apply it to deep-tissue imaging of actual biological targets.

Acknowledgements This work was supported in part by JSPS KAKENHI Grant Numbers JP24681031, JP15H05408, JP15K13337, JP16H00911, JP17K19055, by JST PRESTO, by JST CREST (JPMJCR16F3), by the Research Foundation for Opto-Science and Technology, and by the Nakatani Foundation.

References

1. Iijima S, Ichihashi T (1993) Single-shell carbon nanotubes of 1-nm diameter. Nature 363 (6430):603–605
2. Saito R, Dresselhaus G, Dresselhaus MS (1998) Physical properties of carbon nanotubes. Imperial College Press, London
3. Ando T (1997) Excitons in carbon nanotubes. J Phys Soc Jpn 66(4):1066–1073
4. Wang F, Dukovic G, Brus LE, Heinz TF (2005) The optical resonances in carbon nanotubes arise from excitons. Science 308(5723):838–841
5. O'Connell MJ, Bachilo SM, Huffman CB, Moore VC, Strano MS, Haroz EH, Rialon KL, Boul PJ, Noon WH, Kittrell C, Ma J, Hauge RH, Weisman RB, Smalley RE (2002) Band gap fluorescence from individual single-walled carbon nanotubes. Science 297(5581):593–596
6. Miyauchi Y (2013) Photoluminescence studies on exciton photophysics in carbon nanotubes. J Mater Chem C 1(40):6499–6521
7. Welsher K, Liu Z, Sherlock SP, Robinson JT, Chen Z, Daranciang D, Dai H (2009) A route to brightly fluorescent carbon nanotubes for near-infrared imaging in mice. Nat Nanotechnol 4 (11):773–780
8. Welsher K, Sherlock SP, Dai H (2011) Deep-tissue anatomical imaging of mice using carbon nanotube fluorophores in the second near-infrared window. Proc Natl Acad Sci USA 108 (22):8943–8948
9. Yomogida Y, Tanaka T, Zhang M, Yudasaka M, Wei X, Kataura H (2016) Industrial-scale separation of high-purity single-chirality single-wall carbon nanotubes for biological imaging. Nat Commun 7:12056
10. Yudasaka M, Yomogida Y, Zhang M, Tanaka T, Nakahara M, Kobayashi N, Okamatsu-Ogura Y, Machida K, Ishihara K, Saeki K, Kataura H (2017) Near-infrared photoluminescent carbon nanotubes for imaging of brown fat. Sci Rep 7:44760
11. Hong G, Diao S, Chang J, Antaris AL, Chen C, Zhang B, Zhao S, Atochin DN, Huang PL, Andreasson KI, Kuo CJ, Dai H (2014) Through-skull fluorescence imaging of the brain in a new near-infrared window. Nat Photonics 8(9):723–730
12. Iverson NM, Barone PW, Shandell M, Trudel LJ, Sen S, Sen F, Ivanov V, Atolia E, Farias E, McNicholas TP, Reuel N, Parry NMA, Wogan GN, Strano MS (2013) In vivo biosensing via tissue-localizable near-infrared-fluorescent single-walled carbon nanotubes. Nat Nanotechnol 8(11):873–880
13. Godin AG, Varela JA, Gao Z, Danné N, Dupuis JP, Lounis B, Groc L, Cognet L (2017) Single-nanotube tracking reveals the nanoscale organization of the extracellular space in the live brain. Nat Nanotechnol 12(3):238–243
14. Stokes GG (1852) On the change of refrangibility of light. Philos Trans R Soc Lond 142: 463–562
15. Auzel F (2004) Up-conversion and anti-stokes processes with f and d ions in solids. Chem Rev 104(1):139–174

16. Baluschev S, Miteva T, Yakutkin V, Nelles G, Yasuda A, Wegner G (2006) Up-conversion fluorescence: noncoherent excitation by sunlight. Phys Rev Lett 97(14):143903
17. Zhao J, Ji S, Guo H (2011) Triplet-triplet annihilation based up-conversion: from triplet sensitizers and triplet acceptors to up-conversion quantum yields. RSC Adv 1(6):937–950
18. Sheik-Bahae M, Epstein RI (2007) Optical refrigeration. Nat Photonics 1(12):693–699
19. Rakovich YP, Donegan JF (2008) Anti-stokes photoluminescence in semiconductor nanocrystal quantum dots. In: Rogach AL (ed) Semiconductor nanocrystal quantum dots. Springer, Vienna, pp 257–275. https://doi.org/10.1007/978-3-211-75237-1_9
20. Murakami Y (2011) Photochemical photon up-converters with ionic liquids. Chem Phys Lett 516(1):56–61
21. Dou QQ, Guo HC, Ye E (2014) Near-infrared up-conversion nanoparticles for bio-applications. Mater Sci Eng C 45:635–643
22. Downing E, Hesselink L, Ralston J, Macfarlane R (1996) A three-color, solid-state, three-dimensional display. Science 273(5279):1185–1189
23. He GS, Markowicz PP, Lin T-C, Prasad PN (2002) Observation of stimulated emission by direct three-photon excitation. Nature 415(6873):767–770
24. Epstein RI, Buchwald MI, Edwards BC, Gosnell TR, Mungan CE (1995) Observation of laser-induced fluorescent cooling of a solid. Nature 377(6549):500–503
25. Akizuki N, Aota S, Mouri S, Matsuda K, Miyauchi Y (2015) Efficient near-infrared up-conversion photoluminescence in carbon nanotubes. Nat Commun 6:8920
26. Aota S, Akizuki N, Mouri S, Matsuda K, Miyauchi Y (2016) Up-conversion photoluminescence imaging and spectroscopy of individual single-walled carbon nanotubes. Appl Phys Express 9(4):045103
27. Ghosh S, Bachilo SM, Simonette RA, Beckingham KM, Weisman RB (2010) Oxygen doping modifies near-infrared band gaps in fluorescent single-walled carbon nanotubes. Science 330 (6011):1656–1659
28. Kilina S, Ramirez J, Tretiak S (2012) Brightening of the lowest exciton in carbon nanotubes via chemical functionalization. Nano Lett 12(5):2306–2312
29. Miyauchi Y, Iwamura M, Mouri S, Kawazoe T, Ohtsu M, Matsuda K (2013) Brightening of excitons in carbon nanotubes on dimensionality modification. Nat Photonics 7(9):715–719
30. Piao Y, Meany B, Powell LR, Valley N, Kwon H, Schatz GC, Wang Y (2013) Brightening of carbon nanotube photoluminescence through the incorporation of sp^3 defects. Nat Chem 5 (10):840–845
31. Ma X, Adamska L, Yamaguchi H, Yalcin SE, Tretiak S, Doorn SK, Htoon H (2014) Electronic structure and chemical nature of oxygen dopant states in carbon nanotubes. ACS Nano 8(10):10782–10789
32. Iwamura M, Akizuki N, Miyauchi Y, Mouri S, Shaver J, Gao Z, Cognet L, Lounis B, Matsuda K (2014) Nonlinear photoluminescence spectroscopy of carbon nanotubes with localized exciton states. ACS Nano 8(11):11254–11260
33. Maeda Y, Takehana Y, Yamada M, Suzuki M, Murakami T (2015) Control of the photoluminescence properties of single-walled carbon nanotubes by alkylation and subsequent thermal treatment. Chem Commun 51(70):13462–13465
34. Shiraishi T, Juhász G, Shiraki T, Akizuki N, Miyauchi Y, Matsuda K, Nakashima N (2016) Determination of precise redox properties of oxygen-doped single-walled carbon nanotubes based on in situ photoluminescence electrochemistry. J Phys Chem C 120(29):15632–15639
35. Shiraki T, Shiraishi T, Juhász G, Nakashima N (2016) Emergence of new red-shifted carbon nanotube photoluminescence based on proximal doped-site design. Sci Rep 6:28393
36. Maeda Y, Minami S, Takehana Y, Dang J-S, Aota S, Matsuda K, Miyauchi Y, Yamada M, Suzuki M, Zhao R-S, Zhao X, Nagase S (2016) Tuning of the photoluminescence and up-conversion photoluminescence properties of single-walled carbon nanotubes by chemical functionalization. Nanoscale 8(38):16916–16921
37. Kim M, Adamska L, Hartmann NF, Kwon H, Liu J, Velizhanin KA, Piao Y, Powell LR, Meany B, Doorn SK, Tretiak S, Wang Y (2016) Fluorescent carbon nanotube defects manifest substantial vibrational reorganization. J Phys Chem C 120(20):11268–11276

38. Avouris P, Freitag M, Perebeinos V (2008) Carbon-nanotube photonics and optoelectronics. Nat Photonics 2(6):341–350
39. Endo T, Ishi-Hayase J, Maki H (2015) Photon antibunching in single-walled carbon nanotubes at telecommunication wavelengths and room temperature. Appl Phys Lett 106 (11):113106
40. Jiang M, Kumamoto Y, Ishii A, Yoshida M, Shimada T, Kato YK (2015) Gate-controlled generation of optical pulse trains using individual carbon nanotubes. Nat Commun 6:6335
41. Ma X, Hartmann NF, BaldwinJon KS, Doorn SK, Htoon H (2015) Room-temperature single-photon generation from solitary dopants of carbon nanotubes. Nat Nanotechnol 10 (8):671–675
42. Ishii A, Uda T, Kato YK (2017) Room-temperature single-photon emission from micrometer-long air-suspended carbon nanotubes. Phys Rev Appl 8(5):054039
43. Heller DA, Jeng ES, Yeung T-K, Martinez BM, Moll AE, Gastala JB, Strano MS (2006) Optical detection of DNA conformational polymorphism on single-walled carbon nanotubes. Science 311(5760):508–511
44. Wei X, Tanaka T, Akizuki N, Miyauchi Y, Matsuda K, Ohfuchi M, Kataura H (2016) Single-chirality separation and optical properties of (5,4) single-wall carbon nanotubes. J Phys Chem C 120(19):10705–10710

Chapter 22
Carbon Nanotube-Based Thermoelectric Devices

Tsuyohiko Fujigaya

22.1 Introduction

Thermoelectric (TE) conversion is one of the key technologies to realize a sustainable society since large quantities of energy [1, 2] have been wasted as heat [3]; therefore recovery of heat into electricity via TE technology is quite attractive. A number of semiconducting materials including inorganic [4, 5] and organic materials [6–8] function as TE materials that directly convert heat into electricity via a carrier movement in the materials based on the Seebeck effect. Especially, inorganic materials such as bismuth telluride (Bi_2Te_3), Si/Ge, and PbTe have been intensively studied due to their high power factors that originated from the large Seebeck coefficients ranging from 100 to 45,000 μV K^{-1} [4, 5] and high carrier conductivity. Recently, TE system is also attracting strong interests for flexible batteries with increasing needs for the wearable devices using our body as a heat source [9–15]. For such applications, especially, transition metal dicharcogenite [16], conducting polymers [6–8, 17–21] including poly(3,4-ethylenedioxythiphene) (PEDOT) doped with poly(styrenesulfonate) (PSS) [26] and acid-doped polyanilines (PANi), and carbon materials such as carbon nanotubes (CNTs) [22–25] together with their composites [26–30] are extensively studied due to their flexibility, lightness, nontoxicity to our skin, material abundance, production scalability, and so on. Since the conversion efficiency of TE materials can be evaluated by the dimensionless figure of merit, $ZT = S^2\sigma T/\kappa$, where S, σ, T, and κ denote the Seebeck coefficient, electrical conductivity, absolute temperature, and thermal conductivity, respectively [6, 7], the materials with a high Seebeck coefficient, high electrical conductivity, and low thermal conductivity are desired for an effective TE conversion [31]. In the conventional inorganic TE materials such as Bi_2Te_3, ZT

T. Fujigaya (✉)
Department of Applied Chemistry, Kyushu University, 744 Motooka, Fukuoka 819-0395, Japan
e-mail: fujigaya.tsuyohiko.948@m.kyushu-u.ac.jp

N. Nakashima (ed.), *Nanocarbons for Energy Conversion: Supramolecular Approaches*, Nanostructure Science and Technology,
https://doi.org/10.1007/978-3-319-92917-0_22

value often exceeds 1 near room temperature [32]. Among these candidates, CNTs are considered as the most promising materials primally because of their high electrical conductivity (10^4–10^5 S/cm) even without doping. Although thermal conductivity of single CNT is quite high (more than 3000 W/m K), the films made of many CNTs, so-called buckypapers, have a rather low thermal conductivity due to the thermal resistance at the CNT–CNT contact, thus the CNT sheets have been used for the TE studies.

In the CNT family including single-walled CNTs (SWNTs), especially semiconducting-type SWNTs, are quite promising due to their large Seebeck coefficient [33–36]. Indeed, Maniwa et al. reported that a sorted semiconducting SWNT film shows a very large Seebeck coefficient (170 $\mu V\ K^{-1}$ at 300 K) [37]. So far, most of the reported polymer-based and CNT-based TE materials have a positive Seebeck coefficient, i.e., p-type TE materials, while fewer reports have been published on the n-type TE materials since the n-type polymer- and CNT-based materials are usually unstable and easily react with oxygen and/or water in air. The CNTs shows p-type nature in atmospheric condition because of oxygen-doping and, therefore, needs electron doping with electron donors. Up to date, electron donors such as alkali metal [38], hydrazine [39–41], and sodium borohydride ($NaBH_4$) [42, 43], polyethyleneimine (PEI) [44–48], coenzyme [49], viologen [50], 1*H*-benzoimidazole derivatives [51], triphenylphosphine [17], amino-substituted Rylene dimide [52], n-type Bi_2Te_3 [53], and silver telluride nanowires [54] were reported to form the n-type CNTs [55, 56]. However, air stability of the n-type SWNTs had been insufficient and was an obstacle to develop SWNT-based TE devices [57–59]. In order to solve such a problem, several approaches including polymer wrapping [42, 60] and deposition of inorganic layer [59] were tried but the stability was only shorter than a month under atmospheric condition.

22.2 N-Doping of SWNT by Cobaltocene and Benzimidazole Derivative

In our study, we chose two types of dopants, cobaltocene ($CoCp_2$), Fig. 22.1a) [61, 62] and 2-aryl-1,3-dimethyl-2,3-dihydro-1*H*-benzo[*d*]imidazole (*o*-MeO-DMBI, Fig. 22.1b). We chose $CoCp_2$ since it was reported that the stability of the molecules was improved by encapsulation inside SWNTs [63] and *o*-MeO-DMBI was chosen since n-doping of the SWNT by *o*-MeO-DMBI has already been reported [64–67]. The $CoCp_2$@SWNTs were synthesized according to a previous method [61], the non-doped SWNT film was dipped in an ethanol solution of *o*-MeO-DMBI (50.0 mM) for 10 min, followed by vacuum drying at room temperature for 12 h. The encapsulation of $CoCp_2$ and the doping of *o*-MeO-DMBI were confirmed by X-ray photoelectron spectroscopy (XPS) measurements. We clearly observed peaks at around 780 and 795 eV assignable to Co^{3+}, while no such peak was recognized in the empty SWNTs. For *o*-MeO-DMBI-doped, a peak at 401.5 eV ascribable to N^+ [68] was detected in N 1s region, indicating the *o*-MeO-DMBI formed cation species after the SWNT doping as proposed by Bao et al. [29]. The non-doped SWNT film showed a positive (45.3 $\mu V\ K^{-1}$ at 320 K) Seebeck coefficient due to the p-type nature of the pristine SWNT film under atmospheric conditions [69]. On the other hand, the $CoCp_2$@SWNT film and *o*-MeO-DMBI-doped SWNT film showed a negative Seebeck coefficient of ca. −41.8 and −43.0 $\mu V\ K^{-1}$ at 320 K, respectively, which was maintained in the temperature range of 30–90 °C, indicating the successful n-doping of the film. On the other hand, electrical conductivity was increased to 43,200 $S\ m^{-1}$ and 62,000 $S\ m^{-1}$ for $CoCp_2$@SWNT and *o*-MeO-DMBI SWNT, respectively, at 30 °C These values allowed us to calculate the power factor ($S^2\sigma$ [$W\ m^{-1}\ K^{-2}$]). The power factor of the $CoCp_2$@SWNT and *o*-MeO-DMBI SWNT were 75.4 and 129.0 $\mu W\ m^{-1}\ K^{-2}$ at 320 K, which is much larger than that of the SWNT film due to its higher electrical conductivity than that of the empty SWNT film, and the value was larger than any other reported n-type SWNT-based materials.

22.3 Air Stability of N-doped SWNT Films

In order to evaluate the air stability of the n-type nature, the Seebeck coefficient of the film was monitored. As plotted in Fig. 22.2 (red dots), the Seebeck coefficient of the *o*-MeO-DMBI-doped SWNTs shows a constant negative value of ca. −43 $\mu V\ K^{-1}$ during the stability test [70], while that of the $CoCp_2$@SWNT gradually turned to p-type Fig. 22.2 (blue dots) [71]. We assumed that remarkable air stability of *o*-MeO-DMBI-doped SWNT film was due to the stable coverage of the negatively charged n-doped SWNT by the air-stable *o*-MeO-DMBI cation through a strong Coulomb interaction to suppress the access of oxygen. On the other hand, in the case of $CoCp_2$@SWNT, oxygen molecules can be adsorbed on the SWNT surface to oxidize the n-doped SWNT gradually. DMBI derivatives are known to act as a reducing agent via either hydride, hydrogen atom + 1 electron or 2 electron + proton transfer pathways. In addition, DMBI derivatives themselves are stable under atmospheric conditions and the cationic form of DMBI derivatives after the above transfers are also stable under atmospheric conditions [64]. Owing to their unique reactivity and stability, DMBI derivatives have been used as a reducing agent for various organic syntheses [72, 73] such as hydrogen evolution [74], artificial photosynthesis [75], organic semiconductor [76], and solar cell [77].

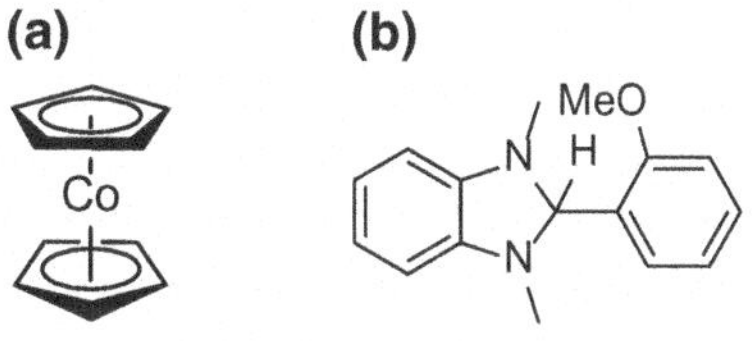

Fig. 22.1 Chemical structures of **a** $CoCp_2$ and **b** *o*-MeO-DMBI

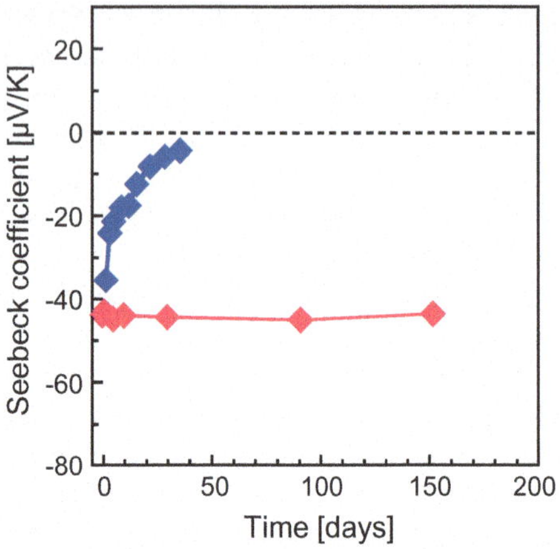

Fig. 22.2 Air stability of Seebeck coefficient of *o*-MeO-DMBI-doped SWNT (red) and $CoCp_2$@SWNT (blue) films

We believe that the DMBI derivatives are promising candidates for SWNT TE application.

Before our report, Nonoguchi and Kawai et al. reported the pioneering work of the air-stable SWNTs based on the doping by salt anions, such as Cl^-, BH_4^-, and OH^-, in which the counter cations were stabilized by a tetraalkylammonium (R4 N^+) or crown ether [78]. They pointed to the importance of the size of the counter cation to compensate the negatively charged SWNTs to realize a good air stability [78]. They found that KOH-doped SWNT stabilized by benzo-18-crown ether provided stable n-type nature even under 100 °C for over 600 h. In this case, hydroxide donated electron to SWNTs to provide negative charge to the SWNT and this negative charge was then stabilized with the cationic crown ether complex through electrostatic attraction. Therefore, it is considered that larger crown ether complexes likely block adsorption of atmospheric O_2 effectively improving the n-type lifetime of the CNTs either through electrostatic rejection by the oxygen in the crown ether or by simple physical blocking. After our report [70], they also reported air-stable n-type SWNT (over an 800 h) using triphenylmethane carbinol base (TPM-CB) generated from malachite green. Although malachite green also possesses a positive charge that is able to resonate throughout its structure, they demonstrated the advantage of the TPM-CB over regular malachite green. It was found that while malachite green did partially n-type dope the SWNTs (only 2.5 $\mu V\ K^{-1}$), doping with TPM-CB drastically altered the Seebeck coefficient to $-59\ \mu V\ K^{-1}$. In the pH-controlled basic solution, insoluble TPM-CB nanoparticles were formed and the stability of the cationic TPM-CB nanoparticles contributed to the n-type air stability [79].

Finally, in order to evaluate the in-plane *ZT* value (*ZT*) of the n-doped SWNTs, we determined the in-plane thermal conductivity ($\kappa = C_p \alpha \rho$) by measuring the in-plane thermal diffusivity (α), specific heat capacity (C_p), and density (ρ). The thermal diffusivity of organic TE materials has been measured in the through-plane direction, since the precise measurement of the thermal conductivity of thin films in the in-plane direction is not easy, and ZT of the organic film has been traditionally calculated using the Seebeck coefficient and electrical conductivity in an in-plane direction and the thermal conductivity in a through-plane direction. Recently, it was pointed out that conducting polymer such as PEDOT–PSS films [38] and SWNT films are known to show a large anisotropy of the thermal conductivity and electrical conductivity between in-plane and through-plane directions, sometimes over 100 times [80]. Therefore, traditional calculation contained overestimation of the ZT value. Indeed, we reported the κ values of through-plane direction as 0.15 W m^{-1} K^{-1} for $CoCp_2$@SWNT film and *ZT* value as 0.157 at 320 K, which was largely overestimated. Recently, after establishing the measurement condition of in-plane thermal diffusivity, we measured in-plane thermal conductivity to be 20.7, which was much larger than the typical through-plane thermal conductivity of SWNT films (0.1–0.2 W m^{-1} K^{-1}) [17, 81] and found that the in-plane ZT value of $CoCp_2$@SWNT was 1.17×10^{-3}. The value was quite similar to the value for *o*-MeO-DMBI-doped SWNT film (1.70×10^{-3}) and the other n-doped SWNT film [78]. Our studies revealed that the ZT values of the SWNT film were smaller than that of inorganic TE materials and need to be improved.

22.4 Perspective

Recently, to improve the ZT value of the SWNT films, many approaches to improve Seebeck coefficient and electrical conductivity and to suppress thermal conductivity have been investigated both experimentally and theoretically. To improve the Seebeck coefficient, an enrichment of semiconducting SWNT [59], use of SWNT with smaller diameter [34, 58, 59, 82], and tuning the doping level are studied. On the other hand, to enhance the electrical conductivity, removal of dispersant [59], control of bundle size [83–85] were known as the promising strategy. Moreover, to lower the thermal conductivity, controlling of junction structure [86, 87], polymer wrapping [88, 89], and carrier doping [34] are now studied. These researches will improve the ZT values of SWNT-based TE materials and many efforts for an effective assembly of these materials will further improve the efficiency of the device [15].

References

1. Hochbaum AI, Chen RK, Delgado RD, Liang WJ, Garnett EC, Najarian M, Majumdar A, Yang PD (2008) Enhanced thermoelectric performance of rough silicon nanowires. Nature 451:163–167
2. Wang Q, Yao Q, Chang J, Chen L (2012) Enhanced thermoelectric properties of CNT/PANI composite nanofibers by highly orienting the arrangement of polymer chains. J Mater Chem 22:17612–17618

3. Zebarjadi M, Esfarjani K, Dresselhaus MS, Ren ZF, Chen G (2012) Perspectives on thermoelectrics: from fundamentals to device applications. Energy Environ Sci 5:5147–5162
4. Vineis CJ, Shakouri A, Majumdar A, Kanatzidis MG (2010) Nanostructured thermoelectrics: big efficiency gains from small features. Adv Mater 22:3970–3980
5. Sootsman JR, Chung DY, Kanatzidis MG (2009) New and old concepts in thermoelectric materials. Angew Chem Int Ed 48:8616–8639
6. Dubey N, Leclerc M (2011) Conducting polymers: efficient thermoelectric materials. J Polym Sci Part B: Polym Phys 49:467–475
7. Poehler TO, Katz HE (2012) Prospects for polymer-based thermoelectrics: state of the art and theoretical analysis. Energy Environ Sci 5:8110–8115
8. Zhang Q, Sun Y, Xu W, Zhu D (2014) Organic thermoelectric materials: emerging green energy materials converting heat to electricity directly and efficiently. Adv Mater. https://doi.org/10.1002/adma.201305371
9. Chortos A, Bao Z (2014) Skin-inspired electronic devices. Mater Today 17:321–331
10. Cornell BA, Braach-Maksvytis VLB, King LG, Osman PDJ, Raguse B, Wieczorek L, Pace RJ (1997) A biosensor that uses ion-channel switches. Nature 387:580–583
11. Bahk J-H, Fang H, Yazawa K, Shakouri A (2015) Flexible thermoelectric materials and device optimization for wearable energy harvesting. J Mater Chem C 3:10362–10374
12. Kim SJ, We JH, Cho BJ (2014) A wearable thermoelectric generator fabricated on a glass fabric. Energy Environ Sci 7:1959–1965
13. DiSalvo FJ (1999) Thermoelectric cooling and power generation. Science 285:703–706
14. Liu W, Yan X, Chen G, Ren Z (2012) Recent advances in thermoelectric nanocomposites. Nano Energy 1:42–56
15. Siddique ARM, Mahmud S, Heyst BV (2017) A review of the state of the science on wearable thermoelectric power generators (TEGs) and their existing challenges. Renew Sustain Energy Rev 73:730–744
16. Wan C, Gu X, Dang F, Itoh T, Wang Y, Sasaki H, Kondo M, Koga K, Yabuki K, Snyder GJ, Yang R, Koumoto K (2015) Flexible n-type thermoelectric materials by organic intercalation of layered transition metal dichalcogenide TiS2. Nat Mater 14:622–627
17. Nonoguchi Y, Ohashi K, Kanazawa R, Ashiba K, Hata K, Nakagawa T, Adachi C, Tanase T, Kawai T (2013) Systematic conversion of single walled carbon nanotubes into n-type thermoelectric materials by molecular dopants. Sci Rep 3:3344
18. Suemori K, Hoshino S, Kamata T (2013) Flexible and lightweight thermoelectric generators composed of carbon nanotube-polystyrene composites printed on film substrate. Appl Phys Lett 103:153902
19. Pang H, Xu L, Yan D-X, Li Z-M (2014) Conductive polymer composites with segregated structures. Prog Polym Sci 39:1908–1933
20. He M, Qiu F, Lin Z (2013) Towards high-performance polymer-based thermoelectric materials. Energy Environ Sci 6:1352–1361
21. Wei Q, Mukaida M, Kirihara K, Naitoh Y, Ishida T (2015) Recent progress on PEDOT-based thermoelectric materials. Materials 8:732–750
22. Meng CZ, Liu CH, Fan SS (2010) A promising approach to enhanced thermoelectric properties using carbon nanotube networks. Adv Mater 22:535–539
23. Kim D, Kim Y, Choi K, Grunlan JC, Yu CH (2010) Improved thermoelectric behavior of nanotube-filled polymer composites with poly(3,4-ethylenedioxythiophene) poly(styrenesulfonate). ACS Nano 4:513–523
24. Moriarty GP, De S, King PJ, Khan U, Via M, King JA, Coleman JN, Grunlan JC (2013) Thermoelectric behavior of organic thin film nanocomposites. J Polym Sci Part B: Polym Phys 51:119–123
25. Bounioux C, Diaz-Chao P, Campoy-Quiles M, Martin-Gonzalez MS, Goni AR, Yerushalmi-Rozene R, Muller C (2013) Thermoelectric composites of poly (3-hexylthiophene) and carbon nanotubes with a large power factor. Energy Environ Sci 6:918–925
26. Baibarac M, Baltog I, Lefrant S (2010) In: Nanostructured conductive polymers, pp 209–260

27. Yao Q, Chen L, Zhang W, Liufu S, Chen X (2010) Enhanced thermoelectric performance of single-walled carbon nanotubes/polyaniline hybrid nanocomposites. ACS Nano 4:2445–2451
28. Meng C, Liu C, Fan S (2010) A promising approach to enhanced thermoelectric properties using carbon nanotube networks. Adv Mater 22:535–539
29. Liu J, Sun J, Gao L (2011) Flexible single-walled carbon nanotubes/polyaniline composite films and their enhanced thermoelectric properties. Nanoscale 3:3616–3619
30. Yao Q, Wang Q, Wang L, Chen L (2014) Abnormally enhanced thermoelectric transport properties of SWNT/PANI hybrid films by the strengthened PANI molecular ordering. Energy Envison Sci 7:3801–3807
31. Snyder GJ, Toberer ES (2008) Complex thermoelectric materials. Nat Mater 7:105–114
32. Clingman WH (1961) Entropy production and optimum device design. Adv Energy Convers 1:61–79
33. Zhang Q, Sun Y, Xu W, Zhu D (2014) Organic thermoelectric materials: emerging green energy materials converting heat to electricity directly and efficiently. Adv Mater 26:6829–6851
34. Avery AD, Zhou BH, Lee J, Lee E-S, Miller EM, Ihly R, Wesenberg D, Mistry KS, Guillot SL, Zink BL, Kim Y-H, Blackburn JL, Ferguson AJ (2016) Tailored semiconducting carbon nanotube networks with enhanced thermoelectric properties. Nat Energy 1:16033
35. Nakai Y, Honda K, Yanagi K, Kataura K, Kato T, Yamamoto T, Maniwa Y (2014) Giant Seebeck coefficient in semiconducting single-wall carbon nanotube film. Appl Phys Express 7:025103
36. Dey A, Bajpai OP, Sikder AK, Chattopadhyay S, Shafeeuulla Khan MA (2016) Recent advances in CNT/graphene based thermoelectric polymer nanocomposite: a proficient move towards waste energy harvesting. Renew Sustain Energy Rev 53:653–671
37. Nakai Y, Honda K, Yanagi K, Kataura H, Kato T, Yamamoto T, Maniwa Y (2014) Giant Seebeck coefficient in semiconducting single-wall carbon nanotube film. Appl Phys Express 7:025103
38. Lee RS, Kim HJ, Fischer JE, Thess A, Smalley RE (1997) Conductivity enhancement in single-walled carbon nanotube bundles doped with K and Br. Nature 388:255–257
39. Klinke C, Chen J, Afzali A, Avouris P (2005) Charge transfer induced polarity switching in carbon nanotube transistors. Nano Lett 5:555–558
40. Barnes TM, Blackburn JL, van de Lagemaat J, Coutts TJ, Heben MJ (2008) Reversibility, dopant desorption, and tunneling in the temperature-dependent conductivity of type-separated, conductive carbon nanotube networks. ACS Nano 2:1968–1976
41. Yu M, Tian WQ, Jayanthi CS, Wu SY (2011) The effect of humidity on the adsorption of the hydrazine on single-wall carbon nanotubes: first-principles electronic structure calculations. Chem Phys Lett 518:93–98
42. Yu C, Murali A, Choi K, Ryu Y (2012) Air-stable fabric thermoelectric modules made of N- and P-type carbon nanotubes. Energy Environ Sci 5:9481–9486
43. Kim SL, Choi K, Tazebay A, Yu C (2014) Flexible power fabrics made of carbon nanotubes for harvesting thermoelectricity. ACS Nano 8:2377–2386
44. Shimizu R, Matsuzaki S, Yanagi K, Takenobu T (2012) Optical signature of charge transfer in n-type carbon nanotube transistors doped with printable organic molecules. Appl Phys Express 5:125102
45. Hewitt CA, Montgomery DS, Barbalace RL, Carlson RD, Carroll DL (2014) Improved thermoelectric power output from multilayered polyethylenimine doped carbon nanotube based organic composites. J Appl Phys 115:184502
46. Piao M, Joo MK, Choi JH, Shin JM, Moon YS, Kim GT, Dettlaff-Weglikowska U (2015) Evaluation of power generated by thermoelectric modules comprising a p-type and n-type single walled carbon nanotube composite paper. RSC Adv 5:78099–78103
47. Bark H, Lee W, Lee H (2015) Enhanced thermoelectric performance of CNT thin film p/n junctions doped with N-containing organic molecules. Macromol Res 23:795–801

48. Montgomery DS, Hewitt CA, Barbalace R, Jones T, Carroll DL (2016) Spray doping method to create a low-profile high-density carbon nanotube thermoelectric generator. Carbon 96:778–781
49. Kang BR, Yu WJ, Kim KK, Park HK, Kim SM, Park Y, Kim G, Shin HJ, Kim UJ, Lee EH, Choi JY, Lee YH (2009) Restorable type conversion of carbon nanotube transistor using pyrolytically controlled antioxidizing photosynthesis coenzyme. Adv Funct Mater 19:2553–2559
50. Kim SM, Jang JH, Kim KK, Park HK, Bae JJ, Yu WJ, Lee IH, Kim G, Loc DD, Kim UJ, Lee EH, Shin HJ, Choi JY, Lee YH (2009) Reduction-controlled viologen in bisolvent as an environmentally stable n-type dopant for carbon nanotubes. J Am Chem Soc 131:327–331
51. Wang HL, Wei P, Li YX, Han J, Lee HR, Naab BD, Liu N, Wang CG, Adijanto E, Tee BCK, Morishita S, Li QC, Gao YL, Cui Y, Bao ZN (2014) Tuning the threshold voltage of carbon nanotube transistors by n-type molecular doping for robust and flexible complementary circuits. Proc Natl Acad Sci USA 111:4776–4781
52. Wu G, Zhang Z-G, Li Y, Gao C, Wang X, Chen G (2017) Exploring high-performance n-type thermoelectric composites using amino-substituted rylene dimides and carbon nanotubes. ACS Nano 11:5746–5752
53. Xia D, Jiang S, Liu C, Fan S, Chen L (2015) A demo solar thermoelectric conversion device based on Bi_2Te_3 and carbon nanotubes. Sol Energy Mater 141:331–336
54. Zhao W, Tan HT, Tan LP, Fan S, Hng HH, Boey YCF, Beloborodov I, Yan Q (2014) N-type carbon nanotubes/silver telluride nanohybrid buckypaper with a high-thermoelectric figure of merit. ACS Appl Mater Interfaces 6:4940–4946
55. Heeger AJ (2001) Semiconducting and metallic polymers: the fourth generation of polymeric materials (nobel lecture). Angew Chem Int Ed 40:2591–2611
56. Mai C-K, Russ B, Fronk SL, Hu N, Chan-Park MB, Urban JJ, Segalman RA, Chabinyc ML, Bazan GC (2015) Varying the ionic functionalities of conjugated polyelectrolytes leads to both p- and n-type carbon nanotube composites for flexible thermoelectrics. Energy Environ Sci 8:2341–2346
57. Savage T, Bhattacharya S, Sadanadan B, Gaillard J, Tritt TM, Sun YP, Wu Y, Nayak S, Car R, Marzari N, Ajayan PM, Rao AM (2003) Photoinduced oxidation of carbon nanotubes. J Phys: Condens Matter 15:5915
58. Geier ML, McMorrow JJ, Xu W, Zhu J, Kim CH, Marks TJ, Hersam MC (2015) Solution-processed carbon nanotube thin-film complementary static random access memory. Nat Nanotechol 10:944
59. MacLeod BA, Stanton NJ, Gould IE, Wesenberg D, Ihly R, Owczarczyk ZR, Hurst KE, Fewox CS, Folmar CN, Holman Hughes K, Zink BL, Blackburn JL, Ferguson AJ (2017) Large n- and p-type thermoelectric power factors from doped semiconducting single-walled carbon nanotube thin films. Energy Environ Sci 10:2168–2179
60. Shim M, Javey A, Shi Kam NW, Dai H (2001) Polymer functionalization for air-stable n-type carbon nanotube field-effect transistors. J Am Chem Soc 123:11512–11513
61. Li LJ, Khlobystov AN, Wiltshire JG, Briggs GAD, Nicholas RJ (2005) Diameter-selective encapsulation of metallocenes in single-walled carbon nanotubes. Nat Mater 4:481–485
62. Li XK, Guard LM, Jiang J, Sakimoto K, Huang JS, Wu JG, Li JY, Yu LQ, Pokhrel R, Brudvig GW, Ismail-Beigi S, Hazari N, Taylor AD (2014) Controlled doping of carbon nanotubes with metallocenes for application in hybrid carbon nanotube/Si Solar cells. Nano Lett 14:3388–3394
63. Takenobu T, Takano T, Shiraishi M, Murakami Y, Ata M, Kataura H, Achiba Y, Iwasa Y (2003) Stable and controlled amphoteric doping by encapsulation of organic molecules inside carbon nanotubes. Nat Mater 2:683–688
64. Zhu X-Q, Zhang M-T, Yu A, Wang C-H, Cheng J-P (2008) Hydride, hydrogen atom, proton, and electron transfer driving forces of various five-membered heterocyclic organic hydrides and their reaction intermediates in acetonitrile. J Am Chem Soc 130:2501–2516

65. Wei P, Menke T, Naab BD, Leo K, Riede M, Bao Z (2012) 2-(2-methoxyphenyl)-1,3-dimethyl-1H-benzoimidazol-3-ium iodide as a new air-stable n-type dopant for vacuum-processed organic semiconductor thin films. J Am Chem Soc 134:3999–4002
66. Wei P, Liu N, Lee HR, Adijanto E, Ci L, Naab BD, Zhong JQ, Park J, Chen W, Cui Y, Bao Z (2013) Tuning the dirac point in CVD-grown graphene through solution processed n-type doping with 2-(2-methoxyphenyl)-1,3-dimethyl-2,3-dihydro-1H-benzoimidazole. Nano Lett 13:1890–1897
67. Wang H, Wei P, Li Y, Han J, Lee HR, Naab BD, Liu N, Wang C, Adijanto E, Tee BC-K, Morishita S, Li Q, Gao Y, Cui Y, Bao Z (2014) Tuning the threshold voltage of carbon nanotube transistors by n-type molecular doping for robust and flexible complementary circuits. Proc Natl Acad Sci USA 111:4776–4781
68. Byon HR, Suntivich J, Shao-Horn Y (2011) Graphene-based non-noble-metal catalysts for oxygen reduction reaction in acid. Chem Mater 23:3421–3428
69. Grigorian L, Sumanasekera GU, Loper AL, Fang S, Allen JL, Eklund PC (1998) Transport properties of alkali-metal-doped single-wall carbon nanotubes. Phys Rev B 58:R4195–R4198
70. Nakashima Y, Nakashima N, Fujigaya T (2017) Development of air-stable n-type single-walled carbon nanotubes by doping with 2-(2-methoxyphenyl)-1,3-dimethyl-2,3-dihydro-1H-benzo[d]imidazole and their thermoelectric properties. Synth Met 225:76–80
71. Fukumaru T, Fujigaya T, Nakashima N (2015) Development of n-type cobaltocene-encapsulated carbon nanotubes with remarkable thermoelectric property. Sci Rep 5
72. Chikashita H, Itoh K (1986) $AlCl_3$-promoted conjugate reduction of αβ unsaturated carbonyl compounds with 1,3-dimethyl-2-phenylbenzimidazoline. Bull Chem Soc Jpn 59:1747–1752
73. Chikashita H, Ide H, Itoh K (1986) 1,3-dimethyl-2-phenylbenzimidazoline as a novel and efficient reagent for mild reductive dehalogenation of alpha.-halo carbonyl compounds and acid chlorides. J Org Chem 51:5400–5405
74. Schwarz DE, Cameron TM, Hay PJ, Scott BL, Tumas W, Thorn DL (2005) Hydrogen evolution from organic "hydrides". Chem Commun 5919–5921
75. Tamaki Y, Koike K, Morimoto T, Ishitani O (2013) Substantial improvement in the efficiency and durability of a photocatalyst for carbon dioxide reduction using a benzoimidazole derivative as an electron donor. J Catal 304:22–28
76. Wei P, Oh JH, Dong G, Bao Z (2010) Use of a 1H-benzoimidazole derivative as an n-type dopant and to enable air-stable solution-processed n-channel organic thin-film transistors. J Am Chem Soc 132:8852–8853
77. Kim SS, Bae S, Jo WH (2015) Performance enhancement of planar heterojunction perovskite solar cells by n-doping of the electron transporting layer. Chem Commun 51:17413–17416
78. Nonoguchi Y, Nakano M, Murayama T, Hagino H, Hama S, Miyazaki K, Matsubara R, Nakamura M, Kawai T (2016) Simple salt-coordinated n-type nanocarbon materials stable in air. Adv Funct Mater 26:3021–3028
79. Nonoguchi Y, Tani A, Ikeda T, Goto C, Tanifuji N, Uda R M, Kawai T (2017) Water-processable, air-stable organic nanoparticle–carbon nanotube nanocomposites exhibiting n-type thermoelectric properties. Small 13:1603420-n/a
80. Sahoo S, Chitturi VR, Agarwal R, Jiang J-W, Katiyar RS (2014) Thermal conductivity of freestanding single wall carbon nanotube sheet by Raman spectroscopy. ACS Appl Mater Interfaces 6:19958–19965
81. Prasher RS, Hu XJ, Chalopin Y, Mingo N, Lofgreen K, Volz S, Cleri F, Keblinski P (2009) Turning carbon nanotubes from exceptional heat conductors into insulators. Phys Rev Lett 102:105901
82. Hung NT, Nugraha ART, Hasdeo EH, Dresselhaus MS, Saito R (2015) Diameter dependence of thermoelectric power of semiconducting carbon nanotubes. Phys Rev B 92:165426

83. Lyons PE, De S, Blighe F, Nicolosi V, Pereira LFC, Ferreira MS, Coleman JN (2008) The relationship between network morphology and conductivity in nanotube films. J Appl Phys 104:044302
84. Nirmalraj PN, Lyons PE, De S, Coleman JN, Boland JJ (2009) Electrical connectivity in single-walled carbon nanotube networks. Nano Lett 9:3890–3895
85. Mustonen K, Laiho P, Kaskela A, Susi T, Nasibulin AG, Kauppinen EI (2015) Uncovering the ultimate performance of single-walled carbon nanotube films as transparent conductors. Appl Phys Lett 107:143113
86. Mitsuhiro I, Naofumi O, Ryo A, Hirotaka K, Ryosuke M, Ichiro Y, Masakazu N (2014) Enhancement of thermoelectric properties of carbon nanotube composites by inserting biomolecules at nanotube junctions. Appl Phys Express 7:065102
87. Yu C, Choi K, Yin L, Grunlan JC (2011) Light-weight flexible carbon nanotube based organic composites with large thermoelectric power factors. ACS Nano 5:7885–7892
88. An CJ, Kang YH, Lee AY, Jang K-S, Jeong Y, Cho SY (2016) Foldable thermoelectric materials: improvement of the thermoelectric performance of directly spun CNT webs by individual control of electrical and thermal conductivity. ACS Appl Mater Interfaces 8:22142–22150
89. Hida S, Hori T, Shiga T, Elliott J, Shiomi J (2013) Thermal resistance and phonon scattering at the interface between carbon nanotube and amorphous polyethylene. Int J Heat Mass Transf 67:1024–1029

Index

N. Nakashima (ed.), *Nanocarbons for Energy Conversion: Supramolecular Approaches*, Nanostructure Science and Technology,
https://doi.org/10.1007/978-3-319-92917-0

The manufacturer's authorised representative in the EU is Springer Nature Customer Service Centre GmbH, Europaplatz 3, 69115 Heidelberg, Germany. If you have any concerns regarding our products, please contact ProductSafety@springernature.com

Printed and bound by CPI Group (UK) Ltd, Croydon, CR0 4YY
05/10/2026
02228122-0018